TRANSACTIONS

OF THE

AMERICAN PHILOSOPHICAL SOCIETY

HELD AT PHILADELPHIA
FOR PROMOTING USEFUL KNOWLEDGE

NEW SERIES—VOLUME 62, PART 7
1972

BURMESE EARTHWORMS

An Introduction to the Systematics and Biology of Megadrile Oligochaetes
with Special Reference to Southeast Asia

G. E. GATES

University of Maine

THE AMERICAN PHILOSOPHICAL SOCIETY
INDEPENDENCE SQUARE
PHILADELPHIA

November, 1972

IN MEMORY OF

K. N. Bahl

W. Michaelsen

J. Stephenson

PREFACE

This opus is the war-induced abortion of a project parthenogenetically conceived in the mind of an unsophisticated M.A., ignorant of systematics, who had to identify Rangoon earthworms for use in laboratory sections of the first biology course of university grade to be taught in Burma. Although the only pertinent literature available in the country at that time was a set of *Nature*. Stephenson's "Oligochaeta" (*Fauna of British India*, 1923) shortly appeared and enabled identification of a few Rangoon species. The diversity of structure shown by taxa of the four families that were represented in the college compound proved to be so interesting, especially to one hitherto familiar only with peregrine lumbricids, that the author rashly decided to find out how many and what species were present in Burma as a preliminary to beginning "more interesting" studies of subjects such as parasitology, ecology, and regeneration. Little exploration was necessary to compel recognition of the fact that an answer to a supposedly simple question would not easily or quickly be obtained. However, as years rolled by, hope persisted that the original question could be answered, at least by a fairly good estimate, in a monograph to be written after retirement from the classroom. That plan, too, may well have been markedly over-optimistic but in any case was not to be followed as World War II not only resulted in destruction of the author's collections, library and most of his records as well as various manuscripts but also in the termination of collecting in Burma. A John Simon Guggenheim Foundation fellowship enabled preparation for publication of such of the author's records as had been put aboard the last passenger boat leaving Rangoon before arrival of invading armies and the writing of a monograph entitled "Burmese Earthworms."

To earthworms, political boundaries are nonexistent. Time and again, the fact became clear that questions posed by Burmese taxa could be answered only outside of the country and finally that, in so far as megadriles were concerned, Burma was a *via media* into and sometimes through which, in former times, came forms that had arisen elsewhere. In hope of answering some of those questions study was continued as material became available during research financed by the Rockefeller and the National Science Foundations, more especially by the latter. Such work had to be mainly concerned with species that are exotic in Burma and adjacent regions. So little was known about megadrile faunas of adjacent territories that few additional species could be added by extending scope of the book to include neighboring portions of Assam and Yunnan as well as Thailand, along with the Andaman and Nicobar islands. Indeed, all of

Thailand at first required inclusion herein of only one species not found in Burma. So much ignorance required use, in the subtitle, of the qualifying "Introduction" and it is probable, even after nearly twenty years of collecting in Burma, that many forms remain to be found in the almost unexplored areas of the western and northern mountain walls as well as the forests of upper Burma between the Chindwin and the Irrawaddy rivers. Much certainly still remains to be recorded about the megascolecid fauna of the Shan Plateau because of the junking of numerous collections comprising mainly endemic species of *Pheretima* and *Tonoscolex*.

Collecting rarely was possible for the author whose pedagogical duties required his presence in laboratory and lecture room during the few weeks when the worms were at that stage in their life cycles best for systematics. Most of the Burmese material was secured by unsupervised collectors. The regions in which they worked were exactly those that were accessible conveniently during the monsoon rains, from mule and footpaths, cart roads, motor roads, railways, and the waterways along which the Irrawaddy Flotilla Company maintained passenger services. This means no oligochaetes ever were collected in much of the country. Unsurveyed areas, densely covered with rain forest, heavily infested with vicious little leeches, *Haemadipsa silvestris*, are just the places where the more interesting, and perhaps haemerophobic, endemics are to be found. While it is possible to state that 23 genera and 174 taxa thought to be of specific rank are present within the political boundaries of Burma, no estimate as to the number remaining to be discovered will be ventured.

Physical difficulties imposed by terrain, climate, distances, lack of time, were by no means all of the obstacles that hindered obtaining an answer to the original question. Greatest, was lack of a criterion that would enable recognition of interspecific boundaries. The classical system itself provided no assistance in that connection, as the very characters by which species (as well as other taxa) were defined often were found to be just those varying the most in more or less closely related populations. Certainly the writer must admit that he cannot claim to be qualified, in Tate Regan's sense, to define species or delineate interspecific boundaries. But even such qualifications would have been of little help inasmuch as large sections of various Burmese populations cannot belong to any species according to several authorities on taxonomic theory. A pragmatic approach to the problems involved had to be adopted, based primarily on a study of individual variation. Any particular species then was assumed to be composed

3

of individuals with very considerable anatomical similarity, in fact with so much that a recent common ancestry was conceivable. That did not, however, provide criteria for assigning populations to sub-species, species, super-species levels. As is demonstrated hereinafter, inability to provide firm formal decisions on such matters also requires the "Introduction" in the title.

As is instanced below, certain preconceptions biased the classical system, and, although the author has attempted to be more objective, it is necessary to admit that he has acted on the following beliefs. Species so closely related as to belong in the same genus should demonstrate a single ancestry by characters universally present in all populations and, similarly, that genera related closely enough to be in one family should also show that relationship by some common anatomy though of course less than is to be expected for species of a genus. Those familiar with the classical system will recall the lack therein of any such ideology.

Although the writer long has known that large sections of the classical system require more or less drastic revision, such action usually has not been feasible and often, the following pages present only predictions as to how the changes now seem likely to be made. Nevertheless, the author has not hesitated to abandon classical family and subfamily groupings for the sake of some anatomical uniformity even though the present arrangements are not expected to be more than interim to the classification that a vastly increased knowledge of anatomy will permit.

Writing the manuscript of the present opus was begun in 1952 and hitherto has been kept up to date at cost of frequent changes. Doubtless some portions of the text, even in spite of the rather widely spread derogation of systematics and especially of megadriles, will be out of date before appearance in printed form.

Many people assisted in one way or another, during the last forty years, the struggle to obtain the data on which the present opus is based. Too many, in fact, to be listed now, and so, contrary to the usual custom in such circumstances, all names of individuals are omitted. Nevertheless, all such assistance is appreciated and whenever possible in the past has been acknowledged. For the financial support provided by the John Simon Guggenheim Foundation that enabled writing of the original draft and for a continuation of such support by the Rockefeller Foundation and the National Science Foundation during a twelve-year period of revision and addition, the author is deeply grateful.

Since the final draft of the manuscript was typed in the fall of 1964, only one name had to be changed, that of a species exotic in Burma. The only records that had to be added were of several peregrine species recently reported from the Andaman and Nicobar Islands.

251 Silver Road   G. E. G.
Bangor, Maine 04401
April, 1972

BURMESE EARTHWORMS

An Introduction to the Systematics and Biology of Megadrile Oligochaetes with Special Reference to Southeast Asia

G. E. GATES

CONTENTS

GEOGRAPHY AND BURMESE MEGADRILES

Great Asiatic rivers[1] originate in the "roof of the world," where mountains reach nearly 30,000 feet into the sky, and they are not far apart in southeastern Tibet, the Kam or land of great canyons. There, each river flows eastward at the bottom of a deep gorge and is separated from its neighbors by sharp ranges of peaks (the "hump" with crests of 19,000 feet over which World War II pilots flew supplies to

[1] Including in addition to those mentioned above, the Hwang Ho or Yellow River, and the Yangtze.

Chungking for the Chinese). One of those rivers, the Brahmaputra breaks through the Himalayas into Assam where it flows westward until past the Khasi-Jaintia-Garo Hills and then turns south to empty into the Bay of Bengal.

The Irrawaddy, with headwaters from Himalayan glaciers, runs south to Mandalay near which it supposedly was diverted from its original course to flow west through a gap in the middle hills only to turn south again after being joined by the Chindwin that has tributaries in the foothills of the Himalayas. The former valley of the Irrawaddy below Mandalay,

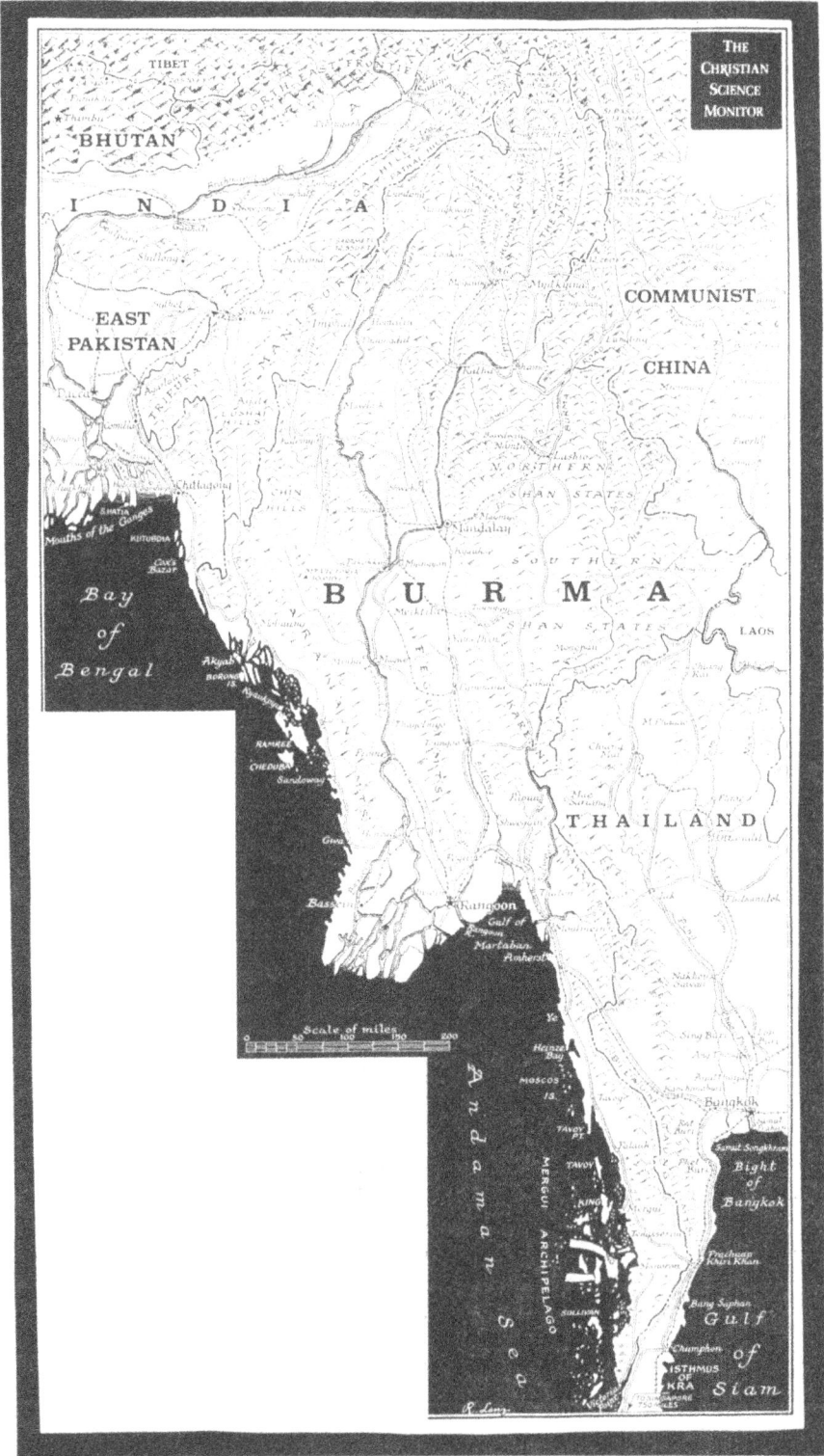

FIG. 1. Important physical features of Burma and immediately adjacent countries.
Lenz in *The Christian Science Monitor* © TCSPS.

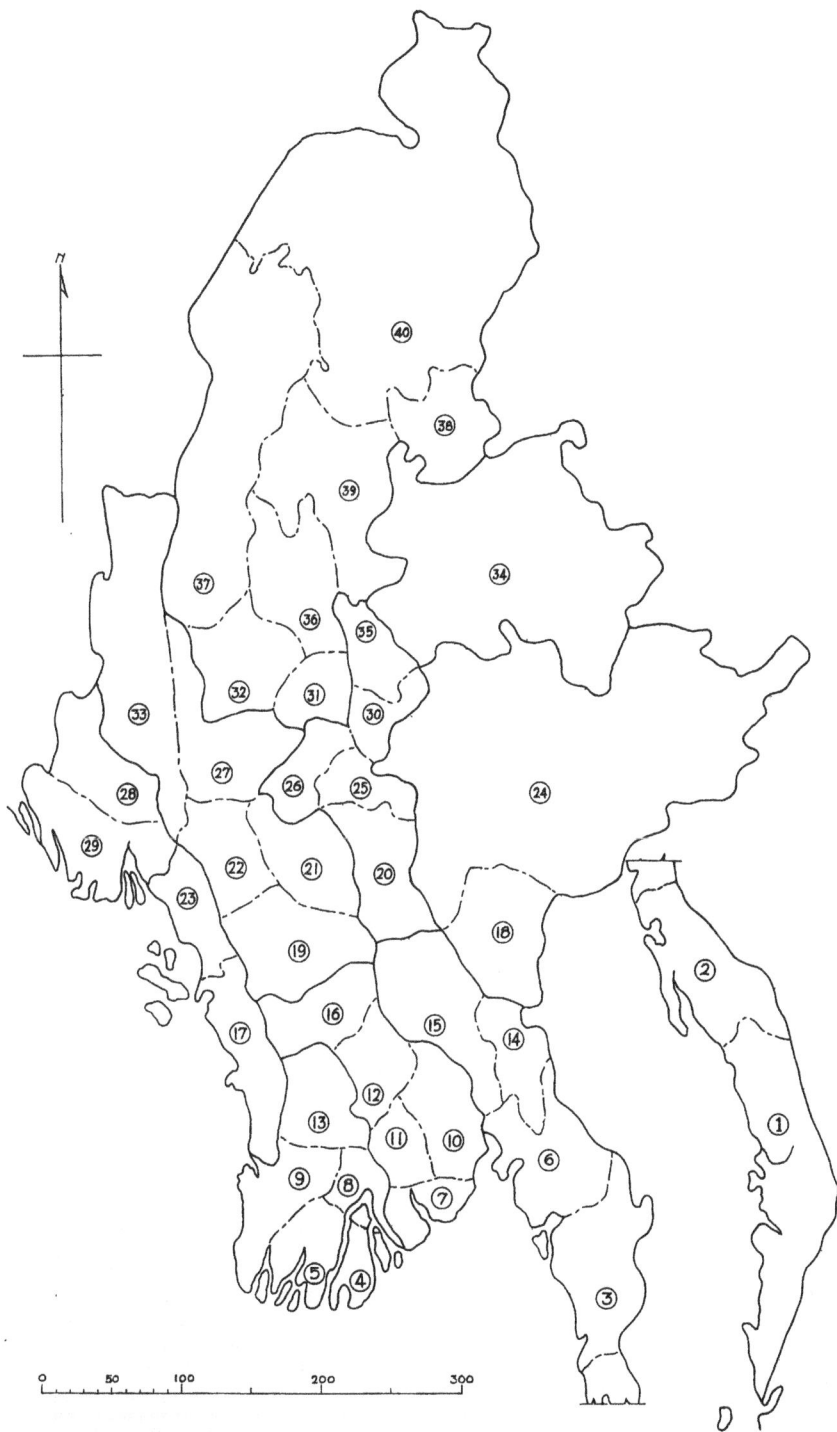

FIG. 2. Political subdivisions of Burma: 1. Upper Chindwin. 2. Myitkyina. 3. Bhamo. 4. Katha. 5. Shwebo. 6. Northern Shan States. 7. Chin Hills. 8. Lower Chindwin. 9. Sagaing. 10. Mandalay. 11. Kyaukse. 12. Pakokku. 13. Arakan Hill Tracts. 14. Myingyan. 15. Meiktila. 16. Southern Shan States. 17. Akyab. 18. Minbu. 19. Magwe. 20. Yamethin. 21. Karenni. 22. Kyaukpyu. 23. Thayetmyo. 24. Toungoo. 25. Prome. 26. Sandoway. 27. Menzada. 28. Tharrawaddy. 29. Pegu. 30. Salween. 31. Bassein. 32. Myaungmya. 33. Maubin. 34. Insein. 35. Pyapon. 36. Hanthawaddy. 37. Thaton. 38. Amherst. 39. Tavoy. 40. Mergui.

7

today is drained by the Sittang, a river that is not navigable because of a high tidal bore. Middle Burma thus has two north-south riverine valleys or axes, the Chindwin-Irrawaddy and the Irrawaddy-Sittang. The Salween emerges from Tibet into the Yunnan Plateau and there turns south to pass through the Shan Plateau in a gorge three to six thousand feet deep and flows into the Bay of Bengal near Moulmein. The Mekong River has a similar course to the east as it emerges from the Yunnan Plateau to form for some distance the western boundary between Burma as well as Thailand and the former French Colonies of Indo-China. A shorter river, the Menam, with tributaries in the Shan Plateau runs south through Thailand and empties into the Gulf of Siam.

The Himalayas, in an eastern sector of the Tibetan highlands, turn almost at right angles to run south much as do Burmese rivers. The western mountain wall of Burma, with successive sections known as the Patkai, Naga, Chin, Lushai Hills,[2] in the far north comprises parallel ranges but in the south narrows to a single range, the Arakan Yoma. A spur to the west, terminating with the Garo Hills at the Rajmehal Gap may have been, in past ages, of considerable zoogeographical importance. The western wall is continuous with a submerged chain that emerges in volcanic Narcondam, the 204 Andaman Islands, the 19 Nicobar Islands as well as in Sumatra and Java.

Through middle Burma, the Central Basin region, between the western mountain wall and the Shan Plateau, north-south ranges partly determine drainage patterns of the two riverine valleys. The Pegu Yomas, in the southern half, beginning with the 5,000-foot Mount Popa, gradually lose height towards the sea and peter out on reaching the vicinity of Rangoon. Similarly, in the northern sector, about which little is known, heights decrease in a southerly direction from a maximum of 19+ thousand feet down to those of the Sagaing Hills which are across the river from Mandalay almost at the center of Burma. Above 27° latitude traces of former glaciers can be recognized but the snowline now is at 20,000 feet and so is found today only north of Burma.

The Shan Plateau, with an escarpment rising 3,000 feet from the Irrawaddy and Sittang Valleys and with an average elevation between three and four thousand feet, is a continuation of Yunnan Plateau. The gorge of the Salween separates off a higher portion to the east. The highest peak is 8,700+ feet. Below the region of Toungoo and Karenni, the plateau character is lost in a series of parallel rows of high peaks, with steep slopes, that passes down through peninsular Burma and Thailand. Southward through Mergui district the elevations decrease to 2,000 feet. The Burma-Thailand boundary follows the watershed

divide from the Isthmus of Kra in the south to the Mekong drainage basin in the north.

Political boundaries of course have no significance for earthworms unless they coincide with physiographic features that act effectively as barriers. Such information as now is available shows that genera and even endemic species are present on both sides of high mountain ranges, deep gorges, and broad rivers. Those physiographic features in Burma, at first thought, seem not to be effective barriers to megadrile migration. Time, or more precisely insufficient time to go beyond present limits, often seems to have been the factor delimiting species ranges but only in certain directions. Perhaps geologists will be able eventually to provide explanations for some of the distribution, but data next to be considered require the limiting adjective.

The area of Burma is 261,000 square miles and so the country is somewhat larger than France and nearly as large as Texas. However, the maximum width is only 575 miles though the maximum length is 1,200 miles of which 500 are a southward extension into the Malay Peninsula. Burma thus extends from above the Tropic of Cancer well down toward the equator. A map of Burma, drawn on one of North America would reach from New Orleans to Fort William, Ontario, and from Cincinnati to Topeka, with Rangoon near St. Louis and Mandalay near Madison, Wisconsin.[3]

Distributions of species belonging to the octochaetid *Eutyphoeus* that has a western origin, as well as to the megascolecid *Pheretima* with an eastern origin, sometimes show a horizontal or transverse banding across Burma. A little evidence is available to suggest that such north-south stratification may be continued, in case of *Pheretima*, into Thailand, for *Eutyphoeus*, into India and East Pakistan. Physical barriers, of the sort that divide Burma vertically, have not been found between those closely related and geographically neighboring species. Possibly temperature differences at different latitudes during Quaternary glaciation may have been involved.

The past geological history of Burma, reluctantly and for various reasons, is ignored completely herein. Data *re* megadrile distribution and the soils of Burma no longer are available. However, mention is permissible of the fact that certain forms seemed to be about equally at home in fertile alluvial deposits, in red soils (leached of their lime) that are thought to be especially characteristic of much of the Shan Plateau, in sandy soils of the dry zone, as well as in occasional sites with more than usual amounts of organic matter. Tropical soils usually are said to

---

[2] Ranges usually are known in Burma as yomas or hills. Only a few isolated peaks are called mountains.

[3] Surrounded on three sides by rugged mountains and on the fourth by the sea, Burma has no railway connection with her neighbors and the only all-weather motor highway from the country is the famous Burma road of World War II into Yunnan. Again at first thought, earthworms would seem to have done better in crossing barriers than man with all his modern machines.

have little humus as organic matter is quickly oxidized during the long and hot dry season. Yet, various species, in considerable sections of Burma, grow to "giant" size.

Perhaps of greater importance to Burmese megadriles than physical barriers and the nature of the soils is the monsoon climate. Trade winds blow from the southwest during May–October and, coming from the ocean, they bring rain, 200+ inches[4] to the Arakan and Tenasserim coastal areas, 100 inches to Rangoon and some of the deltas region, 70 inches to the Shan Plateau, the Chin Hills, northern portions of middle Burma, 20–40 inches in a central part of middle Burma bounded by the 40 inch isohyet, which accordingly is known as the dry zone. A small core, with only 20–25 inches, is semi-arid and with a xerophytic flora. The northeast monsoon, during the remainder of the year, as it comes overland, brings no rain. The earthworm year accordingly is divided throughout most of the country[5] into rainy and dry seasons, to the first of which earthworm activity, with few exceptions,[6] is confined. The length of that season varies from one part of the country to another and may be five months or more in some parts of the Shan Plateau (perhaps also in other hill areas?) but only three or four months in portions of the dry zone. Evidence of megadrile activity often is not obvious in areas with much rain until the monsoon is well underway but in the dry zone, towerlike castings almost seem to spring out of the ground over night after the first real rainfall. Some of such differences may be associated with soil types, a porous and rather sandy soil perhaps enabling, with the same amount of precipitation, earlier activity than alluvial clays that are penetrated more slowly.

At the beginning of the season the worms quite generally are immature and for most species breeding comes toward the end of the rains, August–September in dry zone, September–October rather widely, perhaps later still on some of the trans-Salween heights and various other elevations. Well before the end of the rainy season, earthworms begin to disappear rather generally from the upper layers of the soil. Amounts of daily rainfall that are sufficient to enable initiation of activity early in the monsoon now seem inadequate. Probably involved is the increase in number of sunny hours without cloud cover that results in hardening of the surface soil, followed, at least in alluvial deposits, by shrinking and cracking. Where the cocoons are deposited, and in what circumstances, as well as what happens therein during the long dry season are unknown.

The breeding season may end in at least three ways. One, perhaps the more common, involves burrowing down through the subsoil to any level where conditions will permit survival during the long drought, presumably in some sort of a resting state. No report on that state for Burmese endemics is possible. Whether something more than mere inactivity, for instance like the deep diapause of certain lumbricids, is involved remains to be learned. The terms "estivation" and "hibernation" are inapplicable to whatever the state may be as the dry period of monsoon Burma comprises two seasons, the cool from November into February, and the hot from mid-February through May.[7] A portmanteau word, "hibernestivation" a splendidly impressive characterization for the depths of our ignorance of the subject, has been used.

Drawida grandis, a moniligastrid "giant," could be obtained during May, even on the Nilgiri Hills of South India (Bourne, 1894), only at depths of 9–10 feet. The single Burmese record in the literature (Gates, 1927: p. 185) is of the reported finding by a Burman, at a depth of 10 feet in a pit he was digging, of a ball of worms some 6 to 10 inches in diameter. The observer was unwilling to venture a guess as to whether more than one species was involved. A mass of worms placed in a laboratory tray with only a little water has been observed, on many occasions, to gather into a ball shape. At Pang-Yang, on the Shan Plateau at an elevation of *ca.* 5,000 feet, a missionary, Harold Young, reported (*in lit.* to the author) that "Numbers of earthworms were found about ten feet below the surface, in February, while digging a well. Each worm was coiled up into a sort of ball and there were several kinds. Water was obtained at twelve feet." These few instances suggest that the worms may go straight down until a level with sufficient moisture is reached somewhere near the water table. The individual balling observed on the Shan Plateau also suggests possibility of a diapause like that of certain lumbricids. Aggregation into a single ball (depth not mentioned), during an unfavorable season, has been recorded for unknown species of the Lumbricidae but without subsequent confirmation.

A phenomenon characterized as "mortal wanderings" provides a second way of ending a breeding period. Some days after the last rain, both dates, as well as the interval between them, seemingly unpredictable, mature earthworms are found in great numbers after sunrise moving aimlessly about on the surface of the ground. The wanderings appear to end always in desiccation, and natives in various sections of the Shan Plateau and the Tenasserim yomas where the phenomenon has been observed agree that the worms come to the surface *in order to die*, i.e., to commit suicide on a mass scale. The species involved,

---

[4] Numbers mentioned are approximations and amounts vary even in the areas mentioned, e.g. the average annual rainfall at Rangoon is 98.99 inches, at Toungoo, 78.05, at Mandalay 32.63.

[5] Throughout the year some rain falls in a northernmost portion of Burma where the megadrile fauna is almost unknown.

[6] See page 10 for exceptions other than those of note 5.

[7] Months called summer in many other parts of the northern hemisphere are less hot in Burma, at least in part because of the cloud cover.

according to the author's own observations, are of medium size but not of giant length, the largest being *Desmogaster doriae*. An American after many years of residence and travel through the Shan and Yunnan Plateau provided (V. Young, *in lit.*) the following information.

Many earthworms come to the surface in and around Bana (Mong Lem State) towards the close of the rainy season. The worms are of all sizes from one to twelve inches long. Many of them at least, if not all, apparently die shortly after coming to the surface. In some places there are so many that they cause a terrible stench. I have seen several hundred near each other when walking along jungle paths. There were stretches where none were seen but every few yards one is sure to come to places where there are large numbers on the surface. The Lahu people claim that the earthworms come to the surface, during the last showers of the rainy season, to die.

Bana is in Yunnan province of China where the wanderings appear to be associated with somewhat different conditions from those in Burma. Whether more than one species was involved, Mr. Young was unable to tell.

Possibly of a quite different category is a phenomenon called "mass migrations." Oral and written reports to this author by American missionaries and British officials who lived in or traveled through the Chin Hills, agreed in the general tenor of their content and none included any mention of, or even hint of, suicidal aspects though several did speak about the crushing of many specimens by mules of the pack trains. A typical report (Gates, 1933; p. 415) was as follows.

In the early morning on certain days in October and November, at the beginning of the cold season[8] the road is almost covered with worms, one can see worms tumbling down from the banks above onto the road. In the evening not a worm is to be found. I have always assumed that the worms were moving down-hill perhaps in search of water. In this region the mountains are covered with a thick undergrowth of mosses and ferns.

Later, after a request for more information and a sizable sample from a migration the following statements were received. "All the migrating worms are of the same kind. Both on the slope above the road and the slope below the road the worms were going down; where they followed the road they were going down hill as well." Each specimen of a fairly large sample from one such migration was of a species of *Perionyx* but was unidentifiable further because of immaturity.

Mass migrations were recorded at least once in India. There the movement was observed in an early portion of the rainy season, and the worms were all going up hill.

---

[8] At elevations above three thousand feet almost everywhere in Burma winter temperatures are low enough to warrant the "cold season" characterization.

A third eventuality, after completion of breeding, is continuation of activity. That is, of course, possible only in restricted areas that remain wet long after soils elsewhere are dry. Such places are near or around hydrants or other devices for supplying piped water to the public, wells, tanks, buffalo wallows, ponds, lakes, rivers, irrigation ditches, odd spots where household wastes are emptied onto or into the ground, earth of potted plants in houses or on verandas. Metabolic activity, in favorable circumstances, may be continued at some of those sites throughout the drought. However, sex organs regress and specifically distinguishing external markings become unrecognizable or almost so. A majority of the individuals and of the species remaining active after the rains in such isolated areas (Gates, 1926, 1961) are exotic.

Much of that dry season activity obviously was made possible by man and the opportunities thus provided doubtless have continually increased during recent centuries as he sought, especially in the deltas region and in the dry zone, to satisfy the needs of his own rapidly growing population. No history of the canal systems, constructed in the days of the Burmese kings, was available, and no attempt was made to estimate how megadrile faunas were modified in the dry zone as a result of the increase in available water.

Nowhere in the world has any real study been made of the role of man in the megadrile environment and for the Burmese region comment now is warranted only for two other aspects of the problem.

A common kind of cultivation in eastern as well as western hills of Burma is called "taungya." Trees are cut down, allowed to dry through the rainless season and then burned to provide fertilizer for some hand-sown crop. The clearing is cultivated until fertility is exhausted and then is allowed to revert to jungle. Long experience has taught the hill folk that different soils have various grades of fertility so that deforesting for taungya crops is profitable only once in a period of seven, thirteen, seventeen, or even more years, as the case may be. Taungya cultivation may have been responsible for the inability to find certain endemic megadriles near villages and in areas where that method of cropping is practiced. After exhaustion of the soil, villages sometimes are moved and then may or may not take the old name to a new site. Thus, places at which some of Fea's collections were made no longer can be pinpointed on a map. Effects of burning on the soil fauna have not been studied in Burma. No difference in numbers of towerlike castings was apparent to casual observers in the Kalaw Forest Reserve where ground cover had been burned each year towards the end of the dry season.

For more than a century earthworms were known to have been taken across oceans in earth around roots of live plants. Nevertheless, little effort has been made to ascertain the significance of such transport for native megadrile faunas in various parts of the world. As one result of a twenty-year study of some aspects of the problem, the author (1958) learned: that even a small handful of fertile, unsterilized soil surrounding plant roots is liable to contain cocoons and/or the almost threadlike, just-hatched juveniles, that larger earth-samples often contained one or more adults even of fair-sized species, that in spite of long-standing quarantine regulations megadriles still are frequently transported. Wherever live plants have gone in the past, and they have gone along with Europeans almost everywhere, earthworms sooner or later also have gone.

Records of transport of plants to Burma prior to the British occupation have not as yet been found, possibly because the right sources were not investigated. Live plants certainly were brought into Burma on very many occasions during British rule. They were grown and distributed from sources such as botanical gardens, the Agri-Horticultural Gardens of Rangoon, and also were divided and passed from one householder to another. Plants in pots of various sizes as well as in the ubiquitous kerosene tins frequently were seen, in prewar days, on oxcarts, buses, trains, Irrawaddy timber rafts, boats. The plants were taken from the plains to the hills, from hill stations and summer resorts to the plains, and to dak and other government bungalows in every part of the country reached by officials. Earthworms often were found in such pots during the author's twenty-year survey of the Burmese fauna, but few of the records appeared in print. Among some that did, the most interesting were those of two species obtained from pots on the author's veranda, neither of which was secured again, in spite of careful searches at the Agri-Horticultural Gardens and other places from which the plants could have been brought.

Various Burmese[9] species obviously are peregrine, i.e., in the original meaning of the word, foreign or imported. They are as follows.

Eudrilidae. This family, with endemics confined to tropical Africa, is widely represented in extra-African tropics by *Eudrilus eugeniae* now familiar to millions of Americans as the African night crawler. Arrival of the species in Asia may have been later than that of *P. corethrurus*.

Glossoscolecidae. This family, with endemics confined to tropical America and the West Indies is represented elsewhere by a single species, *Pontoscolex corethrurus*, now widely distributed in extra-American tropics and even with numerical dominance in conditions such as are provided by rubber plantations of Lower Burma and the Malay Peninsula. Arrival in Asia must have been after A.D. 1500.

Lumbricidae. Endemics of this family are confined, perhaps by temperature barriers, to temperate-zone North America (eastern portion only) and Eurasia.

Although the southern border of the lubricid domain has been repeatedly said to be at Calcutta there now is no reason to believe that any of the family is indigenous in Asia south of the Tian Shan and Altai Mountains, Mongolia and Manchuria. Twelve of the 16 European species that colonized successfully throughout temperate[10] North and South America, South Africa, Australia and New Zealand during the last five centuries, have been reported from southeastern Asia. Some of those reports have not been confirmed, and if colonies actually were founded they probably long since became extinct. Eleven of those 12 European species are known to be established in India from where there also has been reported another European species that is much less widely distributed throughout the world. Four other markedly anthropochorous lumbricids are liable to turn up at any time in India. The single lumbricid of mainland Burma, also present in India, always has been thought to have originated in eastern North America. Domicile has been acquired by lumbricids in southeastern Asia only at elevations with climatic conditions similar to those of more northern sections of the hemisphere from which the worms originally were taken. Outside the Indo-Gangetic Valleys and Kashmir, lumbricids are known to be established only at hill resorts where Europeans have resided in the past. Whether lumbricids could have been brought to western India along the trade routes in use during the reigns of Alexander the Great, his Persian predecessors, and later Greco-Bactrian rulers, is unknown but this author has had little opportunity to investigate documents that might provide evidence one way or another. To the hill resorts, some of which were developed by the British, lumbricids may have been taken within the last 150 years.

Ocnerodrilidae. Ocnerodrilinae. All ocnerodriline species of southeastern Asia came originally from America or Africa and presumably since A.D. 1500. Malabariinae. All species of this subfamily are of Asiatic origin and all genera probably evolved in the Indian peninsula. One of the two Burmese species is exotic in Hainan. Presence of both in Burma may be a result of transportation from an Indian home but if not, the species presumably reached Burma across what is now the Rajmehal gap.

Three of the four remaining families of southeastern Asia have species now known to be peregrine in some parts of the world.

Acanthodrilidae. Two taxa, of *Microscolex*, long thought to be species, originally from somewhere in temperate-zone South America, may be only parthenogenetic morphs isolated from a single biparental population. The species of *Pontodrilus* obviously is peregrine in much of its circumtropical range, but where

---

[9] Burmese, also Indian, clearly can mean no more than existing in Burma or India.

[10] Temperate is here understood to refer also to elevated tropical areas with temperature conditions similar to those of the northern zone.

it originated and how it was taken around the world is unknown. Post-transportation colonization is unknown in the other acanthodrilid genus.

Megascolecidae. *Lampito* has endemic species in a small hill section of South India. The single peregrine, *L. mauritii*, so named because it was first found in Mauritius, was not recorded from its homeland until some twenty years later and then under another name. Transportation of the species toward Africa by man, even before the Christian era, is not impossible as apes, ivory, and peacocks are known, from profane as well as sacred sources, to have gone west to Egypt and eventually to Tyre from Tamil kingdoms. The species seems to be able to maintain itself within rather close quarters, in relatively large numbers, and was found on many occasions, both in Burma and in India, in the earth of plant pots. The little known *Lennoscolex* is represented in Burma by two species with closest relatives in the extreme south of India and in Ceylon. Presence in Burma, and also of one of the taxa in Java, resulted from transportation perhaps centuries before arrival of the Portuguese in the Indian Ocean. One species of *Perionyx*, *P. excavatus*, with established colonies in the West Indies and Hawaii, clearly is peregrine. The following species of *Pheretima*, *alexandri*, *anomala*, *bicincta*, *birmanica*, *californica*,[11] *diffringens*, *elongata*,[12] *hawayana*, *houlleti*, *minima*, *morrisi*, *papulosa*, *peguana*, *planata*, *posthuma*, *robusta*,[13] *rodericensis*, have overland or oceanic discontinuities in their distribution. Success in colonizing seems comparable to that of some lumbricid species.

Moniligastridae. Five species of *Drawida* have been transported across oceans or seas and so are peregrine. Records of other species outside the generic range are valueless as no one knows what was involved. One species of *Desmogaster* may be exotic in China.

Octochaetidae. *Octochaetona* and *Ramiella* are represented in Burma only by peregrine species that could have been taken to Burma before the arrival of the Portuguese. Each species of *Dichogaster* came to India from another continent.

Presence of a species such as *P. corethrurus* in many parts of the world admittedly has no more zoogeographical significance than that of rabbits or horses in Australia. Attachment of the peregrine label to a

---

[11] Types were collected at San Francisco in 1852 but the species was not recognized in its probable homeland until 1936 prior to which it had been described under several other specific names from widely separated sites where it also was exotic.

[12] Erected on a specimen from Peru.

[13] Erected on specimens from Mauritius and supposedly also from Manila. The species was not reported from its probable homeland, China, until 1930, prior to which it also had been described under several specific names. Another instance, *Perionyx sansibaricus* Michaelsen, 1891, erected as the name indicates on specimens from the island of Zanzibar, was not reported from its homeland in India until 1909.

species seems to have outcasted the taxon from further consideration by classical systematists. As one result, the original homes of species such as the anthropochorous lumbricids, the microscolexes, *E. eugeniae*, *P. corethrurus* can be indicated, even today, only quite generally on a map or by imprecise statements such as those above. Even the continent in which the oriental dichogasters arose is unknown. The most widely distributed anthropochore must have originated somewhere. Some portions of its present range must have been self-acquired. Why determination of that area of endemism was of no interest, the writer has been unable to understand. In such cases he has been seeking, though often unsuccessfully, to learn in what portions of a range endemism is likely and, in so far as is feasible, has distinguished in subsequent pages between what perhaps can appropriately be called a "donated" portion of a species distribution and that portion which is self-acquired. Thus, for instance, the original home of *P. elongata* now should be sought in a region including Borneo, Celebes, and the Philippines. Absence of *P. posthuma* in the eastern as well as the western and northern mountain walls of Burma seems to indicate that origin is likely to have been east of the Menam River if not still further away. One Burmese species of *Octochaetona* now is known to have arisen in a northeastern portion of peninsular India. If sufficient time had been available that taxon presumably could have got into Assam and then Burma across the Rajmahal gap (as some malabariines and perhaps also the ancestors of two lennogasters may have done). However, the similarity between worms of the three areas seems to be such as to require recent transportation. The second Burmese species of *Octochaetona* must have arisen somewhere in the Indian Peninsula, but more definite pinpointing is at present impossible because of the paucity of our knowledge of the megadrile faunas of so much of that great block of stable land. One Burmese species of *Ramiella* is likely to have arisen in a northern (and perhaps western?) portion of the Deccan.

L. mauritii presumably arose somewhere in the vicinity of, if not on, the Cardamon and Palni Hills of South India. Its range includes most of the subcontinent of India and it is, even though the species seems to be limited to the plains and a few hills that are under 2,500 feet, much greater than that of most autochthones. The only obvious discontinuities are oceanic.

Marine and/or terrestrial discontinuities in the ranges of *Perionyx excavatus*, *Pheretima alexandri*, *anomala*, *peguana*, *planata*, testify to the fact that those megadriles are indeed exotic in certain regions outside of Burma. In that country, the distribution of each is greater than those of most untransported species (as now known), and also is without obvious discontinuities. *P. excavatus* quite possibly arose in the Himalayas and the pheretimas are now thought to have evolved further to the east, perhaps in or near Thailand, but how much and what parts of the distributions in Burma are self-acquired is undecided.[14]

Megadriles, because of inherent limitations, and of course after exclusion of the peregrine forms, were thought (cf. Stephenson, 1930: p. 667) to be capable of yielding results for paleogeography second to no other group of animals. The various intercontinental bridges that were assumed to explain distributions based on erroneous assumptions of the classical system now need no consideration and discussion of wegenerian relationships is unlikely to be rewarding until much more of the classical system has been drastically revised. However, the slow-moving megadriles may be able, when much more information is available, to contribute significantly and in unexpected ways to zoogeographical discussions.

Nelloscolex, about which little is known, was found only on the western side of the Indo-Burman mountain wall. Somewhere in that region, the genus may have arisen too recently to have had opportunity to go further. *Tonoscolex* may have arisen in or just above northernmost Burma long enough ago to have had time to spread to the south, as far as Karenni in the Shan Plateau, and well down in the upper half of middle Burma, to the west along the outer Himalayas as far as the vicinity of Darjiling. Although as yet unknown on the western mountain wall, species are to be expected there, at least in a northern portion, to link with *T. horai* of the Garo Hills section in the western spur. The two genera seemingly are unrelated to other megascolecids, whether Asiatic or extra-Asiatic, and they may be the last survivors of a lineage which, before becoming extinct in the northern half of the Indian peninsula, also gave rise to a branch that evolved into the megascolecids of south India and Ceylon.

Perionyx, in a quite different megascolecid lineage, is believed to have evolved in the Indian Peninsula so long ago that time was available for penetration into Ceylon, Burma (presumably by way of the Rajmahal gap), and the eastern Himalayas. The genus usually seems to be rather unimportant, confined to a few scattered sites with unusual amounts of organic matter. In the eastern Himalayas, however, what might be called a "species explosion" seems to have occurred and to have convinced Stephenson (1923: p. 27) that in these heights "is the great focus of evolution of species of *Perionyx*." Certainly distinguishing many of the Himalayan taxa that were named is difficult and perhaps, in a region with considerable and regular rainfall, probably with more organic matter in the soil than in the tropical plains, a kind of generic rejuvenation really had gotten under way.

Eutyphoeus, along with *Scolioscolides* and the wholly Indian *Bahlia*, are the culmination of one of the major octochaetid lines that evolved (perhaps in the northern half of) the Indian Peninsula. The center of evolution for *Eutypheus* was thought (Gates, 1958c p. 218) to be in what is now the Indo-Burman mountain wall. From that center a migrant wave passed through the southern half of Burma at least. During that time or after reaching the foot of the eastern escarpment, the wave was stratified horizontally into species of the "*levis* group." Horizontal stratification of a subsequent wave appears to have been completed in northern parts of middle Burma but in the south seemingly is still under way. A final wave was stratified even before getting across the lower Irrawaddy. Several of those horizontal ranges (cf. 1958c) are surprisingly narrow.

Plutellus (cf. p. 37) is a huge classical congeries in need of drastic revision. Asiatic plutelli, all endemic, have no extra-continental relationships. Many taxa are known only from single sites. Somatic sizes are small. Because of the wide distribution of the genus in India as well as in Burma, and the paucity of distributional as well as morphological data, origin on either side of the Bay of Bengal could be assumed. The author is inclined to favor evolution from a malabariine-like ancestry in peninsular India. For either alternative, the only migration route that now can be suggested is the one across the Rajmahal gap where the alluvium brought down by the Ganges and the Brahmaputra rivers is shallow.[15]

Glyphidrilus certainly does not belong in any of the families otherwise represented in the Orient and by classical specialists was placed in the Microchaetidae which requires for the genus an African origin. Because of the strictly limicolous habits,[16] all species were assumed by classical specialists to have no zoogeographical interest. An initial study of neglected internal organization (Gates, 1958b) provided reasons for believing that all species are endemic in the areas where they live.[17] If of African origin, *Glyphidrilus* must have arrived in Asia (from tropical east Africa in Tanganyika?) so long ago that there has been time for passage down through all of India to Ceylon, through all of Burma and the Malay Peninsula to various islands of the Malay Archipelago as well as to Hainan. Unless habitat restrictions have been completely changed for each species after arrival in its present domain, those migrations must have been under fresh water.[18]

The Moniligastridae presumably arose well to the east of Burma in a region now occupied by the Malay

---

[14] Also compare with *L. mauritii* above.

[15] Thickness of the alluvium in the ante-Himalayan, synclinal trough is said to be between six and ten thousand feet.

[16] No record is available of any specimen having been seen alive outside of water-covered mud.

[17] A Burmese species has been recorded from the island of Hainan but is even less likely to be there than at Lucknow from which an endemic also was misidentified as *G. papillatus*.

[18] Perhaps well below the surface. *G. gangeticus* Gates 1958 was secured along banks of a river at a depth of about 18 inches.

Archipelago. There, *Drawida* originated long enough ago to have had time to get into Japan, east Siberia, Korea, China, through Burma (and over the Rajmahal Gap?), down to the southern tip of India, perhaps also into Ceylon. Whether any megadrile genus really has a self-acquired range of greater expanse now seems doubtful. Little is known about the distribution of nearly all of the species but the range of one, *D. caerulea*, seems to be in horizontal belts of greater north-south width than is known (as yet!) in other megadrile genera. The primitive *D. c. caerulea* is in the southern half of middle Burma and at an unknown altitude above Mandalay "shades" into *D. c. rasilis* which is expected to "shade" into *D. decourcyi*, as yet known only from the Abor Hills in front of the Assam Himalayas.

One desmogaster may have nearly reached the Tenasserim seacoast in the vicinity of the Ye River but, about at the latitude of Toungoo, is across the Salween River. Two other desmogasters, in the region of the same parallel have nearly reached the western escarpment of the Shan Plateau but without getting down into the Sittang Valley. Still further north, another desmogaster not only passed the Salween but also the Irrawaddy above the latitude of Myitkyina. Two hastirogasters got across the Salween. One has not been found west of Lashio, the terminus of the Burma Railways, but the other was secured near the left bank of the Chindwin, the furthest west to be reached by a moniligastrid since *Drawida* passed on into India. Presence of the various forms on both sides of mountains and river gorges that might seem capable of acting as barriers suggests that some moniligastrid migrations have not reached further west mainly because of lack of time. Certainly none of the present physiographic features in the west are likely to be more effective as barriers.

Species of *Pheretima*, with distributions that seemingly are self-acquired, also have western distributional boundaries that do not coincide with physiographic features that might act as barriers. Two taxa of the *andersoni* complex are present in the Pegu Yomas of the southern half of middle Burma and there, are in the same sort of horizontally striated ranges as is most of the remainder of the complex on the Shan Plateau. Crossing of the Sittang Valley may have been in the vicinity of Toungoo but no vestiges of the passage were found. Another of the *andersoni* complex is well across the Irrawaddy in Myitkyina district, but in northern Burma, ranges of the complex now seem more likely to be vertically stratified.

To summarize this portion of the discussion. Two megascolecid and two octochaetid genera originated in or close to Burma where distributions are likely to be horizontally bandlike and without visible physiographic features as boundaries. Burma's importance in the megadrile world accordingly seems to lie more

in its having served as a bridge for passage from Malaysia of the oldest moniligastrid into India as well as for passage of the supposedly African *Glyphidrilus* to Malaysia.

The Andaman and Nicobar Islands belong politically to India but geographically to Burma, both of which relationships are indicated by those megadriles that were taken there by man. If the anthropochorous peregrines are omitted from consideration, the islands have no species in common with the Asiatic mainland. Endemics all are of the genus *Pheretima*. Relationships presumably will prove to be closest to presently unknown Sumatran species. Separation of the islands from the Burmese mainland must have been prior to that from Sumatra and before *Eutyphoeus* could have gone so far south.

The role of man in the transport of earthworms has not been appreciated. Considerable evidence on the subject has been accumulated during the last twenty years, but only a portion of it is in print. Transport of megadriles for bait would not have been suspected in Burma where, at the end of the rains, water is bailed out of ponds, tanks and channels until the fish can be scooped up by hand. During his travels throughout much of Burma, the author never saw anyone fishing with hook and line to say nothing of using angleworms as the bait. Nevertheless, he is reminded by one of his own experiences[19] that, even in Burma, anglers may have had a role in distribution of earthworms. Whenever a discontinuity in distribution, or an unusually large continuous range, is to be explained, the first factor to be taken into consideration now must be man. Various other agents have indeed been suggested; cattle—mud with their hooves[20], birds—mud on feet and legs, floating logs, natural rafts—especially those washed out to sea. Undoubtedly transport of earthworms by the latter two agents is possible but no evidence, based on actual observations, has ever been recorded.

Occasionally, however, postulating man as the agent in distributing a species seems inadvisable. One such instance is provided by *Pheretima feai*, which is

---

[19] Late in a hot-season afternoon, the author was digging in the sand along the left bank of the Irrawaddy, almost the only place where worms could be found at that time of the year. Two species had been secured, one never seen before, also never again, and another that is obtainable almost anywhere in the plains all the way down to Rangoon. Perspiring profusely, the digger and his clothes were thoroughly dirty. A high government official, out for a "cool-of-the-afternoon stroll," put his cane under his arm, picked up the can, tossed out everything except *Pheretima posthuma*, handed back the remainder with, "There, my good man, this is the only kind that is any good," and then departed hence presumably investing himself with a halo because of the good deed done that day for one obviously in much need of wisdom.

[20] To check on that possibility, megadriles were sought in cowyards of central Maine farms, but were found there only under stones, logs, or boards whence cocoons are unlikely to have been picked up.

known only from the Dawna Hills of the Tenasserim Yomas and from Mount Kambaiti, at 2,300 meters, in a northern portion of Mytkyina district. A range large enough to include both of those regions would be most unusual and an overland gap of such magnitude in the range of an edemic megadrile would be even more amazing. Cogitation on the facts involved, almost inevitably results in raising the question as to whether water, *per se* and without assistance of rafts, logs, or any other floating material can transport the animals. The author himself never witnessed anything of the sort, but inquiries did reveal that balls of earthworms a foot in diameter had been seen[21] in the deltas region at times of unusually high floods when entire villages had to be abandoned. Corroboration of those reports was obtained from an American missionary, a keen and reliable observer, who had traveled on various occasions by rowboat in flooded areas while on relief missions. Moreover, earthworms in shallow water in the laboratory on very many occasions were seen aggregated into a clump that immediately after a slight jar, took on the shape of a ball and so could be rolled about with slight pressure. Conceivably then, floods may have had a part in the distribution of earthworms in the alluvium deposited in the deltas region during the last several or perhaps even more centuries. Flood plains also flank the Chindwin and doubtless some of the lesser rivers. Riverine transport does not, however, need to be invoked to explain distributions in deltas region as all species present there, with the single exception of the mysterious *P. bermudensis*, are known to have been transported, as cocoons, juveniles or adults in earth with potted plants.

Perhaps some of the species of *Eutyphoeus* that line, for short distances, both sides of the Irrawaddy, got across in flood water, but if so, would not colonies on the left bank be further south than on the west shore? Specimens of southern members of the *andersoni* complex presumably could have been brought down into the Sittang Valley from the Plateau but they could not have been washed up by any such floods to their present sites on the Pegu Yomas. Moreover, no vestiges of the crossing have been found in the intervening valley. Even if floods do assist in the spreading of some species, such a method of transport is of little or no value in helping to understand the horizontal banding that now seems to be so characteristic of endemic distributions in Burma, nor the increasing specialization of *D. caerulea* to the north. *Perionyx excavatus*, with a presumed origin in the eastern Himalayas, seems to be the most logical candidate for riverine distributions, but its presence in Hawaii and the West Indies could be thought to obviate need for postulating any other transporting

agent than man. However, if the huge migrating populations of immature perionyxes, that were seen so often in the Chin Hills, should prove to be of *excavatus*, man is unlikely to have been responsible for presence of the species there.

Nothing of importance can now be said about correlation of earthworm distributions with any of the various series of ecological or "natural" regions supposedly existing in Burma.

Two, possibly 3, of 8 families are represented in Burma solely by imported species. Only 11 of the 23 Burmese genera have species that are endemic in that country. Even some pheretimas that may be indigenous in an eastern region including a part of the Shan Plateau are known to have been distributed throughout much of the remainder of Burma by man. The fact that 30–35 of the 170-odd species recorded from the country are more or less markedly anthropochorous provides little more than a hint as to what faunal changes may have resulted from all those introductions. The last war obviated profitable discussion of the subject. However, three conclusions that had been reached prior to Pearl Harbor are expected to be just as true 25 years later. Riverine sandbanks of Burmese lowlands are occupied exclusively, or almost so, by *Pheretima posthuma*. *Pontoscolex corethrurus*, in lower and peninsular Burma as well as further south, dominates, almost to the exclusion of other forms, cultivated areas of several kinds. *Perionyx excavatus* dominates habitats characterized by unusual amounts of organic matter and moisture.

Finally, some explanation regarding the frequently mentioned Rajmahal Gap now is appropriate. Rajmahal is at the edge of hills spurring out in a northeastern direction from peninsular India. The gap is the lowland between those hills and the Garo spur from the western mountain wall of Burma. In that gap, the Pleistocene alluvium that fills the Himalayan foredeep is shallow and through that gap the Brahmaputra as well as the Ganges now flows into the Bay of Bengal. When Tethys covered the sites of Indo-Gangetic Valleys and the Himalayas, a Rajmahal spur of some sort was continuous, for a time, with the Garo Hills and provided the land bridge by which octochaetids from peninsular India may have passed into Burma and along which moniligastrids may have migrated from eastern Asia into peninsular India. Attempts to interest various authorities in collecting through and around that gap were futile perhaps because the costs of a thorough survey would have been so small as not to justify consideration of the project! Nevertheless, the pauper fauna of the Gangetic Valley[22] shows that species from the Himalayas, or from the still older peninsular block, have not had enough time to colonize widely in the more recent Gangetic alluvium.

[21] Whether the balls were thought to be floating or to be merely rolled along perhaps by a swift current in shallow water annot now be stated because of wartime loss of records.

[22] To be discussed, circumstances permitting, in a subsequent publication.

REFERENCES

BOURNE, A. G.   1894.   "On *Moniligaster grandis* A. G. B. from the Nilgiris, S. India." *Quart. Jour. Micros. Sci.* **36**: pp. 307–360.

CHHIBBER, H. L.   1933.   *The Physiography of Burma* (London)
—— 1934.   *The Geology of Burma* (London).

GATES, G. E.   1926.   "Notes on the Seasonal Occurrence of Rangoon Earthworms." *Jour. Bombay Nat. Hist. Soc.* **31**: pp. 180–185.
—— 1930.   "The Earthworms of Burma. I." *Rec. Indian Mus.* **32**: pp. 257–356.
—— 1933.   "The Earthworms of Burma. IV." *Rec. Indian Mus.* **35**: pp. 413–606.
—— 1961.   "The Ecology of Some Earthworms with Special Reference to Seasonal Activity." *American Midland Nat.* **66**: pp. 61–86.

KARMANOV, I. I.   1961.   "Soils of the Rice Fields of Lower Burma and of Certain other Regions of the Burmese Union." *Soviet Soil Sci.* 1960, **8**: pp. 828–833.

STAMP, L. D.   1925.   *The Vegetation of Burma from an Ecological Standpoint* (Calcutta).
—— 1930.   "Burma, an Undeveloped Monsoon Country." *Geogr. Rev.* **20**: pp. 86–109.
—— 1938.   *Asia, a Regional and Economic Geography* (New York).

TERRA, HELLMUT DE.   1944.   "Component Geographic Factors of the Natural Regions of Burma." *Ann. Assoc. American Geogr.* **34**: pp. 67–96.

MEGADRILE REPRODUCTION

Basic in classical thinking was a belief that the only evolutionary modifications of an ancestral holandric-hologynous state, during all of the eons since megadriles arose perhaps "early in the Tertiary," were eliminations of one or the other pair of ovaries along with, or without, deletion of one or another of the two ancestral pairs of testes. Implicit, though perhaps never so stated before, were beliefs that the sex of each pair in the octogonadal battery was immutable and that reproduction was obligatorily biparental. So deeply entrenched were those ideas that evidence to the contrary was ignored, explained as the result of massive parasitic infestations, or attributed to evolution of dioecism.

Absence of mature sperm, throughout a short monsoon breeding season in each of several consecutive years, in Burmese forms of the megascolecid genus *Pheretima*, early in the thirties was recorded by this author, as well as absence of spermathecae (in which to receive sperm) or of organs by which sperm could be passed to the exterior. Later, Kobayashi (1937) obtained in the laboratory from anarsenosomphic individuals of *Pheretima hilgendorfi* (Michaelsen, 1892) uniparental reproduction. Meanwhile, that same kind of reproduction, also secured in the laboratory, was recorded for the first time (Gavrilov, 1935) and for two lumbricid species *Eiseniella tetraedra*, *Eisenia foetida*. Later additions to the list (Gavrilov, 1939; Evans & Guild, 1948), all of them lumbricid, were 2 species each of *Dendrobaena* and *Octolasion*, 1 of *Eisenia*, and 1 then in *Bimastos*. Subsequent confirmation is lacking only for *E. foetida*.

Then, in contributions appearing through 1951–1957, Muldal and Omodeo reported on the chromosome numbers of several markedly anthropochorous lumbricids, some of them being species already known to breed successfully when individuals are raised to maturity in isolation.

Polyploidy was recorded for the first time from oligochaetes and only for taxa with uniparental reproduction, through some even-numbered ploids were believed to be amphimictic. Ability of chromosome studies "to remove systematic confusion" in earthworms was not substantiated by the contribution in which the claim was made. Thus, for instance, synonymization of *Bimastos tenuis* with *Dendrobaena rubida* was not enabled, nor was recognition of the closer relationship of *A. chlorotica* to Levinsen's *eiseni* than to the unrelated species of *Allolobophora* with which it was thought by its chromosomes to be affiliated. Moreover, some generalizations extrapolated from the data *re* chromosomes of "not less than ten specimens" of a taxon certainly are suspect. The statement that "within their genera the polyploids are all the most successful and widespread species," whatever it may have been intended to mean, is supported by very little evidence as: (1) "the various polyploid mutations in one and the same species do not show any significant morphological difference from each other" or indeed even from the diploids, i.e., they cannot be distinguished by characters recognizable on examination of externalia or of the anatomy as displayed in dissection. (2) Although the same lumbricid species may be represented in foreign areas by polyploids, their chromosomes have not been studied in Africa, the Americas, Australia, New Zealand, or even most of Europe.

Parthenogenesis was said to be important because it makes retention of polyploidy possible and also favors the spread of polyploid forms into new areas, since even a single individual may establish a population there. Advantages of parthenogenesis probably were overrated as a single individual of a hermaphroditic species with obligatory biparental reproduction can breed, in favorable conditions, for several months, perhaps five or six, after copulation. Polyploidy is not essential for parthenogenesis as some male sterile diploids are known.

Although the promised cytology of parthenogenesis has not been forthcoming, knowledge of the chromosome studies fortunately did eventuate in a partial relaxation (*cf.* Gates, 1954, 1956) of a taboo against publication, in absence of cytological proofs, on that method of reproduction or on the male sterility of megadriles. Parthenogenesis often can be recognized in a variety of ways, some of which were cited by Muldal and Omodeo in support of conclusions supposedly derived from their chromosome studies. Male sterility, one of the more important proofs, may be variously evidenced: (1) By retention through adult life of testes in a juvenile state. (2) By the juvenile state in adults, of seminal vesicles and absence therein of sperm. (3) By absence at maturity of the iridescence on male funnels that demonstrates aggregation of mature sperm. (4) By absence of similar iridescence in male ducts and, even more important, in all of the spermathecae. (5) By lack of externally adhesive spermatophores or, if present, absence therein of spermatozoa. None of those conditions alone, variously combined, or all of them together simultaneously, even when clitellar tumescence is maximal and ovarian egg strings are long, guarantees, for any solitary individual, more than male sterility which can result from interference with normal embryonic or subsequent growth. The larger the number of specimens with similar conditions, the greater becomes the probability that the sterility is genetically controlled or inheritable. When all specimens from a locality, throughout a short monsoon (or other) breeding season, year after year, show the same condition, further proof of parthenogenesis is unnecessary.

Testes, seminal vesicles and reservoirs, male ducts, copulatory chambers, ejaculatory bulbs, intromittent organs of various sorts and penes, prostates, GM and various other kinds of glands, spermathecae, copulatory, genital and penial setae, no longer are essential when reproduction is parthenogenetic. Each set of those organs, various combinations as well as the totality of all such sets, this author has found, may be lacking in male sterile morphs. Only the clitellum, ovaries, oviducts, and perhaps in some cases ovisacs also, are essential, in megadriles, for perpetuation of the taxon. Absence of any set of deletable organs may be indicative of parthenogenesis, except perhaps when a substitute for the lost organs is available, as for instance the spermatophores of the athecal but supposedly amphimictic species *Criodrilus lacuum*. Furthermore, the author's studies also have shown that male sterility may be associated sometimes, not so much with organ absence as with organ modification. Accordingly, anatomical degradation as well as organ deletions may be indicative of parthenogenesis and here again probability that the condition is not an individual aberration increases with the number of specimens available. Deletion and degradation may well prove to be especially important when maturation of sperm might lead to expectation of biparental reproduction.

Sparse maturation is indeed sometimes associated with parthenogenesis which then may be optional though such a possibility remains to be proved. Profuse maturation is known to be associated, in one lumbricid species with facultative parthenogenesis, and a similar option is anticipated for one species at least of *Pheretima*. One lumbricid, supposedly amphimictic because development had been seen to follow penetration of the ova by sperm, subsequently was found to be parthenogenetic. The sperm merely initiated development (perhaps prematurely?), without contributing to heredity, a condition called pseudogamy.

Absence of spermathecal sperm sometimes may be of little value as an indicator of anything except lack of opportunity for copulating with another individual but also could result from copulation with a male sterile partner. Absence of iridescence on male funnels, especially of aclitellate immatures, may mean only that spermatogenesis had not yet been completed even though spermathecal sperm prove copulation already had taken place. Sperm in spermathecae of male sterile individuals also require an answer to the questions as to whether amphimixis is optional in such circumstances. Copulating individuals caught in the act need not be assumed to be amphimictic. Pseudogamy may be involved or even male sterility as post mortems have demonstrated occasionally for both partners of a copulating pair. Absence of, or inability to find the testes of, fully mature individuals often may mean only that male gonads regress after spermatogenesis (which is completed in the vesiculae seminalis) is well under way. A rare individual variation, aberration, or some result of cephalic regeneration, also can be misinterpreted because of mimicking, as it were, results of parthenogenetically permitted deletions or modifications.

The list of deletable genital organs at the beginning of a previous paragraph includes those structures by which, in the classical system, species and genera most often were defined, identified, and distinguished from each other. Instances of closely related species, differing from one another mainly in their reproductive anatomy, increasingly are being found. Genital organs are the very ones that are lacking or modified in male sterile taxa. Parthenogenesis was said above to be associated with such organ deletions and modifications, but sufficient evidence now is available to support an additional generalization, to wit, those very changes are permitted or enabled by such a method of reproduction. Parthenogenetically allowed changes may exactly parallel past evolutionary modifications made, in other lineages, without loss of amphimixis. Thus, for example, three closely related but distinct (i.e., not interbreeding) species, a-c, may differ from each other mainly as to number of spermathecae, the apertures of which in the three taxa could be at 5/6–6/7, 5/6–7/8, 5/6–8/9, respectively. Parthenogenesis in species c, enabling deletion of the last pair or the last two pairs of spermathecae would provide individuals c_b and c_a mimicking worms of species b and a, respectively. All individuals with the c_b spermathecal battery can be referred to collectively as a morph which, if necessary for sake of simplifying future discussions, can be named in various ways, for instance as a sexthecal c or as the c_b morph. In the same way, individuals lacking four spermathecae could be characterized collectively as a quadrithecal c or the c_a morph. Italicized, Latin names, however, should not be provided. If now, species c differs from species b and a only by the number of spermathecae, the c_b and c_a morphs will be distinguishable only by their male sterility or perhaps also by retention of seminal vesicles in a juvenile state at maturity. If the parthenogenesis is associated with profuse maturation of sperm or pseudogamy, distinguishing a mimicking morph from the mimicked species might be possible only by a cytological study which usually cannot be made satisfactorily on ordinary museum material.

Parthenogenesis theoretically can be of considerable importance in megadrile systematics but of how great actual significance is as yet unknown as study of the problem only recently was begun. The situation is considerably complicated by the fact that the changes allowed or enabled by parthenogenesis sometimes amount to what might be called mutational explosions. These result in the production of what is termed genital or parthenogenetic polymorphism. Several instances are considered hereinafter at greater length but for the present attention is directed to the fact that although dozens, perhaps hundreds, of morphs are theoretically possible, many can be classified (Gates, 1956) into several major sorts that have arisen time and again in widely divergent taxa. Some common morphs are athecal, and in each species they are, in the author's terminology, designated as "A." Another common sort lacks prostates and is conveniently characterized as aprostatic. If the remainder of the male terminalia also is lacking, such an anarsenosomphic morph is designated "R." Morphs without spermathecae and male terminalia, known already in several species, are referred to as "AR." Morphs without testes, and of course also without seminal vesicles and testis sacs, are designated "Z." One end of morph evolution seemingly is provided by the AZR morphs that now have been found in species of two unrelated megadrile families. Such morphs are at the same time most advanced and most degraded or degenerate.

Just how many "species" were erected on individuals of more or less degraded morphs has not been determined, but the number certainly is greater than several, and this writer is by no means the only one who has sinned in that respect. Synonymizing some of those specific names now seems likely to be forever impossible, but intensive collecting at appropriate sites has enabled illuminating solutions to several of the "systematist's nightmares." Clues were provided in some instances by individuals in which deletion of an organ set had not yet been completed. Thus, for instance, an individual with a single spermatheca could, by the characteristics of that organ, show relationship of athecal morphs to another with a complete and normal battery. Similarly, one normal spermatheca, of a set in which all other organs had lost usual conformations, enabled recognition of rela-

tionship with a morph having a complete set of normal structure. Retention of a single characteristic GM gland and its associated marking also has allowed recognition of affiliations. Individuals providing such clues sometimes are common and then are referred to collectively as intermediate morphs because they do show stages intermediate between normal and degraded morphs or between the less degraded and those that are more advanced. An "I," when associated with an appropriate subscript (Gates, 1956, or cf. the glossary below), provides a convenient way of designating between which major morphs certain common morphs actually are intermediate.

Intermediate morphs, in several instances (for examples see *P. anomala* and *houlleti* below) enabled a solution to a dilemma presented oligochaetologists by the New Systematics. Definition of a species as an interbreeding population, even with one or more of the proposed qualifications, would leave (Gates, 1960: p. 281) considerable proportions of the earthworms in various regions of the world in an "ataxic limbo." Fortunately, certain parthenogenetic morphs could be linked through intermediates to a population that can be regarded with some degree of confidence as amphimictic. In such a case, the species is understood to include not only the interbreeding population but also all recently evolved uniparental strains, clones, or morphs that clearly are affiliated with it. If an interbreeding population is as yet unknown, all parthenogenetic strains, clones, or morphs that can be similarly linked together by intermediates are considered to be of one species. The hope of course is that further collecting will reveal an interbreeding population from which the morphs have been isolated by their parthenogenesis. In regions where the Quaternary glaciation probably exterminated large sections of various genera, the search may be futile. However, even in such cases accumulating data may enable affiliation of morphs having a common origin and, hopefully, even a fairly good characterization of the extinct population.

Evolutionary changes allowed or enabled by parthenogenesis may be characterized as negative and positive. Among the former, by far the more common, are the following: Deletion of some part of an organ, as the diverticulum of a spermatheca or the main axis ental to the diverticular junction. Deletion of an entire organ on one side or another of a segment. Deletion of both organs of a segment as well as various other aysmmetrical or unilateral deletions, in a battery of spermathecae or genital markings. Deletion of an entire organ set. Deformation of structure, e.g., abnormal differential growth resulting in hypertrophied main spermathecal axes and dwarfed diverticula, or stunted main spermathecal axes along with normal or enlarged diverticula. Positive changes sometimes can be characterized as reversionary or predictive. Addition of pairs of organs to existing sets, may belong to one or the other class according to past history. Thus, presence of gonads in ix and xii, as well as in other segments (v-viii in *P. anomala*), represents reversion to various more or less ancient ancestral states. Presence of a third pair of spermathecae at the posterior end of the series in morphs of *E. rosea* is called predictive as that addition has not yet been made in the ancestral biparental population though it can be anticipated in the future. However, if the additional pair had been in front of the normal series, reversion would have been involved. Positive changes that may be neither predictive nor reversionary include modification of testes into hermaphroditic gonads or even completely into ovaries. Histological changes, in certain circumstances, may be difficult to classify. Thus, in one pheretima, spermathecal ducts are slender and with little muscularity or are thick and massively muscularized. Which of those conditions parthenogenesis may have been responsible for is unknown.

Somatic anatomy has remained unchanged, in three species of two families, during the entire period (however long or short it may have been) that was required for deletion of all genital organs (*cf.* list in paragraph above) except ovaries, (ovisacs in 1 species?), oviducts and clitellum. Somatic changes also have not been recognized in any of the numerous species of five families in which evolution of parthenogenetic polymorphism is less advanced. In so far then as data thus obtained can provide reliable proofs, somatic anatomy clearly must be recognized as more resistant to evolutionary modification than the genitalia. Quite possibly then genital evolution proceeds much more rapidly than the somatic. In either case, the facts are of profound significance for the classification of megadriles.

Much remains to be learned about incidence of parthenogenesis, primarily because the evidence for it has not been sought and also because few endemics have been studied carefully since megadrile male sterility became scientifically respectable. Although further proof of one sort or another may be desired by those with various special interests, evidence indicative of parthenogenetic reproduction has been found in species of 4 acanthodrilid genera, 3 of the Glossoscolecidae, 6 of the Lumbricidae, 3 of the Megascolecidae, and 4 of the Ocnerodrilidae. Very probably other instances await investigation. Geographically, parthenogenesis also is widespread as it characterizes taxa endemic in North and South America, Europe, Africa, southeast Asia, China, Japan, and Australia.

Parthenogenetic individuals seemingly can support, according to considerable evidence accumulated in Burma but which no longer is available, a much greater parasitic fauna (especially of gregarine Sporozoa), without detriment to themselves or to their ability to reproduce at end of the annual breeding period, than amphimictic individuals of the same species.

More data *re* megadrile parthenogenesis, proven, probable or possible, may have been published in the last few years than appeared *re* biparental reproduction in more than a century. Copulation has been observed in a few species, perhaps more than a dozen, but has been studied carefully only in three or four. The literature seemingly indicates that clitellate individuals alone copulate, but spermathecal sperm occasionally are found (obviously as a result of copulation with another worm) in individuals with clitellar tumescence lacking or only just beginning to be macroscopically recognizable.

Several earthworm species usually live together in the same habitat, and the number of taxa often is greater in sites with extra amounts of organic matter. Numerous attempts have been made to secure hybrid crosses merely by pairing individuals of different species. All reliable records are of copulation only between worms of the same species. Many times during the short breeding season, in Burma and India, a bucket full of nothing but mixed earthworms was brought into the laboratory just as they were dug up. In spite of the fact that other species of the same genus were present, in breeding condition, and in intimate contact, intraspecific copulation alone was observed. Presumably then individuals, in heat as it were, have some way of distinguishing, among a mixture of 3 to 12 or more taxa, potential mates in the same condition and of their own species.

REFERENCES

EVANS, A. C., and W. J. M. GUILD. 1948. "Some Notes on Reproduction in British Earthworms." *Ann. Mag. Nat. Hist.*, Ser. 11, 14: pp. 654–659.

GATES, G. E. 1954. "On the Evolution of an Oriental Earthworm Species, *Pheretima anomala* Michaelsen, 1907." *Breviora, Mus. Comp. Zool. Harvard College*, No. 37: pp. 1–8.

—— 1956. "Reproductive Organ Polymorphism in Earthworms of the Oriental Megascolecine Genus *Pheretima* Kinberg 1867." *Evolution* 10: pp. 213–227.

—— 1960. "On Burmese Earthworms of the Family Megascolecidae." *Bull. Mus. Comp. Zool. Harvard College* 123: pp. 203–282.

GAVRILOV, K. 1935. "Contribution a l'étude de l'autofecondation chez les Oligochetes." *Acta Zool. Stockholm* 16: pp. 111–115.

—— 1939. "Sur la reproduction de *Eiseniella tetraedra* (Sav.) forma *typica.*" *Ibid.* 20: pp. 439–464.

—— 1948. "Sobre la reproduccion uni y biparental de los Oligoquetos." *Acta Zool. Lilloana, Tucuman* 5: pp. 221–311.

KOBAYASHI, S. 1937. "On the Breeding Habit of the Earthworms without Male Pores. I. Isolating Experiments in *Pheretima hilgendorfi* (Michaelsen)." *Sci. Rept. Tohoku Univ.*, Ser. 4, 11: pp. 473–485.

MULDAL, S. 1952. "The Chromosomes of the Earthworms. I. The Evolution of Polyploidy." *Heredity* 6: pp. 55–76.

OMODEO, P. 1952. "Cariologia di Lumbricidae." *Caryologia* 4: pp. 173–275.

—— 1955. "Cariologia dei Lumbricidae. II Contributo." *Caryologia* 8: pp. 137–178.

—— 1957. "Lumbricidae and Lumbriculidae of Greenland." *Meddel. Om Gronland* 24, 96: pp. 1–27.

CEPHALIZATION

Segmentation in oligochaetes sometimes has been said to be homonomous, meaning that all metameres are similar which, as high school and college biology students know, does not connote identical likeness even if the peristomium and the pygomere are excluded as non-segmental. Uniformity perhaps is lacking most of all in the digestive system where organs such as the following have been differentiated; crops, gizzards, calciferous and supra-intestinal glands, enterosegmental organs, typhlosoles and intestinal caeca, as well as the almost universally present pharynx with which special glands sometimes are associated. Even in the family supposedly with the simplest alimentary tract, behind the buccal cavity there is a region thought to combine functions of a pharynx and a gizzard. Genital organs usually are confined to a short anterior region along with the brain and a subpharyngeal ganglion, one or both of which may contain, in certain areas, neurosecretory cells. There is, accordingly, a concentration of special organs in an anterior region which, because of the presence of the brain, appropriately can be called "head." Some zoologists believed the anterior portion of the body that could be exactly regenerated after experimental deletions should be called "head." Five or fewer were the numbers of segments usually obtained from those earlier operations on lumbricids and so the head was said to comprise the first five (or perhaps six) metameres. However, more than six subsequently were obtained, even from the most used species, *Eisenia foetida*, and two to three times the supposed head number easily are replaced by species of *Criodrilus* and *Perionyx*, the only nonlumbricid genera in which regeneration has been studied. Inasmuch as "head" cannot be defined satisfactorily from results of regeneration nor from embryology, the author believes it to be convenient to regard the entire pre-intestinal region as cephalic. In favorable circumstances a head can regenerate—but the product is only a heteromorphic copy, and the monstrosity is doomed to die of constipation. However, if even a very few intestinal segments, perhaps only one or two, are associated with an intact pre-intestinal region, production of a homomorphic tail may be possible. A tail of appropriate length, on the contrary, can develop even at levels well back toward the anal region,[1] a head containing all special organs of the anterior region except prostates, which in normal, unregenerate animals usually are located in the intestinal region.[2]

---

[1] Although prostates are not formed in epimorphic regenerates, they sometimes are differentiated morphallactically by intestinal segments of the substrate.

[2] During one of the two Guggenheim fellowships the manuscript of a monograph on megadrile regeneration was written. The present contribution was submitted first because little interest was shown in the other.

Although haplotaxid heads provide one of the two instances of incompatability with the author's head definition, their excretory and vascular systems may well show homonomous segmentation better than in any other megadriles. Presence in each metamere of a ganglionic swelling of the nerve cord and three pairs of nerves needs no further consideration as such organization now appears to be common to all megadriles. Each segment is bounded internally as well as externally and contains a pair of blood vessels connecting the dorsal and ventral vascular trunks as well as two nephridia that open anteriorly into the coelom through preseptal funnels and externally by pores slightly behind the posterior intersegmental furrow of the segment containing the excretory organs.

Modifications of those conditions primarily have been in the minds of those speaking of cephalization. Little is known about that subject. Septa 1/2–3/4 or 4/5 very frequently are aborted during embryonic development. In *Octochaetoides* 1/2–3/4 seemingly are aborted but 4/5 is thickly muscularized and during evolution of various species the number of aborted septa, beginning with 5/6, has increased until even 9/10 may be represented only by a ventral rudiment. Interpretation of some of the cephalizations in the septal system is not easy, especially when the partitions are very delicate and parietal insertions are not what they at first seem to be. Morphologists as well as systematists have placed organs in the wrong segments because some one or more of the septa had been misidentified.

Cephalization in the vascular system most often involves abortion of parts or all of segmental commissures connecting dorsal and ventral trunks, especially in ii–iii or iv, less frequently in v–vi. Additionally the first pair of vascular arches, and then also some portion of the dorsal trunk in the first few to eight segments, may be aborted. Even in the ventral trunk of at least one species, a gap has appeared. Segmental commissures in some of metameres v–xiii usually become muscularized and moniliform, in which case they are called hearts. Some of the hearts may acquire connections with a third longitudinal trunk. Connections of hearts with the longitudinal trunks now appear to be of considerable systematic importance but unfortunately were ignored usually in the past and, when mentioned, hearts frequently were so carelessly characterized that the "information" is useless systematically.

Cephalization in the excretory system may involve abortion of the nephridia in the anterior segments of a short to a long portion, if not all, of the head. Perhaps more frequent is the acquisition by nephridia, in meroic genera, of openings into the pharynx. These enteroic organs were studied in several species by Bahl, but much more information is needed about them in other taxa. Even in holoic genera, nephridia of the head may, at least in some anterior portion, differ significantly from those behind (*cf.* for example, Bahl's account of those organs in *P. corethrurus*).

Also sometimes involved in cephalization is a gradual but quite definite anteroposterior increase in segment size within the head behind which sizes may be very much smaller. Setal specializations are not considered as cephalizations. With the single exception of copulatory setae, other kinds of modified setae also are found in the intestinal region or there only. Abortion of setal follicles may be confined to the head or a portion thereof (small or large) but may be continued into the intestinal region or even throughout the body. Thus, setae (at least of adults) are lacking; in ii–iii of *Desmogaster albalabia* and *planata*, in ii–v or vi of *D. doriae*, in ii–viii, ix or x of *D. ferina* and *Hastirogaster browni*, in the first 9–23 segments of *H. livida*, throughout the entire body of *Desmogaster sinensis* even in juveniles. Absence of setae may be associated with inconspicuous or invisible nephropores and secondary or tertiary annulation as well as absence of dorsal pores. Then, as in various large forms, determination of segmental boundaries externally may be difficult. Abortion or displacement of anterior septa can considerably complicate the problem.

Perhaps the most interesting kind of cephalization is what Beddard so dramatically characterized as a worm swallowing its own head. The process is now under way in *P. corethrurus* and already has been completed in other glossoscolecids[3] as well as in two megasolecid genera[4].

"METHODS OF EXAMINATION: SYSTEMATIC DESCRIPTION"

The heading above is that of the first chapter of Stephenson's Indian "Oligochaeta." His advice to beginners need not be repeated here as his monograph (1923) still is available at a reasonable price.[1] Stephenson's views about dissection are corroborated by this author's detection of mistakes that previous specialists almost certainly would not have made if they had relied more on pinned-open specimens. Such a dissection, for many purposes and especially for the novice or when specimens are scarce, seems preferable to any halving of the anterior end by a safety razor. Dissections have been avoided by some, and doubtless there always will be those who want to make an identification from a hasty glance at the exterior. All such should remember that generic identification of megadriles, almost without exception, is decided by internal anatomy.

Even some of the simplest characters long-used in megadrile systematics need more than the mere

listing they usually have had in the past. Measurements of size, for instance, often are of little value because they were made on bloated, macerated or relaxed as well as on strongly contracted, markedly homoeotic or even amputated individuals. Length of relaxed earthworms can be two, three or even more times greater than that of strongly contracted specimens. The latter condition is preferred by the author because uniform contraction often is more easily secured than uniform relaxation of the entire soma. Determination of certain characters, for example clitellar extent, becomes increasingly difficult the greater the relaxation until resort to microtome sections may be necessary. Gigantism and dwarfism, when labeled as such, are of interest but what should be known are the limits of individual variation in normal members of a population. Number of segments likewise should be determined from normal individuals which is not always possible as sometimes every specimen, even of a large sample, lacks a portion of the tail. Screening out of worms with tail or head regenerates also is recommended at least until more is known about mery at the levels and directions involved. Here again limits of normal variation would seem to be of greater interest than averages, means, medians, or modes of unscreened samples. An average or a mean at present is of little help in identification of a difficult specimen. Statements *re* segment number, as well as size, in the precis below, do not always include recorded minima and maxima, especially if there is reason for believing that aberration, regeneration or maceration was involved. A statement such as "Segments, 97–120" means that number was counted on two specimens or that the figures indicate minimum and maximum of several records. Occasionally sufficient observations enable characterization of the range shown by 51 per cent to a majority which then is preceded by "usually."

Recounting segments of types sometimes provides a number different from that originally recorded. Differences may be unimportant. The pygomere may not have been believed to be a segment, or may have been withdrawn into the hind end of the body and so missed, or slight rectal eversion may have been taken for a last metamere. Abnormality in the metamerism, not uncommon in the intestinal region, may be unrecognizable on the side of the body from which the count was made and if recognized can be variously interpreted.

Color characterizations of the past often are of little or no importance today. An unpigmented lumbricid species, *Octolasium cyaneum*, by its name provides one bit of supporting evidence. Another lumbricid example is furnished by *Eisenia rosea* which usually appears to be unpigmented though minute yellowish or brownish flecks often are recognizable under the binocular in older individuals. Colors sometimes mentioned, or occasionally shown in plates,

---

[3] *Thamnodriloides* Gates, 1968, and *Estherella* Gates, 1970, for example.

[4] *Tonoscolex* and *Nelloscolex* (see below).

[1] Or Ljungstrom, 1970, *Pedobiologia* 10: pp. 265–285 may be consulted.

now are attributable to cuticular iridescence or refraction of light by the epidermis (as in *O. cyaneum*), to blood (as in *E. rosea*), even to ingesta within the intestine or (especially at the posterior end of the body) to accumulations of coelomic corpuscles along with other debris. A green color, apparently characteristic of a few species, cannot be traced to discrete particles. The yellow color, golden yellow, straw, lemon, yellowish brown, or brownish recorded for preserved specimens often is now attributable to different degrees of alcoholic browning, the final stage being a dark brown without optical differentiation throughout the internal tissues. A red color of the clitellum in many drawidas is "developed" after preservation by formalin and perhaps also by other chemical preservatives. The fine granules responsible for that red are in the outermost portions of the epidermal cells. Similarly located granules may be responsible for a striking orange or fiery red coloration of the clitellum in live lumbricids with maximal clitellar tumescence. After preservation, and especially after clitellar regression, the pigment flecks appear to be yellowish or brownish. Rather generally, color is limited to the dorsum, so frequently indeed that the fact often is not definitely mentioned. Extension into the ventrum, as well as patterning, of course is to be indicated.

Pigment, in material studied by the author, usually has been in or associated with circular muscle layer of the body wall but sometimes does extend into the longitudinal layer at the anterior end and dorsally. Presence of red, yellowish, or brown granules in the body wall does not necessarily enable a similar characterization of color as seen externally. Thus, when deposition is dense, the color may be slate or even almost black. Chemical differences in the pigments, when known, are expected to be of importance in megadrile systematics.

Small differences in length of prostomial tongue, indicated in classical writings by fractions such as $\frac{1}{2}$, $\frac{3}{4}$, are unimportant but whether the tongue approximates or reaches the first intersegmental furrow sometimes (as in the Lumbricidae) is of interest. Closure of the posterior end of the tongue if a prostomium is epilobous may be recorded but the characters "open" and "closed" are unlikely to be of any value in difficult cases because of occasional individual variation. Of much more interest are differences in the prostomium itself, concerning rudimentation, elimination, separation from the first segment (universal in the Moniligastridae), replacement by a protrusible proboscis. As an important external organ, perhaps a worm's equivalent of an elephant's trunk, mistaken identity seems unlikely. Nevertheless, when rudimentary or absent, some fortuitous lobe of the buccal cavity occasionally has been thought to be prostomial. Eversion of the buccal cavity may hamper recognition of the prostomium, but the protuberance usually can

be distinguished, in formalin preservation, by its flaccidity as well as color differences though the latter may be slight in unpigmented species. The anterior end of a proboscis, just visible in the buccal cavity, also has been mistaken for a prostomium.

Location of the first dorsal pore often is not determinable because of inability to distinguish between functional apertures and non-functioning but more or less pore-like markings. Bending the anterior end appropriately, sometimes enables one to see fluid ooze or even squirt out but without certainty that a weak spot in the body wall was not ruptured by the pressure. Location of the first pore also may be subject to more or less individual variation the specific limits of which almost never have been learned.

Nephropores, not mentioned by Stephenson, usually are unrecognizable macroscopically if the excretory system is meroic, and if nephridia are holoic may be distinguishable only on relaxed individuals or after preservation by some special method. Easily seen pores are called obvious but the structure that is lacking, or perhaps only much less developed when pores are inconspicuous is as yet uncertain. Location and number of nephropore ranks now are systematically important (*cf.* lumbricid species below), and, when setae are not in regular ranks, sometimes provide the only meridians of longitude. Positions of holonephric apertures cannot be determined, as sometimes was thought, from the level at which nephridial ducts enter the body wall, as in many species, emergence therefrom many be far laterally or even dorsally.

Characters of the setae, or those indirectly provided by them, long have been regarded as of major systematic importance. Genera and families, in the past, have been defined by one classical pair, lumbricine and perichaetine (now spelled without the "e") indicating presence per segment of eight or more. Shedding, at least into the coelom where they are incorporated into brown bodies, probably also to the exterior through follicle apertures, is assumed to be followed usually by a reserve shaft becoming functional. Little more than that is known about a phenomenon which, in perichaetin species, may be continuous or spasmodic, involving only a few widely separated follicles or those of small areas and then frequent enough to hamper attempts to determine limits of individual variation in number (when perichaetin) at definite levels along the anteroposterior axis. Rarely, shedding may be complete or very nearly so, as was true of a considerable number of large juveniles of three species of *Tonoscolex* obtained during the rains. The same species, without individual exceptions, at the same locality, during a later breeding period were normally setigerous. Whether a seasonal and normal or a pathological and inconstant phenomenon was involved remains to be learned. Each individual of a Chinese species of *Desmogaster* that has been seen was completely asetal.

Slight markings in the longitudinal musculature as seen from the coelomic face, located about where setal gaps could be expected, suggest that setae once were present. Whether shedding takes place before or after hatching also is unknown. Four or six longitudinal ranks of setae are completely lacking in some haplotaxids, but whether loss is ontogenetic is not indicated by the literature. Ontogenetic dehiscence, without replacement, in a varying number of the anteriormost segments, is one of the aspects of cephalization in some species of several families. Also noteworthy is occasional retention of juvenile size in a particular segment when size in adjacent segments is markedly greater.

The setal shape rather generally is sigmoid, i.e., with a slight double curvature. Somewhat ectal to its middle the shaft has a slight thickening called the modulus. The ental end usually is more or less bluntly rounded and an ectalmost portion usually is slightly tapered. Divergence from that shape, except in local regions next to be considered, is rare, though a sickle-shape has been recorded from certain haplotaxids. In a varying number of preclitellar segments or towards the hind end setae may be considerably enlarged but without marked change in shape, often with some sculpturing or ornamentation—near the ectal end so universally that mention of its location on the shaft often is omitted. Change of shape usually is restricted to setae believed to have some function in reproduction (perhaps in copulation) and that are associated more or less closely with certain genital organs. Setae of follicles opening through genital tumescences are called genital. Those associated with sperm receptacles are called spermathecal or copulatory and others associated with male pores, copulatory chambers or prostates, are called penial. The systematic importance claimed for differences of shape, sculpturing, and ornamentation in those kinds of setae in the classical system, may have been too great. Penial setae may show considerable intraspecific variation or may be similar if not the same in several species. Copulatory setae with a rather claw-shaped tip and ornamented with longitudinal rows of characteristic gouges, have been recorded from ten species of *Octochaetona*, two species of *Lennogaster*, one each of *Bahlia* and *Calebiella*, all in the Indian section of the Octochaetidae, and from genera of the unrelated South American Glossoscolecidae. Longitudinal grooving on genital setae, according to several specialists, shows that *Criodrilus* is related to lumbricids and, for some, so closely as to warrant transfer into that family. Genital, as well as somatic anatomy, in both families proves that the setal sculpturing is but another instance of convergence.

Perhaps just as important, if not more so, than any character of the setae themselves, are the reference points that apertures of the setal follicles provide for describing locations of genital pores, grooves, and markings. Such situations now are most concisely stated by reference to meridians coincident with the longitudinal ranks of setae and indicated by italicizing and capitalizing letters that long have designated particular setae. Thus *BC* indicates the space between the ranks of the *b* and the *c* setae. Similarly useful are circumferential meridians provided by apertures of the setal follicles. These levels almost universally are equatorial in each segment but a preequatorial location of the setae has been recorded, in which case pre- and post-setal have a meaning different from that which is usually understood.

Genital apertures so generally are in the ventrum that location there is to be uuderstood unless presence in the dorsum is definitely mentioned. The position of the pores long has been recorded but for those of the spermathecae not always with precision because of failure to distinguish between a circumferential groove and the therein contained fine line that is the real boundary. Much more than their positions, as the author's studies have shown, now must be known about genital openings. Female pores, except in the Eudrilidae, always are of a size characterized herein as minute. Primarily (as well as primitively?) other genital apertures are not markedly larger and at maturity often can be characterized by the same adjective. Spermathecal pores usually are minute but some species do appear to have variously enlarged apertures that remain superficial. Primary, with reference to male pores, refers to apertures (apparently always minute) at ectal ends of the sperm ducts (sometimes called vasa deferentia). Primary male and spermathecal pores may be superficial, concealed within slight clefts or fissures, invaginated into chambers that are confined within the body wall or that protrude more or less markedly into coelomic cavities. Chamber apertures on the surface, never minute, though commonly called male or spermathecal, of course actually are secondary and each or both may be withdrawn into some room the external opening of which must be tertiary. Further complications will be found in the male terminalia as the vasa deferentia may unite in various ways with the prostatic ducts and before or after union with ducts of other glands or with follicles of penial setae. Male chambers may be eversible into characteristic porophores and may contain protrusible penes or other, and often specifically characteristic, protuberances called penial bodies because of the presence thereon of the male pores. Glands also may open into spermathecal ducts or into chambers containing the primary pores. Such glands, just like those associated with external genital markings, may be of several quite different kinds. Thus, genital systems often can provide systematically useful characters that were unknown to or were ignored by classical authorities.

Everted parietal and coelomic chambers may be difficult to distinguish, especially with alcoholic preservation, from more or less marked protuberances of uninvaginate male pore areas. The problem sometimes is complicated when a particular method of preservation always results in protrusion or eversion. Because of the resulting inadequate, if not also inaccurate, characterization of the male terminalia, identifications have been hampered or prevented and unnecessary species names have been provided.

Some methods of preservation markedly decrease cuticle transparency and may have been responsible in the past for some imprecision, in particular with reference to situation of spermathecal pores. Microscopic examination of already loose cuticle sometimes reveals important characters not determinable macroscopically.

Most organs of megadriles, except the prostomium, gizzard, brain, nerve cord (which, like the dorsal blood vessel, is of double origin), and some of the major vascular trunks, are paired and are to be so understood except when the contrary is definitely indicated.

Although calciferous glands were allowed a small role in the classical system, descriptions of those organs, even after publication of Michaelsen's "Lumbricidae" (1918), often were so careless or perfunctory as to be of little or no use. Even without resort to microtome sections, careful observation reveals, as is shown hereinafter as well as on previous occasions, various distinguishing characters. Although necessity for recording segment of intestinal origin was mentioned by Stephenson (1923: p. 7), the esophageal valve that marks the posterior boundary of the esophagus seems rarely to have been looked for and the gizzard, even in the most common lumbricids has been said to be at the end of the esophagus or at the beginning of the intestine, neither of which is correct. Intestinal caeca also were listed by Stephenson, but the frequent lack of any mention of those organs in species descriptions certainly does not mean that a search for the organs had been futile. Other parts of the digestive system now known to have systematic importance are typhlosoles and supra-intestinal glands.

The vascular system was almost completely derogated by classical authorities who did sometimes mention the segment containing the last pair of hearts though not always correctly. Megadrile hearts are of at least three kinds and for nearly three-quarters of a century occasionally have been precisely characterized; as lateral if connecting dorsal and ventral vessels only, as esophageal if joining the supra-esophageal and the ventral trunks, but latero-esophageal if opening above into both the supra-esophageal and the dorsal vessels. However, "latero-esophageal" often was not used even in circumstances justifying a suspicion that stated characterizations of "esophageal" and "lateral" provide none of the needed information.

The dorsal and ventral trunks now seem likely to

be present in all megadriles but, even so, do provide systematically useful characters when the dorsal vessel is doubled, in various ways and in diverse portions of the anteroposterior axis, or when a portion of either trunk is ontogenetically aborted. The subneural trunk not only provides present and absent characters of considerable importance but when present may be within the muscular sheath of the nerve cord or completely free of the cord and adherent to the parietes. Very little is known as yet about lateroneural trunks, but sufficient evidence already has been accumulated to show the necessity of recording presence or absence as well as connections with other trunks of vessels such as the following; extra-esophageals, one or two pairs of anterior lateroparietal trunks, posterior lateroparietal trunks perhaps often present when a subneural is lacking. Long portions or even all of a major trunk may be almost or definitely unrecognizable when empty and so necessitate dissection of a number of specimens, perhaps also varying methods of preservation.

Bahl's studies of the excretory system in certain Indian megadriles provided the first demonstration, though not at once so recognized, of the folly involved in limiting systematic characterizations to the two classical key characters. Diversion of Bahl from his patiently detailed morphological studies and the lack of a successor in his major field are greatly to be regretted. The finer details of excretory anatomy, especially in meroic systems, will not easily or quickly be learned, but determining presence or absence of a bladder and its shape, when conditon of holoic organs is favorable, ought not to be too difficult. Also worthy of record is concealment of nephridial loopings by storage, possibly of reserve metabolites, in enlarged cells of the peritoneal investment. Such accumulations may characterize certain meroic as well as holoic organs.

Genital organs, at least those of the classical system, instead of being "the most important of all for systematic purposes" now are known to warrant no such valuation. However, previously ignored or unknown genital characters increasingly are proving their usefulness as various examples herein show. Probably by far the most important are those having to do with the shape and the egg strings of ovaries. Those organs are in xiii, and the female pores are in xiv, so very generally that repetition for each species often has seemed unnecessary, and presumably, presence in other situations always has been recorded. GM glands, perhaps also various sorts of glands associated with setal follicles, especially when microscopic structures has been elucidated, now seem to promise being of some interest.

Even with regard to organs emphasized in classical speculations, more precision in characterization often was needed as testis sacs were mistaken for seminal vesicles, clefts for primary male pores, GM glands for spermathecal atria, pseudovesicles for ovisacs. Another instance of rather common imprecision is provided by statements indicating that seminal vesicles of xii, in common lumbricids, extend through xiii or xiii–xiv when the organs actually are confined to xii though portions are contained within posterior pockets of 13/14.

One of the major defects in the classical system, as already is shown above, was the basic restriction of evolutionary significance and of systematic usefulness to a very limited number of phylogenetic changes that could be indicated by simple key characters. The system used in this work is better than the classical because it takes into consideration much more of megadrile anatomy and will be improved as our knowledge of that organization and of other attributes of the animals increases.

NOMENCLATURE

Spellings of certain generic names will be unfamiliar to many. "*Bimastos*," "*Bothrioneuron*," and "*Octolasion*," as originally published were changed by Michaelsen in his Tierreich monograph (1900) to *Bimastus, Bothrioneurum*, and *Octolasium*. Dr. Curtis W. Sabrosky, of the International Zoological Commission, states (courtesy of Dr. Fenner A. Chace, Jr.) that there is nothing in the International Code of Zoological Nomenclature requiring an author to use either the Greek *-os* and *-on* endings or the Latin *-us* and *-um* equivalents. Furthermore, there also is nothing in the Code permitting a subsequent author to change a "correct original spelling" of any name. Accordingly, if the Code is to be followed, the three generic names must be spelled, *Bimastos, Bothrioneuron*, and *Octolasion*, as by their authors.

One application of the correct spelling principle, *Pheretima omtrekensis* Cognetti, 1911, especially amuses the Dutch. Its author thought Omtrek on a label was a place name instead of a preposition meaning inside. When the mistake was discovered the author changed the patently ridiculous name to *Pheretima homoeotrocha* Cognetti 1914 which, according to the Code, had to be a synonym of *P. omtrekensis*.

Names of infrasubspecific taxa, such as variety (var.), forma (f.), mutant (mut.), are excluded from zoological nomenclature by article 1 of the Code. Accordingly there no longer is any justification for italicizing them. Portentous names, such as *Dendrobaena* or *Eisenia veneta* (Rosa) var. *hibernica* (Friend, 1892) f. *dendroidea* (Friend, 1909) are suspected by some of having a length approximating the depth of ignorance about the relationships involved. Specimens that the author once believed to be referable to *dendroidea* are now known to be of *E. hortensis*.

The words "typicus" and "typica," that long have been common in oligochaete literature, do not convey any such meaning as usual, normal, or customary.

They refer only to the fact that the variety so characterized was the first of two or more to be described. In *E. tetraedra*, var. "*typica*" is an aggregate of markedly aberrant parthenogenetic morphs. Other parthenogenetic morphs, perhaps with but slight or even no morphological divergence from lumbricid norms, were referred to as var. *hercynia*.

Parthenogenetic morphs have not been provided by the author with Latin names, since megadrile male sterility became scientifically respectable. In some species there are many easily recognizable morphs, in fact too many to be provided with names. A system of designating morphs according to the degree of morphological degradation, originated by this author, is explained hereinafter. Nevertheless convenience sometimes is promoted by use of a former varietal or even specific name without italics in some such manner as, "the hercynian morphs."

TYPES

The type of a species, according to the Code,[1] is a single specimen that, for distinction from all other kinds of types, is called the holotype. When splitting of a long-known species becomes necessary, the type functions as a nomenifer (name bearer) and supposedly determines which taxon retains the old name. An investigator wishing some such assistance with a nomenclatural problem in the Oligochaeta, will immediately encounter a considerable series of problems. The first of which is provided by the long-continued practice of omitting from systematic literature any mention of the fate of types. There is no easy way of learning where a type may be or even if it is still in existence. This author for twenty years has been trying to compile a list of types and the institutions at which they can be consulted. Although the British, Dutch, Genoa, Indian, Paris, and U. S. National museums, as well as several other institutions, have cooperated, the list is nowhere near completion. Before World War II, the largest collection of earthworms, and of types, was that amassed by Michaelsen during some fifty years of publication on oligochaetes. However, the Hamburg Museum has been unable to provide a list of those types that did survive the war, even without any indications as to condition.

Megadrile types that were in Burma, China, Korea, Japan were destroyed during the war, as were those that were in Hungary during the rebellion. Eisen's types were lost, along with much unstudied material, in the San Francisco earthquake of 1906. Other types very probably have disappeared. Many of Beddard's and some of Friend's types were lost. How many other additions will have to be made to the list of

---

[1] The International Code of Zoological Nomenclature, as adopted by the XV International Congress of Zoology, and published by the International Trust for Zoological Nomenclature, London, 1961.

lost types cannot now be estimated but material was lost by "stable institutions" as well as from individual collections.

Extant types, when finally located, often will be found to be of little if any value for solution of nomenclatural or systematic problems. Species are known to have been erected, thirty to eighty or more years ago, at a time when exotic material was much less easily obtainable, on partially rotted material, usually more delicately characterized as macerated. The decay often eventuated from crowding worms into much too small containers with far too little preservative. Most of the material that comes to museums or directly to specialists for identification today still is poorly to badly preserved. Even zoologists sometimes seem to have thought that dropping specimens into preservative, regardless of strength and amount or of dirt and slime, is good enough for any one so stupid as to waste time on "mere identification." Worms in stable institutions also have become hard, brittle, and brown because alcohol was too strong, had evaporated or was drunk. Even with the best of curatorial care, earthworms deteriorate with age. In formalin, specimens gradually soften and presumably will begin to disintegrate eventually. In spirits, worms may gradually brown until nearly all or even all optical differentiation is lost in the tissues. Only the setae, in the last stages of somatic devaluation, now seem likely to be of any interest to a systematist and nothing is known about their resistance to change during a century and more. A recent conference[2] recommended that journals publish descriptions of new species only if and when the types are available in stable institutions. That recommendation undoubtedly was made in the interest of advancing knowledge but it is in stable institutions that megadrile types are slowly deteriorating, often because methods are exactly the same as a hundred years ago. Nowhere, so far as this author could discover, is any research under way on how to prevent earthworm types from losing all of those values that justify occupation of space on museum shelves[3].

Neither Beddard, Michaelsen, nor Stephenson, seems to have indulged in the practice of designating

---

[2] *Science* 131 (1960): pp. 937; and 132 (1960): p. 832.

[3] Experiments set up by the author early in the twenties to provide information about methods of preservation and fixation were terminated in 1943 by the Japanese army of occupation when the worms, mostly in hermetically sealed glass containers, were thrown out with the "rubbish."

Tuberculata pubertatis and genital tumescences were not distinguishable from the clitellum in some recently received alcoholic lumbricids that otherwise appeared to be well preserved.

Formalin, used in one way or another, especially immediately *post mortem*, often does seem to enable as much or more optical differentiation than in the best of alcoholic material.

Strongly contracted worms, especially if the intestine is completely filled by the ingesta, are less likely to be damaged during shipment and do stand better the handling involved in a careful study of external characters.

a holotype. That kind of a type, accordingly, is available only when the species was erected on a single specimen. Michaelsen customarily kept one or more specimens for his own collection. Often, one of them was sectioned.[4] Remaining worms of the species, not always from a single locality, were returned to the owner. As a result material in both the Hamburg and Indian museums is labeled "Type." Theoretically, in selecting a holotype, one should choose a dissected or sectioned specimen, inasmuch as generic and family identifications can be made only from internal anatomy. However, in the very act of making the dissections or the sections, some of the value of the individual as a type is likely to be lost. Hence an undissected syntype may be of greater systematic value but only if conspecificity with the dissected or sectioned type is certain. The problem thus posed cannot always be solved. Michaelsen once mentioned that Beddard sometimes labeled contents of a whole jar from the top specimens only. Some sections of types that were examined by this author in various museums now are almost if not wholly useless. Stains have faded. Gaps are present in the series, occasionally at the most interesting level. Sections sometimes are torn or folded back on themselves.

Another factor that may need to be taken into consideration before deciding whether search for and study of a type will justify the requisite expenditure of money and time. A type often is not typical. It may be markedly variant from a species norm in several ways. It may even be abnormal because of interference with development, defective cephalic regeneration, or parthenogenetic degradation. The lost types of several "species" of *Pheretima* were of such degraded morphs and the descriptions were so inadequate that any one of two or more species may have been involved. The type of *Allolobophora relictus*, probably also lost, was so abnormal as to warrant the characterization of monstrosity and the name could be placed in a synonymy[5] only after considerable study of lumbricid aberrations.

If there is no type material, resort must then be to a search of the type locality. Unfortunately, the source of the specimens for some of the earlier-named species is unknown or was stated as West Indies, New Zealand, and in other ways that provide little basis for mapping an itinerary. In many areas of the tropics, much travel can be futile because a native village, near which a type had been obtained, was moved, perhaps on various occasions, with or without taking the old name along. The niches from which

---

[4] The writer cannot now state whether Michaelsen ever dissected any of his specimens. Microtome sections probably were made routinely. If the German authority relied solely on the sections, that may have been in part responsible for his lack of interest in the vascular and excretory systems the anatomy of which is much better studied in dissections.

[5] Gates, 1956, *Ann. Mag. Nat. Hist.*, Ser. 12, 9: pp. 369–373.

the types were obtained may have been destroyed during recent years by building of dams, draining of swamps, cutting down of jungles, or by various other activities that change the face of nature as civilization advances. Yet, in spite of those difficulties, careful collecting in and near the type locality may provide information of greater importance than can be obtained from inspection of a holotype, especially if that is the only specimen available and which, accordingly, must be handled with caution so as not to bring about further damage.

Even a long series of specimens from a type locality occasionally may be quite unable to provide a satisfactory species characterization. That is so in some cases where the type locality is in a region to which the species was introduced and more especially if the colony is parthenogenetic. Only in areas of species endemicity can one expect to secure the necessary data. Yet, the original homes of various anthropochorous species, in *Pheretima* and several other genera, are unknown.

Although holotypes of the author's Burmese species were thrown out with the rubbish during the Japanese occupation of Rangoon, paratypes (other specimens from the same locality, secured at the same time as the holotype) or specimens collected at type localities, were deposited, in prewar days, in one or more of the British, Indian, and U. S. National museums.[6] The author disclaims responsibility for any characterization as to kind of type that may be mentioned on museum labels. Practice has differed from one institution to another and over a period of years may have varied even in an institution.

SYSTEMATICS

Phylum ANNELIDA

Class CLITELLATA

Hermaphroditic annelids, without parapodia but with metameric segmentation internally as well as externally, with gonads very few and in definite segmental as well as intrasegmental locations, with special ducts for discharge of genital products, at maturity with a definitely located tumescence of the epidermis (clitellum) to secrete a cocoon in which eggs are deposited to be fertilized and develop therein without a free larval stage, with spermathecae for storage of gametes received during exchange of sperm with a copulatory partner.

Remarks: Departures from what now appears to be the norm of structure in the Clitellata, as characterized above, are known. The African leech genus *Marsupiobdella* Goddard & Malan, 1912, as its name suggests, needs no cocoon (or clitellum to secrete one) because of the brood pouch in which development takes place. Leeches with hypodermic impregnation, Rhynchobdellae and Pharyngobdellae, get along without spermathecae. The megadrile *Criodrilus lacuum* Hoffmeister, 1945 also has no spermathecae though reproduction still may be biparental. Parthenogenetic morphs that evolved in various megadrile species have lost not only spermathecae but also various other genital organs and in some microdriles that reproduce asexually, genital organs never have been seen.

Oligochaetes are more closely related to the leeches than to the polychaetes, as was recognized by Avel (1959), Pickford (1948, 1964) and Stephenson (1930). To indicate that relationship, Michaelsen (1919, 1928) combined the two orders, Oligochaeta and Hirudinea, in a class that he named the Clitellata. Acceptance of that grouping requires the Oligochaeta to be a subclass or an order rather than a class, as ranked by some authors, unless the Clitellata is to become a subphylum of the Annelida.

In view of the author's agreement with Stephenson as to the value of Michaelsen's suborders, the taxon Oligochaeta, in continuation of previous practice of monographers Beddard (1895), Michaelsen (1900), and Stephenson (1930), is retained as an order. Michaelsen's Clitellata, is retained, as a class, because its demonstration of closer relationship between oligochaetes and leeches is considered to be of more importance than whatever "convenience" may result from giving the Polychaeta, Oligochaeta and Hirudinea the appearance of being equally related to each other because of the same hierarchical rank.

Order OLIGOCHAETA

Annelids with spacious coelomic cavities containing coelomocytes of various sorts, with a closed vascular system comprising at least a dorsal and a ventral trunk, with setae typically in each segment except the peristomium and the periproct. Ovaries behind testis segments.

Remarks: Here also there are departures from what now seems to be the norm of oligochaete structure as characterized above. The branchiobdellids not only

---

[6] The species involved are Acanthodrilidae, *Plutellus compositus, pandus.* Megascolecidae, *Perionyx viridis. Pheretima aculeata, analecta, austrina, canaliculata, exigua, immerita, insolita, labosa, maculosa, manicata, mendosa, ornata, papilio, pauxillula, planata, rimosa, rufula, rugosa, tenellula, terrigena, velata, youngi. Tonoscolex birmanicus, depressus, ferinus, lunatus, montanus, scutatus, triquetrus.* Moniligastridae, *Desmogaster albalabia. Drawida abscisa, caerulea, constricta, flexa, fucosa, lacertosa, longatria, montana, peguana, rangoonensis, rara, sepulta, spissata, tecta, tumida, vulgaris.* Octochaetidae, *Eutyphoeus annulatus, compositus, bifovis, bullatus, cochlearis, constrictus, excavatus, falcifer, hastatus, longiseta, marmoreus, peguanus, planatus, rarus, sejunctus, strigosus. Octochaetus birmanicus, lunatus.*

Types now in the Indian Musem are as follows: Acanthodrilidae, *Plutellus exilis, himalayanus.* Megascolecidae, *Perionyx miniatus. Nelloscolex strigosus. Tonoscolex montanus.* Moniligastridae, *Drawida exilis, limella.* Ocnerodrilidae, *Malabaria sulcata, Thatonia parva.* Octochaetidae, *Bahlia albida, Calebiella parva, Eudichogaster barailanus, Lennogaster elongatus, Pellogaster isabellae, Ramiella nainiana, Octochaetus albidus, comptus, parvus, pearsoni.*

have lost their setae but have developed a sucker posteriorly as well as certain other leechlike characters. Family status within the Oligochaeta has been questioned by Holt (1963) who prefers ordinal rank within the Clitellata.

Megadriles occasionally may lose a few, more or all of their setae, temporarily or even permanently. However, *Acanthobdella peledina*, which was included in a single volume of Kükenthal and Krumbach's massive *Handbuch der Zoologie* both as an oligochaete and as a leech, retains the ancestral setae anteriorly and in a typical lumbricid arrangement, along with more leechlike anatomy than in the Branchiobdellidae. Ordinal rank for a taxon Acanthobdelliformes, in the Clitellata, also is preferred (1963) by Holt.

Between order and families, in his later systems, Michaelsen interposed two (1921), then three (1928), and finally four (1930) suborders as well as various series. Involved in definitions of some of those taxa are vast extrapolations from very little base. Number of cells in the upper lip of nephridial funnels, for instance, when nephridia of only a very few species have been examined. So much elaboration of the classification, in agreement with Stephenson (1930: p. 719) seems unwarranted by the present state of our knowledge which is as yet inadequate to enable recognition of more than three or four monophyletic families.

A need for a single dichotomy in the classification of the Oligochaeta was evidenced by Avel (1959), Pickford (1948), as well as Stephenson (1930). Accordingly, the author's long-established practice recently (1962) was modified to meet that need with two old terms. Thus earthworms are now characterized as megadrile and other oligochaetes or non-earthworms as microdrile. That procedure is, for the present, regarded merely as a semantic device to facilitate discussion and, as such, needs no definition for either term. However, it so happened that the author's megadriles proved to be the exact equivalent of Michaelsen's suborder Opisthopora as well as of Yamaguchi's (1953) Opisthopora diplotesticulata.

Somatic anatomy of microdriles may have been neglected as much as that of megadriles. However, attention is directed to the distribution of certain characters in hope that those who are better informed may be stimulated to contribute their observations. Megadrile setae, except as specially modified in local areas for reproductive purposes, are sigmoid, with a nodulus, and almost universally[1] with a single tip. Single-tipped, sigmoid setae are not, however, confined to the megadriles but are shared with some lumbriculid and enchytraeid taxa. Three major

vascular trunks, extra-esophageals and the subneural, in so far as could be determined from the available literature, are lacking in the Microdrili. Not all megadriles do have those trunks but evidence is accumulating to support a belief that in absence of one or both of those vessels, lateroparietal trunks, also lacking in microdriles, are present. Megadriles quite generally seem to have three pairs of nerves per segment and information *re* conditions in the Microdrili might be of interest.

After the preceding paragraphs were typed *The Aquatic Oligochaeta of the USSR* was received. Chekanovskaya also sensed a need for a single dichotomy in the Oligochaeta and in that monograph (1962) divided the Oligochaeta, as a class, into two orders. The Naidomorpha comprises six microdile families, Aeolosomatidae, Naididae, Opisthocystidae, Enchytraeidae, Tubificidae, Phreodrilidae. The Lumbricomorpha comprises the Lumbriculidae, Lycodrilidae, Branchiobdellidae and six megadrile families, Haplotaxidae, Alluroididae, Moniligastridae, Megascolecidae, Eudrilidae, Glossoscolecidae, Lumbricidae. Megadrile families, excepting the Haplotaxidae (*q. v.*), presumably are as in Stephenson's final defense (1930) of the classical system. The supposedly new names (Chekanovskaya, 1962: pp. 105, 141) already had been used though with somewhat different content and at subclass level in 1890 by Benham. Naidomorpha subsequently (Beddard, 1895: p. 275) was a family name for a group comprising *Chaetogaster, Amphichaeta, Nais, Bohemilla, Pristina, Uncinais, Chaetobranchus,* and *Dero*. Subclass Lumbricomorpha was divided by Benham into two orders, Microdrili and Megadrili. Families of the latter were divided among two branches, Meganephrica and Plectonephrica.

Other oligochaete dichotomies were as follows. Naidea and Lumbricina (Grube, 1850). Gemmiferes and Agemmes (Udekem, 1855). Oligochetes limicoles, and Oligochetes terricoles (Claparede, 1862). Naidina and Lumbricina in a group Scoloces which also included, as in some previous classifications, the polychaete Capitellidae (Johnston, 1865). Naidea and Lumbricina (including, Lumbricina propria, Enchytraeina, Vaillant, 1868). Because of his doubts about relationships of the Aeolosomatidae with the Oligochaeta and the Polychaeta, Beddard (1895: p. 160) recognized for the family a group Aphaneura distinguished, as the name suggests, from microdriles as well as megadriles by a supposed absence of the nerve cord.

In a classification that Stephenson (1930: p. 716) considered "a triumph of arrangement which brought order into confusion and constituted a remarkable advance in our understanding of the group" Michaelsen (1900), divided the Oligochaeta directly into families. Stephenson's opinion, stated and supported in two much-used monographs, may well have been

---

[1] Exceptions; penial setae of *Pheretima andamanensis*, some of the clitellar setae in certain morphs of *P. houlleti*, some setae of the glossoscolecid *Periscolex fuhrmanni* Michaelsen, 1913, one seta of each couple (in each segment?) of the haplotaxid *Heterochaetella glandularis* Yamaguchi, 1953.

in part responsible for Michaelsen's system having remained, with relatively unimportant modifications, in use much longer than any other classification of the oligochaetes.

MEGADRILE OLIGOCHAETES

Megadriles seem to fall naturally into two groups. One, comprising the Alluroididae, Haplotaxidae, and Moniligastridae, supposedly has seminal vesicles that are simple, unpartitioned, tubular pockets posteriorly directed from the septa, ovisacs of similar structure, large ova with yolk, and unilayered clitellum. However, egg- and/or sperm-sacs may be lacking in some alluroidids; sperm sacs never are present in moniligastrids. A unilayered state of the moniligastrid clitellum has been definitely denied and further observations are required. Moniligastrids may be intermediate between families with microdrile characteristics and families of the next group. This, comprising all other megadriles, supposedly has seminal vesicles that are lobed, trabeculate, and initially at least not posteriorly directed, ovisacs (when present) that are of about the same structure but always much smaller than the vesicles, eggs without yolk, and a multilayered clitellum. Possibly more significant, because of fewer exceptions, than any of the above-mentioned characters is the difference in male pore locations, in front of the female pores (a microdrile character) in the first group, behind the female pores in the second group. A single exception does have to be noted, that of a lumbricid species (*cf.* below) in which parthenogenesis has enabled reversion presumably to a long-lost microdrile condition. Because of known exceptions, the necessity for confirmation in some taxa of certain supposedly universal characters, as well as our ignorance of much anatomy, megadriles herewith are divided directly into families. When convenience requires, reference is made to Group I and to Group II families.

Juliana de Berners, prioress of a nunnery, in one of the very first of printed English books, *Fysshynge with an Angle* (1496), mentioned several kinds of earthworms. Izaak Walton also knew several kinds of angleworms and in his *Compleat Angler* (1653) told which were good bait for fish and which were not. Nevertheless, *Lumbricus* of Linnaeus, in 1785, comprised *L. terrestris* (earthworms) and *L. marinus* (sea worms). By few, except anglers, were earthworms seemingly considered to be worthy of observation and study until into the nineteenth century. Valid species names for earthworms were provided only in 1826 when Savigny astonished French savants by demonstrating that a dozen or so species existed in Paris and immediate vicinity. Savigny defined his taxa by external as well as internal, microscopic as well as macroscopic, somatic as well as genital characters. The standard set by Savigny was not long maintained and already by 1866 newly erected genera as well as

species were being characterized by reference only to external characters. Because of a museum rule forbidding dissection of types, internal anatomy of some species still is unknown and as one result the species cannot, of course, be referred to any genus. In the latter half of the nineteenth century, English and European zoologists became aware of the existence of different earthworms in other parts of the world and, neglecting the inadequately characterized faunas of their own backyards, concentrated on the exotic. Almost inevitably it would seem, the genitalia, primarily internal, because of their unfamiliar diversity, macroscopic size and ease of observation, received the most attention. Already by 1889, Beddard (*Trans. Roy. Soc. Edinburgh* **35**, 1: p. 636) was stating that "Generic and family distinctions are chiefly based upon variations of the reproductive system." Ever since, genitalia have dominated oligochaete systematics, probably even more completely than one is likely to deduce from Stephenson's dictum (1923: p. 7), "The sexual organs are the most important of all for systematic purposes." That idea is part of the esotery on which the classical system really was based. To understand the system a basic esotery must be known. The more important beliefs, they might almost be called axioms, of the esotery can be briefly stated as follows:

1. The genital system is extremely conservative, i.e., resistant to evolutionary modification, e.g., during the millions of years since the Jurassic or Cretaceous the only changes made in the megadrile battery of gonads are elimination of ovaries in xii sometimes along with disappearance of the testes in one or the other of x and xi.

2. Somatic systems are much more liable to evolutionary modification, especially the digestive because (Stephenson, 1930: p. 720) of a "well-known dependence of the conformation of the alimentary tract on food and environment."

3. Morphological changes that mean anything from the evolutionary point of view are few. Their phylogenetic sequence is known. The organs or systems, the changes with evolutionary meaning and the systematically useful characters they provide (in parentheses) are as follows: Gonads, paired testes in x and in xi, paired ovaries in xii and in xiii (holandry, hologyny). Changes, elimination of the ovaries in xii (metagyny) sometimes along with abortion of one or the other pair of testes (proandry, metandry). Male terminalia, sperm ducts opening to the exterior in the middle of three segments, a pair of prostates opening to the exterior through each of the anterior and the posterior metameres (acanthodrilin). Changes, elimination of the anterior (balantin) or of the posterior (microscolecin) pair of prostates, dislocation of the male pores from the middle of the three segments to the one containing the remaining pair of prostates, or

dislocation of the remaining pair of prostates into the middle segment (megascolecin). Union of the sperm and prostatic ducts to open to the exterior by one or two common apertures. Prostates, originally tubular, the only change being to a racemose condition through various intermediate stages. Spermathecae had considerable importance in classical definitions of genera and families but without the explanations that were provided for other organs or systems. The adiverticulate state may have been considered primitive and the diverticulate state secondary. The original battery may have been thought to comprise four sacs opening to the exterior at 7/8 and 8/9. The characters, aside from diverticulate and adiverticulate, were number of organs and location of the pores especially with reference to segments containing testes and male pores. Setae, a single change, from eight per segment (lumbricin) to many in an equatorial circle (perichaetin). Excretory system, changes from two units per segment (meganephridia) to more, usually many more and up to hundreds in "forests" (micronephridia). Gizzard and calciferous glands, absence or presence, the other characters employed usually were number and axial location.

4. The characters mentioned above are sufficient to define genera as well as subfamilies and families.

5. Genera so defined can be arranged in phylogenetic sequences comparable to the *Eohippus-Equus* lineage except that in the megadrile Oligochaeta each genus while giving birth to a daughter genus has remained unchanged, ancestral taxa surviving to the present alongside daughter, granddaughter and even great-granddaughter genera, to provide a "living paleontology."

6. Proof that phylogenies are correct is provided by transitional forms between mother and daughter genera.

7. Phylogenetic sequences with a common ancestral genus constitute a subfamily.

8. Subfamilies that evolved from a common and still extant ancestor comprise a family.

9. Families, just like genera and subfamilies, can be arranged in phylogenetic series.

10. Convergence is common and polyphyly has been detected. Some additional instances of polyphyly doubtless will be detected but possibly, if not probably, many more will be forever unrecognizable.

11. In a phylogenetic classification, boundary lines, and, in particular, those between genera, are "bound to be merely arbitrary" (Stephenson, 1923: p. 193). Accordingly, "since all such lines are arbitrary interruptions in the record of a continuous process" (*idem,* 1930: p. 833), it does not matter very much where they are drawn.

Evolutionary sequences in a system based on so few and such simple key-characters can, in absence of fossils, begin in several different ways and the fact that other seriations were not suggested must be regarded as a tribute to the author of the system, Wilhelm Michaelsen, as well as to the system's major proponent, John Stephenson. Their presentations were so effective that for long no one seems to have thought of questioning the basic esotery. Not even when those preconceptions compelled placing an Indian species, *Perichaeta pellucida* Bourne, 1894, in an Australian genus in spite of morphological and geographical evidence demonstrating (Stephenson, 1923: p. 317) that the Indian taxon "is not phyletically related to the Australian species." The significance of other instances of admitted and supposedly inherent and unavoidable polyphyly, of disagreements by specialists as to placement of dividing lines across the phylogenetic lineages, of Michaelsen's various shifts of his own boundary lines, was unrecognized. A growing awareness that the classical system, as defined and defended by Stephenson (1930), was less than perfect, resulted in several neoclassical revisions. None were based on new insights or evidenced understanding of what really was wrong with the system. Little more than reshufflings of the generic and specific cards was involved and in some instances polyphyly and morphological heterogeneity were greatly increased. Also misunderstood or ignored was evidence occasionally appearing in the literature that had indicated importance of previously unused characters. Thus, by 1917, Michaelsen already had employed (an isolated instance) microscopic structure of calciferous glands in addition to the classical key characters throughout a revision of the South American glossoscolecid genera. Pickford (1937) also had shown that hitherto neglected characters of organs in two somatic systems could be used similarly.

However, the first demolition of a major portion of the classical system eventuated from the author's study (1937) of a megascolecine, Himalayan form, *Megascolides bergtheili* Michaelsen, 1907. The species was found to differ so little, in so much of its somatic anatomy from the octochaetine *Eutyphoeus* that erection therein of a subgenus for Michaelsen's species had to be considered. The single anatomical character by which the classical Megascolecinae and Octochaetinae were distinguished from each other had been found in both subfamilies and later was to be found in more than one genus of others.[1] Moreover the heterogeneity of somatic anatomy in each of the Megascolecinae and the Octochaetinae was so great that neither could be defined without drastic revision. A revision such as obviously was required could not, however, be undertaken then because of the vast ignorance of megadrile morphology. A century of neglect could not be mitigated in a few years.

Somatic anatomy it was that enabled demonstration of the real relationships of *bergtheili*. Somatic char-

---

[1] Also note inclusion of *Diplotrema* in the classical Megascolecinae.

acters subsequently were found by the author in case of *Eutyphoeus* and one genus after another to enable satisfactory characterization of the taxa, often even without mention of genitalia. Just one kind of somatic organ, the calciferous portion of the gut, without any attempt to study microscopic anatomy, by use only of macroscopically recognizable characters (Gates, 1958), enabled keying of nearly all Oriental octochaetine genera. Microscopic anatomy of the glands almost certainly would have provided additional useful characters. Somatic characters have been used below in construction of one key to Oriental species of the Lumbricidae. A constantly growing mass of evidence for almost half a century proves that somatic anatomy, in megadrile classification at least, can be ignored only at risk of compounding prevailing systematic confusion.

The classical system was based on inadequately tested preconceptions and its replacement by a more natural system was prevented by refusal to reexamine those concepts. Frequent mention of necessity for adequate characterization of somatic anatomy previously, as well as herein, should not be misunderstood to indicate a personal bias on the part of this author. To the contrary, as is shown below, he long ago recognized that taxa of one whole family, at generic as well as specific level, can be defined almost only by genitalia. Furthermore, preliminary study of the Lumbricidae already has revealed various genital characters that will have important systematic uses, and at more than one level. Prerequisite to any satisfactory revision of megadrile systematics, according to the author's experience of some thirty-odd years, is intensive study of intraspecific variation in all genital as well as in all somatic structure. Such research will enable definition of a taxon according to what may be called ordinary or normal, individual and geographic variation. Species characterizations then only will be free of current distortions based on type-specimen idolatry, temporary pathological or parasitological changes, aberrations arising during embryonic and regenerative development or resulting from parthenogenesis. Even more important, at least for the megadrile oligochaetes, should be the invariation[2] revealed by such studies which will enable definition of each species by characters universally valid through all of its populations. Invariant characters common to increasingly larger groups of species then could serve to define taxa at various hierarchical levels. The process obviously will be slow but, as data accumulate, extrapolations will be less liable to error. As more aspects of the organisms, not merely the grossly structural, are taken into consideration, the system will become more natural.

Having abandoned the classical method of fitting systematic units, defined only by a few, simple, "key" characters, to preconceived evolutionary schemes, the author saw his problem to be that of finding invariant characters-in-common by which individuals of species, species of a genus, genera of a family, are to be recognized. The study of variation accordingly, has been, in one sense, a search for invariation. Our knowledge of individual and geographic variation, as well as of much anatomy and physiology that also may prove to be significant, is incomplete for all megadriles. Accordingly, each summary of such known data as now seem to be systematically useful at species level, is referred to as a précis rather than a definition. That procedure seems appropriate since content of earthworm faunas has been determined, in all of Asia, only at Allahabad, Lahore, and Rangoon. Even at each of those places an occasional addition to the local list from more recent importation will not be surprising.

KEY TO FAMILIES OF MEGADRILE OLIGOCHAETES

1. Testes and male funnels, intraseptal
                                        Moniligastridae
   Testes and male funnels, not intraseptal..... 2
2. Male pores, in front of female pores......... 3
   Male pores, behind female pores........... 5
3. Intestinal gizzard, present...Lumbricidae (part)[a]
   Intestinal gizzard, lacking................. 4
4. Male pores two pairs, prostates lacking
                                        Haplotaxidae
   Male pores one pair, prostates or prostate-like
      glands present.................Alluroididae
5. Prostates[b] with muscular ducts, generally,[c]
      present............................... 6
   Glands of that sort, not generally present.. 10
6. Spermathecal pores, in front of testis segments
      or mostly so located[d].................... 7
   Spermathecal pores, in or behind first testis
      segment........................Eudrilidae
7. Last hearts, in xi..............Ocnerodrilidae
   Last hearts, behind xi or a homoeoetic equiva-
      lent................................. 8
8. Prostates, racemose, of mesodermal origin
                                        Megascolecidae
   Prostates, tubular, of ectodermal origin...... 9
9. Nephridia, holoic.............Acanthodrilidae
   Nephridia, meroic..............Octochaetidae
10. Dorsal pores, present............Lumbricidae[e]
    Dorsal pores, absent...................... 11
11. Spermathecae, lacking................... 12
    Spermathecae, present................... 13
12. Gizzard and calciferous glands, lacking
                                        Criodrilidae
    Gizzard and extramural calciferous glands,
       present.............Glossoscolecidae (part)[f]
13. Spermathecal pores, in front of testis segments
       or mostly so[g]............................ 14

---

[2] Absence of significant invariance characterized most units of the classical system. Classical taxa that can be retained unaltered in a modern system had such invariance, as it were, by accident and almost without conscious recognition.

[a] This entry is required by widely distributed, common, parthenogenetic morphs in which male pores may be in the same segment as the female pores or even in the metamere next in front. Species without parthenogenetic degradation of organization, so far as is now known, key out to 10.

[b] Prostate, in this work and contrary to practice of some authors, refers only to glands, reaching well into coelomic cavities, that usually are provided with easily distinguishable ducts, and that are in the male pore segment or one or both of those next to it. Similar glands associated with genital markings are, in two families, characterized as GM glands. Other glands that some specialists called prostates are designated by different names e.g., atrial in the Lumbricidae.

[c] Here again parthenogenetic morphs of three families can cause trouble to students unfamiliar with the anatomy of those groups. Sometimes an aprostatic aberration will be evidenced as such by the presence of glandless prostatic ducts that may be of nearly normal size when not more or less rudimentary. If male terminalia are lacking, worms with preclitellar female pores and adiverticulate spermathecae are likely to be lumbricid but those with intraclitellar female pores and diverticulate spermathecae are most likely to be megascolecid. However, acanthodrilid as well as ocnerodrilid species (possibly even octochaetid?) may be involved and spermathecae may have lost their diverticula or may themselves be lacking.

[d] The qualifying alternative is required by plutelli that have additional spermathecae in some or all of x-xii.

[e] Absence of dorsal pores, reported from an occasional rare lumbricid, requires confirmation.

[f] This entry is required by species of the athecal and classical Glossoscolex. Little or nothing is known about the reproduction of those forms. Whether parthenogenesis is involved, as in so many athecal morphs of other families, or whether some adaptation (such as the spermatophores of Criodrilus) enables continuation of biparental reproduction remains to be determined.

[g] The qualification of the key characters is necessitated, first, by species such as those of the glossoscolecid genera, Thamnodrilus and Rhinodrilus, that have additional spermathecae in one to five of segments x-xiv, and, second, by a species of the microchaetid Glyphidrilus that has rudimentary (and presumably functionless spermathecae) in x, xi.

REFERENCES NOT IN THE BURMESE BIBLIOGRAPHY

AVEL, M. 1959. "Classe des Annelides Oligochates." In: Grasse, P. P. (ed.), Traité de Zool. 5, 1: pp. 224–270.

CHEKANOVSKAYA, O. V. 1962. (The Aquatic Oligochaeta of the USSR) (Moscow), pp. 441.

CLAPAREDE, E. 1862. "Recherches anatomiques sur les Oligochetes." Mem. Soc. Phys. Hist. Nat. Geneve 16: pp. 71–164.

GATES, G. E. 1959. "On a Taxonomic Puzzle and the Classification of the Earthworms." Bull. Mus. Comp. Zool. Harvard College 121: pp. 229–261.

GRUBE, A. E. 1850. "Die Familien der Anneliden mit Angabe ihrer Gattungen und Arten." Arch. Naturgesch. 16, 1: vide p. 281.

HOLT, P. C. 1963. "The Systematic Position of the Branchiobdellidae." American Zool. 3: pp. 522–523.

JOHNSTON, G. 1865. A Catalogue of British non-parasitical Worms in the Collection of the British Museum (London).

MICHAELSEN, W. 1919. "Ueber die Beziehungen der Hirudineen zu den Oligochäten." Mitt. Naturhist. Mus. Hamburg 36: pp. 131–153.

—— 1921. "Zur Stammesgeschichte und Systematik der Oligochäten." Arch. Naturgesch. 86, A-8: pp. 130–141.

—— 1928–1930. Oligochaeta. In: Kuekenthal & Krumbach, Handbuch der Zool. 2, 2-8: pp. 1–118.

—— 1929. "Zur Stammesgeschichte der Oligochäten." Zeitschr. Wissensch. Zool. 134: pp. 693–716.

PERRIER, E. 1872. "Recherches pour servir a l'histoire des Lombriciens terrestres." Nouv. Arch. Mus. Hist. Nat. Paris 8: pp. 5–198.

PICKFORD, G. 1948. "Annelida." Encyclopedia Britannica (14th ed.) 1: pp. 998–1006.

VAILLANT, L. 1868. "Note sur l'anatomie de deux especes du genre Perichaeta, et essai de classification des Annelides lombriciens." Ann. Sci. Nat., Ser. 5, 10: pp. 225–256.

VEJDOVSKY, F. 1884. System und Morphologie der Oligochaeten (Prag).

YAMAGUCHI, H. 1953. "Studies on the aquatic Oligochaeta of Japan. VI." Jour. Fac. Sci. Hokkaido Univ. (6, Zool.) 11: pp. 277–342.

ACANTHODRILIDAE

1880. Acanthodrilidae (part[1]) Claus, Grundzüge der Zoologie (ed. 4) 1: p. 479.

1884. Acanthodrilidae (part[1]) + Pontodrilidae + Plutellidae, Vejdovsky, System und Morphologie der Oligochaeten (Prag), p. 63.

1888. Acanthodrilidae (part[1]) + Microscolex, Photodrilus, Pontodrilus, Plutellus (Eudrilidae), Rosa, Bull. Mus. Zool. Univ. Torino 3, 41: p. 9.

1890. Acanthodrilidae (part[1]), + Pontodrilus, Photodrilus, Microscolex, Rhododrilus, Plutellus (Eudrilidae), Benham, Quart. Jour. Micros. Sci. 31: pp. 220, 221.

1895. Acanthodrilidae (part[1]), + Pontodrilus, Plutellus, Microscolex, Microdrilus, Fletcherodrilus (Cyrptodrilidae), Beddard, A Monogr. of the Order of Oligochaeta (Oxford), pp. 443, 468, 452, 459, 505, 480.

1900. Acanthodrilinae (Megascolecidae), part (excluding Maheina) + Plutellus, Fletcherodrilus, Pontodrilus and Diporochaeta (Megascolecinae) + Diplocardia and Zapotecia (Diplocardiinae), Michaelsen, 1900, Das Tierreich 10: pp. 122, 163, 178, 179, 199, 324, 329.

1907. Acanthodrilinae (Megascolecidae) + Plutellus, Pontodrilus, Fletcherodrilus and Diporochaeta (Megascolecinae), Michaelsen, Fauna Südwest Australiens 1: pp. 138, 156.

1922. Acanthodrilinae (Acanthodrilidae), part, Michaelsen, Mitt. Zool. Mus. Hamburg 38: p. 58. (Excluding Howascolex.)

1923. Acanthodrilinae (Megascolecidae) + Diplotrema, Plutellus, Pontodrilus and Diporochaeta (Megascolecinae) + Diplocardia + Zapotecia (Diplocardiinae), Stephenson, (Fauna of British India), Oligochaeta, pp. 163, 165, 469.

1928. Acanthodrilinae (Acanthodrilidae) + Diplocardia and Zapotecia (Diplocardiinae) + Diplotrema + Plutellus + Diporochaeta (Megascolecidae), Michaelsen, Handbuch der Zoologie, Berlin 2, 2-8: pp. 109–110.

1930. Acanthodrilinae (Megascolecidae) + Diplocardia + Zapotecia (Diplocardiinae, Megascolecidae) + Plutellus, Diplotrema, Pontodrilus, Diporochaeta (Megascolecinae), Stephenson, The Oligochaeta (Oxford): pp. 820, 831, 833, 840.

1937. Acanthodrilinae (Megascolecidae), Pickford, A Monogr. of the Acanthodriline Earthworms of South Africa (Cambridge), p. 98.

[1] Excluding species with meroic excretory systems.

1959. Acanthodrilidae, Gates, *Bull. Mus. Comp. Zool. Harvard College* 121: p. 255. Acanthodrilinae (part[1], Megascolecidae), Lee, *The Earthworm Fauna of New Zealand* (Wellington), p. 32.

Digestive system, with an intestinal origin behind xiii. Vascular system, with hearts behind xi. Excretory system of holoic nephridia.

Spermathecae,[2] diverticulate. Clitellum, multilayered, including female pore segment. Ovaries fanshaped and with several egg-strings (?). (Ovisacs, small and lobed?) Ova, not yolky. Seminal vesicles, trabeculate. Prostates, tubular and of ectodermal origin.

Distribution: Burma. Australia, Tasmania, New Caledonia. New Zealand, Auckland, Chatham and subantarctic islands. United States, Mexico. Central America, southern South America. South Africa, Madagascar. Ceylon, India.

Remarks: Genera of the classical Octochaetinae, along with certain megacolecid taxa, were recently (Lee, 1959) transferred to the classical Acanthodrilinae. The enlarged subfamily was then defined by location of prostatic and male pores as well as by prostatic structure. A glance at the definition shows that no uniquely diagnostic character is mentioned and that no single character is common to all of the included genera and species. A taxon composed of organisms for which no morphological or other character in common is stated seems to be a negation of classification and no improvement over the phyletically interesting but outmoded older system.

The Acanthodrilidae certainly appear to be more closely related to the Ocnerodrilidae and the Octochaetidae than to the Megascolecidae. That affinity long has been recognized. Michaelsen (1921, 1929) included the Ocnerodrilidae, Acanthodrilidae, and Octochaetidae as subfamilies in his enlarged Acanthodrilidae but, unable to escape the handicaps imposed by his own phylogenetic assumptions, left genera such as *Plutellus* in his Megascolecidae. A monophyletic origin of the Acanthodrilidae probably is little if any more likely than the same sort of ancestry for the Ocnerodrilidae, Acanthodrilidae, and Octochaetidae together.

The Acanthodrilidae as now recognized is defined morphologically by characters common to all included species and genera. The grouping meets and promises to continue to meet Stephenson's criterion of convenience until such time in the more or less distant future when slowly accumulating knowledge of somatic anatomy has become sufficient to permit a better arrangement.

KEY TO ORIENTAL GENERA OF ACANTHODRILIDAE

1. Nephridia with terminal vesicles......*Microscolex*
   Nephridia avesiculate........................ 2
2. Nephridia present in some preclitellar segments,
   not littoral......................*Plutellus*

[2] In front of the testis segments (except in 1 Burmese species of *Plutellus*?).

Nephridia lacking in preclitellar segments,
littoral..........................*Pontodrilus*

Microscolex

1887. *Microscolex* Rosa, *Boll. Mus. Zool. Univ. Torino* 2, 19: p. 1. (Type species, *M. modestus* Rosa 1887 = *Lumbricus phosphoreus* Duges 1837).

1937. *Microscolex*, Pickford, *A Monograph of the Acanthodriline Earthworms of South Africa* (Cambridge, England), p. 424.

1962. *Microscolex*, Gates, *Proc. Louisiana Acad. Sci.* 25: p. 7.

Digestive system (with an intestinal origin behind xv?), without a strong gizzard, calciferous and supraintestinal glands, intestinal caeca and typhlosoles. Vascular system, with unpaired dorsal, supra-esophageal and ventral trunks but no subneural, with paired extra-esophageals median to the hearts and united posteriorly at mV on ventral face of gut (associated with posterior lateroparietal trunks?), with lateroesophageal hearts in x–xii. Nephridia, holoic, present from ii, vesiculate, posterior bladders ocarina-shaped with pointed ends mesially and with short ducts from ventral side laterally (provided with parietal sphincters?). Nephropores (obvious?), in a single longitudinal rank on each side of the body. Setae, eight per segment. Septa all present from 5/6. (Dorsal pores, lacking?)

Holandric. Spermathecae, diverticulate. Ovisacs, present. Ovaries, fan-shaped and with several egg strings.

Distribution: Of the classical taxon, according to Stephenson (1930), South Patagonia, Tierra del Fuego, west Argentina, Darien(?), Falkland Islands, South Georgia, Kerguelen, Marion Island, Cape Colony, Crozet, Campbell, Auckland, Antipodes and Macquarie Islands. Not included were distributions of two species (*cf.* below) that had been carried around the world by man.

Systematics: Acanthodrilin and microscolecin, two of the standard characters by which genera were defined in the classical system, fifty years ago no longer could define the taxa from which those adjectives were derived. *Acanthodrilus* and *Microscolex*, as defined by Stephenson (1930: p. 824), even after considerable discussion,[4] were distinguishable from each other, solely by presence or absence of a well-developed gizzard. Validity of gizzard characters at generic level was questioned subsequently as it also had been previously. Some of the reasons are unacceptable today and need not be repeated here.[5] Although use

[4] That discussion was summarized by Pickford, 1937, in her *Acanthodriline Earthworms*, p. 76. Or, see Stephenson, 1930: p. 822.

[5] Some of the discussion also seems to have been tinged with a belief in or a search for absolutes equally valid throughout the system. More recent work of course shows that a character universally present in a group of species so closely related as to belong together in one genus, need not, and frequently is not, of similar or even of any systematic usefulness for other genera, sometimes even at species level.

of certain gizzard characters sometimes may force, as was claimed, an unnatural separation of related species, attention with respect to that organ is invited for generic definitions in the present monograph.

A significant contribution toward clarification of the prevailing confusion with regard to the *Acanthodrilus-Eodrilus-Microscolex* complex was provided by Pickford who defined *Microscolex* (1937: p. 424), in part, by the following characters-in-common:

Nephridia vesiculate. Nephropores, in a single series on each side (of the body). Setae, eight per segment.

Her gizzard character, "reduced or rudimentary" can be better stated in Stephenson's wording "Gut, without 'a well developed gizzard.'" Such a definition lacks the alternatives and exceptions so typical of the classical system and, which is especially noteworthy, comprises only somatic characters.

The generic précis of the present author adds to the somatic skeleton of Pickford's definition, further characters shared by the only species that are well known, i.e., the exotics. Those characters, or similar ones, elsewhere have been found to be sometimes valid at generic level. Those enclosed by parentheses may be more liable than the others to intrageneric variation. Whatever future study may reveal, the revision that now seems to be needed may result in a much smaller and more nearly continuous generic range than that of the classical system. Indeed, the portion of the *Microscolex* distribution (of Stephenson, 1930) that was self-acquired is unknown.

Remarks: One anthropochorous taxon already has been recorded from India and another can be expected to turn up sooner or later in southeast Asia.

KEY TO THE WIDELY DISTRIBUTED TAXA OF *Microscolex*

Spermathecae, present.................*phosphoreus*
Spermathecae, absent.....................*dubius*

Microscolex dubius

1887. *"Eudrilus?" dubius* Fletcher, *Proc. Linnean Soc. New South Wales*, Ser. 2, **2**: p. 378. (Type, locality, Sydney, New South Wales, Australia. Types, none.)
1962. *Microscolex dubius*, Gates, *Proc. Louisiana Acad. Sci.* **25**: p. 7.
GAMETOGENESIS: Omodeo, 1952, *Caryologia* **4**: p. 359.

Athecal. Male pores, minute, along with prostate pores, and usually also with apertures of penisetal follicles, in a small invagination opening just lateral to *A* at eq/xvii. Female pores, slightly anteromedian to *a*. Clitellum, annular, xiii, xiii/n–xvi, xvii/n, xvii. Setae, present from ii, widely paired but *AB* gradually narrowed through clitellum and into xviii, gradually widening through xix–xxv, $AB < CD < BC$ ca. $= AA$, $DD < \frac{1}{2}C$, *a,b*/xvii penial. Nephropores, obvious, except in ii–v about at mL. Prostomium, epilobous, tongue open. Dorsal pores and pigment, lacking.

Segments, to 117 but usually 111–116. Size, 35–90 by 2–5 mm.

Septa, 6/7–13/14 slightly strengthened. Digestive system, with a small and weak gizzard in v, esophagus widened and moniliform in x–xiii, intestinal origin in xvi, but without caeca, typhlosoles, calciferous and supra-intestinal glands. Vascular system, with dorsal (single), ventral and supra-esophageal (in x–xii) trunks, extra-esophageals median to segmental connectives and uniting at mV on gut in ix to pass back into region of xii–xiii, but without a subneural. Hearts, of ix and anteriorly lateral, of x–xii latero-esophageal. Nephridial vesicles, ocarina-shaped posteriorly, with pointed ends mesially and with short ducts from ventral side laterally (with well-developed parietal sphincters?). Holandric. Seminal vesicles, in xi, xii. Ovisacs, small, lobed. Ovaries, fan-shaped and with several egg strings.

Reproduction: Externally adhesive spermatophores are unknown. In absence of such spermatophores and of spermathecae, reproduction is unlikely to be biparental. Iridescence on male funnels, when recognizable, usually has been such as to indicate sparse sperm maturation. Male sterility for many individuals and some morphs is anticipated, in which case reproduction must be parthenogenetic. Polyploidy presumably is not involved as Omodeo (1952) found gametogenesis to be normal, the number of chromosomes approximating those of diploid lumbricids.

Distribution: Australia, New Zealand, Norfolk I. Oregon, California, Louisiana, North and South Carolina, Florida, Mexico. Bolivia, Chile, Argentina, Uruguay, Paraguay.

France, Portugal, Spain, Italy, Balearic Islands, Corsica, Sardinia, Sicily, Rhodes, Crete. Madeira, Canary Islands, St. Helena. Tunis, South Africa.

The original home of *M. dubius* supposedly is somewhere in a southern portion of South America. Much of the distribution, accordingly, must be due to transportation and presumably by man. Interception by U. S. Bureau of Plant Quarantine, of 15 specimens of *M. dubius* on 4 occasions, as well as of 29 specimens of *M. phosphoreus* on 17 occasions, shows that transoceanic transportations still are under way.

Habitats: Soil, in plant pots, greenhouses (Oregon), gardens. Under stones, "am Meeresstrande," in a leaf pile.

Biology: The species is unlikely to have been introduced only to those countries from which it has been recorded. Restriction to subtropical and warmer temperate zone areas seemingly is indicated.

Much of the ingesta examined was humus or undecayed vegetable matter sometimes including fibers up to 10 mm. long associated with a few very fine sand grains. These worms may be discriminating feeders.

Luminescence has not been recorded from this species, and was said not to be evoked by alcohol and ammonia.

Variation and Abnormality: An unusually short range of variation in segment number was indicated (1962) by American collections from which obvious and probable posterior amputees had been screened out. For Australian juveniles, presumably referable to *M. dubius* the number was said to be 113 (with a standard deviation of 5).

Abnormalities, such as absence of a heart in xii (1962: No. 4), presumably result from developmental accidents. Others are of the degradational kinds permitted by parthenogenesis, such as absence of male terminalia on one side (1962: No. 3), or reversionary such as presence of a gonad (presumably male) in ix (1962: No. 7). Of much more interest perhaps is divergence, from what seems to be normal, of a kind that is characterized as predictive, such as presence of one-half of a heart or of a complete heart on one side of xiii. However, if hearts had been present in xiii of some ancestral stage, the conditions mentioned (1962: Nos. 1, 2) would be merely one further illustration of parthenogenetically permitted reversion.

Polymorphism: Prostates clearly are disappearing in many strains, the glands being confined to xvii in part of the author's material, sometimes straight and too short to reach beyond m*BC*. Absence of male terminalia on one side of a single specimen, warrants anticipation of AR morphs. Of similar interest is the megascolecin union of male and prostate ducts just below the gland that was shown by specimens from California and Louisiana. A similar union is reached in the oriental section of the genus *Plutellus* only by the more advanced species. Morphs with ovaries in x, xi, even in x–xiii would not be unexpected in view of probable presence of ova in gonads that should have been testes. As so many morphs have been found in the few specimens that were carefully dissected, polymorphism may be as rampant as in *D. octaedra, E. rosea, P. houlleti*, etc.

Regeneration: Only one regenerate, caudal, has been recognized by the author and that was morphallactic, at 100/101, as yet without metameric differentiation though original nephropores and sites of former follicle apertures still were visible.

Parasites: INSECTA: *Onesia accepta* (Malloch, 1927), Australia, first and second instars under the epidermis, third in the coelom (Fuller, 1933, *Parasitology* 27).

Systematics: The classical *M. dubius* now appears to be a congeries of parthenogenetic morphs. The amphimictic population from which those morphs were isolated is unknown. Less degraded intermediate morphs that might provide clues to a proper characterization of the biparental population are unknown unless they are provided by another classical taxon, *M. phosphoreus*. At very least, the interbreeding ancestral population is likely to be bithecal and biprostatic which is just the state now characterizing most of the *phosphoreus* morphs. Comparing the

précis of the two taxa will show that there are almost no really good contraindications to the suggested relationship. Size, shape, and ornamentation of penial setae can, of course, be cited against the relationship but to little purpose as parthenogenesis also permits degradation of penisetal structure, indeed so much so that characterization was omitted above. In that connection, an abnormal specimen (1962: No. 3), is of especial interest as it shows, that in absence of a prostate, setae do not become penial. Differences between penial and genital setae of the two taxa may then be a result of differential degradation or only an expression of different stages reached in a single series of changes. Differences in size and segment number are interesting but both characters are liable to considerable variation in biparental populations. The anal region of all individuals of both taxa that were seen by the author, except after posterior amputation, had two pairs of nephropores. Metameric differentiation of two segments and a pygomere never had been completed. In view of what is now known about changes permitted by parthenogenesis, it is not unthinkable that inhibition of further metameric differentiation can become effective at different levels in different morphs. This discussion is continued on a subsequent page after consideration of the morphs of *M. phosphoreus*.

Remarks: Unless the genital tumescences and associated genital setae as well as the GS glands of a single one of the author's specimens are a result of predictive mutations, the ancestral biparental population may have had such structures. Their characters, not excluding shape and ornamentation of the setal shaft, are similar to those of certain species of *Lumbricus*.

If male terminalia have been degraded during evolution of the *dubius* morphs, male pores of the biparental population may have been in fairly large parietal invaginations or even in coelomic chambers, perhaps on some sort of an intromittent organ. Differences in characterization of the male terminalia by various authors scarcely need reconciliation because any and all of the conditions, as well as others as yet unrecorded, are allowed by the parthenogenesis.

Posterior lateroparietal trunks are believed to be present in *M. phosphoreus*. If confirmed, the same trunks can be expected in *M. dubius*.

Microscolex phosphoreus

1837. *Lumbricus phosphoreus* Duges, *Ann. Sci. Nat.*, Ser. 2, 8: pp. 17, 24. (Type locality, greenhouses, Jardin des Plantes, Montpellier, France. Types, None.)

1914. *Microscolex phosphoreus*, Stephenson, *Rec. Indian Mus.* 10: p. 388.

1937. *Microscolex, phosphoreus*, Pickford, *A Monograph of the Acanthodiline Earthworms of South Africa* (Cambridge, England), p. 433. GAMETOGENESIS: Omodeo, 1952, *Caryologia* 4: p. 259. LUMINESCENCE: Pierantoni, 1923, *Boll. Soc. Nat. Napoli* 36. Skowron, 1926, *Biol. Bull.* 51: 1928, *ibid.* 54.

Bithecal, pores minute, each often on a very small tubercle, at or just lateral to A, at 8/9. Male pores, minute, between apertures of a and b follicles at eq/xvii. Prostatic pores, common opening of a prostate and of its adjacent b follicle, lateral to male pore in xvii. Female pores, at or slightly median to A, about halfway between 13/14 and eq/xiii or slightly nearer the latter. Genital markings, circular to shortly elliptical and then transversely placed, unpaired, median or laterally, each with a grayish translucent center. Clitellum, xiii, xiii/n, xiv–xvi/n, xvi, xvii/n, xvii, annular but often lacking ventrally in xiii and/or xvii, sometimes even in a short section of xvi. Setae, present from ii in which d already is above mL, widely paired posteriorly where $AB < CD$, $AA < BC$, $DD < \frac{1}{2}C$, a,b/xvii penial. Nephropores, obvious, in ii–iv at or near D, elsewhere at or somewhat below C. Dorsal pores, usually lacking(?). Prostomium, epilobous, tongue open. Pigment, none. Segments, to 90 but usually 73–88. Size, 10–35 by 1–1.5 mm.

Septa, 7/8–10/11 or 12/13 or 8/9–13/14 somewhat thickened. Digestive and vascular systems, as in *M. dubius*. Nephridial vesicles, small anteriorly and at or above D, posteriorly large, ocarina-shaped with pointed end continued mesially by a slender cord, bluntly rounded end laterally, funnel-shaped floor narrowing ventrally to parietes (with well-developed parietal sphincters?), tubular portion of the nephridium joining vesicle dorsally and mesially. Holandric. Seminal vesicles, acinous, in xi, xii. Spermathecae, small and subesophageal, duct short to nearly as long as ampulla, with two equisized or slightly subequal diverticula, each about as long as its ellipsoidal to ovoidal seminal chamber, united ectally to open into anterior face of duct close to parietes. Ovisacs, small, lobed. Ovaries, more nearly oblong than fan-shaped, with few egg strings, usually four. (GM glands, none.)

Reproduction: Probably parthenogenetic in many morphs because of male sterility. Sperm maturation seemingly is sparse in some morphs and spermathecal seminal chambers often have no spermatozoal iridescence. If individuals fail to copulate, reproduction could be uniparental in spite of profuse sperm maturation. Polyploidy presumably is not involved as Omodeo (1952) found gametogenesis to be normal, the number of chromosomes approximating those of diploid lumbricids.

Distribution: Japan. New Zealand. Hawaii,[6] California, Illinois, Washington (D. C.), North Carolina, Florida, Mexico.[6] El Salvador (at 1,000 m.). Ecuador, Chile, South Patagonia, Elisabeth Island, Argentina, Paraguay, Brazil.

Poland, Germany, Jersey, France, Switzerland, Jugoslavia,[6] Italy, Sardinia, Bulgaria, Greece.

Algeria. Canary Islands. Southwest Africa, South Africa (at 8,700 feet in Basutoland [Lesotho]). Turkey, Israel, West Pakistan.

The original home of the species long has been supposed to be in some part of southern South America but without any more precise delimitation. In France, as late as 1898, the species was said to be present only in the vicinity of gardens with recently introduced, rare plants.

Habitats: Soil, in plant pots (Hawaii, Poland), greenhouses (North Dakota, Maine, France, Germany), botanical gardens (Israel, Sardinia), gardens, under stones, brook banks. Under manure heap, leaf mold around tree stump, sawdust pile kept moist by water from spring (central Illinois, the only record for the state). Abandoned coal mines (Poland) where "great quantities" were present in a passage 230 m. below the surface. Caves (Italy).

Biology: The luminescence of *M. phosphoreus*, according to one author, is produced by symbiotic bacteria. A luciferin-luciferase-like reaction, according to others, is involved. Granules supposedly responsible for giving off the light are liberated from coelomic corpuscles in a slime discharged from the mouth and/or the anus. Preformed openings by which the corpuscles pass from the coelomic cavities into the enteron have been postulated but probably have not been seen. The most recent discussion of the subject still is that of Stephenson's (1930) "The Oligochaeta."

The distribution, as in case of *M. dubius*, suggests a climatic restriction to subtropical and warmer temperate-zone areas.

Variation and Abnormality: Differences as to manner of opening of prostatic ducts, as recorded by several authors, are such as now can be expected when reproduction is parthenogenetic.

Sizes, larger than those mentioned in the précis above, to 55 by 3 mm., are recorded but may have been measured on macerated, relaxed, or bloated worms. All specimens available to the author were within the smaller range.

Dorsal pores usually were believed to be absent. However, porelike markings at mD sometimes are recognizable, more often just in front of and just behind the clitellum than elsewhere. Similar markings sometimes have been visible near the hind end. Patent apertures have been seen at 19/20 and 21/22. Coelomic coagulum of one individual protruded to the exterior through a pore at 11/12. Whether progression or regression is involved remains to be determined.

Genital markings, previously not mentioned in definitions, were recognized on more than a hundred individuals. Markings were present only in x, xi, xii, xvi, xvii or across 16/17. Positions varied from median to as far laterally as in AB.

---

[6] Based on specimens intercepted by the U. S. Bureau of Plant Quarantine. Ten intercepted specimens from Israel confirm a previous record for that country that might also have been of intercepted worms.

Abnormalities, such as absence of one heart in xii, presumably resulted from some sort of developmental accident. Other aberrations, such as presence of gonads in xii are regarded as reversions to ancestral conditions, permitted by the parthenogenesis. A graded series of spermathecal abnormalities seemingly can be read in either of two ways, fusing of two originally discrete diverticula, or splitting of an originally single diverticulum into two except for the common junction with the duct.

Polymorphism: Prostates, of xvii, clearly are disappearing in some strains but an R morph has not yet been found. Degradation of spermathecal structure seemingly is under way in some morphs, regardless of which way the evolutionary changes are trending. An ovary frequently is present in xii and morphs with hologyny can be expected. Gonads of x, xi sometimes have about the same shape as the ovaries of xiii and do contain, instead of sperm, bodies that look more like *phosphoreus* ova than parasites.

Few individuals have been examined, and other morphs are likely to be found when the taxon is more carefully investigated than in the past.

Systematics: The classical *M. phosphoreus* now appears to be a congeries of parthenogenetic morphs. The amphimictic population from which those morphs were isolated is unknown. A less degraded ancestral population could very well have been quadriprostatic and quadrithecal. Two morphs, of *Microscolex georgianus* (Michaelsen, 1888), distinguishable from each other only by presence or absence of the posterior prostates, afford a partial parallel. Spermathecae are just as easily and perhaps just as often lost as prostates after intervention of parthenogenesis. If then, a quadriprostatic and quadrithecal taxon is ancestral, *Microscolex luisae* (Michaelsen, 1899), supposedly endemic in a small coastal sector of South Africa, needs careful consideration. Genital markings, hitherto unmentioned in definitions of the peregrine taxa, often are present in *phosphoreus* morphs and are like those of *M. luisae*. Differences from *M. phosphoreus*, excepting number of spermathecae and prostates, are small and most now appear to be unimportant in view of the parthenogenesis. Some consideration perhaps should be given to situation of nephropores in m*BC* rather than just below *C* but the supposed difference may be more a matter of variation in intersetal intervals than in location of the pores. Nothing is known about reproduction of the few individuals referred to *M. luisae*, but the rudimentary state of seminal vesicles in ix (when present) suggests that even in the quadrithecal and quadriprostatic taxon some degradation of structure already has taken place. Indeed, instead of being endemic as long assumed, *M. luisae* may have been introduced to Africa along with *dubius* and *phosphoreus* morphs.

M. dubius, phosphoreus, and even *luisae,* may have been segregated from a common ancestral interbreeding population. *M. luisae,* in that case, should approximate most closely to the biparental morphology. So little is known about the results of parthenogenesis in the Acanthodrilidae that any of the suggested synonymizations may be erroneous. Prerequisite to a solution of some of the problems seems to be a search for the biparental population in the presumed South American homeland of all of the morphs.

Remarks: Posterior lateroparietal trunks are believed to be present in *M. phosphoreus* though they have been filled with blood, and accordingly visible, only in xviii to xiv. The trunks are continuous in xiii with posterior bifurcations of the extraesophageal.

If both anthropochorous microscolexes evolved from the same amphimictic stock, degradation of the male terminalia has gone further in *phosphoreus* than in *dubius* morphs.

Whether GM glands can be eliminated while retaining associated markings is not known.

Plutellus

1873. *Plutellus* Perrier, Arch. Zool. Exp. Gen. 2: p. 250. (Type species, *Plutellus heteroporus* Perrier, 1873.)

Digestive system, with one esophageal gizzard in region of v–vii. Setae, eight per segment.

Biprostatic. Metagynous.

Distribution: Ceylon. India. Burma. Tasmania, Australia. New Caledonia. New Zealand, Stewart and Auckland Islands. Queen Charlotte Island and the Pacific coastal strip of the United States. Central America and northern South America.

Systematics: The definition above, with omission of family characters, is essentially that of the classical system. If, however, the usual classical practice of including certain genital characters in a definition is followed, several additional statements can be added.

Male pores, in xviii, xix, or xx. Holandric, pro-andric or metandric. Sperm ducts open to exterior near prostatic ducts or pass into prostatic ducts in the parietes, just above the parietes, in the middle or the ental end of the prostatic duct or even into the prostate ental to the duct. Spermathecae, 1, 2, 3, 4, 5, 6, or 7 pairs in some or all of vi–xii.

A definition of *Plutellus* with reference only to a very limited number of simple somatic characters would have to read somewhat as follows.

Digestive system, agiceriate, or with a vestigial or a well developed gizzard, in v, vi, or vii, without calciferous glands or with such glands (and without any consideration of internal structure) as follows; one pair intramural in xvi or extramural in xvii, two pairs in xiv–xv or in xv–xvi, three pairs in x–xii or in xi–xiii, four pairs in x–xiii, xii–xv or in xiii–xvi, five pairs in ix–xiii. Intestinal origin, in xiv, xv, xvi, xvii, or xviii, etc. Vascular system, with or without a subneural trunk, with dorsal blood vessel double or single, with all hearts lateral or some esophageal or latero-esophageal.

So much heterogeneity of conservative somatic structure, along with the vast distributional gaps, guarantees that the classical *Plutellus*, with or without *Diplotrema* and *Pontodrilus*, is a massive polyphyletic congeries.

Plutellus includes, according to Michaelsen, *Argilophilus* Eisen, 1893, *Fletcherodrilus* Michaelsen, 1891, *Diplotrema* Spencer, 1900, *Pontodrilus* Perrier, 1874. Stephenson, in 1923 as well as 1930, allowed *Diplotrema* and *Pontodrilus* to retain generic rank.

The type species of *Plutellus*, is known only from Perrier's description of two softened worms that had been in the Paris Museum since 1822 and which were "meles dans le meme bocal avec cinq ou six veritables Lombrics provenant de la Pennsylvanie" (p. 249). The type locality long has been assumed to be Pennsylvania. Such an origin is now known to be impossible even if the types did get to Paris via that state or along with worms from that state. American specialists never found anything that might be *heteroporus* in the Pacific coastal strip and it now seems as if the original source might be Australia.

P. heteroporus has three pairs of stalked, kidney-shaped, extramural calciferous glands as well as a subneural blood vessel. *Plutellus* now must be defined, in part at least, by characters of the calciferous glands of the type species as well as by characters of the vascular system. Such a definition excludes Eisen's *Argilophilus* which has no calciferous glands at all. *Argilophilus* in so far as can be determined from the literature, is to be defined somewhat as follows.

Digestive system, with a large gizzard (in v?), an intestinal origin behind xvi, an intestinal typhlosole but without intestinal caeca, calciferous and super-intestinal glands. Vascular system, with complete dorsal (single) and ventral trunks, a supra-esophageal, extra-esophageals median to the hearts and passing onto the gut posteriorly (in region of?), posterior latero-parietals from anal region passing in xiii–xiv onto the gut (and to the supra- and/or extra-esophageals), but without a subneural trunk. Hearts, from x posteriorly latero-esophageal. Excretory system, holoic and of holoic, avesiculate nephridia. Septa, present from 4/5. Setae, eight per segment. Dorsal pores, present behind the clitellum.

Metagynous, ovaries fan-shaped and with numerous egg strings. Ovisacs, in xiv.

The situation of the inconspicuous nephropores, irregularly alternating and with asymmetry between several positions, also was thought (Eisen, 1896) to be generically characteristic. Distribution, at the same time, was said to be from the San Joaquin Valley in California to Vancouver Island in British Columbia but in 1900 Eisen included *hyalinus*, erected on (and known to this day only from) a single macerated individual. Whether that Guatemalan worm has enough anatomy-in-common to belong in the same genus with the northern forms cannot be decided from the very brief description. Spermathecae of North American species originally may have been adiverticulate. Seminal chambers within the wall of the spermathecal duct, in some lines, seemingly are being aggregated into groups that in later evolutionary stages may well be distinctly protuberant as polyloculate diverticula. An evolutionary sequence of increasingly higher calciferous lamellae, somewhat as Smith's series (1924) in the genus *Diplocardia*, perhaps will be associated with so much anatomy-in-common that generic separation with reference to presence or absence of high calciferous lamellae (in a rather primitive sort of gland?) may be impossible. However, presence of intestinal caeca and ventral branching of the intestinal typhlosole, in certain plutelli, warrant a search for additional characters that could enable generic separation from *Argilophilus*. Species with extramural calciferous glands, like *P. kincaidi* Altman, 1936, very probably will have to go into another genus than *Argilophilus*.

Burmese if not also all oriental species can go better into *Argilophilus* than any genus of which *heteroporus* is the type species. Known morphological differences, presence of dorsal pores from the region of 7/8–8/9, presence on the spermathecae of uniloculate diverticula, may seem rather unimportant. An intestinal typhlosole usually is lacking but has appeared independently in two or three species. Funnels never were seen on most of the preclitellar nephridia, perhaps because of the poor preservation, but it does seem possible that the excretory system of the oriental worms has undergone more modification (cephalization) than in North American worms. Such is the evidence, together with the vast oceanic gap, that contraindicates congeneric affinity. Inasmuch as their excretory systems cannot be adequately characterized, oriental plutelli probably should remain for the present, or until such time as a sufficiency of properly preserved material has provided the necessary information, in the classical congeries.

Characters that now seem likely to be common to the oriental plutelli and that were not mentioned for *Argilophilus* are as follows. Digestive system, with an intestinal origin behind xiii. Extra-esophageals, passing onto gut in region of viii–ix (as in at least one Californian species). Clitellum, saddle-shaped, intersegmental furrows obliterated and dorsal pores occluded. Genital apertures, all minute and superficial. Pigment, lacking.

Phylogeny: A hypothetical form (Gates, 1961), from which all Burmese species could have been derived, differs from a hypothetical ancestral ocnerodrile by the greater length of the esophagus which ended with a valve in xiv rather than in xii, and also by the presence of diverticula from spermathecal ducts.

Among the earliest steps in evolution of the oriental plutelli would seem to have been acquisition of dorsal pores, addition of a pair of hearts in xii and reduction of prostates to a single pair which became able to attract into themselves during ontogenetic development the posteriorly growing sperm ducts. The various differences in level at which male and prostatic ducts unite was believed (Gates, 1961) to be due to the different states of development of the prostatic anlage at the moment they were met, presumably at parietal level, by the sperm ducts. Another early development in the oriental line may have resulted in unknown modifications of excretory tubules in preclitellar segments. Subsequent evolutionary changes in the group, are numerous (for a list *cf.* Gates, 1961) but none are unique. They have been made in various genera of different families and apparently have appeared independently, at different times and places, in the oriental plutelli.

Ancestors of that group presumably arose in peninsular India long enough ago to have allowed migration into Ceylon, up into the Himalayas of eastern India from Darjiling district into the Abor Country of Assam, in Burma down almost to the sea centrally and to the east at least to the foot of the Shan Plateau.

Remarks: The classical *Plutellus*, like its supposed descendant *Megascolides*, was believed to have reached North America from a region centered in Australia, possibly over an Angara continent (Stephenson, 1930, p. 682) or (Omodeo, 1963: p. 141) to and along the eastern coast of Asia, around the Pacific presumably in a region where Siberia and Alaska nearly meet, and then down into the United States, leaving behind no evidence of its former presence[7] outside of Australia and New Zealand. Now that the real nature of the classical system and of the esoteric phylogeny on which it was based has been exposed, further consideration of such overland migrations clearly is needless.

KEY TO ORIENTAL SPECIES OF *Plutellus*

1. Length more than 200 mm., ventral setae of preclitellar segments very obviously enlarged
 longus
 Length less than 200 mm., ventral setae not so obviously enlarged......................... 2
2. Seminal vesicles present in ix............... 3
 Seminal vesicles lacking in ix 10
3. Intestinal origin in front of xv[a].............. 4
 Intestinal origin behind xiv................. 8
4. Quadrithecal............................. 5
 Decathecal........... ....*sikkimensis*
5. Spermathecal pores at 7/8–8/9.............. 6
 Spermathecal pores in front of 7/8–8/9......*macer*
6. Spermathecal pores at or median to *A*....... 7
 Spermathecal pores at or lateral to *B*.....*subtilis*
7. Spermathecal pores at *A*.................*rudis*
 Spermathecal pores median to *A*.......*montanus*
8. Anterior spermathecal pores at 7/8......... 9
 Anterior spermathecal pores at or behind eq/viii...............................*indicus*
9. Last hearts in xi, spermathecal diverticulum from duct ectally...............*singhalensis*

Last hearts in xii, spermathecal diverticulum from duct entally................*campsiaulus*
10. Holandic............................... 11
 Metandric.............................. 27
11. Last hearts in xii[b]........................ 12
 Last hearts in xiii...................... 22
12. Intestinal origin in xiv[c].................... 13
 Intestinal origin behind xiv................ 19
13. Bithecal.............................. 14
 Not bithecal........................... 15
14. Spermathecae unidiverticulate, penial setae lacking............................*comptus*
 Spermathecae bidiverticulate, penial setae present...........................*geminatus*
15. Quadrithecal........................... 16
 Not quadrithecal........................ 17
16. Male pores in xviii......................*exilis*
 Male pores in xx.....................*leucaspis*
17. Sexthecal..........................*pauxillulus*
 Octothecal............................. 18
18. Spermathecal pores median to *A*......*compositus*
 Spermathecal pores lateral to *A*.........*inflexus*
19. Intestinal origin in xv..................... 20
 Intestinal origin in xvii..................*rallus*
20. Bithecal.......................*himalayanus*
 Quadrithecal........................... 21
21. Penial setae present................*pratensis*
 Penial setae absent..................*aquatilis*
22. Intestinal origin in front of xvi[d]............ 23
 Intestinal origin behind xvi............... 25
23. Octothecal............................. 24
 12–14 thecal.........................*pandus*
24. Intestinal origin in xiv..............*thanbulanus*
 Intestinal origin in xv................*ambiguus*
25. Quadrithecal........................... 26
 Bithecal or athecal..................*variabilis*
26. Typhlosole lacking.....................*tenuis*
 Typhlosole present.....................*kempi*
27. Spermathecal pores paired, at or lateral to *A*
 macrochaetus
 Spermathecal pores median to A, unpaired at mV..........................*palniensis*

---

[a] *P. halyi* (Michaelsen, 1898) drops out here. If intestinal origin is in xiv this species is distinguished from the *subtilis* group by the unpaired and median female pore. The same character will also distinguish *halyi* from Indian species with an intestinal origin behind xiv.

[b] *P. timidus* Cognetti, 1911, drops out here. The species is distinguished from other oriental plutelli by location of the male pores at m*BC*.

[c] *P. dubariensis* Michaelsen, 1921, drops out here. It would seem to be distinguishable from other oriental plutelli by the large, diagonally slitlike and mesially convergent male pores. However minute and superficial male pores are so typical that some further characterization probably is needed. The supposed pores may be only deep seminal grooves such as are present in *variabilis*.

[d] *P. aborensis* Stephenson, 1914, falls out here. If intestinal origin is in xvi or anteriorly the species is distinguished from the *inflexus* group by the quadrithecal battery. If intestinal origin is behind xvi, *aborensis* may be distinguished from *tenuis* and

---

[7] Presence on Queen Charlotte Island, of *Plutellus perrieri* Benham, 1892, has not been confirmed. Various searches there turned up only anthropochorous lumbricids.

kempi by the peculiar spermathecae. The unique type may have been abnormal or a *lapsus calami* may have been responsible for the anomalous location of the seminal vesicles.

Note: Only one Burmese species, *P. pandus*, could be secured at will during the active season but information as to niches inhabited by that taxon no longer is available. Absence of plutelli in so many collections in part may have been due to the small size and difficulty of distinguishing unpigmented and perhaps sluggish individuals, if not also to lack of the requisite patience and perseverance on part of the collectors. However, only one specimen of the largest Burmese species was secured, in spite of repeated searches at the type locality, by the most efficient collector.

Plutellus aborensis

1914. *Plutellus aborensis* Stephenson, *Rec. Indian Mus.* 8: p. 384. (Type locality, Rotung, Abor Country, eastern Himalayas. Holotype, in the Indian Mus.)
1923. *Plutellus aborensis*, Stephenson (*Fauna of British India*), *Oligochaeta*, p. 171.

Quadrithecal, pores in *AB*, at 7/8–8/9. Male pores, in xviii, in small porophores restricted to *AB*. (Genital markings? Clitellum?). Setae, small, lacking or unrecognizable in ii–x, *a,b*/xviii penial, $DD = \frac{1}{3}C$. (Nephropores, inconspicuous?) First dorsal pore, at 9/10. Prostomium, small, prolobous. Color, none (? unpigmented?). Segments, 385. Size, 100 by 3 mm. (to $1\frac{1}{2}$ mm. posteriorly).

Gizzard, in v. (Intestinal origin?) Last hearts, in xiii. Holandric. Seminal vesicles, in xi, xii. (Junction of male and prostatic ducts?) Penial setae, 0.88 mm. long, 1 μ thick, shaft with a gentle wavy curve ectally, tip narrowing to a sharp point, ornamentation lacking. Spermathecae, ampulla sausage-shaped, transversely placed and sessile on the parietes, diverticulum a shortly pear-shaped protuberance from the median end of the ampulla.

Distribution: Rotung, Abor Country, eastern Himalayas at 1,300 feet, Assam.

Remarks: The species is known only from the original description which stated that the seminal vesicles were attached to the anterior faces of 11/12 and 12/13. Stephenson's comment (1923) is "The situation of the seminal vesicles seems peculiar—one would have expected vesicles in xi and xii to be attached to the posterior faces of 10/11 and 11/12." Without having had an opportunity to examine the type, the author is inclined to believe that a *lapsus calami* may have been involved. Many instances of such slips probably can be found in the literature on oligochaetes, and this author has found it very difficult to catch them during the boredom of proofreading.

As unusual as the recorded location of the seminal vesicles, is the position of the diverticulum, supposedly an outgrowth of the spermathecal ampulla rather than of the duct. Also unusual is the segment number. Only one of the other oriental plutelli has more than

205 segments, *P. palniensis* Michaelsen, 1907, which has 240–260.

The specimen from which the species is alone known was immature and accordingly there is a possibility that genital markings may appear at full maturity.

Plutellus ambiguus

1931. *Plutellus ambiguus*, Gates, *Rec. Indian Mus.* **33**: p. 357. (Type locality, unknown, Types, none.)
1932. *Plutellus ambiguus*, Gates, *ibid.* **34**: p. 362.
1955. *Plutellus ambiguus*, Gates, *ibid.* **52**: p. 60.

Octothecal, pores minute and superficial, at or just median to *B*, at 5/6–8/9. Male pores, in xviii, in *AB* in a field extending from postsetal secondary furrow of xvii to presetal secondary of xix, at widest-close to 17/18-reaching just beyond *B*. Genital markings, unpaired, median, transversely elliptical, in *AA*, across 11/12, 12/13, smaller and paired, in *AB*, across 17/18. (Clitellum?) Setae, $AB < CD < BC < AA$, *a,b*/xviii penial (follicles without discrete apertures?). Nephropores, inconspicuous (?), at *D* (regularly?). First dorsal pore, at 7/8. Prostomium, prolobous. Color, none (unpigmented). (Segments?) Size, 85 by 1 mm.

Gizzard, large, in v. Intestinal origin, in xv. (Typhlosole, none but gut roof thickened slightly at mD.) Last hearts, in xiii. (Nephridia, avesiculate?) Holandric. Seminal vesicles, in xi, xii. Sperm ducts, into ental ends of prostatic ducts (?). Penial setae, 0.41 mm. long, 5 μ thick, ornamentation ectally of several fine spines. (GM glands, none, epidermis thickened only.) Spermathecae, small, duct much shorter than ampulla and almost confined to parietes, diverticulum from median face of duct, shorter than main axis, club-shaped.

Distribution: Chindwin Valley (possibly from a northern portion).

Remarks: This species is known only from an aclitellate type. Immaturity may explain absence of a ventral curvature in the clitellar region. A slight greenish tinge that characterized the worm when first studied may have been acquired after preservation. Nothing of the sort has been reported for other oriental plutelli. The only important difference from *P. inflexus* now known is the presence of hearts in xiii.

Plutellus compositus

1933. *Plutellus inflexus* var. *compositus* Gates, *Rec. Indian Mus.* **35**: p. 477. (Type locality, Tonbo. Types, None.)
1955. *Plutellus compositus*, Gates, *ibid.* **52**: p. 61.
1961. *Plutellus compositus*, Gates, *Ann. Mag. Nat. Hist.*, Ser. 13, **4**: p. 417.

Octothecal, pores minute and superficial, median to *A*, at 5/6–8/9. Male pores, in xviii, at or close to *A*, in an indistinctly delimited field reaching laterally to *B* and slightly(?) into xvii and xix. Genital markings, unpaired, median, transverse areas in *AA* across 11/12, 12/13 (17/18? 18/19?). Clitellum, saddle-shaped, xiii–xviii, occasionally reaching into xii and xix. Setae, $AB < CD$, CC ca. = $\frac{1}{2}C$, a,b/xviii penial (follicle apertures not discrete?). Nephropores, inconspicuous(?), at or just above D (throughout?). First dorsal pore, at 7/8. Prostomium, prolobous, marked off by a transverse furrow from a small triangular, posteriorly pointed tongue in i. Color, none (unpigmented). Segments, 149. Size, 50–87 by 2 mm.

Gizzard, in v. Intestinal origin, in xiv. (Typhlosole, none.) Last hearts, in xii. (Nephridia, avesiculate?) Holandric. Seminal vesicles, acinous, in xi, xii. Sperm ducts, into prostate glands near ectal ends (?). Penial setae, 0.4–0.58 mm. long, 8–12 μ thick, ornamentation ectally of 3–5 circles of rather elongately triangular spines, tip sharply pointed and flattened. (GM glands, none, epidermis thickened.) Spermathecae, small, duct shorter than ampulla and almost confined to parietes, diverticulum from median face of duct shorter than main axis, with an ovoidal to ellipsoidal seminal chamber and short stalk.

Distribution: Tonbo (Mandalay).

Variation and Abnormality: The median distance between pores of a pair decreases posteriorly in some speciments and on one worm (of 131) there was only a single pore at mV and 8/9. The ducts of the posterior pair of spermathecae, in that specimen, united as they entered the body wall.

Remarks: Relationships now appear to be closest with the *inflexus* group of species. Rank as a southern subspecies of *P. inflexus* may eventually be found to represent relationships better.

Plutellus comptus

1955. *Plutellus comptus* Gates, *Rec. Indian Mus.* **52:** p. 62. (Type locality, Hlegu. Types, none.)

Bithecal, pores minute and superficial, at B and 8/9. Male pores, in xviii, at or just lateral to B, in transversely elliptical porophores (male field indistinct or lacking?). Genital markings, small, transversely elliptical, unpaired, median, and postsetal in AA of viii, ix, xviii, paired in region of AB, presetal in xviii or across 17/18, pre- and postsetal in viii, ix. Clitellum, saddle-shaped, xiv–xviii. Setae, a,b/xviii lacking, in xx $AB < $ or = CD, $BC < $ or = AA, Nephropores, inconspicuous(?), at B (?). First dorsal pore, at 7/8, 8/9. Prostomium, epilobous. Color, none (unpigmented). (Segments?) Size, 24–57 by 1–2 mm.

Gizzard, in v. Intestinal origin, in xiv. (Typhlosole, none.) Last hearts, in xii. (Nephridia, avesiculate?) Holandric. Seminal vesicles, in xi, xii. Sperm ducts, into prostatic ducts close to parietes(?). (Penial setae, none.) (GM glands, none, epidermis thickened only?) Spermathecae, medium-sized, duct shorter than ampulla and nearly as thick except in parietes, diverticulum from median face of duct-sessile

or with a very short stalk and an ellipsoidal seminal chamber.

Reproduction: Presumably biparental as sperm are exchanged during copulation.

Distribution: Twante (Hanthawaddy). Pegu (Pegu). Hlegu (Insein). Paukkaung (Prome).

Remarks: The clitellum originally was assumed to have regressed in the single worm having spermatozoa in a spermathecal seminal chamber. The assumption may have been incorrect as on various subsequent occasions spermathecal sperm have been found in aclitellate individuals showing no recognizable indications of having ever had a clitellar tumescence. A ventral curvature in the clitellar region of the single clitellate specimen apparently was lacking.

The relation to *P. geminatus*, suggested by the number of spermathecae, seems not to be close.

Plutellus geminatus

1961. *Plutellus geminatus* Gates, *Ann. Mag. Nat. Hist.*, Ser. 13, 4: p. 418. (Type locality, Thazi. Types, at TTRS).

Bithecal, pores superficial but not minute, transverse slits between B and mBC at 8/9. Male pores, in xviii, at B, each at center of a small circular protuberance from a (usually indistinct?) male field. Genital markings, transversely elliptical, presetal, in AB of xiii (xviii and xix?). Clitellum, saddle-shaped, xiii–xviii. Setae, $AB < $ or ca. = CD, $DD < \frac{1}{2}C$, a,b/xviii penial (follicle apertures discrete?). (Nephropores, inconspicuous and at? Dorsal pores, lacking?) Prostomium, epilobous, tongue open. Color, none (unpigmented). Segments, 65–98. Size, 23–36 by 1 mm.

Gizzard, weak, in v. Intestinal origin, in xiv. Typhlosole, none, though roof slightly thickened at mD. Last hearts, in xii. (Nephridia, avesiculate?) Holandric. Seminal vesicles, in xii. (Junction of sperm and prostatic ducts?) Penial setae, ornamented ectally by two longitudinal rows of triangular teeth. (GM glands, none, epidermis thickened only?) Spermathecae, large, duct barrel-shaped and shorter than ampulla, diverticula (2) from median and lateral faces of duct slightly nearer ental end.

Reproduction: Presumably biparental as sperm are exchanged during copulation.

Distribution: Thazi (Meiktila).

Abnormality: No. 1. Right spermatheca, lacking.

Remarks: Known only from seven clitellate types.

Dorsal pores, if present, should have been recognizable at the posterior end where the body was strongly bent ventrally.

Two pairs of vesicles are to be expected in holandric species with biparental reproduction. None were present in ix and failure to recognize a pair in xi may have resulted from concealment of organs in a very sticky coelomic coagulum.

Relationships with *P. comptus*, suggested by the number of spermathecae, seem not to be close.

Plutellus himalayanus

1945. *Plutellus himalayanus* Gates, *Jour. Roy. Asiatic Soc. Bengal*, (*Sci.*) **11**: p. 71. (Type locality, Changu, Sikkim. Type, in the Indian Mus.)

Bithecal, pores (minute and superficial?) at or slightly median to C, at 8/9. Male pores, in xviii, minute and superficial, each within a small, circular, translucent area included in a conspicuously raised, shortly elliptical porophore lateral to a deep median depression. Genital markings, small, nearly circular, distinctly delimited, in transversely placed rows of three or four, each row within a distinctly demarcated, transversely rectangular field that reaches B, markings across 12/13 and 22/23 (also 21/22?). Clitellum, saddle-shaped, without dorsal pores and intersegmental furrows, from 17/18 into a posterior portion of xiii. Setae, behind xxv $AB < CD < BC < AA, DD < \frac{1}{2}C$, a,b/xviii lacking(?). Nephropores, inconspicuous(?). Dorsal pores, at least in front of clitellum, unrecognizable. Prostomium, slightly epilobous. Color, none (unpigmented). (Segments?) Size, 70 by 2 mm.

Gizzard, in vi. Intestinal origin, in xv. (Typhlosole, probably none but on roof at mD a low ridge of triangular section.) Last hearts, in xii. (Nephridia, avesiculate?) Holandric. Seminal vesicles, small, in xi, xii. (Junction of sperm and prostatic ducts? GM glands, none.) Spermathecae, fairly large, ducts almost confined to parietes, diverticula slenderly club-shaped, one laterally, one mesially and one anteriorly from each duct.

Reproduction: Presumably biparental as sperm are exchanged during copulation.

Distribution: Changu, Sikkim.

Remarks: *P. himalayanus* is known only from the original description of a single specimen. The species is distinguished from all other oriental plutelli by location of the single pair of spermathecal pores at 8/9 and by the tridiverticulate spermathecae.

Male porophores and GM fields are unlikely to be protuberant *in vivo* but whether, in that state, porophores are withdrawn into some sort of a median depression is unknown.

Relationships of *P. himalayanus* presumably are with unknown species in the large region, extending from the Himalayas to south India, where plutelli have not yet been collected.

Plutellus inflexus

1931. *Plutellus inflexus* Stephenson, *Rec. Indian Mus.* **33**: p. 179. (Type locality, Kalewa. Types, in the British Museum.)
1933. *Plutellus inflexus* var. *typicus*, Gates, *idem* **35**: p. 477.
1955. *Plutellus inflexus*, Gates, *ibid.* **52**: p. 64. (After examination of types returned to this author by Stephenson.)

Octothecal, pores at B, at 5/6–8/9. Male pores, in xviii, slightly lateral to A, in small indefinite papillae. Genital markings, unpaired, median, in AA, across 11/12, 12/13. Clitellum, saddle-shaped, xiii–xviii,

xix/n, xix/eq. Setae, $AA > BC, CD = 2AB, DD = \frac{1}{2}C$, a,b/xviii penial (follicle apertures not discrete?). Nephropores, inconspicuous (?), at D (?). First dorsal pore, at 7/8. Prostomium, prolobous (?). Color, none (unpigmented). Segments, 150. Size, 40–55 by 1.5–2 mm.

Gizzard, in v. Intestinal origin, in xiv. Typhlosole, none though gut roof thickened slightly at mD. Last hearts, in xii. (Nephridia, avesiculate?) Holandric. Seminal vesicles, in xi, xii. Sperm ducts, into middle (?) of prostatic ducts. Penial setae, 0.4–0.45 mm. long, 4 μ thick at midshaft, 5 μ at base, bowed slightly with considerable basal curve, ornamentation ectally of several conspicuous spines, tip tapering gently to a point. GM glands, none (? genital markings only areas of slight epidermal thinning and thickening?). Spermathecae, small, duct not clearly marked off from ampulla (and almost confined to parietes?), diverticulum from median face of duct, much shorter than main axis, narrowed ectally.

Reproduction: Presumably biparental as sperm are exchanged during copulation.

Distribution: Kalewa (Lower Chindwin).

Remarks: A short transverse vessel in each of segments, vi, vii, and viii connects the two extra-esophageals with each other. That condition has not been reported from other oriental species.

P. inflexus is related to *P. ambiguus, compositus, thanbulanus* and *pandus* but more especially to *compositus* which still has the intestinal origin in xiv and no hearts in xiii. North, perhaps of the *inflexus-compositus* range, *ambiguus* has pushed back the intestinal origin into xv and has acquired hearts in xiii. To the south of the *inflexus-compositus* range, *thanbulanus* with the intestinal origin still in xiv has acquired hearts in xiii. Still farther south, and in the deltas region, *pandus* not only has hearts in xiii and an intestinal origin in xv but also an extra pair or two of spermathecae in such very unusual locations as x and xi.

Plutellus leucaspis

1955. *Plutellus leucaspis* Gates, *Rec. Indian Mus.* **62**: p. 65. (Type locality, Yandoon. Types, none.)

Quadrithecal, pores minute and superficial, at B, at 7/8–8/9. Male pores, in xx, at B, near lateral margins of a transverse male field. Genital markings, small, transversely elliptical, one postsetal pair in male field, one pair in xv within a presetal field similar to that of xx, a postsetal pair in AA of vii, a postsetal median in viii. Clitellum, saddle-shaped, xiv–xviii. Setae, ventral setae enlarged anteriorly, in xxii AB slightly $< CD, AA = 2BC, a,b$/xx penial (follicle apertures not discrete?). Nephropores, inconspicuous(?), at or near D (?). First dorsal pore, at 7/8, 8/9. Prostomium, epilobous, tongue open. Color, none (unpigmented.) (Segments?) Size, 25 by 2 mm.

Gizzard, rudimentary, in v. Intestinal origin, in xiv. (Typhlosole, none.) Last hearts, in xii. (Nephridia, avesiculate?) Holandric. Seminal vesicles, in xi, xii. Sperm ducts, into prostatic ducts close to parietes. Penial setae, 0.36–0.37 mm. long, 9–10 μ thick, shaft nearly straight, ornamentation ectally of 3–5 rows of triangular teeth, tip flat and truncate. (GM glands, none, epidermis thickened only?) Spermathecae, medium-sized, duct slightly shorter than ampulla and widened entally, diverticulum from median face of duct just below ampulla, shorter than main axis, with ellipsoidal seminal chamber and a short stalk.

Reproduction: Presumably biparental as sperm are exchanged during copulation.

Distribution: Yandoon (Maubin).

Remarks: Known only from the holotype, a posterior amputee that was aclitellate (in October) and presumably postsexual.

P. leucaspis is much like an Indian species, *P. exilis* Gates, 1945, known only from Allahabad and Naini in the Gangetic Plains. The major difference is location of the male pores in xx instead of the usual metamere. Probability that such homoeosis is a result of regeneration seems to be quite small. If male pores should prove to be normally in xviii in the Burmese population, possibility of synonymy with *exilis* and of transportation of the species from one region to another should be considered.

Plutellus longus

1933. *Plutellus* sp., Gates, *Rec. Indian Mus.* **35**: p. 484.
1955. *Plutellus longus* Gates, *ibid.* **52**: p. 66. (Type locality, Chaukan near Pegu. Types, none.)

(Quadrithecal, pores at 7/8–8/9?) Male pores, in xviii, at *A*. (Genital markings, lacking? Clitellum? Dorsal pores? Nephropores? Segments? Prostomium?) Setae, of ii–vii much enlarged. Color, none (unpigmented). Size, 250 by 2 mm.

(Gizzard, in v?) Intestinal origin, in xvii. (Typhlosole?) Last hearts, in xiii. (Nephridia? Holandric? Seminal vesicles? Junction of sperm and prostatic ducts? Penial setae? Spermathecae?)

Distribution: Chaukan (Pegu).

Remarks: Known only from an aclitellate specimen. The type was secured by the laboratory steward K. John who, during subsequent searches for the species, was able to find only *D. gracilis*.

The markedly enlarged anterior setae, with their black tips, give the anterior end in this species an appearance much like that characterizing contracted specimens of *Drawida gracilis*.

The species is distinguished from all known oriental plutelli by the much greater length. That length together with an intestinal origin in xvii and the unusually obvious enlargement of the setae in ii–vii, a combination of three rare characters, guarantees recognition of the species when further material is obtained.

Plutellus macer

1955. *Plutellus macer* Gates, *Rec. Indian Mus.* **52**: p. 67. (Type locality, Lashio. Types, none.)

Quadrithecal, pores superficial, large, transverse slits centered at *C*, about halfway between equators and posterior boundaries of vii–viii. Male pores, in xviii, at *B*, in small circular porophores extending from *A* into *BC*. (Genital markings, none.) Clitellum, saddle-shaped, reaching into xiii and xviii. Setae, in xxi *AB* < *CD* < *BC* < *AA*, *a,b*/xviii penial (follicle apertures not discrete?). (Nephropores, inconspicuous? Dorsal pores? Segments?) Prostomium, epilobous, tongue open. Color, none (unpigmented). Size, 32 by 1 mm.

Gizzard, none. Intestinal origin, in xiv. (Typhlosole, none.) Last hearts, in xii. (Nephridia, avesiculate?) Holandric. Seminal vesicles, in ix and xii. Sperm ducts, into prostatic ducts entally near glands (?). Penial setae, 0.32–0.37 mm. long, 19 μ thick, unornamented (?), tip sharply pointed. Spermathecae, large, duct longer than and nearly as thick as ampulla, diverticulum spheroidal and sessile on median face of duct just below ampulla.

Reproduction: Presumably biparental as sperm are exchanged during copulation.

Distribution: Lashio (Shan Plateau).

Remarks: Known only from the holotype, the single plutellus obtained in all of the Shan Plateau. Relationships were thought to be with *P. subtilis*, one of the Thaton district plutelli.

Plutellus montanus

1961. *Plutellus montanus* Gates, *Ann. Mag. Nat. Hist.*, Ser. 13, **4**: p. 419. (Type locality, Mount Popa. Types, none.)

Quadrithecal, pores in *AA* but slightly nearer to *A* than to each other, at 7/8–8/9. Male pores, in xviii, slightly median to *B* (male field?). Genital markings, transversely elliptical, unpaired, median, in *AA*, presetal in vii (xxi, xxii?). Clitellum, saddle-shaped, xiii–xviii. Setae, *AB* < *CD*, *DD* < ½*C* and posteriorly may = *AA* as well as *BC* and *CD*, *a,b*/xviii penial (apertures of *a* follicles discrete). Nephropores, inconspicuous(?). (Dorsal pores, none.) Prostomium, epilobous, tongue open. Color, none (unpigmented). Segments, 77–82. Size, 23–32 by 1 mm.

Gizzard, weak or rudimentary, in v. Intestinal origin, in xiv. (Typhlosole, none.) Last hearts, in xii. (Nephridia avesiculate?) Holandric. Seminal vesicles, in ix and xii. Sperm ducts, into prostatic ducts close to parietes. Penial setae, slender. (GM glands, none?) Spermathecae, large, duct shorter than ampulla, diverticulum from lateral face of duct close to parietes, shorter than main axis, with a shortly ellipsoidal seminal chamber and a slender stalk.

Reproduction: Presumably biparental as sperm are exchanged during copulation.

Distribution: Mount Popa (Myingyan).

Remarks: Known only from three clitellate types. *P. montanus* from a mountain in the semidesert central part of Burma appears to be closely related to *P. subtilis* which was secured from a lowland locality in a northern portion of peninsular Burma where rainfall is much greater.

Plutellus pandus

1933. *Plutellus pandus* Gates, *Rec. Indian Mus.* **35**: p. 480.
(Type locality, Rangoon. Types, none.)
1955. *Plutellus pandus*, Gates, *ibid.* **52**: p. 68.
1960. *Plutellus pandus*, Gates, *American Midland Nat.* **66**: p. 62.

Duodecathecal (rarely 14-thecal), pores at or slightly median to A, at 5/6–10/11 (11/12). Male pores, in xviii, at A, in an indistinctly delimited field that reaches lateral to A (and into xvii, xix?). Genital markings, median, unpaired, transverse, in BB, across some or all of 9/10–12/13. Clitellum, saddle-shaped, xiii–xviii, occasionally extending into xii and xix. Setae, $AB < CD, c,d$, longer and thicker than ventral setae, a,b/xviii penial (follicles without discrete apertures?). Nephropores, inconspicuous(?). First dorsal pore, at 7/8. Prostomium, prolobous, often with a small triangular tongue demarcated in i so that apex points posteriorly. Color, none (unpigmented). Segments, 185–205. Size, 57–115 by 2 mm.

Gizzard, in v. Intestinal origin, in xv. (Typhlosole, none.) Last hearts, in xiii. Nephridia, avesiculate(?). Holandric. Seminal vesicles, in xi, xii. Sperm ducts, into middle(?) of prostatic ducts. Penial setae, 0.33–0.60 mm. long, 5–11 μ thick at base, 6–7 μ at midshaft, ornamentation ectally of 15–20 irregular circles of triangular teeth, tip flattened and bluntly rounded or narrowed to a point. (GM glands, none, epidermis thickened only.) Spermathecae, small, duct slightly shorter than ampulla, diverticulum from median face of duct, shorter than main axis, with short stalk and ovoidal to ellipsoidal seminal chamber.

Reproduction: Presumably biparental as sperm are exchanged during copulation. Spermathecae of xi and xii, it should be noted, contained sperm.

Breeding season, in Rangoon, may be earlier than for most endemics, from last of June through August, sometimes into early portion of September.

Distribution: Twante, Rangoon (Hanthawaddy). Yandoon (Maubin). Pegu (Pegu). Hmawbi, Wanetchaung (Insein). Henzada (Henzada).

Remarks: All preserved adults with a certain amount of clitellar development are "hooked,"[8] strongly bent ventrally in the clitellar region. Younger specimens, preserved in the same way, are straight.

The spermathecal battery of this species can be regarded as primitive and for most individuals of the species already shortened and perhaps from one formerly much more extensive. A more usual view is

[8] Other plutelli for which a similar hooking was recorded are *compositus* and *inflexus*.

that a quadrithecal battery was primitive. In that case *pandus* may be still adding to a battery already duodecathecal.

Relationships presumably are closest with octothecal species of the *inflexus* group. Possibly the youngest of that group, *pandus* seemingly is restricted to the deltas region geologically the youngest part of Burma. In spite of the differences, *pandus* may be closest to *thanbulanus* for geographical reasons.

Plutellus pauxillulus

1955. *Plutellus pauxillulus* Gates, *Rec. Indian Mus.* **52**: p. 70.
(Type locality, Augsaing. Types, none.)

Sexthecal, pores minute and superficial, at or just lateral to B, at 6/7–8/9. Male pores, in xix, at B, in circular porophores. (Male field? Genital markings?) Clitellum, saddle-shaped, extending into xiii and xix. Setae, in xxi $AB < CD < BC < AA$, posteriorly d progressively near to mD until $DD < \frac{1}{4}C$, a,b/xix lacking. Nephropores, inconspicuous(?). First dorsal pore, at 8/9. Prostomium, prolobous. Color, none (unpigmented). (Segments?) Size, 28 by 1 mm.

(Gizzard?) Intestinal origin, in xiv. (Typhlosole, none.) Last hearts, in xii. (Nephridia, avesiculate?) Holandric. Seminal vesicles, in (xi?) xii. Sperm ducts, into prostatic ducts close to parietes. (Penial setae, none.) Spermathecae, large, duct slender and much shorter than ampulla, diverticulum from median face of duct close to parietes, about as long as main axis and club-shaped.

Reproduction: Possibly biparental as sperm are matured. However, no evidence of exchange during copulation was found.

Distribution: Aungsaing (Thaton).

Abnormality and Regeneration: The type is abnormal. Anterior to the female pore metamere which should be xiv there are only 11 normal segments. A short region between ii and iii is marked off by a circumferential furrow into two parts without usual evidence of metameric differentiation (setae and nephropores). The abnormalities involved are known to arise during regeneration. Amputation may have been in region of 4/5–5/6. Except for this one instance, cephalic regeneration has not been recorded for oriental plutelli.

Remarks: Known only from the holotype which is abnormal anteriorly. Probability that homoeosis of the male pores is a result of regeneration seems to be small (*cf.* reorganization in perionyxes where regenerative capacity almost certainly is much greater than in any of the oriental plutelli).

Plutellus pratensis

1932. *Plutellus pratensis* Gates, *Rec. Indian Mus.* **34**: p. 363.
(Type locality, Toungoo. Types, none.)
1955. *Plutellus pratensis*, Gates, *ibid.* **52**: p. 71.

Quadrithecal, pores minute and superficial, at or close to B, near 7/8–8/9 in viii, ix. Male pores, in xviii, at or just lateral to B, in small circular tubercles on lateral walls of a median depression. Genital markings, transversely elliptical areas centered in AB, small and presetal in viii–x, larger and across 16/17, 17/18, (18/19?) and 19/20. Clitellum, saddle-shaped, xiii–xviii. Setae, $AB < CD$, $BC <$ or $ca. = AA$, a,b/xviii penial (follicle apertures not discrete?). Nephropores, inconspicuous (?), at C(?) or nearer B (? alternated?). First dorsal pore, at 7/8. Prostomium, prolobous. Color, none (unpigmented). Segments, 167. Size, 55–85 by 3–4 mm.

Gizzard, in v. Intestinal origin, in xv. Typhlosole, rudimentary, present from xxiv, interrupted midsegmentally. Last hearts, in xii. Nephridia, avesiculate(?). Holandric. Seminal vesicles, in xi, xii. Sperm ducts, into prostatic ducts slightly below glands (?). Penial setae, 0.44–0.65 mm. long, 10–11 μ thick, ornamentation ectally of 6–11 circles of fine spines, tip narrowing to a terminal filament(?). (GM glands, none, epidermis thickened only.) Spermathecae, medium sized, duct much shorter than ampulla, diverticulum from median face of duct close to parietes, shorter than main axis and club-shaped.

Reproduction: Presumably biparental as sperm are exchanged during copulation.

Distribution: Toungoo, Pegu Yomas to the west of Toungoo (Toungoo).

Remarks: Known only from 4 specimens secured at three sites some 7 miles or more apart.

Plutellus rudis

1961. *Plutellus rudis* Gates, *Ann. Mag. Nat. Hist.*, Ser. 13, 4: p. 420. (Type locality, unknown. Types, none.)

Quadrithecal, pores superficial but not minute, at A and 7/8–8/9. Male pores, in xviii, at A (?), in a pitlike depression along with apertures of penisetal follicles. Genital markings, transversely elliptical, presetal, in region of AB of x, xxii, xxiii, unpaired and median in AA of xvi, xvii. Clitellum, saddle-shaped, xiii–xviii. Setae, $AB < CD$ throughout, DD becoming smaller posteriorly until $ca. = AA$, a,b/xviii penial (apertures of follicles seemingly discrete). Nephropores, inconspicuous (?), at C(?). (Dorsal pores, none.) Prostomium, epilobous, tongue open. Color, none (unpigmented). Segments, 90–104. Size, 40–85 by 1.5–2 mm.

Gizzard, small, in v. Intestinal origin, in xiv. (Typhlosole, none.) Last hearts, in xii. Nephridia, lacking in i–xi(?), avesiculate(?). Holandric. Seminal vesicles, in ix and xii. Sperm ducts, into prostatic ducts close to parietes. Penial setae, ornamented ectally with two (?) longitudinal rows of triangular teeth. (GM glands, none?) Spermathecae, medium-sized, duct short and rather stout, diverticulum from duct nearer ental end, shorter than main axis.

Reproduction: Presumably biparental as sperm are exchanged during copulation.

Distribution: Unknown.

Regeneration: Tail regenerates, of 15 segments at 80/81 and of 8 segments of 81/82, were recorded from the type series which probably contained several other individuals with regenerated tails. Other records of caudal regeneration by oriental plutelli are of *P. tenuis.*

Remarks: All of the 24 specimens that were available were from one site presumably in Burma.

The species seems to be close to *P. montanus* and *subtilis*, the locations of the spermathecal pores are intermediate between those of the other 2 species.

Plutellus sikkimensis

1907. *Plutellus sikkimensis* Michaelsen, *Mitt. Naturhist. Mus. Hamburg* **24**: p. 147. (Type locality, Sandakphu, Darjiling district, eastern Himalayas. Types, 11, in the Indian Museum. One or more may still be in the Hamburg Museum.)

1909. *Plutellus sikkimensis*, Michaelsen, *Mem. Indian Mus.* 1: p. 155.

1916. *Plutellus sikkimensis*, Michaelsen, *Svenska Vetenks. Akad. Handl.* **52**: 13, p. 43.

1923. *Plutellus sikkimensis*, Stephenson, (*Fauna of British India*), *Oligochaeta*, p. 177.

Decathecal(?), pores just median to B and at 4/5–8/9. Male pores, in xviii, at B, within a median field that reaches into xvii and xix. Genital markings, paired, transversely elliptical, in AB and across 12/13 (sometimes connected with each other by a glandular median area). (Clitellum? Nephropores inconspicuous?) Setae, rather stout, widely spaced, $DD < \frac{1}{2}C$, a,b/xviii penial. First dorsal pore, at 6/7. Prostomium, epilobous, tongue open. Color, none (unpigmented). Segments, 90. Size, 42 by $ca.$ 1 mm.

Gizzard, small, in v. Intestinal origin, in xiv. Last hearts, in xii. Holandric. Seminal vesicles, "apparently" in ix, xi, xii. (Junction of male and prostatic ducts?) Penial setae, $ca.$ 0.33 mm. long, 9 μ thick at middle, ectal portion of shaft bent at an obtuse angle and somewhat tapering, ending in a sharply pointed, slightly recurved, slender tip, ornamentation ectally of 9 oblique circlets each of about 9 very large teeth. (Spermathecae, diverticulate? GM glands?)

Distribution: Sandakphu, Bengal Province, India.

Remarks: The species is known only from immature specimens. Spermathecae were solid, still within the body wall and presumably too juvenile for specific characteristics to be recognizable. Location of the spermathecal pores, probably estimated from sections, should be confirmed. Genital markings, at maturity, may have a quite different appearance.

Supposed presence of seminal vesicles in ix allows an assumption that a more primitive condition was octovesiculate as in certain lumbricids today.

One other species, *P. pandus*, has more than four pairs of spermathecae but number of those organs at present seems unlikely to have the systematic significance allowed in the classical system.

Plutellus subtilis

1955. *Plutellus subtilis* Gates, *Rec. Indian Mus.* **52**: p. 73. (Type locality, Kyaikto. Types, none.)

Quadrithecal, pores at or just lateral to B, at 7/8–8/9. Male pores, in xviii, at B, laterally in transversely oval porophores. Genital markings, in region of AB, presetal in xviii. Clitellum, saddle-shaped, xiv–xvii. Setae, $AB < CD$, $AA < BC$, $a,b/$ xviii penial, (a follicles at least with discrete apertures). Nephropores, inconspicuous(?), at C(?). (First dorsal pore? Prostomium?) Color, none (unpigmented). (Segments?) Size, 40 by 1 mm.

Gizzard, rudimentary, in v. Intestinal origin, in xiv. (Typhlosole, none.) Last hearts, in xii. Nephridia, avesiculate(?). Holandric. Seminal vesicles, in ix and xii. Sperm ducts, into prostatic ducts slightly ectal to their midpoints. Penial setae, 0.37–0.38 mm. long, 5 μ thick, slightly curved ectally, (ornamentation?), tip bluntly rounded. Spermathecae, large, duct longer than ampulla, bulbous or spindle-shaped, diverticulum from middle of duct laterally, shorter than the duct, with a spheroidal to ellipsoidal seminal chamber and a slightly longer stalk. (GM glands none?)

Reproduction: Presumably biparental as sperm are exchanged during copulation.

Distribution: Kyaikto (Thaton).

Remarks: The species is known only from the clitellate type and was thought to be related to *P. macer*, the single plutellus recorded from all of the Shan Plateau.

Plutellus tenuis

1961. *Plutellus tenuis* Gates, *Ann. Mag. Nat. Hist.*, Ser. 13, **4**; p. 422. (Type locality, Tharrawaddy. Types, at TTRS.)

Quadrithecal, pores minute and superficial, in AA, slightly nearer A than each other, at or slightly behind equators of viii and ix. Male pores, in xviii, at A (?), within a transversely elliptical field. (Genital markings, none.) Clitellum, saddle-shaped (?), xiii–xviii. Setae, $AB < CD$, $DD < \frac{1}{2}C$, $a,b/$xviii lacking. Nephropores, inconspicuous(?), at C(?). First dorsal pore, at 10/11, 12/13. Prostomium, epilobous, tongue open or tapering to a posterior point. Color, none (unpigmented). Segments, 109–130. Size, 50–70 by *ca.* 1.5 mm.

Gizzard, large and long, in v. Intestinal origin, in xix. (Typhlosole, none.) Last hearts, in xiii. Nephridia, avesiculate(?). Holandric. Seminal vesicles, in xii. Sperm ducts, into prostate glands, near ectal ends. (Penial setae, none.) Spermathecae, small, duct shorter than ampulla, diverticulum from median face of duct close to parietes, shorter than main axis.

Reproduction: Possibly biparental as sperm are matured. No proof that sperm had been exchanged during copulation was found and the various abnormalities recorded for this species are of kinds often associated with parthenogenesis.

Distribution: Tharrawaddy (Tharrawaddy).

Abnormality and Polymorphism: Abnormal No. 1. Right spermatheca of ix with two shortly digitiform diverticula from median and lateral faces of the duct close to parietes. No. 2. Two spermathecae doubled, ducts united in the parietes. The third spermatheca has three diverticula. No. 3. Two pairs of spermathecae in viii, one pair just about as far in front of the segmental equator as the other is behind, all spermathecal pores in two regular longitudinal ranks. No. 4. Bithecal, the anterior pair lacking.

Each of the above-mentioned characters is abnormal only because markedly divergent from normal—normal being that condition prevailing through a majority of individuals. Each of those deviations has, however, appeared in the evolution of various species in several families. Some at least of the changes appear in uniparental lineages after the establishment of parthenogenesis. Retention of spermathecae in a juvenile condition at maturity is incompatible with biparental reproduction.

Regeneration: A tail regenerate at 106/107 had 15 segments. A tail regenerate at 118/119 had 8 segments but metameric differentiation was incomplete.

Remarks: The species is known only from a type series of 92 specimens in which spermathecae appeared to be juvenile even when spermatozoal iridescence was recognizable on male funnels.

P. tenuis, except for absence of a typhlosole, is much like *P. kempi* Stephenson, 1924, from the Palni Hills of South India. Convergence presumably is involved as both species are believed to be autochthonous, one near the southwestern and the other near the southeastern boundary of the range (as presently known) of the oriental plutelli.

Plutellus thanbulanus

1955. *Plutellus thanbulanus* Gates, *Rec. Indian Mus.* **52**: p. 74. (Type locality, Thanbula. Types, none.)

Octothecal, pores minute and superficial, at B and 5/6–8/9. Male pores, in xviii, at B, in small indistinctly delimited porophores extending from A slightly into BC. (Genital markings, none? Clitellum? Setae? Nephropores? Dorsal pores? Prostomium? Segments? Size?) Color, none (unpigmented).

Gizzard, in v (?). Intestinal origin, in xiv. (Typhlosole, none.) Last hearts, in xiii. Nephridia, avesiculate(?). Holandric. Seminal vesicles, in (xi?), xii. (Sperm ducts, into prostatic ducts entally? Penial setae?) Spermathecae, duct almost confined to parietes, diverticulum from median face of duct, with short stalk and a spheroidal to ellipsoidal seminal chamber.

Reproduction: Presumably biparental as sperm are exchanged during copulation.

Distribution: Thanbula (Thayetmyo).

Remarks: The species is known from four juvenile and aclitellate types. Presence of sperm in spermathecae of two individuals with no trace of present or past clitellar tumescence again suggests exchange of sperm before the clitellum appears.

Relationships, as indicated by the spermathecal battery, are with the *inflexus* group.

Pontodrilus

1874. *Pontodrilus* Perrier, *Compt. Rend. Acad. Sci. Paris* **78**: p. 1582. (Type species, *P. marionis* Perrier, 1874 = *Lumbricus litoralis* Grube, 1855.)

Digestive system, with an intestinal origin behind xiii, but without calciferous and supra-intestinal glands, intestinal caeca and typhlosoles. Vascular system, with complete dorsal (single) and ventral trunks but without a subneural, with paired extra-esophageals passing onto gut in xii–xiii and uniting mesially in xiv–xv, (a short supra-esophageal?), hearts of v–ix slender, of x–xiii much thicker, all lateral(?). Excretory system, of holoic (avesiculate?) nephridia, lacking in preclitellar segments. Septa, all present from 4/5. Setae, 8 per segment. Nephropores, inconspicuous. Pigment, lacking.

Biprostatic, male pores (common openings of sperm and prostatic ducts), in xviii. Female pores, in xiv. Spermathecal pores, at 7/8–8/9. Metagynous, ovaries fan-shaped (? and with several egg-strings? Ovisacs?).

Distribution: Circumglobal, on seashores in the tropics and warmer parts of temperate zones in both hemispheres.

Nothing is known as to the original home of the species (one guess, in region of Australia-New Zealand) or as to how species have spread around the world.

Systematics: *Pontodrilus* has had a checkered history. Michaelsen has, at various times, regarded it as a distinct genus or as a subgenus of *Plutellus*. Stephenson (1923, 1930) retained generic rank and included a terrestrial Ceylon species (*agnesae* Stephenson, 1915) as well as a lacustrine New Zealand species (*Plutellus lacustris* Benham, 1904) with 8 spermathecae and with penial setae.

Perrier's genus is retained but only for the littoral species that quite obviously are physiologically distinct. The genus is distinguished morphologically from nearly all species of the phylogenetically defined *Plutellus* by absence of nephridia in preclitellar segments.

Three littoral species were recognized in the last revision of the genus (Michaelsen, 1910) but they were so variable as to require recognition of a number of "formae" mostly distinguished by length and shape of prostatic ducts. Other structure, there is some reason to believe, will provide better characters for species or subspecies delimitation (*cf.* for example,

Pontodrilus gracilis Gates, 1943 and *Cryptodrilus insularis* Rosa, 1891).

Presently recognized as belonging to *P. bermudensis* are only such individuals as are known to have the structure indicated in the species précis. Doubtless such conservatism has resulted in a considerable derogation of the distribution.

Species currently excluded from *Pontodrilus*, *agnesae* and *lacustris*, are returned to the phylogenetically defined *Plutellus* where they will not significantly increase the confusion already prevailing therein.

Remarks: Most specimens available to the author for study were not statisfactorily preserved to enable adequate characterization of vascular and excretory systems. As so often, a major trunk, such as the supra-esophageal, may be completely unrecognizable in dissections if empty. When the supra-esophageal was filled with blood no trace of branches to it from the hearts of x–xiii was recognized. The hearts of x–xiii were easily traced to the dorsal trunk. If such hearts really are lateral, that characterization may well distinguish *Pontodrilus* from *Plutellus*. If the pontodriles do have connections from the hearts of x–xiii to the supra-esophageal trunk there would then seem to be some physiological difference at least from conditions characterizing preserved specimens of most other genera and species (in which the bifurcation to the dorsal trunk usually is empty and difficult to find while the bifurcation to the supra-esophageal is distended with blood).

Whether pontodriles are agiceriate or with a rudimentary gizzard seems not to have been determined.

Pontodrilus bermudensis

1891. *Pontodrilus bermudensis* Beddard, *Ann. Mag. Nat. Hist.*, Ser. 6, **7**: p. 96. (Type locality, Bermuda. Types none.)

1926. *Pontodrilus bermudensis*, Gates, *Rec. Indian Mus.* **28**: p. 150.

1961. *Pontodrilus bermudensis*, Gates, *Burma Res. Soc. 50th Anniv. Pub. No. 1*: p. 57.

1970. *Pontodrilus bermudensis*, Soota & Julka, *Proc. Zool. Soc. Calcutta* **23**: p. 202.

?

1936. *Pontodrilus* sp., Gates, *Rec. Indian Mus.* **38**: p. 379.

Spermathecal pores at or just lateral to *B*. Male pores, minute, at *B*, on small papillae on lateral walls of longitudinal depressions median to longitudinal ridges(?). Genital markings, unpaired, median, transversely elliptical, across 19/20 and less often 12/13, 13/14. Clitellum, saddle-shaped, xiii–xvii, xviii. Setae, ornamented ectally, *a,b*–xviii lacking, *AB* < *CD*, *AA* and *BC* may *ca.* = *CD*, *DD* < ½C. Nephropores, inconspicuous(?), at *C* (?). (Dorsal pores, none.) Prostomium, epilobous. Color, none (unpigmented). Segments, 78–120. Size, 32–120 by 2–4 mm.

Septa, 5/6–12/13 muscular. Gizzard, none (or very rudimentary? and then in v?). Intestinal origin, in xvii. Nephridia, discoidal, avesiculate (?), lacking in i–xii and xiv, small in xiii, larger from xv, cells of peritoneal investment hypertrophied. Holandric. Seminal vesicles, acinous, in xi, xii. Sperm ducts, into prostatic ducts at parietes (? or more entally?). Prostatic ducts, to 2 mm. long, curved into a crescent shape, with muscular sheen, narrowed at each end. (Penial setae, none.) (GM glands, none, epidermis thickened.) Spermathecae, long enough to reach dorsal parietes, duct shorter than ampulla and emerging from parietes only at equators of viii, ix, diverticulum digitiform to club-shaped, from duct within parietes and emerging into coelom just behind 7/8, 8/9.

Reproduction: Presumably biparental as sperm are exchanged during copulation.

The breeding period is unknown. It may be short. Various collections, from a Texas shore, contained only immatures.

Habitats: Under logs on sandy soil away from seashore (Burma). Under algae thrown up by waves on wet beach (Palmyra). In mud with high content of organic matter and of salt, salinity in lagoon 10 per cent (Laysan). Under seaweed on beach (St. Croix, Virgin Islands).

Worms were in quantity, according to the collector, in the salty Laysan mud.

Distribution: Kadonkani (Pyapon), Burma. Andaman Islands. Car Nicobar. Vietnam, Hainan, Java, Christmas Island, Borneo, Celebes, Aru, New Guinea, Australia. Palmyra Atoll, Fannin Island, Laysan. Texas, Mississippi, Louisiana (new records), Florida, Virginia. Mexico, Bermuda, Bahamas, Dry Tortugas, Haiti, Jamaica, Virgin Islands, Mona Island. Colombia, Brazil. Congo, Angola, South Africa, Tanzania, Madagascar. Laccadive Islands, India (east and west coasts). Maldive Islands, Ceylon.

Biology: The gut sometimes contained a gelatinous slime with sparsely scattered reddish brown flecks. Often the gut lumen was filled with bits of dead coral, shells of larval gastropods and of two foraminiferas, *Amphistegina lessonii* and *Marginopera vertebralis*. Length of such objects may be as much as 1½ mm. A bit of coral or of limestone, of appropriate size and bulging gut wall in v, vi, or vii, can produce an appearance of a strong gizzard with marked muscular sheen. Many of these worms presumably subsist on such microscopic life as exists in the slime on and in the crevices of the inorganic materials so frequently ingested.

A terminal portion on each of 21 Palmyra specimens, without recognizable evidence of amputation, was a growth zone. Segments, in that zone, became increasingly shorter posteriorly. Setae first became unrecognizable. Then nephridia no longer were visible through the transparent body wall. Finally intersegmental furrows became indistinct or unrecognizable. Number of countable segments ranged from 78 to 119. Production of new segments after hatching is suspected.

Abnormality: No. 1. Male terminalia lacking on the right side where there is no trace of any depression or parietal thickening.

Parasites: Ciliata: *Anoplophrya macronucleata, Hysterocineta pontodrila, Maupasella leptas* Wichterman, 1942, gut. Dry Tortugas.

Remarks: Circumtropical distributions of species that are not especially primitive seem to require transportation from the original homes wherever those may have been. The agent involved in this instance seems unlikely to have been man. One method usually suggested in such circumstances is carriage of cocoons in mud on birds' feet. If megadrile cocoons ever were seen on birds' feet no records to substantiate the find got into scientific journals. Those who have thought most about the ornithological solution believe cocoons are likely to be too deeply buried and that getting onto birds' feet and staying there viably during transoceanic flights is improbable. Absence of geographical variation does suggest that transportation may have been recent.

The "well known dependence of the conformation of the alimentary on food and environment" (Stephenson, 1930: p. 740) seems to be contraindicated by the gut of *P. bermudensis.* Beddard, who found the gut in that same species to be always filled with coral debris (1895: p. 55), concluded, "It is evidently therefore not safe to lay down any such general statement about the cause of the presence or absence of the gizzard."

Hearts of xiii, in three specimens examined after completion of the manuscript, did seem to be lateroesophageal but again certainty was not possible.

ALLUROIDIDAE

1900. Alluroididae Michaelsen, *Das Tierreich* 10: p. 106.
1903. Alluroididae, Michaelsen, *Die geographische Verbreitung der Oligochaeten* (Berlin), p. 65.
1919. Syngenodrilinae (Moniligastridae) Smith & Green, *Proc. U. S. Natl. Mus.* 55: p. 155.
1921. Alluroididae + Syngenodrilidae, Michaelsen, *Arch. Naturgesch.* 86, A: p. 141.
1928. Alluroididae + Syngenodrilidae (Phreoryctina), Michaelsen, *Handbuch der Zoologie* (Berlin) 2, 2-8: p. 106.
1930. Alluroididae + Syngenodrilinae (Moniligastridae), Stephenson, *The Oligochaeta* (Oxford), pp. 806 and 811.
1945. Syngenodrilinae (Alluroididae?), Gates, *Jour. Washington Acad. Sci.* 35: p. 396. Snygenodrilidae + Alluroididae, Pickford, *ibid.*, p. 399.
1959. Alluroididae, Gates, *Bull. Mus. Comp. Zool. Harvard College* 121: p. 255.
1964. Alluroididae, Brinkhurst, *Proc. Zool. Soc. London* 142: p. 534.
1968. Alluroididae, Jamieson, *Jour. Zool. London* 155: p. 72.

Digestive system, with pre-intestinal gut in more than 10 segments, without caeca, typhlosoles, calciferous and supra-intestinal glands. (Vascular system?) Nephridia, holoic. Setae, sigmoid, singly pointed, 4 pairs per segment. Dorsal pores, lacking.

Male pores, one pair only, in front of female pores. Clitellum, unilayered, including male and female pore segments. Seminal vesicles, simple. Spermathecae, anterior to testis segments, adiverticulate. Ovaries, in xiii, (shape?). Ova, yolky. Ovisacs, simple.

Distribution: Africa, Argentina.

Remarks: *Alluroides* and *Syngenodrilus*, as monogeneric subfamilies or families, were distinguished from the Haplotaxidae, in the classical system, by location of male pores behind 12/13, situation of female pores in xiv, and by presence of prostates (*Alluroides*) or of prostate-like glands (*Syngenodrilus*). *Alluroides*, being proandric, is more advanced as regards gonads, than *Syngenodrilus* which has retained the supposedly ancestral holandry. However, both sperm ducts of a side in *Syngenodrilus* join to open by one male pore, a condition that characterizes holandric megadriles, except several haplotaxids, moniligastrids, and the quadriprostatic species of a small, Indian, octochaetid genus, *Hoplochaetella* Michaelsen, 1900. As that junction is so very common, it does seem possible that it was acquired long ago and rather early in megadrile evolution. Nevertheless, in spite of the union, male pores of *Syngenodrilus* still are in front of the female pores. Such a stage accordingly is intermediate between the haplotaxid state and that characterizing all other megadriles, in which male pores are behind the female and at various levels from xv to xxii.

As one result of derogation of somatic anatomy by classical oligochaetologists, little is known about such structure and in particular about the vascular system of *Alluroides*. Presence of dorsal and ventral trunks, as well as direct or indirect connectives between them in each segment of course can be assumed. If other longitudinal trunks are indeed absent, such conditions would show closer relationships to *Haplotaxis* than to the other genus of its own family. *Syngenodrilus*, on the contrary, is said to have hearts and the location of its last pair in the second segment in front of the ovaries is a condition shared not only with the Moniligastridae but also the Ocnerodrilidae and Lumbricidae. Even more interesting and perhaps of much greater systematic importance is presence in *Syngenodrilus* of additional longitudinal trunks. These extra-esophageal vessels are lateral to the hearts as also in the Moniligastridae and the Sparganophilidae. All other megadriles in which location is known have their extra-esophageals median to the hearts.

Both dorsal and ventral trunks are distributing vessels anterior to the hearts, according to Stephenson (1930: p. 171), and so the function of the extra-esophageals becomes "draining the anterior part of the body" and carrying blood to the section of the gut where calciferous tissues do develop in many genera. Some other somatic anatomy, as for instance presence of two well-developed esophageal gizzards,

also suggests that *Syngenodrilus* is much more advanced than *Alluroides*.

CRIODRILIDAE

1884. Criodrilidae Vejdovsky, *System und Morphologie der Oligochaeten* (Prag), p. 57.
1885. Criodrilina, Oerley, *Ertek, Term. Magyar Akad.* 15, 18: p. 8.
1886. Criodrilinae (Lumbricidae), Rosa, *Boll. Mus. Zool. Univ. Torino* 1, 15: p. 2.
1887. Criodrilina (Lumbricidae), Rosa, *Mem. Accad. Sci. Torino*, Ser. 2, 38: p. 179.
1895. Microchaetinae (part), Beddard, *A Monogr. of the Order of Oligochaeta* (Oxford), p. 665. (Including *Criodrilus* only.)
1900. Criodrilinae (Glossoscolecidae), part, Michaelsen, *Das Tierreich* 10: p. 463. (Excluding, *Alma, Sparganophilus*, and South American species).
1903. Criodrilinae (Glossoscolecidae), part, Michaelsen, *Die geographische Verbreitung der Oligochaeten* (Berlin), p. 150. (Exclusions as in 1900.)
1918. Criodrilinae (Lumbricidae), Michaelsen, *Zool. Jahrb. Syst.* 41: p. 372.
1921. Criodrilinae (Lumbricina), *Arch. Naturgesch.* 86, A: p. 141.
1923. Criodrilinae (Lumbricidae), Stephenson, (*Fauna of British India*), *Oligochaeta*, p. 495.
1928. Criodrilidae (Lumbricina), Michaelsen, *Handbuch der Zoologie* (Berlin) 2, 2-8: p. 108.
1930. Criodrilidae (part), Cernosvitov, *Zool. Jahrb. Anat.* 52: p. 494. (Excluding *Alma*). Criodrilinae (Glossoscolecidae), Stephenson, *The Oligochaeta* (Oxford), p. 903.
1950. Lumbricidae (part), Pop, *An. Acad. Romane, Sect. Sti. Geol. Geogr. Biol.*, Ser. A, 1, 9: p. 48. (Excluding all but *Criodrilus*.)
1959. Criodrilidae, Gates, *Bull. Mus. Comp. Zool. Harvard College* 121: p. 255.

Digestive system, with an esophageal crop (? in?), an intestinal origin in xv (?), but without gizzards, caeca, calciferous and supra-intestinal glands. Vascular system, with complete dorsal, ventral, and subneural trunks, the latter adherent (along with lateroneural trunks?) to the nerve cord, extra-esophageal trunks that are lateral to the hearts and join the dorsal trunk in xii, but without a supra-esophageal (?). Hearts, moniliform, in (vi?) vii–xi, lateral. Nephridia, holoic (present from? vesicles?), ducts passing into parietes at *B* (throughout?). Setae, sigmoid, single-pointed, four pairs per segment, sexual setae longitudinally grooved ectally. Nephropores, at *B* (throughout?). Dorsal pores, lacking. Anus, dorsal.

Reproductive apertures, all minute, equatorial, female in xiv, male in xv. Clitellum, multilayered, behind genital apertures. Holandric. Seminal vesicles, in ix, x, xi, xii, trabeculate. Ovaries, in xiii, fan-shaped and not terminating in a single egg string. Ova, not yolky. Ovisacs, in xiv, small and lobed. (Athecal.)

Distribution: Italy, Germany, Austria, Hungary, Latvia, Russia, Moldavia, Syria, Lebanon, Israel, Amur River region of Siberia. (India, Japan, England?)

The history of specimens supposedly collected in

America is unknown. They presumably were introduced from Europe but without acquisition of domicile.

Systematics: The English genus *Anagaster* Friend 1921, attributed by its author to the Criodrilinae, like its single species that also is dubious, has not been placed. The family now appears to be monogeneric. Immature specimens from Indian localities that Stephenson identified as *Criodrilus lacuum* probably were glyphidriles. The Japanese species referred to *Criodrilus* differs enough from the European as to cause doubts, that cannot be resolved from published information, about the generic identification. Atrial glands were associated in xiii with the male pores, a condition that characterizes certain parthenogenetic morphs of *Eiseniella tetraedra*.

Criodrilus, except for the dubious Japanese form, always has been monospecific. The species was erected on German worms that supposedly are atyphlosolate like those from Bohemia. Italian specimens, or at least some of them, have a typhlosole that nearly fills the intestinal lumen. Whether species or geographical races can be distinguished in Europe remains to be determined. A range that extends from central Germany to the Pacific coast of Siberia is so unusually large for any autochthonous megadrile as to require more exacting investigation of somatic anatomy than yet has been made.

Longitudinal grooving of the ectal end of the genital setae has been thought to prove that *Criodrilus* is closely related to the Lumbricidae in which Pop (1950) did place the genus. Another sort of setal ornamentation that was thought, at one time or another, to be of considerable systematic significance, is now known to characterize some of the species in each of several genera in different families. Accordingly, the longitudinal grooving of genital setae now seems likely to be another of the systematically unimportant convergences that are common throughout the megadrile Oligochaeta. Extra-esophageal trunks of the vascular system, lateral to the hearts, rather than median to them as in the Lumbricidae, require consideration of relationships with the Moniligastridae and Alluroididae if not also the Haplotaxidae and the Sparganophilidae. Finally, the ovaries, which seem of all organs of the body to be the most resistant to evolutionary modification, contraindicate close relationships between the Criodrilidae and the Lumbricidae.

Remarks: As long ago as 1887 an English (Benham) and a Hungarian (Oerley) zoologist failed to find in criodrile ovaries evidence for close relationship with the Lumbricidae. Nevertheless, location segmentally was the only ovarian character allowed importance and use in the classical system.

Presence of more than one species in Europe is anticipated.

REFERENCES

BENHAM, W. B. 1887. "Studies on earthworms. No. 3." *Quart. Jour. Micros. Sci.* 27: pp. 561–572.

OERLEY, L. 1887. "Morphological and Biological Observations on *Criodrilus lacuum.*" *Quart. Jour. Micros. Sci.* 27: pp. 551–560.

POP, V. 1950. "Lumbricidele din Romania." *An. Acad. Romane, Sect. Sti. Geol. Geogr. Biol.* Ser. A, 1, 9: pp. 1–129.

EUDRILIDAE

1880. Eudrilidae (part), Claus, *Grundzüge der Zoologie* (Marburg, 4th ed.) 1: p. 479. (Excluding all genera except *Eudrilus.*)

1884. Eudrilidae (part), Vejdovsky, *System und Morph. der Oligochaeten* (Prag), p. 63. (Excluding all genera except *Eudrilus.*)

1888. Eudrilidae (part), Rosa, *Boll. Mus. Zool. Univ. Torino,* Ser. 3, 41: p. 9. (Excluding all genera except *Eudrilus.*)

1890. Eudrilidae, Beddard, *Proc. Roy. Phys. Soc. Edinburgh* 10: p. 256. Eudrilidae (part), Benham, *Quart. Jour. Micros. Sci.* 31: p. 221. (Excluding all genera except *Eudrilus* and *Teleudrilus.*)

1891. Eudrilinae (part, Megascolicidae), Rosa, *Ann. Naturhist. Hofmus. Wien* 6: p. 379.

1895. Eudrilidae, Beddard, *A Monogr. of the Order of Oligochaeta* (Oxford), p. 573.

1900. Eudrilinae (Megascolecidae), Michaelsen, *Das Tierreich* 109: p. 387.

1921. Eudrilidae, Michaelsen, *Arch. Naturgesch.* 86A: p. 141.

1923. Eudrilinae (Megascolecidae), Stephenson, (*The Fauna of British India*), *Oligochaeta,* p. 485.

1928. Eudrilidae (Megascolecina), Michaelsen, in Kükenthal & Krumbach, *Handbuch der Zoologie* (Berlin) 2, 2-8: p. 109.

1930. Eudrilidae, Stephenson, *The Oligochaeta* (Oxford), p. 863.

1959. Eudrilidae, Gates, *Bull. Mus. Comp. Zool. Harvard College* 121: p. 255.

Digestive system, giceriate. Vascular system, with dorsal, ventral and supra-esophageal, and extra-esophageal trunks. Excretory system, of holoic nephridia. Setae, 8 per segment.

Sperm ducts, pass into euprostates, digitiform glands with muscular walls. Seminal vesicles, trabeculate. Spermathecae, behind testis segments. Clitellum, multilayered. Metagynous.

Distribution: Tropical and subtropical Africa. One species has been widely transported by man.

Remarks: The euprostates, for long seemingly uniquely diagnostic of the family, according to Stephenson (1930: p. 372), "are actually loops of the male duct, the ectal or terminal limb being much thickened, and more or less extensively united with the thin inner (ental) limb." Little seems to be known about these glands and for further information Stephenson refers the reader to a short discussion in Beddard's monograph (1895: pp. 111–112). Presence in *Rosadrilus* Cognetti, 1907, of euprostates unconnected with any sperm ducts proves that the glands cannot be mere gonoduct thickenings. Presumably then eudrilid prostates, just as in so many megadriles, develop from parietal invaginations. Organs, superficially like the euprostates were found very recently in an ocnerodrile but whether the similarity extends to microscopic structure remains to be learned. The "muscular" prostates of some moniligastrids also may require consideration along with the ocnerodrile gland.

SYSTEMATICS

All outgrowths from the eudrilid esophagus except the fat-body-like appendages were chylustaschen to Michaelsen who seems to have believed that their function was absorption rather than secretion. The paired glands were recognized as calciferous by Stephenson (1930) but he seems to have had reservations about the ventromedian glands which were termed esophageal sacs. As calcareous granules have been found in median as well as paired glands, all will be called calciferous until found to have other function.

Female reproductive systems, of eudrilids, as those of almost all amphimictic megadriles, are provided with "spermathecae." Some of those organs, arising in several genera as ingrowths through the body wall late in ontogeny, are homologous with similar, simple receptacles in many other megadrile families. Other eudrilid structures, known for more than seventy years to develop by a closing off of various coelomic spaces (cf. Beddard, 1895: p. 134), have been called spermathecae except by Sims (1964, *Proc. Zool. Soc. London* **143**: pp. 587–608). Nevertheless, he continued to call certain structures spermathecal pores, spermathecal atria, spermathecal ducts, and frequently spoke of spermathecal systems in contrast with female systems. Some confusion may be avoided by provision of different but appropriate terms for structures that are not homologous. The chamber into which the "spermathecal pore" opens in various eudrilids, e.g., *Eudrilus eugeniae*, does function as a vagina and herein is so called. Its external opening then is a vaginal aperture.

Dorsal pores, as in the Ocnerodrilidae, usually are lacking but have been said to be present in some species of *Platydrilus* and *Eudriloides*.

The Eudrilidae comprises species groups of unknown relationships to each other and we do not know, according to Stephenson (1930: p. 873), "whether the groups are to be considered as subgenera of a larger genus, as separate but nearly related genera, or as genera which are only distantly related. The groups themselves, however, are mostly well defined, and for the present they may simply be placed side by side and regarded as genera." That, of course, is what is to be expected of classical systematics, in part because of its disregard or derogation of vascular and excretory anatomy. Of the 32 eudrilid "genera" recognized by Stephenson (1930), 22 have three or fewer species, and they are described, defined, and keyed primarily by genital characters some of which can be expected, from our knowledge of other genera, to be subject to intrageneric if not also to individual variation.

Eudrilus

1871. *Eudrilus* Perrier, *Compt. Rend. Acad. Sci. Paris* **73**: p. 1175. (Type species, *Eudrilus decipiens* Perrier, 1871, = *Lumbricus eugeniae* Kinberg, 1867).
1900. *Eudrilus*, Michaelsen, *Das Tierreich* **10**: p. 401.

1930. *Eudrilus*, Stephenson, 1930, *The Oligochaeta* (Oxford), p. 881.

Digestive system, with a gizzard (in v?), a ventromedian calciferous gland in each of x–xi, in xii a pair of rather reniform and lobed calciferous glands each opening by a very short and slender duct into gut midlaterally at region of insertion of 12/13 (intestinal origin in region of insertion of 14/15? intestinal caeca and typhlosole lacking? but supra-intestinal glands present?). Vascular system, with unpaired dorsal and ventral trunks, paired supra-esophageals in vii–xiv, a complete subneural adherent to the parietes, paired extra-esophageals median to the hearts and with branches to median calciferous glands. Hearts, in vii–xi. Nephridia, holoic, present from iv (bladders?). Setae, closely paired, ventral couples of xvii lacking. Nephropores, present from iv, obvious. Dorsal pores, lacking.

Vaginal apertures, well anterior to male pores. Holandric. Seminal vesicles, in xi, xii. Female organs, completely paired.

Distribution: Tropical west Africa.

The proper range as thus indicated has been extended throughout the tropics, presumably as a result of transportation of one species by man since 1500 A.D.

Remarks: Most of the uncertainties about generic definition are due to lack of information about somatic anatomy of the two endemic species.

KEY TO SPECIES OF *Eudrilus*

1. Copulatory chambers lacking.............*simplex*
   Copulatory chambers present................ 2
2. A *Y*-shaped gland opens through own porophore into copulatory chamber....*eugeniae*
   Y-shaped gland and porophore lacking....*pallidus*

Location of the gizzard in *E. simplex* Michaelsen, 1913, and in *E. pallidus* Michaelsen, 1891, is unknown.

Eudrilus eugeniae

1867. *Lumbricus eugeniae* Kinberg, 1867, Öfvers. K. Vetensk. Akad. Förhandl. Stockholm **23**: p. 98. (Type locality, St. Helena Island, South Atlantic. Types, in the Stockholm Mus.)

ANATOMY: Eisen, 1900, *Proc. California Acad. Sci.*, Ser. 3, 2: p. 135. Gates 1942, *Bull. Mus. Comp. Zool. Harvard College* **89**: p. 137. Vidyavati, 1945, *Proc. Natl. Inst. Sci. India* **11**: p. 345. ECOLOGY: Madge, 1969, *Pedobiologia* **9**: p. 190.

Copulatory chamber apertures, transverse slits centered at or lateral to *B*, slightly in front of 17/18. Vaginal apertures, transverse slits centered at or median to *C*, presetal in xiv. (Genital markings, none.) Clitellum, intersegmental furrows faintly indicated and setae retained, xiii, xiv–xviii. Setae, closely paired, AA ca. $= BC$, DD slightly $> \frac{1}{2}C$, a,b/xvii lacking. Nephropores, just lateral to *C*. Prostomium, epilobous, tongue open. Color, restricted to dorsum, red. Segments, to 211 (+ ?). Size, 90–185 by 5–8 mm.

Septa, all present from 4/5, 6/7 and several subsequent septa slightly strengthened. Pigment, red, in circular muscle layer. Gizzard, in v. Intestinal origin, close to 14/15. Typhlosole and caeca, none. Supra-intestinal glands, small, paired, postseptal, in 8–42 consecutive segments of lxii–cxxxii. Dorsal blood vessel, aborted in front of hearts of vii. Hearts, of vii lateral, of viii–xi latero-esophageal. Testis sacs, unpaired, ventral. Sperm ducts, discrete to entrance into ectal portion of prostate, widened between small male funnels (without plications) and their septa. Prostates, to 8 mm. long, ducts short and slender but muscular. Copulatory chamber, large, containing penis and Y-gland porophore, a groove from Y-gland pore passing up the porophore and then down nearly to the tip of the penis.

Reproduction: Presumably biparental as sperm are exchanged during copulation.

Distribution: Upper Guinea, Sierra Leone, Liberia, Ivory Coast, Togo, Nigeria, Cameroons. That portion of Africa presumably comprises the original home, or some part thereof, that was self-acquired by *E. eugeniae*. Presence elsewhere, as indicated in the next paragraph, presumably is due to oversea transport by man.

Madagascar, Great Comoro, Anjouan, Moheli, Mayotte, Ceylon, India, New Caledonia, New Zealand, Panama, Colombia, Venezuela, Guyana, Surinam, French Guiana, Brazil. Trinidad, Martinique, St. Thomas, St. Croix, Puerto Rico, Haiti, Cuba, Bahamas, Bermuda. St. Pierre-Miquelon. St. Helena, Cape Verde, Fernando Po and São Thome Islands.

Although very many specimens, possibly millions, have been distributed throughout North America during the last fifteen or more years, (*cf.* Economic Importance below) no information as yet has become available to indicate successful colonization north of Panama.

E. eugeniae "colonizes much of the United States" (Omodeo, 1963: p. 142) and so supposedly provided a second instance of extraordinary adaptability unparalleled by other invertebrates. Actually in spite of the very wide distribution by human agents there is no published record of finding the species in a natural habitat in any state. Those who tried to establish it in natural environments found that the "worms do very well until the temperature drops below 40° F at which time they die" (*in lit.* from R. N. Henson).

Ecology: Plentiful in coastal, shaded grasslands of West Africa. Castings deposited in mounds of pellets with a base diameter of 30–50 mm. and a height usually under 30 mm., during rainy season only. Castings, with higher organic content than parent soil, may amount to 70 tons per acre per annum.

Worms had a water content of 83.9 per cent. They survived in mud under water but not in water alone and remained quiescent in solutions with a pH of 5.6–9.2. They emerged onto surface of soil only when driven out by ants.

Economic Importance: *Eudrilus eugeniae* has been cultured by many earthworm farmers in the United States and even by some in Canada. The species has been shipped into every one of the United States, even including Hawaii, as well as several Canadian provinces, for use as bait. The species also is recommended as food for fish, birds, etc., that refuse or are harmed by the more commonly cultured *Eisenia foetida*.

Remarks: Characterization of the female genitalia is omitted from the précis as well as from the generic definition because of uncertainty about homology of various parts. Whether or not hologyny was indicated, Miss Vidyavati maintained the species has two pairs of ovaries, rudimentary in xiii and functional in xiv. Her specimens, from Ceylon, seemed to have paired testis sacs but, in absence of coagulum, walls of a middle portion of the sac usually are apposed to each other. Unless the delicate, transparent membranes are carefully separated the unpaired condition is not likely to be recognized.

Male funnels of *eugeniae* are small and sperm do not aggregate thereon but in the vesicular widening of the sperm duct between the funnel and the funnel septum. Aggregation there may prove to be characteristic of all eudrilids with calciferous glands.

Although hearts of viii, ix definitely bifurcate dorsally no blood has been seen in the anterior bifurcations of the many specimens examined.

Extra-esophageal trunks have been traced into xii. Median unions, or transverse connectives, between the two supra-esophageals, have been seen in any one, two, or in all three of segments x–xii.

Brown bodies, small, each formed around a single seta were seen commonly in the coelomic cavities of various segments from vii posteriorly. The bodies were found so frequently as to raise a question as to whether all shedding from setal follicles is internal rather than external.

GLOSSOSCOLECIDAE

1900. Glossoscolecinae (Glossoscoledicae), Michaelsen, *Das Tierreich* 10: p. 420.
1918. Glossoscolecinae (Lumbricidae), Michaelsen, *Zool. Jahrb. Syst.* 41: p. 54.
1921. Glossoscolecinae, Michaelsen, *Arch. Naturg.* 86, A: p. 141.
1923. Glossoscolecinae (Lumbricidae), Stephenson (*The Fauna of British India*), *Oligochaeta*, p. 488.
1928. Glossoscolecinae (Lumbricina), Michaelsen, "Oligochaeta." In: Kükenthal & Krumbach, *Handbuch der Zool.* 2, 2-8: p. 107.
1930. Glossoscolecinae (Glossoscolecidae), Stephenson, *The Oligochaeta*, p. 891.
1945. Glossoscolecinae (Glossoscolecidae), Cordero, *Comm. Zool. Mus. Hist. Nat. Montevideo* 1, 22: p. 1.
1959. Glossoscolecidae, Gates, *Bull. Mus. Comp. Zool. Harvard College* 121: p. 255.

Digestive system, with one esophageal gizzard in vi (or its homoeotic equivalent), with paired, extramural calciferous glands in some of segments vii–xiv (or their homoeotic equivalents). Vascular system, with dorsal and ventral trunks, a supra-esophageal, paired extra-esophageals median to the hearts, and a subneural adherent to the parietes (?). Nephridia, macroic, in intestinal region holoic, vesiculate (?). Setae, sigmoid, simply pointed. Dorsal pores, none.

Spermathecae, adiverticulate.[1] (Ovaries, in xiii? shape? ovisacs?) Male pores, behind female pores. Clitellum, multilayered. Ova, not yolky. (Prostates with muscular ducts, none. Metagynous?)

Distribution: Central and South America, Trinidad, West Indies. One species has been widely transported.

Systematics: Beddard, in his monograph (1895: p. 623) admitted that there was "no one character found in all the Geoscolecidae which is absolutely characteristic of the family." Nevertheless, there has been no disagreement since 1900 as to the content of the taxon but only as to its hierarchical status, phylogeny, and relations to other taxa. The range, restricted to a definite portion of the Americas (save for one markedly anthropochorous species), rather than the anatomy, presumably provides the explanation for a situation so unusual in the classical system. The only characters without exceptions or qualifications in definitions of the group by Stephenson, (1923, 1930) were: (1) A single esophageal gizzard. (2) Calciferous glands present. Both characters, it is noteworthy, are solely somatic. Neither character, nor a combination of the two, is diagnostic, being present alone or in combination in other families. A third character, spermathecae adiverticulate, was included by Beddard (1895) and Michaelsen (1900) in their definitions but is equally applicable to various other families.

Although by 1918 Michaelsen had realized that internal structure as well as number and location of calciferous glands was of some importance in defining genera, he continued to ignore vascular and excretory systems. Nephridia of one species only have been studied. Accordingly, the tentative definition above, involves considerable extrapolation from the little that is known about somatic anatomy of the family.

Remarks: Our knowledge of an important somatic system and its significance in the obsolete phylogenies of the classical system is indicated by a casual comment in the latest monograph on "The Oligochaeta" (Stephenson, 1930: p. 892). "Cognetti gives some details concerning the distribution of the vessels in *Aptodrilus festae.* The same author also notes the number of hearts, and the situation of the last heart, in several genera." Even today, the characterization of the vascular system above had to be based on data from a very few species.

Certain glands in some genera in the past have been called prostates, but none have muscular ducts.

Examples of the "homoeotic equivalent" (pre-

viously unrecognized in the family) of the above précis will be found in two of the author's recent publications: *Megadrilogica* 1, 1 (1968), and *Breviora, Mus. Comp. Zool. Harvard,* No. 356 (1969).

Pontoscolex

1861. *Pontoscolex* Schmarda, *Neue wirbellose Thiere, Leipsig* 1, 2: p. 11. (Type species, *P. arenicola* Schmarda 1861 = *Lumbricus corethrurus* Müller 1856.)
1895. *Pontoscolex,* Beddard, *A Monogr. of the Order of Oligochaeta* (Oxford), p. 653.
1900. *Pontoscolex,* Michaelsen, *Das Tierreich* 10: p. 424.
1918. *Pontoscolex,* Michaelsen, *Zool. Jahrb. Syst.* 41: p. 233.
1923. *Pontoscolex,* Stephenson (*The Fauna of British India*), *Oligochaeta,* p. 489.
1930. *Pontoscolex,* Stephenson, *The Oligochaeta* (Oxford), p. 895.
1945. *Pontoscolex,* Cordero, *Com. Zool. Mus. Hist. Nat. Montevideo* 1, 22: p. 7.

Digestive system, with paired, solid, "paniceled tubular" calciferous glands in vii–ix, each gland with a more or less rudimentary distal appendage and a duct from ventral or median end passing to gut dorsolaterally just in front of a septal insertion, with a well-developed intestinal typhlosole but without caeca and supra-intestinal glands. Vascular system, with complete dorsal (single), ventral, and subneural trunks, the latter adherent to the parietes and the dorsal trunk markedly moniliform in several preintestinal segments, a supra-esophageal trunk in vi–xiv, paired extra-esophageal trunks (median to the hearts) united on gut at mV (from vii to?) with connectives to supra-esophageal and subneural, hearts of vi–ix lateral, of x, xi (latero?)-esophageal. Nephridia, macroic, vesiculate, in intestinal segments holoic, the transversely placed bladders elongately ocarina-shaped and opening to the exterior through short, thick-walled and rather conical ducts from the ventral side, anteriorly bladders elongately sausage-shaped and opening to the exterior through terminal ducts. Nephropores, obvious. Pigment, none. Septa, all present at least from 5/6.

Metandric (?). Seminal vesicles, long, extending from xii back through several segments. Metagynous (?), ovaries band-shaped (? and with egg-strings?). Male pores and tubercula pubertatis, in the clitellar region.

Distribution: Surinam, Guyana, Guatemala, (Panama?). The generic range as thus indicated by the autochthonous species has been vastly extended by the recent transportation of one species.

Systematics: Two species were recognized in the last revision of the genus (Michaelsen, 1918), *corethrurus* (*cf.* below) and *lilljeborgi* Eisen, 1896. Subsequently *hingstoni* Stephenson, 1931, and *vandersleeni* Michaelsen, 1933, were erected. Little is known about the somatic anatomy of the three autochthonous species. The definition above rests on assumptions that certain characters of the better-known type species, of kinds now known to be uniform throughout

---

[1] With but few exceptions anterior to the testis segments.

certain genera, also are definitive in *Pontoscolex*. As elsewhere, certain characters now included in generic definitions may eventually be transferable to definitions of higher categories. Although almost every megadrile specialist has, at one time or another, studied the type species more or less carefully, no certainty yet has been reached about several important and possibly definitive characters.

Relationships presumably are closest with *Meroscolex* Cernosvitov, 1934, *Onychochaeta* Beddard, 1891, *Opisthodrilus* Rosa, 1895, *Rhinodrilus* Perrier, 1872, which also have in vii–ix calciferous glands with dichotomously branched longitudinal canals. As defined in the classical system (*cf.* Stephenson, 1930) *Rhinodrilus* and *Onychochaeta* are distinguished from *Pontoscolex* by the holandry as well as by the shorter seminal vesicles while *Opisthodrilus* is additionally distinguished by location of male pores and tubercula pubertatis behind the clitellum. Andry and vesicle length, to say nothing of length of clitella, location of male pores and of tubercula pubertatis, all are subject to rapid evolutionary modification and genera so distinguished must be considered dubius until defined by more stable and presumably somatic characters. *Meroscolex* supposedly was distinguished from *Pontoscolex* by regularity of its setal ranks a character which is, however, shared with two species of *Pontoscolex* (*cf.* key below). *Diachaeta* Benham, 1886, with only low ridges on inner walls of the calciferous glands presumably is somewhat less closely related.

KEY TO SPECIES OF *Pontoscolex*

1. Setae of a posterior part of the body in the "quincunx" arrangement.................... 2
   Setae not in the quincunx arrangement......... 3
2. Sexthecal, pores at 6/7–8/9............*corethrurus*
   Octothecal, pores at 6/7–9/10........*vandersleeni*
3. Sphermathecal pores at 7/8–9/10, *B* and *D* lines slightly irregular....................*hingstoni*
   Spermathecal pores at 8/9–10/11, *A* and *C* lines slightly irregular.................*lilljeborgi*[a]

[a] Structure of the calciferous glands is unknown. Accordingly the species may not belong in *Pontoscolex*.

Note. *P. corethrurus*, long thought to have a prostomium, now is known to have none though buccal tissue may be turgescent so as to simulate a prostomium or even a proboscis. *P. vandersleeni* and *hingstoni* were thought to have a prostomium that was prolobous in the first species and pro-epilobous in the second. The tip of a proboscis just visible in the buccal cavity could simulate a prolobous prostomium like that of moniligastrids. *P. lilljeborgi* supposedly has a "very long" prostomium but a figure (1, pl. 1 of Eisen, 1896) suggests a proboscis much more than a prostomium.

Pontoscolex corethrurus

1856. *Lumbricus corethrurus* Müller, *Abhandl. Naturgesch. Ges. Halle* **4**: p. 26. (Type locality, Itajahy, Brazil, Types, probably none.)
1916. *Pontoscolex corethrurus*, Stephenson, *Rec. Indian Mus.* **12**: p. 349.
1923. *Pontoscolex corethrurus*, Stephenson (*The Fauna of British India*), *Oligochaeta*, p. 489.
1925. *Pontoscolex corethrurus*, Stephenson, *Rec. Indian Mus.* **27**: p. 73.
1926. *Pontoscolex corethrurus*, Gates, *Ann. Mag. Nat. Hist.*, Ser. 9, **17**: p. 472; *Jour. Burma Res. Soc.* **15**: p. 217; *Jour. Bombay Nat. Hist. Soc.* **31**: p. 183; *Rec. Indian Mus.* **28**: p. 169.
1930. *Pontoscolex corethrurus*, Gates, *Rec. Indian Mus.* **32**: p. 351.
1931. *Pontoscolex corethrurus*, Gates, *ibid*, **33**: p. 431.
1933. *Pontoscolex corethrurus*, Gates, *ibid.* **35**: p. 601.
1936. *Pontoscolex corethrurus*, Hla Kyaw & Gates, *Jour. Roy. Asiatic Soc. Bengal*, (*Sci.*) **2**: p. 166.
1939. *Pontoscolex corethrurus*, Gates, *Jour. Thailand Res. Soc. Nat. Hist. Sup.* **12**: p. 110.
1955. *Pontoscolex corethrurus*, Gates. *Rec. Indian Mus.* **52**: p. 92.
1961. *Pontoscolex corethrurus*, Gates. *American Midland Nat.* **66**: p. 62. *Burma Res. Soc. 50th Anniv. Pub. No. 1*: p. 57.
1970. *Pontoscolex corethrurus*, Soota & Julka, *Proc. Zool. Soc. Calcutta* **23**: p. 205.

ANATOMY: Perrier, 1874, *Arch. Zool. Exp.* **3**: p. 379. Beddard, 1888, *Quart. Jour. Micros. Sci.* **29**: p. 234; 1890, *ibid.* **31**: p. 473. Eisen, 1900, *Proc. California Acad. Sci.*, Ser. 3, **2**: p. 87. Gates, 1943, *Ohio Jour. Sci.* **43**: p. 92; 1954, *Bull. Mus. Comp. Zool. Harvard College* **111**: p. 219. BIOLOGY: Vanucci, 1953, *Dusenia* **4**: p. 287. NEPHRIDIA: Bahl, 1942, *Quart. Jour. Micros. Sci.* **84**: p. 1. PARASITES (BACTERIA): Knop, 1926, *Zeitschr. Morph. Ökol. Tiere* **6**: p. 608. (PROTOZOA): Boisson, 1957, *Ann. Sci. Nat. Paris* **19**: p. 84. REGENERATION: Gates, 1927, *Biol. Bull.* **53**, p. 352. REPRODUCTION: Gates, 1958, *American Mus. Novitates*, No. 1886: p. 8; 1962, *Proc. Biol. Soc. Washington* **75**; p. 140. SPERMATOZOAL DEGENERATION: Cernosvitov, 1930, *Zool. Jahrb. Anat.* **52**: p. 494. TYPHLOSOLE: Hertling, 1923, *Zeitschr. Wiss. Zool.* **120**: p. 201.

Sexthecal, pores minute and superficial, at *C* and at or close to 6/7–8/9. Female pore, on left side, a minute transverse slit in or close to *AB* and slightly in front of 14/15. Male pores, lateral to *B* (at or near 20/21?). Clitellum, saddle-shaped, intersegmental furrows slightly indicated at maximal tumescence, setae retained(?), down to region of *B* or *A*, xv, xv/n, xvi–xxii, xxiii. Tubercula pubertatis, translucent longitudinal bands, just lateral to *B*, xix/n–xxi, xxii/n. Genital tumescences, around *a,b*, or *a* or *b*, (xiv, xviii)xix–xxi(xxii). Setae usually present from i–ii in which they are very closely paired, *AB* and *CD* gradually wider from iii, one rank after another becoming more and more irregular until the "quincunx" arrangement is attained, towards the posterior end enlarged and ornamented ectally by transverse rows of fine teeth, one or both setae of some of xiv–xxii genital and ornamented ectally with longitudinal rows of gouges. Nephropores, obvious, present or absent in ii–iii, in a single longitudinal rank on each side, about at *C* anteriorly and remaining about at mL throughout.

Peristomium, flaccid, usually withdrawn into anterior end of body, not recognizably delimited from ii. (Prostomium and proboscis, lacking.) Pygomere, often withdrawn into posterior end of body. Color, none. Segments, to 212. Size, 60–120 by 4–6 mm.

Septa, 5/6 membranous, 6/7–13/14 (at least) funnel-shaped and posteriorly directed, 6/7–9/10 thickly muscular and displaced posteriorly, 9/10 apparently inserted on parietes over site of intersegmental furrow 10/11, septum 10/11 and the next few close together (incomplete dorsally?), and apparently with parietal insertions displaced posteriorly. (Pigment, none. Intestinal origin, in xiv or xv?) Typhlosole, beginning in region of xxi–xxv, lamelliform, height greater than width of gut lumen, rolled up like a scroll, ending abruptly in the region between 107th and 131st segments. Male funnels, large, plicate, facing dorsally within testis sacs between nerve cord and ventral blood vessel. Male gonoducts, without epididymis, slender, passing anteroventrally to the parietes and thence posteriorly lateral to B under peritoneum. Seminal vesicles, one pair, lateromesially flattened, more or less deeply incised by the septa, extending posteriorly through 8–10 segments. TP and GS glands, acinous and supraparietal. (Progynous or metagynous?) Ovaries, small, under testis sacs, narrowly bandlike (eggstrings?). (Oviducal funnels?) Oviduct, of left side long and straight, passing anteroventrally to parietes. Right oviduct, slenderer than the left, attenuating at or within the parietes or without a functional aperture. Spermathecae, long, ducts slender.

Reproduction: A population with normally biparental reproduction perhaps is to be expected in the original home of the species but that home has not yet been found. Elsewhere, reproduction usually if not always is parthenogenetic. Absence of mature male gametes in all specimens of frequent samples from sites where the species is common shows that sperm are not even necessary for initiation of embryonic development.

Distribution: Ross Island, Aberdeen, Port Blair, Mount Harriet, Minnie Bay, Wrightmyo (Andaman Islands). Victoria Point, Mergui, Nyaungbinkwin, Kala Island, Labaw, Wuzinok (Mergui). Tavoy, Mindat, Myittha, San Hlan, Maungmagaun, Kanyindaung, Sitpye, Myinmaw, various localities in the vicinity of the Kamaungthwe River, the Myaungdonle, Mayan and Posoe Chaungs (Tavoy). Ye, Moulmein, Martaban, Chaungson, Kya In, Kawkareik, Kyaikmaraw, Mupun (Amherst). Thaton, Sittang, Boyagyi, Duyinzeik, Bilin, Kyaikto, Kyaiktiyo, Kumingyaung, Taungzun, Aungsaing (Thaton). Rangoon, Syriam, Pazunmyaung, Kyauktan, Thongwa, Kungyangon (Hanthawaddy). Pyapon, Thameintaw, Dedaye, Kyaiklat (Pyapon). Myaungmya, Thanchitaw, Myohaung (Myaungmya). Bassein, Coomzamu, Thinbawgyin (Bassein). Pegu, Thanatpin, Chaukan, Nyaunglebin (Pegu). Insein, Taikkyi (Insein). Thonze, Tharrawaddy (Tharrawaddy). Henzada, Letpadan (Henzada). Toungoo, Tantabin, Thandaung, Leiktho Circle (Toungoo). Prome (Prome). Thayetmyo (Thayetmyo). Sandoway, Taungup, Kyauktaga, Tanyagyi, Patle (Sandoway). Pyinmana (Yamethin). Kyaukpyu (Kyaukpyu). Kalaw, Taungyi, Maymyo (Shan States). Thazi, Meiktila (Meiktila). Kyaukpadaung,Toungoo, Mount Popa (Myingyan). Akyab, Buthidaung, Maungdaw, Kyauktaw (Akyab). Mandalay (Mandalay). Sagaing (Sagaing). Paungbyin (Lower Chindwin). Hsiwpaw, Kyaukme (northern Shan States). Mogok, Indaw Lake (Katha). Bhamo (Bhamo).

Ko Chang, Chiengmai (Thailand). Cambodia. Hong Kong.

Penang, Malay Peninsula, Pulau Berhala (Malacca Straits). Sumatra. Java, Christmas Island. Borneo. Celebes. Moluccas. Philippine Islands. Salibaboe and Saonek Islands. Riouw Archipelago. Engano. Ternate. New Caledonia. Australia. Loyalty, Solomon, New Hebrides, Fiji, Tonga, Tahiti, Samoa, Marquesas, Hawaiian and Gorgona Islands.

United States (greenhouses at Chatham, New Jersey, and Pana, Illinois). Mexico, Guatemala, Swan Island, Honduras, Salvador, Nicaragua, Costa Rica, Panama, Colombia, Ecuador, Brazil, Fernando Noronha, Paraguay, French Guiana, Surinam, Guyana, Venezuela, Trinidad, Bonaco, Martinique, Dominica, St. Thomas, Puerto Rico, Haiti.

Cape Verde Islands, Algeria, Liberia, Belgian Congo, Tanzania, Natal, Kenya, Madagascar, Mauritius, Comoro Islands, Mahe, Seychelles. St. Helena.

Persia, Pakistan, India, Ceylon.

In India and Burma present as yet only in the lowlands. Reported from elevations of 100–1,800 m. in El Salvador. Although recorded from elevations of 6,200 feet above sea level, the species appears to belong primarily in the lowlands. Although originally from the tropics, neither Cancer nor Capricorn now bounds its acquired range.

P. corethrurus has been said to be "the most widely distributed earthworm" though that honor also has been claimed for several lumbricid species. Supposed restriction to coastal areas (Stephenson, 1930) probably no longer is true in many regions. In the Indian peninsula the species has now gone about as far from the coast as is possible.

In many places reportedly more common than the endemic earthworms, as in various sections of Burma and the Malay Peninsula. The species, according to Vanucci, "is undoubtedly the most widespread and, in most places, the commonest worm in Brazil."

The extensive distribution is attributed primarily to transportation by man. The species was about the commonest among worms brought to Kew Gardens in Wardian cases according to Beddard (*Proc.*

Zool. Soc. London **1893** (1894): p. 737) who earlier (*Quart. Jour. Micros. Sci.* **34** (1893); p. 258) had characterized the species as so ubiquitous that it can hardly be safely assumed that its presence in tropical parts of the Old World is due to man's agency.

Habitats: Primary jungle soil. Agricultural and cultivated clay soils. Red and black earths. Sandy soil (near seashore?). Just above high-water mark (presumably at seashore). Gardens, coconut plantations (South Sea islands), rubber plantations (Burma, Malay Peninsula). Manure and compost heaps. Earth of plant benches in green houses (United States), earth around roots of potted plants (Burma, Ceylon). In termite galleries. Under bark of trees (Malay Peninsula), under rotting tree trunks, under dead wood, in rotting plantain trunks, under stones, in fresh water (Celebes).

Common down to a depth of 200 mm. (Vanucci) in Brazil.

Biology: Clitellate worms at a breeding stage were common in Burma from July to March and probably are obtainable during April–June at sites where moisture and other requirements are met. Cocoons were found from September till end of March and were deposited in the laboratory during December–January. Cocoons were secured during April in the Malay Peninsula and are extremely common all the year round in Brazil (Vanucci) wherever adults are present. Accordingly, year-round activity and breeding are possible for the species in favorable conditions. Little is known about individual life histories. Less moisture probably is required than by any other Burmese species and it withstands drought (Vanucci) better than excess moisture.

Cocoons, 4–6 by 3–4 mm., are suspended each by a thread in a slightly larger chamber lined with hardened mucus. Development, of one individual only per capsule, is completed (Vanucci) in the laboratory at 15–20°C, in 34–42 days, presumably more slowly in cooler soil. Diapause, entered after gut has been emptied, in a chamber lined with hardened mucus where the worm is coiled into a tight ball, is deep. On interruption worms do not at once become active. Two worms in a chamber, with anterior ends parallel, were found on several occasions by Vanucci in laboratory cultures.

Worms are geophagous but course bits of plant tissues occasionally are found in the gut. Nothing is known about castings, and copulation has not been observed.

The average daily rate of segment production during embryonic development appears to be between 3+ and 8.7+.

Variation and Abnormality: Segments of the smallest available juveniles, 16–54 by 1½-2 mm., were 136–297, of somewhat larger juveniles were 136–281. No evidence of post-hatching production of metameres (except of course in tail regenerates) was recognized.

Juveniles with smaller numbers of segments, 136 and up, already may have lost considerable tail portions though no evidence indicative of amputation, other than the fewer metameres, could be distinguished. Frequency of posterior amputation, with or without tail regeneration, hampers determination of normal segment number for adults.

Length of segment iii usually is less than ½ mm. but the length of the portion of the body in front of 2/3 varies considerably and may be 2 mm. or more when the anterior end is everted. That end usually is flaccid and preserves poorly thus hampering determination of extent of variation in characters such as number and location of nephropores. Presence of a prostomium has been recorded at least twice but if *corethrurus* was involved the observations almost certainly were erroneous.

Except for the usual deviations in metamerism, little is known about aberrant structure. Any discussion of abnormality most likely would be futile until anatomical norms (of amphimictic populations) are known.

Polymorphism: Only one quadrithecal individual, and then with pores at 5/6–6/7, was found among the many hundreds of specimens examined by this author. The spermathecal battery may be unusually resistant to mutational elimination of its members. Yet, the species was long defined as quadrithecal.

The parthenogenetic degradation of genital structure that is obvious in genera of various other families is much less readily recognized in *P. corethrurus*. "A" morphs should have been readily recognized and may not yet have evolved. R morphs have been seen but can be recognized only in tedious dissections of favorable material or in really good microtome sections. Recognition of some other morphs that now can be anticipated will be just as difficult, if not more so, and must be preceded by discovery of norms of andry and gyny as well as of several other characters of the ancestral amphimictic population that has not yet been found. Avesiculate morphs and those with vestigial vesicles can of course be recognized easily in dissections. More common morphs lack tubercula pubertatis and associated TP glands as well as genital setae and GS glands. Presence of those structures, in an occasional rare specimen, allowed the prediction that they will be found to characterize the biparental population from which the morphs were isolated by parthenogenesis. The protandry, consecutive hermaphroditism, dioecism, masculine and feminine phases, invoked at various times in the past to explain misunderstood conditions probably involving parthenogenetic polymorphism, are lacking. Involved may well have been some or all of the following; anandry, progyny, metagyny, hologyny, hypergyny, as well as morphs with hermaphroditic gonads or with completely feminized testes.

Regeneration: Cephalic regeneration at several

anterior levels probably is possible but was so slow in the first experiments that further study of the species was abandoned. Tail regeneration is not rare in nature, but little information now is available except that replacement is possible at levels from 98/99 posteriorly.

Growth, in tail regenerates, during the Brazilian summer in the laboratory (Vanucci), was "5 mm. every 30 days."

Setae, in all of the numerous tail regenerates observed by this author, were in the usual quincunx arrangement of unregenerate postclitellar metameres. Supposed tail regenerates with paired setae, mentioned by several authors, must now be regarded as of some other species just the same as unregenerate individuals with regularly paired setae that were on various occasions referred to *P. corethrurus*. One of the author's specimens did have a posterior portion obviously with all of the stigmata of recent regeneration and with closely paired setae. Dissection alone revealed that a juvenile of quite another family had crawled part way into the tightly fitting rectum!

Parasites: SPOROZOA: *Monocystis radiata, biacuminata, Nematocystis glossoscoleci* Boisson 1957, seminal vesicles, Cochin China (now part of Vietnam).

Systematics: Worms with setae in eight regular longitudinal ranks through the body, on several occasions were referred to *P. corethrurus*. Regularity of setal ranks also was assumed to be a reversion to an ancestral condition in tail regenerates. That explanation no longer is tenable.

Quadrithecal Jamaican individuals could have been of another species unless they belonged to a very rare parthenogenetic morph of *P. corethrurus*.

Eisen's var. *mexicana* supposedly was distinguished by an intrasegmental location of the spermathecal pores. Eisen may not have made allowances for seeming insertion of the septa on the parietes behind the usual intersegmental levels.

Relationships of *corethrurus* scarcely warrant consideration in the present state of our ignorance of so much structure in its congeners.

Excretory system: Though distinguished specialists from Perrier to Bahl studied this system, nephridia, at least in anterior segments of *corethrurus*, still need accurate characterization, primarily perhaps with reference to the segments in which they belong morphologically. Fusion of peristomium and second segment was not recognized, and each of Bahl's (1942) septal or segmental numbers must be increased by one. An apparent insertion of septum 8/9 (Bahl's numbering) over intersegmental level 9/10 also was not recognized, and the ninth segment does have a pair of nephridia.

The six discrete and supposedly holoic nephridia in segment v, each with a preseptal funnel on the anterior face of 4/5 and looped tubules just behind that septum, were said to belong to segments iii, iv

and v but to "have become telescoped together" (*ibid.*, p. 13) because of absence of septa 2/3 and 3/4. Such telescoping would seem to require union of septa 2/3–4/5 for which the only evidence was the presence, side by side, of the funnels. Ducts, in other genera, not uncommonly open to the exterior or into the pharynx some distance from the metamere in which the nephridia are contained. Indeed, meronephridia of one segment may open, as Bahl showed in a subsequent page of the same journal number, to the exterior in two or more segments well in front of that containing the tubules.

If septa 2/3–4/5 are not united, each of the six discrete nephridia in segment v would appear, regardless of location of nephropores, to be truly meroic, and perhaps originating in the same way as in *Perionyx* "*dubius.*" Bahl, however, calls the single pair of nephridia in each of the next few segments meronephric because of the presence of several short twisted loops opening independently into a glandular portion of the organ. Those short loops, contrary perhaps to Bahl's anticipation, never became discrete anywhere in the Glossoscolecidae; they are branches of a single nephridium with a single nephrostome, nephropore and (presumably also) single embryonic origin. Such a unit ought to retain the holonephric characterization in spite of the complexity if for no other reason than to obviate future confusion such as was associated in the past with classical terms such as meganephridial and micronephridial.

The large nephridial mass in front of 4/5 comprising, according to Bahl, 80–100 meronephridia also has a single nephrostome and a single duct. Characterization as a much branched holonephridium seems preferable. The nephrostome of that nephridium supposedly is located on septum 1/2 (really 2/3) but in the figure (*ibid.*, p. 12) is shown as passing to the nerve cord behind the ganglion of segment iv. Because of the marked cephalization that *P. corethrurus* has undergone, septum 4/5 (really 5/6) is much more likely to be present than 1/2 or 2/3. Ducts of the anterior four pairs of nephridia, as shown in Bahl's figure 10, are free throughout their course. In specimens examined by the author, the ducts are free only rarely but usually pass into tissues of or associated with the pharyngeal bulb. From those tissues some ducts do emerge to acquire epidermal openings. Others, however, do not emerge. In such cases a corresponding epidermal pore is lacking and the nephridium presumably does open into gut lumen as Stephenson believed. Although past disagreements as to manner of terminating might then be attributable to individual variation another explanation may be preferable. The cephalization that has been evolving in *P. corethrurus* seems to be directed toward the complete elimination of the first two segments as portions of the body exposed to the external environment while conserving the nephridia which are "swallowed" so as to open into gut possibly now with some quite different function. The swallowing process then is further along in some morphs than in the others which still have the nephropores externally.

Economic Importance: Friable and loose soil in plant benches of greenhouses infested by *P. corethrurus* (Gates, 1954, *Bull. Mus. Comp. Zool. Harvard College* 112; p. 122) became hard, and the species was regarded as a pest by those who had experience with it. The species, along with *P. elongata*, in India (Puttarudriah & Sastry, 1961, *Mysore Agric. Jour.* **36**: p. 4) was found to have rendered a soil compact, hard, cloddy, thus hampering digging, preparation of the soil, rooting of vegetable crops, normal percolation. Reference was to subsurface casting. Since, *P. elongata* does cast at least in part on the surface, *P. corethrurus* may have been more responsible for the soil changes. Perhaps the most important word in connection with earthworm activity in the Mysore field was "overpopulation"!

Remarks: Although some progress has been made in clearing up the confusion regarding the anatomy of this species, definitive characterization of important structure still is impossible.

As already indicated, andry and gyny of the species remains to be determined. Attribution of ovaries as well as of testes to xii–xiii (Beddard) and of hermaphroditic gonads to x, xi (Vanucci) is presently unacceptable as previous authors were unaware of the anomalous parietal insertion of septum 9/10 which was thought to be 10/11. Septum 10/11 probably is often unrecognizable because of adherence to 9/10. Indeed, this author has been sure about 10/11 only once when it was possible to peel off the delicate membrane from the hearts as well as from 9/10. Ovaries are very small and in the writer's experience often were not found in the closely crowded septal and connective tissues. When seen, the ovaries were immediately under the testis sacs (presumably of xi), and no intervening segment was recognized.

Setae of *P. corethrurus*, according to Beddard (1895), are bifid distally, but no such termination has been seen by the author, and confirmation or more accurate regional localization is needed.

Because of the diversity in accounts of the vascular system by different authors, more study even of the larger trunks and hearts still is needed.

Pulsations of calciferous glands in developing embryos (Vanucci) may be like those in the calciferous glands of adults of *O. occidentalis* but nothing is known about their occurrence in adults of *P. corethrurus* or their function in either species.

Infestation of *corethrurus* cocoons by enchytraeids, according to Vanucci, resulted in sterility.

Amino acids of the soil are not depleted (Dubash & Ganti, 1964, *Current Sci., Bangalore* **33**; p. 219) by passage through the gut of *P. corethrurus*. On the contrary, cystine is deposited and arginine is increased.

HAPLOTAXIDAE

1880. Phreoryctidae Claus, *Grundzüge der Zool.* (4th ed.) 1: p. 482.
1875. Phreoryctidae, Vejdovsky, *Sitzber. Böhmischen Ges. Wissensch. Prag* 1875; p. 198.
1884. Phreoryctidae, Vejdovsky, *System und Morphologie der Oligochaeten* (Prag), p. 16.
1888. Phreoryctidae, Beddard, *Ann. Mag. Nat. Hist.*, Ser. 6, 1: p. 394.
1890. Phreoryctidae, Forbes, *Bull. Illinois State Lab. Nat. Hist.* 3: p. 108.
1895. Phreoryctidae, Beddard, *A Monogr. of the Order of Oligochaeta* (Oxford), p. 187.
1900. Haplotaxidae, Michaelsen, *Das Tierreich* **10**: p. 107.
1903. Haplotaxidae, *Die geographische Verbreitung der Oligochaeten* (Berlin), p. 64.
1913. Haplotaxidae, Piguet & Bretscher, *Catalogue des invertebres de la Suisse*, No. 7: p. 161.
1921. Phreoryctidae, Michaelsen, *Arch. Naturgesch.* **86**, A: p. 141.
1928. Phreoryctidae (Phreoryctina), Michaelsen, *Handbuch der Zool. Berlin* 2, 2-8: p. 106.
1930. Haplotaxidae, Stephenson, *The Oligochaeta* (Oxford), p. 802.
1953. Haplotaxidae, Yamaguchi, *Jour. Fac. Sci. Hokkaido Univ. (6-Zool)* 11: p. 304.
1958. Haplotaxidae (part only?), Omodeo, *Mem. Inst. Français Afrique Noire*, No. 53: p. 17.
1959. Haplotaxidae, Gates, *Bull. Mus. Comp. Zool. Harvard College* 121: p. 255.
1962. Haplotaxidae (Lumbricomorpha), Chekanovskaya, *The Aquatic Oligochaeta of the USSR* (Moscow), p. 320. (In part only?)
1966. Haplotaxidae, Brinkhurst, *Jour. Zool. London* **150**: p. 32.

Digestive system, agiceriate(?), without caeca, typhlosoles, calciferous and supra-intestinal glands, pre-intestinal gut in fewer than 10 segments. Vascular system, with complete dorsal and ventral trunks connected in each(?) segment by a pair of long and looped lateral vessels (without other longitudinal trunks?). Hearts, none. Nephridia, holoic, lacking in some anterior segments (avesiculate?). Setae, sigmoid(?), single-pointed(?). Dorsal pores, lacking.

Reproductive apertures, all minute and superficial, male pores (each an external opening of a single duct from one male funnel) behind spermathecal but anterior to female pores. Clitellum, unilayered, including male and female pore segments. Seminal vesicles, simple. (Prostates, prostate-like glands and atria lacking?) Spermathecae, adiverticulate. (Ovaries, shape?) Ova, yolky(?). Ovisacs, simple.

Distribution: United States (Washington, California, Illinois, Indiana, Ohio, Virginia, New York). Denmark, Germany, Poland, Bohemia, Switzerland, France, Belgium, England, Corsica, Italy, Jugoslavia, Hungary, Bulgaria, Russia, Siberia, Japan.

Poeloe Berhala (near Sumatra), Australia, New Zealand, Auckland and Campbell Islands, Argentina, Paraguay, Peru, South and equatorial Africa.

The range of the family now seems to comprise most of the habitable world with exception of Central America, West Indies, Iberian Peninsula, north Africa, China, Asia south of the Caucasus and Himalayan mountains. Little significance is allowed those exceptions because of the paucity of our knowledge about the oligochaete faunas.

The Haplotaxid range is so much greater than that of any other megadrile family as to require careful consideration. Increasing knowledge of oligochaete transportation by man during the last two to three thousand years might be thought to suggest possibility of that method of carriage especially since the aquatic *Criodrilus lacuum* is now known to have been carried from Europe to the United States. Seemingly favoring human transport is the range of *Haplotaxis gordioides* (Hartmann, 1821). That species has been reported from 12 European countries, Russia, Siberia, Japan, as well as Washington, Michigan and New York states in North America. No other megadrile has a range anywhere near as large, except as a result of much assistance from men. However, much of the information about the biology of these animals seems to contraindicate likelihood of frequent transportation. *Haplotaxis gordioides* may be another classical congeries masquerading as a species. It is known, in many areas from which it has been reported, only from immature individuals.

Biology: Haplotaxids usually are considered to be aquatic and as inhabiting primarily subterranean waters or mud at bottom of deep lakes. Species, in certain areas, have been found only in spring fed pools, in water pumped from deep wells or in mud brought up from lake bottoms at depths such as the following: *Haplotaxis ascaridioides* Michaelsen 1905, Lake Baikal, Siberia, 7–1,300 m. *Haplotaxis gordioides* (Hartmann 1821), Telezkischen See, Russia, 15–319 m.; Lake Leman, Switzerland, 309 m.; Luganer See, 270 m.; Other Swiss Lakes, 30–150 m. *Haplotaxis heterogyne Benham*, 1903, New Zealand, 550 feet. *Pelodrilus ignatovi* Michaelsen, 1903, Telezkischen See, 38–319 m.

H. gordioides was found to be abundant in Alsatian soil at depths of 0.8–2.0 m. (observed, during May, 1915, June, 1916, from World War I trenches), and other species were found in habitats such as, under stone in very wet earth (Corsica), marshy soil, wet sand at river edge, rich wet soil, under moss below water at edge of tarn, forest pool. Moreover, Thienemann (1912) maintains that *H. gordioides* really is terrestrial and breeds only in earth. The same author (1913) also characterizes *gordioides* as haloxene, the worms having been found in Westphalia in water with 9.193–13.98 g. salt per liter, "in einer Salzquelle von Sassendorf leben."

Systematics and Phylogeny: The family, from 1895 until recently, seemed to be fairly well defined though it did comprise only two genera distinguished from each other mainly by setal characters and presence or absence of a gizzard. Such somatic characters recently were said (Omodeo, 1958) not to distinguish *Haplotaxis* Hoffmeister 1843 and *Pelodrilus* Beddard 1891 because of guts having been found in inter-

mediate conditions and because of possible convergences in the setal system.

The so-called gizzard of certain species of *Haplotaxis* probably needs better histological characterization. The organ does not seem to be a gizzard in the usual megadrile sense, i.e., of a greatly increased thickness of circular muscle layer in a definite segment of the esophagus or of the intestine, but rather a muscularization of the pharynx in which glandularity has been retained. Any such specialization should be investigated in as much detail as is necessary to be of use in defining taxa that are systematically difficult because of considerable morphological simplicity.

Setae, in classical definitions, were sigmoid and simply pointed (i.e., characteristically megadrile). *Heterochaetella* Yamaguchi, 1953, was placed by its author in the Haplotaxidae, though one in each of the 4 setal couples per segment has a bifid tip, a condition that is common in the Naididae, Tubificidae, some phreodrilids, one enchytraeid genus, and 4 lumbriculid genera. The Japanese genus is known only from a description of the type species which was erected on immature individuals. Location of male and female pores is unknown and nothing was recorded about the sperm ducts. *Adenodrilus* Chekanovskaya, 1959, likewise placed by its author in the Haplotaxidae, not only has bifid setal tips but also fine bristles in the crotch, a condition that is known elsewhere only from Microdrili. The genus is monospecific.

One of Michaelsen's species of the year 1914, provides further proof of the need for much more information about supposedly primitive megadriles and also about possible closely related microdriles. The species, *kraepelini*, erected on 3 immature worms from Congo, was placed by its author originally in a tubificid genus *Lycodrilus* Grube, 1873 but was subsequently transferred by Michaelsen himself to a lumbriculid genus, *Eclipidrilus* Eisen, 1881, erected on a Californian worm. A year later, after a study of microtome sections, Hrabe (1933) transferred the species *to Pelodrilus*. However, for Grube's genus and its Russian species, Svetlov (in Chekanovskaya, 1962) erected a new family, the Lycodrilidae. Previously Omodeo (1958) had transferred the Congo species to "*Haplolumbriculus* Omodeo (*in litt.*)" which remains to this day a nomen nudem as do two species names, *Haplolumbriculus insectivorus* and *Pelodrilus carnivorus*, of the same publication. *Pelodrilus falcifer* Omodeo 1958 has simple setal tips and clearly seems to belong in the classical *Pelodrilus*.

Absence of hearts, if universal in the family, would seem to be uniquely diagnostic. Accordingly, it is unfortunate that so little has been recorded about vascular structure. In *Pelodrilus bureschi* Michaelsen, 1925, for instance, the circulatory system was not even mentioned in either of the descriptions but it is in just that species, because of the unusual size, 5 mm.

thick, that hearts seem most likely to be needed. The length (and perhaps also a necessarily associated looping?) of the lateral connectives between dorsal and ventral trunks may prove to be characteristically haplotaxid. Whether a pair of such connectives is present in "each" body segment, as sometimes has been assumed, remains to be determined. The vessels were recognizable, in the types of *Haplotaxis gastrochaetus* Yamaguchi, 1953, only in ii-xx. Differences in number and location might be systematically useful, but considerable care will be required in determining such characters because of the invisibility of thin-walled oligochaete blood vessels when empty.

Cells in the peritoneal investment of the nephridia are known to be enlarged and filled with a fatlike material in some species. Information is lacking that would enable use of presence or absence characters even at species level.

Characters provided by the male ducts and their apertures, allowed greatest systematic importance in the classical system, are less typical of the family. Length of the ducts is not a character of much systematic importance as the moniligastrids conclusively show. Of significance is location of the male pore with reference to the septum bearing the male funnel. Male pores of haplotaxids, in so far as they have been recorded, are somewhat behind the intersegmental level next posterior to that of the septum bearing the appropriate male funnel. Male pores in the Moniligastridae often are still nearer the funnel-bearing septum though sperm ducts may always be longer. Perhaps of more systematic importance is independence of the sperm ducts, i.e., acquisition by each duct, during growth, of its own external aperture. The character is shared by haplotaxids with the holandric moniligastrids as well as with biprostatic species of an octochaetid genus, *Hoplochaetella* Michaelsen, 1900, meroandric species of all megadrile families, meroandric and other parthenogenetic morphs of several families. The character in the Haplotaxidae and Moniligastridae presumably is primitive but elsewhere is a secondary simplification commonly and seemingly rather easily acquired. Location of the male pores in front of the female pores, presumably primitive, is shared by the Haplotaxidae with the Alluroididae, Moniligastridae, and parthenogenetic morphs of a Lumbricid species, in the latter of which the uniparental method of reproduction had enabled reversion to an Alluroidid stage.

The Haplotaxidae, in the classical system, is primitive and also is the great-great-grandparent of all megadrile families, having maintained its Haplotaxid nature to this very day while some of its descendants were evolving into all the other megadrile genera and families, just as if (*cf.* Stephenson, 1930) *Eohippus* had lived on alongside all of its evolutionary derivatives including *Equus*. The near cosmopolitan distribution does now seem to support greater age than for any other megadrile family.

However, if ovaries of *Haplotaxis forbesi* Smith 1918 are in xv, xvi because gonads of segments xii–xiv in a long primitive series have disappeared, as the author maintains, instead of being displaced posteriorly from xii–xiii, as in classical thought, the genital organization of haplotaxids already has undergone considerable simplification. What now needs to be determined is whether and how much of the simplicity in haplotaxid somatic structure is secondary and perhaps independently acquired in various areas separated from each other by wide oceanic distances.

REFERENCES

CERNOSVITOV, L. 1945. "Oligochaeta from Windermere and the Lake District." *Proc. Zool. Soc. London* 114: p. 523–548.
CHEKANOVSKAYA, O. V. 1962. (*The Aquatic Oligochaeta of the USSR*) (Moscow), pp. 412.
HRABE, S. 1933. "Zur Kenntnis des *Pelodrilus kraepelini* Michaelsen." *Zool. Anz.* 104: pp. 225–228.
OMODEO, P. 1958. "La reserve naturelle integrale du Mont Nimba. I. Oligochetes." *Mem. Inst. Français Afrique Noire*, No. 53: pp. 109.
THIENEMANN, A. 1912. "Der Bergbach des Sauerlands, Teil I: Die Organismen des Mitteldeutschen Bergbaches." *Internatl. Rev. Hydrobiol. Suppl.* 4, 2: pp. 1–125.
—— 1912. "*Haplotaxis gordioides* G. L. Hartmann als terrestrischer Wurm." *Arch. Hydrobiol.* 19: p. 377.
—— 1913. "Die Salzwassertierwelt Westfalens." *Verhandl. Deutschen Zool. Ges.* 23: pp. 55–68.

HORMOGASTRIDAE

1900. Hormogastrinae (Glossoscolecidae) Michaelsen, 1900, *Das Tierreich* 10: p. 446.
1903. Hormogastrinae (Glossoscolecidae), Michaelsen, *Die geographische Verbreitung der Oligochaeten* (Berlin), p. 134.
1918. Hormogastrinae (Lumbricidae), Michaelsen, *Zool. Jahrb. Syst.* 41: p. 376.
1921. Hormogastridae, Michaelsen, *Arch. Naturgesch.* 86, A: p. 141.
1928. Hormogastridae (Lumbricina), Michaelsen, *Handbuch der Zoologie, Berlin* 2, 2-8: p. 108.
1930. Hormogastridae (Glossoscolecidae), Stephenson, *The Oligochaeta* (Oxford), p. 903.
1956. Hormogastrinae (Microchaetidae), Omodeo, *Arch. Bot. Biogeogr. Italiano*, Ser. 4, 1: p. 177.
1959. Hormogastridae, Gates, *Bull. Mus. Comp. Zool. Harvard College* 121: p. 255.

Digestive system, with three strong gizzards in vi–viii, with an intramural, nearly circumferential, esophageal(?) calciferous gland (in xx? Intestinal origin?), and a large typhlosole divided anteriorly into longitudinal lamellae. Vascular system, with complete dorsal and ventral trunks (subneural and lateroneurals?), a supra-esophageal free from gut in region of vi–xi (extra-esophageals?), but without lateroparietal trunks. Hearts, in vi–xi, lateral. Nephridia, holoic, vesiculate, each vesicle with a caecum(?), (ducts passing into parietes at mL?). Nephropores, obvious, in two longitudinal ranks, about at mL. Setae, eight per segment, in genital tumescences elongated but slender and longitudinally grooved ectally. Dorsal pores, lacking.

Female pores, post-equatorial in xiv. Males pores, in region of 15/16. Spermathecae, adiverticulate, in front of ovarian segment. Clitellum, multilayered including male and female pore segments. Holandric. Ovaries, in xiii, (shape?). Ovisacs, in xiv, small and lobed. Ova, not yolky. Prostates with muscular ducts, none.

Distribution: Italy, Sicily, Sardinia, Corsica, Spain. (Tunisia, Algeria?).

Presence of the family in north Africa now seems likely to have resulted from transport by man. For the single recorded instance of overseas transport (Gates, 1954, *Bull. Mus. Comp. Zool. Harvard College* 111: p. 248) there must have been very many others in the last two thousand years of Mediterranean traffic. Also noteworthy is the fact that transport to America was of clitellate adults and not of cocoons or recently hatched juveniles.

Remarks: The family is monogeneric and presently with but 2 species. Doubts have been expressed as to the distinctness of those species, in each of which varieties, formae, or subspecies have been named. Individual variation seems never to have been studied from a long series at any single locality.

Our ignorance of so much that would be of interest regarding these European forms may have resulted from their rarity. If they are haemerophobic they could have been disappearing from much of their autochthonous realm because of millennia-long disturbances of the soil by agricultural and other human activities. *Hormogaster*, perhaps like *Criodrilus*, may be the only survivor of a large family that was decimated during the ice ages.

Longitudinal grooving in ectal portions of the elongated but slenderized genital setae, as well as other genital characters of no more systematic importance, once were believed to prove such close relationships as to require at most no more than subfamily status in the Lumbricidae. Hearts seemingly are lateral, as in the lumbricids, which does seem rather surprising in view of presence of a supra-esophageal trunk so unusually free from the gut. Somatic anatomy, of course, was given little or no consideration previously. Even gizzards were allowed little systematic value, presumably because of a belief in "the well known conformity of the digestive tract to diet." Three esophageal gizzards, distinctly demarcated metamerically, instead of one intestinal gizzard possibly without metameric delimitation posteriorly, along with other known differences in conservative somatic anatomy, including the calciferous section of the gut, contraindicate specially close relationship with the Lumbricidae. So little is known about somatic anatomy of the Microchaetidae and the Glossoscolecidae to which the European genus at one time or another has been thought to be affiliated

that any consideration of such views at the present time is likely to be fruitless.

LUMBRICIDAE

1876. Lumbricidae (part), Claus, *Grundzüge der Zool.* (ed. 3) 1: p. 416. (Including only *Lumbricus* and *Helodrilus*.)
1880. Lumbricidae (part), Claus, *ibid.* (ed. 4) 1: p. 478. (Excluding *Criodrilus* and *Pontoscolex*.)
1884. Lumbricidae, Vejdovsky, *System und Morphologie der Oligochaeten* (Prag), p. 59.
1885. Lumbricina, Oerley, *Ertek. Term. Magyar Akad.* 15, 18: p. 10.
1887. Lumbricina (Lumbricidae), Rosa, *Mem. R. Acad. Sci. Torino* 38: p. 179.
1888. Lumbricidae (part), Rosa, *Boll. Mus. Zool. Univ. Torino* 3, 41: p. 8. (Excluding *Criodrilus*.)
1890. Lumbricidae (part), Benham, *Quart. Jour. Micros. Sci.* 31: p. 222. (Excluding *Criodrilus*.)
1893. Lumbricidae (part), Rosa, *Mem. Accad. Sci. Torino* 43: p. 403. (Excluding *Criodrilus*.)
1895. Lumbricidae, Beddard, *A Monogr. of the Order of Oligochaeta* (Oxford), p. 687.
1897. Lumbricini (Lumbricidae), Michaelsen, *Verhandl. Naturwiss. Ver. Hamburg*, Ser. 3, 4: p. 25.
1900. Lumbricidae, Michaelsen, *Das Tierreich* 10: p. 470.
1903. Lumbricidae, Michaelsen, *Die geographische Verbreitung der Oligochaeten* (Berlin), p. 135.
1910. Lumbricidae, Michaelsen, *Ann. Mus. Zool. St. Petersburg* 15: p. 2.
1913. Neolumbriciden, *Szüts, Allat. Közl.* 12: p. 55.
1918. Lumbricinae (Lumbricidae), Michaelsen, *Zool. Jahrb Syst.* 41: p. 379.
1921. Lumbricidae, Michaelsen, *Arch. Naturgesch.* 86, A: p. 141.
1923. Lumbricinae (Lumbricidae), Stephenson (*Fauna of British India*), *Oligochaeta*, p. 496.
1928. Lumbricidae (Lumbricina), Michaelsen, *Handbuch der Zoologie* 2, 2-8: p. 108.
1930. Lumbricidae, Stephenson, *The Oligochaeta* (Oxford), p. 905.
1950. Lumbricidae (part), Pop, *An. Acad. Romane, Sect. Sti. Geol. Geogr. Biol.* (A), 1, 9: p. 48. (Excluding *Criodrilus*.)
1956. Lumbricidae, Omodeo, *Arch. Zool. Italiano* 41: p. 170.
1959. Lumbricidae, Gates, *Bull. Mus. Comp. Zool. Harvard College* 121: p. 255.

Digestive system, with an intramural calciferous gland[1] comprising longitudinal chambers that open at their anterior ends into the esophageal lumen, a terminal esophageal valve reaching into xv, an intestine beginning with a "crop" followed by a gizzard,[2] a sacculated as well as an unsacculated portion and ending in an atyphlosolate region,[3] but without intestinal caeca and supra-intestinal glands. Vascular system, with complete dorsal, ventral, subneural (and lateroneural?) trunks, the latter adherent to nerve cord, extra-esophageal trunks that are median to the hearts and which pass to dorsal trunk in region of x–xii, without supra-esophageal and lateroparietal trunks. Hearts, lateral, the last pair anterior to segment xii.

---

[1] Always in the region of x–xiv?

[2] Always mainly in xvii of anthropochorous species. In the remainder of the family?

[3] Sometimes called a rectum or a proctodeum. Correct morphological characterization remains to be determined. Definitely implied, of course, is presence of an intestinal typhlosole anteriorly.

TABLE 1.

Species	Number of times intercepted	Number of specimens intercepted
Allolobophora chlorotica	58	107
Allolobophora longa	4	10
Allolobophora trapezoides	20	160
Allolobophora tuberculata	11	51
Allolobophora turgida	25	55
Bimastos parvus	20	31
Dendrobaena octaedra	26	42
Dendrobaena rubida	53	104
Eisenia foetida	51	61
Eisenia rosea	25	368
Eiseniella tetraedra	9	14
Lumbricus castaneus	10	44
Lumbricus rubellus	31	49
Lumbricus terrestris	21	31
Octolasium cyaneum	2	2
Octolasium lacteum	14	58
(?)eiseni	1	1
Totals	379	1,182
Unidentified lumbricids probably mostly of the above species	212	410
Grand Total		1,592

Nephridia, holoic, vesiculate, ducts passing into parietes in region of *B*. Setae, sigmoid and single pointed, eight per segment, in regular longitudinal ranks, in genital tumescences elongated but slender and longitudinally grooved ectally. Dorsal pores, present.

Reproductive apertures, all minute, female (always in xiv?) and male pores in that anteroposterior order,[4] equatorial and anterior to the multilayered[5] clitellum which is always behind xvii(?). Spermathecae, adiverticulate, pores at intersegmental levels. Ovaries, in xiii, bandlike, each terminating distally in a single eggstring. Ovisacs, in xiv, small, lobed. Ova, not yolky. Prostates with muscular ducts, none.

Distribution: Southeastern United States below the southern limit of glaciation and west from New Jersey (perhaps to the Mississippi River or somewhat beyond?), Europe below the southern limit of glaciation, Asia Minor, Siberia, Japan. Northern Africa

[4] Except in certain parthenogenetic morphs in which the aberrant method of reproduction has enabled return of male ducts to an alluroidid stage or some approximation thereto. Presence of male pores in any of xii–xiv and xvi, when somite deletion and fragmentation is not involved, now appears to be among the aberrations that are established after reproduction becomes parthenogenetic. If presence of male pores behind xvi resulted similarly, location of those pores in xv would be invariant for amphimictic taxa as presence of female pores in xiv now seems to be. Male pores in atypical locations are always (?) equatorial.

[5] And always saddle-shaped (except in parthenogenetic morphs?).

has been thought by some to belong in the range but proof is lacking.

The size of the range, greater than that of any megadrile family except the Haplotaxidae, disproves a long-held classical assumption that the family was recently evolved.

North of that range, megadrile populations were exterminated by the sheering, grinding, plucking, crushing pressures exerted by slowly flowing ice sheets thousands of feet thick. Even in unglaciated adjacent regions, subarctic temperatures probably destroyed many populations. Certainly, few if any megadriles were able to advance northward close to the receding ice. In fact, American earthworms still were far from having reached as far north as the Canadian border by the time Europeans began to settle that country. All earthworms in Canada today are exotic and, perhaps with one exception, anthropochorous. The natural family range, i.e., one comprising the totality of all ranges self-acquired by endemic species, has been greatly extended by man's transportation of a dozen or more taxa into every area where European man has gone to live. Proof of that transportation is provided by presence of European lumbricids on oceanic islands such as the Azores, St. Helena and Ascension, St. Paul's Rock, the Hawaiian Islands, and others that can be mentioned.

Accidental importations, into more advanced countries across national boundaries, were considerably reduced during the last seventy years, as a result of increasingly effective enforcement by government agencies of rules controlling importation of plants and soils. Nevertheless, during a fifteen-year period ending early in 1964, inspectors of the U. S. Bureau of Plant Quarantine intercepted earthworms on 1,244 occasions. Among the 3,430 specimens were a number of lumbricids mainly of anthropochorous species. Unpublished data as to hereinafter included species are presented in table 1.

Transportation of the markedly anthropochorous species, only into unoccupied glaciated regions of the Holarctic, is unlikely but no attempt seems to have been made to determine what happens following introductions into territory occupied by endemic and possibly haemerophobic lumbricids.

The distribution of anthropochorous lumbricids has been much misunderstood. The finding of a species at some thirty sites in California late in the previous century was almost all the justification that can be detected for including "ganz Nord America" (Michaelsen, 1900, *Das Tierreich* **10**: p. 486) in the distribution of *A. chlorotica*. A later list, in a 180-page examination of oligochaete distribution, correctly mentioned only North Carolina (one site), California, Vancouver, Mexico (Mexico City), and Guatemala. The later publication is, however, less frequently consulted than the earlier which still is the most recent systematic treatment of all of the Oligochaeta.

A single morph of *D. rubida*, according to Cernosvitov and Evans (1947: p. 21) is "widely distributed, known from almost every country and can be regarded as one of the commonest species" but the quotation contains a typographical error, "country" of course should be "county" since the statement is immediately preceded by "British records." That statement, carelessly read, presumably was responsible five years later for another zoologist declaiming that the morph had been recorded "from more places all over the world than any other species." By 1963, the legend was stated in greater precision, "*Dendrobaena rubida* lives on the island of Disko off Greenland, on the Himalayas and throughout the tropical Malayan peninsula" (*cf.* correct listing of recorded distribution of *D. rubida* below).

Most peregrine lumbricids often are said to be cosmopolitan. The first meaning of that word, according to the *Oxford Universal Dictionary*, is "belonging to all parts of the world" or in natural history, "found in all countries." Alternate meanings, in the same source, are, "not restricted to any one country" or "found in many countries." In its first meaning, cosmopolitan, as is shown below, cannot characterize correctly any single megadrile species. Unfortunately, just that primary meaning usually has been understood erroneously in the past. In its secondary meanings, cosmopolitan certainly conveys no more information than "eurytopic" or the equivalent phrase "widely distributed," each of which is less likely to be misunderstood. The phrase "holoarctic species" characterizes lumbricid anthropochores no more accurately than it would the Colorado potato beetle, *Leptinotarsa decemlineata* (Say, 1824) now spread from Central America into Canada and through much of Europe.

Because of the unfortunate prevalence of so many misconceptions about lumbricids some emphasis, at this point, seems necessary on one fact. No lumbricid species is known to have colonized tropical lowlands permanently, anywhere. Deliberate introductions, by the hundreds if not thousands of individuals, failed. The evidence now available indicates that every anthropochorous species is climatically restricted, either to cooler areas such as are found in tropical highlands and in the temperate zone or to warmer areas of the same zone like those around the Mediterranean. This disposes at once of a classical belief regarding lumbricids that recently has been stated in these words, "earthworm species have often a latitude of adaptation unparalleled by any other invertebrate." What anthropochorous lumbricids do seem to have is an ability to live in many situations that provide adequate moisture and metabolic necessities but only within certain climatic boundaries determined by specific genotypes. Additionally, genotypes of peregrine lumbricids express themselves in ways as yet not understood but which enable survival

in environments more or less constantly disturbed by man, and so closely to him that the worms are unintentionally taken on very many of his travels about the world. The factors thus involved, included in the characterization haemerophilic, effectively set apart a very few lumbricids from the bulk of the family, the endemic species which appear to be haemerophobic (averse to culture and human disturbances of the environment). Anthropochorous forms, accordingly, may already have replaced less well-adapted congeners in considerable portions of their own ranges and especially in habitats influenced during the last two millennia by more or less continuous cultivation. Determination of self-acquired species ranges in Eurasia, especially for lumbricid peregrines, may prove to be very difficult, perhaps impossible.

Systematics: Hierarchical rank allowed the taxon has varied from time to time and by Michaelsen was changed more than once. Content of the taxon, except for occasional inclusion of *Criodrilus*, has not varied significantly for some eighty years, a fact that has some hitherto unrecognized significance.

An early definition of the family (Beddard, 1895) mentioned characters of five somatic and four genital organs. Unfortunately, inexactness then, as on many subsequent occasions, reduced the value of some characters, more particularly those that were somatic. The gizzard, for instance, never is at the end of the esophagus nor is it at the beginning of the intestine. Nephridia indeed are paired but they were not "all similar" whatever that may have been intended to mean.

The much longer definition of the next monograph (Michaelsen, 1900) did include a statement that calciferous "glands" are present, which was a slight advance in the right direction even though failure to mention any anatomical distinctions from the glands of various other families allowed the character almost no systematic value. The forward step was coupled with a large step backward by depriving the circulatory system of systematic usefulness. The single vascular character of Beddard's definition obviously has a double significance and now seems likely to be of considerable assistance in determination of interfamily relationships. Beddard's excretory character was replaced by "meganephridial" which Bahl and others long ago recognized as too inexact to have any real systematic usefulness. Later monographs usually repeated the 1900 definition, with or without unimportant emendations. One recent revision of the Lumbricidae provided no definition of the family.

The author's examination of thousands of specimens showed that certain previously ignored, somatic characters are invariant, at least in the American endemics as well as in peregrine European species of seven classical genera. A calciferous gland, for example, always was present and always had, even in the very most aberrant individual, the characteristic anatomy. One genital character, never used in the classical system, rather unexpectedly, also, was found to be invariant, as well as to be promising aid in determining interfamily relationships.

Although the same invariableness was found in a few short series of European endemics, only a small portion of the family has been studied and the bulk of the material has been of anthropochorous species that admittedly are exceptional, at least in their physiology. Accordingly, the definition above, though believed to represent a "giant step forward" in lumbricid taxonomy, may be liable to considerable emendation. Some characters that could have been included on the basis of the material studied are indicated in the footnotes.

Parthenogenesis handicaps the defining of lumbricids just as in other families. The clitellum, for instance, was at first thought to be saddle-shaped (Beddard, 1895) but in the next monograph the character was qualified by "meist" and "usually" has been retained hitherto. Exceptions that were thought to require "usually" now appear to be the annularity of certain parthenogenetic morphs. The male terminalia provide similar problems. Male pores, in the author's material were in xv, except in instances of developmental, regenerative or parthenogenetic abnormality, and then were in one or two of segments ix–xvi. Aberrant conditions arising during regeneration or because of some interference with normal embryonic development, presumably need no mention in a systematic definition, unless perhaps uniquely confined to a single taxon. But, should parthenogenetically allowed, aberrant characters also be excluded? If so, and if situation of male pores behind xvi is similarly associated, the Lumbricidae then can be defined by location of male terminalia in xv as well as by other characters not now included.

The Lumbricidae, as defined above and in spite of the several uncertainties, now is about as clearly distinguished from other megadrile families, including all those of Michaelsen's Lumbricina, as is the Moniligastridae. Such clearcut distinctions at least hint at long phylogenetic isolation from other evolutionary lines and support the author's deductions as to family age based on the distribution.

Subdivision of the family into genera, hitherto has been, as Michaelsen noted (1910: p. 2) in his last major contribution to the problem, one of the "schwierigen Aufgaben der Oligochätologie." Difficult, according to Beddard (1895: p. 687), because the range of variation in lumbricid structure is "very small." Difficult, according to others, because variation is too great and also because of the existence of intermediates (cf. Stephenson, 1930: p. 906–909) with the supposedly distinguishing characters of two or more genera. Little more was to be expected, it is

now clear, because of the classical restriction of systematic usefulness to a very few, simple, key characters.

During examination of considerable material of the type species of seven classical lumbricid genera, several characters were found to be intraspecifically invariable. The same characters were found to be equally invariable in all available specimens of other species of Lumbricus. That genus, accordingly is now defined primarily by those characters. Similar characters of the type species of four lumbricid genera, on the contrary, differed significantly from those of other species in the same classical genus. Such somatic characters, in other families, have enabled resolution of classical congeries into natural groupings of species at generic level. There is reason for believing that use of appropriate somatic characters can enable resolution of the classical congeries that have masqueraded as genera under such names as Allolobophora, Bimastos, Dendrobaena, and Eisenia. Each of those genera now can be split into two, definable primarily by somatic characters of the sort that are under consideration. Information as to those characters, unfortunately, is available almost only for common anthropochores. Little or nothing is known about the calciferous glands, the excretory and vascular systems of the European endemics that constitute a major portion of the family. As recently as 1967, the description of a newly erected species contained no mention of excretory organs and the calciferous gland was dismissed with a statement to the effect that a pair is in xi (of glands or of terminal sacs?). Accordingly, and inasmuch as the minority anthropochores are exceptional at least in their physiology, as well as for other reasons, lumbricid genera, with the exception of Bimastos, are not revised herein. Instead, some of those somatic characters by which other lumbricid genera now seem most likely to be defined in the future are merely indicated for consideration of those who have access to material of the European endemics.

Adequate somatic characterization of megadrile families now seems likely to provide a much better basis for phylogenetic and zoogeographical speculation than the genitalia. A first and tentative attempt at a lumbricid phylogeny, based on somatic structure, is presented below.

Phylogeny: Primitively, calciferous tissues were within longitudinal folds of an inner portion of the esophageal wall stretching through ix–xiv. Then, as in the beautiful evolutionary series found (Smith, 1924) in the acanthodrilid American genus Diplocardia, the folds became higher, more thinly lamelliform and eventually united mesially so as to leave longitudinal tunnels in the esophageal wall. The process, whether or not it began posteriorly, remains incomplete anteriorly so that the tunnels open into the gut lumen through a circumferential circle of small pores. That is the condition retained to the present in one major lineage of the Lumbricidae which includes *Dendrobaena*, *Eisenia*, and *Helodrilus* (of Hoffmeister 1845, a genus long forgotten and still misunderstood). In that sacless line, apertures into the esophageal lumen sometimes were retained in x (*Helodrilus*). Calciferous lamellae, in that line, gradually narrow anteriorly in the segment within which the tunnels are to open into the gut lumen. So much decrease is avoided by a second major lineage in which calciferous lamellae end midsegmentally in x. Associated therewith, even if not causally, the wall of the esophagus is bulged out on each side to form a lateral pocket opening widely and vertically into the gut lumen. In such lateral sacs, calciferous lamellae are on the posterior wall which is thick. The anterior wall of the sac is at least as thin, if not thinner, than that of the esophagus anteriorly.

Calciferous lamellae of x, in a third evolutionary line, were eliminated as they had been in the ancestry of *Dendrobaena* and of *Eisenia*. Subsequently, however, in the third line, sacs were evaginated anteriorly from the floor of the gut near insertion of septum 10/11. Into those sacs calciferous lamellae are continued but on the lateral wall. Other modes of communication with esophageal lumen will be evidenced by the more rarely studied species.

Some special widening of calciferous lamellae through one or two segments, in certain taxa of each major lineage, results in a more or less marked moniliform shape of the gut, possibly associated with increased activity of the gland. On the contrary, lamellae are narrowed in *E. tetraedra* and its gland has been thought to have little function and perhaps to be disappearing.

Easily recognized evolutionary changes in the vascular system involve the hearts, and the extra-esophageal trunks. All of the six (or seven or more?) pairs of hearts, sometimes called aortic arches or pseudohearts, that develop in the embryo usually are not retained. Those eliminated, as is shown by species considered below, may be at one or the other end or at both ends of the series. The primitive character of the extra-esophageal trunks is believed to be that of opening into the dorsal trunk posteriorly in xii. Corroborative evidence is supplied in an occasional individual of species in which the junction now is in region of insertion of 10/11, by retention into maturity of the posterior part of the trunk as a diminutive but patent tube or as a functionless rudiment.

From a primitive nephridial bladder, with a shortly ellipsoidal or digitiform shape, easily can be derived the other major kinds of lumbricid vesicles. The ocarina form requires little more than depression of the floor of the primitive kind in an asymmetrically funnel-shaped manner. The *J*- to *U*-shaped sort requires little more than elongation of the ancestral bladder, perhaps in circumstances that require the growing organ to turn completely back on itself in

order to function best in available space. Penetration of nephridial ducts straight through the body wall to open by a single rank of pores on each side is assumed to be primitive. Then, the indirect manner, involving growth of some ducts well laterally within the parietes before opening through a second rank of nephropores on each side, is secondary. Obvious nephropores are associated with something that inconspicuous pores lack (or have in a much less developed state?) and that something, perhaps only a much enlarged sphincter, is regarded as secondary.

One classical assumption, that four pairs of seminal vesicles are primitive, does not need to be questioned at present. Nor assumptions that absence of vesicles, in x – – – the "*Dendrobaena* stage," in ix, x "the *Bimastos* or *Eophila* stage," are secondary. Inasmuch as those same changes are now known to be made within limits of infrasubspecific parthenogenetic morphs there is no longer any justification for regarding such changes as intrinsically of generic significance or distinction. Newer assumptions are: The primitive spermathecal battery comprised two pairs of organs opening at intersegmental levels 7/8–8/9. Derivative states, as in other families, arose by increase in number of spermathecae within a segment or by addition of new pairs of organs at either end of the series as well as by deletions of one or more pairs of organs at either end of the series.

With reference to genitalia that were ignored in the classical system, the ancestral lumbricid had its male pores within atria deeply invaginated into coelomic cavities. Associated with those invaginations were well-developed, supraparietal atrial glands. Deep coelomic invaginations have been retained almost only in one of the relict American species, *Bimastos palustris* Moore 1895. Elsewhere the atrium usually has been reduced to a transverse cleft (often called a male pore) of variable depth within a male tumescence, though the atrial gland itself may still be well developed. Subterminal and terminal stages of that evolutionary sequence are shown by *Lumbricus* and, in particular, by one species which now has superficial male pores without trace of atrium, atrial cleft or atrial gland.

Recent anatomical changes found in various infrasubspecific, parthenogenetic morphs show that the absence of TP and GS glands, as well as of the associated tubercula pubertatis and genital tumescences, is secondary. Accordingly then, the ancestor of most if not all lumbricids had those organs as well as, and for exactly the same reasons, the atria and atrial glands.

Anatomical changes made during evolution of an ancestral lumbricid from the postulated protomegadrilid were listed (Gates, 1956) in an earlier publication. Those not already mentioned above are as follows. Acquisition of dorsal pores at intersegmental levels. Dislocation laterally of spermathecal pores from region of *AB* to *CD*. Elimination of an esophageal gizzard, extension of the esophagus into xv, appearance of an intestinal crop, gizzard, and typhlosole. Differentiation of a subneural trunk and, median to the hearts, extra-esophageal trunks. Elongation of male ducts to open externally in *BC* at the equator of xv.

Terminology: Calciferous tissues of the lumbricidae have been said to be in a single gland or in paired glands and then in one or two pairs. Either of the last two specifications usually (always?) is erroneous as the lamellae extend one or two segments behind the one or two moniliform gut widenings that are involved in the statement as to number. Basic in all lumbricid glands of known structure is a longitudinal series of channels closed posteriorly and opening anteriorly into the gut directly or indirectly through some sort of a sac. As the channels are uninterrupted i.e., are continuous from one end to the other, and as there is no real segregation of those channels into right and left fractions of a cylinder, "gland" seems a more accurate characterization than the plural and accordingly is preferred.

The crop, at the beginning of the intestine, is directly followed by a gizzard that usually is said to be in xvii–xix or occasionally even farther behind. Merely cutting off a dorsal portion of the gut behind the crop shows the marked thickening of muscular layers that distinguishes a gizzard seemingly ending in xviii anteriorly. The exact level at which a marked thickening of a muscular layer becomes macroscopically unrecognizable is not easily determined. Septum 17/18 is membranous and its insertion on the gizzard usually is unrecognizable, perhaps often because of presence of a large fenestra that may not have been produced as the worm was being opened and pinned out. Whether that portion of the gut with a superficial external appearance of being a gizzard is the sphincter or valve that has been reported, at least for one species, is uncertain. Determination of the real posterior end of the gizzard may provide systematically useful information as is suggested in each lumbricid précis.

Remarks: To be common is to be misunderstood, almost seems to be proved by the anthropochorous lumbricids. About them there probably are more misconceptions than about all other megadriles together, and in spite of the frequency of use in laboratory instruction and experimentation. Attention has been called at various times by others to errors in textbooks. As recently as 1963, treatment of the alimentary tract of *L. terrestris*, even in advanced and modern texts, was said not only to be inadequate with reference to functional anatomy but also to contain contradictory statements. Errors and inadequacies likewise are to be found in manuals written for those beginning work on the megadriles. Most deplorable, however, are those perpetuated more or less carelessly by those of us, who, because of special studies, may be thought to be speaking with more unimpeachable authority than is warranted. Some instances, more especially of misconceptions as to distribution, are mentioned elsewhere in this opus. Certain others are as follows.

The position of the clitellum and the number of segments occupied by it are not constant in each species and, as shown below, intraspecific variations in the characters often are not such as to be considered slight. Clitellar characters, like those of the tubercula pubertatis, often are useful aids to a tentative specific identification but their systematic value in the Lumbricidae frequently has been grossly overrated.

Erroneous records of calciferous sacs being absent presumably are due to concealment by connective tissues that sometimes bind the sacs, not only to the esophagus but also to both of the adjacent septa, and to a failure to carefully complete the dissection.

The writer is not the only one who has said that hearts are present in xii. The vessels involved, in each case that has been carefully checked, are posterior portions of extra-esophageal trunks. Location of the last pair of hearts in xi also may characterize the Glossoscolecidae, about the vascular system of which little is known. The situation in the Moniligastridae is various but actually is, in each genus, just as in the Lumbricidae and the Glossoscolecidae, i.e., in the second segment in front of the ovarian metamere. Hearts have been added in xii (other families) but in the ovarian segment only in some megascolecids, acanthodrilids, octochaetids, and eudrilids.

Considerable caution is required in connection with "replace." Almost certainly, peregrine lumbricids have replaced endemics in California, South Africa, southern South America, Australia, New Zealand, and the literature (perhaps because of the very considerable gaps in our knowledge of the native species) indicates possibility of a similar change even in unglaciated portions of Europe. The faunal modification often is thought to involve extermination of the endemics during a competitive struggle with the anthropochores, e.g., introduction of exotic lumbricids "frequently causes the disappearance of the endemic earthworm fauna" (Stephenson, 1930: p. 905). However, there is at present little evidence to support that conclusion. Much more careful surveys than ever before are required to show just how extensive are the areas of faunal change and the nature of the habitats in which the change has been made, e.g., whether limited to the immediate vicinity of cities, towns, villages, or if cultivated lands as well as pastures were affected. Whether the replacement involved much more than occupation of niches already vacant, the endemics already having been exterminated therein by human activities often may no longer be determinable.

"All the polyploid worms show obligate parthenogenesis. They are highly successful species, peregrine and cosmopolitan, and exhibit less variation than do the sexual species." To one word, peregrine, there can be little objection! "Lumbricidae can withstand any climate which allows herbs and moss to grow," and "able to establish themselves in any climate" (*cf.* distributions of species below.)

KEYS

Two keys to the oriental lumbricids are included. The first is based primarily on external and easily determined internal conditions but omitting, because of parthenogenetic organ eliminations, some of the hitherto commonly used genital characters.

Construction of the second key was begun in order to learn how many of the species could be included without mention of genitalia. Rather unexpectedly, certain characters hitherto unused in keys or specific definitions enabled completion and provide yet another demonstration of the systematic importance of somatic anatomy.

Use of both keys is recommended to any who may be identifying earthworms for the first time, followed by a careful check of the appropriate specific definition. Earthworms, for some time yet, certainly should not be identified merely by use of one or even two keys.

KEY TO INDO-BURMESE SPECIES OF LUMBRICIDAE

I

1. Male pores on xiii.......... *Eiseniella tetraedra*
   Male pores behind xiii.................... 2
2. Athecal................................. 3
   Thecal.................................. 6
3. Pigmented............................. 4
   Unpigmented..........morphs of *Eisenia rosea*
4. Setae widely paired
                     morphs of *Dendrobaena rubida*
   Setae not or not all widely paired.......... 5
5. Prostomium, epilobous.........*Bimastos parvus*
   Prostomium, tanylobous......."*Bimastos*" *eiseni*
6. Spermathecal pores near mD.............. 7
   Spermathecal pores in region of *CD*........ 9
7. Calciferous sacs, present in x
                         thecal morphs of *E. rosea*
   Calciferous sacs, lacking in x.............. 8
8. Setae closely paired, atrial and TP glands
   lacking.....................*Eisenia foetida*
   Setae widely paired, atrial and TP glands
   present...................*Eisenia hortensis*
9. Setae closely paired..................... 10
   Setae widely paired or separated........... 16
10. Prostomium tanylobous, pigment red, seminal
    vesicles three pairs..................... 11
    Prostomium epilobous, no red pigment, seminal
    vesicles four pairs...................... 13

11. Clitellum, begins behind xxx.. *Lumbricus terrestris*
    Clitellum, begins in front of xxx........... 12
12. Clitellum, in xxvi, xxvii–xxx, xxxi, xxxii
                         Lumbricus rubellus
    Clitellum, in xxviii–xxxiii....*Lumbricus castaneus*
13. Sexthecal, pores at 8/9–10/11
                         Allolobophora chlorotica
    Quadrithecal, pores at 9/10–10/11.......... 14
14. Unpigmented............................ 15
    Pigmented (*Cf.* 2nd note below)
                         Allolobophora trapezoides
15. Clitellum usually begins in front of 27/28, no
    genital tumescences in xxxiii
                         Allolobophora tuberculata
    Clitellum usually begins behind 27/28, genital
    tumescences present in xxxiii (*cf.* note below)
                         Allolobophora turgida
16. Pigment red........................... 17
    Unpigmented or without red pigment....... 18
17. Sexthecal, pores at 9/10–11/12, clitellum xxix,
    xxix/n–xxxiv/n, xxxiv....*Dendrobaena octaedra*
    Quadrithecal, pores at 9/10–10/11, clitellum
    does not reach xxxiii but extends well in
    front of xxix.............*Dendrobaena rubida*
18. Clitellum in xxix–xxxiv......*Octolasium cyaneum*
    Clitellum in xxx–xxxv......*Octolasium tyrtaeum*

Note: *Allolobophora jassyensis* Michaelsen, 1891, was reported once (on two specimens) from the Murree subdivision of the Punjab, but the record has not been confirmed. Nor has agreement yet been reached as to how the species should be defined.

Note: *A. longa* has not yet been recognized in India but may be distinguished from *A. trapezoides* by location of the tubercula in xxxii–xxxiv instead of xxxi–xxxiiii as well as by the greater number of segments.

II

1. Calciferous sacs, lacking.................. 2
   Calciferous sacs, present.................. 4
2. Nephridial vesicles, ocarina-shaped...*D. octaedra*
   Nephridial vesicles, not ocarina-shaped...... 3
3. Setae, closely paired................*E. foetida*
   Setae, not closely paired...........*E. hortensis*
4. Calciferous sacs, vertical and opening into gut
   mesially............................. 5
   Calciferous sacs, horizontal and opening into
   gut posteriorly........................ 12
5. Pigment, red........................*B. parvus*
   Pigment, lacking or if present not red....... 6
6. Nephropores, inconspicuous and in two ranks
   on each side of body, setae closely paired.. 7
   Nephropores, obvious and in one rank on each
   side of body, setae not closely paired..... 11
7. Pigment, present........................ 8
   Pigment, lacking........................ 9

8. Segments, 130–167, typhlosole ends in the
93rd–114th segments leaving 25–57
atyphlosolate.................*A. trapezoides*
Segments, 145–220, typhlosole ends in 115th–
130th leaving 56–88 atyphlosolate.....*A. longa*

9. Tail end usually truncate, body section behind
the clitellum circular to squarish......*E. rosea*
Tail end not normally truncate, body section
behind the clitellum transversely elliptical
to rectangular.......................... 10

10. Size 35–85 by 1½–4½ mm., segments 125–168,
typhlosole ends in 90th–108th segments
A. turgida
Size 50–150 by 4–6 mm., segments 146–194,
typhlosole ends in 102nd–125th segments
A. tuberculata

11. Typhlosole, ends in 88th–110th segments
leaving 22–23 atyphlosolate........*O. tyrtaeum*
Typhlosole, ends in 115th–126th segments
leaving 32–37 atyphlosolate.......*O. cyaneum*

12. Setae, closely paired...................... 13
Setae, not closely paired................. 17

13. Extra-esophageals pass up to dorsal trunk
in xii................................. 14
Extra-esophageals pass to dorsal trunk in
vicinity of 9/10........................ 15

14. Longitudinal musculature pinnate (?), prostomium tanylobous, nephropores obvious,
typhlosole without longitudinal ridges on its
ventral face.................*Levinsen's eiseni*
Longitudinal musculature fasciculate, prostomium epilobous, nephropores inconspicuous, typhlosole with 3–5 longitudinal
ridges on its ventral face.........*A. chlorotica*

15. Soma large, to 350 by 12 mm., typhlosole ends
in 110th–116th segments leaving 27–44
atyphlosolate...................*L. terrestris*
Soma smaller, typhlosole ends more anteriorly
and leaves fewer atyphlosolate............ 16

16. Size 30–45 by 3–5 mm., segments 69–99,
typhosole ends in 70th–77th segments leaving 14–20 atyphlosolate... ......*L. castaneus*
Size 60–150 by 4–6 mm., segments 100–123,
typhlosole ends in 78th–98th segments leaving 20–26 atyphlosolate...........*L. rubellus*

17. Pigment, red, typhlosole ends in 79th–100th
segments, nephridial vesicles *U*-shaped
D. rubida
Pigment not red, typhlosole ends in 46th–75th
segments, nephridial vesicles sausage-shaped
E. tetraedra

Note: Distribution is of major zoogeographical
interest only when known from self-acquired areas of
endemic taxa. Until lumbricid genera receive the
drastic revision that is so much needed their distribution needs no consideration. Because of much
past and current misunderstanding, present dis-

tributions of lumbricid anthropochores are rather fully
stated.

Allolobophora

SYSTEMATICS: Eisen, 1874, *Ofvers. Vetensk.-Akad. Förhandl.
Stockholm* **30**, 8. Michaelsen, 1900, Oligochaeta, *Das Tierreich*
10: p. 480; 1910, *Ann. Mus. Zool. St. Petersburg* **15**. Omodeo,
1956, *Arch. Zool. Italiano* **41**. Pop, 1941, *Zool. Jahrb. Syst.* **74**.
Stephenson, 1930, *The Oligochaeta* (Oxford), pp. 905–908.

Nomenclature and Systematics: *Allolobophora* was
erected by Eisen, without designation of a type species,
for *riparia, turgida, mucosa, norvegica, arborea, foetida,
subrubicunda*. Two names, *riparia* and *mucosa*, respectively, are synonyms of *chlorotica* and *rosea*
(Savigny, 1826). Three, *norvegica, arborea* and *subrubicunda*, are synonyms of *Dendrobaena rubida*
(Savigny, 1826). Another, *turgida*, was placed by
Beddard (1895) in the synonymy of *caliginosa*. One
species, *foetida*, became the type of *Eisenia*. Thus,
when Michaelsen was revising the Lumbricidae for
his Tierreich monograph, *chlorotica, rosea*, or *turgida*
(= *caliginosa*) could have been designated as the
type. However, primarily because of location of
spermathecal pores near mD, *rosea* was transferred
from *Allolobophora* to the genus *Eisenia*. Then, faced
with the necessity of choosing between two so different species as *caliginosa*, already known to be polymorphic, and *chlorotica*, Michaelsen, contrary to his
usual practice of citing or designating a type, omitted
any mention of the matter.

Some of the problems that would have arisen from
a designation of either species doubtless were known
to Michaelsen but what was not to be appreciated,
perhaps for seventy years, was the fact that a system
based on a few pairs of simple key characters, with
those of the reproductive system given by far the most
importance, had failed, even before publication, a
crucial test.

Designation of a type for *Allolobophora* should have
been left (*cf. chlorotica*) until such time as availability
of necessary data could have permitted, in place of
the recent neoclassical rearrangements, the really
drastic revision that was needed. Unless a recent
designation can be annulled, or Oerley's *Aporrectodea*
can be resurrected with *chlorotica* as its type, some
of the better-known lumbricid species must have a
new generic name. At the present moment, such a
genus, with one of the widely spread anthropochorous
species as a type, would seem to be definable from
somatic anatomy as follows.

Calciferous gland, opening into gut through a pair of
vertical sacs in x. Gizzard, mostly in xvii. Extra-
esophageal vessels, passing to dorsal trunk in xii. Hearts,
in vi–xi. Nephridial bladders, *J*-shaped, closed end
laterally, ducts passing into parietes near *B*. Longitudinal
musculature, pinnate. Nephropores, inconspicuous, behind the clitellum irregularly alternating, and with
asymmetry, between levels slightly above *B* and above *D*.
Seta, paired.

Anthropochorous species of *Allolobophora*, with exception of *chlorotica* (q.v.), can be characterized additionally by certain characters of the genital system as follow.

Holandric, testes free. Seminal vesicles, 4 pairs, in ix–xii. Atrial, TP and supraparietal GS glands, lacking. Spermathecal pores, in region of *CD*. Genital tumescences, around *a,b* only, in some of viii–xiii, xvi–xvii, xxv–xxxviii.

In a group so defined, can be included in addition to the widely distributed *longa, trapezoides, tuberculata*, and *turgida*, the less-markedly peregrine *limicola* Michaelsen, 1890, *nocturna* Evans, 1946. Doubtless other species, presently in *Allolobophora*, should be included. Among them, might be *bashkirica* Malevitch, 1950, *ictericum* Savigny, 1826, *moebii* Michaelsen, 1895, *roseum* Savigny 1826, etc. Inclusion of *roseum* would require elimination from the definition of spermathecal and genital tumescence characters, whereas inclusion of *ictericum* and *moebii* would only require deletion of the segmental restrictions of genital tumescences. Quite possibly some taxa now in *Eophila* (especially if the quadrivesiculate state is associated with parthenogenesis) will be found to have so much of the same somatic anatomy as to require elimination of any specification as to number of seminal vesicles. Whatever future study may show, almost certainly delimiting and defining lumbricid genera will be decided henceforth primarily by somatic rather than genital anatomy.

Nevertheless, some reproductive organs may have systematic values not even suspected by specialists of the classical and neoclassical schools. One such is the ovarian character now moved up into the family definition where it seems likely to aid in estimating interfamily relationships. Atrial, TP and supraparietal GS glands promise to be of systematic use at supraspecific levels. Even patterns of genital tumescence location, when adequate information as to individual variation has been obtained, may aid in determination of interspecific relationships especially, perhaps, when affinities are close. Patterns of six species are indicated in table 2. For the last five of those forms considerable data as to individual variation already are available. The anterior tumescences, those of ix–xi, often are very resistant to intraspecific deletions. An almost complete absence of individual variation with respect to five of the species suggests close relationship (conditions in xii then may indicate less close affinities of *A. nocturna* and *longa* to the other three, and in x of *limicola* perhaps still more remote affinities to all of them).

Allolobophora, however defined in the future, is palearctic, except as transportation has been involved.

Tumescences in the clitellar region, on the contrary, suggest a closer relation to each other of *nocturna, trapezoides*, and *turgida*, than to *tuberculata*. Conditions in segment xxxi suggest that *longa* and

TABLE 2

PATTERNS OF GENITAL TUMESCENCE LOCATIONS IN *Allolobophora*

Segment	Nocturna	Trapezoides	Tuberculota	Turgida	Longa	Limicola
8			R			
9	U	U	U	U	U	U
10	U	U	U	U	U	R
11	U	U	U	U	U	U
12	U		R		R	U
13						R
14						
16						R
17						R
18–24						
25			R	R	R	
26			C	R	R	R
27		R	R	C	R	R
28	R	O	R	R	R	R
29	C	R	R	R	R	C
30	U	U	U	U	R	U
31			R	R	U	U
32	U	U	U	U		U
33	U	U	R	U	U	
34	U	U	U	U	U	
35		O	R	R	R	
36			R			C
37						O
38						R

U: Usual to almost always or, with exception of bilateral asymmetry, even always (amputation and developmental aberrations of course excepted).

C: Common but always in less than 50% of specimens examined.

O: Occasional.

R: Rare, sometimes indicative of a single specimen only.

limicola may be more closely related to each other than either is to any of the *trapezoides* group.

Remarks : Persistence of hearts in vi, restriction of the gizzard mainly to xvii, presence of 8 seminal vesicles, are believed to be primitive characters. Calciferous sacs of the sort characterizing species except *chlorotica*, apparently are in an early stage of being constricted off from the gut. Total absence of atrial, TP and supraparietal GS glands, on the contrary, seems to represent considerable specialization.

Allolobophora chlorotica

1826. *Enterion chloroticum* Savigny, *Mem. Acad. Sci. Inst. France*, **5** (Hist. Acct.): p. 183. (Type locality, Paris. Types, in the Mus. Hist. Nat. Paris.)

1956. *Allolobophora chlorotica*, Omodeo, *Arch. Zool. Italiano* **41**: pp. 180, 181. (Designated the type species of the genus and a subgenus.)

BIOLOGY: Kollmansperger, 1934, *Die Oligochaeten des Bellinchengebietes, Köln*. Evans & Guild, 1948, *Ann. Applied Biol.* **35**. BACTERIA, in the nephridia: Knop, 1926, *Zeitschr. Morph. Oekol. Tiere* **6**: p. 600. CALCIFEROUS GLAND: Smith, 1924, *Illinois Biol. Monogr.* 9, 1: p. 33. CELLULASE, CHITINASE: Tracey, 1951, *Nature* **157**: p. 777. CHROMOSOMES: Muldal, 1952, *Heredity* **6**: p. 59. CHLORAGOGEN: Liebmann, 1927, *Zool. Jahrb. Allg. Zool. Physiol.* **44**. COCOONS: Evans & Guild, 1948, *Ann. Mag. Nat. Hist.*, Ser. 11, 14; COE-

LOMIC CORPUSCLES & PHAGOCYTOSIS: Cameron, 1932, *Jour. Pathol. Bact.* 35: p. 248. (EMBRYOLOGY: Bergh, 1890, *Zeitschr. Wiss. Zool.* 50: p. 474). ESTIVATION: Keilin, 1915, *Bull. Sci. France Belgique* 49. GROWTH: Michon, 1954, Thesis, Poitiers. HEREDITY: Kalmus *et al.*, 1955, *Ann. Mag. Nat. Hist.* 8. LEARNING: Heck, 1921, *Lotos, Prag*, 67–68: p. 185. MUSCULATURE: Pool, 1937, *Acta Zool.* 18: p. 80. NEUROSECRETORY CELLS: Michon & Alaphilippe, 1959, *Compt. Rend. Acad. Sci. Paris* 249, p. 835. PARASITES, PROTOZOA: Hesse, 1909, *Arch. Zool. Exp. Gen.*, Ser. 5, 3. Cepede, 1910, *Ibid.* Berlin, 1924, *Arch. Protistenk.* 48. Williams, 1942, *Jour. Morph.* 70. Meier, 1954, *Arch. Protistenk.* 100. Puytorac, 1954, *Ann. Sci. Nat. (Zool).* 16. Loubatières, 1955, *ibid.* 17. Meier, 1956, *Arch. Protistenk.* 101. Puytorac, 1957, *Arch. Zool. Exp. Gen.* 94, N. & R. 2. Rees, 1961, *Parasitology* 51. NEMATODA: Voelk, 1950, *Zool. Jahrb. Syst.* 79. Dunn, 1955, *British Vet. Jour.* 111. INSECTA: Keilin, 1909, *Compt. Rend. Soc. Biol. Paris* 67. Webb & Hutchinson, 1916, *Proc. Entomol. Soc. Washintgon* 18. Kirchberg, 1961. *Anz. Schaedlinks.* 34. ARACHNIDA: Hughes & Jackson, 1958, *Virginia Jour. Sci.* 9. PHARYNGEAL (septal) GLANDS: Keilin, 1920, *Quart. Jour. Micros. Sci.* 65. PHYSIOLOGY: Laverack, 1963, *The Physiology of Earthworms* (Pergamon Press). SPERMATOGENESIS: Tuzet, 1945, *Arch. Zool. Exp. Gen.* 84, N. & R. TYPHLOSOLE: Hertling, 1923, *Zeitschr. Wiss. Zool.* 120.

Sexthecal, pores at *C* or *D* and 8/9–10/11. Clitellum, saddle-shaped, down to m*BC*, xxx–xxxvi, usually reaching into xxix and/or xxxvii, rarely reaching both 28/29 and 37/38. Tubercula pubertatis, slightly lateral to *B*, transversely and shortly elliptical to almost circular, after certain preservations suckerlike, in xxxi, xxxiii, xxxv, occasionally elsewhere. Male pores, each near m*BC* and laterally in a deep cleft almost confined to median half of *BC*. Male tumescences, extending from *B* well toward *C* and into xiv as well as xvi, obliterating 14/15, 15/16, in median half of *BC*. Female pores, in xiv, just above *B*. Genital tumescences, including some of the following, *a,b*/ix, x, xxvii–xxxviii, *c,d*/ix, x. Setae, closely paired, behind the clitellum, *AB ca.* = *CD*, *BC* < *AA*, *DD* < ½*C*, follicles opening through genital tumescences thickened but not much more protuberant into coelom than other ventral follicles. Nephropores, inconspicuous, behind the clitellum irregularly alternating, and with asymmetry, between levels slightly above *B* and above *D*. First dorsal pore, usually at 4/5 or 5/6. Prostomium, epilobous, tongue usually open. Compression of body behind clitellum such that ratio of right–left to dorsoventral thickness at most is 2:1. Color, green, yellowish, pink, reddish, red, slate. Segments, 104–143, usually 110–119. Size, 35–70 by 4–5 mm.

Septa, 6/7–9/10 somewhat muscularized, 10/11–13/14 less so. Pigment: red, rarely present, in circular muscle layer, a green color present throughout the body wall but greatest concentration in longitudinal musculature of the preclitellar portion of the body, color almost instantly leached out from a freshly killed worm by acetone and not associated with granules. Longitudinal musculature, fasciculate. Calciferous sacs, in x, digitiform to pyriform, anteriorly (and then often in contact with 9/10),

anterolaterally or even dorsally directed, opening into gut ventrally about at level of insertion of anterior lamella of septum 10/11, calciferous lamellae continued to anterior ends of the sacs. Gut lumen, vertically slitlike in xi–xii, wider behind an internal constriction just in front of 12/13. Intestinal origin, in xv. Gizzard, mostly in xvii (ending at ?). Gut, narrowed and valvular in xix or at insertion of 18/19. Typhlosole, beginning in region of xxi–xxiii, anteriorly compressed dorsoventrally so as to have an inverted *T*-shape in cross section, with deep longitudinal grooves marking off 3–5 ridges on ventral face, ridges gradually disappearing posteriorly but *T*-shape in section retained much further back, ending abruptly in lxxxvi–xcv. Extra-esophageal vessels, joining dorsal trunk in xii. Hearts, in vii–xi. Nephridial bladders, *J*-shaped to almost *U*-shaped, closed end laterally, longer ectal limb posteriorly or ventrally. Holandric. Seminal vesicles, in ix, x, xi, xii. Sperm ducts, without epididymis. Atrial glands, extending into xiv and to 16/17. (TP and GS glands, none.) Spermathecae, in ix–xi, with shortly ellpitical ampullae, coelomic portion of duct to as long as ampulla.

Reproduction. Biparental and obligatorily so. Individuals isolated since hatching occasionally did deposit infertile cocoons.

Chromosomes, $2n = 32$.

Distribution: New Zealand.

British Columbia, Washington, Oregon, California, Colorado, Missouri, Wisconsin, Illinois, Ontario, Michigan, Indiana, Ohio, West Virginia, Virginia, North Carolina, Washington, D. C., Maryland, Pennsylvania, New York, Quebec, Vermont, Connecticut, Massachusetts, Maine, New Brunswick, Nova Scotia.

Mexico, Guatemala, Bermuda, Peru, Chile, Uruguay.

Greenland, Iceland.

Norway, Sweden, Finland, Russia, Latvia, Lithuania, Poland, Germany, Denmark, Isle of May, Scotland, Hebrides, Clare Island, Ireland, Isle of Man, England, Wales, Alderney, Guernsey, Herm, and Jersey of the Channel Islands, Netherlands, Belgium, France, Switzerland, Czechoslovakia, Austria, Hungary, Romania, Bulgaria, Greece, Corfu, Jugoslavia, Sicily, Italy, Spain, Portugal.

Azores, Madeira, Canary Islands, St. Helena (new record). Algeria.

Syria, Lebanon, Iran (interception).

At elevations of 300 m. in the Alps, of 2,300 ft., in Wales. Very few elevations were recorded but Mexican, Guatemalan and Peruvian records are not for low lands.

Before 1900, *A. chlorotica* already had been secured from 29 places in California and something of that sort must have been responsible for Michaelsen's citation (1900: p. 486) as "ganz Nord America" when actually the species had been recorded elsewhere in that continent only from Vancouver, Mexico, and Guatemala. However, at about the same time, individuals of *chlorotica* from an unknown but probably eastern site in North America, possibly at or near Baltimore where specimens now are known to have been collected prior to 1900, got to Hamburg in soil with fern roots.

A dead specimen was found in Central Illinois, September 12, 1918, but the species was not seen alive until 1921 and by 1928 Frank Smith had had only ten specimens as a result of collecting in the vicinity of the University of Illinois. The species still seems to be scarce in the region as Harman, during his survey (thesis, 1960), found individuals at only 6 sites of 4 counties. From the area where Smith first found *A. chlorotica* the species already had disappeared. Although not widely distributed *A. chlorotica* was found (*idem.*), in large colonies, the one mentioned extending 200 feet along a stream. Just such discontinuities are to be expected because of the accidental, fortuitous, random ways in which earthworms are distributed by man. Only when such activities are taken into adequate consideration can the distribution of earthworm species in areas of Pleistocene glaciation be understood.

The specimens intercepted at American ports were from 17 countries in America, Europe, Africa, Asia.

Habitats: Soils, with a *p*H of 4.5–8.00, with much plant matter, of house gardens, of fields, pastures and forests, steppes, wet organic soil, stiff alluvial clay, very clayey soil, peaty soil, earth of seedling beds, *Corylus* groves with rich soil, earth with carrots in ships stores, sandy and stony shores of lakes, banks of stream polluted with sewage, estuarine flats (near Belfast docks), just below high-tide level of seashore (Bermuda), mud, among roots of *Littorella* and *Phragmites* spp. to a distance of 20 m. from the shore and under permanent water, under stones, wet moss, rotting leaves, rotting straw, humus, dead wood, rubbish heap, manure pile, damp soil with horse manure.

Caves (West Virginia, Virginia, England, Belgium, France, Italy, Hungary, Germany). Mines (Nancy, France). Botanical gardens (California, England, Portugal). Greenhouses (France, Switzerland, Poland, Finland, Maine, Massachusetts, New York).

Active specimens have been found in soil from close to the surface down to depths of 300 mm. Coiled specimens, in a quiescent state, were found down to depths of 600 mm.

Samples of a stiff alluvial clay soil provided 50 worms per square meter; in Michigan, other samples provided 100 per square meter.

Biology: Activity, including breeding, is possible year round in appropriate circumstances. However, in many places, only two periods of activity and breeding are allowed climatically, in the spring and early summer, in the fall. In the northern part of the range, activity may be restricted to a single period in the summer and elsewhere climatic fluctuations may reduce activity to six or fewer months of the year. Copulation and defecation are beneath the surface. Laying, in England, 25–27 cocoons per worm per annum but in Germany, at 15° Celsius, an average of 12 cocoons already had been lain between February 1 and March 31. One hatchling emerged from each of 142 English cocoons but emergence of 2 from a cocoon was seen on another occasion at which time embryonic fission was thought to have been involved. Growth can be completed without a diapause, a condition which Omodeo calls "homodynamic" but which Michon calls "amphodynamic" because diapause is facultative (but presumably required by climatic conditions?). Activity in England is resumed as soon as soil moisture becomes sufficient which, in some Hungarian observations, was found to be 40 per cent. Repeated immersion in water was found to be necessary to restore Italian worms to activity from a summer diapause. Estivation in the upper 0–20 cm. of some Hungarian soils was believed to result fatally if drought was extended. Prolonged submersion of Italian worms brought on regression of sex characters, torpidity, and high mortality but in England *chlorotica* was found to be abundant in purely limnic habitats, indeed at Lake Windermere in the *Littorella* zone up to 20 m. from the shore. In Wicken Fen the species was abundant among undisturbed sedge roots where clitellates were rare in August but abundant in May and October, possibly indicative of seasonal rhythm even in a continuously moist environment.

Incubation, at 15°C in Germany, 30 days, in England 12½ weeks. Maturity, in the German laboratory, at 120–130 days, in England at 29–42 weeks. Mean duration of life, at 18°C, in a French laboratory (Michon, 1954) 463 days, at 9° 447 days, at 25° 536 days (an increase that was paralleled only by *E. foetida*). Intermittent alimentation at 18° raised the average to 554 days, diapause in the preclitellar and clitellar phases raised the average to 587 days.

Male clefts are present as shallow grooves in which male pores are not distinguishable, and intersegmental furrows in median parts of *BC* are obliterated, before any evidence of differentiation of clitellum or of tubercula was macroscopically recognizable. Tubercula, when first visible, appeared as slightly postequatorial patches.

Water content may reach as much as 84.3 per cent of body weight.

A. chlorotica often has been thought to be sluggish, prone to roll into a spiral when alarmed or to coil up "au moindre danger."

The species has been characterized as shallow-burrowing which may be correct for a considerable portion of the life of many individuals but active specimens have been found at various depths down to 300 mm. below the surface.

Intolerance of acid soils supposedly is characteristic.

Cultivation, according to Evans & Guild (1948), does not reduce numbers of the species. That, together with the frequent occurrence in gardens (England where *A. chlorotica* often has been called the garden worm), perhaps may warrant the characterization haemerophile. Further evidence is provided by the data *re* presence in soil imported to the United States during the last fifteen years: 58 interceptions, 107 specimens from 18 countries including one (Iran) from which *A. chlorotica* hitherto had not been reported.

Taste presumably is unpleasant as in England moles (*Talpa europaea*) do not store *A, chlorotica*, a species which anglers have found no use as bait.

Variation and Abnormality: Color of a Californian worm was so dark as to seem black. Pigment, red, was dense in circular musculature of dorsum and even in longitudinal muscles of the first four segments. No trace of usual green color was recognized in the less-common pigmented individuals. A yellow color has been attributed to coelomic corpuscles and the chloragogen, presumably as seen through a translucent or transparent body wall of live specimens, but in pickled material could have resulted from early stages of alcoholic browning. Populations of *A. chlorotica* in English woodlands and gardens are predominantly or entirely "pink," but in permanent grassland the green form predominates. Experiments (Kalmus) provided evidence indicating that recessive green color is controlled by a single Mendelian factor.

Largest number of segments in the author's specimens from some thirty-odd countries was 133. At Göttingen, Hertling found segments were 102–135 with 16–33 atyphlosolate (as a proctodeum?) as compared with 21–36 in the author's series. From Italy, Omodeo recorded a maximum of 142 segments.

Rudimentary gonads, presumably female(?), have been seen in xii.

Abnormal No. 1. Location of the female pore in xiii, and of the male pore in xiv, on the left side of one of Friend's worms, presumably resulted from suppression of one mesoblastic somite. If data *re* other divergence from normal had been provided an estimate as to which somite was deleted might have been possible.

No. 2. Left female pore, equatorial in xiv but, as in a specimen of *E. tetraedra* also seen by the writer, at m*AB*. This probably represents a partial return to a former ancestral location in *AA*. No. 3. Left posterior seminal vesicle, in xiii but with a slender stalk passing to posterior face of 11/12. No. 4. Atrial glands, lacking, no macroscopically recognizable rudiments. Male sterility, not involved, spermatozoal iridescence on male funnels marked even though the individual was aclitellate. No. 5. Male

pore, of one side in xvi. A mesoblastic somite probably had been halved at some level in front of the fifteenth but here again the description lacked data that might have provided a clue as to somite that had been split.

Regeneration: A tail was regenerated, according to the first published record, only nine months after amputation of a posterior half of the body, but no information was provided as to number of segments or size of the regenerate.

Absence of left skewness in a curve of segment number was attributed (Evans, 1946, *Nature* **158**: p. 98) to mitigation by caudal regeneration of the effect of predation. Another explanation of Evans normal curve for *chlorotica* now seems possible, absence of predation (perhaps because of a taste that is unpleasant to birds, moles, and other enemies). Certainly little has been recorded about the subject and the author has yet to see a *chlorotica* regenerate. Statements such as, no posterior regeneration as long as animals remain active and well fed, need not necessarily connote tail regeneration in other conditions as similar claims have been made for species that do not regenerate posteriorly. However, formation in 19 days, of 26 segments, after amputation of the last 34, has been recorded (Michon, 1954) as well as of 24, after removal of the last 38. Absence of information *re* cephalic regeneration, even when predation is absent or insignificant, hints that a head is less liable to be lost than a tail or else that the worms are much more likely to survive after loss of a caudal portion.

Parasites.[6] CILIATA: *Anoplophrya lumbrici*[7] (Schrank, 1803), Germany. *A. striata* (Dujardin, 1845), France. *Maupasella nova* Cepede, 1910, France, Germany. *M. cepedei* Puytorac, 1954, France. *Metaradiophrya chlorotica*[8] Williams, 1942, West Virginia. *M. falcifera* (Stein, 1861) (West

---

[6] Ciliate parasites are restricted so frequently to the gut lumen and monocystids so often to the vesiculae seminalis that those sites are hereinafter to be understood in absence of indication to the contrary. Exceptional sites, when preceded by an "and," are additional to those that are understood. Helminthologists often failed to mention organs or tissues inhabited by the parasites, also the geographical locality from which hosts were obtained. North America and/or Europe at various places below means nothing more than a required assumption that research recorded in an American journal made use of North American material. A similar assumption sometimes is required *re* work published in Great Britain, Germany, Russia. Final decisions *re* status of various species names have not been reached. Nor have proposed synonymies always been acceptable. Much obviously remains to be done on previously studied megadrile parasites.

According to some workers in the field; *Monocystis cristata* Loubatières, 1949, is preoccupied by *M. cristata* Schmidt, 1854 = *M. lumbrici* Henle, 1845. *Monocystis cuneiformis* Loubatières, 1949 is preoccupied by *M. cuneiformis* Ruschhaupt, 1885 = *Rhynchocystis pilosa* (Cuenot, 1901).

[7] May belong in *Metaradiophrya*.

[8] May be a synonym of *M. falcifera*.

Virginia?). *M. hovassei* Puytorac, 1954, Wales. *M. lumbrici* (Dujardin, 1841), Germany.

SPOROZOA: *Apolocystis herculea* (Bosanquet, 1894), France. *A. perfida* Rees, 1963, coelom midbody, Wales. *A. pertusa* Loubatières, 1955, France. *A. spinosa* Rees, 1963, Wales. *Dirhynchocystis elongata* Loubatières, 1955, France. *Monocystis cognettii*[9] Hesse, 1909, France, Germany, Sweden. *M. densa* Berlin, 1923, Sweden. *M. hederacea* Loubatières, 1955, France. *M. herculea* Bosanquet, 1894, Sweden. *M. proteiformis* Loubatières, 1955, France. *M. securiformis* Loubatières, 1955, France. *Nematocystis claviformis* Loubatières, 1955, France. *N. sinuosa* Loubatières, 1955, France. *Rhynchocystis ovata* Loubatières, 1955, France. *Zygocystis cometa* Stein, 1848, Germany, Sweden. *Z. legeri* Hesse, 1909, France, Germany, Sweden.

CESTODA: *Amoebotaenia cuneata* (Linstow, 1872) Cohn, 1900, France, said to be cosmopolitan in chickens (perhaps also in earthworms?). *A. sphenoides* (Railliet, 1892), France.

NEMATODA: *Diplogaster eurycephalus = Diplogasteritus eurycephalus* (Völk, 1950) Goodey, 1963, Germany. *D. stöckherti = Diplogasteritus stoeckerti* (Völk, 1950) Paramonov & Sobolev in Skryabin *et al.,* 1954, Germany. *Diploscapter lycostoma* Völk, 1950, Germany. *Metastrongylus* sp., *Rhabditis dolichura = Caenorhabditis dolichura* (Schneider, 1866) Dougherty, 1955, Germany. *Rhabditis* (*Rhabditella*) *axei* (Cobbold, 1884) Dougherty, 1955, recorded from the earthworms as *R. elongata* (Schneider, 1866) Bütschli, 1876, Germany. *Rhabditis* (*Cephaloboides*) *oxycerca* de Man 1895-subgen. by Dougherty 1955, recorded from the earthworms as *Rhabditis oerleyi* Völk, 1950, Germany. *Pelodera* (*Pelodera*) *strongyloides* (Schneider, 1860) Schneider, 1866, recorded from the earthworms as *Rhabditis strongyloides*, Germany. *Pelodera* (*Pelodera*) *teres* Schneider, 1866, recorded from the earthworms as *Rhabditis teres*, Germany. *Rhabditis* (*Rhabditis*) *terricola* Dujardin, 1845, recorded from the earthworms as *Rhabditis aspera* Bütschli, 1873, Germany. *Rhabditolaimus leptosoma* Völk, 1950, Germany.

INSECTA: *Onesia sepulchralis*, Meigen = *O. floralis* Robinaux & Desdoity 1830, phagocytosed, Europe. *Pollenia rudis* (Fabricius, 1786), Europe, North America, larvae that devour their host. *Sarcophaga carnaria* (Linnaeus, 1758), Europe.

ARACHNIDA: *Histiosoma murchiei* Hughes & Jackson, 1958 (cocoons from which no hatchlings emerge), Michigan. In three counties 45 per cent of the *chlorotica* cocoons were thought to be infested.

Systematics and Nomenclature: Calciferous sacs of *A. chlorotica*, as in *E. tetraedra*, Levinsen's *eiseni* and the genus *Lumbricus*, are horizontal opening into

[9] Not found in out-of-doors hosts but only in those from greenhouses and so was thought to be exotic in France (= *Z. legeri*?).

the oesophagus posteriorly in the region where septum 10/11 is inserted. Such sacs appear to be so different in their development, ontogenetically as well as phylogenetically, from the lateral bulges that characterize other well-known species of *Allolobophora*, as to contraindicate inclusion in the same genus. The color, when green, may be unique in the family. The species accordingly is markedly divergent from others of its genus that are at all well known and should not have been designated the type of the genus.

The calciferous sacs and the nephridial bladders require consideration of relationships with the genus *Lumbricus*. Whether *A. chlorotica* should be included in *Lumbricus* depends upon the value to be attributed to such differences as fasciculate versus pinnate longitudinal muscles, epilobous versus tanylobous prostomium, absence versus presence of testis sacs, absence versus presence of seminal vesicles in x, presence versus absence of a pair of spermathecae opening at 8/9. Perhaps also some other differences such as in the typhlosoles.

Whether or not *A. chlorotica* goes into *Lumbricus*, the species no longer should be in the same genus as *A. turgida, longa, trapezoides, tuberculata* and others like those just cited. A new generic name accordingly is required for widely distributed species fairly common throughout much of the world unless there is some way of annulling the designation of *chlorotica* as the type of *Allolobophora*.

A. chlorotica may be closely related to *A. georgii* Michaelsen 1890 and to *A. cupulifera* Tetry 1937 both of which also have suckerlike *tubercula*. Nothing recorded about those two species contraindicates the relationship but too little is known about their somatic anatomy at present to warrant including them in a genus of which *chlorotica* is the type.

Spermathecal pores of some specimens of *chlorotica*, according to Omodeo (1956, *Arch. Zool. Italiano* 41: p. 182), are at 9/10–11/12 which, if no typographical error is involved, would seem to suggest a possibility of existence of still another taxon with the suckerlike *tubercula*.

Remarks: Speleologists have called the species troglophile. A characterization of "euro-americain" seems inadvisable for a species of European origin that was taken to America by man since 1492.

The periods of *chlorotica* quiescence are in need of further investigation.

Allolobophora longa

1885. *Allolobophora* longa Ude, *Zeitschr. Wiss. Zool.* 43: p. 134, etc. (Type locality, garden, Göttingen, Germany. Types, nothing has been learned about them.)
1958. *Allolobophora longa*, Gates, *Ann. Mag. Nat. Hist.*, Ser. 13, 1: p. 38.

ABNORMALITY: Woodward, 1893, *Proc. Zool. Soc. London* 1893. BIOLOGY: Avel, 1929, *Bull. Biol. France Belgique* 63. Evans, 1947, *Jour. Board Greenkeeping Res.* 7. Evans & Guild,

1947, *Ann. Appl. Biol.* **34**. Jefferson, 1958, *Jour. Sports Turf Res. Inst.* **9**. Michon, 1954, "Contribution experimentale a l'etude de la Biologie des Lumbricidae" (thesis, Poitiers, France). CHROMOSOMES.: Muldal, 1952, *Heredity* 6. COMMENSALISM: Saussey, 1956, *Bull. Soc. Zool. France* 81. PARASITES: (PROTOZOA) Loubatières, 1955, *Ann. Sci. Nat. Zool.* **17**, 1. Meier, 1956, *Arch. Protistenk.* **101**. PHYSIOLOGY: Laverack, 1963, *The Physiology of Earthworms* (Pergamon Press).

Quadrithecal, pores in *CD*, at 9/10, 10/11. Clitellum, saddle-shaped, (xxvi, xxvi/n), xxvii, xxvii/n–xxxiv(once), xxxv/n(once), xxxv, (xxxvi/n, xxxvi, occasionally). Tubercula pubertatis, longitudinal bands just lateral to *B*, xxxii–xxxiv, occasionally extending into xxxi, very rarely through xxxi. Male pores, each in a cleft. Male tumescences, extending into xiv and/or xvi. Female pores, in xiv, just above *B*. Genital tumescences, around *a,b* only, ix, x, xi (almost always), xxxi, xxxiii, xxxiv (usually), occasionally or rarely in one or more of xii, xxv–xxviii, xxx, xxxii, xxxv. Setae, closely paired, *a* and *b*/xv even more closely, *AB ca.* = *CD*, *BC* < *AA*, *DD* < ½*C*. Nephropores, inconspicuous, alternating irregularly and with asymmetry between levels slightly above *B* and above *D*. First dorsal pore, usually at 12/13. Prostomium, epilobous, tongue usually closed. Dorsoventral compression of body, in anal region, slight to so marked as to produce a trapezoidal section as in *L. terrestris* or *E. tetraedra*. Color, light yellowish brown, reddish brown, dark brown, slate, albinism rare. Segments, 145–220. Size, 120–160 by 6–9 mm.

Septa, none especially muscular (previous authors), muscularity increasing from 5/6 to 8/9, from 9/10 to 14/15 muscularity decreasing, 15/16 membranous. Pigment, brown in circular muscle layer. Longitudinal musculature, pinnate. Calciferous sacs, vertical, in x. Gut lumen, vertically slitlike in xi–xii, wider from xiii. Intestinal origin, in xv. Gizzard, mostly in xvii (ending at ?). Typhlosole, beginning in region of xxiii–xxiv, thicker ventrally, at first with regular transverse grooves and one median longitudinal ridge on the ventral face, posteriorly with a deep longitudinal groove on each side of the median ridge or with an appearance of three longitudinal ridges, ending abruptly (unamputated individuals) in region of 115th–130th segments, leaving an atyphlosolate section (rectum?) extending through the last 56 to 88 metameres. Extra-esophageal vessels, joining dorsal trunk in xii. Hearts, in vi–xi. Nephridial bladders, *J*-shaped, closed end laterally, reaching on parietes beyond *D*. Holandric. Seminal vesicles, in ix, x, xi, xii. Sperm ducts, with epididymis, of one long hairpin loop, several shortly *U*-shaped loops, or in a ball of loops or coils. Spermathecae, in ix, x. (Atrial, TP, and suprapaprietal GS glands, none.)

Reproduction: Biparental and obligatorily so, according to Muldal who found chromosome number to be 2*n* = 36. Individuals isolated since hatching, according to Evans, became mature but did not lay.

Distribution: Australia, Kermadec Island, and New Zealand ("throughout the region").

Ontario, Nova Scotia, New Brunswick, Grand Manan. Colorado, Michigan, Indiana, Ohio, North Carolina, Pennsylvania, New Jersey, New York, Massachusetts, Connecticut, Vermont, New Hampshire, Maine. Mexico (new record).

Norway, Sweden, Finland (extreme south only), Russia, (including Urals, Pri-Urals), Latvia, Lithuania, Denmark, Germany, Poland, Czechoslovakia, Netherlands, Belgium, Hebrides, Scotland, Islay Island, Ireland, England, Wales, Jersey, Jerbourg (Channel Islands), France, Switzerland, Austria, Hungary, Bulgaria.

Algeria, South Africa.

Habitats: Earth with roots of potted plants. Greenhouses (several in central Maine). Outside flower beds of greenhouses, gardens. Botanical gardens and arboretums (Germany, England, North America). Lawns, grassland, grass moor, fields, pastures, roadsides, dune basins, swampy stream bank, peat bog (Massachusetts). Under elm logs and stones. Cowyards, chicken yards. Compost, humus, cow and other manure, rotten vegetables, grass clippings. Fresh beech mull on clayey marlaceous ground, sandy soil, sandy loam, clay loam, very acid soil with scattered reeds and rushes, fresh mull soil abounding in humus, soils with a *p*H of 4.5 to 8.0. Forests, steppes.

Caves (France, Britain).

Very common "in every county" of Great Britain. A Scotland population was estimated to be 252,700 per acre. In a 2-year-old pasture at Boghill, the *longa* population was estimated to be 112,300 ± 8,700 per acre. The average, in light soil of alluvial origin in permanent pasture, Durham County, England, was thought to be 149.8 per sq. m.

Biology: Year-round activity is possible if climatic conditions permit. Breeding cannot yet be so characterized and in Maine, laying usually is twice a year, in early spring and late fall, perhaps occasionally with an omission of one period of laying if either season is unusually dry or otherwise unfavorable. Laying periods, in England, comprise mid-March to June or July, and October–November. Reproductive rates are low, annual average of laying being eight cocoons a year. Incubation is 10–10½ weeks. Only one hatchling emerged from each of 89 cocoons. Number of segments, in recently hatched Maine juveniles, was 146–173. Growth, for many individuals presumably involves formation of new segments (perhaps during the summer diapause?). Ages of 5¾ to 10½ years were attained, by operated worms in favorable laboratory conditions. Nevertheless, average life (Michon, 1954), when development was direct, at 18°C, was 928 days, at 9° was 958 days, with intermittent alimentation (at 18°) was 990 days, when interrupted by a diapause in the clitellar and in the preclitellar phase, respectively, was 982 and 976 days.

Critical heat-death temperature (Miles, 1963, *Nature* **199**: p. 826) is 25.7°C.

A. longa has been thought to be heterodynamous, having an internally imposed and internally terminated summer diapause, quite regardless of conditions in the external environment. What happened to such a seasonal diapause "down under" in New Zealand (and Australia?) never has been investigated. Maturity has been said to be reached without a diapause but in central Maine, both juveniles and postsexual aclitellates were found, with empty guts, coiled tightly in closed chambers only 2–3 inches below the surface of an open, unshaded field in a dry September. Diapause seemingly is ended automatically, according to another author, when soil moisture reaches 20 per cent. Excess water together with plenty of food, according to yet another, permanently prevents diapause. Therefore, since growth can be completed without such a period of obligatory quiescence, the species has been also said to be amphodynamous. During cold periods, the worms remain active just under the frost, even when temperature is 1.5 to 0.5. Worms in Jefferson's cages, on the contrary, went into a winter diapause. Moreover, some of the specimens experimentally exposed to temperature of −2 to −5°C for 12 hours, survived.

Feeding, at the surface during the night, seems to be selective and leaves occasionally are dragged into the burrows. Deep burrowing involves swallowing considerable amounts of soil and *A. longa* is believed to be responsible, along with *A. nocturna*,[10] for nearly all of surface casting in England about which Darwin wrote. Nevertheless, Jefferson found that more soil always was voided below ground than on the surface, at maximum activity 40 grams per diem were deposited below to 2 grams cast on the surface.

Small stones, according to Jefferson, are deposited around burrow entrances.

Copulation modestly is beneath the surface.

German authors who call *A. longa* a near-surface-living species believe it is disturbed by cultivation and further state that it was rarely found in soil of cultivated fields at Kiel. The *longa* percentages of populations were higher in well-established pastures of England than in leys and arable fields. Nevertheless, characterization as haemerophobic is contraindicated by frequent reports of occurrence in gardens, flowerpots, and greenhouses.

Variation and Abnormality: Anterior genital tumescences, according to the definition of Cernosvitov & Evans (1947), are in ix–xii. Presence in xii has been found (American collections), only in four of several hundred specimens. Absence of tumescences in any of ix–xi was seen only once and then in ix alone.

[10] *A. nocturna* has a very restricted distribution in England, presumably as a result of rather recent importation from Europe.

Double heads (each of 2 segments) and double tails have been recorded.

Hermaphroditic gonads were recorded from xii, in one of which mature sperm and ova were present. An ovary with eight egg-strings provides the only record (found by the author) of deviation in any lumbricid from the typical single-string condition. Some of these egg-strings passed through 13/14 into xiv perhaps at the site where an ovisac should have been located.

Regeneration: A head can be regenerated at any time during the year and at least at each level from 5/6 anteriorly. The maximum number of segments recorded for head regenerates is 4. Tails supposedly are regenerated only during the summer diapause. Records of caudal regeneration for winter quiescence or dispause are lacking. Tail regeneration, in a posterior direction, is now known to be possible at any level behind 40/41. The largest number of epimorphic segments was 139, in a tail regenerate at 46/47. A regenerate at 100/101 had 100 segments, and another, at 170/171 had 14. Fewer metameres than that, three, were found but once and then in a regenerate (perhaps morphallactic?) at 180/181.

Parasites: CILIATA: *Anoplophrya lumbrici* (Schrank, 1803), gut, Germany. *Maupasella cepedei* Puytorac, 1954, gut, France, CSR. SPOROZOA: *Apolocystis lavernensis* Rees, 1963, ovisacs and coelom of genital segments, Wales. *Monocystis caudata* Berlin, 1923, Sweden. *M. ciliata* Drzewecki, 1907, and coelom, France. *M. crenulata* Hesse, 1909, France. *M. densa* Berlin, 1923, Sweden. *M. oblonga* Berlin, 1923, and coelom, Sweden. *M. wallengrenii* Berlin, 1923, and coelom, Germany, Sweden. *Nematocystis elmassiani* (Hesse, 1909), and coelom, France. *N. lumbricoides* Hesse, 1909, Germany. *N. vermicularis* Hesse, 1909, and coelom, France. *Pleurocystis cuenoti* Hesse, 1909, and male funnels, France, England. *Zygocystis cometa* Stein, 1848, France, Germany. *Z. pilosa* Hesse, 1909, France.

(?) *Polythei helodrili* Leigh-Sharpe, 1924, on a nephridium of an intestinal segment, London.

CESTODA: *Dilepis undula* (Schrank, 1788), cysticercoid larvae, coelom of midbody, Scotland.

NEMATODA: *Metastrongylus* spp. *Rhabditis anomala* Hertwig, 1922, Germany. *R. longicauda* Hertwig, 1922, Germany. *R. maupasi* Seurat, 1919, Germany. *R. pellio* (Schneider, 1966), Germany. *Syngamus trachea* (Montagu, 1811), England.

TREMATODA: Cf. *A. trapezoides.*

Nomenclature and Systematics: Although *Enterion terrestre* Savigny 1826 was still born, at moment of publication, Savigny's name continues to be used, contrary to the international rules, for a species of *Allolobophora* (*A. longa*), and in spite of the fact that attention was again called, by the author (1958) in a well-known journal, to the continued violation of the International Code.

Although worms identified as *A. terrestris* (Savigny, 1826) f. *typica* have had considerable use in French laboratory experimentation, no attempt seems to have been made to determine the systematic status of the population bearing the illegitimate name. Solving that problem might well aid acquiring a better understanding of the way some markedly anthropochorous lumbricids evolved.

Discontinuities in the range of Savigny's taxon may be due in part to human disturbances of the environment and if so the taxon may be disappearing or giving way in its own territory to haemerophile peregrines distributed by man. No evidence is available to indicate that individuals of Savigny's taxon have colonized transoceanic regions and no evidence for such transportation or for interceptions has been found.

Remarks: Until recently, complained Friend in 1926, *A. longa* was "continually confused" with *L. terrestris*. Around 1955, an American university professor gave specimens of *longa* along with *L. terrestris* to his students and published the data for *longa* as of and irrevocably intermingled with that of *terrestris*.

In Ohio *A. longa* was not found by Olson in his survey of the earthworms of that state published in 1928 but this author now has had specimens from there. The species is being, and doubtless for some time has been, distributed with potted plants from greenhouses in America and, though Ohio now has been reached, many of the states where *L. terrestris* probably is domiciled are not known to have been colonized. Presumably, *A. longa* was not brought to America as early or as often as *L. terrestris*. Interceptions tend to support such assumptions as during the last fifteen years *A. longa* was secured only twice by the U. S. Bureau of Plant Quarantine which thereby obtained only two specimens each from England and Ireland. That seems rather surprising in view of the frequent records of presence of *A. longa* in plant pots, gardens, and greenhouses. *A. longa*, reported to be very common in "every county" of Great Britain, may have been taken there earlier and more frequently than *L. terrestris*. Both probably got there much earlier than *A. nocturna*.

A. longa, in spite of its commonness in England, only occasionally was found in the stores of moles. If the taste is not pleasing to moles presumably *A. longa* was not the species served by a famous French chef to his delighted guests.

A. longa is not known to have been sold or used for bait in North America but the species may have occasionally been mistaken for *terrestris*, unless anglers are more careful than university professors, and often being mistaken for *A. trapezoides* seems even more likely.

Total castrates, in some of Avel's beautiful experiments, developed secondary sex characters, just as in the controls, copulated, and laid cocoons containing no eggs. The castrates also remained fully clitellate longer than the controls.

Allolobophora trapezoides

1828. *Lumbricus trapezoides* Duges, *Ann. Sci. Nat.* 15: p. 289, etc. (Type locality, unknown. Types, none.)
1958. *Allolobophora trapezoides*, Gates, *Breviora, Mus. Comp. Zool.* Cambridge, No. 91: p. 2, and *Ann. Mag. Nat. Hist.*, Ser. 13, 1: p. 38.

ABNORMALITY: Gates, 1956, *Ann. Mag. Nat. Hist.*, Ser. 12, 9: p. 369. CHROMOSOMES: Omodeo, 1955, *Caryologia* 8: p. 141, etc. NOMENCLATURE and SYNONYMY: Tetry, 1937, *Bull. Mus. Hist. Nat. Paris*, Ser. 2, 9. Cain, 1955, *Ann. Mag. Nat. Hist.*, Ser. 12, 8.

Quadrithecal, pores at or just above C, at 9/10, 10/11. Clitellum, saddle-shaped, xxvii–xxxiv, occasionally extending slightly into xxvi and/or xxxv, less often to the equators of those segments or even very rarely to 35/36 or 25/26. Tubercula pubertatis, uninterrupted, longitudinal bands of translucence, usually occupying most of BC through xxxi–xxxiii, occasionally with a much narrowed extension into xxx that rarely reaches 29/30. Male pores, at or just above mBC and at lateral end of an equatorial cleft in xv. Male tumescences, usually conspicuous and obliterating 14/15, 15/16, to or beyond eq/xiv, xvi, the tumescences sometimes paired and then variously shaped, sometimes united mesially and then often markedly protuberant throughout CC, the tumescence decreasing only beyond eq/xiv, xvi. Female pores, slightly lateral to B, at eq/xiv. Genital tumescences, often large enough to be united mesially, surrounding a,b only, in ix, x, xi (almost always), usually in xxx, xxxii, with decreasing frequency in xxxiii, xxxiv, one or more of xxiv–xxix, xxxi, xxxv. Setae, closely paired, a,b even more so in xv, ventral setae at least in ix–xxiv larger than the lateral setae, $BC < AA$, $DD < \frac{1}{2}C$. Nephropores, inconspicuous, behind xvi alternating irregularly, with asymmetry (?) between a level just above B and one above D. First dorsal pore, at any of levels from 6/7 to 13/14. Prostomium, epilobous, tongue almost always closed, a longitudinal furrow at mD in the tongue occasionally obvious. Compression of body near anal region, such as to produce a transversely rectangular section with a setal couple at each of the four corners, ratio of left–right to dorsoventral thickness 5:3. Color, in dorsum, especially dark in the first 15 segments, usually lighter behind the clitellum, sometimes darkening again near the hind end, reddish (juveniles only), reddish brown, brown, slate, almost black, albinism rare. Segments, 130–169, usually 132–159. Size, 80–137 by 3–7 mm. (contracted), to 220 by 3–5 mm. (relaxed).

Septa, 5/6 somewhat muscular, 6/7–9/10 increasingly muscular, maximum thickness in 9/10, muscularity decreasing posteriorly through 10/11–14/15, 15/16 membranous. Pigment, in circular muscular layer, reddish brown, or brown, seeming albinos with some slight color recognizable in free hand sections of the body wall. Longitudinal musculature, pinnate. Calciferous sacs, vertical, reaching slightly above and below esophageal level, wide and usually without marked constriction from the esophagus. Calciferous gland, thickest in xi, with a slight external constriction at 10/11, an internal and/or an external constriction near insertion of 11/12, behind which calciferous lamellae are narrower. Gizzard, mostly in xvii (ending at?). Gut, more or less valvular in region of insertion of 19/20 or in xx. Typhlosole, beginning rather gradually in or behind xxi, anteriorly thicker ventrally and broadened, with regular and close transverse grooving interrupted only at mV by a single, longitudinal ridge, ending abruptly in region of 93rd to 114th segments, leaving 27–57 metameres atyphlosolate. Extra-esophageals, passing up to dorsal trunk posteriorly in xii. Hearts, in vi–xi, occasionally unrecognizable in vi (because invisible when empty?). Nephridial bladders, J-shaped, closed end of loop laterally and about at CD though other portions of the organ reach well into DD. Holandric. Seminal vesicles, small to almost rudimentary in ix, x, medium-sized to smaller in xi, xii. Sperm ducts, with an epididymis comprising a single hairpin-loop, shorter loops bound together in a ball, or spirally coiled as in a watch spring. Spermathecae, in x, xi, ducts confined to body wall. (Atrial, TP, and supraparietal GS glands, lacking.)

Reproduction : Parthenogenetic, sometimes with pseudogamy, in Italy where Omodeo found only triploids ($3n = 54$) and tetraploids ($4n = 72$). Sperm never were found by the present author in spermathecae. A transparent watery fluid without solid matter distended ampullae of fully clitellate worms from various parts of the world, including Italy. Copulation either had not taken place or, if attempted, had not resulted in exchange of sperm. Male sterility appears to be very common. The only evidence for maturation of sperm was provided by iridescence in several spots on the male funnels of a South African individual. A slight sheen in ental portions of male ducts, of a few specimens, may have been due to presence of mature sperm but no iridescence was noted in testes. Seminal vesicles contained none of the usual stages of massive spermatogenesis but, in juvenile and pre-reproductive aclitellate, as well as clitellate and postreproductive aclitellate specimens, did contain brown matter, presumably the usual end product of phagocytosis.

Distribution : China (Chen, according to whom the taxon is "fairly common in most parts of China").

South Australia, New South Wales, Tasmania, New Zealand. Oahu, Kauai, and Lanai of the Hawaiian Islands (where the species is not common, only 11 speciments received from the islands during the last ten years).

British Columbia, Alberta (one site only). Washington, Oregon, California, Idaho, Nevada, Montana, Wyoming, Utah, Arizona, Colorado, Oklahoma, Texas, Iowa, Missouri (Stebbings), Arkansas, Louisiana, Wisconsin (driftless section), Illinois (Harman), Michigan,[11] Kentucky, Tennessee, Mississippi, Alabama, South Carolina, North Carolina, Virginia, Ohio (southernmost portion only), Pennsylvania, Massachusetts (single record of a specimen that may have escaped from a neighboring greenhouse).

Mexico[12] northern states of Sonora and Chihuahua (?). Chile, Argentina.[12]

England,[11] Guernsey (Channel Islands), France (Duges),[12] Italy,[12] Jugoslavia,[12] Greece.[12]

St. Helena, Algeria (Cain), South Africa.

Iran,[12] Afghanistan, Baluchistan, Pakistan, and Kashmir.

Many authors who recorded A. "caliginosa" failed to provide data that could enable present recognition of the taxon involved. The distribution above is based mainly on this author's materials including interceptions. Inclusion of New Zealand was permitted only by two specimens although A. "caliginosa," according to Lee (1959), is the most common earthworm in New Zealand.

A. iowana (= trapezoides) was found in Harman's survey of central Illinois (thesis, 1960) to be the second in number of times collected and in number per site.

A. trapezoides probably is the "caliginosa" most frequently recorded from all Mediterranean countries and the islands in that sea. The known distribution in transoceanic areas, colonized after introduction by man, supports an assumption that the species is restricted to climatic conditions similar to those of the Mediterranean region. A. trapezoides may then have acquired its present physiology farther from the great glaciers of the last Ice Age than did A. tuberculata and turgida.

A. trapezoides is one of the five species commonly sold and used for bait in North America. Much of the present distribution on that continent, especially

[11] Two lots of worms definitely identifiable as A. trapezoides were received from Michigan. One was a sample obtained some ten years ago from one of the bait dealers who imported the species from Louisiana for sale, primarily to anglers. Some worms could have been purchased and placed in gardens, lawns, and fields (where megadriles may have been lacking) as a result of the propaganda about earthworms increasing fertility of the soil. The second and later lot was from a garden.

Although A. trapezoides has been recorded, as a variety of A. caliginosa, most of the English records probably are referable to A. turgida if not also to A. tuberculata. However, a record of A. trapezoides by Davies (1960) for the Bristol area is believed to be correct.

[12] Record based on specimens intercepted by U. S. Bureau of Plant Quarantine.

in areas away from cultivation, probably is due to anglers.

Habitats: Earth around roots of plotted plants, including jasmine, fig, apple, rose, ferns, papaya, amaryllis, *Philodendron, Buxus, Echeveria, Cymbopogon* spp., and *Ficus elastica*, greenhouses, gardens, cultivated fields, pastures, along creek and spring banks (Missouri, in which sites it is "extremely common" and in a clitellate state through the summer, according to Stebbings, 1962), moss bank, under rocks and boards, yellow clay loam with rock fragments, soil of bracken covered hillside, black muck near edge of creek, quite dry sandy soil, soil of yellow pine, larch, white-fir forest, banks of drain-off from cess pool, cow manure pile.

Caves (Kentucky, Tennessee, West Virginia, Arkansas, Texas, Afghanistan).

From sea level, in California, to elevations of 5,800 feet. In Arizona at an elevation of 5,000 feet. Culture beds of earthworm farms.

Biology: Activity may be year-round when conditions are favorable but a similar statement cannot yet be made for breeding even though clitellate individuals were said to be common throughout the summer in Missouri stream banks. In South Australia, activity is restricted to 24 weeks each year, from mid-April to mid-October and the rest of the year is passed at bottom of the burrow rolled tightly into a ball within an oval chamber. Laying, in South Africa, is indicated for September–December and for May.

Genital tumescences provide the first externally recognizable indications of approaching maturity though epidermal modification at first is slight and easily overlooked. A bit more obvious is enlargement of included apertures of GS follicles which already are slightly farther apart than the paired openings of unmodified follicles. The genital setae, in the author's specimens, always were retracted deeply into the parietes so as to be externally unrecognizable. The GS follicles already protrude more conspicuously than others into coelomic cavities. The male pores, at first are superficial and without tumescences though a very fine, shallow, equatorial groove extends mesially from each pore. As that furrow deepens the immediate margins become slightly tumescent. The tubercula pubertatis appear before the clitellum and when the earliest rudiment is recognizable already are continuous through each of the three segments, without slightest indication of a bipartite origin. Hence, the tubercula are said to be of tripartite origin. Sites of the tubercula usually are distinguishable, after clitellar regression, by a faint, yellowish or yellowish-brown coloration.

A high proportion of mineral soil is ingested even when organic food is available. Food passes through the gut in 10–20 hours.

Variation and Abnormality: Considerable variation in segment number has been recorded. The maximum number, in an Italian series of 700 individuals, was 152 and the "medio" was 124.7. The number in another sample varied from 90 to 168 with a "media" of 135. In a Baluchistan series, after screening out amputees, number was 93-147, with all except six worms having 110–132. Average number of segments was higher in most of this author's samples, and the numbers cited in the précis are those he found in his own material. Maximum number was 174 in central Illinois where Harman (thesis, 1960), obtained an average of 141.7.

Tubercula pubertatis, in what may have been best preservation, were distinctly demarcated, each by a fine circumferential furrow. A major portion of each tuberculum was grayish translucent bordered by an opaque peripheral band. Tubercula of other specimens, some of which were in very good condition for certain purposes, were indistinguishable from the clitellum, and the central translucence usually was unrecognizable. Although the tuberculum of each immature individual was continuous through the three segments, in some adults a division or indication of a division into two was indicated by a transverse groove. The metabolic condition of the worm at time of preservation, the stage of the life cycle and the manner of preservation seem to be responsible for the differences.

Very rarely was any of the genital tumescences of ix–xi lacking and almost equally rare was presence of an extra tumescence in xii.

Among the rare organ abnormalities seen are the following: Absence of male terminalia on one side of xv. Presence on one side of xvi of an extra male pore cleft and tumescence. Sperm ducts disappeared into parieties in xv and a male pore probably was lacking in xvi. Total doubling of a posterior spermatheca, each organ with its own external aperture.

An unusually extensive homoeosis of an Arizona specimen with two pairs of male pores (Gates, 1956), was found to have resulted from halving of embryonic somites at second to fifteenth levels. The halving involved in that specimen provided well over 210 segments as a tail portion of unknown length had been amputated some time prior to capture. Elucidation of the cause of the anomalies enabled synonymization of *A. relictus* Southern 1909, a species supposedly having Lusitanian connections, with *L. rubellus*.

Polymorphism: Major morphs, of the kinds so often evolved, after reproduction becomes parthenogenetic, have not as yet been recognized in *trapezoides*. Seminal vesicles, and spermathecae of course do not have the same appearance as in fully mature individuals of forms with biparental reproduction but such physiological differences must be expected when animals are male sterile and the organs no longer function. No major morphological degradation was recognized by the author and none seems to have been reported by others. Some reduction of number of genital tumescences behind xxxii seemingly is under way in various clones.

Unless *A. trapezoides* is unusually resistant to evolutionary degradation of structure, acquisition of parthenogenesis would seem likely to have been recent. However, his Italian triploids are believed by Omodeo to be 40 million years old.

The triploids are morphologically indistinguishable from the tetraploids. Some Italian triploid morphs have gonads, supposedly hermaphroditic, in xii. Extra gonads were not found in any of this author's material.

Regeneration. Head regeneration in an anterior direction is possible back to 9/10 or 10/11, at which level regenerates, with metameric anomalies, had 3 or 4 segments. Tail regeneration is not uncommon and in them the number of segments produced decreases as level of amputation moves posteriorly as the following indicate, 76 at 77/78, 70 at 84/85, 45 at 88/89, 42 at 112/113, 34 at 132, 27 at 138/139. One worm had two consecutive tail regenerates, at 63/64 and at 83/84, the total number of segments 137.

Tails were said not to be regenerated when the worms are active and well fed. Perhaps, as in certain other species, tails are regenerated only during periods of quiescence or diapause.

A typhlosole may be present in proximal portions of long tail regenerates, ending at 104th, 97th, and 93rd segments when regeneration was at 77/78, 84/85, 88/89, respectively.

Parasites: TREMATODA: A specimen identified by the Beltsville Parasitological Laboratory as *Mehlisia* sp., was found in lumbricid fragments taken from stomachs of Bandicoots, *Perameles gunni* Gray, 1838, at Wiltshire Junction, Tasmania. The fragments probably were mostly of *A. trapezoides* but *A. longa* also could have been involved. Neither host species has been recorded hitherto from Tasmania.

Nomenclature and Systematics: Types of Duges's species never were seen by any of his successors and his *trapezoides* was for many decades in the synonymy of *A. caliginosa* though often recognized, for erroneous reasons, as a "variety" or a subspecies. Duges did define his taxon by at least one character that now has been found to be always lacking in other species also mistakenly thought to be synonyms of *A. caliginosa*. A taxon, *trapezoides*, even without any types, now can be satisfactorily characterized and adequately distinguished from all others of the congeries so long known as *A. caliginosa*. The questions still unanswered as to Duges's taxon[13] are no longer nomenclatural but biological.

Is the population, many samples of which now have

been correctly referred to *A. trapezoides*, a species? The answer, according to some of the criteria of modern systematics, is no, because reproduction, in so far as now known is parthenogenetic or at most, pseudogamic. An amphimictic population from which the known *trapezoides* morphs could have been isolated by their uniparental reproduction has not been found and may no longer be extant. If so, the species may have been able to survive the rigors of subarctic climates during the last glacial period only by becoming uniparental as seems also to have happened to many of the endemic American lumbricids. Whether that preglacial and amphimictic, ancestral population was specifically distinct from related populations that were able to retain biparental reproduction as well as distinct specific status down to the present time seems unlikely to be determined in the near future.

Some of the *trapezoides* morphs seen by the author appear to have undergone little or no morphological degradation. Such morphs may differ from the biparental morph only by functional differences attendant on failure to mature sperm. If, by a reversal of evolutionary process, such morphs again became amphimictic they would not ordinarily interbreed with biparental species of the *caliginosa* complex because of differences in distribution. Occasional interbreeding where ranges overlap presumably would be of no more significance than when "good" species of birds interbreed at common range boundaries. Whatever the distant future may decide, for the present and the near future, recognition of distinctness from others of the *caliginosa* complex seems necessary and, in the circumstances, most conveniently at species level.

Remarks: Looking down on the ventral face of the typhlosole, the appearance of the latter anteriorly suggests a row of on-edge coins close together and with a string along the middle of the row.

Allolobophora tuberculata

1874. *Allolobophora turgida* f. *tuberculata* Eisen, *Öfvers, Vetensk. Akad. Förhandl. Stockholm* 31, 2: p. 43. (Holotype designation, none. Syntypes, from Mt. Lebanon, presumably secured on the Massachusetts side, and at Niagara, Canada, in the U. S. Natl. Mus.)

1910. *Allolobophora similis* Friend, *Gardener's Chronicle*, Ser. 3, 48: p. 99, fig. 35. (Type locality, Kew, England. Types, in the British Mus.)

1911. *Aporrectodea similis*, Friend, *Naturalist* 1911: p. 396, fig. 1–3A.

1917. *Helodrilus (Allolobophora) caliginosa* f. *typica (part)*, + f. *trapezoides (part)*, Smith, *Proc. U. S. Natl. Mus.* 52: p. 166.

1930. *Allolobophora turgida* + *A. trapezoides* (major part?), Bornebusch, *Forst. Forsogs. Danmark* 11: pp. 30, 33, 94, 96, etc.

1942. "*Aporrectodea similis*" and *Allolobophora caliginosa* "*forma?*" Cernosvitov, *Proc. Zool. Soc. London*, B, 111: pp. 241, 273.

---

[13] To meet requirements of the International Commission, worms characterized as in the précis above should be collected at Montpellier, preserved properly and deposited in some institution where they can be examined by any who are interested in the problem.

1947. *Allolobophora caliginosa* f. *typica* (part), Cernosvitov & Evans, *Linnean Soc. London, Synopses British Fauna*, No. 6: p. 13.
1948. *Allolobophora caliginosa* (part only?), Evans, *Ann. Mag. Nat. Hist.*, Ser. 11, **14**: p. 515.
1951. *Allolobophora caliginosa* (part), Gates, *Proc. Natl. Acad. Sci. India*, B, **21**: p. 19.
1952. *Allolobophora arnoldi* Gates, *Mus. Comp. Zool. Harvard College, Breviora*, No. 9: p. 1; No. 10: p. 2. (Type locality, Arnold Arboretum, Boston, Mass. Type, in Mus. Comp. Zool., No. 4441.)
1953. *Allolobophora arnoldi*, Gates, *Bull. Mus. Comp. Zool. Harvard College* **107**: p. 510; *Breviora*, No. 15: pp. 2, 9.
1954. *Allolobophora arnoldi*, Davies, *Breviora*, No. 26: p. 1; *Ann. Mag. Nat. Hist.*, Ser. 12, **7**: p. 351.
1955. *Allolobophora arnoldi*, Jefferson, *Jour. Sports Turf Res. Inst.* **9**, 31: p. 13, footnote.
1958. *Allolobophora tuberculata*, Gates, *Mus. Comp. Zool. Harvard College, Breviora*, No. 91: p. 4.
1964. *Allolobophora tuberculata*, Gerard, *Linnean Soc. London Synopses British Fauna*, No. 6 (2nd ed.): p. 30.

ABNORMALITY: Gates, 1957, *Ann. Mag. Nat. Hist.*, Ser. 12, **10**: p. 204. ACTIVITIES: Jefferson, 1958, *Jour. Sports Turf Res. Inst.* **9**: pp. 442, 444, 449, 450. CHROMOSOMES: Omodeo, 1955, *Caryologia* **8**: pp. 140–142. ECOLOGY: Gates, 1961, *American Midland Nat.* **66**: p. 73, etc. HOMOEOSIS: Gates, 1957, *Ann. Mag. Nat. Hist.*, Ser. 12, **10**: pp. 204–208. REGENERATION: Gates, 1961, *American Midland Nat.* **65**: p. 42.

Quadrithecal, pores at *C*, at 9/10, 10/11. Clitellum, saddle-shaped, 7–9 (very rarely 10) segments, xxvi, xxvi/n, xxvii, xxvii/n, xxviii–xxxiv, xxxv/n, xxxv. Tubercula pubertatis, of double origin, in xxxi–xxxiii, occasionally reaching into or through xxx and/or xxxiv but very rarely through both, nearer *B* than *C* but occupying much of *BC*, median margin concave, concavity deepest at eq/xxxii or a transverse furrow at eq/xxxii demarcating the two halves from each other. Male pores, at bottom near lateral end of an equatorial cleft almost reaching *B* but not so close to *C*. Male tumescences, paired and reaching median to *A* and nearly to *C* as well as into xiv and xvi, obliterating 14/15 and 15/16, variously shaped, or unpaired, deeply depressed or markedly protuberant. Female pores, just lateral to *B* at eq/xiv. Genital tumescences, surrounding *a,b* only, six pairs, in ix–xi, xxx, xxxii, xxxiv, occasionally in xxvi. Setae, closely paired, *a,b*/xv even more closely, *a,b*/xvi–xxix larger than *c,d* of the same segments, follicle apertures in genital tumescences more widely separated than elsewhere, *BC*<*AA*. Nephropores, inconspicuous, behind xvi, alternating irregularly, with asymmetry, between a level just above *B* and others at or above *C*. First dorsal pore, at any of levels 9/10–12/13. Prostomium, epilobous, tongue open or closed. Compression of body dorsoventrally near hind end, slight. Color, white (preserved), grayish, brownish, as well as other seeming colors of live specimens due to ingesta. Segments, 145–194 but usually < 181. Size, 50–150 by 4–6 mm.

Septa, muscular anteriorly, thickness increasing from 5/6 to 8/9, 9/10 or 10/11, decreasing posteriorly

to 12/13 or 13/14. Pigment, none of usual sort present, sparse yellow flecks externally recognizable in older individuals under the binocular perhaps respresenting remains of eleocytes. Longitudinal musculature, pinnate. Calciferous sacs, vertical, reaching slightly above and below the esophageal level, usually opening widely into the gut without any marked constriction. Calciferous gland, thickest in xi, with a slight constriction at 10/11, an internal and/or an external constriction near insertion of 11/12, behind which calciferous lamellae are narrower. Gizzard, mostly in xvii (ending at ?), circular muscle layer markedly narrowed anteriorly in xviii. Gut, more or less valvular in region of 19/20 or xx. Typhlosole, beginning rather gradually in region of xxi–xxiv, anteriorly thickened ventrally and broadened, for a short distance with a low honeycomblike ridging except where interrupted mesially by a low longitudinal ridge, posteriorly with three longitudinal ridges on the ventral face, ending abruptly in region of 102nd to 125th segments, leaving 34–76 segments atyphlosolate. Extra-esophageals, passing up to the dorsal trunk posteriorly in xii. Hearts, in vi–xi. Nephridial bladders, *U*-shaped, in *BC*, with closed end laterally. Holandric. Sperm ducts, with an epididymis of short irregular loops that may be closely bound together in a flat disc or in a ball. Seminal vesicles, 4 pairs, in ix–xii. Spermathecae, usually in x, xi, ducts confined or almost so to parietes. (Atrial, TP and supraparietal GS glands, none.)

Reproduction: Mature individuals almost always have spermatozoal iridescence on male funnels and, if copulation has been allowed, also in the spermathecae. Chromosome number, $2n = 36$ (Omodeo). Reproduction seemingly is biparental and perhaps obligatorily so.

Distribution: New South Wales, Australia.

Queen Charlotte Island, British Columbia, Alberta, Saskatchewan, Manitoba, Ontario, Newfoundland, New Brunswick, Nova Scotia.

Oregon, Colorado, Iowa, Wisconsin (driftless area), Michigan, Florida (*cf.* Habitats), Ohio, Pennsylvania, New Jersey, New York, Connecticut, Massachusetts, New Hampshire, Maine.

Mexico.[14] Chile.[14]

Norway,[14] Sweden, Finland,[14] Germany,[14] Denmark,[14] England[14] (4 localities only), Ireland,[14] Netherlands.[14]

Iran,[14] India, western Himalayas in region of Simla, at elevations of 6,500–8,000 feet.

Manchurian specimens referred[15] to f. *typica* of *A. caliginosa* did have genital tumescences in the *tuberculata* locations but were of a uniformly light

---

[14] Based on intercepted specimens, except the new record for Finland, and as yet without confirmation from local collections in case of Mexico, Chile, Ireland, and Iran.

[15] Kobayashi, 1940, *Sci. Rept. Tohoku Univ.* **15**: p. 295.

slate color. If that color was due to gut contents or some postmortem chemical change rather than to presence of a pigment, *A. tuberculata* may have been involved. Much less information is available as to a single Samarcand worm, referred[16] to f. *trapezoides*, that may have had genital tumescences in the *tuberculata* locations. *A. tuberculata* probably is the species that already had become common in Milwaukee, Wisconsin, before 1885.

In central Maine, *tuberculata* is the most widely distributed and probably the most common megadrile. A similar characterization for other New England states, and for eastern provinces of Canada, may be possible when their faunas are studied. Dominance of a similar sort seemingly is to be expected in Denmark, if not also in Norway and Sweden, but evidence for the Netherlands and Germany, as well as other areas is inadequate. The spotty distribution in England, where it has been found only at four places, suggests a rather recent introduction from the continent. As the North American distribution has been attained since 1500 A.D., much of the spread through the glaciated areas of Europe may also be due to transport by man. Certainly the centuries before 1500 A.D. must have provided many opportunities for such carriage. If *A. tuberculata* did not actually arise in southern Europe, its present physiology with reference to climate must have evolved in a cool if not subarctic climate below the ice sheets.

A. tuberculata, being so common in gardens, fields, and pastures, is the species most likely to be dug up for bait in much of Canada as well as in the northeastern part of the United States, and undoubtedly has been widely distributed through the uninhabited or heavily forested areas to which American anglers are known to have resorted for more than a century.

Habitats: Earth around roots of plotted plants, greenhouses (frequently in Central Maine), outside beds of greenhouses, turf, lawns, gardens, in and under short grass cuttings, compost, in and under decaying leaves, manure pile (once), cowyard, henyard, under cow pats in pastures, cultivated fields, pastures, peat bog (along with *A. limicola*), in a load of peat (once).

Under stones in sand-gravel beside river. Aggregated silty clay. Sandy soil under brush of wasteland. Saturated soil of stream banks. Leaves and mud at bottom of a spring. Under logs in a stream bed. Rich soil under horn-beam-alder mixture. Sand-humus mixture over clay. Soils with a *p*H of 4.8 to 7.5. Under elm logs. On roads and sidewalks after heavy rain in fall and spring.

Culture beds on an earthworm farm, Florida (the only record for that state).

Biology: Activity is year-round when conditions permit. Breeding also can be year-round when the

climate, as in greenhouses or in permanently moist sites, is favorable. However, estivation and hibernation are climatically imposed at most sites in central Maine, probably also throughout most of New England, New York, Canada, where breeding generally is in March or April–May or June, October or November–December. State of the animals during the periods of quiescence twice a year is unknown but some estivating individuals, with gut empty or nearly so, were coiled up tightly, each in a closed cell, conditions suggestive more of diapause than mere quiescence.

Genital tumescences along with very fine rudiments of male clefts provide the first externally recognizable indications of approaching maturity. Tubercula pubertatis first appear as a pair of subcircular tumescences in each of xxxi and xxxiii. Growth results in obliteration of 31/32 and 32/33, invasion of xxxii, until eq/xxxii is almost reached. Then, the two on each side may unite, leaving only a concavity of the median margin, deepest at eq/xxxii, to indicate the double origin, or, in absence of union, a transverse groove appears at eq/xxxii between the two markings of a side. (Concavity, equatorial groove, as well as tubercula often are unrecognizable in markedly relaxed individuals.) Prior to appearance of the clitellum, genital tumescences of preserved material may be markedly protuberant, much more so than the tubercula. Such protuberances still characterized clitellar segments of the types of *similis* and, mistaken by Friend for *tubercula*, were thought to be similar to those of *chlorotica*. At close of the breeding period, the clitellum is yellow or (in preserved material) almost red. After disappearance of the clitellar tumescence a faint yellowish to brownish discoloration may demonstrate that the animal once had been able to lay. Male tumescences and tubercula also regress during periods of quiescence and may be unrecognizable or represented only by a faint yellowish discoloration. Genital tumescences may become indistinguishable except by the slight larger size of the GS follicle apertures. Any remaining spermatozoal iridescence on male funnels and in spermathecal ampullae disappears during quiescence, perhaps in the spermathecae as a result of lysis, elsewhere perhaps in part as a result of phagocytosis by coelomocytes. A brilliant yellow material had been deposited, during winter quiescence, under the peritoneum of the ventrum in clitellar and anterior segments, also in the GS follicles. Seminal vesicles, however, contained only granular brown debris or aggregates of such material into "brown bodies," a number of which usually were present in coelomic cavities of ix–xx. Spermatozoa already were producing iridescence on male funnels of worms very recently emerged from hibernation but none were as yet present in spermathecae. However, sperm frequently were found in spermathecae of individuals still aclitellate, or at

---

[16] Michaelsen, 1910, *Ann. Mus. Zool. Acad. Sci. St. Petersburg* **15**: p. 55.

most only with a very slight thickening of clitellar epidermis distinguishable, as well as of individuals that had not yet matured enough gametes to have any iridescence on their male funnels. Copulation, accordingly may not always result in an exchange of sperm between two animals. What happens to the defrauded partner in such transactions is unknown.

Much mineral soil is ingested but individuals may feed selectively as large sections of the gut frequently were filled with organic matter or long bundles of indigestible plant fibers. Feces were deposited on the surface, according to Jefferson, in every type of soil examined. The casts were bulky and cohesive even when the soils were light.

Preferences and behavior, also according to Jefferson, were so different from those of *A. caliginosa* f. *typica* (probably reference was to *A. turgida*) as to warrant separation from that taxon of *tuberculata* at species level.

Variation and Abnormality: Individual variation with regard to some characters of the clitellum and tubercula pubertatis already has been indicated. An unusually long or short clitellum, especially if tubercula extended through xxx–xxxiii, xxxi–xxxiv or even xxx–xxxv, could have led to misidentification or even unnecessary Latin-named taxa, because of the paucity of information as to intraspecific variation and the importance erroneously allowed those two organs in the classical system. Among variant specimens "difficult" for the hasty identifier are those with *trapezoides*-like tubercula just as occasionally concave median margins sometimes associated with a transverse groove give *trapezoides* tubercula an appearance like that usually characteristic of *A. tuberculata*.

Absence of genital tumescences in either of ix–xi is very rare as also is presence of tumescences in viii or xii–xiv. Pattern of tumescence location in the region anterior to xv is identical in *A. trapezoides*, *tuberculata*, *turgida*, and in each species also seems to be equally resistant to modification. The pattern posteriorly is subject to more individual variation. The alternate spacing of the tumescences in xxx–xxxiv does seem to be characteristic of *A. tuberculata* though finding it in a rare individual of *A. nocturna*, *turgida* or even *trapezoides* would not be surprising. Appearance of tumescences in xxxiii or xxxi of *A. tuberculata* is less likely. Deletion of the tumescences in xxxiv occasionally results in a pattern like that of certain *trapezoides* morphs. Any one of the clitellar tumescences of *A. tuberculata* may fail to develop. As all clitellar tumescences of one side sometimes are lacking, occasional absence on both sides presumably should be anticipated.

DD in the anteriormost portion of the body may about equal ½C but in the clitellar region of presexual and postsexual aclitellate individuals (after preservation) may be greater than ½C. Posteriorly, as the transverse section changes in shape from subcircular to transversely elliptical, then followed by flattening of the ventrum and the lateral sides, the dorsal side remaining slightly convex, *DD* becomes smaller than ½C. Dorsoventral compression near the hind end, except in case of certain posterior amputees, almost always was recognizable but never seemed to be as marked as in *A. trapezoides* and *L. terrestris*.

Regeneration: Some of the experimental results attributed to *A. caliginosa* in the past are likely to have been furnished by *A. tuberculata*. Hypermeric head regeneration is probable at some levels in front of 8/9. Head regeneration, in an anterior direction, should be possible at all levels back to and including 47/48 even though the posteriormost record for homomorphic head regeneration yet recorded is 23/24 (for *E. foetida*, *q.v.*). The proof is provided (Gates, 1961: p. 42) by a purely cephalic regenerate in a posterior direction at 47/48. An extensive region of bipotential regenerative capacity, as in *E. foetida*, is anticipated.

Tail regenerates, all homomorphic, were found at various levels from 36/37 back to 163/164. Number of new segments, rather generally, declines as level of amputation becomes more posterior. However, 195 segments were present in one individual that had regenerated at 86/87 even though 187 was the highest number counted on any of the author's unamputated specimens. Average number of segments in seven regenerates at 101/102–102/103, was 69 (*cf.* with *A. turgida*).

Regenerative capacity may prove to be of about the same order, if not greater, as in the often studied *E. foetida*. If, as in some lumbricid species, caudal regeneration is possible only during diapause, maximal results might well be unobtainable in conditions inhibiting or unfavorable to such periods of quiescence.

Individuals with more or less marked organ or regional homoeosis (Gates, 1957) because of deletion or splitting of embryonic somites are not rare. Origin of some organ homoeoses such as male terminalia of one side of the body opening to the exterior just lateral to the female pore, an extra male pore cleft with some marginal tumescence just lateral to one of the female pores, is unexplained.

Systematics and Nomenclature: *A. caliginosa*, according to Cernosvitov and Evans,[17] is "cosmopolitan, imported by man into many parts of the world, where it follows development of agriculture and usually replaces the indigenous species (India, Australia, New Zealand). Must be regarded as the commonest species of earthworm in the world." Because of that worldwide distribution, dominance or near dominance in many areas, frequent use in laboratory experiments, an economically important

---

[17] Cernosvitov & Evans, 1947 "Lumbricidae," *Linnean Soc. London Synopses British Fauna* (London) *cf.* p. 14.

role in soil faunas increasingly being studied almost everywhere except in the Americas,[18] solution of long-standing nomenclatural, systematic and biological problems is desirable.

E. caliginosum, as defined by its author, has paired setae, a clitellum in segments xxvii–xxxiv, four pairs of seminal vesicles, and probably, though not certainly, would have been referable to the genus *Allolobophora* as long defined. Tuberculata pubertatis, according to Tetry,[19] who certainly is much better qualified than this author to put Savigny's characterizations into modern terminology, are in xxxi–xxxii but according to Cain,[20] they are in xxxi and xxxiii. Neither of those specifications accurately characterizes most of the specimens from various parts of the world that for a century and more were referred to *A. caliginosa*. Savigny's successors never saw any specimens he had identified as *E. caliginosum*. Absence in the Paris museum of any such material may be because the French scientist had only one specimen, perhaps at most two or three, or a unique type may have been a more or less aberrant individual of some common taxon. The species involved, owing to absence of types, never will be known, as the information supplied by Savigny is insufficient to enable distinction from several related forms also present in the type locality.

Another of Savigny's species, *E. carneum*, differed from *E. caliginosum* in two ways; the clitellum is in xxviii–xxxiv and one pair of vesicles, presumably that in x, is lacking. Specimens in the Paris museum, supposedly identified by Savigny as *E. carneum*, were found on dissection to be octovesiculate rather than sexvesiculate and were assumed to be of *E. caliginosum*. Tetry[19] thought that Savigny erred in recording the number of vesicles in *E. carneum* but that assumption seems gratuitous unless a parthenogenetic morph was involved (in which case original specimens most likely would have been of *trapezoides*). If a mistake was made, the more likely one now would seem to have been identifying worms by external characters only when internal anatomy was necessarily involved.

Two of the Paris museum specimens of *E. carneum* were believed by Cain[20] to be of a taxon commonly known in England as *A. caliginosa* var. *typica*. The identification may be correct but cannot be certain because condition of the worms is too poor to permit determination of all of the distinguishing characters. A third specimen of *E. carneum* was designated the "lectotype" of *A. caliginosa* (p. 491) because of an assumption that the three worms were conspecific. Unfortunately, there is good reason for suspecting

that the lectotype is of a species distinct from that of the other two specimens though here again certainty is lacking because of poor condition and perhaps also because of age (either not fully mature or perhaps postsexual?). The lectotype now appears to be less likely to function satisfactorily as a nomenifer than the other worms. Accordingly it seems as if the International Commission could rule that the names *caliginosa* and *carnea* are to be suppressed because of absence of types and the inadequacy of Savigny's characterizations.

Chronologically the next form requiring mention is Duges's *L. trapezoides* which already has been considered above. Further comment here is unnecessary except to emphasize that even today, in some unfavorable cricumstances, specimens of *A. trapezoides* cannot be distinguished with certainty from those of related species and this possibility[21] provides another one of the additional reasons that can be cited for suppression of the names *caliginosa* and *carnea*.

Ever since Savigny, oligochaetologists seem to have felt that the taxon known as *A. caliginosa* really was dimorphic. Hoffmeister thought his German forms were discrete and Michaelsen believed (as now appears correctly) a northern race could be distinguished from a southern though his criteria for doing so were worse than useless. Cernosvitov, after examination of the types of *A. similis*, was unwilling to treat the name as a mere synonym but thought a "form" in some way distinguishable from the common English race (= *turgida*) might have to be recognized. Evans, who studied common lumbricid species much more intensively than ever before, found certain American specimens from the type locality of his *A. iowana* (= *trapezoides*) that differed "significantly" from specimens of *A. caliginosa* (= *turgida*) he himself had bred at Rothamsted and which he thought probably were of a new species. Additional specimens from that Iowa site were found by this author, to be of the same species as the types of Friend's *A. similis* and some of the types of Eisen's f. *tuberculata*. Eisen's original material still is available and its condition when examined by this author several years ago was good enough to permit adequate characterization of *A. tuberculata*. Selection of one of Eisen's specimens from collections in the U. S. National Museum or the Swedish museum at Stockholm should enable stabilization of the name. Mount Lebanon, perhaps the better choice for a type locality, originally was said to be in New England but later was thought by another to be in New York. The mountain actually is in both

[18] Barley, 1961, *Advances in Agronomy* 13: p. 266.

[19] Tetry, A., 1937–1938, "Revision des Lombriciens de la collection de Savigny," *Bull. Mus. Natl. Hist. Nat. Paris*, Ser. 2, 9: pp. 140–155 (see p. 146) and 10: pp. 72–81.

[20] Cain, A. J., 1955, "The Taxonomic Status of *Allolobophora iowana* Evans, 1948 (Oligochaeta, Lumbricidae)," *Ann. Mag. Nat. Hist.*, Ser. 12, 8: pp. 481–497.

[21] No variation in number of seminal vesicles has been found in thousands of normal specimens of three amphimictic species probably closely related to whatever Savigny believed *caliginosum* and *carneum* to be. If a parthenogenetic morph was involved it most likely would have been of *trapezoides* in male sterile morphs of which vesicles of x may be small to vestigial.

New York and Massachusetts and may well have been approached from its eastern side in New England. Presence there of the species, prior to 1874, in such numbers as to be found by a passing traveler, indicates once more the rapidity with which species can spread with human assistance.

Parthenogenesis effectively isolates *A. trapezoides*, as it is known today, from *A. tuberculata*. The latter now appears to be reproductively isolated from its closest relatives, *A. turgida*, *nocturna*, and *longa*, species probably with obligatory biparental reproduction, as effectively as from *A. trapezoides*.

Remarks: The conspicuousness of the genital tumescences, often as obvious as on those specimens of *E. rosea* referred to "var. *macedonica*," as well as the regularity of the intervals between them in the clitellar region, enables easy identification in the field, without a glass, of many individuals *in vivo*. A large percentage of properly preserved specimens of *A. tuberculata* probably can be correctly identified, at least in North America, merely by the characteristic spacing of the clitellar tumescences.[22] Other specific characteristics should enable identification of most variant and aberrant individuals. Distinguishing some alcoholic and softened amputees from *A. turgida* or even *A. trapezoides* is likely to be impossible.

Clitellar tumescences may become conspicuous before appearance of the tubercula pubertatis and they were mistaken by Friend, on the types of *A. similis*, for tubercula and so were responsible for a comparison of the species with *A. chlorotica*. However, Friend did note a similarity in crawling with that of *A. caliginosa* (= *A. turgida*).

Specimens used in a recent electron-microscope study of cytology were identified as *Lumbricus terrestris*(!), which certainly is difficult to understand, though Stephenson (1930, p. xi) may have had the explanation.

Allolobophora turgida

1874. *Allolobophora turgida* Eisen, *Öfvers. Vetensk.-Akad. För-handl. Stockholm* 30, 8: p. 46. (Type locality, unknown, but in Sweden. Types, no designation.)
1964. *Allolobophora caliginosa* (in toto?) Gerard, 1964, *Linnean Soc. London Synopses British Fauna*, No. 6: p. 27.

Quadrithecal, pores at or just above *C*, at 9/10–10/11. Clitellum, saddle-shaped, 5½–8 segments (xxvii/n?), xxviii, xxviii/n, xxix–xxxiv, xxxv/n, xxxv. Tubercula pubertatis, of double origin, in xxxi–xxxiii, occasionally reaching into or through xxx or xxxiv, median margin concave, concavity deepest at eq/xxxii, or a transverse

---

[22] A rare specimen of *A. turgida*, or even of *A. trapezoides* may have the same arrangement of clitellar tumescences. Tubercula not uncommonly are indistinguishable from the clitellum when preservation is alcoholic, another reason for some of the past difficulties with the *caliginosa* complex now known to have comprised at least four species, *A. trapezoides*, *tuberculata*, *turgida*, and *nocturna* Evans 1946, if not also *A. limicola* Michaelsen, 1890, and perhaps at times *A. longa*.

furrow at eq/xxxii demarcating the two parts from each other. Male pores, at bottom near lateral end of an equatorial cleft almost reaching *B* but not so close to *C*, at or lateral to m*BC*. Male tumescences, paired and reaching median to *A* and nearly to *C* as well as into xiv and xvi, obliterating 14/15 and 15/16, longitudinally elliptical, transversely oval with pointed ends laterally, or otherwise shaped, the ventrum in *AA* finely wrinkled, or depressed and with translucent epidermis, or male tumescences united mesially so as to show no evidence of double origin. Female pores, just lateral to *B* at eq/xiv. Genital tumescences, surrounding *a,b* only, in ix–xi, xxx, xxxii–xxxiv, often also in xxvii. Setae, closely paired, *a,b*/xv even more closely, follicle apertures in genital tumescences more widely separated than elsewhere, *BC* < *AA*. Nephropores, inconspicuous, usually just above *B* in xv–xvi, behind xvi alternating irregularly, with asymmetry, between a level just above *B* and others at or above *D*. First dorsal pore, at any of levels 11/12–14/15. Prostomium, epilobous, tongue open. Color, white (preserved), pinkish, grayish, brownish or otherwise (alive) depending on content of gut, capillary relaxation in parietes, accumulation of particulate matter in coelomic cavities. Segments, 125–168, usually 131–152. Size, 35–85 by 1.5–4.5 mm.

Septa, 5/6–9/10 muscular, but 6/7–7/8 though not massively muscular slightly thicker than the others, 10/11–13/14 strengthened. Pigment, of usual sorts lacking, sparse yellow flecks externally recognizable under the binocular in older individuals possibly representing, as also internally, remains of eleocytes. Longitudinal musculature, pinnate. Calciferous sacs, in x, vertical, reaching slightly above and below the esophageal levels, usually opening widely into the gut without any marked constriction. Esophagus, thickest in xi, with a slight constriction at 10/11, an internal and/or an external constriction near insertion of 11/12, behind which calciferous lamellae gradually narrow. Gizzard, mostly in xvii (ending at?). Gut, more or less valvular in region of 19/20 or xx. Typhlosole, beginning rather gradually in region of xxi–xxiv, anteriorly thickened ventrally and broadened, the ventral face with three or five longitudinal ridges but without transverse grooving, becoming vertically lamelliform posteriorly, ending abruptly in region of 90th to 108th segments leaving the last 30–65 metameres atyphlosolate. Extra-esophageals, passing up to the dorsal trunk posteriorly in xii. Hearts, in vi–xi. Nephridial bladders, *J*-shaped, closed end laterally, one limb about half the length of the other. Holandric. Sperm ducts, usually with an epididymis comprising one long hairpin loop. Seminal vesicles, 4 pairs, in ix–xii. Spermathecae, usually in x, xi, ducts confined or almost so to the parietes. (Atrial, TP and supraparietal GS glands, none.)

Reproduction: Mature individuals almost always

have spermatozal iridescence on male funnels, and, if copulation has been allowed, also in the spermathecae. Chromosome number, probably $2n = 36$. Reproduction seemingly is biparental and perhaps obligatorily so.

Distribution: New South Wales, Australia.

British Columbia, Ontario, New Brunswick, Nova Scotia, Newfoundland.

Alaska, Oregon, California, Idaho, Iowa, Michigan, Ohio, Pennsylvania, Virginia (at elevations of *ca.* 3,500 feet), Connecticut, Massachusetts, New Hampshire, Maine (not common but somewhat more so than *A. chlorotica* and much less so than *A. tuberculata*). Argentina.

Norway, Sweden, Finland, Poland, Germany, Denmark, Scotland, Ireland, England, Guernsey (Channel Islands), Netherlands, Switzerland, Italy (northern).

South Africa, from the Cape into Orange Free State.

Baluchistan (localities unknown), western Pakistan. India, Simla, also several localities in Himachal Pradesh in the western Himalayas, at 7,000–8,000 feet where soil is under snow for 3–4 winter months.

A. turgida was found to be abundant wherever collections were made (Davies, 1960) in the Bristol area of England. Definitions and figures of *A. caliginosa* in both editions of the "Lumbricidae" in Linnean Synopses (Cernosvitov & Evans, 1947, Gerard, 1964) indicate that *A. turgida* is the component of the *caliginosa* complex that is most common and widespread in the British Isles. Of the other components, the paucity of reports *re A. trapezoides*, *tuberculata* and *nocturna* suggest for each of them a much more recent introduction to the isles. The last two taxa may be common in those sites from which it is now known but *A. trapezoides* is not known to be common anywhere in Britain or Ireland.

In North America, *A. turgida* now appears to be much less common than *A. tuberculata*. In central Maine, in one locality only, was *A. turgida* present in larger numbers than *A. tuberculata* and then only at a single site. Although quite inadequate for a proper estimate of commonness, fewer individuals of *A. turgida* (than of *tuberculata*) have been available to the author from Denmark, and Finland.

Intercepted specimens were from 13 countries (including Greece) and from the Azores.

Habitats: Earth around roots of potted plants, greenhouses, outside beds of greenhouses, botanic gardens (South Africa), flower beds elsewhere, gardens, cultivated fields, meadows, pastures, turf, humus in forest, compost, leaf litter, banks of springs, streams, rivers, abandoned grape arbor, wasteland, city dump, cave (West Virginia), on tarred sidewalks and roads after heavy rain.

Biology: Activity is year-round when conditions permit. Breeding also probably can be year-round when the climate, as in greenhouses or in permanently

moist sites, is favorable. However, estivation and hibernation are climatically imposed at most sites in central Maine, probably also in most North American areas where it is present. State of hibernating individuals is unknown but those found estivating (central Maine) were in a condition usually characterized as diapause.

Post-embryonic development, and especially of the tubercula, is about the same as *A. tuberculata*.

Feeding, seemingly is entirely below the surface of the soil where the casts are deposited.

Variation and Abnormality: Individual variation with regard to some taxonomically important characters already has been indicated in the précis. Percentage of specimens with genital tumescences in xxvii, varies widely in different series, 20, 33, 39, 50, 75, 80 per cent. Additional tumescences, rarely present, were in one of xxvi, xxviii, xxix, xxxi, xxxv, xxxvi, even more rarely in two of them.

Variations in the tubercula that may cause an overhasty identifier trouble, dumbbell-shape, band-shape without evidence of bipartite origin, extension into or even through xxx and/or xxxiv. Although "difficult" specimens are not especially rare, identification of properly preserved and unamputated individuals usually is possible.

Individuals with regional homoeosis (symmetrical or asymmetrical) are not especially rare and as in *A. tuberculata* seemingly have arisen from deletion or splitting of embryonic somites.

Seeming absent of hearts of vi may be due to inability to recognize those vessels when empty and without any red color.

Regeneration: Head regeneration has not been seen. Tail regenerates, all homomorphic, were found at various levels from 52/53 posteriorly. Number of new segments, declines as level of amputation becomes more posterior. The largest number of new metameres was 90 in a regenerate at 52/53 and the smallest was 17 at 128/129. Average number of metameres in six regenerates at 101/102–102/103, was 43 (*cf.* with *A. tuberculata*).

Systematics: The means by which *A. turgida* is isolated reproductively from *A. tuberculata* are unknown. Although both species often are reported from the same area, presence together at same site is not common. When together in one site, growth stage often seems to be slightly different.

Economic Importance: Within a period of twenty-five years (according to K. S. P. Rao, 1959, *in lit.*) many acres of land in Himachal Pradesh had gone out of cultivation as a result of the appearance of earthworms that made fertile soil cloddy and unproductive. The worms were said to be present to as many as 250 per square feet of surface down to a depth of 10 inches. A sample of the accused worms, sent to the author for identification, comprised 1 juvenile of

Perionyx sp., 1 of *A. trapezoides*, and 7 specimens of *A. turgida*.

Remarks: Specimens that had been preserved in a relaxed state occasionally were seen. In some of them, clitellar boundaries were not certainly determinable, male clefts had been obliterated, intersegmental furrows were unrecognizable and the tubercula were indistinguishable from the clitellum. The latter condition, however, has been noted also in strongly contracted alcoholic material. Relaxed individuals may be as long as 200 mm.

The few ventral setae of xv that were examined showed the ectal sculpturing characterizing setae associated with genital tumescences and so also can be called genital.

Bimastos

1893. *Bimastos* Moore, *Zool. Anz.* 16: p. 333.
1895. *Bimastos*, Moore, *Jour. Morph.* 10: p. 473. (Type species, *B. palustris* Moore, 1895.)
1968. *Bimastos*, Gates, *Jour. Nat. Hist. London* 9: p. 306. (Redefined to exclude degenerate parthenogenetic morphs of various unrelated European species.)

Calciferous gland, without marked widening in xi–xii, opening into gut in x through paired, vertical sacs. Gizzard, mainly in xvii (ending at?). Extra-esophageal trunks, joining dorsal trunk in xii. Hearts, in vii–xi. Nephridial bladders, U-shaped, closed ends laterally, ducts passing into parietes near *B*. Nephropores, inconspicuous, irregularly alternating, and with asymmetry, between levels somewhat above *B* and well above *D*. Setae, closely paired. Dorsal pores, present from region of 5/6. Prostomium, epilobous. Pigment, red.

Male pores, equatorial in xv, in atrial chambers invaginated deeply into the coelom and bearing acinous glands. Female pores, equatorial in xiv, shortly above *B*. Holandric, seminal vesicles in xi–xii. (Spermathecae, tubercula pubertatis and TP glands, lacking.)

Distribution: United States.

Much remains to be learned about the range of the genus, in particular as to where species are endemic (presumably only in unglaciated sections) and where exotic (presumably introduced by man since the glaciers retreated). Such information as now is available does seem to allow the following suggestions for future consideration.

Bimastos originated in North America where it presumably acquired a wide range in which numerous species evolved. During the glacial epoch most of the species were exterminated and those that survived the rigors of arctic or subarctic climates, perhaps with one exception, became parthenogenetic. The possible exception, *B. palustris* (the generic type), lost its spermathecae, tubercula pubertatis and TP glands, just as did the surviving parthenogenetic morphs. If *palustris* still is biparental, that must be because

of some specializations associated with adhesive spermatophores. An apt comparison may be with *Criodrilus lacuum* for which an explanation of how an athecal species retains amphimixis also is incomplete.

Systematics: *Bimastos* was not recognized in 1895, presumably because Beddard knew only Moore's preliminary contribution. The genus already had become, in 1900, a systematic wastebasket in which were dumped quadrivesiculate and athecal morphs of then unknown relationships. Some, *Allolobophora tenuis* Eisen, 1874, *A. constricta* Rosa, 1884, and others are now known to be anatomically degraded, parthenogenetic morphs of *Dendrobaena rubida*. Two foreign morphs included in Michaelsen's *Bimastos* (1900) are *Lumbricus eiseni* and *Allolobophora syriaca* Rosa, 1893. The latter, seemingly known only from the now inadequate original description, was transferred (Omodeo, 1856) to *Dendrobaena*. Levinsen's *eiseni* certainly does not belong in *Bimastos*. With those deletions *Bimastos*, as recognized by Michaelsen in 1900, would have become American.

Bimastos was, however, suppressed by Pop (1941) in the synonymy of *Eisenia*. Even more unfortunate (*cf.* below) was the placement of *L. eiseni* in the synonymy of Eisen's *parvus*. Elimination of *Bimastos* was not acceptable to Omodeo (1956) who further diversified the congeries by including in it the following rarely seen and inadequately characterized taxa, *Eophila antiqua* Cernosvitov 1938, *Allolobophora* (*Bimastus*) *icenorum* Pickford, 1926, *Allolobophora minuscula* Rosa, 1906, *Bimastus muldali* Omodeo, 1956. Those certainly are not of American origin and insofar as their anatomy is known do not belong in *Bimastos* as now defined.

All American taxa, perhaps with one exception, now appear to be morphs isolated by their parthenogenesis from unknown biparental populations probably no longer extant. Information now available is insufficient to enable certain determinations of those that could have had a common origin in a single biparental population which, in the modern systematics, would be considered a good species. *B. parvus, beddardi* (Michaelsen, 1894) and *longicinctus* Smith & Gittins, 1915, intergrade without known ways of delimiting each from the others. Accordingly, isolation of each of those morphs from a common biparental population does seem to be possible.

Phylogeny: European lumbricids and their parthenogenetic morphs, perhaps more so than in America, show various stages in simplification and final elimination of the male terminalia. However, the American *B. palustris* still shows the best development yet recorded, of the organs. In that presumably ancient and primitive state, the minute male pore is invaginated deeply into an atrium that protrudes conspicuously into the coelomic cavity and bears on its coelomic surface a large, acinous, atrial gland of unknown function. The atrium opens to the exterior

SYSTEMATICS

by a transverse slit usually called the male pore but which it is, only in a secondary sense.

Extinct species of *Bimastos* presumably were thecal and if the battery was as primitive as the male terminalia, 4 spermathecae may have been present with pores at 7/8–8/9 perhaps even in region of *AB* or slightly laterally, at least in some of the forms. Tubercula pubertatis and associated TP glands probably also were present as well as genital setae, perhaps of the common European sort, along with GS glands. Species may have differed from each other by location of a 5 to 8 segment clitellum in various parts of a region comprising xx–xl, by the segments in which genital setae and their associated glands were present, and presumably also by size and segment number.

Remarks: *Bimastos*, much better than any other Oligochaete genus now known, testifies to the zoological devastation wrought by the Pleistocene glaciation.

Bimastos parvus

1874. *Allolobophora parva* Eisen, *Öfvers. K. Vetens. Akad. Förhandl.* **31**, 2 : p. 46. (Type locality, Mount Lebanon, New York-New England. Types, in the U. S. Natl. Mus.)
1930. *Bimastus parvus*, Gates, *Rec. Indian Mus.* **32** : p. 352.
1931. *Bimastus parvus*, Gates, *ibid*, **33** : p. 433. Stephenson, *ibid.* **33** : p. 201.
1933. *Bimastus parvus*, Gates, *ibid.* **35** : p. 604.
1961. *Bimastos parvus*, Gates, *Burma Res. Soc. 50th Anniv. Publ.* No. 1 : p. 57.
1956. *Bimastus parvus*, Gates, *Bull. Mus. Comp. Zool. Harvard College* **115** : p. 6.

BACTEROIDS: Knop, 1926, *Zeitschr. Morph. Oekol. Tiere* **6** : p. 601. BEHAVIOR: Swartz, 1929, *Jour. Comp. Psychol.* **9**. CALCIFEROUS GLANDS: Stephenson & Prashad, 1919, *Trans. Roy. Soc. Edinburgh* **52** : pp. 457, 474. Smith, 1924, *Illinois Biol. Monogr.* **9** : p. 31 CASTINGS: Teotia *et al.*, 1950, *Res. Bull. Nebraska Agric. Exp. Sta.*, No. 165 : pp. 2, 7, 12, 14. COCOON FORMATION AND DEPOSITION: Parshad, 1916, *Jour. Bombay Nat. Hist. Soc.* **23** : p. 494. PHARYNGEAL (septal) GLANDS: Stephenson, 1917, *Quart. Jour. Micros. Sci.* **62** : p. 274. PARASITES.: Scott, 1913, *Science* **38** : p. 672. POPULATIONS: Teotia *et al.*, 1950, *Res. Bull. Nebraska Agric. Exp. Sta.*, No. 165 : p. 2.

Clitellum, saddle-shaped, xxiv-xxx. Male tumescences, in a median portion of *BC*, restricted to xv or extending into xiv and/or xvi. (Genital tumescences and setae, as well as GS glands, lacking.) Setae, present from ii, *AB* slightly > *CD*, *AA* somewhat > *BC*, *DD* ca. = or slightly > ½*C*. Prostomial tongue, open. Segments, 85–124. Size, 23–46 by 2–3 mm.

Septa, none markedly thickened, muscularity decreasing through 6/7–13/14. Typhlosole, simply lamelliform. Seminal vesicles, small. Epididymis, lacking or represented only by sinuosities or by a single *U*-shaped loop just behind funnel septum. Atrial glands, supraparietal and extending into xiv and xvi or rudimentary and then confined, or almost so, to the parietes.

Reproduction: Sperm rarely are present on male funnels and most specimens appear to be male sterile. Spermatophores are unknown. Whether reproduction always is parthenogenetic or whether there is occasionally some possibility of amphimixis remains to be determined.

Distribution: Thandaung, Kalaw, Tanungyi, Maymyo, all on the Shan Plateau.

India, the Himalayas from Darjiling west into Kashmir and Northwestern Frontier Province of Pakistan, down into Rajputana. Allahabad in the Gangetic Plain, Kodaikanal in the Palni Hills of South India. Cameron Highlands, Malaya.

Tibet, China (Szechuan, Chekiang, Kiangsi, Kiangsu provinces), Manchukuo, Korea, Japan. Java (Malan). Australia. Tahiti (first record, elevation unknown). Hawaii.

United States; Washington, California, Montana, Nebraska, Kansas, Texas, Arkansas, Louisiana, Illinois, Michigan, Ohio, Maryland, New York, Massachusetts. Mexico, Guatemala, Costa Rica. Brazil (Nova Teutonia). Argentina.

Iceland, Denmark, Germany, England, Wales, Portugal, Spain, Switzerland, Italy, Corsica, Rhodes, Hungary, Bulgaria, Romania, Russia.

Southwest Africa, South Africa, Mauritius. St. Helena (new record). St. Paul (Indian Ocean).

Afghanistan. Kazakstan and Central Asia. Siberia.

Each of the Burmese sites, at which the species was secured during twenty years of collecting, is a hill station (as also in case of Malaya) where domiciled Europeans had introduced ornamental and other cultivated plants. In that same way, *B. parvus* was taken by man, even to the isolated rock of St. Paul. Elevations at which the species have been recorded, Simla Hills—5,000–6,000 ft., Naini Tal—7,000 ft., Tibet—11,000 ft., Kashmir—9,000 ft., Costa Rica—2,100 m., and its absence from tropical lowlands to which it must also have been introduced frequently, testify to a physiology evolved in subarctic climates and since unchanged.

North America long has been thought to be the original home of the species, but no area of endemicity has yet been recognized. Nor in that continent is any biparental population known from which *parvus* could have been segregated by its parthenogenesis.

Specimens intercepted at American ports were from Taiwan, Japan, Australia, Mexico, England, and Italy.

Habitats: Soil, with *p*H of 7.5 and wetted by waste effluents from human habitations, near water, of wooded areas, gardens, fields. Under logs and decaying grass. Litter in caves. Humus, moss, manure, dumps. Sapling beds of nurseries, greenhouses, earth around roots of potted plants.

Biology: Hints as to some physiology of the species are provided by the distribution and the elevations at which individuals have been taken. Some part of the distribution may have been obtained by aboveground wanderings, rather than by migration through the soil. However, the only record found in the literature is of an individual crawling along garden walls in rainy weather.

Activity, presumably including breeding, seems to be possible year-round in favorable conditions. Cocoons may contain as many as four eggs. If more than one hatchling emerges from a concoon, dispersal through a new area presumably will be more rapid.

Castings: Small, size according to Teotia *et al.* (1950), 0.5–1.0 mm., usually spheroidal to spindle-shaped. Average per diem production, in Nebraska (Teotia), 0.27 gram, by worms with an average weight of 0.49 gram. The castings were found to be several times more stable than control soil.

Parasites: NEMATODA: *Ascaridium perspicillum* (Rudolphi, 1803) Dujardin, 1845.

Remarks: Slight but probably unimportant differences in appearance of the epidermis near ventral margin of a saddle-shaped clitellum have been responsible for records of tubercula pubertatis.

As already indicated above, *B. parvus* is not a species but an aggregate of morphs or clones. Until recently records of *B. parvus* could be assumed to refer to worms with a clitellum limited to xxiv–xxx as the taxon was so defined by Michaelsen (1900), Smith (1917) and Stephenson (1923) though the latter recognized occasional slight invasion of xxxi as well as absence in xxiv. Subsequently, *B. eiseni*, with a clitellum extending through xxxii or xxxiii, was mistakenly suppressed as a synonym of *parvus* which was transferred (Pop, 1941, 1949) to *Eisenia*. Some recent reports of *B. parvus*, in absence of evidence as to generic affiliation, accordingly require confirmation.

Closest relationships of *B. parvus* are with *B. beddardi* (Michaelsen, 1894) and *B. longicinctus* Smith & Gittins, 1915, hitherto distinguished from each other by location of the clitellum. That organ extends in *B. beddardi*, as defined by Michaelsen (1900), through xxiv or xxv–xxxi or xxxii and in *B. longicinctus* through xxiii, xxiv–xxxii, xxxiii. However, English specimens referred to *B. beddardi* have had locations as follows: xxiii, xxiv, xxv–xxxi, xxxii. The clitellum of a Tibetan specimen referred by Michaelsen to *B. beddardi* was in xxiii–xxxii exactly as in *B. longicinctus* but of a Japanese specimen doubtfully referred to *B. parvus* by Yamaguchi (1953) was in xxii–xxxi and so was one segment further forward than even in *B. longicinctus*. Pakistan specimens of *Bimastos* (Gates, MS) have the clitellum in xxiii–xxix, xxiii–xxx,[1] xxiv–xxx, xxiv–xxx/n. Distinguishing *B. parvus*, *beddardi* and *longicinctus* from each other does not seem possible from information now available and all three of the taxa may be morphs isolated by their parthenogenesis from a single biparental population no longer extant. The précis above, however, characterizes *B. parvus*

[1] Also one specimen from a Texas cave.

as it was found by the author in India and Burma. The distribution, on the contrary, includes all records of *B. beddardi* as well as of *B. parvus*.

Dendrobaena

1874. *Dendrobaena* Eisen, *Öfvers.-Akad. Förhandl. Stockholm* **30**, 8: p. 53. (Type and only species, *D. Boeckii* n. sp. = *Enterion octaedrum* Savigny 1826.)

Somatic characters of the type species that may prove to be definitive at a generic level are as follows:

Calciferous gland, without sacs opening into gut at vicinity of 10/11, markedly moniliform in xi–xii. Gizzard, mainly in xvii. Extra-esophageal trunks, passing to dorsal trunk in vicinity of 9/10. Hearts, in vii–ix. Nephridial bladders, ocarina-shaped, with bluntly rounded end laterally and pointed end mesially, ventral side funnel-shaped and narrowing to pass into parietes at *B*. Nephropores, obvious, behind first few segments in one rank on each side, just about *B*. Setae, not closely paired. Longitudinal musculature, pinnate. Pigment, red.

Genital characters of *D. octaedra* and related species that might in part characterize the group at generic level:

Holandric, testes free. Seminal vesicles, three pairs, in ix, xi, xii. Atrial, TP and supraparietal GS glands, present. Spermathecal pores, in region of *CD*.

Also included in *Dendrobaena* for many years is another of Savigny's species, *rubida*. Characters of the taxon that may prove to be definitive at a generic level are as follows:

Calciferous sacs, in x, high, lateromesially flattened and bound against the esophagus (which there has a vertically slitlike lumen), opening into gut posteriorly close to insertion of 10/11. Calciferous gland, not markedly moniliform in xi–xii where its lumen is vertically slitlike but then widened in xiii. Gizzard, mainly in xvii. Extra-esophageal trunks, passing to dorsal trunk posteriorly in xii. Hearts, in vii–xi. Nephridial bladders, *U*-shaped, open mesially. Nephropores, inconspicuous, behind first few segments irregularly alternating with asymmetry between two levels on each side, one just above *B*, the other above *D*. Setae, not closely paired. Longitudinal musculature, pinnate. Pigment, red.

Genital characters, of course, must be assumed for the present to be much the same as for a radically revised *Dendrobaena*, inasmuch as both *octaedra* and *rubida* presently are congeneric almost only because of similar genitalia.

Location of nephropores in a single longitudinal row on each side of the soma was found, by Pickford, as long ago as 1937, to characterize all South African Acanthodrilids. In all of certain lumbricid material (*D. octaedra* and *Octolasion* spp.) examined by the present author, no individual variation was found with respect to the same character. Inconspicuous nephropores in two longitudinal rows were associated, in Pickford's South African acanthodrilids, with absence of nephric vesicles but, in the present author's collections of *D. rubida* invariably were associated with 4 longitudinal rows (2 on each side of the soma) as well as with existence of U-shaped vesicles. Accordingly, even the nephropores alone, sometimes do provide characters of real systematic usefulness.

Dendrobaena octaedra

1826. *Enterion octaedrum* Savigny, *Mem. Acad. Sci. Inst. France* **5**, (Hist. Acct.): p. 183. (Type locality, Paris. Types, in the Mus. Hist. Nat. Paris.)

1958. *Dendrobaena octaedra*, Gates, *Breviora, Mus. Comp. Zool. Harvard*, No. 91, p. 5.

CALCIFEROUS GLANDS, etc.: Smith, 1924, *Illinois Biol. Monogr.* 9, 1; 1928, *Rept. Sci. Results Norwegian Exped. Novaya Zemlya* 1921, No. 41. COCOONS: Svendsen, 1956, *Ann. Mag. Nat. Hist.*, Ser. 12, 9. CHROMOSOMES: Omodeo, 1952, *Arch. Zool. Italiano* **37**; *Mem. Mus. Sto. Nat. Verona* **4**; 1957, *Meddel. Gronland*, **124**, 6. DESSICATION RESISTANCE: Ferroniere, 1901, *Bull. Soc. Sci. Nat. Ouest France*, Ser. 2, 1. ECONOMIC IMPORTANCE: Ribaucourt & Combault, 1906, *Rev. Gen. Agron.* 1. ENZYMES: Nielsen, 1962, *Oikos* **13**. EVOLUTION: Omodeo, 1956, *Proc. XIV Internatl. Congr. Zool.*, p. 152. MUSCLES: Pool, 1937, *Acta Zool.* 18. SEGMENT NUMBER: Backlund, 1949, *Zool. of Iceland* **2**, 20; 1951, *Acta Univ. Lund* **46**, 2. Baltzer, 1956, *Zool. Jahrb. Syst.* **84**. Harman, 1960, thesis. Omodeo, 1955, *Atti. Soc. Sci. Nat. Pisa. Mem.* **62**, B; 1961, *Mem. Mus. Civ. Sto. Nat. Verona* 9. Plisko, 1961, *Fragmenta Faunistica, Warsaw* **8**. Pop, 1947, *An. Acad. Romane Mem. Sect. Sti.*, Ser. 3, **22**, 3. SYSTEMATICS: Beddard, 1895, *A Monogr. of the Order of Oligochaeta (Oxford)*. Bretscher, 1896, *Rev. Suisse Zool.* **3**. Cernosvitov & Evans, 1947, *Linnean Soc. London, Synposes British Fauna*, No. 6. Michaelsen, 1900, *Das Tierreich* 10. Olson, 1933, *Ohio Jour. Sci.* **23**. Ribaucourt, 1896, *Rev. Suisse Zool.* **4**. Rosa, 1893, *Mem. Acad. Sci. Torino* **43**. Smith, 1917, *Proc. U. S. Natl. Mus.* **52**. Tetry, 1938, *Contribution a l'étude de la fauna de l'est de la France (Lorraine)* (Nancy). RESPIRATION (rate): Byzova, 1965, *Rev. Ecol. Biol. Sol* 2.

Sexthecal, pores, at or near D, at 9/10, 10/11, 11/12. Clitellum, saddle-shaped, 5–6 segments, xxix, xxix/n–xxxiii/n–xxxiii, xxxiv/n, xxxiv usually 5 segments, xxix–xxxiii. Tubercula pubertatis, longitudinal bands of translucence slightly depressed, just lateral to B, in xxxi–xxxiii. Male pores, at bottom of equatorial clefts, at or near mBC. Male tumescences, large, protuberant, confined to xv and BC but often slightly dislocating 14/15 and 15/16. Female pores, at eq/xiv just lateral to B. Genital tumescences, usually around a and around b setae, in xvi, some or all of xxix–xxxiii, increasingly rarely in one of xxviii, xxii, xxiii or around c in one of ix–xii. Setae, not paired, present from ii, B $ca.$ at mL, D just lateral to level of furrows bounding prostomial tongue, BC smaller than other intervals, DD behind clitellum often smaller than AA. Nephropores, obvious, in iii–iv, v, or vi, at or above C posteriorly in two longitudinal ranks, each somewhat above B. First dorsal pore, at any of 4/5–6/7 but usually at 5/6. Body section, behind clitellum almost circular to nearly octagonal. Color, red, dark red to slate. Segments, 59–109, usually 90–101. Size, 20–60 by 2.75–5 mm.

Septa, all delicate. Pigment, red, mostly associated with circular muscle layer but in i–x, xi often also in an outer portion of the inner muscular layer. Longitudinal musculature, pinnate. Blister organs, in CD, ix–xiii. Calciferous gland, without sacs, opening into gut near insertion of 10/11, lamellae widest in xi–xii and behind an internal equatorial constriction in xiii gradually narrowing. Esophagus widened and markedly moniliform, with vertically slitlike lumen in xi–xii. Intestinal origin, in xv. Gizzard, mostly in xvii (ending at?). Intestine, valvular in xix or region of insertion of 19/20. Typhlosole, beginning in region of xx–xxiii, rather high, anteriorly slightly widened ventrally so as to have a sectional shape transversely of an inverted T, with a flat ventral face (and then sometimes with a median groove) or merely concave, ending abruptly in region of 73rd to 90th segments, leaving 11–17 metameres atyphlosolate. (Dorsal trunk, aborted in front of 6/7?). Extra-esophageal trunks, passing up to dorsal vessel in region of insertion of 9/10. Hearts, in vii–ix. Nephridial bladders, vertically ocarina-shaped, in AD or reaching beyond D, pointed end laterally, bluntly rounded end mesially. Holandric. Seminal vesicles, 3 pairs, in ix, xi, xii. Sperm ducts, without epididymis. Spermathecae, stalked, free in ix, x, xi. TP, GS and atrial glands, present.

Reproduction: Phagocytosis of sperm, by lymphocytes, in seminal vesicles and/or the coelom was described by Cernosvitov (1930) in addition to degeneration of sperm in cytophores and resorption of sperm in spermathecal epithelium. Two of the worms isolated by Janda and Gavrilov (1939) before sexual maturity laid viable cocoons. Sperms were lacking in the spermathecae. Mature (and presumably normal) sperm were found in x–xi of one worm. Whether self-fertilization or parthenogenesis had been involved was not determined.

Specimens from the Dolomites, Italy, were found (Omodeo, 1952, 1953) to be male sterile, with abortive spermatogenesis or with sparse maturation of male gametes varying in size (and abnormal?). Chromosome numbers, 120–130, regarded as hexaploid or heptaploid ($7 \times 18 = 126$), in the Dolomites, 108, hexaploid, in Greenland. Lesser "ploids" are unknown.

Sperm occasionally have been seen in spermathecae and were there presumably as a result of copulation. One such fertilized individual had 108 mitotic chromosomes (Omodeo, 1957) and pseudogamy was suggested.

Parthenogenesis clearly is widespread in *D. octaedra*. Whether any individuals characterized as in the above précis have biparental reproduction remains to be proved.

Distribution: Alaska, British Columbia, Alberta, New Brunswick, Labrador, Newfoundland.

Washington, California, Colorado, Nebraska, Arkansas, Wisconsin (driftless section), Illinois, Indiana, Michigan, Ohio, Virginia, North Carolina (Asheville. At $ca.$ 2,000 ft.) Pennsylvania, New Jersey, New York

(Lake Placid only), Vermont, Massachusetts, Connecticut, New Hampshire, Maine.

Mexico (Mexico City. At *ca.* 7,800 ft.?) Colombia, Bogotá, at 2,300–2,800 m.

Greenland, Iceland and Grimsey Island. Norway, Sweden, Lapland, Finland. Russia, Kolguyev, Nova Zemlya, Murmansk, Archangelsk south to Moldavia, through the Caucasus and into Transcaucasus, east to the Urals, Pri-Urals, including Byelorussia as well as Ukraine and Crimea, Estonia, Latvia, Lithuania. Poland, Germany, Denmark, Hebrides, Scotland, Ireland and Clare Island, England, Jersey (Channel Islands), Netherlands, Belgium, Czechoslovakia, Romania, Bulgaria, Jugoslavia, Hungary, Austria, Switzerland, France, Italy (northern part but excluding the peninsula), Corsica, Balearic Islands, Portugal, Madeira.

Siberia, Tomsk to Lake Baikal, Transcaucasus, Kazakstan, Chuvash and Tatar (Kirgiziya) Republics. (Afghanistan?) Pakistan, Baluchistan. India, Kodaikanal in Palni Hills of South India (at 6,000–7,000 feet).

Characterization of the distribution (Cernosvitov & Evans, 1947) as "whole of Europe, . . . North and South America" obviously is incorrect even for Europe. Such abbreviated statements presumably are responsible for some of the misunderstandings about lumbricid ranges. "Euro-american" does indicate the present distribution but is just as applicable to the Colorado potato beetle.

Elevations not mentioned above are: 900–2,800 m. Italy. 890–925 m. Corsica. 1579 m. Austria and Hungary. 1,200–2,600 m. Switzerland. 2,000–3,000 m. at Valais. To 2,510 feet on Clare Island, to 1,900 m. in the Austrian Alps, to 1,400 m. in the Bavarian Alps, to 1,940 m. in Bulgaria, 950–1,000 m. in mid-arctic grass heaths, at 950–1,200 m. in high altitude heaths without continuous plant cover, Sweden.

D. octaedra has been found in more localities of Finland than any other megadrile. The species is said to be the dominant earthworm on the Hebrides and is the only megadrile in the northern parts of Sweden and Finland.

Restriction to the northern hemisphere, among the better known peregrine lumbricids, is unique. Much of the range, on various occasions, has been attributed to spontaneous distribution. Presence in Greenland, at sites said to be far away from present settlements, as well as in districts inhabited by the Norse ten centuries ago, was thought to support migration from Europe to America across a land bridge that no longer exists. A supposition that *octaedra* is haemerophobic may also have predisposed thinking toward spontaneous distribution.

Eurytopic, or widely distributed, in the northern hemisphere, as a characterization of the present *octaedra* range, seems less likely to be misunderstood than "cosmopolitan." The original home, as for other anthropochores, is unknown.

Interception, during the last fifteen years, at American ports, of 50 specimens of *D. octaedra* with plants from Norway, Sweden, Denmark, Germany, Russia, Czeckoslovakia, Netherlands, Belgium, England, Ireland, Spain (no previous record of the species from that country), Portugal, Canada, Mexico, shows that the species is not as haemerophobic as may have been thought. Presence in Mexico, Colombia, and a southern portion of the Indian peninsula, in absence of proof for any other method of carriage, is to be regarded as result of transportation by man. Even more likely, is human introduction to various parts of the Holarctic, perhaps originally from a region including the Caucasus where a center of Lumbricid evolution has been thought to have existed.

D. octaedra, whether it survived Pleistocene ice ages in Greenland or but the last millennium or only the last century or two, clearly is uniquely adapted among megadriles for conditions prevailing in Nova Zemlya, Greenland, arctic portions of Finland, Sweden, Norway, and by that very adaptation seemingly is inhibited from colonizing tropical lowlands. Adaptation to arctic and subarctic climates presumably was not acquired in a region with tropical or subtropical climate.

Habitats: Under stones, logs, bark, roots, moss on rocks and on trees. Buried in moss. In peat moss, sphagnum, rotten stumps, logs and wood, fallen tree trunks, dead leaves, compost, raw humus of deciduous and coniferous forests, pine litter, drift, wrack along seashore, manure, dung pats. Raw and natural mull soils, sandy soil, podsolized soil, soil so saturated that water could be squeezed out from a handful, soil with a *p*H of 3.0–7.7 (6.8 said to be optimal). Taiga, forests, steppes, oak, pine, ash, linden, spruce, mixed forests "of all sorts," orchards. Swamps, stream banks, trout streams, "torrents." Various sites little affected by culture, including meadow flood plains, hill pasture (Scotland). Cultivated fields rarely, Kiel. High mountains, Scandinavia, where there is no continuous plant cover. Mountain tops, France, where no other megadrile lives. Caves, Italy, Slovakia.

Garden (Bogotá). Botanical gardens, Kew, Berlin. Arboretums (Boston, San Francisco). From soil or other material with potted plants: *Azalea indica*, *Gerusta lusitanica*, *Ligustrum ovalifolium*, *Phillodendron amurense*, *Humulus* sp., *Viburnum* sp., *Acer* sp., *Ribes* sp., spruce, ash, roses, geraniums, begonia cuttings, lily bulbs, holly, heather, evergreen seedlings.

Restricted to the Deschampsietum, where it shares dominance with *O. lacteum*, Kama floodplain of Gouv. Perm, Russia. Sometimes sharing dominance with *L. rubellus* in untilled sandy loam, sea meadow, sharing dominance with *L. rubellus* and Levinsen's *eiseni* in a grass-moor and a heather-moor community. Dominance is shared with *Eisenia nordenskiöldi* (Eisen, 1879) in meadow saline and meadow marshy saline soils of the Troitsk district of the Urals and is second only to *E. nordenskiöldi* in the A_0 and A_1 horizons of alkaline tchernozem. On the Pennines, with Levinsen's *eiseni*, forms the bulk of the population in moor and bog, giving way to *D. rubida* and *L. rubellus* on "better soils." *D. octaedra* also has been found in grazed as well as ungrazed maritime grassland.

Presence of individuals, near Saratov on a branch of the Volga, at a depth of 10 meters was later said to be the result of an accident. However, Russians did find that *D. octaedra* withstands waterlogging and in that respect was compared to *E. tetraedra*.

The "least exacting" of the Iceland megadriles often is said to be characterized by great ecological valence or able "to adapt itself to all conditions of life" and has been called "the pioneer worm in unfavorable sites." Probability of *D. octaedra* being taken indoors would seem to be nearly as great as in case of *D. rubida* but like that taxon never has been recorded from greenhouses, the temperature of which may be too high.

Biology: Activity, and presumably breeding also, is year-round in appropriate conditions. However, in central Maine, as well probably as in most North American colonies, imposition of a winter and of a summer rest breaks the earthworm year into two breeding periods. In arctic and subarctic conditions, a single, short summer breeding season, and a long hibernation are expected. Nothing is known about the nature of the inactive state.

The species often is said to be surface living, i.e., inhabiting the upper layers of soil, in the Netherlands mostly in the litter layer, less in the H layer and very rarely in the sand layer. Upper layers supposedly are abandoned only to go down deeper for estivation or hibernation in a special mold-lined chamber, but aboveground wanderings perhaps should be anticipated. Feeding is selective, little or no sand and rock particles being found in the intestine. Beech leaves are consumed except for stalks and veins but undigestible plant fibers, up to 15 mm. long, have been seen in the gut.

Variation and Abnormality: Number of segments, was given as 95 by Beddard (1895) though Rosa (1893) had indicated a range of 80–95 which was copied by Ribaucourt and by Bretscher (1896), Michaelsen (1900), Tetry (1938), Cernosvitov & Evans (1947), Smith (1917), Olson (1933). Counts up to 100 however had been made by Friend, and recently the limits have been still further extended presumably as a result of actual counts of material from a number of sources, by Pop (1947) of 65–100 and 54–100, by Backlund (1949, 1951) of 59–100 and 55–97, by Baltzer (1956) of 63–109, by Harman (1960) of 61–103, by Omodeo (1955, 1957, 1961) of 64–109, 61–108, by Plisko (1963) of 70–104. Bimodality of Omodeo's curves (1955), at least in part, probably is due to posterior amputation which is frequent. Unfortunately, distinguishing old posterior amputees often is impossible

without a dissection and that too may yield no clues. Halving of mesoblastic somites, without obvious indications externally, does sometimes provide abnormally high numbers.

Much less is known about variation in other characters. Strong contraction can produce an appearance of the tanyloby that occasionally was reported.

Abnormalities that must often have been seen, once said to be "innombrables," and that occasionally were recorded, doubtless were mostly of the kinds that now must be expected to appear after reproduction becomes uniparental. Presence of one or two hearts in x, recorded by Smith (1928) who may have been the first to mention the blister organs, clearly is a partial reversion to a more or less distant ancestral condition, but whether loss of hearts in x–xi preceded or followed establishment of parthenogenesis is unknown.

Polymorphism: Parthenogenetic polymorphism is rampant in *D. octaedra*, probably more so than in any other lumbricid if not also than in any other megadrile. The morphs already found are too many even for a mere listing here and further consideration of this subject is left to a subsequent occasion.

Regeneration: Head regeneration is possible at each level back to 9/10, perhaps also to 12/13. The maximum number of segments was 5, in regenerates at 6/7 and 7/8. Damage to the buccal region of a New Jersey specimen had resulted in a bifid anterior end, with two symmetrical peristomia, each with a normal epilobous prostomium.

Tail regenerates have been seen at several levels behind 37/38. Maximum number of segments was three, in regenerates at 44/45 and 76/77. Formation of the pygomere may have been epimorphic but the other one or two segments, in each regenerate, probably were morphallactic. Few of the posterior amputees had regenerated and some of the brevicaudate individuals (frequently found) had the clitellum in the middle of the body or even in the posterior half.

Parasites: CILIATA: None were found in German material. An undentified species of *Anoplophrya* was found in the gut at Prag. SPOROZOA: Usually were not found and the monocystids seen on a single occasion were not identified. CESTODA: *Cysticercus dendrobaenae* Tetry, 1938, coelomic cavities, France. Cysts, found by the present author in coelom, in or on wall of gut, were identified as *Cysticercus* sp. Some were believed (according to Dr. Allen McIntosh, *in lit.*) to be of *Southwellia ransomi* Chapin 1926, a parasite of the robin, *Turdus migratorius*.

NEMATODA: *Rhabditis pellio* Schneider, 1866, Germany (according to Völk, 1950). *Capillaria caudinflata* (Molin, 1858) Travassos, 1915.

Economic Importance: Individuals allowed to crawl (Rabaucourt & Combault, 1906) on bare mountain rocks, left a viscous trail, the site of which was recognized a year later by the lichens growing thereon, thus supposedly initiating a lichen-moss-vascular plant succession.

D. *octaedra* is believed to be chiefly responsible for decomposition of oak leaves in the Kursk region (Russia), where it is regarded as an important converter of leaf substances.

The species was not stored by *Talpa europaea* (meadow soil in vicinity of Cracow) though present in the soil.

Systematics: The clitellum of *D. octaedra*, according to Savigny (1826, 1828) occupies five segments, xxix–xxxiii but according to Beddard (1895) it may also extend through xxxiv. One of the cotypes more recently was found to have the clitellum in xxix–xxxiv. Clitellar glandularity in xxvii and xxviii had been reported and those segments are included by Michaelsen (1900) in his definition of the species. Subsequent authors, Smith, (1917) Tetry (1938), Cernosvitov & Evans, (1947), Pop (1950), Gerard (1964), seemingly without further investigation followed Michaelsen in their definitions of *D. octaedra*. All specimens with the clitellum extending through xxviii seen by the author were quadrithecal and appeared to be referable to *D. attemsi* Michaelsen 1902.

Relationships of *D. octaedra* to *D. attemsi* as well as of each of them to other taxa never have been determined. A biparental, diploid population of *D. octaedra*, as already indicated, is unknown. Parthenogenesis, even in all male-fertile morphs of *D. octaedra*, may be optional. If so, that method of reproduction may have permitted, just as in *E. rosea*, segregation of a sexthecal population from one that was quadrithecal, perhaps much like *D. attemsi*. Tubercula of *D. attemsi* are in xxx–xxxii, a location occasionally recorded for *D. octaedra*. The clitellum of *D. attemsi* has been said to occupy xxviii–xxxiii, xxviii–xxxiv, xxix–xxxiii, xxix–xxxiv, a range perhaps approximating that to be expected in an amphimictic population. Spermathecal pores of *D. attemsi* usually were said to be in 9/10–10/11 but for worms from Turkey the location recently was recorded as 10/11–11/12. Such a variant condition now seems more likely to have arisen by deletion of the anterior pair of *D. octaedra* spermathecae than by a transfer posteriorly of the entire battery of *D. attemsi*. Until much more is known about individual variation and reproduction in *D. attemsi*, further consideration of relationships is unlikely to be rewarding.

Origin in the Miocene has been suggested (Omodeo, 1956) as well as in age of 25 million years for the parthenogenetic mutation.

For further comment cf. *D. rubida* (Systematics).

Remarks: Among the species that now seem likely to belong in a somatically defined *Dendrobaena* are *mariupoliensis* Wyssotsky, 1898, *pygmaea* (Savigny, 1826), *samarigera* (Rosa, 1893) and some of the *byblica* complex (Rosa, 1893).

Dendrobaena rubida

1826. *Enterion rubidum* Savigny, *Mem. Acad. Sci. Inst. France* 5, (Hist. Acct.): p. 182. (Type locality, Paris. Types, in the Mus. Hist. Nat. Paris.)
1958. *Dendrobaena rubida*, Gates, *Breviora, Mus. Comp. Zool.* Cambridge, No. 91: p. 6.

ABNORMALITY: Smith, 1922, *Trans. American Micros. Soc.* 41; 1925, *ibid.* 44. CALCIFEROUS GLANDS: Smith, 1924, *Illinois, Biol. Monogr.* 9. CHROMOSOMES: Muldal, 1949, *Proc. Linnean Soc. London* 161; 1952, *Heredity* 6. Omodeo, 1955, *Caryologia* 8; 1957, *Meddels. Grönland* 124, 6. COCOONS: Evans & Guild, 1948, *Ann. Mag. Nat. Hist.*, Ser. 11, 14. Svendsen, 1956, *Ann. Mag. Nat. Hist.*, Ser. 12, 9. DEVELOPMENT: Menzi, 1919, *Rev. Suisse Zool.* 27. Svetlov, 1928, *Trav. Lab. Zool. Sta. Biol. Sebastopol*, Ser. 2, 11–13. ENZYMES: Tracey, 1951, *Nature* 167. HISTOLOGY: Ude, 1885, *Zeitschr. Wiss. Zool.* 43. LOCOMOTION: Roots, 1956, *Nature* 177. MUSCLES: Pool, 1937, *Acta Zool.* 18. Pop & Dragos, 1959, *Stud. Univ. Babes-Bolyai*, Ser. 2, 2 (Biol.). PARASITES: Meier, 1956, *Arch. Protistenk.* 101. REGENERATION: Gates, *Breviora*, No. 61. REPRODUCTION: Evans & Guild, 1848, *Ann. Applied Biol.* 35; *Ann. Mag. Nat. Hist.*, Ser. 11, 14. Ribaucourt, 1897, *Bull. Sci. France, Belgique* 30. RESISTANCE TO FREEZING: Sekera, 1896, *Zool. Anz.* 19. TYPHLOSOLE: Hertling, 1923, *Zeitschr. Wiss. Zool.* 120. WATER RELATIONSHIPS: Roots, 1956, *Jour. Exp. Biol.* 33.

Quadrithecal, pores at or near *C*, at 9/10, 10/11. Clitellum, saddle-shaped, xxv, xxv/n, xxvi–xxxi, xxxii/2, xxxii. Tubercula pubertatis, longitudinal bands just lateral to *B*, usually with 1 or 2 longitudinal grooves therein, in (xxvii very rarely), xxviii–xxx. Male pores, at bottom of shallow equatorial clefts, at or near m*BC*. Male tumescences, confined to xv, usually *U*-shaped and open laterally, reaching nearly to *B* but not so close to *C*. Female pores, at eq/xiv just lateral to *B*. Genital tumescences, around *a,b* only, most frequently in xvi, xxiv, xxvii, xxxi, xxxiii–xxxiv or xxxv, but occasionally in one or more of viii–xi, xvii–xxiii, xxv, xxvi, xxviii, even xxix and xxxii though but very rarely. Setae, present from ii, paired, *AB* ca = or < *CD* < *BC* < or = *AA* < *DD* < ½*C*, *d* setae usually above m*L* throughout. Nephropores, inconspicuous, of iii–v at or close to *D*, posteriorly alternating irregularly and with asymmetry between two levels, just above *B* and above *D*. First dorsal pore, usually at 5/6, rarely at 4/5 (3/4?) or 6/7. Prostomium, epilobous, tongue open. Body section, slightly compressed dorso-ventrally behind the clitellum, from nearly transversely elliptical but with ventrum slightly more flat, to almost oblong. Color, light to dark red, slate, some segments in front of the pygomere yellowish (owing to massive accumulations of coelomocytes). Segments, 74–110, usually 87–107. Size, 15–50 by 1½–5 mm.

Septa, none thickly muscular, 5/6–12/13 or 13/14 somewhat strengthened, (17/18 fenestrated?). Pigment, red, associated with circular muscle layer, lacking immediately underneath intersegmental furrows. Longitudinal muscle, pinnate. Calciferous sacs, in x, high, latero-mesially flattened and bound against esophagus, opening into gut posteriorly, close to insertion of 10/11. Esophagus, not markedly moniliform in xi–xii where lumen is vertically slitlike as also in x, lamellae gradually narrowing behind xii. Intestinal origin, in xv. Gizzard, mostly in xvii (ending at?). Intestine, rather valvular at 18/19 or in xix. Typhlosole, beginning in or behind xx somewhat thicker ventrally, with a median longitudinal groove on the ventral face, ending abruptly in 66th to 100th segments, leaving 8–14 segments atyphlosolate. Extraesophageal trunks, passing up to dorsal trunk in xii. Hearts, in vii–xi. Nephridial bladders, each in a hairpin-shaped loop, open mesially, slightly shorter ectal limb often posterior. Holandric. Seminal vesicles, in ix, xi, xii, the last pair apparently extending through xiii but actually within pockets of 12/13. Sperm ducts, with an epididymis of one long hairpin loop or 2–3 shortly *U*-shaped loops. Spermathecae, usually in ix, x, ampulla medium sized and nearly spheroidal or ovoidal, large and vertically ellipsoidal, duct slender, shorter than ampulla but coelomic or at least above the parietes and within the associated septum. TP, GS and atrial glands, present, the latter extending through xiv, xvi, constricted by 14/15, 15/16.

Reproduction: Individuals reared in isolation to sexual maturity (Evans & Guild, 1948), of athecal as well as of thecal morphs, laid viable cocoons, but only nine months after attaining maturity in case of thecal individuals.

Polyploidy is common, perhaps more so than diploidy which has been reported for specimens from Iceland, Italy, and Turkey. Chromosome numbers, 34 in diploids, 48 hypotriploid (−3) England, 102 hexaploid Greenland, Iceland and Italian Alps, 115–120 hypo-octoploid Italy. Reproduction in diploids and some hexaploids is said to be amphigonic or in case of some Iceland hexaploids pseudogamic. Maturation and exchange of sperm no longer can be regarded always as proof that reproduction is biparental.

Male sterility and absence of spermathecae is common and in many of such morphs parthenogenesis must be obligatory. Parthenogenesis is said to be optional, even for the large thecal and tetraploid morphs.

Distribution: New South Wales, Queensland, southwest Australia, west Australia. Auckland, Campbell, Kermadec, Stewart, Sunday Islands and New Zealand.

Hawaii.

Alaska, British Columbia including Queen Charlotte Island, Alberta, Saskatchewan (new record, on an interception), Manitoba, Ontario, Quebec, New Brunswick, Nova Scotia, Labrador (new record). Newfoundland.

Washington, Oregon, California, Utah, Colorado, Missouri, Arkansas, Louisiana, Illinois, Michigan, Indiana, Alabama, Kentucky, Tennessee, Ohio, West Virginia, Virginia, Pennsylvania, New Jersey, New York, Vermont, Massachusetts, Connecticut, New Hampshire, Maine.

Mexico, Guatemala, Colombia (Medelín and Bogotá, 1,600–3,600 m), Ecuador, (Río Bamba, Loja), Brazil (Petropolis and Rio de Janeiro), Chile (numerous localities) and Juan Fernandez Island, Tierra del Fuego, Falkland Islands, Argentina, Tristan da Cunha, Uruguay.

Greenland and Disko Island, Iceland and Grimsey Island. Lofoten Island, Norway, Sweden, Lapland, Finland, Russia, from the shores of the White Sea through Ukraine into Moldavia and Caucasus, including Urals, Pri-urals, Byelorussia, as well as Crimea and the Ukraine. Estonia, Latvia, Lithuania, Poland, Germany, Borkum Island, Denmark, Hebrides, Islay Island, Scotland, Ireland, Lambay and Clare Island, England, Wales, Jersey and Gurnsey (Channel Islands), Netherlands, Belgium, Czechoslovakia, Romania, Bulgaria, Jugoslavia, Hungary, Austria, Switzerland, Andorra, France, Corsica, Portugal, Spain, Italy, Sardinia, Sicily, Greece, Balearic Islands, Azores, Madeira.

South Africa from the Cape into Natal, Southwest Africa, Madagascar, Reunion.

Siberia, Tomsk to Lake Baikal, Kamchatka. Manchuria, Korea, Hokkaido (Japan). Turkey, Transcaucasus, Turkestan, Kazakstan, Chuvash and Tartar (Kirgiziya) Republics. Pakistan, India, Naini Tal (6,400 ft.), Simla (7,500 ft.), Junga (4,000 ft.), Almora, (6,560 ft.), Matiana and Mashobra (8,000 ft.), Simla Hills in the western Himalayas, Sandakphu, Phallut, Darjiling district, in the eastern Himalayas, Fern Hill, Ootacamund, Kodaikanal, at 5,000–7,000 ft. on the Nilgiri and the Palni Hills well south in the Indian Peninsula.

Kerguelen and St. Paul.

Elevations not mentioned above are: 540–2,600 m. Switzerland, 0–2.500 m. Italian Alps, 1,649 m. Croatia, 1,800 m. Bulgaria, 2,000–2,600 m. Spain, to 2,500 ft. in Scotland, 2,000–3,000 ft. White Mountains of New Hampshire.

D. rubida has been said to be present in Sardinia, the Faeroes, Kashmir, and throughout the tropical Malay Peninsula. Records for some of those places have not been found by the author and the species almost certainly is lacking throughout the Malay Peninsula.

Presence of *D. rubida* on oceanic islands such as the Hawaiians, Juan Fernandez, Tristan da Cunha, Kerguelen, St. Paul, obviously resulted from transportation. Presence in Australia, New Zealand and its islands, South America, South and Southwest Africa, Madagascar, Reunion, just as undeniably, must be explained by transportation. Acquisition of so much of the distribution presumably involved a rather considerable number of introductions as there is no certainty, even when reproduction is parthenogenetic, that every instance ends in permanent colonization. Accordingly, the problem now requiring solution is to determine how much of the present Eurasian-American distribution is due to transportation.

During the last fifteen years, and in spite of efforts to prevent importation of unsterilized soil, inspectors of the U. S. Bureau of Plant Quarantine, intercepted some 130 specimens[1] belonging to several morphs of *D. rubida*. The worms were found in sphagnum as well as wood duff but mostly in earth along with plants[2] from the following 21 regions or countries: Australia, Hawaii, Saskatchewan, Newfoundland, Mexico, Canal Zone (new record for the species), Brazil, Germany, Poland, Scotland, Ireland, England, Netherlands, Belgium, France, Greece, Italy, Spain, Portugal, Madeira, South Africa. In absence of enforced quarantine rules, the number of transported specimens almost certainly would have been much greater. Clearly then, man has been transporting *D. rubida* with his plants, probably for many years. The worms have been found only where man has gone and usually where he has settled. No other agent has been shown to be involved in transportation of earthworms. There has been sufficient time since 1500 A.D. for the species to have become established in each of the areas involving transoceanic travel. Postulation of preglacial presence in North America, and survival on probably barren nunataks, no longer is necessary. Similarly, glaciated regions of Europe must have been entered since the geologically recent recession of the Pleistocene ice sheets. At present only one agent is known to be involved in such penetration and that is man. Some 400–500 years was sufficient for *D. rubida* to be taken to and through the presently populated regions of North and South America, Australia, New Zealand, and South Africa. More time presumably was available for a similar transportation through glaciated portions of Europe.

Rio de Janeiro and Petropolis are just within the tropics. The elevations at which the worms actually were obtained, as in case of Ecuador, Guatemala, and the Canal Zone, are unknown. All tropical localities for which elevations were recorded are above the four thousand foot level. Extensive colonization of tropical lowlands anywhere now seems highly unlikely. Although *D. rubida* may have "a latitude of adaptation unparalleled by any other invertebrate" (Omodeo, 1963; p. 141), that latitude has been inadequate to permit colonization of the tropical lowlands to which it must have been taken often. The species presumably acquired its present physiology in a climate not too different from that presently prevailing in arctic and subarctic regions. That deduction from the distribu-

tion is not contra-indicated by recovery alive of 6 individuals from natural ice (Sekera, 1896) and such temperature data (Michon, 1945) as are available.

Habitats: Under stones, boards, bark of stumps or of standing and down trees. In rotting oak bark, rotting wood and straw, pine stumps (common), pine litter, compost, peat, moss mats, fucus beds, dumps, sawdust, corn-cob piles, compost, final settling tank of sewage plant (U. S.), sides of a sewage tank (England), manure. Banks of springs, black earth of brook and stream banks, in trout stream "torrents," among moss in running water, springs, wells (but not as common there as *E. tetraedra*), wandering on streets after rain. Gardens, cultivated fields (rarely at Kiel but perhaps more commonly at 8,000 feet in the Himalayas). Wet soil with more or less sewage, damp clay soil, loamy brown soil, in soil with pH of 5.0–6.2. Beech raw humus, coniferous forest mull and in forests of various kinds. In flood plains, meadows, and, in the Urals in meadow saline and meadow marshy saline soils but rarely in lightly leached chernozems. In peaty humus soils (Vologda Oblast), as well as in birch and spruce forest, mown meadows, podzol of a spruce-bilberry patch.

Caves, Germany, Belgium, France, Switzerland, Italy, Austria, Hungary, Jugoslavia, Bulgaria, Greece, Portugal, Pennsylvania, Virginia, West Virginia, Kentucky, Tennessee, Alabama, Missouri. Mines, England, France, Pennsylvania. Earth with potted plants: Greenhouses, Berlin, Mark Brandenburg, Poznan (Poland), Berne, Rome, Naples, Maine. Melon frames, old hot beds, culture beds of earthworm farms. Botanical gardens, Copenhagen, Oxford, Rome, Naples, Palermo.

Dominant in pure dung (Troitsk district of the Urals) and in substrates having a high concentration of organic matter as well as in rotten trees.

Biology: Activity is year-round when conditions permit. Breeding presumably also can be year-round in appropriate circumstances, but in central Maine and in many of the North American areas, a winter rest, if not also one during summer, is climatically imposed. A single, short, summer season of breeding presumably is characteristic in subarctic and arctic areas. Estivation is spent tightly coiled and immobile in a small mucus-lined cell, activity being resumed as soon as moisture reaches appropriate levels, according to one author, but Michon (1954) said the species is homodynamous and so never has a diapause.

Copulation, as in most megadriles, has not been studied but the couple seen by Ribaucourt (1895) was in a position otherwise unrecorded for megadriles, of ventral apposition to each other from one end to the other, with head to head and tail to tail. A copulating pair was seen at 10.5°C by Baltzer who gave no further information except that a slime tube was thin. Cocoon production (England) was said to be low, 42 per year, but in Germany 40 cocoons, on the average, were

[1] Not included are 39 juveniles of *Dendrobaena* spp. probably mostly of *D. rubida* and *octaedra*.

[2] Entire plants, cuttings, bulbs, or tubers. *Antherium* sp., *Astilbe* sp., *Cymbidium* sp., *Dicentra spectabilis*, *Ficus* sp., *Fragaria* sp., *Humulus lupolus*, *Laburnum vossi*, *Rheum* sp., *Ribes*, sp., *Trollius* sp., azaleas, begonias, cacti, cedar, dahlias, geraniums, heather, hydrangeas, jasmine, lilies, mountain ash, oleanders, peonies, orchids, roses, spruce, as well unidentified plants possibly of other kinds. Other worms were on celeriac, cabbages and lettuce of ships stores.

laid between February 1 and March 31, the annual average estimated at 95. Number of eggs in the cocoons is one-third more than the number of hatchlings. Only one hatchling emerged from 62–75 per cent of the cocoons but 2–4 do sometimes hatch out. Incubation has been said to be 12–15 days in summer, $3\frac{1}{2}$ weeks in fall and winter or, according to another author, even $8\frac{1}{2}$ weeks. For growth, the optimum temperature is 18°C. At 25°, mortality is high. Suboptimal temperatures are less harmful than supraoptimal—worms have been recovered alive (Sekera) from natural ice. Maturity sometimes is reached in 22–42 weeks. Life expectancy has been said to be less than a year; however, the average life in a French laboratory at 18° (Michon, 1954) was 547 days or at 9° was somewhat higher, 595 days. The average was reduced (at 18°) to 526 days when isolation prevented copulation. Individuals totally submerged (Roots, 1956) survived almost a year.

The worms live in the upper soil layers which they have been said to leave only to hibernate, but it is known that they aggregate under sheep dung on mineral soil as a result of movement "at or near the surface." Although the latter alternative is not to be ruled out, the first probably is often involved as individuals have been seen wandering on the soil surface or even climbing trees during wet nights. Above-ground migration now seems likely to be more frequent than the few past records have indicated.

Individuals can fling themselves around by several rapid, side-to-side movements of the body.

The species was thought to show itself as "exacting" in Iceland but with the athecal morphs less so.

A population of 36,000 per acre, with dominance, was found in a very acid, untreated, hill pasture in Scotland. Dominance, in a field recently brought into pasture was with 55.5–59.2 percent of the population. Numbers rose from 28.6 to 129.7 thousands per acre in fields treated with lime and fertilizers. Further improvements of the soil was thought to have resulted in an increase in numbers of individuals of pasture species but with decrease in number of *D. rubida*.

Soil *p*H was not often recorded for *D. rubida* which has however been characterized as "ubiquitous" with respect to it.

The stomadeum originally develops in i–iv but later extends through i–vi. How much of the atyphlosolate portion of the intestine is proctodeal, as for other megadriles, is unknown.

D. rubida, though present in soils of the vicinity of Cracow, like *D. octaedra*, is not stored by the moles (*Talpa europaea*).

Variation and Abnormality: Segments to 115, to 123, and in Göttingen to 134 (Hertling), clitella beginning at 23/24, 27/28, and ending at 33/34, tubercula pubertatis in xxix–xxxi, 25–29 atyphlosolate segments of the intestine, as well as other divergences from an assumed norm, were recorded but without the data that are needed to enable decisions as to what was involved. Atyphlosolate intestinal segments at Göttingen (Hertling) were 4–29 but only three of those worms had more than the maximum of 14 segments (25, 28, 29) found by the present author in a series of 100+ specimens. In such circumstances, one is now entitled to suspect abnormality of some kind or presence of a second species. Many of the worms, with fewer than 8 atyphlosolate intestinal segments in this author's series were posterior amputees and all others are justifiably suspected of having suffered similarly. Seeming absence of hearts of vii often may be due merely to inability to recognize smaller vessels when empty. Lengths of 90–120 mm., if not otherwise explainable may have been measured on macerated material.

Some of the variations recorded may be characteristic of markedly divergent morphs or clones. Other differences from usual norms are now known not to be abnormal in the ordinary sense but of the very kinds that often appear after parthenogenesis is established. Thus, shortening of the tubercula so as to extend only through xxix–xxx probably has arisen since reproduction became uniparental and is an early stage in an evolutionary sequence that ends with complete deletion as well as disappearance of the associated glands.

Presence of male pores in xiv, recorded some years ago without pertinent data, could have resulted from deletion of a pair of embryonic somites, from hypomeric head regeneration, or from postparthenogenetic mutation. Similarly, presence of rudimentary gonads in xii, recorded as long ago as 1894, may or may not have been associated with parthenogenesis.

Polymorphism: Morphs are numerous. A biparental and diploid H morph that has been recorded from several places needs more adequate characterization especially if amphimixis is obligatory. H_p morphs probably are numerous. First order intermediate morphs also are numerous, differing mainly from the H_p in the absence of any 1, 2, or 3 of the spermathecae, or with such spermathecae as are present being juvenile, rudimentary, defective or abnormal. When rudimentation or deletion of seminal vesicles, tubercula and TP glands as well as GS and atrial glands, also are taken into consideration, along with possibly clonal limitations of size, segment number, clitellar extent, the number of more or less readily recognizable morphs becomes so great as to make provision of Latin names for all of them ridiculous even with omission of anatomically indistinguishable morphs of different ploidy. Various athecal, quadrivesiculate morphs, long regarded as species of *Bimastos* with the names of *constrictus* and *tenuis* often still are referred to as subspecies, varieties, or formae of *D. rubida*. Morphs with tubercula restricted to xxix–xxx then are referred to collectively at one or another hierarchical level as "typica" and those with tubercula extending through xxviii–xxx as *D. subrubicunda*.

Atrial, TP, and GS glands are being reduced and eliminated, or already had been eliminated, in many of the morphs represented in the material available to the author. GS glands, at least as objects recognizable from the coelom, apparently have been deleted more often than the other two kinds of supraparietal glands.

Regeneration: Homomorphic head regeneration (Gates, 1956) can be expected, in optimal conditions, at all levels back to 17/18. Equimery has not been seen and is not expected behind 5/6.

Homomorphic tail regeneration, from a growth zone of rapid segment formation, seemingly is not to be expected ordinarily, if at all. Instead a terminal region of the substrate rarely may be reorganized (morphallaxis) into a pygomere and one or two segments with some of the stigmata of regeneration, the reorganization possibly being more drastic and more rapid in the middle of the body than at more posterior levels. A regenerate at 71/72, of same length as segments lxix-lxxii, was metamerically undifferentiated. Whether epimorphosis or morphallaxis was involved could not be determined. Regeneration of a pygomere may be, perhaps often is, epimorphic. Posterior amputation seems to be rather common but determination of subsequent events becomes more difficult as time passes.

Parasites: CILIATA: *Anoplophrya lumbrici* (Schrank, 1803), Czechoslovakia, Germany. *Maupasella nova* Cepede, 1910, Germany. *Metaradiophrya lumbrici* (Dujardin, 1895), Germany. SPOROZOA: *Apolocystis herculea* (Bosanquet, 1894), France, Germany, Sweden. *A. lumbrici-olidi* (Schmidt, 1854), France, Germany. *Monocystis herculea* Bosanquet, 1894, Germany, Sweden. *Nematocystis elmassiani* (Hesse, 1909), France, Germany. *N. lumbricoides* Hesse, 1909, Germany. NEMATODA: *Choerostrongylus pudendotectus* (Wostokow, 1905), larvae, Europe. *Dicelis dendrobaenae* Timm, 1962, Pennsylvania. *Dicelis* sp., coelom, x, xi, Central Maine. *Metastrongylus elongatus* (Dujardin, 1845), Europe. *Rhabditis pellio* Bütschli 1876, non Schneider 1866, nephridia, England.

Systematics: The species précis above is as much of a characterization of an assumed original, diploid population with obligatory amphimixis as could be deduced from such material as was available to the author from more than a dozen countries.

Somatic anatomy of *D. rubida*, as already indicated, differs so much from that of *D. octaedra* as to suggest advisability of transfer to another genus which, according to the evidence provided by the calciferous sacs would be closer to *Lumbricus*, and to genera containing *A. chlorotica* and Levinsen's *eiseni*, than to *Dendrobaena*.

Further consideration of relationships of *D. rubida* still is contra-indicated by lack of properly preserved material of other species as well as by absence, in the literature, of pertinent information re somatic anatomy.

Remarks: Calciferous sacs are bound by connective tissue against the esophagus in x and usually are unrecognizable as such merely on pinning back an opened specimen. Prior to some further dissection one sees only a seemingly moniliform widening of the gut. The thin-walled section of the gut in x is compressed by the sacs so that the lumen is vertically slitlike.

Eisenia

1877. *Eisenia* Malm, *Ofvers. Sällsk. Hortikult. Vänn. Förh. Goteborg* 1: p. 45. (Type species, none designated.)
1968. *Eisenia*, Gates, *Jour. Nat. Hist. London* 9: 305. (Redefined to eliminate American endemics transferred to a new genus.)

Eisenia was erected for three species, *Enterion foetidum* Savigny 1826, *Allolobophora norvegica* and *subrubicunda* Eisen 1874. The last two were found to be synonyms of *Dendrobaena rubida*, thus leaving *foetidum* to be the type species.

Characters of the type species that may prove to be definitive at a generic level are as follows:

Calciferous gland, without sacs, opening into gut behind insertion of 10/11 through a circumferential circle of small pores. Gizzard, mostly in xvii. (Extra-esophageal trunks, or a major branch of each, passing to dorsal trunk in region of insertion of 9/10?) Hearts, in vii–xi. Nephridial bladders, sausage-shaped or digitiform, transversely placed. Nephropores, inconspicuous, in two ranks on each side, alternating irregularly and with asymmetry between a level just above *B* and one above *D*. Setae, closely paired. Longitudinal musculature, pinnate. Pigment, red.

Another of Savigny's (1826) species, *Enterion roseum*, has been in *Eisenia* since 1900, primarily because of location of spermathecal pores at the *foetida* level. Somatic anatomy, however, differs from that of *foetida* and characters of *rosea* that may prove to be definitive of a group at generic level are as follows:

Calciferous sacs, in x, large, lateral, opening vertically and widely into esophagus. Gizzard, mostly in xvii. Extra-esophageals, passing up to dorsal trunk posteriorly in xii. Hearts, in vi–xi. Nephridial bladders, *U*-shaped, transversely placed. Nephropores, inconspicuous, alternating irregularly and with asymmetry, between a level somewhat above *B* and another above *D*. Setae, closely paired. Longitudinal musculature, pinnate. Pigment, lacking.

Because of some long-known differences from *foetida*, Omodeo and Pop have transferred *rosea* to *Allolobophora*. However, somatic anatomy contra-indicates placing *rosea* in any genus of which Savigny's *chloroticum* is the type. For that reason, and to avoid an extra change in the nomenclature, the author has felt constrained to follow much current usage, and retain the generic name of *Eisenia* for *rosea* until such time as some of the biological and nomenclatural problems that are involved have been solved. This will avoid at least one of those changes that many biologists resent.

Eisenia foetida

1826. *Enterion fetidum* Savigny, *Mem. Acad. Sci. Inst. France* 5 (Hist. Acct.): p. 182. (Type locality, Paris. Types, in Mus. Hist. Nat. Paris.)
1891. *Allolobophora foetida*, Rosa, *Ann. Hofmus. Wien* 6: p. 381.
1900. *Eisenia foetida*, Michaelsen, *Das Tierreich* 10: p. 475.
1903. *Eisenia foetida*, Michaelsen, *Die geographische Verbreitung der Oligochaeten* (Berlin), p. 136, etc.
1923. *Allolobophora (Eisenia) foetida*, Stephenson, (*The Fauna of British India*), *Oligochaeta*, p. 499.
1958. *Eisenia foetida*, Gates, *Breviora, Mus. Comp. Zool. Harvard*, No. 91: p. 6.

ABNORMALITY: Gates, 1958, *American Midland Nat.* 59. ARGINASE: Kruger, 1952, *Comp. Biochem. Physiol.* 5. CALCIFEROUS GLAND: Guardabassi, 1957, *Zeitschr. Zellforsch. Mikr. Anat.* 46. CHLORAGOGEN: Semal van Gansen, *Bull. Biol. France Belgique* 90. COPULATION (of 3 individuals): Wilcke, 1957, *Zool. Anz.* 158. DEVELOPMENT, GROWTH, REPRODUCTION: André, 1957, *Compt. Rend. Acad. Sci. Paris*, 245; 1962, *ibid.* 253. *Compt. Rend. Soc. Biol. Paris* 156; 1963, *Bull. Soc. Zool. France* 87 and *Bull. Biol. France* 97. Devries, 1964, *Bull. Soc. Zool. France* 89. Evans, 1947, *Jour. Board Greenkeeping Res.* 7. Evans & Guild, 1948, *Ann. Applied Biol.* 35 and *Ann. Mag. Nat. Hist.*, Ser. 11, 14. Graff, 1953, *Regenwürmer Deutschlands*, Hannover. Herlant-Meewis, 1954–1967, *Ann. Soc. Zool. Belgique* 85, 87, 89, 96; 1962, *Gen. & Comp. Endocrinol.* 2, Janda & Gavrilov, 1939, *Vest. Cesk. Zool. Spol. Praze* 6-7. Michon, 1954, *Compt. Rend. Acad. Sci. Paris* 238; 1957, Année Biol. 33. Stinauer, 1951, Life Histories and Culture of Earthworms," Thesis, Michigan State Univ. DIGESTIVE SYSTEM (structure and function): Van Gansen, 1962, "Structures et fonctions de tube digestif du lombricien *Eisenia foetida* Savigny," Thesis, Brussels. ENDOCRINOLOGY: Herlant-Meewis, 1962, *Mem. Soc. Endocrinol.* 12. Rude & Linder, 1964, *American Zool.* 4. ENZYMES: Fujimoto & Adams, 1965, *Biochim. Biophys. Acta* 105. DIET: Miles, 1963, *Soil. Sci.* 95. FREEZING: Korschelt, 1914, *Zool. Anz.* 45. Kobayashi, 1938, *Sci. Rept. Tohoku Univ.* 13. HEREDITY: André, 1963, *Bull. Soc. Zool. France* 87. MUSCULATURE: Harman, 1960, "Studies on the Taxonomy and Musculature of the Earthworms of Central Illinois," Thesis, Univ. Illinois. Kawaguti & Ikemoto, 1957. *Biol. Jour. Okayama Univ.* 3. Pop & Dragos, 1959. *Stud. Univ. Babes-Bolyai*, Ser. 2, 2 (Biol.). NERVE CORD STRUCTURE: Guardabassi, 1959, *Zeitschr. Zellforsch.* 50 and 1963, *Arch. Zool. Italiano* 48. PARASITES: Heidenreich, 1935, *Arch. Protistenk.* 84. Loubatières, 1955, *Ann. Sci. Nat. (Zool. 11)* 17. Meier, 1956, *Arch. Protistenk.* 101. Miles, 1962, *Jour. Protozool.* 9. Rees, 1961, *Parasitology* 51; 1962, *ibid.* 52. PHARYNGEAL GLANDS: Keilin, 1920, *Quart. Jour. Micros. Sci.* 65. PHYSIOLOGY: Laverack, 1963, *The Physiology of Earthworms* (London). REGENERATION: Avel, 1929, *Bull. Biol. France Belgique* 63 and *Année Biol.*, Ser. 3, 26. Gates, 1949–1950, *Biol. Bull.* 96, 98, 99. RESPIRATION: Byzova, 1965, *Rev. Ecol. Biol. Sol.* 2.

Quadrithecal, pores at 9/10, 10/11, slightly lateral to mD. Clitellum, saddle-shaped, of 6–8 segments (but usually 7?), xxiv, xxv, xxv/n, xxvi, xxvi/n, xxvii–xxxi, xxxii/n, xxxii, xxxiii/n, xxxiii, xxxiv, usually beginning at 24/25 or 25/26 and ending at 31/32 or 32/33. Tubercula pubertatis, uninterrupted longitudinal bands, just lateral to B, in 3–6 segments but usually 3, xxvii, xxviii, xxviii/eq–xxx, xxxi/eq, xxxi, xxxii/eq. Male pores, at or near mBC usually in a slight depression with little resemblance to an atrial cleft. Male tumescences, confined to xv, usually slight. Female pores, just lateral to B at eq/xiv. Genital tumescences, including any setal couple of viii–xii, ventral couples only in any of xvii–xxxv, most often in some of xxii, xxiii and xxvii–xxxii. Setae, closely paired, AB ca. = or $> CD$, $BC < AA < DD$ posteriorly $< \frac{1}{2}C$. Nephropores, inconspicuous, in two ranks on each side, alternating irregularly and with asymmetry between a level just above B and above D, the dorsal rank quite irregular. First dorsal pore, at 4/5 or 5/6, usually the former. Prostomium, epilobous, tongue open. Body section, behind clitellum dorso-ventrally compressed, setal couples at the four corners, the ventrum narrower than the dorsum and flat or slightly concave, the dorsum slightly convex, the sides diagonal and slanted towards mV. Color, purple, red, dark red, slate, brownish red or reddish brown, brownish, usually in transverse, mid-segmental bands alternating with white or yellowish bands that include the intersegmental furrows. Segments, 80–131, usually 90–110. Size, 27–125 by 3–5 mm.

Septa, none thickly muscular. Pigment, red, in circular muscle layer and, when dense, through entire thickness of that layer down to B. Longitudinal musculature, pinnate. Blister organs, present but without definite boundaries and regular locations (?). Calciferous sacs, none. Esophagus gradually widened through x–xi, widest in xii, gradually narrowing through xiii–xiv, lumen vertically slitlike in xii, widening in xiii, calciferous chambers, ca. 60. Intestinal origin, in xv. Gizzard, mostly in xvii (ending at ?). Gut, valvular in region of insertion of 19/20. Typhlosole, beginning in region of xx–xxiii, rather thick and vertically oblong in transverse section or slightly flanged ventrally, usually with a median longitudinal furrow on ventral face, ending abruptly in region of 60th to 109th segments but usually in 75th–108th, leaving 8–15 metameres atyphlosolate. Extra-esophageals, passing to dorsal trunk in region of insertion of 9/10. Hearts, in vii–xi. Nephridial bladders, sausage-shaped or digitiform, transversely placed and when straight reaching beyond D. Holandric. Seminal vesicles, 4 pairs, in ix–xii. Sperm ducts, with an epididymis in various loops or coils. Spermathecae, usually in ix, x, ducts slender, short and almost confined to the parietes. TP and atrial glands, absent (unless rudimentary and wholly within parietes). GS glands, lacking (unless represented by tissue under peritoneal blisters).

Reproduction: Amphimixis has been thought to be obligatory. Isolated individuals reared to sexual maturity (Evans & Guild, 1948) did not lay or, in case of 8 worms, produced only a few infertile cocoons. Worms that were mated 10–14 months after isolation laid an average of 5 viable cocoons in the first three weeks. Nevertheless, three individuals isolated before maturity (Janda & Gavrilov, 1939) deposited cocoons from which 5 hatchlings emerged. Uniparental reproduction clearly is possible, even if only rarely, but whether self-fertilization or parthenogenesis had been involved was not determined. Castrates, after removal of segments ix–xi (André, 1957), laid eggs that developed, as haploids or diploids, until the blastula stage. None lived longer than 48 hours after initiation of development. Autofertility was proved (André, 1961) by two sets of experiments involving; (1) exchange of male segments between individuals of a pair, or (2) exchange of male segments and then, after healing, excision of male organs only from one of each couple. Subsequent experiments (André, 1962) were said to prove that self-fertilization is possible only by the sperm that had been within spermathecae, presumably as a result of self-copulation. Controls involved germ cells from seminal vesicles but seemingly not any sperm that had aggregated on the male funnels or passed down into the sperm ducts.

The brilliant spermatozoal iridescence on large and polyplicate male funnels of nearly all adults from most colonies and a similar brilliance of material distending spermathecal ampullae almost always testifies to possibility of amphimictic reproduction. Male sterility has been suspected by this author only in some of the markedly aberrant regional homoeotics.

Chromosome number, $2n = 22$.

Fertilization is secured with ease artificially (André, 1962) by adding spermathecal sperm to coelomic fluid containing ova removed from the ovisac, but prevention of polyspermy is difficult.

Distribution: Nicobar Islands.

New South Wales, Victoria, southwest Australia. New Zealand and Sunday or Raoul Island (Kermadecs).

(Hawaii?)

British Columbia, Ontario, Quebec, New Brunswick, Nova Scotia, Cape Breton Island.

Washington, Oregon, California,[1] Oklahoma, Texas (EF?),[2] Minnesota, Missouri, Arkansas, Louisiana, Illinois, Michigan, Indiana, Kentucky (also EF), Tennessee, Alabama (EF?), Georgia (EF), Ohio, Virginia (also EF), District of Columbia, Pennsylvania, Maryland, New York, New Jersey, Vermont, Massachusetts, Connecticut, New Hampshire (also EF), Maine (also EF).

Mexico.[3] Guatemala. Colombia (Bogotá 2,600 m.), Peru, Chile, Argentina, Uruguay, Porto Alegre and Rio Grande do Sul, Brazil.

Bermuda.

Greenland. Iceland. Norway, Spitzbergen, Sweden, Finland, Russia, from the north to Moldavian SSR, Black Sea, Caucasus, Transcaucasus, including Crimea, Ukraine, Byelorussia, Kuban, Urals and

Pri-Urals. Latvia, Lithuania, Poland, Germany, Denmark, Faroes, Hebrides, Scotland, Ireland, England, Wales, Jersey, Guernsey (Channel Islands), Belgium, Czechoslovakia, Romania, Bulgaria, Jugoslavia, Hungary, Austria, Switzerland, France, Portugal, Spain, Balearic Islands, Corsica, Sardinia, Italy, Sicily.

Azores, Madeira, Canary Islands, St. Helena (1,000–2,000 ft.).

Morocco, South Africa.

Siberia, from the Yenesei and Tomsk to Lake Baikal. Manchuria, Korea, Japan. Caucasus, Transcaucasus, Turkey, Lebanon, Turkmenistan, Kazakstan, Chuvash and Tatar Republics. Afghanistan. Kasauli, Simla (7,500 ft.), Simla Hill States (6,000 ft.), western Himalayas. Kurseong (5,000 ft.), Darjiling (7,000 ft.), Sikkim, eastern Himalayas. Fern Hill (7,500 ft.), Ootacamund (6,700–8000 ft.), Coonoor (2,000 m.), Ponmudi (750 m.), Travancore, Shembanagur, Kodaikanal (7,000 ft.), Palni Hills, South India.

Elevations not already mentioned above, to 1,200 m. Italian Alps, to 1,000 m. Croatia, 200–700 m. Switzerland, 1,000–1,600 m. Carpathians.

Specimens intercepted at American ports during the last fifteen years, 86, were from 22 countries or regions, as follows: Canada, Mexico, Argentina, Norway, Sweden, Poland, Germany, Denmark, Scotland, Ireland, England, Wales, France, Belgium, Netherlands, Switzerland, Italy, Greece, Spain, Portugal, Azores, Madeira, India, Japan, Korea, Australia, New Zealand. In the literature, no records were found by author for Netherlands and Greece though there is no reason to doubt presence of the species in those two countries.

At least two writers have mentioned Burma when discussing the range of *E. foetida*, presumably because of the Nicobar record. The species never was found on the Burmese mainland nor in more recent collections from the islands. Europeans have lived there, and Rosa's specimens actually could have been obtained from earth with recently imported plants. If still on the islands, *E. foetida* is likely to be present only at elevations above a thousand feet, as is the case on the South Atlantic island of St. Helena.

Nearly seventy years ago, *E. foetida* was said (Beddard, 1895) to be "universally distributed." A little more caution was evinced by "Durch Verschleppung nahezu kosmopolitisch" in the next oligochaete monograph (Michaelsen, 1900) though there the range was said to include "ganz Nord-and Zentral-Amerika." Such statements, unfortunately accepted more or less widely without question, presumably convinced Smith (North American Lumbricidae, 1917) that even mention of those states in which he had found the species to be present was needless. Nevertheless, in "Die geographische Verbreitung der Oligochaeten" (1903) Michaelsen could

---

[1] Date of first collection in the state, July, 1877.

[2] EF: Earthworm farms or from dealers selling worms raised in culture beds of the farms.

[3] Date of first collection in the country, April 8, 1881, at Guanajuato, Recorded from Mexico City in 1900.

list only three states, one Canadian province, one Central American republic and for those areas, except California (with 14) and Guatemala (with 2), only a single locality each had been recorded. Also responsible for much misunderstanding was the failure for many years to record pertinent data, especially as to elevation above sea level.

E. foetida clearly is of European origin (*cf.* Systematics, below). The transoceanic portions of the presently known range obviously were not self-acquired, and there is no need to postulate any other mode of carriage than by man. Moreover, since we know the species has been carried around the world in plant pots and has been distributed locally from new centers in greenhouses, arboretums, and botanical gardens for two or more centuries, the important problem now is that of determining just what part of the Eurasian distribution was self acquired and where the species originated.

In 1896 *E. foetida*, though common around Lake Leman, was "rarissime" in central Switzerland and the Jura bernois. That area accordingly provides an interesting opportunity to discover what has happened to the species to the present.

Various introductions to tropical lowlands are known to have failed. Extinction of the species in any region where colonies once had been well established has not been demonstrated as yet. Nevertheless, collections made during the last fifteen years or so in St. Helena and Hawaii suggest that *foetida* now provides an insignificant fraction of the megadrile population of the islands, in spite of repeated recent introductions.

Habitats: Between outer leaves of cabbages and of lettuce in markets and ships' stores, in axils of plantain leaves, in debris within tree-trunk cavities, under moss on beech stumps, under bark of rotten tree stumps, in decaying fallen logs, under elm logs, fermenting vegetable matter, leaf mould, compost, manure, coarse gravel in bed of an intermittent raw sewage stream, final settling tank of sewage plant (Illinois, but not as numerous there as *E. tetraedra*), taiga, forests, and steppes (Russia).

Earth of plant pots and cold frames. Gardens, meadows (Germany) among roots of weeds at water's edge, banks of forest creek, wandering on streets after rain (Switzerland, Australia, New Zealand).

Botanic gardens, Christiania, Upsala, Copenhagen, Berlin, Oxford, Naples, Palermo, Coimbra (Portugal), South Africa, Philadelphia. Arboretums, California, Massachusetts. Greenhouses, Sweden, Germany, Poland, Wales, central Maine, Massachusetts, Maryland, Illinois, often also from outdoor beds around the greenhouses (including Maine, Connecticut, New York). Caves, Germany, Russia, Caucasus, Jugoslavia, Czechoslovakia, France, England, Portugal, Kentucky. Mines, England, France.

The few records of *p*H are, in the range 6.8–7.6 but specimens survived for 29 days in experimental earth with a *p*H of 5.4.

In Scandinavia, *E. foetida* has been thought to be a haemerobiont, entirely "dependent on culture." The first Australian record for the species is of gardens to which plants had been brought from other parts of the world and *E. foetida* still is among the forms most frequently encountered by gardeners of the Sidney district. The first New Zealand record also was of presence in gardens. The literature provides other instances; common in gardens of Scotland (1910), abundant in Montevideo gardens as well as in gardens of Chile, "readily obtained in any market garden" of an eastern part of the United States (1920). The Colombia worms were from a garden. A more recent report of presence in a cultivated area of Arkansas may mean little more than that an "Organic Gardener" had introduced the worms into a area heavily fertilized with organic matter, a practice highly praised and widely recommended throughout the country to practitioners of that kind of gardening. In the New England states, in the early part of the present century, *E. foetida* could be counted on to be present in almost every farm manure pile. Similar sites, throughout much of Europe may well have been similarly populated. The species is dominant throughout Raoul Island in forest litter and the writer has found specimens in leaf litter of well-shaded portions of a Massachusetts arboretum. The first record for St. Helena, "earth in moist situation on the high land" was associated with a comment that at certain times of the year the species may be seen lying in dead and dying states on the hard surface of roads.

E. foetida need not however be confined to gardens, manure piles, or man-made accumulations of leaf litter and other organic matter, as Russian records of presence in taiga, forests, and steppes seemingly prove. Already by 1887, in New Zealand, the species was said to be five times more numerous than any other, abounding in gardens, pastures and in almost every quality of land from peaty soil on low flats to dry friable loam several thousand feet up on slopes of the range. Hundreds of specimens sometimes were seen dead in ditches or paddocks where water stood after rain. That distribution may not have resulted entirely from human transport as there are good reasons now for believing that individuals wander widely and frequently on the surface although some of such travels obviously end in death.

Adequate moisture and plenty of slowly decaying organic matter were thought to favor or to enable dominance on Raoul. Similar conditions, early in the rainy season, presumably also enabled dominance at Darjiling (71 per cent of 1,012 specimens collected there) and subdominance at Simla. The habitat records certainly suggest a strong preference if not also necessity for gross organic matter, and examination of gut contents shows little if any soil.

Some recent writers have found E. *foetida* to be less prevalent in the United States than had been assumed from the frequency with which the worm was used experimentally in this country (*cf.* Stephenson, 1930: p. 912). Thus, Murchie (1954) maintains that the species is quite uncommon in many parts of North America and is surprisingly so in natural environments of Michigan because of the use of the species for bait. He also cites Guild as authority for never finding the species in natural field populations of Scotland, with the exception of a single specimen (which doubtless had wandered away from its own niche nearby). Except in barnyards and active compost heaps, almost continuous introduction into an area, Murchie feels, is necessary in order to maintain the species.

A marked decrease of number in New Zealand seemingly is indicated by Benham who stated in 1915 that E. *foetida* was "still fairly common." Latest information on New Zealand was provided by Lee (1959) who credits the species with being "very common . . . in compost heaps, manure heaps and similar concentrations of organic matter. It is rarely found in pasture and garden soils."[4] Presumably following introduction, E. *foetida* was widely distributed by man, with dispersal accelerated by a wandering propensity of the individuals. Eventually an exploding population peaked. The decline may have been hastened by reduction, through activities of the worms themselves, of organic matter in soils to levels inadequate for the species. An exploding population may also have peaked in New England during early years of the nineteenth century when there were many complaints about contamination of wells by hordes of earthworms. Unfortunately, except for Smith's studies in Central Illinois, no surveys of earthworm faunas were made until into the twentieth century, and whether at the turn of the century E. *foetida* was largely restricted to its present habitats in New England is unknown. Searches in numerous manure piles of several New England states since 1946 have shown that such habitats are much less frequently occupied than forty-five years ago. Abandonment of farms, replacement of horsepower in the flesh by horsepower in engines, substitution of inorganic fertilizers for organic manures, decrease in livestock as well as horses, in New England, together have resulted in suitable *foetida* habitats often being too far from those already occupied to be reached by random wanderings. Nevertheless, the species does still sometimes appear in organic accumulations when presence in the immediate neighborhood is or at least seems unlikely.

4 Aside from mention of the fact E. *foetida* is not in litter of native forest, and the statement quoted above, Lee unfortunately provides no indication as to just how universally the species now is found, in appropriate habitats, throughout the mainland islands. Such information might well have been helpful in estimating rate of spreading there as well as elsewhere.

Biology: Copulation, which takes place beneath the surface, has been mentioned by various authors since 1845 and has been observed more often than in any other megadrile species. Even an abnormal process, involving three individuals simultaneously, was recorded.

Laying begins 48 hours after copulation and 1 cocoon is deposited by unisolated worms every 5 days throughout the year without diapause. Individuals isolated after copulation laid once every 3–4 days for $4\frac{1}{2}$ months, sperm having been seen in the spermatheca for 4 months after copulation. Repeated copulation, accordingly reduces fertility! Hatching, at end of 21–28 days, usually 26 days, at 19°C. Sexual maturity, at end of 3 months. Growth ceases after copulation begins (Herlant-Meewis, 1954).

Up to 20 eggs are laid in a cocoon from which 1–8, but usually 2, hatchlings emerge. Laying, uninterrupted until October; 11 cocoons per year. Incubation, 11 weeks. Maturity, at 47–74 weeks. Diapause, none during growth (Evans).

Laying, continued throughout the year, in appropriate circumstances. Average interval between laying, 5.3 days at 70°F, 1.9 days at 75°F. Incubation, 22–33 days at 70°F. Maturity, $3\frac{1}{2}$–$4\frac{1}{2}$ months after hatching (Stinauer).

Laying, 26 cocoons per individual between February 1 and March 31. Average number of hatchlings, 2.5 per cocoon. Hatching, after 16 days at 25° Celsius (Graff, 1953), "puberty" 70–80 days later.

For growth, the optimal temperature is 28°C. A temperature of 32° from hatching inhibits growth but after normal growth at 28° for a month or more development can continue at 32°C (Michon, 1954). Critical heat-death temperature (Miles, 1963, *Nature* 199: p. 826), 33.3°C. Individuals that had been gradually acclimated were still active after 11 days at 95–100°F (Stinauer). Laying stops at 10–3°C, but sex organs do not regress, copulation and laying being resumed at once in warmth (Herlant-Meewis 1954). Some of the specimens exposed (Korschelt 1914), for about 12 hours, to temperatures of −2 to −5°C, survived. On January 11, at a depth of 110 cm., Kobayashi (1938) found 63 specimens when the temperature was −12.3°C. Presumably present physiology was acquired in arctic or subarctic climates, but in view of the experimentally indicated acclimation to temperatures that are fairly high for megadriles, the failure of the species to maintain itself in tropical lowlands to which it was introduced during the last fifteen years, does seem surprising.

Individuals have survived total immersion for periods up to 6 months. Maximum life expectancy has been found to be $4\frac{1}{2}$–5 years (Herlant-Meewis, 1967) but average life (Michon 1954) at 28° was 594 days, at 18° was 589 days. The averages dropped to 503 days (at 28°) and 519 (at 18°) for isolated individuals that could not copulate.

Rest periods sometimes are involved in the life cycle but little is known about them. Thus, estivation has been said to involve being tightly coiled, in an immobile state, within a small, mucus-lined cell and so apparently in a state usually called diapause. Activity was said to be resumed as soon as moisture again became adequate. Winter rest seemingly is mentioned only once in the literature and then some 78 years ago in a New Zealand article which indicated that "half grown" individuals occasionally hibernate in pots even when conditions supposedly were favorable for activity. Michon (1954), on the contrary, called the species homodynamous, i.e., without any diapause in its life history. Activity, in favorable conditions can be continuous throughout the year.

Feeding, at least usually, is selective in so far as ingestion of earth is concerned. The diet, according to results obtained from feeding experiments (Miles, 1963), must contain soil protozoa (but *cf.* Van Gansen, 1962). Ingesta may reach xl in ten minutes (Van Gansen), cx in 4 hours, the first defecation 8–15 hours after beginning of feeding.

Though present in the soil at Cracow, *E. foetida* is not stored by moles, *Talpa europaea*. Birds are said to refrain from eating the species. An Australian curlew, tricked into taking one of the worms instantly rejected the tidbit, then showed its disgust, presumably because of an unpleasant taste, by wiping off its bill.

Variation and Abnormality: Color restriction to transverse bands has been regarded as characteristic since 1845. The bands tend to be longer (anteroposteriorly) in culture beds of numerous earthworm farms perhaps as a result of much selection for many generations. Even when the bands are longest, pigment is lacking immediately under intersegmental furrows. Uniform coloration (in dorsum only?) was thought, prior to 1910, to justify recognition of a distinct variety that was later abandoned as appearance of uniformity was found to be due to strong contraction and perhaps more so, because of considerable variation in length of the bands. Recently, "uniformly" colored French worms have been referred to as f., var. or even subspecies *unicolor* (*cf.* Systematics, below). Banding, in America, usually is confined to, or is much more marked in, the dorsum. However, denser deposition often is associated with extension into the ventrum and, when banding is unrecognizable dorsally, even the ventrum sometimes has a uniformly dark pigmentation from one end of the body to another. Totally albinized individuals were not found by the author though on a few freshly preserved individuals the color was so faint that the bands were recognized only under a binocular. Pigment, in a few other specimens, was so sparse in postclitellar segments that the red color was visible only close to mD.

The recorded maximum for segment number is 131 but in the author's collections the largest number was 123. Relictuslike, abnormal individuals have had as many as 138–142 or 149–152 segments depending on which side of the body the count was made. Splitting of mesoblastic somites was involved in such cases and could have been involved, even when evidence enabling its recognition is lacking, in specimens with more than 123 segments.

The variation indicated in the précis with respect to extent and location of clitellum and tubercula pubertatis is based on individuals without recognizable indications of somite splitting during embryonic development.

Variation in number of chambers in the calciferous glands is likely to be greater than has been recorded. Variation in level of the calciferous opening into the esophagus may not be of special interest though considerable shortening of calciferous lamellae (formerly extending into, if not also through, x or perhaps even into ix) may have been involved in the ancestry of the genus *Eisenia*.

Atyphlosolate segments of the gut, in the author's worms from a wide variety of sources, were 8–15, and the number 15 was found only once though 16 and 22 were recorded (Hertling) for two German worms. Somatic segments, in those two worms, 107 and 119 respectively, were well within the *foetida* range, but other data that could have provided evidence for aberration or for a different taxon were not provided. Complicating the situation are possibilities of typhlosolar regression after posterior amputation (frequent) or of extension of the typhlosole into a caudal regenerate in appropriate conditions.

Almost no variation in number of hearts was found, but one heart of ix had no connection at all, though filled with blood, to the ventral trunk but instead, well down in the coelomic cavity, turned anteriorly to the posterior face of 8/9. One extra-esophageal trunk once was seen to pass into xii from whence it continued into xiii after giving off a slender branch to the dorsal vessel. Such instances almost certainly are to be regarded as reversions to an ancestral condition perhaps no more remote than origin of *E. foetida* as a species. Not now regarded as of any phylogenetic interest is adherence of the subneural trunk, through about one-fourth of the length of a single individual, to the parietes instead of to the nerve cord as along the remainder of the axis. The extra-esophageal of one side in an Italian worm did turn up in xii and pass to the dorsal trunk, without a recognizable connection thereto in the region of 9/10, or a continuation into xiii.

Transverse serrations ornamenting ectal ends of setae are sometimes thought to provide a character of sufficient systematic importance to be included even in generic and subfamily definitions. Little seems to have been recorded about the character in *E. foetida* and in particular as to which setae of which regions of the body are so modified. The ornamentation was not visible on setae in a number of *foetida* follicles randomly selected by the author but was easily recognized on larger setae from species of other genera.

Transversely elliptical areas of slight epidermal thickening, with a distinctively brilliant opacity like that supposedly distinguishing "f. *macedonica*" in *Eisenia rosea* and likewise including apertures of the *a* and *b* follicles, occasionally were observed. Blisters, with a quite compact coagulum separating the peritoneum from the longitudinal musculature, in a number of worms, extended through ix–xi from spermathecal pore levels to D.

Other individual variations in the genitalia are of the kinds now to be expected, penetration of a spermatheca into the coelomic cavity behind or in front of the septum with which it is associated, doubling of the spermathecal ampulla, doubling of an entire spermatheca so that each may have its own external aperture, absence of a single spermatheca or a seminal vesicle. Perhaps unique are the extra ovisacs, in some Italian worms with uniform pigmentation, located on the anterior face of 13/14 and connected with the female funnels and the oviducts.

Aberrations such as metameric anomalies and more or less marked regional as well as individual organ homoeoses (Gates, 1958) are rather common.

Regeneration: Head regeneration, in an anterior direction, is possible at each intersegmental level back to and including 23/24. Equimery can be expected from 9/10 forward and hypermery at any more anterior level. Heteromorphic tails may be regenerated at any of levels 20/21–54/55.

Tail regeneration in a posterior direction is possible back from 20/21, without a zero level posteriorly, and, as in case of head regeneration, throughout the entire year. Heteromorphic heads can be obtained, in unknown circumstances, at levels from 15/16 to 34/35.

Regenerative capacity (Gates, 1949, 1950) varies along the anteroposterior axis. Three regions, (1) of cephalic regeneration only regardless of direction, (2) of cephalic and caudal regeneration, presumably according to internally controlled circumstances, in either direction at each level, (3) caudal regeneration only. Boundaries of those regions and of possible subregions have not been precisely determined.

Gonads are not regenerated.

Results of beautiful operations, mostly on *E. foetida*, will be found in contributions by Avel and his school.

Parasites: CILIATA: *Anoplophrya lumbrici* (Schrank, 1803), Germany, *A. striata* Cepede, 1910, France. *A. vulgaris* Puytorac, 1954, France. *Maupasella cepedei* Puytorac, 1954, France. *M. nova* Cepede, 1910, Germany. *Metaradiophrya gardneri* Rees, 1961, Wales. *M. lumbrici* (Dujardin, 1847), Pennsylvania, Germany. *M. varians* Puytorac, 1964, France. SPOROZOA: *Apolocystis elongata* Phillips & MacKinnon, 1946, England. *A. gigantea* Troisi,

1933, New Jersey, Pennsylvania. *A. herculea* (Bosanquet, 1894), France, Germany. *A. lumbriciolidi* (Schmidt, 1854), France, Germany. *Cephalocystis singularis* Rees, 1962, coelomic cavities throughout the body, Wales. *Monocystis agilis* Stein, 1848, Germany, Sweden. *M. arcuata* Boldt, 1910, Germany, Sweden. *M. cambrensis* Rees, 1961, intestinal wall and coelom, Wales. *M. cristata* Loubatieres 1955, France. *M. cuneiformis* Loubatieres 1955, France. *M. densa* Berlin, 1923, Sweden. *M. elongata* (Phillips & MacKinnon, 1946), England, Wales. *M. lopadiformis* Loubatieres, 1955, France. *M. hispida* Loubatieres, 1955, France. *M. lumbriciolidi* Henle, 1845, France, Germany, New Jersey, Pennsylvania. *M. suecica* Berlin, 1923, Sweden. *M. turbo* Hesse, 1909, France, Germany, Sweden. *M. ventrosa* Berlin, 1923, Sweden. *Nematocystis elmassiani* (Hesse, 1909), France, also coelom, New England to Virginia including New Jersey, Pennsylvania, Germany. *N. lumbricoides* Hesse, 1909, Germany. *N. plurikaryosomata* Bhatia & Chatterjee, 1925, Kasauli, India (possibly = *elmassiani*, according to Rees, *in lit.*). *Rhynchocystis pilosa* (Cuenot, 1901), England, France, Sweden, New Jersey, Pennsylvania. *R. piriformis* Berlin, 1923, Sweden. *R. porrecta* (Schmidt, 1854), England, Germany, New Jersey, Pennsylvania. *Zygocystis cometa* Stein, 1848, France, Germany. *Z. eiseniae* Loubatieres, 1955, France. *Z. suecica* Berlin, 1923, France, Sweden. *Z. wenrichi* Troisi, 1933, New Jersey, Pennsylvania.

CESTODA: *Taenia cuneata* (Linstow, 1872) Cohn, 1900, cysticercoid, Italy.

NEMATODA: *Capillaria annulata* (Molin, 1858) Cram, 1926, larvae, longitudinal muscle preference, North America. *C. caudinflata* (Molin, 1858) Travassos, 1915, said to be cosmopolitan in chickens (in earthworms?). *C. contorta* (Creplin, 1839) Chapin, 1925, said to be cosmopolitan in ducks, gulls and fowls (in earthworms?). *Choerostrongylus pudendotectus* (Wostokow, 1905), larvae to infective stage, Europe, North America. *Dicelis filaria* Dujardin, 1845, sexual stages in seminal vesicles and coelom of genital segments, *Heterakis gallinarum* Schrank, 1788, which is a vector for *Histomonas meleagridis* (Smith, 1895) Tyzzer, 1920, the causative agent of blackhead of chickens and turkeys. *Metastrongylus apri* (Gmelin, 1790) Vostokov, 1905, after experimental exposure, England. *M. elongatus* (Dujardin, 1845), infective stages, North America, England. *Porrocaecum* sp., England. *Rhabditis pellio* Bütschli 1876 non Schneider 1866, nephridia, England. *Stephanurus dentatus* Diesing, 1839, Europe. *Syngamus trachea* (Montagu, 1811), larvae, England, North America, said to be cosmopolitan in turkeys. *Tetrameres fissispina* (Diesing, 1861) Travassos 1914, Africa. Asia, Europe, Oceania, South America (in earthworms?).

INSECTA: *Onesia sepulchralis* Meigen = *O. floralis* Robinaux & Desdoity, 1830, larvae phagocytized.

Hybridization: Popular literature during the last twenty-five years has contained much about hybrid earthworms, supposedly of superior qualities as bait and for enriching soils of organic gardeners. Lots purchased from the originator of the "soilution hybrid" and from dealers who advertised them for sale have been examined by three zoologists. Most of the specimens quite obviously were *E. foetida*, and occasional exceptions always were of well-known species.

Economic importance: More than a century ago an Englishman claimed that "all anglers praise the Brandling, and some have sung it in immortal verse." *E. foetida* now is being or in the immediate past has been raised or offered for sale in every state of the United States, possibly excepting Alaska and Hawaii (for which no information is available), and in several provinces of Canada. In very many places in each of those states or provinces, the brandling has been used for bait. Some estimate of the number of people who have purchased the worms in hope of enriching the soil of their gardens and fields might be possible if data were available as to number of organic-gardening books and magazines that have been sold since that sort of culture became a fad. Dealers often advertise for sale the worms under their own trade names, such as Soilution Worms, Egyptian Reds, Red Hybrids, California Hybrids, etc. Castings, accumulated in culture beds, also have been packaged and sold as fertilizer for potted plants. An earthworm oil advertised for relief of headaches and other human ills by a "Dr. Oliver" was analyzed by a chemist who found nothing likely to have come from earthworms.

Systematics: Cross breeding of banded individuals with unicolored French specimens (André, 1963), though with considerable difficulty, was accomplished. F_1 hybrids, with a pigment pattern intermediate between those of the two parents, were sterile (no hatching from 220 cocoons laid by four couples). Unicolored and banded forms, even though otherwise not morphologically distinguishable from each other, were regarded as "especes jumelles." Little is known about unicolored worms, but recorded variation in characters of major importance in the classical system, such as extent and location of clitellum as well as tubercula, does seem unusually high in the banded worms.

Closest relationships of *E. foetida* now appear to be with *E. nordenskiöldi* (Eisen, 1879). Previously identified specimens of both taxa that have been available to the writer are indistinguishable specifically from each other by any of the characters of the classical system. Indeed, the only difference that could be recognized in Siberian specimens of *nordenskiöldi* was a definitely greater number of atyphlosolate segments at the hind end of the body.

E. foetida, having close relationships only to European species, having originated in or close to Europe should be characterized, if the term is to have any more than a mere anthropochorous connotation, as European and not as holarctic.

Remarks: The species name indicates truthfully that much has been said about the odor supposedly given off by the worms. Having handled many thousands of specimens alive, the author has found no reason to complain of the odor which never seemed to be offensive. Moreover, a feminine zoologist who also has handled live specimens, compared the odor to that of a delicately fragant American flower. Nor has either of us experienced any difficulty in removing from the hands any stains produced by the yellowish material ejected through the dorsal pores.

One of the interesting rewards of a search through medical literature was the finding of a record of a specimen of *foetida* (identified by the Italian specialist Daniele Rosa) that was passed in the urine of a female patient.

Although damned by its scientific name, obviously banded individuals of *E. foetida* have evoked admiration in words such as "belle espece," "a really beautiful species" from systematists or "a very beautiful worm" from an angler (Bowlker, 1786, *The Art of Angling*).

A population of 1,000 individuals, according to Dr. A. W. Khan, increased in about six months to 456,380! The habitat to which the worms were introduced was shaded, irrigated plots covered with cow manure and leaves to a height of 2 feet. At the end of the period the worms were found in actual soil to a depth of as much as $1\frac{1}{2}$ feet. (*6th Ann. Progr. Rept. Scheme Res. Soil Zool., Lahore, Pakistan.*)

Eisenia hortensis

1890. *Allolobophora subrubicunda* f. *hortensis* Michaelsen, *Mitt. Naturhist. Mus. Hamburg* 7: p. 15. (Type locality, Hamburg, Germany. Types, originally in the Hamburg Museum but present condition unknown.)

1951. *Eisenia veneta* f. *hortensis*, Gates, *Proc. Natl. Acad. Sci. India*, B, 21: pp. 19, 21.

1958. *Eisenia hortensis*, Gates, *Breviora, Mus. Comp. Zool. Cambridge*, No. 91: p. 6.

1968. *Eisenia hortensis*, Gates, *ibid.* No. 300, p. 1.

BACTERIA, of nephridia: Knop, 1926, *Zeitschr. Morph. Oekol. Tiere*, 6; p. 601. CALCIFEROUS GLAND. Smith, 1924, *Illinois Biol. Monogr.* 9, 1: p. 26. Kreutz, 1936, *Zeitschr. Morph. Oekol. Tiere* 30: p. 792. CHROMOSOMES: 1952, Muldal, *Heredity* 6: p. 59 and Omodeo, *Caryologia* 4: p. 188.

Quadrithecal, pores at 9/10–10/11, close to mD. Males pores, in xv, near m*BC* and in a slight equatorial cleft almost confined to median half of *BC*. Male tumescences, extending into xiv and xvi or dislocating 14/15 and 15/16. Female pores, in xiv, just above *B*. Genital tumescences, some of the following, *a*, *b*/xi, xii, xvi, xxix, xxx, xxxi, xxxii, xxxiii, *c*, *d*/xi.

Clitellum, saddle-shaped, down to or nearly to B, (xxvi?), xxvii, xxviii–xxxii, xxxiii. Tubercula pubertatis, just lateral to B, longitudinal, two, uninterrupted bands of translucence, in xxx–xxxi sometimes invading xxxii slightly, or four, of subcircular outline, 30/31 uninterrupted. Setae, not closely paired, behind the clitellum $CD =$ or slightly $< AB < BC < AA < DD$ which is $<\frac{1}{2}C$, c setae often at or close to mL, follicles opening through genital tumescences obviously thickened but those of the a setae much shorter than those of other genital setae. Nephropores, inconspicuous, alternating irregularly, and with asymmetry, between three levels, slightly above B, slightly above D, much further dorsally. First dorsal pore, at 5/6. Prostomium, epilobous, tongue usually open. Color, reddish. Segments, 42–130 but usually (?) between 95–105. Size, 22–50 by $1\frac{1}{2}$–3 mm.

Septa, none thickly muscular or 13/14–14/15 fairly muscular. Pigment, red, in circular muscle layer. (Longitudinal musculature?) Calciferous gland, without sacs and moniliform widening, opening into gut in xi midway between insertions of 10/11 and 11/12 or at more anterior levels to one slightly behind 10/11. Gizzard, mostly in xvii (ending at?). Gut, narrowed and valvular in region of 19/20. Typhlosole, simply lamelliform, beginning in region of 20–25th segments and ending (unamputated individuals) in 80th–90th. Extra-esophageal vessels, joining dorsal trunk in xii (always?). Hearts, in vii–xi. Nephridial bladders, sausage-shaped, reaching laterally nearly to C or well beyond D. Holandric. Seminal vesicles, in ix, xi, xii. Sperm ducts, without epididymis. Atrial, TP, GS glands, present and supraparietal. Spermathecae, usually in ix and x, ducts slender, coelomic, shorter than to as long as the ampullae.

Reproduction: Sperm are matured, profusely as indicated by the brilliance of the iridescence on male funnels, and are exchanged during copulation, as shown by iridescence in the spermathecae. Spermatogenesis, according to Omodeo, is normal. Chromosome number, diploid, is 36. Reproduction formerly was assumed to be biparental. Nevertheless, parthenogenesis is now anticipated (*cf.* Polymorphism, also Gates, 1968).

Distribution: Simla, western Himalayas (probably at about the same elevation as at Darjiling), Darjiling, at 6,250 feet, eastern Himalayas.

Oregon, California, Arkansas, Illinois, Ohio, Virginia, New York, Maine (greenhouses only). Chile, Argentina.

Iceland, Germany, Russia steppes and Caucasus, Slovakia, Hungary, Greece (interception), Albania, Switzerland, Italy, Portugal, France, England, Ireland (interception).

Azores, South Africa.

Habitats: Earth of plant beds with a pH of 8.1,

"fetter Erde," rich organic soil, botanic gardens (England and Germany), arboretum (San Francisco), caves (Westphalia), greenhouses (Maine), "hot beds," temperature 80–85°F. (New York). Blenheim Palace Gardens, bank of stream polluted with sewage, chicken yard, earth receiving kitchen drainage, old cow manure, friable black soil saturated with septic-tank effluvium (Oregon). Under wet cardboard carton on a back porch of a farmhouse.

The species may have been secured in manure and/or compost more often than published records indicate. The Ohio record is of a single specimen, found in a toilet bowl, that came there in the city water. Paucity of records, according to McKey-Fender (*in lit.*), may be due to "the filthy places it inhabits. It takes some fortitude to collect it."

Variation: Number of segments of some Swiss worms, 50–80 by 4–5 mm., identified as *E. hortensis* during the nineteenth century, was 120–155. What may have been involved can only be guessed. Much of the abnormality seen by this author is now believed to be associated with genital polymorphism.

Polymorphism: Even in sperm maturing and exchanging populations, morphs can be recognized by absence, of atrial glands, of TP glands, of GS glands, of seminal vesicles of ix, each alone or in various combinations, with other deletions.

An athecal, metandric morph with one pair of seminal vesicles in xii and with vertical testis sacs is known from a single Irish aclitellate which had normally developed atrial and TP glands. Another morph, also known only from one specimen, has an extra pair of atrial glands and male pore clefts (without male pores!), in xvi.

Remarks: Two independent studies of the chromosomes seemingly provided no data to suggest possibility of parthenogenesis.

Eisenia rosea

1826. *Enterion roseum* Savigny, *Mem. Acad. Sci. Inst. France* **5**, (Hist. Acct.): p. 182. (Type locality, Paris. Types, in the Mus. Hist. Nat. Paris.)

1958. *Eisenia rosea*, Gates, Breviora, *Mus. Comp. Zool. Cambridge*, No. 91: p. 6.

CHROMOSOMES. Omodeo, 1952, *Caryologia* 4. COELOMOCYTES: Thermes, 1960, *Bull. Zool.* 27. GROWTH & REPRODUCTION: Barley, 1959, *Australian Jour. Agric. Res.* 10. Cernosvitov, 1930, *Zool. Jahrb. Anat.* 52. Evans & Guild, 1948, *Ann. Applied Biol.* 35, and *Ann. Mag. Nat. Hist.* ser. 11, 14. Graff, 1953, *Regenwürmer Deutschlands.* Michon, 1954, Thesis, Poitiers. Tuzet, 1946, *Arch. Zool. Exp. Gen.* 84. MUSCLES: Pop & Dragos, 1959, Stud. Univ. Babes-Bolyai, Ser. 2, 2 (Biol.). PARASITES: Loubatières, 1955, *Ann. Sci. Nat.* (Zool.) 17. RESPIRATION (rates): Byzova, 1965, *Rev. Ecol. Biol. Sol.* 2.

Quadrithecal, pores near mD, at 9/10–10/11. Clitellum, saddle-shaped, intersegmental furrows not obliterated, in 7–11 segments, xxiii, xxiv, xxv, xxvi, xxvii–xxxi, xxxii, xxxiii, usually xxv or xxvi–xxxi or

xxxii. Tubercula pubertatis, longitudinal bands just lateral to B, in xxviii/n, xxix–xxx, xxxi, but usually in xxix–xxxi. Male pores, at or below mBC, in an equatorial cleft. Male tumescences, reaching well into xiv and xvi, obliterating 14/15 and 15/16, of variable shape. Female pores, margin usually opposed so that pore can be found only by traction on adjacent epidermis, at eq/xiv, just lateral to B. Genital tumescences, including apertures of any couple of setal follicles in ix–xiii, including only apertures of ventral couples in some of xvi, xxiv, xxvi–xxxiii. Setae, present from ii, closely paired, AB a trifle $> CD$, $BC < AA < DD$, DD ca. $=$ or $< \frac{1}{2}C$, a, b/xv enlarged and with transverse serration ectally. Nephropores, inconspicuous, behind xvi usually alternating irregularly and with asymmetry between a level somewhat above B and another above D, the dorsal rank on each side often quite irregular. First dorsal pore, at or in front of 7/8 but usually at 4/5. Body section, behind the clitellum shortly elliptical to almost square, with a setal couple at each corner, the dorsum slightly convex, the ventrum flat, the lateral sides sloped toward mV. Prostomium, epilobous, tongue open. Color, alive-rosy or grayish and with orange or red clitellum; preserved-white with clitellum some shade of red or also white. Segments, 112–176, usually 120–140. Size, 23–90 by $2\frac{1}{2}$–5 mm.

Septa, increasingly thickened to 9/10 and then decreasingly so to 13/14 or 14/15 but none very thickly muscular. Pigment, none, yellowish or brownish flecks occasionally recognizable externally under high powers of the binocular presumably representing remains of defunct coelomocytes. Longitudinal muscles, pinnate. Calciferous sacs, in x, large, vertical, reaching into contact with 9/10 as well as 10/11, projecting well above and also slightly below the esophagus but scarcely constricted off therefrom and opening widely therein, lamellae on the posterior wall. Esophagus, often seemingly much widened in x because connective tissue conceals the sacs, slightly constricted at insertions of 10/11, 11/12, 12/13, with a vertically slitlike lumen in xi–xii, with an internal constriction at or behind insertion of 12/13 behind which the lamellae gradually narrow. Intestinal origin, in xv. Gizzard, mostly in xvii, (ending at?). Typhlosole, beginning in region of xxi–xxiii, anteriorly compressed dorsoventrally so that ventral face may be almost 1 mm. wide and with two longitudinal grooves thereon, posteriorly the shape in transverse section that of an inverted T which may be retained to the end located in region of 84th or 135th segments, leaving 19–37 metameres atyphlosolate. Extra-esophageal trunks, passing up to dorsal vessel in xii. Hearts, in vi–xi. Nephridial bladders, U-shaped, transversely placed, reaching into AA as well as beyond D. Holandric. Seminal vesicles, 4 pairs in ix–xii. Sperm ducts, with a long variously looped or coiled but not discoidal epididymis. Spermathecae, usually in x, xi, ampullae spheroidal to ovoidal or ellipsoidal, duct slender, as long as ampulla, within the coelom or associated septum. TP, GS and atrial glands, present.

Reproduction: Degeneration of sperm in cytophores, phagocytosis of sperm in seminal vesicles and coelom as well as resorption in spermathecal epithelia were described by Cernosvitov in 1930. Atypical spermatogenesis later recorded (Tuzet, 1946), resulted in amitoses, polyvalents, and giants with short tails. Individuals reared in isolation to sexual maturity (Evans & Guild, 1948) laid fertile cocoons. Individuals of some morphs clearly are male sterile and do not copulate. Individuals of other morphs perhaps may mature sperm that aggregate on male funnels but none have been found in spermathecae of any of the numerous specimens examined by this author. Biparental reproduction, in any of the anthropochorous morphs, is unknown.

Chromosome numbers, 53, 54, 72, 90, 108, 160–174.

Distribution: New South Wales, South Australia. North, South and Chatham Islands, New Zealand.

Hawaii.

Alberta, Ontario, Quebec, New Brunswick, Nova Scotia.

Washington, Oregon, California, Arizona, Oklahoma, Texas, Iowa, Missouri, Arkansas, Louisiana, Illinois, Michigan, Indiana, Georgia, Ohio, Virginia, North Carolina, Pennsylvania, New Jersey, New York, Vermont, Massachusetts, Connecticut, New Hampshire, Maine.

Mexico, including Lower California.

Colombia (at 2,600 m.), Peru (elevation not recorded), Chile, Argentina, Uruguay, Porto Alegre and Rio Grande do Sul, Brazil.

Iceland. Norway, Sweden, Finland. Russia, from the north through Ukraine and Crimea into Caucasus and the Moldavian SSR, including Urals and Pri-Urals, Byelorussia, Latvia, Lithuania. Poland, Germany, Denmark. Hebrides, Scotland, Clare Island, Lambay Island, England, Wales, Jersey (Channel Islands), Netherlands, Belgium, Czechoslavakia, Romania, Bulgaria, Jugoslavia, Hungary, Austria, Switzerland, France, Portugal, Spain, Italy, Greece, Rodi, Crete, Sicily, Corsica, Sardinia, Balearic Islands, Azores, Canary Islands, St. Helena, Tristan da Cunha.

Morocco, Algeria, Tunis, Libya and Egypt including oases, Eritrea, Cape Province to Orange Free State, South Africa.

Siberia, Manchuria, Korea. Transcaucasus, Turkey, Lebanon, Israel, Turkmenia, Tajikistan, Turkestan, Kazakstan, Kirgizia, Chuvash and Tatar Republics, Iran, Afghanistan. Chitral, Pakistan. Kashmir. Murree and Dhar, Simla, Darjiling, Calcutta (once), Kodaikanal, Palni Hills of South India.

Elevations not mentioned above; 6,000 feet and

above, Kashmir, Simla, Darjiling, Kodaikanal. 0–1,900 m. in Italy, 200–2,600 m. in Switzerland, 1,000–1,500 feet in Scotland, 1,900–2,100 feet in St. Helena.

Although *E. rosea* has been recorded from Hawaii, no specimens were obtained on any of the islands in various collections of the last dozen years. On the contrary, the species had not hitherto been recorded from St. Helena but appears to be not uncommon there as it was taken at five places during the last six years.

■During the last fifteen years some 360 specimens, intercepted by inspectors of the U. S. Bureau of Plant Quarantine, were from 15 countries including Spain, Jugoslavia, Tunisia, Mexico, and Iran. Although 13 worms were from Iran, no previous record for that country has been found.

Absence in tropical lowlands, certainly would not be apparent to anyone from a statement such as "introduced by man into all parts of the world and therefore cosmopolitan" (Cernosvitov and Evans, 1947: p. 23).

Habitats: Under stones, elm logs, boards, leaves, in and under grass cuttings, sawdust piles, compost, in slime on banks of fresh water, among sedge roots of undisturbed fen, along banks of swamps, brooks, streams, roadside ditches. Barnyards, gardens, cultivated fields, alfalfa fields, lawns, pastures, flood plains, taiga, forests, and steppes. Mineral soil, clay, silty clay loam, forest soils, soils rich in mull. Eichen-Mecurialis-mull, forestless areas on mountain tops (Caucasus), soils of a stand of *Picea* sp. and of a grove of *Corylus* sp.

Caves, England, Germany, Belgium, France, Italy, Austria, Jugoslavia, Hungary, Turkey, Bulgaria, Portugal, Afghanistan, Kentucky, Virginia. Mines, France. Strip mines, Illinois. Botanical gardens, Coimbra (Portugal), Oxford, Chelsea, Kew, Palermo. Greenhouses, Naples, Oklahoma, Maine (8 sets). Arboretum, Massachusetts. Wandering on streets after rain, Switzerland, Maine.

The only species widely distributed in virgin steppes of Russia, the only species found abundantly on forestless summits of mountains in the Caucasus. From soils with a *p*H of 4.9–8.0.

Found often enough in conditions that were thought to justify characterization as "amphibious."

A population with an average of 89.9 per square meter was observed in light alluvial soil of permanent pasture, Durham County, England.

Biology: Activity, including breeding, is possible year-round in appropriate circumstances but in much of the holarctic portion of the range summer drought and winter cold are passed in a resting stage. Estivation as well as hibernation is spent tightly coiled into a small pink ball. Activity supposedly is resumed as soon as moisture becomes sufficient. *E. rosea*, in central Maine, becomes active before the frost is out of the ground. That something more than

drought and temperature may be involved in diapause is suggested by two records. In Manchuria, at temperatures of 1.5–2.4°C, and at depths of 50–70 cm., some individuals were active though others were coiled into balls. The other record is of finding, during the summer, at depths of 25–50 cm., active individuals along with those in the coiled state.

Enforced duration of submersion in laboratory water has been thought to result in cessation of sex activity, regression of external sex characters and eventually many deaths. Individuals, however, are frequently found, in natural habitats where the soil is saturated.

Incubation, 60 days, puberty 150–180 days later, in Germany at 12°C. An average of 6 cocoons per worm was laid between February 1 and March 31. Average number of hatchlings per cocoon, 1.0. In England, Evans & Guild (1948) found incubation to be 17½ weeks with maturity being achieved in 50–62 weeks. Average laying was only 8 cocoons per year. In the vicinity of Adelaide (Australia) where *E. rosea* constitutes 55–75 per cent of the populations in permanent pastures and cultivated fields, the breeding period is of ten weeks in the spring. Duration of the clitellar phase of the life history, in France (Michon, 1953), was found to be 215–240 days following a preclitellar phase of 104 days. Average life at 18° (Michon, 1954) was 662 days for animals that did not diapause. The average increased to 776 days for animals that diapaused in the clitellar phase, to 795 days for those with the diapause in the preclitellar phase.

Unpleasant substances causing rejection of species such as *D. octaedra* and *E. foetida* presumably are lacking in *E. rosea* which is stored in Poland by moles (*Talpa europaea*).

Polymorphism: Spermathecae may be 4, of normal size or digitiform, only slightly protuberant into the coelom, the ampulla represented only by the translucent, thinner-walled ental portion, or 2 and then in x or in xi, or 6 with the third pair of pores at 11/12, or none at all. Spermathecal pores may become more lateral, as much as 0.5 mm. or more from mD in a partial reversion to a past ancestral state.

Of the seminal vesicles, the pair in ix, the pairs in ix, x may be eliminated, one or more pairs may be vestigialized or nearly so. Those of ix and x rudimentary, of xi only slightly larger, those of xii small, all functionless, is the state characterizing many individuals.

Male funnels may be more or less reduced in size though without loss of plications.

Atrial glands may extend into xiv and xvi, be confined to xv not even reaching to 14/15 and 15/16, be confined to the parietes except for strands of glandular material just visible between longitudinal muscle bundles, be wholly unrecognizable from the coelom though bulging body wall internally, or even unrecognizable within the parietes. Associated of course with such changes are those involving male tumescences and male pore clefts which also are decreased in size until little or no trace of either is recognizable, sometimes only a slight concavity containing the male pore or a circular patch of epidermal translucence. Male tumescences, along with the clefts (and the male pores?) may be doubled, the extra pair in xiv or in xvi.

Tubercula may be shortened and confined to xxix–xxx and, with loss of the TP glands, may become indistinct.

GS glands may be reduced in size as well as number and confined to the parietes or perhaps be completely aborted.

Only one set of organs may be affected by the parthenogenetic degradation, or two or more sets. If segment number, clitellar extent, and location are also taken into consideration, many morphs can be recognized. Attempts to subdivide the species into a few major varieties, according to number of spermathecae, number of seminal vesicles, location of genital tumescences, are of little use because of the great variation characterizing each of such taxa.

Variation and Abnormality: *E. rosea*, about the smallest of megadriles living outdoors in central Maine, seems to be of similar size from Mexico into Canada. American specimens, preserved in a relaxed condition have sometimes been as long as 145 mm. but of course were slenderer. Diameters of 5–8 mm. in the clitellar region are regarded as abnormal, that region of the body sometimes becomes bloated because of fluid accumulating in coelomic cavities. Reported lengths of more than 100 mm., if of contracted states, require confirmation and specification of other characteristics. Individuals with more than 150 segments in American collections were abnormal, the greater number of segments due to considerable splitting of embryonic somites. The range of variation in number of body segments, of material studied hitherto, seems to be greater than in the number of atyphlosolate intestinal metameres.

Although pigment is said to be absent, a faint yellowish to brownish equatorial striping of postclitellar segments sometimes can be recognized under the binocular, perhaps as a result of local accumulations of defunct eleocytes.

Transverse sections of the postclitellar portion of the body vary from transversely oblong to almost squarish, perhaps in accordance with methods of preservation, but without loss of slight convexity of the dorsum, the flatness of the ventrum, and the slope of the lateral sides toward mV. The hind end usually has a characteristically truncate shape.

Divergence of spermathecal pores from a location close to mD, not often reported, may represent a tendency to revert to some more lateral, or even ventral, ancestral site or only a phylogenetic lag in reaching the usual position. Sites of spermathecal pores, often almost undetectable, sometimes are advertised by the presence of small, tuberclelike porophores.

Complete doubling of a single spermatheca has been noticed. Other abnormalities that have been seen are attributable to splitting of embryonic somites or to the parthenogenetically permitted category.

Regeneration: Very little information is available in the literature. Recently, three regenerates were found. (1) A homomorphic head of 2 segments (–2 hypomery) at 4/5. (2) A homorphic but abnormal cephalic regenerate, of 5 or 6 segments (metamerism abnormal), with a second head comprising prostomium and peristomium (both normal) on the right side of the third segment. (3) A heteromorphic regenerate of 32 segments (metamerism normal) at the anterior end of a South African tail comprising 80 metameres of which 26 are atyphlosolate.

Hybridization: According to Muldal (1948, *39th Ann. Rept. Innes Hortic. Inst.*, p. 21), "the haploid number of 29 in *Eisenia rosea* is the sum of the numbers in two other species of the genus. It therefore seems not only that species with these different numbers can cross but also that polyploidy can follow the hybridization"! The species referred to are, *E. foetida* and *veneta* (Rosa, 1886). The latter is a congeries and by some is placed in *Dendrobaena*. Successful experimental interspecific hybridizations are unknown (*cf.* p. 117) among megadriles.

Parasites: CILIATA: *Anoplophrya lumbrici* (Schrank, 1803), Germany. SPOROZOA: *Apolocystis pertusa* Loubatières, 1955, France (= *N. elmassiani?*). *Dirhynchocystis elongata* Loubatières, 1955, France. *Monocystis hederacea* Loubatières, 1955, France. *M. proteiformis* Loubatières, 1955, France. *Nematocystis elmassiani* (Hesse, 1909), Germany. *N. lumbricoides* Hesse, 1909, Germany. *N. sinuosa* Loubatières, 1955, France. *Rhynchocystis ovata* Loubatières, 1955, France. NEMATODA: *Choerostrongylus pudendotectus* (Wostokow, 1905) larvae, Lithuania. *Dicelis filaria* Dujardin, 1845, sexual stages. *Metastrongylus elongatus* (Dujardin, 1845), Lithuania. *Rhabditis craspedocerca* Völk, 1955, Germany. *R. maupasi* Seurat in Maupas, 1919, Germany. *R. pellio* Schneider, 1866, England, Germany. INSECTA: *Pollenia rudis* (Fabricius, 1786), Europe. *Sarcophaga carnaria* (Linnaeus, 1758), Europe. ARACHNIDA: *Histiosoma murchiei* Hughes & Jackson, 1958, cocoons, Michigan.

Systematics: *E. rosea* already has a number of synonyms and more probably are to be added to the list.

Helodrilus (Bimastus) indicus Michaelsen 1907 is known only from the original description of five types that already were "weakened," i.e., macerated and now are unlikely to have any systematic usefulness.

The description of those types, in view of the condition of the specimens, provides no evidence for specific distinctness from *E. rosea*. This author's synonymization of Michaelsen's *indicus* has not been approved by some, but regardless of what Michaelsen may have had, the taxon no longer can be thought to have been endemic in India. The types, if not secured at the botanical gardens in Sibpur to which live plants long had been brought, could have been found in earth of plant pots brought from any one of a number of hill stations or even from Kashmir. Another not impossible alternative source could have been fresh vegetables obtained from ships stores.

Recent synonymizations and the present knowledge of lumbricid ecology allow no lumbricid endemicity in Asia south of the Hindu Kush and Karakorum ranges and from Baluchistan west to the Pacific. Indeed any endemicity of lumbricids south of the Tian Shan and Altai Mountains, Mongolia and Manchuria, will be quite unexpected.

E. rosea, according to several authors, belongs in the genus *Allolobophora*. The problem is discussed on a previous page (p. 96).

The species précis above omits mention of degraded anatomy of parthenogenetic morphs but does indicate characteristics to be expected in an ancestral amphimictic population.

Eiseniella

1900. *Eiseniella* Michaelsen, *Das Tierreich* 10: p. 471. (Type species, *Enterion tetraedrum* Savigny 1826.)

SYSTEMATICS: Michaelsen, 1932, *Jenaische Zeitschr. Naturwiss.* 67. Omodeo, 1956, *Arch. Zool. Italiano* 41. Pool, 1937, *Acta Zool.* 18. Pop, 1952, *Stud. Cerc. Sti. Cluj* 3, 3–4.

Characters of the type species that may prove to be definitive at a generic level are as follows:

Calciferous sacs, in x, digitiform, opening posteriorly into the gut ventrally in region of insertion of 10/11. Esophagus of nearly uniform width through xi–xiv, calciferous channels narrow, lamellae low. Intestinal origin, in xv. Gizzard, in xvii, weak, 17/18 not fenestrated(?). Typhlosole, simply lamelliform. Extra-esophageal trunks, joining dorsal vessel in xii. Hearts, in vii–xi. Nephridial bladder, shortly sausage-shaped. Nephropores, inconspicuous, behind xv alternating irregularly and with asymmetry between a level just above *B* and one above *D*. Setae, not closely paired behind the clitellum. Longitudinal musculature, pinnate.

Too little is known about the somatic anatomy of those species included at one time or another in the classical *Eiseniella* to warrant a guess as to whether any can be congeneric with *E. tetraedra*.

Eiseniella tetraedra

1826. *Enterion tetraedrum* Savigny, *Mem. Acad. Sci. Inst. France* 5 (Hist. Acct.): p. 184. (Type locality, Paris. Types, in the Mus. Hist. Nat. Paris.)
1937. *Eiseniella tetraedra* f. *typica*, Cernosvitov, *Rec. Indian Mus.* 39: p. 107.

ANATOMY: Hesse, 1894, *Zeitschr. Wiss. Zool.* 58: p. 69, etc. Pool, 1936, *Acta Zool. Stockholm* 18. BACTERIA and BACTEROIDS: Knopf, 1926, *Zeitschr. Morph. Oekol. Tiere* 6: p. 600. CALCIFEROUS GLAND: Smith, 1924, *Illinois Biol. Monogr.* 9, 1. CELLULASE: Tracey, 1951, *Nature* 167: p. 777. CHROMOSOMES: Muldal, 1952, *Heredity* 6. Omodeo, 1952, *Caryologia* 4: p. 200; 1955, *ibid.* 8: p. 162. COCOONS: Evans & Guild, 1948, *Ann. Applied Biol.* 35, *Ann. Mag. Nat. Hist.*, ser. 11, 14: p. 714. DIAPAUSE and QUIESCENCE: Omodeo, 1948, *Boll. Zool.* 15: p. 12. GROWTH. Michon, 1953, *Compt. Rend. Acad. Sci. Paris* 236: p. 2347; and 1954, *ibid.* 238: pp. 2200, 2457, and 1957, *Année Biol.* 33; 367–376. MUSCULATURE: Pop, 1941, *Zool. Jahrb. Syst.* 74: p. 521. OXYGEN RELATIONSHIPS: Knoz, 1957, *Vest. Ceskoslovenske Spol. Zool.* 21: p. 203. PARASITES: Berlin, 1924, *Arch. Protistenk.* 48: p. 77. Malevitch, 1940, *Bull. Soc. Nat. Moscow,* (*Biol.*) 49: p. 36. Meier, 1954, *Arch. Protistenk* 100; 1956, *ibid.* 101. Oliver, 1962, *Jour. Parasitol.* 48. Puytorac, 1957, *Arch. Zool. Exp. Gen.* 94, N & R, 2: p. 118. Völk, 1950, *Zool. Jahrb. Syst.* 79. REACTIONS (to temperature and atmospheric humidity): REGENERATION: Gay, 1963, *Bull. Soc. Zool. France* 88. Heimburger, *Ecology* 5: p. 276. REPRODUCTION: Cernosvitov, 1930, *Zool. Jahrb. Anat.* 52: p. 493. Gavrilov, 1939, *Acta Zool. Stockholm* 20. TYPHLOSOLE: Hertling, 1923, *Zeitschr. Wiss. Zool.* 120: p. 210. REGENERATION AND NEUROSECRETION: Michon *et al.*, 1964, *Compt. Rend. Acad. Sci. Paris* 259. RESPIRATION (rates): Byzova, 1965, *Rev. Biol. Ecol. Sol.* 2.

Quadrithecal, pores at 9/10–10/11, at one or another of several levels in *DD*. Male pores, in xv, at or somewhat below *C*, each at lateral end of a deep transverse cleft. Male tumescenses, slight, involving only immediate margins of the male pore cleft. Female pore, in xiv, median to *A*. Genital tumescenses, including *a*, *b* in any of viii–xii, xiv–xv, xviii–xxv, (rarely including *c*, *d*, in ix–xi, xii). Clitellum, saddle-shaped and then down nearly to *A*, or mostly annular but not so well developed ventrally, xxii, xxii/n, xxiii–xxvi, xxvii/n, xxvii. Tubercula pubertatis, longitudinally bandlike, uninterrupted by intersegmental furrows, with greyish translucence centrally, usually between eq/xxiii and eq/xxvi, but subject to shortening or even some elongation. Setae, paired, *AB* ca. = *CD*, *BC* < *AA*, *DD* < ½*C*, *a*, *b* of male pore metamere more closely paired than elsewhere, more or less sigmoid but ornamented near tip by irregularly interrupted circles of fine serrations, (setae of genital tumescenses?). Nephropores, inconspicuous, behind xv, irregularly alternating, and with asymmetry, usually between levels above *B* and above *D* (?). First dorsal pore, at (3/4, 4/5), 5/6, or even more posteriorly. Prostomium, epilobous, tongue open. Body, in section nearly circular in fron of the clitellum, posteriorly almost squarish and with a pair of setae at each corner, anus terminal. Color, brownish, brown, yellowish or reddish brown, sometimes with a golden tinge, "golden yellow." Segments, 50–92 but usually 74–78. Size, 30–58 by 2–4 mm.

Septa, none thickly muscular. Pigment, in circular muscle layer. Longitudinal musculature, pinnate. Peritoneum, blistered away from the musculature in *BC* of ix–xii. Calciferous sacs, in x, digitiform, anteriorly or anterolaterally directed and then reaching or connected by a cord to 9/10, or dorsally directed, opening into gut ventrally and just in front of the insertion of 10/11. Esophagus, of uniform width or nearly so through xi–xiv. Gizzard, in xvii weak, 17/18 not fenestrated. Typhlosole, beginning in region of xx–xxii, simply lamelliform, ending abruptly (unamputated individuals) in 68th–78th segment. Extra-esophageal vessels, joining dorsal vessel in xii. Hearts, in vii–xi. Nephridial bladders, shortly sausage-shaped transversely placed, not reaching C. Holandric. Seminal vesicles, in xi, xii. Sperm ducts, with well-developed epididymis. Spermathecae, with short coelomic stalks. Atrial, TP and GS glands, present and supraparietal.

Reproduction: Atypical mitoses in the testes, as long ago as 1894, were recognized. Sperm, according to Cernosvitov (1930), are phagocytized by lymphocytes in the seminal vesicles and/or the coelom. Uniparental reproduction was experimentally established (Gavrilov, 1939), for the first time for any megadrile, in E. tetraedra. Chromosome number, in oogenesis (not less than ten specimens, Muldal, 1952), is 72 and "presumably tetraploid." Reproduction, according to the same author, is obligatory parthenogenesis but his promised "cytology of parthenogenesis in Lumbricidae" was not published. Chromosome numbers of 50 and 54 (Omodeo, 1955), regarded as triploid, were found in Italian worms for which reproduction was also said to be obligatory parthenogenesis.

Sperm never were found in the spermathecae of the author's specimens from six states and from several provinces of Canada, as well as worms from India, South Africa, Mexico, Italy, Madeira, and the Channel Islands. Nevertheless, many individuals are not male sterile, as shown by aggregation of mature sperm on the male funnels and their presence at least in the ental half of the sperm ducts. However, the spots of iridescence on the male funnels usually were peripheral and such as to suggest sparse maturation. Spermatophores rarely were found, and one reason for uniparental reproduction by sperm-maturing individuals may be simply that they do not copulate. However, on French worms spermatophores several times were found, and some of them did contain sperm.

Distribution: Naini Tal in the western Himalayas and Ootacamund in the Nilgiri Hills of a southern part of the Indian Peninsula.

Southwest Australia and New South Wales, Australia. Macquairie Island and New Zealand.

British Columbia, Ontario, Quebec, New Brunswick, Nova Scotia (Canada). Washington, Oregon, California, Colorado, Missouri, Arkansas, Wisconsin (driftless section), Illinois, Michigan, Indiana, Ohio, North Carolina, Virginia, District of Columbia, Pennsylvania, New Jersey, New York, Vermont, Massachusetts, New Hampshire, and Maine.

Mexico, Peru (at an elevation of 4,300 m.). Chile and Juan Fernandez Island.

Iceland, Norway, Sweden, Lapland, Finland, Russia, including Carpathians, Urals and Pri-Urals, Ukraine, Crimea and Caucasus, Estonia, Latvia, Lithuania, Denmark, Germany, Poland, Czechoslovakia, Netherlands, Belgium, Faroe Islands, Hebrides Islands, Scotland, Ireland, Isle of May, England, Wales, Jersey, Guernsey and Alderney of the Channel Islands, France, Switzerland, Austria, Hungary, Romania, Bulgaria, Jugoslavia, Albania, Italy, Spain, Portugal (Balearic Islands?), Elba, Corsica, Sardinia, Sicily, Crete, Rhodes, Greece.

Azores, Madeira, Canary Islands, Cape Verde, St. Helena (first record), Tristan de Cunha, South Africa, Morocco, Libya.

Transcaucasia, Turkey, Syria, Lebanon, Israel, Iran, Afghanistan, Turkestan, Kazakstan, Chuvash, and Tartar Republics. Tajikistan (where it was found in virgin, meadow-marshy soil of the Hissar Valley).

The distribution is typical in several respects and provides a standard for subsequent comparisons. The list of habitats below provides little reason for suspecting that E. tetraedra is more likely to be carried around than other earthworms. In fact, during the last fifteen years, only fourteen specimens (from five countries including Mexico, herewith recorded for the first time) were intercepted by the U. S. Bureau of Plant Quarantine. Yet, the distribution could have been achieved only by frequent (and often repeated?) transportation. Also noteworthy is the fact that colonization by the earthworm has largely (a rather surprising exception of course is Hawaii) paralleled colonization by Europeans, hill resorts of India, Australia, New Zealand, but not including China and Japan. Islands such as Tristan da Cunha, Juan Fernandez, the Faroes, and the Hebrides were colonized but not Hainan, Formosa, and the Andamans. That the species was taken only to just those islands and countries that are mentioned above seems unlikely. Absence of E. tetraedra in tropical Africa and America and presence in glaciated areas of the Holarctic except in much of the more recently settled Canadian prairies also seem significant.

Accordingly, there are some reasons, in addition to others mentioned below, for thinking that E. tetraedra attained its present state in arctic or subarctic climates just below the Pleistocene glaciers of Europe. From that particular area the species has been distributed by man to the north, east, and west, as well as to the south, since recession of the glaciers. In such circumstances the species should be considered European and not Holarctic.

Habitats: Wells, springs, underground waters, mountain torrents, banks of hillside rills, of ditches, brooks, streams, canals, rivers, pools, ponds, lakes, at depths of 0.5–0.6 m. in Peruvian mountain-lakes, under stones and moss in running water, among plant roots in brooks, streams, and swamps, under dead leaves and rotting vegetation on brook or stream beds, under flood refuse, in wet rubbish, in leaves of forest ditches, in peat moor, swampy soil under alder near spring, in swampy fir forest, wrack bank soaked with fresh water, in watercress of airplane stores, slime, mud, highly organic mud, moist sand, wet gravel, damp clay soil, warm earth near hot spring, under bark of rotten log, in compost, in settling tanks of sewage plants, under dung pats, in "pure dung," in horse and sheep dung, in a "nearly saline" meadow, in a drained lake bed (Virginia) where soil was not especially moist. In soil: in rather dry abandoned hayfield (once, Maine), of pasture some distance from nearest water (once, Scotland), with pH of 7.4, with a pH from 6.8 to 8.5, a moisture content of 25–35 per cent and an organic content of 4–5 per cent (22 Ohio counties). Lightly leached chernozem, alkaline chernozems. Forests, steppes, taiga (Russia). Caves, in Austria, Belgium, Britain (the commonest megadrile in English caves), Crete, Czechoslovakia, France, Germany, Greece, Hungary, Italy, Jugoslavia, South Africa, Switzerland. Mines, France.

Earth with potted plants. Greenhouses (Maine, Indiana, Finland). Outside flower beds of greenhouses (Maine). Botanic Gardens, Oxford. Arboretums, Boston. Gardens, Chile, Maine.

From usually unspecified sites at elevations of 1,600 m. (Carpathians), 1,800 m. and 2,000 m. (Bulgaria), 1,836 m. and 1,900 m. (Switzerland), 2,500 m. (Italy), 2,100 m. (Asia Minor), 2,300–4,300 m. (Peru).

In dense moss of swift streams, one of the dominant animals. In marshes of the Hebrides, the dominant species.

Biology: Activity, and presumably also breeding, is year-round if conditions are favorable. Quiescence, however, is required by climatic conditions in many areas, during summer as well as winter. Estivation, involving immobility and tight coiling in a small mucus-lined cavity, ends as soon as moisture reaches the appropriate level. Because of this the species has been called homodynamous and quiescence has been said to be facultative (rather than obligatory as when imposed by internal conditions). Actually, of course, estivation seems to be no less obligatory because it is externally imposed than if, as it were, self-required. Whether hibernation involves quiescence or diapause is unknown. Until that has been learned the characterization of homodynamous remains tentative.

Populations of 20–150 per square meter were reported.

Hatching is after $3\frac{1}{2}$ weeks. Usually only one hatchling emerges from a cocoon but in a series of 198 cocoons, two wormlings hatched from $4\frac{1}{2}$ per cent of the cocoons. Average duration of life (Michon, 1954),

at 18°C, was 239 days, but at 9° was 279. Intermittent alimentation, at 18° raised the average to 306 days, but the average fell to 221 for individuals isolated to prevent copulation.

Optimal pH for the species has been said to be 6.5. Optimum temperature for growth is said to be 18°. Suboptimal temperatures are less harmful than supraoptimal.

Food, at the Endgebieten der Schlei, was found to be diatoms. Individuals of *tetraedra* are not stored by moles in Poland.

Variation and Abnormality: Little information is as yet available in the literature with regard to individual variation. Some 125+ specimens from six countries, numerous states and provinces, as well as four oceanic islands, provided most of the data on which the précis was based. For instance, calciferous sacs were characteristically digitiform in every specimen though, when dorsally directed, the sacs at first appeared to be of the vertical sort present in species of *Allolobophora*.

Abnormalities often are mentioned in the literature but usually without the information required to decide whether embryonic aberration, regenerative abnormality, or parthenogenetic and genetically determined degradation or modification was involved.

Instances of asymmetrical and symmetrical homoeoses resulting from halving or suppression of mesoblastic somites on one or both sides of the main axis have been seen though all such records are as yet unpublished.

Regeneration: Incidence of anterior and of posterior amputation usually seems not to be low. Autotomy of tail portions may be easily induced but anterior autotomy is unknown. Tail regenerates have been found by the writer occasionally but none of the records have been published.

Head regeneration had followed anterior amputation except in two of the author's worms which had lost the first eight and nine segments. Regenerates had one to five segments, and were at various levels back to 12/13 at which a head of three segments was abnormal.

Polymorphism: Rampant polymorphism characterizes *E. tetraedra*. Perhaps almost as many parthenogenetic morphs are to be recorded as in case of *D. octaedra*. Published characterizations in the past almost always have provided far too little information to enable distinguishing parthenogenetic morphs from those with embryonic aberrations and others with regenerative abnormalities.

Morphs have been recorded with male pores in ix, x, xi, xii, xiii, xiv, xiv and xv, xv, but without even such scraps of necessary information as location of female pores or ovaries. Individuals with male pores in ix–xi, possibly also xii, now seem more likely to have resulted from hypomeric head regeneration. Location of female pores, or of ovaries, alone would have enabled determination not only of the origin of the abnormality but also, in some cases, identification of the parthenogenetic morph.

Uniparental morphs may be athecal, quadrithecal, or with one or more of the battery lacking, or with some or all spermathecae more or less rudimentary. A hyperthecal individual, with paired spermathecae in each of segments v–viii, had resulted from regeneration of a five-segment head at 8–9. This is the first time that spermathecae were found in a lumbricid head regenerate. Spermathecae may be in the segment in front of or behind the level at which they open. Often a spermathecae may be in two segments and more or less deeply constricted by the intervening septum.

Seminal vesicles often are four pairs but they may be quite rudimentary in x or ix, x, or unrecognizable in x (but represented by microscopic vestiges?). Vesicles of xi, xii often are constricted into dorsal and ventral portions that are unequally developed. When the vesicles of xii are of maximal size they appear to reach into xiv but actually are contained within posteriorly directed pockets of septa 12/13 and 13/14. Such pockets had been formed in a number of individuals in which vesicles of xii were small and confined to that segment!

The epididymis, at best development, is a disc of closely crowded loops but may be represented by several unaggregated, free loops, a single long U-loop, several short loops, slight sinuosities, or may be wholly lacking.

Hypergynous and hyperandric morphs were lacking in the author's collections but, in some German locality, five-sixth of the hatchlings had gonads, presumably ovarian, in xiv (hypergyny).

Atrial, TP and GS glands can be reduced in size or even completely eliminated. Presumably a next step after disappearance of the GS glands is elimination of the genital setae.

As each of these series of organ degradations and eliminations can be associated with other series in various ways, as well as with considerable diversity in number and location of genital tumescences, it is obvious that a long list of morphs is possible. Many already have been seen though not as yet the hyper- and hypo-andry and gyny that are anticipated.

Since the above was first written an athecal hypergynous morph has been seen. Extra ovaries were in xiv rather than xii.

Systematics and Phylogeny: An amphimictic population of *tetraedra* is unknown and, as in case of *B. parvus* and Levinsen's *eiseni*, its characterization must be deduced from least degraded structure of parthenogenetic morphs. This is not difficult for some of the *tetraedra* anatomy. The ancestral biparental population very probably was quadrithecal, with pores at 9/10 and 10/11, holandric and with four pairs of seminal vesicles, with a well-developed

epididymis in each male duct and with functioning atrial, TP and GS glands. The male pores probably were, as is usual throughout biparental lumbricids, in xv. The clitellum may well have been saddle-shaped—some data hint that annularity has appeared in certain lumbricids since establishment of parthenogenesis.

Spermathecal pores now usually are at one or the other of two locations, midway between D and mD, or nearer D. The more ventral location presumably is more ancient phylogenetically. Then, does location of the pores near D represent retention of a primitive character by some less-advanced morphs or does it represent a parthenogenetically permitted reversion to an ancestral condition? Both locations theoretically could have been attained as a result of parthenogenetically permitted shifts in different morphs from a still more ventral location CD.

The female pores provide similar difficulties. Those apertures presumably are located, in most lumbricids, lateral to B. Then, is the *tetraedra* location, median to A to be regarded as a reversion to an ancestral condition enabled by the parthenogenesis or as an ancestral character retained long after it had been lost by other lumbricids? In either case, what is the significance of the rarely recorded location of the female pores in AB? A halfway step toward complete reversion or a halfway prediction of a future evolutionary change seems equally possible. Because of the almost complete absence of reported locations elsewhere than median to A, that position is now assumed to characterize the biparental population even though there is little reason for believing that the character under consideration has been carefully studied in the past.

Male pore locations had considerable importance in the classical system. Morphs of various sorts but with male pores in xiii were referred to as subspecies, variety, or forma *typica*. Morphs, closest to the ancestral biparental population because of presence of male pores in xv, were referred to as subspecies, variety or forma *hercynia!* Some morphs have two pairs of male pores, in xiv and in xv, each pore in association with normal terminalia including cleft, tumescence, and atrial gland. Male ducts of a side, almost universally throughout Group 2 megadrile families, come into contact shortly behind the last male-funnel septum and continue back together so as to open, after union, to the exterior by a single primary male pore. That near universality permits a deduction that presence of a single pair of male pores is a character acquired long ago and early in evolution of Group 2 megadrile families. Parthenogenesis, then, almost as if by a single macromutation, had enabled elimination of a condition characterizing Group 2 megadriles since their first appearance. Independent opening to the exterior of the sperm ducts now is known from parthenogenetic morphs of two lumbricid genera but has been reported elsewhere only from supposedly biparental species belonging to the quadriprostatic section of one Octochaetid genus (*Hoplochaetella*).

Male pores of the most common *tetraedra* morphs are in front of the female pores, a situation unknown elsewhere in Group 2 megadrile families. Parthenogenesis, in such morphs, enabled elimination of a character otherwise universal in Group 2 families but as yet only with a near approximation of return to the microdrile condition that characterizes Group 1 megadrile families, for male ducts of a side still unite and still open more than one segment behind the last male funnel-septum. Quadriporal morphs show that male terminalia can be doubled so as to allow one pair of male ducts to acquire external pores independently of the other pair but that they still have to do so at segmental equators. Is, then, elimination of the posterior atrial glands, reunion of sperm ducts of a side to open by a single pore in xiv, subsequently allowed by parthenogenesis? Some morphs, inadequately characterized in the literature, presumably could have arisen in that way, though two different processes, independently of the parthenogenesis, can bring about the same result. Whether male terminalia got into xiii from xv by a single step, by two steps one of which involved transfer from xv into xiv, or otherwise, is unknown and as yet intermediate morphs that might throw some light on the problems seem not to have been recognized.

The species was defined, Michaelsen (1900) and Cernosvitov (1947), as having closely paired setae, but Omodeo (1956) omitted the "closely." From the coelom the follicles certainly appear to be closely paired but external appearances sometimes, perhaps because of the poor preservation, seemed to belie the adverb.

Characterization, in the précis, of the nephridial bladders must be regarded as tentative. Preservation often was too poor to allow certainty as to the shape of the bladder which may perhaps have been shortly *J*-shaped.

Friend's *E. macrura* has 160 segments and "wide apart setae," two characters (unless developmental abnormality was responsible for each) that at very least require re-examination of the synonymization with *E. tetraedra*.

Other taxa, possibly not parthenogenetic morphs of *E. tetraedra*, have been placed as species in *Eiseniella* but too little is known about their somatic anatomy to enable a decision as to whether they really do belong there. Only when the requisite data *re* somatic systems are available will it be possible to be certain as to how many of the characters in the précis of the type species can go up into a generic definition.

Relationships, as indicated by the calciferous sacs, appear now to be closest with *Lumbricus*. If nephridial bladders of *E. tetraedra* also are *J*-shaped, they will provide further and perhaps equally important evidence for closeness to *Lumbricus*.

Remarks: *E. tetraedra* is markedly anthropochorus, highly polymorphic, and by spelunkers is thought to be troglophile. Characterization as homodynamous may or may not be equally justified. The species also has been said to be amphibious, limicole, hygrophile, stenobiote.

Tips of clitellar setae once were said to be hastate, a characterization that should be confirmed. Further information might be systematically useful.

Absence of widening of the gut in xi–xii of *E. tetraedra* is not associated with marked reduction in size of the central lumen. Calciferous lamellae accordingly are small, squarish in x, narrowly rectangular in xi, xii, as was obvious in somewhat macerated individuals in which much of the intestinal epithelium at least mesially had disappeared. Lamellar size evoked comments by Smith (1924) who thought "something akin to retrogression of the organ is taking place in some specimens at least." Perhaps differences in lamellar sizes are functional and can be correlated with existence in aquatic conditions and in certain kinds of soils.

Text and figures (Smith, 1924) provide no hint that the characteristic shape and origin of the calciferous sacs were recognized, presumably because of reliance solely on microtome sections. Careful dissections scarcely could have failed to reveal digitiform shape of the sacs.

Especially noteworthy in connection with *E. tetraedra* is the fact that organs, evolved as or very soon after polychaetes became oligochaetes, and present in almost every biparental species, have completely eliminated without vestiges, during a period of time so short as to permit little or no evolutionary modification of somatic anatomy.

The surprisingly little variation in segment number at a single locality, that was mentioned in a recent publication, may have been due to a clonal origin from a single parthenogenetic individual accidentally introduced by man.

Parasites: CILIATA: *Anoplophrya alluri* Cepede, 1907, (=*A. lumbrici?*), France. *A. nodulata* Meier, 1954, Germany. *Maupasella mucronata* (Cepede, 1910), France. *M. nova* Cepede, 1910, Germany. *Protoptychostomum simplex* (André, 1915) Raabe 1949, Switzerland. *Schultzellina mucronata* Cepede, 1910, France. SPOROZOA: *Apolocystis vivax* Berlin, 1923, Germany, Sweden. NEMATODA: *Hystrichis tricolor* Dujardin, 1845, Ukraine or Moldavia (also Asia?) *Rhabditis pellio* Schneider, 1866, Germany. ARACHNIDA: *Histiosoma murchiei* Hughes & Jackson, 1958, found in 15% of 85 Michigan cocoons, perhaps should not be called a parasite or even a predator but embryos in the infested cocoons did not survive.

Hesse's comment *re* his examinations of *E. tetraedra*

was "Jamais de Gregarines." However, cysts filled the seminal vesicles of Maine worms and also were present in hypertrophied testes as well as in coelomic cavities. Believed to be a pentastomid, by R. E. Crabhill, was an incomplete and mutilated parasite from a hole in the intestinal wall behind the gizzard of an Alderney (Channel Islands) specimen.

E. tetraedra can be infected experimentally by species of the nematode genus *Metastrongylus* but natural infestations have not as yet been recorded.

Lumbricus

1758. Lumbricus (part Linnaeus, *Syst. Nat.* (ed. 10), p. 657. (Type species, supposedly *L. terrestris* Linnaeus 1758, but certainly in part only, without a formal designation that is satisfactory to the International Commission of Zoological Nomenclature.)

ANATOMY: Hesse, 1894, *Zeitschr. Wiss. Zool.* 58. BACTERIAL SYMBIOSIS: Knop, 1926, *Zeitschr. Morph. Ökol. Tiere* 6. CALCIFEROUS GLANDS: Kreutz, 1936, *ibid.* 30. Smith, 1924, *Illinois Biol. Monogr.* 9, 1. CHROMOSOMES: Muldal, 1949, *Proc. Linnean Soc. London* 161; 1952, *Heredity* 6. COCOONS: Evans & Guild, 1948, *Ann. Mag. Nat. Hist.*, Ser. 11, 14. GROWTH and REPRODUCTION: Evans & Guild, 1948, *Ann. Applied Biol.* 35 and *Ann. Mag. Nat. Hist.*, Ser. 11, 14. Janda & Gavrilov, 1939, *Vest. Cesk. Zool. Prase* 6-7. Kollmannsperger, 1934, *Die Oligochaeten des Bellinchengebietes* (Köln). Michon, 1954, "Contribution experimentale a l'etude de la biologie du Lumbricidae," Thesis, Poitiers. PARASITES: Berlin, 1924, *Arch. Protistenk* 48. Meier, 1954 and 1956, *ibid.* 100 and 101. Troisi, 1933, *Trans. American Micros. Soc.* 52. PHYSIOLOGY: Laverack, 1963, *The Physiology of Earthworms* (Pergamon Press). REGENERATION: Gates, 1953, *American Midland Nat.* 50. TYPHLOSOLE: Hertling, 1923, *Zeitschr. Wissensch. Zool.* 120.

Calciferous sacs, in x, digitiform to pyriform, opening into gut posteriorly and ventrally in region of insertion of 10/11, lamellae on lateral walls. Esophagus, widened and markedly moniliform in xi–xii, in those segments with a vertically slitlike lumen which widens as lamellae gradually narrow behind 12/13. Intestinal origin, in xv. Gizzard, mainly in xvii, (ending at?). Typhlosole, high, rather thick and nearly oblong vertically in cross section, lateral faces deeply grooved vertically, grooves not continuous across ventral face. Extra-esophageal trunks, joining dorsal trunk in region of ix–x (posterior continuation?).[1] Hearts, in vii–xi. Nephridial bladder, *J*-shaped, closed end laterally, duct passing into parietes near *B*. Nephropores, obvious, behind xv irregularly alternating, with asymmetry, between levels just above *B* and well above *D*. Setae, closely paired. Longitudinal musculature, pinnate. First dorsal pore, anterior to 10/11. Prostomium, tanylobous. Pigment, lacking immediately underneath intersegmental

---

[1] Junction with dorsal trunk in region of 9/10 has been found in four species. Posteriorly conditions seem to be subject to considerable variation in one species, but whether incomplete vestigialization, reversion, presence or absence of blood in a delicate vessel, presence or total absence of a vessel is involved was not determined.

furrows, in circular muscle layer. Body compressed dorso-ventrally behind clitellum and with a more or less trapezoidal transverse section.

Quadrithecal, pores in region of *CD*, at 9/10–10/11. Clitellum, saddle-shaped. Tubercula pubertatis, lateral to *B*, continuous longitudinal bands of epidermal thinness. Male pores, in *BC*. Female pores, above *B*. Genital tumescences, only around *a*, *b* setae. Holandric, testes in unpaired sacs, (between nerve cord and ventral vessel?). Seminal vesicles, in ix, xi, xii (the latter contained within posteriorly directed pockets of 12/13). Spermathecae, in ix and x. (Atrial and TP glands, lacking.)

Reproduction: Each species that was studied cytologically (Muldal, 1952) is diploid, $n = 18$, $2n = 36$, with biparental reproduction obligatory.

No evidence suggesting possibility of parthenogenesis was found by the author during study of his material. As there now is little reason (also *cf.* Janda & Gavrilov, 1938, and Evans & Guild, 1948) to believe that amphimixis is not obligatory throughout the genus, a seemingly generic character need not be repeated for each of the species considered below. The single exception, still unexplained, is provided by the fully developed young worms that emerged from 3 of the 1,704 cocoons laid by individuals of *L. castaneus* that had been reared in isolation. Unfortunately, nothing is known about the origin and post-hatching history of the supposedly uniparental progeny.

Distribution: Areas of species endemism presumably are to be sought below the southern limit of glaciation.

The genus is small (relict?), definitely European, and most of the information available about it was derived from studies of the three markedly anthropochorous species. A fourth species, *L. festivus* (Savigny, 1826), has colonized at least one of the maritime provinces of eastern Canada, the British Isles, Germany and some of Scandinavia, presumably after introduction by man. A fifth, *L. friendi* Cognetti 1904, was said to be common in Uruguay, presumably also after introduction by man. A sixth species, *L. meliboeus* Rosa 1884, has been taken to Swedish Lapland, and by what other agency than man? The original home of none of the nine species is known, and information as to somatic anatomy of five species, all except *festivus* and the common anthropochores, is insufficient to guarantee, in the classical system, necessity for retention in the genus *Lumbricus* as it now must be defined.

Remarks: Quoting from Fuchs's (1907) account of the circulatory system in *Lumbricus*, Stephenson (1930: p. 136), says that the extra-esophageal trunks are on the esophagus "between the peritoneal and muscular layers" a condition which the writer has not seen in any species of the family Lumbricidae. On the contrary, the trunks are free from gut throughout much of their extent.

GS glands, if present in any of the anthropochores, must be within the parietes and there (in a more or less rudimentary state?) perhaps responsible for the rather common blisters.

The Type Species: The genus *Lumbricus* of Linnaeus's tenth edition (1758) comprised two species, *L. terrestris* and *L. marinus.* As the latter was not an earthworm (no marine megadriles), the type species must be *L. terrestris* which was formally declared to be so in Opinion 75 of the International Commission on Zoological Nomenclature. Although the name "terrestris" was intended to include all earthworms it must be retained for the type species and that must be a Swedish form with which Linnaeus was familiar. Included in the 1758 definition of *L. terrestris* was the phrase "adscendit noctu." Of the Swedish lumbricids there is only one that habitually feeds, when conditions are favorable, on the surface at night. That species is the one that almost universally is known today as *L. terrestris. Allolobophora longa,* also present in Sweden, sometimes has been thought to feed at the surface during the night. However, among the hundreds of thousands of earthworms that are "picked" commercially while active on the surface at night, *A. longa,* is only very rarely represented although also present in the same soil.

Linnaeus later lengthened his definition of *L. terrestris* and in the twelfth edition of his system (1767) included, among other emendations, the following, "exit supra terram tempore nocturno pro copula." The lumbricid almost universally known today as *L. terrestris,* is the only one that, as was long ago remarked, exhibits its "unabashed venery" by coming up to the surface of the ground to copulate. Linnaeus obviously knew more about the habits of *L. terrestris* than he did about its externalia and its internal organization. There is no longer any reason for use of any other name for the species.

Lumbricus castaneus

1826. *Enterion castaneum* Savigny, *Mem. Acad. Sci. Inst. France* 5 (Hist. Acct.): p. 180. (Type locality, Paris. Types, in the Mus. Hist. Nat. Paris.)

REGENERATION: Gates, 1949, *Science* 110; p. 567. RESPIRATION (rates): Byzova, 1965, *Rev. Ecol. Biol. Sol.* 2.

Clitellum, usually including all of xxviii–xxxiii, occasionally slightly shortened in xxviii and/or xxxiii, very rarely not reaching beyond 32/33. Tubercula pubertatis, usually extending through xxix–xxxii, often with narrower extensions into xxviii, very rarely reaching 27/28. Male tumescences, lacking, male pore cleft lacking or so small as to be almost indistinguishable from the pore itself. Genital tumescences, in any of ix–xiv (but rarely lacking in x), xxvi–xxxiii. Setae, enlarged in some of the preclitellar segments and ornamented ectally with transverse serrations, $CD < AB$, BC ca. = AA, DD anteriorly $ca. = \frac{1}{2}C$ (but posteriorly $< \frac{1}{2}C$. First

dorsal pore, usually at one of levels 5/6–8/9. Segments, 69–99 but usually 85–94. Size, 30–45 by 3–5 mm.

Septa, delicate, none thickly muscular. Typhlosole, beginning in region of xxi–xxii, ending usually in region of 70th to 77th segments, usually leaving 14–20 segments atyphlosolate, transverse ridges continued laterally off the typhlosole onto roof of gut. GS glands, if present not supraparietal, possibly represented by thumb-shaped peritoneal protuberances long enough to reach C and containing, in addition to setal follicles, a delicate, almost flocculent tissue. (Testis sacs, annular ? seminal vesicles excluded.) Sperm ducts, without epidymis.

Distribution: New Zealand.

Ontario, Nova Scotia, Newfoundland. Oregon, Pennsylvania, New York, Massachusetts, Vermont, Maine (where it is much more frequently found than *rubellus*). Mexico.

Norway, Sweden, Finland, Russia (from the polar part of the Usi Basin through Byelorussia and Ukraine Bukowina). Latvia, Lithuania, Denmark, Germany, Poland, Czechoslovakia, Faroes, Hebrides, Scotland, Clare Island, Ireland, England, Wales, Jersey, Guernsey, Alderney and Sark of the Channel Islands, France, Switzerland, Austria, Hungary, Romania, Jugoslavia, Italy, Corsica.

St. Helena.

Ten interceptions, of 44 specimens from 4 countries during the last 15 years, suggest that *L. castaneus* is less likely to be transported than *L. rubellus.*

Although widely distributed in the British Isles, cannot be regarded (Cernosvitov & Evans, 1947) "as a very common species, and is certainly more rare than *L. rubellus.*" Nevertheless, a population in Durham County provided a figure of 54.6 individuals per square meter, and another author found the species common in gardens and pastures of the Bristol area. In central Maine, *L. rubellus* is very rare and *L. castaneus* is rather common. The latter may now be less common in New Zealand than at some time in the previous century. An Iceland record of the species has been questioned. In such a case introduction may have failed to result in permanent colonization.

Habitats: Earth with roots of potted plants. Greenhouses (Poland, Finland, and Maine). Outside beds of greenhouses (Maine). Lettuce wrapped in cellophane purchased at a supermarket in winter. Cabbage heads in ship's stores. Botanical Gardens (Berlin, Portici, Oxford, and America). Gardens, richly manured garden sites, sandy loam, soils with a pH of 4.6–8.0, cultivated fields, pastures (Scotland, common), old forests, taiga, forests and steppes, banks by water, wet and mossy ledge, under stones, under timber, under cow pats, leaf layers, rotting leaves, leaf litter, compost, peat compost, straw heap, matted leaves including pine needles, grass trimmings.

Caves (Belgium, Italy). Mines (France).

Exposed ridges with snow 2½–3 feet deep on either side and with frost at 2–7 inches below surface of the ridge (specimens that had copulated).

In Switzerland down to depths of 400–600 mm., in England down to 3–5 feet.

At elevations up to 2,800 m. in Gran Paradiso Natl. Park, to 1,600 m. in the Alps, to 3,250 feet in Scotland.

Although definite records are few, *castaneus* is said to be "ubiquitous" with reference to *p*H.

Biology: Copulation, which may not involve secretion of a slime tube, is subterranean. Laying, 65 cocoons per year (Evans & Guild, 1948). As many as 8 eggs are passed into a cocoon, of which only 3 (Graff) develop though, according to Evans and Guild (1948), only one hatchling emerged from each of 182 cocoons. Incubation, 14 weeks, maturity at 18–25 weeks and average life (England) less than a year. Average duration of life in Michon's experiments was 698 days at 18° but at 9° was 722. Interrupted alimentation raised the average to 757 days but the average fell to 634 days for worms isolated to prevent copulation. Optimal growth was obtained at 18°C. At 25° mortality was high and development defective. Suboptimal temperatures were less harmful than supra-optimal. Diapause, during growth or in unfavorable seasons of the year, is lacking according to all who studied the problem. Individuals were found, in unfavorable seasons, at depths of 3–5 feet.

Isolated individuals did not lay for Michon or for Janda and Gavrilov (1939) but for Evans & Guild laid 1,704 cocoons from 3 of which, fully developed hatchlings emerged. No information has been vouchsafed as to fate of the hatchlings nor any explanation as to what may have been involved, especially in a genus supposedly with obligatory amphimixis.

Rudiments of male tumescences and of the tubercula are recognizable at maturity before clitellar modifications of the epidermis are externally distinguishable.

Variation and Abnormality: An extra spermatheca (two specimens) in xiii of two worms had its pore at 13/14. A preseptal location of the organ seems to be usual in the genus, but opening of a supernumerary organ at 12/13 would have seemed more probable for an otherwise normal individual.

Male pores once were found in xvi and on another worm an extra male pore was present in xiv.

Regeneration: A 6-segment head at 7/8 (Gates, 1949) was believed to indicate possibility of equimeric cephalic regeneration at 6/7 and all more anterior levels. Epimorphic tail regeneration, as in other species of the genus, is unknown. However a pygomere can be restored.

Parasites: CILIATA: *Anoplophrya* sp., Prague. *Metaradiophrya* sp., Prague. SPOROZOA: *Apolocystis herculea* (Bosanquet, 1894), New Jersey, Pennsylvania. *A. minuta* Troisi, 1933, New Jersey,

Pennsylvania. *Monocystis acuta* Berlin, 1923, Sweden. *M. agilis* Stein, 1848, France, Germany, Sweden. *M. arcuata* Boldt, 1910, Germany. *M. carlgrenii* Berlin, 1924, Sweden. *M. caudata* Berlin, 1924, Sweden. *M. densa* Berlin, 1924, Sweden. *M. hessei* Berlin, 1924, Sweden. *M. hirsuta* Hesse, 1909, France, Germany, Sweden. *M. lumbrici* (Henle, 1845), France, Germany, Sweden. *M. oblonga* Berlin, 1924, and coelom, Sweden. *M. tubiformis* Berlin, 1923, and coelom, Sweden. *M. ventrosa* Berlin, 1924, Sweden. *Nematocystis anguillula* var. *gracilis* Berlin, 1924, Sweden. *Rhynchocystis pilosa* (Cuenot, 1901), France, Germany, Sweden, New Jersey, Pennsylvania. *R. porrecta* (Schmidt, 1854), France, Germany, Sweden. NEMATODA: *Metastrongylus* sp., larvae, Russia.

Nomenclature: *Enterion castaneum* Savigny, 1826, and *Lumbricus castaneus* Risso, 1826, may require action of the International Commission of Zoological Nomenclature.

Systematics: *L. castaneus* is so close to *L. rubellus* as almost to require more evidence for retention of each as species. The two have not been obtained at the same site by this author and even small juveniles, when normal and complete, have been referable to one or the other without trouble.

Hybridization of *L. rubellus* and *L. castaneus* might even be more easily obtainable than the claimed *rubellus/festivus* cross.

Remarks: *L. castaneus*, in the northern Pennines, was found to aggregate in sheep dung as a result "of movement at or near the surface." At night, on a soil surface wet with rain, the species once was seen by the same author and above-ground wanderings probably are more frequent than that single record would indicate.

L. c. f. disjunctus Tetry, 1936, with a size of 50–82 by 4.5–5.5 mm., 93–111 segments, tubercula pubertatis bilobed ventrally because of widenings in region of 29/30 and 31/32, genital tumescences in viii–xiii but rarely in x, was not found in the same locality as worms identified as *L. c. f. typica*. If there is no interbreeding modern systematics would seem to require species status for *disjunctus*.

Lumbricus rubellus

1843. *Lumbricus rubellus* Hoffmeister, *Arch. Naturgesch. 9*: p. 187. (Type locality, unknown. Types, none(?).)

1891. *Lumbricus rubellus*, Rosa, *Ann. Hofmus. Wien 6*: p. 381.

1895. *Lumbricus rubellus*, Beddard, *A Monogr. of the Order of Oligochaeta* (Oxford), p. 722.

1900. *Lumbricus rubellus*, Michaelsen, *Das Tierreich* 10: p. 509.

1903. *Lumbricus rubellus*, Michaelsen, *Die geographische Verbreitung der Oligochaeten* (Berlin), p. 144, etc.

1923. *Lumbricus rubellus*, Stephenson, (*The Fauna of British India*), *Oligochaeta*, p. 508.

1958. *Lumbricus rubellus*, Gates, *Breviora, Mus. Comp. Zool.* Cambridge, No. 91: p. 8.

Clitellum, usually including all of xxvii–xxxii, occasionally slightly shortened in xxvii and/or xxxii, rarely extending into or even through xxvi or xxxiii. Tubercula pubertatis, usually extending through xxviii–xxxi, often with narrower extensions into xxvii and/or xxxii, shortened along with the clitellum, elongated to xxxii/eq or even to 32/33 if the clitellum reaches xxxiii/eq or 33/34. Male pores, in median half of BC but near mBC. Male tumescences, annular or half annular and then median to the pores, too small at best to reach 14/15 or 15/16, often lacking. Genital tumescences, in any of viii–xii (very much less frequently in x than in the other segments), xx–xxiii, xxvi–xxxvi. Setae, enlarged in some of the preclitellar segments and ornamented ectally with transverse serrations, $CD < AB$, $BC < AA$, DD anteriorly $ca. = \frac{1}{2}C$ but posteriorly $< \frac{1}{2}C$. First dorsal pore, usually at one of levels 5/6–8/9. Segments, 70–123 but usually 106–117. Size, 60–150 by 4–6 mm.

Septa, 5/6–12/13 with variable muscularity but never very thick. Typhlosole, beginning in region of xxi–xxii, ending usually in region of the 78th to 98th segments, usually leaving 20–26 segments atyphlosolate, transverse ridges coming into contact with each other so as to form a honeycomb pattern. GS glands, if present not supraparietal, possibly represented by the tissue underneath peritoneal blisters.

Testis sacs, annular (? including hearts, seminal vesicles of xi and spermathecae of x). Sperm ducts, without epididymis.

Distribution: Nicobar Island.

New Zealand (widely distributed in soils of pastures and cultivated lands throughout the mainland, also in some of the outlying islands, Chatham and the Kermadecs).

Alaska (1 site only, at Juneau). British Columbia, Ontario, Quebec, New Brunswick, Nova Scotia, Labrador, Newfoundland. Washington, Oregon, California, Missouri, Arkansas, Illinois, Michigan, Indiana, Ohio, Virginia, District of Columbia, Pennsylvania, New Jersey, New York, Massachusetts, New Hampshire, Maine.

Mexico.[2]

Greenland (recorded from a garden), Iceland, Norway, Sweden, Finland, Russia (from the White Sea and middle Russia through the Urals and Pri-Urals, Byelorussia, Ukraine, Carpathians, Crimea and Caucasus), Estonia, Latvia, Lithuania, Denmark, Germany, Poland, Czechoslovakia, Netherlands, Belgium, Faroes, Hebrides, Scotland, Clare Island, Ireland, England, Wales, Guernsey, Jersey and Sark of the Channel Islands, France, Switzerland, Austria, Hungary, Romania, Bulgaria, Jugoslavia, Italy, Portugal,[2] Balearic Islands, Corsica, Lemnos in the Aegean, Greece.

[2] Based on interceptions.

Madeira,[2] Canary Islands, Tristan da Cunha. South Africa. Turkey, Transcaucasus, Black Sea region, Turkestan, Kazakstan, Iran, Afghanistan, Siberia (Lake Baikal, the mouth of the Lena, etc.), and the Far East.

Specimens intercepted at American ports were from 14 countries, including Iran, as well as Madeira.

The commonest species of *Lumbricus* in the British fauna, subdominant in loamy and sandy soils of Stirling, Scotland. The most common and widely spread of the earthworms in Scotland from sea level to 3,000 m. In Washington, according to Altman (1936, *Univ. Washington Pub. Biol.* **4**–1: p 95), *L. rubellus* is the "most common endemic earthworm west of the Cascade Mountains. It is found in almost every location in this section where any earthworms are found." This is by no means the only instance of endemism being assumed merely because of commonness in the area.

Habitats: Earth with roots of potted plants. Greenhouses (Poland, Naples, Oregon, Maine). Outside beds of greenhouses (Maine). Culture beds of earthworm farms (a number of states). Lettuce heads in stores on an airplane. Cabbage heads in ships' stores. Vegetable gardens, cultivated fields, pastures. Botanic gardens (Oxford, South Africa, America). Flood plains. Under flat stones, slabs, logs, old bagging, moist cow-manure pats, below top six inches of sewage beds, pine needles. Gravel soaked with raw sewage effluent (a "tremendous concentration"). Rubbish heaps, rotting sedge heaps. Moss on peat moors. Peat bog. Litter of beech forest floor, taiga, forest, and steppes. Wrack bank. Springs, banks of polluted streams. Straw heap in field. Damp and highly organic forest soils. Moss on trunk of oak tree. Natural soils with pH of 3.8–8.0. A very acid soil, supporting only scattered reeds and rushes, in which the species is dominant, with a per-acre average of 96,000.

Caves (Germany, Hungary, Jugoslavia, Greece, Italy, Belgium, England). Mines (France). Unspecified sites at various elevations in the Alps up to 2,600 m.

An explanation of its absence in Hawaii and St. Helena to which it presumably had been transported on various occasions is lacking.

The flood plain habitat was so characteristic of the species in central Illinois that Harman (thesis, 1960) thought destruction of the sites, presumably by continuation of human activities already under way, would exclude the species from the state.

Those who have studied the species have expressed the following opinions of *L. rubellus.* Least exacting species of *Lumbricus.* A true follower of culture in the Scandinavian mountains (haemerophilic?) but often found in littoral habitats, one of the commonest species in the skerries. With respect to pH

"ubiquitous." Lives in surface litter or burrows in the soil according to circumstances, but also has been said to be "non-burrowing, shallow living." Probably the most adaptable of all Danish earthworms, occurring, as it does, not only everywhere in mull soils of fields, pastures and gardens, but also everywhere in mully deciduous and coniferous forest soils as well in beech raw humus (in great numbers). Spelunkers call the species "trogloxene," a word which is not in any of the dictionaries available as this is being written but which may mean foreign to caves. Other characterizations, stenotopes, humustier.

Wandering on the surface, especially at night, after rain, has been recorded several times.

Biology: Activity, including breeding, is year-round, in favorable circumstances. Copulation, seemingly without a slime-tube, modestly like defecation, is under ground. Laying also is uninterrupted in appropriate climatic conditions, 79–106 per year. Incubation 16 weeks, 1 juvenile only hatched from each of 105 cocoons, but two were seen on another occasion. Maturity was reached, in the English experiments, at 37 weeks. Corresponding data from German experiments (Graff, 1953), 33 cocoons per worm between February 1 and March 31, at 15–18°C, incubation 45 days, maturity at 135–160 days. Average duration of life (Michon, 1954), at 18°C, was 719 days but at 9°C was 749 days. Interrupted ailmentation had little effect on the average, but that was reduced to 682 days by the isolation that prevented copulation. Diapause is lacking during growth and unfavorable climatic periods (Evans & Guild as well as Michon), but condition of the worms was not indicated for any of the periods of quiescence. At Bellinchen (Kollmannsperger, 1934) individuals were rolled up in a ball (presumably in dispause?) at a depth of 45 cm. and a temperature of 1.3°C but at depths of 20 cm. in the summer were active. Optimal temperature for growth (Michon) is 18°C. Suboptimal temperatures are less harmful than supraoptimal. Indeed, a specimen was recovered alive in the previous century from natural ice, and some of the specimens, exposed by Korschelt (1914) to temperatures of − 2 to − 5°C for about 12 hours, survived, presumably without experimental preconditioning. Feeding probably is selective and food passes through the gut in 16–48 hours.

Populations of 100–275 per square meter have been found in well-ripened and well-aerated clay soil, of 370,000 per acre in pastures on Bardsey Island.

Variation and Abnormality: The author of *L. rubellus* recognized, in his material, two varieties. The smaller never had more than 120 segments. The larger had 140–150 segments. Ever since, in almost every characterization of the species, the number of segments has been said to be 85 or 90 to 145 or 150. Recent counts, however, provided an average for 37 Icelandic specimens of 109.5 or showed

ranges from 110 to 136 (Turkey), 81–130 (Italy). One Hungarian specimen, recently reported to have 156 segments, like Hungarian worms of the previous century that had 130–150, may have been of Hoffmeister's larger variety. All American material seen by the author had 123 or fewer segments. Harman (thesis, 1960) had an average of 124.40 and may have had some of the larger form.

The clitellum, of Hoffmeister's species, after addition of one for the peristomium, not at his time counted as a segment, began with xxv, xxvi, or xxvii and ended with xxx, xxxi, or xxxii. As long ago as 1892 Friend already had learned that in Great Britain clitellar invasion of xxvi was rare. The organ usually has been found to comprise segments xxvii–xxxii and that seems to be the condition on most of the worms examined by the author from various states, provinces, and countries. However, some shortening in one or other or both of xxvii and xxxii occasionally has been noted. Invasion of xxxiii characterized several Channel Island specimens, on one of which 33/34 even was reached. That worm was distinguishable from *castaneus* by presence of clitellum in xxvii as well as by number of segments and the location of the genital tumescences. Segment number, in the author's specimens reached a maximum at 121 (two specimens only), five with 120, five with 119, six with 118, nine with 117. Even the markedly homoeotic type of Southern's supposedly Lusitanian "relictus" (a synonym of *rubellus*) with the male pores on xxviii had only 126.

The information vouchsafed by the various authors is insufficient to warrant any suggestion to as to whether or not they had markedly homoeotic and accordingly abnormal specimens or individuals of a different race or species. Accordingly, the characterization with reference to segment number, in the species precis above, has been based mainly on the author's observations.

Extra gonads were recorded for *L. rubellus* by Woodward (1893) but just what was involved again is uncertain because of absence of pertinent information.

An unusually large Greenland specimen, with only 110 segments, was 175 by 5 mm., but in absence of information as to state of preservation such figures can have little meaning.

Typical dorsoventral flattening of the body is lacking in the species according to Hoffmeister but in the author's specimens has been much more obvious than in *L. castaneus*. Some peculiarity of preservation may have made the flattening unrecognizable.

Hybridization: Individuals reared in isolation to sexual maturity were cross-mated (after having laid demonstrably infertile cocoons? or having failed to lay for how long?) with similar individuals of *L. festivus* (Savigny, 1826) on September 18. Supposedly hybrid hatchlings (Evans & Guild, 1948) emerged December 29. The number of hybrids was not stated, and no further information regarding them was published. That is unfortunate as those authors were the first scientists (*cf. E. foetida* above) to claim making interspecies crosses without aid of surgery. Neither author was able to mention systematically important characters of the supposed hybrids. Nor could they tell, when asked, if the hybrids had reached maturity.

L. rubellus was the donor of the ovaries transplanted (Harms, 1912, *Arch. Entwickmech. Org.* **34**) into "*Helodrilus caliginosus*" to obtain supposedly intergeneric hybrids that did not live to maturity.

Regeneration: The maximum number of segments recorded for head regenerates is 4, obtained at levels 4/5, 5/6, and 7/8. One three-segment head were obtained at 14/15, the posteriormost level at which cephalic regeneration was recorded. Of especial interest, because of experimental results cited to the contrary, are head regenerates obtained by Homan (1936, *Zeitschr. Wiss. Zool.* **148**) at 3/4 and 4/5, in absence of a nerve cord at those levels.

Posterior amputees usually do not regenerate, and the only definite record found in the literature was of one natural tail regenerate with an unrecorded number of segments at an unspecified level. Amputated segments were not restored as long as worms remained active and well fed (Omodeo, 1948, *Boll. Zool.* **15**) does seem to suggest that tail regeneration was involved in other circumstances but no data were provided to support that deduction. Experimental conditions appear to have been unfavorable for the species because of absence of laying.

Economic importance: L. rubellus has been cultured in a number of states and has been widely sold and used for bait. The species, like *D. octaedra*, is regarded as an important converter of leaf substance.

Parasites: CILIATA: *Anoplophrya lumbrici* (Schrank, 1803), Germany. *A. marylandensis* Conklin, 1930,[3] Maryland. *A. striata* Cepede, 1910, France. *Hoplitophrya hamata*[4] Cepede, 1910, France. *H. lumbrici* Cepede, 1910, France. *H. mucronata* Cepede, 1910, France. *Maupasella nova*, Cepede, 1910, Germany, Russia. *Metaradiophrya falcifera* (Stein, 1861), Czechoslovakia, England. *M. lumbrici* (Dujardin, 1841), France, Germany. *Plagiotoma lumbrici* (Dujardin, 1841), Germany, Russia. SPOROZOA: *Apolocystis gigantea* Troisi, 1933, southern New Jersey and/or Philadelphia. *A. herculea* (Bosanquet, 1894), and coelom, France, Germany, Sweden. *A. minuta* Troisi, 1933, southern New Jersey and/or Philadelphia. *Monocystis acuta* Berlin, 1924, Sweden. *M. agilis* Stein, 1848, France, Germany, Sweden. *M. arcuata* Boldt, 1910, Germany. *M. catenata* Mulsow, 1911, coelom, Germany. *M. caudata* Berlin, 1924, Sweden. *M. carlgrenii* Berlin, 1924, Sweden.

M. chlorotica.[5] *M. densa* Berlin, 1924, Sweden. *M. hessei* Berlin, 1924, Sweden. *M. lumbrici* (Henle, 1845),[5] France, Germany, Sweden. *M. oblonga* Berlin, 1924, and coelom, Sweden. *M. polymorpha* Berlin, 1924, and coelom, Sweden. *M. suecica* Berlin, 1923, Sweden. *M. tubiformis* Berlin, 1923, Sweden. *M. turbo* var. *suecica* Berlin, 1923, Sweden. *M. ventrosa* Berlin, 1924, France, Germany, Sweden. *M. wallengrenii* Berlin, 1924, and coelom, Sweden. M. wenrichi.[5] *Nematocystis anguillula* var. *gracilis* Berlin, 1923, Sweden. *N. elmassiani* (Hesse, 1909), France, Germany. *N. pilosa* Tuzet & Loubatieres, 1955, France, Germany, Sweden. *N. vermicularis* Hesse, 1909, Sweden. *Rhynchocystis hessei* Cognetti, 1911, Europe. *R. pilosa* (Cuenot, 1901), France, Germany, Sweden. *R. piriformis* Berlin, 1923, Sweden. *R. porrecta* (Schmidt, 1854) France, Germany, southern New Jersey and/or Philadelphia, Sweden. *Zygocystis wenrichi* Troisi, 1933, southern New Jersey and/or Philadelphia. NEMATODA: *Capillaria plica* (Rudolphi 1819), Russia. *Choerostrongylus pudendotectus* (Wostokow, 1905), larvae, Germany, Lithuania. *Dicelis filaris* Dujardin, 1845, coelom of x–xii, Germany. *Metastrongylus elongatus* (Dujardin, 1845), larvae, Germany, Lithuania, (Russia?). *Rhabditis anomala* Hertwig, 1922, Germany, Georgia (U. S. A.). *R. longicauda* Hertwig, 1922, Germany. *R. maupasi* Seurat, 1919, Germany. *R. pellio* Schnaider, 1866, Germany, Georgia (U. S. A.). *Porrocaecum* sp., larvae. *Thelophania* sp., muscles, France. *Thominx aerophilus* (Creplin, 1839), larvae, Russia. CESTODA: *Amoebotaenia cuneata* (Linstow, 1872) Cohn, 1900, Germany, said to be cosmopolitan in chickens.

Remarks: Occurrence of *L. rubellus* in the Nicobars has not been confirmed. If Rosa's specimens really came from those islands, the worms may have been obtained from soil actually with introduced plants. Certainly there now are no reasons for believing that any colonization was permanent as the species was not obtained in recent collections.

The musculature has been studied by Pop (1957) and the respiration rates by Byzova (1965). The references are on p. 119.

Lumbricus terrestris Linnaeus 1758

1758. *Lumbricus terrestris* Linnaeus, *Syst. Nat.* (ed. 10), p. 647. (Type locality, unknown and none subsequently designated. Types, none.)
1951. *Lumbricus terrestris*, Gates, *Proc. Natl. Acad. Sci. India*, (B) **21**: p. 19.
1958. *Lumbricus terrestris*, Gates, *Breviora, Mus. Comp. Zool.*, No. 91: p. 8.

ABNORMALITY: Gates, 1956, *Ann. Mag. Nat. Hist.*, ser. 12, **9**. ACTIVITIES: Darwin, 1881 and subsequent years, *The Formation of Vegetable Mould Through the Action of Worms, with Observations in their Habits* (London and New York, various

---

[3] This species not recognized by some specialists.
[4] According to Puytorac, 1957, *H. lumbrici and hamata* are synonyms of *Metaradiophrya lumbrici*.

[5] *M. chlorotica* and *wenrichi*, according to Meier (1954) are *M. lumbrici*.

editions). Evans, 1947, *Jour. Board Greenkeeping Res.* **7**. Jefferson, 1958, *Jour. Sports Turf Res. Inst.* **9**. Parley, 1878, *Sci. Gossip* **14**: pp. 121–122 and 154–156. Raw, 1962, *Ann. Appl. Biol.* **50**. ANATOMY: Arthur, 1963, *Proc. Zool. Soc. London* **141** (digestive system). Beams *et al.*, 1957, *Anat. Rec.* **127** (ciliary structure). Mill & Knapp, 1970 *Jour. Cell. Sci.* Nakahara *et al.*, 1969, *Calciferous Tissue Res.* **4** (calciferous gland). BLOOD: Cosgrove & Schwartz, 1965, *Physiol. Zool.* **30**. CHLORAGOGEN: Urich, 1965, *Zeitschr. Vergleich. Physiol.* **50**. ENZYMES: Li & Shetlar, 1965, *Comp. Biochem. Physiol.* (glycosidases). Stolk, 1961, *Experientia* **17** (dehydrogenases). FOSSILS: Bräm, 1956, *Eclogae Geol. Helveticae* **49**. IRON METABOLISM: Delkeskamp, 1964, *Zeitschr. Vergleich. Physiol.* **48**. LOCOMOTION: Seymour, 1969, *Jour. Exp. Biol.* **51**. MUSCLES: Pop & Dragos, 1957, *Stud. Univ. Babes-Bolyai*, ser. 2, **2** (Biol.). OSMOTIC AND IONIC REGULATION: Dietz & Alvarado, 1970, *Biol. Bull.* **138**. PATHOLOGY: Hancock, 1961, *Experientia* **17**; 1965, *ibid.*, **21** and *Nature* **205**. Smirnoff & Heimpel, 1961, *Jour. Insect Pathol.* **3**. PREDATION BY MOLES: Evans, 1948, *Proc. Zool. Soc. London* **118**. Plisko, 1961, *Fragmenta Faunistica, Warsaw* **9**. PHYSIOLOGY: Laverack, 1963, *The Physiology of Earthworms* (New York) (for references not included herein). PSYCHOLOGY: Wyers *et al.*, 1964, *Jour. Comp. Physiol. Psychol.* **57**. RESPIRATION: Byzova, 1965, *Rev. Ecol. Biol. Sol.* **2**. Urich, 1964, *Zeitschr. Vergleich. Physiol.* **48**. TISSUE CULTURE: Janda & Bohuslav, 1934, *Spisy Prir. Fak. Karolvy Univ.* No. 133. VARIATION: Pearl & Fuller, 1905, *Biometrika* **4** (*re* size, segment number, location and extent of clitellum).

Clitellum, usually including all of xxxii–xxxvii, very rarely lacking in xxxvii, very rarely extending through xxxi or xxxviii. Tubercula, pubertatis, usually extending through xxxiii–xxxvi, very rarely through xxxii–xxxv, slightly less rarely through xxxiii–xxxvii. Male pores, at or slightly lateral to mBC, each in an obvious but often rather short cleft. Male tumescences, always present, varying considerably in protuberance after preservation, reaching B and C, confined to xv though 14/15 and 15/16 sometimes are dislocated. Genital tumescences, in any of viii-xiv (but more often in x than elsewhere), xxiv–xxxix (but more often in xxvi and some of the clitellar segments than elsewhere). Setae, enlarged in some of the preclitellar segments and ornamented ectally with transverse serrations, ventral setae of xv with scoop-shaped tips (often? always?), setae of genital tumescences behind xx usually with the typical longitudinal grooves ectally, $CD < AB$, $BC < AA$, anteriorly $DD >$ or *ca.* $= \frac{1}{2}C$ but posteriorly $< \frac{1}{2}C$, AB and CD larger toward ends of the body than elsewhere. First dorsal pore, usually at 7/8. Segments, 127–160 but usually 140–155. Size, to 350 by 12 mm.

Septa, membranous, none thickly muscular. Typhlosole, beginning in region of xxi–xxiii, ending usually in region of 110th to 116th segments, usually leaving 27–44 segments atyphlosolate, the transverse ridges united so as to produce a marked honeycomb appearance continued across the ventral face. Testis sacs, unpaired, subesophageal, spermathecae, hearts, and nephridia excluded. Sperm ducts, with an epididymis of loops loose or aggregated into a compact disc. GS glands, if present not supraparietal, possibly represented by delicate, almost flocculent tissue beneath peritoneal blisters associated with GS follicles.

Distribution: India, Simla (Western Himalayas), at elevations of 6,500–8,000 feet.

Australia. New Zealand (quite common in garden soils at Auckland, Hamilton and probably elsewhere).

British Columbia (possibly including Queen Charlotte Island). (An apparent record of the *L. terrestris* for Saskatoon, Saskatchewan, is based on an erroneous identification.) Ontario, Quebec, New Brunswick, Nova Scotia, Newfoundland. Washington, Oregon, California, Idaho, Minnesota, Iowa, Missouri, Arkansas, Wisconsin (driftless section), Illinois, Michigan, Indiana, Ohio (43 counties), North Carolina, Virginia, Maryland, Pennsylvania, New Jersey, New York, Vermont, Massachusetts, Penikese Island, Connecticut, New Hampshire, Maine.

Mexico. Uruguay. Falkland Islands.

Greenland, Iceland, Norway, Oland Island, Sweden, Finland, Russia (including middle Russia, Byelorussia, Carpathians, Ukraine, Moldavia, Priurals, Urals), Latvia, Lithuania, Denmark, Germany, Poland, Czechoslovakia, Netherlands, Belgium, Hebrides, Scotland, Clare Island, Ireland, England, Wales, Jersey, France, Switzerland, Austria, Hungary, Romania, Bulgaria, Jugoslavia, Italy, Portugal, Balearic Islands, Corsica, Sicily.

Azores, Madeira.

South Africa.

Siberia (including Chuvash and Tatar Republics).

L. terrestris is believed (Harman, 1960, thesis) to be absent below a line from northern Arkansas to Virginia. The only exceptions known to this author, are two localities, at High Hampton (Cashiers) and Elon College in the western mountains of North Carolina to which the species probably was recently introduced, perhaps deliberately.

L. terrestris is known to have been introduced deliberately into the states of Oregon, California, Arkansas (recently), accidentally to Washington in 1909 (also *cf. Economic Importance* below). Angleworms, which may have included *L. terrestris* and/or one more species of *Allolobophora*, already had been introduced to the Lake Superior shores of Michigan before 1830, to Utah by 1877, to Idaho by 1884 but by 1948 still had to be introduced into considerable sections of British Columbia. Seemingly isolated colonies in the hills of western North Carolina may have been rather recently introduced. A Biological Supply Company there still was purchasing preserved material from New England though several specimens already had been found on its own grounds. Absence of *L. terrestris* in the Canadian prairies between Ontario and the Rocky mountains also is significant (an apparent record for Saskatoon rests on a misidentified species of *Allolobophora*).

In southern Illinois, Frank Smith first saw living specimens *ca.* 1896, crawling on the surface during a rain. The species may have arrived at the university in earth with plants brought to the arboretum. Material for dissection in the university had to be purchased from dealers until 1905 at which time some of the needs could be met from the arboretum. By 1927 the species was abundant in Champaign and Urbana.

L. terrestris has been characterized as "euro-americaine" because of a belief that the species reached America by migrating along a North Atlantic bridge instead of being imported to and distributed throughout the continent by man since 1500 A.D.

Juliana de Berners, Izaak Walton, and later English writers seem not to have known *L. terrestris* and as late as 1892 Friend thought the species, for which *A. longa* frequently had been mistaken, was not common in Britain. Protesting characterization as "the common earthworm" in a preface to his monograph (1930) Stephenson said that in vicinity of Edinburgh, *L. terrestris* then was third in order of frequency. Moreover, "the chances are that if a worm is 'the common earthworm' of a locality it is not *Lumbricus terrestris*; and that if it has been correctly identified as *Lumbricus terrestris* it should not be described as the 'common earthworm.'" More recently, and presumably with reference to the Liverpool region, Dunn (1955, *British Vet. Jour.* 111: p. 98) also maintained that *L. terrestris* is not the "common" earthworm by which name it is so frequently known. Arrival of *L. terrestris* in Britain then may have been relatively recent and some time after flooding of the English Channel. Spreading may have been slow because the species was not used as bait by anglers. In North America, the night crawlers are, and long have been, widely used as bait. Spreading through woodlands in America has been said to be not only slow but erratic because mainly dependent on such factors as anglers throwing bait away at end of the day or farmers' dumping of garden refuse.

However, spreading sometimes is rapid unless, of course, presence passed unnoticed until a population achieved considerable density. For instance, the species was seldom seen in Central Ohio in 1918 and in only a few parts of Columbus could it be found. In 1928 the species was present in all parts of the city and had been collected from 43 Ohio counties. A peaking population explosion may have been responsible, late in the eighteenth and early in the nineteenth centuries, for frequent reports of unwanted presence of earthworms in springs and wells of the eastern states.

Habitats: Soils with *p*H of 4.0 to 8.08. Very acid pasture with soil *p*H of 4.0–4.5. Course mull soil in forests and fields, crumbly mull, light soil of alluvial origin in permanent pasture. Golf courses (Maine, Connecticut, Ohio, Illinois, etc.), fields, pastures, gardens, deciduous forests (where it is extremely common in Denmark), rare or lacking in beech forest even in good mull condition (avoids coniferous forests, sandy beech forest soil of the *Oxalis* type and raw humus in Denmark but in the same country shows a particular preference for immediate environments of houses and of manure heaps even when on poor soil). Moist ravines, woodlands, river forests of Michigan, taiga, forests, steppes (Russia). Mud of stream bank, mud flats of small stream, woody peat, inside rotten log, under logs in a stream bed, under boards on the ground, under stones, under cow pats in pastures, compost.

Greenhouses (Maine, Indiana, Illinois, Switzerland). Earth with potted plants. Culture beds of earthworm farms (several states). With celery and cabbage in ships stores. Botanic gardens, England, Austria, Czechoslovakia, Switzerland. Arboretums, Massachusetts, Connecticut, California. Wandering on streets and sidewalks after cold rains in spring or fall, Maine, Ohio, Europe.

Caves, Belgium, Italy.

Unspecified sites in the Alps at elevations up to 2,500 m.

Worms from ships' stores, compost, culture beds, plant pots, culture beds of earthworm farms, mostly were juvenile and rather small. Of the 28 specimens from potted plants that were intercepted during the last 15 years, 24 were juvenile.

An average of 65.1 individuals per square meter was found for light soil of alluvial origin in permanent pasture of Durham County, England.

Golf courses in certain states seem to be able to provide an inexhaustible supply of night crawlers, and concessions are sold for "jacking" the worms at night to those who supply bait dealers and biological supply houses.

Biology: Activity, including copulation and laying, in favorable circumstances, is year round. However, in many areas a period of summer and/or winter rest is climatically imposed. Individuals have been found, in winter, ten feet or more from the surface. Although most authors thought that quiescence in either season did not involve diapause, they failed to characterize the physical state of the resting animals. Somewhat more informative was Jefferson (1958) who, at low temperatures, found the worms to be torpid in sealed-off burrows, but activity was soon resumed with rising temperature. Diapause was reported at least once, but until confirmation is provided, an erroneous identification of the species will seem more probable to some. Copulation takes place on the surface which incited, in the older literature, comment on such unabashed venery.

The interlaying interval, after copulation, in Stinauer's laboratory at room temperature (said to be 70°F) averaged 8.6 days, with incubation requiring 61–80 days, and maturity being reached in $5\frac{1}{2}$–$6\frac{1}{2}$ months. Incubation, in German studies (Graff, 1953) was 10 weeks, maturity being reached in 50 weeks. Average duration of life in the French experiments (Michon), at 18°C (supposedly optimal) was 862 days, but at 9° was 887 days. For isolated individuals unable to copulate, the average fell to 802 days. One worm, after an operation lived $5\frac{1}{2}$ years in the laboratory, and Korschelt's estimate of the life span was 6 years. At 25° mortality was high and development was defective. At 30° survival, even in saturated air was less than 12 hours. Suboptimal temperatures are less harmful than supra-optimal, and individuals are quite active at low temperatures. Indeed, worms lived in melting ice for more than 12 hours, and copulation has been observed on various occasions in melt-water from ice and snowbanks at night. Moreover, some of Korschelt's specimens (1914), exposed for about 12 hours to temperatures of $-2°$ to $-5°C$, survived (without previous acclimitation?). The body temperature of *L. terrestris* was found (Hogben & Kirk, 1944) to be well below that of the surroundings, and comparison was made with an imperfect wet-bulb thermometer. Low thermal death point and rapid dessication in dry air require for the species (and other megadriles?) a cool, moist environment. Such a physiology, as in case of the two markedly anthropochorous congeners, presumably was acquired in subarctic climates of Europe just below the Pleistocene glaciation, and, it is especially noteworthy, without abandonment of amphimixis.

The species has been characterized as ubiquitous with respect to *p*H. Feeding may be both selective and indiscriminate. Burrowing certainly involves swallowing much soil, but feeding on the surface appears to be selective. Bundles of plant (possibly grass) fibers, with very little soil, and *ca.* 40 mm. long, have been found in the gut. Beech leaves in experiments offering various choices, were preferred. Casting has been said to be almost entirely below the surface, hence not involved in Darwin's computations on that activity. This author's observations indicate a possibility of some appreciable surface casting at certain periods or in certain conditions, perhaps mostly by less mature individuals.

Activities unusual for megadriles, that have aroused wonder and that sometimes stimulated investigation, are; lining burrows with small pebbles, cinders or fecal earth, drawing into the burrows of leaves, plugging entrance to burrows with seeds, sticks, straws, feathers, etc. Among the twigs drawn into burrows of the author's plot, was one 230 mm. long, 4 mm. thick at one of the buried nodes, with one end drawn down to a depth of 82 mm. Vertical burrows often are 7-10 mm. wide.

A gradient along the main axis was noted by Perrier (1871, *Compt. Rend. Acad. Sci. Paris*, **73**: p. 386) who found that decomposition advances from behind.

Intelligence tests devised by Darwin and their results are described (1882) in one of his two books that have been said by critics to be really interesting to read.

Variation and Abnormality : Abnormality is not rare but usually has been characterized only by reference to the most obvious and easily recognized aberration so that origin is not determinable. One or both of the male tumescences may be outside of xv. Location of one of the male tumescences in xiv, if the female pore of that side is in xiii, is due to embryonic abortion of a mesoblastic somite in front of the 13th level, but when both male pores are in xiv and both female pores are in xiii, bilateral abortion or hypomeric, post-hatching regeneration could have been involved. Doubling of the male terminalia on one side may have resulted from the same mutation that doubles male terminalia on both sides so often in parthenogenetic forms. However, the doubled male terminalia sometimes have resulted from halving of mesoblastic somites at the 15th level. Absence of male terminalia on one side has been noted, as in certain parthenogenetic morphs of other lumbricids.

Doubling of each spermatheca so that eight discrete spermathecal pores are distinguishable was recognized in one American specimen. A similar increase in spermathecae has been thought to characterize some lumbricids. The condition seemingly can arise in a single macromutational step. Transfer, of ability to develop male terminalia, from one metameric level to another also seems to involve a single step as the terminalia always are equatorial, without intermediate transitional stages.

Prostomia that are not tanylobous probably often resulted from regeneration.

Bifid anteriorly, one individual had two heads each comprising a peristomium and prostomium, the aberration a result of regeneration.

Individuals with a lateral outgrowth of more or less normal, caudal nature are as yet unexplained (see regeneration).

The subneural blood vessel is said to be within the muscular sheath of the nerve cord (the systematic character herein stated as, adherent to cord), but, in one of the specimens examined by the author, was adherent to the parietes from which the cord came easily away leaving the unbroken vessel on the body wall.

Regeneration : Some anterior regeneration has been recorded. Three segments can be replaced at 3/4, but even at that level metamerism often is more or less aberrant. Regenerates at 13/14 and 16/17 contained no gonads, and no more than that was recorded about them. At such levels, in some species, heteromorphic tails, of course without gonads, would be regenerated.

Records of rapid and more extensive head regeneration for this species (1957 *Jour. Tennessee Acad. Sci.* **32**; p. 159) are erroneous. The worms, as so often in the past, were assumed to be of *terrestris* without any real attempt at an identification. Although the situation was explained to the editor, no correction has been made. *Eisenia foetida* very probably was the species.

Tail regeneration never has been found though many thousands of specimens were examined. Supply houses that handle millions have been unable to furnish a single specimen, though many posterior amputees, including markedly brevicaudate individuals, were provided.

The last three metameres of a 60-segment juvenile were unusually small with reference to the adjacent portion of the body and lacked setae and nephropores. The appearance of a regenerate is believed to have resulted from reorganization of one or two segments after the amputation.

Lateral outgrowths, always more or less normally caudal, if developed after hatching would suggest possibility of tail regeneration. Unfortunately, nothing is known as to origin of the growths.

Parasites: CILIATA: *Anoplophrya hamata* Cepede, 1910, France, Russia. *A. lumbrici* (Schrank, 1803), France, Germany. *A. marylandensis* Conklin 1930, (= *A. lumbrici?*) Maryland. *A. oblonga* Puytorac, 1954, France. *A. striata* Cepede, 1910, France. *A.* sp., Russia. *A. hamata* Cepede, 1910, France, Germany, Russia. *Hoplitophrya secans* Stein, 1861, Germany. *Maupasella falcifera* (Stein, 1861), Germany. *M. herculei* Puytorac, 1954, France. *M. nova* Cepede, 1910, France, Germany, Russia. *Mesnillella secans* (Stein, 1861), France. *Metaradiophrya falcifera* (Stein, 1861), Germany. *M. lumbrici* (Dujardin, 1845), England, France, Germany, Czeckoslovakia. *Plagiotoma herculei* Puytorac, 1954, France. *P. lumbrici* (Dujardin, 1841), France, Germany, Russia.

SPOROZOA: *Apolocystis catenata* (Mulsow, 1911), coelom, Russia. *A. gigantea* Troisi, 1933, southern New Jersey and/or Philadelphia. *A. herculea* (Bosanquet, 1894), and coelom posteriorly, England, France, Germany, Sweden. *A. minuta* Troisi, 1933, southern New Jersey and/or Philadelphia, England. *A. pilosa* (Cuenot, 1901), France, Germany. *Dirhynchocystis minuta* Ruston, 1959, England. *Monocystis acuta* Berlin, England, Sweden. *M. agilis* Stein, 1848, England, France, Germany, Sweden. *M. catenata* Mulsow, 1911, France, Germany, Russia. *M. carlgrenii* Berlin, 1923, Sweden. *M. caudata* Berlin, 1923, and coelom posteriorly, England. *M. cristata* Schmidt, 1854, Germany. *M. densa* Berlin, 1923, England, Sweden. *M. hessei* Berlin, 1924, England, Sweden. *M. hispida* Loubatières, 1955, France. *M. lumbrici* (Henle, 1845), England, France, Germany, Sweden. *M. oblonga* Berlin, 1923, and coelom, Sweden. *M. minuta* Ruschhaupt, 1885, Germany. *M. pilosa* Cuenot, 1910, France. *M. rostrata* Mulsow, 1911, Russia. *M. striata* Hesse, 1909, England, France, Germany. *M. suecica* Berlin, 1923, Sweden. *M. ventrosa* Berlin, 1923, Sweden. *M. wallengrenii*

Berlin, 1923, England, Sweden. *Nematocystis anguillula* Hesse, 1909, England. *N. a.* var. *gracilis* Berlin, 1923, Sweden. *N. elmassiani* (Hesse, 1909), England, Sweden, United States (?). *N. caudata* Loubatières, 1955,[6] France. *N. magna* (Schmidt, 1854), and male funnels, France, Germany, Sweden. *N. pilosa* Tuzet & Loubatières, 1955, France. *N. vermicularis* Hesse, 1909, England, Sweden. *Rhynchocystis hessei* Cognetti, 1911, southern New Jersey and/or Philadelphia. *R. pilosa* (Cuenot, 1901), England, France, Germany, Sweden, Southern New Jersey and/or Philadelphia. *R. ovata* Loubatières, 1955, France. *R. piriformis* Berlin, 1923, England, Sweden. *R. porrecta* (Schmidt, 1854), England, France, Germany, Sweden. *Zygocystis cometa* Stein, 1848, and coelom, Germany. *Z. gigantea* Troisi, 1933, southern New Jersey and/or Philadelphia. *Z. suecica* Berlin, 1923, Sweden. *Z. wenrichi* Troisi, 1933, southern New Jersey and/or Philadelphia (= *M. rostrata?*).

CESTODA: *Dilepis undula* (Schrank, 1788), cysticercoid larvae, coelom of midbody, Scotland.

NEMATODA: *Anguillula lumbrici* Gmelin, 1790, Europe. *Capillaria annulata* (Molin, 1858), larvae, longitudinal musculature, North America. *C. caudinflata* (Molin, 1858) Travassos, 1915, said to be cosmopolitan in chickens and likely to be present in other species if lumbricids are lacking. *C. mucronata* (Molin, 1858), Europe. *C. plica* (Rudolphi, 1819), Russia. *Choerostrongylus pudendotectus* (Wostokow, 1905), larvae, vascular system and gut wall, especially in posterior part of the esophagus, Europe, North America. *Cyathostoma lari* Blanchard, 1849, larvae, experimental infection, England. *Dicelis filaria* Dujardin, 1845, coelom of x-xii, Germany. *Heterakis gallinarum* Schrank, 1788 (a vector of the causative agent of blackfeet of turkeys and chickens), United States. *Metastrongylus elongatus* (Dujardin, 1845), larvae, vascular system and gutwall, Europe, North America. *M. salmi* Gedoelst, 1923, larvae, vascular system and gut wall, North America. *Pelodera* (*P.*) *strongyloides* (Schneider, 1860) Schneider, 1866, Germany. *P.* (*P.*) *teres* Schneider, 1866, Germany. *Porrocaecum ensicaudatum* (Zeder, 1800) Baylis, 1920, larvae, ventral trunk and hearts, Illinois, CSR, Germany. *Rhabditis anomala* Hertwig, 1922, Germany, Georgia (U. S. A.). *R. longicauda*, Hertwig, 1922, Germany. *Rhabditis* (*R.*) *maupasi* Seurat in Maupas, 1919, Germany. *R. pellio* (Schneider, 1866), Germany, Georgia (U. S. A.) R. (*R.*) *terricola* Dujardin, 1845, Germany. *Spiroptera turdi* Molin, 1860.[7] *Stephanurus dentatus* Diesing, 1839, larvae, experimental infestation. *Syngamus trachea* (Montague, 1810), larvae, England, North America. *Thominx aerophilus* (Creplin, 1839), larvae, Russia.

---

[6] *N. caudata* = *N. magna?*

[7] *Spiroptera turdi* Molin, 1860 was the name given to larvae now known to develop into *Porrocaecum ensicaudatum*.

(*Viguiera turdi = Spiropters turdi = Porrocaecum ensicaudatum.*)

BACTERIA: *Spirochaeta lumbrici* Cropper, 1912, coelomic epithelium cells of seminal vesicles, England.

Fossils: A study of the external appearance and internal structure of concretions from calciferous glands of *L. terrestris* showed (Bräm) that they were the same as fossils from Late Tertiary and Quaternary beds that had been named *Arion kinkeli* and *hochheimensis.*

Economic Importance: Millions of night crawlers are "jacked" every year, in the dark, on lawns, golf greens and courses, other open areas, for distribution to dealers who sell them for bait or to supply biological houses which sell them to schools and colleges for dissection in laboratory sections of elementary biology or zoology courses. Rights to collect sometimes are auctioned to the highest bidder.

Because of use as bait, *L. terrestris* has been introduced into "many localities" even near the arctic circle in Finland.

Remarks: A belief that almost any earthworm is *Lumbricus terrestris* is not entirely restricted to high-school graduates who have had an elementary course in biology (*cf.* Stephenson, 1930: p. xi). The species used in a recent electron-microscope study of sperm cytology was said to be *L. terrestris* but actually was *Allolobophora tuberculata.* Anatomy of the species probably is widely misunderstood, especially by biologists whose knowledge of the worm was obtained in high-school biology courses. Partly responsible seems to be a trade journal (owned by an outstanding research institute) that is widely distributed to high-school and college teachers. As recently as 1961, an anonymous article said that one of the two exit pores of the nephridium was near the posterior border of the segment and that the dorsal nephropore was connected with the ventral pore by an epithelium lined canal. Although the attention of the editor was called to the misinformation no correction, so far as this author has been able to learn, ever was made.

Because of its habits, *L. terrestris* is in several ways one of the most interesting of all earthworms. Uniquely, it has furnished cells for the only megadrile tissue cultures (Janda & Bohuslav). Cells from intestinal wall lived in cultures for 13 months.

Unlike *A. longa*, *L. terrestris* seems to be a favorite food of moles. In England, 85 to 100 per cent of the individuals in mole stores were of this species, and in Poland the species constituted 74 per cent of the stores of *Talpa europaea* in meadow soil at Cracow. The percentage in the mole stores, in England, was much greater than the percentage of the species in the megadrile population of the area; 25 per cent or less.

A report (Stebbings, 1962, Nature 196, p. 905) that *Lumbricus terrestris* has almost completely destroyed "the zonal horizonation" of the prairie chernozem soils of eastern South Dakota, certainly requires confirmation. The only earthworm species validly recorded from that state, is *E. tetraedra*, a limicolous species unlikely to have been involved.

Octolasion

1885. *Octolasion Oerley*, Ertek, *Term. Magyar Akad.* 15, 18: p. 13. (Type species, *O. lacteum* Oerley 1885 = *Enterion tyrtaeum* Savigny 1826.)

Characters of the type species that may prove to be definitive at generic level are as follows:

Calciferous sacs, in x, large, lateral, communicating vertically and widely with gut lumen though reaching beyond esophagus both dorsally and ventrally. Intestinal origin, in xv. Gizzard, mostly in xvii. Extra-esophageals, passing up to dorsal trunk posteriorly in xii. Hearts, vi–xi. Nephridial bladders, ocarina-shaped. Nephropores, obvious, behind xv in one regular rank on each side, just above *B*. Longitudinal musculature, pinnate. Setae, behind the clitellum not closely paired.

Some mention of testis sacs usually has been included in the generic definition. Inasmuch as in *D. octaedra* testis sacs are present only in certain degenerate morphs, and as all available octolasia also were parthenogenetic, the classical character is suspect and confirmation of universality in amphimictic populations is required. Possibly universally characteristic of a revised genus will be location of spermathecal pores in region of *CD* and presence of four pairs of seminal vesicles, in ix-xii.

Octolasion cyaneum

1826. *Enterion cyaneum* Savigny, *Mem. Acad. Sci. Inst. France* 5 (Hist. Acct.): p. 181. (Type locality, Paris. Types in the Mus. Hist. Nat. Paris.)

1958. *Octolasium cyaneum*, Gates, *Breviora, Mus. Comp. Zool., Harvard*, No. 91: p. 8.

CASTING: Boyd, 1957, *Proc. Roy. Soc. Edinburgh*, B, 66. Jefferson, 1958, *Jour. Sports Turf Res. Inst.* 9. CHROMOSOMES: Muldal, 1948, *39th Ann. Rept. Innes Hort. Inst.*; 1952, *Heredity* 6. DISPERSION: Karppinen & Nurminen, 1964, *Ann. Zool. Fennici* 1. ECOLOGY: Lee, 1951, *Tuatara* 4. GROWTH & REPRODUCTION: Evans & Guild, 1948, *Ann. Applied Biol.* 35 and *Ann. Mag. Nat. Hist.*, ser. 11, 14. Graff, 1953, *Regenwürmer Deutschlands* (Hannover). Michon, 1954, Thesis, Poitiers. PARASITES: Meier, 1954, *Arch. Protistenk.* 100. Rees, 1962, *Parasitology* 53.

Quadrithecal, pores at or slightly above *C*, at 9/10–10/11. Clitellum, saddle-shaped, six segments, xxix–xxxiv. Tubercula pubertatis, uninterrupted longitudinal bands, lateral to *B*, always extending through xxx–xxxiii, often invading xxix and/or xxxiv. Male pores, nearer to *C* than *B*, in lateral portion of an equatorial cleft. Male tumescences, nearly reaching *B* and *C*, confined to xv, but often dislocating 14/15 and 15/16. Female pores, just lateral to *B* at eq/xiv. Genital tumescences, including any one or more of the following, a/x, xvii-xx, xxii, b/x. Setae, paired, widely so posteriorly, $AB < BC > CD$ or $CD < AB < BC < AA < DD$, DD $< \frac{1}{2}C$, a,b/xv enlarged, with transverse serrations ectally. Nephropores, obvious, behind xv in a single rank on each side just above B. First dorsal pore, in region of 9/10–13/14, usually at 11/12. Prostomium, epilobous, tongue open. Color, white, grayish, pinkish, sometimes a few segments in front of the anus bright yellow (owing to coelomocytes that fill the coelomic cavities).[1] Segments, 100–159, usually 140–159. Size, 52–220 (180?), by 5–8 mm.

Septa, 5/6–15/16 strengthened, 6/7–7/8 or 12/13–13/14 more so than the others, none thickly muscular. Pigment, none. Longitudinal musculature, pinnate (?). Calciferous sacs, in x, large, lateral, reaching to or nearly to 9/10, 10/11 and above as well as below the esophagus into which they open widely without constrictions, lamellae on posterior wall only. Esophagus, much widened in xi, with marked external constrictions at insertions of 10/11 and 11/12, lumen vertically slitlike from posterior portion of x into xii, lamellae gradually narrowing behind middle of xii. Intestinal origin, in xv. Gizzard, mostly in xvii (ending at ?). Gut, valvular at or near insertion of 19/20. Typhlosole, beginning behind xix, anteriorly very wide but probably bifid ventrally and with bifurcations curved up to gut roof laterally, on the ventral face a low but distinct longitudinal and median ridge from which closely crowded transverse grooves run laterally, ending in region of 114th–126th segments, leaving 25–37 metameres atyphlosolate. Extra-esophageals, passing to dorsal trunk posteriorly in xii. Hearts, in vi–xi. Nephridial bladders, as in O. tyrtaeum[2](?). Holandric.[3] Seminal vesicles, 4 pairs, in ix–xii. Sperm ducts, with an epididymis of short loops or of coils that may be bound into a disc. Spermathecae, sessile on parietes, usually in x-xi. TP, GS and atrial glands, if at all present, entirely within the parietes and so as to be unrecognizable from the coelom.

Reproduction: One of 5 individuals reared to maturity in isolation by Evans & Guild (1948) layed fertile cocoons 3 months after reaching sexual maturity.

Spermathecae of all specimens seen by the writer were transparent and filled with a watery fluid. Spermatophores never were found. Spermatozoal iridescence on male funnels was seen only twice and then was slight. Copulation is unrecorded and may never have been observed. Many specimens doubtless are male sterile.

Reproduction is parthenogenetic.

Chromosomes, 144, 180, 190 (% 15–20 ?), $n = 19$ (Muldal, 1952).

[1] The specific name refers to iridescence of cuticle or to epidermal refraction of light.
[2] Confirmation required because nephridia of all available specimens were in poor condition.
[3] Testis sacs, supposedly characterizing the genus, were not found.

Distribution: New South Wales, Capital Territory, South Australia. North and South Islands, New Zealand.

British Columbia, Iowa, Massachusetts, Maine.

Argentina, Tuyupara Island in the delta of the Paraná, Uruguay.

Iceland, Sweden, Finland, Russia, Poland, Germany, Denmark, Faroes, Hebrides, Islay Island, Scotland, Ireland, England, Wales, Guernsey, and Jersey (Channel Islands), Netherlands, Belgium, Czechoslovakia, Romania, Hungary, Switzerland, France, Portugal, Spain.

Azores.

Murree, Kufri, at 7,800 feet in the Simla Hill States, Simla at 6,500–8,000 feet, Ootacamund at 6,700–8,000 feet in the Nilgiri Hills of South India.

Elevations not mentioned above, to 2,600 m. Switzerland, 2,000–2,200 m. Dolomites, 1,130–1,600 m. Italy.

Especially noteworthy in view of the misunderstanding about certain lumbricid distributions is the fact that the single tropical record of cyaneum is for elevations of 6,700–8,000 feet.

Only two specimens were intercepted during the last fifteen years at United States ports of entry, one from Italy and one from Denmark.

O. cyaneum is known to have been recently introduced from abroad to a greenhouse in Finland from which the species spread to an adjoining garden and thence to nearby gardens to which plants also had been taken from the greenhouse. This recorded account (Karppinen & Nurminen) of introduction to and distribution within Finland is an example of what has been happening for more than 400 years in North and South America, Australia, Africa, etc., and which probably also has been happening in Eurasia and North Africa for a still longer period.

Habitats: Under stones in water. In moss. Banks of brooks, other limnic habitats. Plowed fields (one of the commonest species there, Scotland). Wet sand, mull type soils, humid and mully hollows in foliferous woods, forest soils, soils with low fertility and then at depths of 4–6 inches or in remnants of native forests (New Zealand). Soils with a pH of 5.2–8.0. Wandering on streets after rains (Switzerland), on surface frequently after rain (Britain).

Greenhouses, central Maine (2 sets only), Finland. Gardens, including those near or supplied from greenhouses. Botanical Garden, Oxford. Arboretum, Boston.

Caves, Belgium, Germany. Mines, France, where the species is second only to D. rubida in number of specimens collected there.

Characterized as "ubiquitous with respect to pH".

Biology: Laying is said (Evans & Guild) to be uninterrupted, presumably year-round in England, with about 13 cocoons a year per individual. The cocoons are small relative to size of the individual producing them. One hatchling apiece emerged from about 90 per cent of the cocoons, 2 each from the other 10 per cent.

Incubation, 60 days. Puberty, at 20°C, attained (Graff) 180–200 days later. Laying, in the German conditions, averaged 20 per individual between February 1 and March 31. Average number of hatchlings per cocoon 0.7. The preclitellar phase, in the French conditions (Michon, 1954), was 260 days, average life being 738 days.

The species is said to be homodynamous, i.e., without diapause. In central Maine summer drought and winter freezing impose two periods of inactivity at each site, except of course those in greenhouses. In appropriate circumstances, activity can be year round.

Casting, in pot and cage experiments (Jefferson, 1958) at St. Ives, always was below the surface, but in the Hebrides, surface casts were said (Boyd) to be higher than the grazed sward. Feces, in cage experiments (Jefferson) were voided into short burrows adjacent to the main run.

Most activity, even at 33°F, is near the surface which in Switzerland is left only for hibernation. Submersion in water for six weeks resulted in no recognizable ill effects. Compact soils were penetrated in five hours. Has been seen on surface, after heavy winter rain, along with *L. terrestris*.

O. cyaneum is stored, in England, by moles.

Populations of *O. cyaneum* in Argentina towards the end of the previous century may have been much larger than elsewhere. New Zealand may now have considerable populations. *O. cyaneum* in those islands (Lee, 1951) usually is the first exotic species to appear after removal of native vegetation for conversion to pasture where it temporarily becomes dominant though eventually replaced by species of *Lumbricus* and of *Allolobophora*. However, if the pasture reverts to a scrub or to forest, dominance again is achieved.

Polymorphism: None of the major anatomical deletions appearing after establishment of parthenogenesis has been recorded for *O. cyaneum*. Vesicles of x, xi often are rudimentary, and the others usually are rather small. Little is known about variation. Seeming absence of hearts in vi–vii of some individuals may be merely because of inability to recognize smaller vessels when empty.

Regeneration: Although *O. cyaneum* was used frequently in transplantation experiments nothing has been recorded hitherto about ability to regenerate after a single transection. Tail regenerates have not been seen. The only cephalic regenerate that was found has a normal prostomium and peristomium, but the brain is small (perhaps development incomplete at time of preservation). Level of regeneration was at or near 15/16, and so in a region from which few lumbricid regenerates have been recorded.

Parasites: CILIATA: *Anoplophrya lumbrici* (Schrank, 1803), Germany. SPOROZOA: *Apolocystis herculea* (Bosanquet, 1894), coelom and intestinal wall, Wales. *A. villosa* (Hesse, 1909), Wales. *Cephalocystis rotaria* Rees, 1962, coelom, Wales.

NEMATODA: *Rhabditis pellio* Bütschli non Schneider, England. Unidentified larval nematodes have been found (England) encysted in brown bodies.

Systematics. Latin names have been given to morphs or clones distinguished by segmental locations of genital tumescences.

Distinctness from *O. tyrtaeum* has been questioned, primarily because the only formerly known differences are in the clitellum. Nearly similar clitellar differences have been found to distinguish two species of *Lumbricus*. Until more is known about variation in individuals now referable to the taxon, and until something is known about the ancestral interbreeding population (if still extant) from which *cyaneum* has been isolated by its parthenogenesis, retaining *cyaneum* at the species level[4] seems preferable as does disregarding the formae and varieties.

Remarks: A record in the medical literature perhaps more interesting than that cited in the chapter on *E. foetida*, is of the specimen of *O. cyaneum* that was discharged from a woman's vagina where it supposedly had been for more than a year.

Absence of *O. cyaneum* in some of the most thoroughly surveyed American areas, and its rarity elsewhere on the continent, contra-indicates characterization of the species as "holarctic."

Octolasion tyrtaeum

1826. *Enterion tyrtaeum* Savigny, *Mem. Acad. Sci. Inst. France* **5** (Hist. Acct.): p. 180. (Type locality, Paris. Types, none.)

1958. *Octolasium lacteum*, Gates, *Breviora, Mus. Comp. Zool. Harvard College*, No. 91: p. 9.

CHROMOSOMES: Muldal, 1952, *Heredity* 6. Omodeo, 1955, *Caryologia* 8. BREEDING, GROWTH, REPRODUCTION: Evans & Guild, 1948, *Ann. Mag. Nat. Hist.*, ser. 11, **14**, and *Ann. Applied Biol.* **35**. Michon, 1954, Thesis, Poitiers. Stinauer, 1951, Thesis, Michigan Sta. Univ. MUSCLES: Pop & Dragos, 1959, *Stud. Univ. Babes-Bolyai*, ser. 2, 2 (Biol.) PARASITES: Meier, 1956, *Arch. Protistenk* 101. NOMENCLATURE. Cernosvitov, 1931, *Zool. Anz.* 96: p. 203. RESPIRATION (rates): Byzova, 1966, *Rev. Biol. Ecol. Sol.* **3**.

Quadrithecal, pores at 9/10, 10/11, at or above *C* but within *CD*. Clitellum, saddle-shaped, dorsal pores occluded, intersegmental furrows obliterated, setae retained, 6 segments, xxx–xxxv. Tubercula pubertatis, uninterrupted longitudinal bands, just lateral to *B*, always in xxxi–xxxiv but usually reaching eq/xxx, xxxiv. Male pores, nearer to *C*, at or near lateral ends of deep equatorial clefts. Male tumescences, of variable shape, obliterating 14/15, 15/16, reaching equators of xiv and xvi. Female pores, just

---

[4] In that connection see new data as to segment number and typhlosole termination in the précis of each species.

lateral to B, at eq/xiv. Genital tumescences, usually including only apertures of a follicles, in one or more of segments viii-xii, xiv, xvi–xxiv, xxvii, xxx, xxxvii, xxxviii. Setae, paired anterior to clitellum, posteriorly widely paired or distant, $a,b/xv$ displaced mesially. Nephropores, obvious, behind xv in one regular rank on each side, just above B. First dorsal pore, in region of 8/9–14/11 usually at 10/11. Prostomium, epilobous, tongue open. Body section, more or less octagonal posteriorly. Color, white, grayish, slate, clittellum in life sometimes orange to red, (also red but of a different shade after formalin preservation). Segments, 76–136, usually 114–136.[5] Size, 30–160 by $2\frac{1}{2}$-8 mm.

Septa, 5/6–14/15 somewhat strengthened, 6/7–8/9 perhaps somewhat more so but none thickly muscular. Pigment, when present yellowish brown and in circular muscle layer. Longitudinal musculature, fasciculate.[6] Calciferous sacs, in x, lateral, large, reaching to or nearly to 9/10 and 10/11, not constricted off from gut though reaching above as well as below and opening widely into it, lamellae on posterior wall of sac wide. Esophagus much widened in xi, with marked external constrictions at insertions of 10/11 and 11/12 and an internal constriction in xii, lumen vertically slitlike from posterior portion of x into xii, lamellae gradually narrowing behing middle of xii. Intestinal origin, in xv. Gizzard, mostly in xvii (ending at ?). Gut, valvular in region of insertion of 19/20. Typhlosole, beginning in region of xx–xxiv, bifid ventrally and with lateral folds curved up against roof of gut, a low but obvious longitudinal ridge at mV on ventral face from which closely spaced, transverse grooves run laterally and sometimes onto intestinal roof, posteriorly the groovings disappear being replaced by three longitudinal ridges, ending abruptly in region of 76th–110th segments, leaving 12–33 metameres atyphlosolate. Extra-esophageals, passing up to dorsal trunk posteriorly in xii. Hearts, 6 pairs, in vi–xi. Nephridial bladders, somewhat ocarina-shaped, transversely placed, joined by the tubule dorsolaterally, a very short duct passing straight down from ventral face to parietes near B. Holandric. Seminal vesicles, 4 pairs, in ix–xii. Sperm ducts, with an epididymis in various loops, coils or in a single hairpin loop. Spermathecae, usually in x, xi, sessile, ducts confined to parietes. TP and GS glands, lacking. Atrial glands, reaching into xiv and xvi.

Reproduction: One of 8 individuals reared in isolation to sexual maturity (Evans & Guild), three months after reaching that stage, laid 4 fertile cocoons. Even when a "fair amount of apparently normal sperm" is produced (Muldal, 1952), parthenogenesis is obliga-

tory. That determination, based on not less than 10 specimens, cannot be challenged as a result of the author's examinations of many times that number. Specimens without any iridescence on male funnels and in sperm ducts clearly were male sterile. Spermathecae, usually were filled with a watery fluid or a pinkish, translucent jelly without trace of any opacity. One copulating couple was seen by this author but on dissection both individuals were found to be male sterile.

Chromosomes, 38, 54, 72, 144, $n = 19$, the highest recorded for megadriles according to Muldal (1952), but Omodeo (1955) found n to be 18.

Distribution: New South Wales, Australia.

British Columbia, Manitoba.

Washington, Oregon, California, Colorado, Nebraska, Iowa, Missouri, Arkansas, Illinois, Michigan, Indiana, Tennessee, Alabama, Ohio, West Virginia, Pennsylvania, New Jersey, New York, Massachusetts, Connecticut, Maine.

Mexico City, Peru, Argentina, Uruguay.

Sweden, Aland Island, Finland, Russia (south at least from Novgorod, Vologda and Perm, through Ukraine and Crimea, the Moldavia SSR, including Urals, Pri-Urals, Byelorussia). Latvia, Lithuania, Poland, Germany, Denmark, Hebrides, Lambay Island, Scotland, Ireland, Clare Island, England, Wales, Guernsey (Channel Islands), Netherlands, Belgium, Czechoslovakia, Romania, Bulgaria, Jugoslavia, Hungary, Austria, Switzerland, France, Corsica, Portugal, Spain, Italy, Greece.

Azores, Madeira, Canary Islands.

Algeria, South Africa.

Siberia. Manchuria. Turkestan, Chuvash and Tatar Republics, Kazakston, Iran.

Dhar and Murree, Mashobra-Simla Hill States, Simila at 7,500 feet, Mussoorie at *ca.* 6,000 feet, Darjiling at 6,250 feet, Ootacamund.

Elevations not mentioned above are 200–2,600 m. Switzerland, to 2,800 m. Italian Alps, to 2,500 m. Dolomites, to 800 m. Romania, to 1,000 m. Bulgaria, to 5,800 ft. California.

Intercepted individuals, U. S. ports during the last fifteen years, 58, from England, France, Germany, Greece, Jugoslavia, Azores, Mexico, Canal Zone. The single specimen from the last area was adult and with orchid roots.

The first Australian record was for a garden to which live plants had been brought from abroad. No information as to date of first introduction to America is available. Although not recorded from America until 1893 by then the species already had become "abundant" in central Illinois as probably also in California. By 1890 the species had become common in the vicinity of Buenos Aires.

Habitats: Under stones (including some that are submerged), logs. In woody peat, leaf mold, compost, decaying wood, forest litter of various kinds,

---

[5] Larger numbers, to 180, have been recorded. Data herein provided are from unpublished studies of the author.

[6] The longitudinal muscle layer also has been said to be pinnate. Specification of level anteroposteriorly as well as dorsoventrally may be required.

debris of different sorts (including that at dumps), hillside detritus. House gardens, cultivated fields, hayfields, pastures. Taiga, forests, steppes, bogs, peat bogs, banks of creeks, brooks, rivers, in springs, "torrents," bottom muds around roots of submerged vegetation, in mud under 50 cm. of water of an Isoëtes stand, clayey beaches, on streets after rain (Switzerland).

Fresh beach mull soil, meadow and meadow-marshy saline soils, fine black mull, matted mor (the only species), damp clay soil, silty clay loam, turfy podzolic, peaty humus and peaty gley soil under spruce (lacking only, one Russian area, under fir-blueberry), sandy soil, soils with pH of 5.5–8.08.

Botanical Gardens, Berlin, Oxford, Torino, Palermo. Arboretum, Massachusetts. Greenhouses, Naples. Caves, Hungary, Italy, Jugoslavia, Tennessee, West Virginia. Mines, France.

Biology: Activity, in appropriate circumstances, is possible throughout the year, but in much of the range summer drought and winter cold necessitates rest. Indications are that the periods may be variously spent by different individuals as in Germany, some of those found in summer at depths of 45–50 cm. were coiled up tightly as if in diapause but others were not. Failure to obtain diapause in his experimental conditons led Michon (1954) to maintain that the species is homodynamous.

Laying (Michigan) averaged once every three days. Incubation, at 70°F (Michigan), 60–76 days, or (England) 21 weeks. Only one hatchling per cocoon was recorded. Maturity was reached, at 70°F (Michigan), in 10–11 months. Average life (Michon 1954), at 18°C, 738 days.

O. tyrtaeum is stored by *Talpa europaea* in meadow soil, vicinity of Cracow, Poland, and in the fall of 1960 constituted 0.7 per cent of the number of worms stored.

O. tyrtaeum has been characterized as sedentary, inhabiting surface layers of the soil, leaving there only to hibernate, and also as a deep burrowing species. It has been said to be stenotopic, eurytopic, as well as amphibious, and the pioneer in Russia of alluvial meadows.

O. tyrtaeum shares dominance, of the Deschampsietum in Russia, with *D. octaedra*, in alluvial grounds of some sections of the same country with *E. nordenskioldi*, in Illinois-Indiana sites near Chicago was found to be the second most abundant.

Variation and Abnormality: Size of one of the author's Mexican specimens with clitellum at maximal tumescence, apparently unamputated posteriorly, was 37 by 2½ mm., the width measured through the clitellum which is of course the thickest portion of the body. That size was recorded because it was so unusual. Much larger sizes, as occasionally were recorded in the past, were not found, and none of the specimens of this species seen by the author approximated size of larger individuals of *cyaneum*.

Considerable variation in number of segments, as indicated in the précis, was recorded in the past. The number, in material from various parts of the world that was examined by the author, almost always was within the range 114–136 if posterior amputees were excluded.

Genital tumescences were found around c and d follicles 5 times and then only in segment x (1 Mexican and 4 Australian worms). Setae of the tumescences on one side of the body (1 worm) were of the usual longitudinally grooved kind. Setae of the other side lacked such grooves but instead were ornamented with irregularly interrupted circles of transverse serrations.

Follicles of genital setae are lateromesially flattened, discoidal and of triangular shape, the apex projecting into the coelom. Supraparietal GS glands never were seen in the coelom, but whether rudiments can still be found within the parietes is unknown.

Variation in the internal organs, of the author's material, was limited to vestigialization of organs such as the seminal vesicles, most advanced in ix, x, and the atrial glands. Whether the TP and GS glands had been eliminated before intervention of parthenogenesis is not known.

Few abnormalities have been recorded or noted. Those seen are of the usual sort, metameric anomalies and homoeoses resulting from some sort of interference with normal development.

Regeneration: Nothing has been published hitherto with regard to regeneration after simple transection. Capacity for tail regeneration, as in *L. terrestris*, may be nil. The single head regenerate seen hitherto was at 3/4. Metamerism was abnormal, but if it had been normal there probably would have been three segments.

Parasites: CILIATA: *Anoplophrya lumbrici* (Schrank, 1803), Germany. *Metaradiophrya falcifera* (Stein, 1861), Germany. SPOROZOA: *Apolocystis herculea* (Bosanquet, 1894), and coelom, France, Germany, Wales. *A. villosa* (Hesse, 1909), France, Germany, Wales. *Monocystis turbo* (Hesse, 1909), France, Germany. *Nematocystis elongata* (Hesse, 1909), male funnels, France. NEMATODA: *Choerostrongylus pudendotectus* (Wostokow, 1905), larvae, vascular system and gut wall, Europe, North America. *Metastrongylus elongata* (Dujardin, 1845), larvae, vascular system and gut wall, Europe, North America. *Porrocaecum ensicaudatum* (Zeder, 1899) Baylis, 1920, larvae, ventral trunk and hearts, Illinois. *Rhabditis maupasi* Seurat in Maupas, 1919, Germany. *R. pellio* Schneider, 1866, Germany.

Economic Importance: At least one dealer in Michigan was known to have sold *O. tyrtaeum* for bait. The castings, numbers per acre in various seasons under different cropping systems are reported by Teotia *et al.* (1950, *Res. Bull. Nebraska Agric. Exp. Sta.* 165).

Introduction of *O. tyrtaeum* into gardens and fields of Alberta and Manitoba was followed some time later by complaints of damage to the soil by excessive caking. Unfortunately, attempts to obtain an objective valuation of the claims were futile, in part because the damage, this author was told, had not yet been great enough to warrant the effort.

Systematics: Spermatophores containing sperm are recorded in the literature. Perhaps then, in or near the original home of this European species (wherever that may be) an amphimictic population may still be able to provide some of the information needed to determine relationships to each other of *tyrtaeum* and *cyaneum*.

Nomenclature: Until the biological problems here involved are solved, nomenclature, just as in various other instances, cannot be stabilized. Now requiring consideration, is the relationship to each other of *cyaneum* and *tyrtaeum*. Close indeed must be the affinity suggested by the fact that the two are distinguishable from each other almost only by a one-segment difference in locations of clitella and of tubercula pubertatis. Two species of *Lumbricus*, that differ from each other by little more than those same characters, long have been thought to be "good." However, the octolasia in question are known only from polyploidal parthenogenetic morphs. In such circumstances, information about the biparental population or populations, from which the uniparental morphs were isolated, appeared to be requisite to a solution of the problem. The name *tyrtaeum*, if only one species is involved, has page priority over *cyaneum* and its use herein, for the first time, in place of *lacteum* requires explanation.

Only three individuals of *Octolasion* (and perhaps a fragment of a fourth, possibly of the type of *tyrtaeum*?), all of *cyanea*, are in Savigny's collection at the Paris Museum. Ribaucourt[7] found no octolasia during a five-year survey (1895–1900) of megadriles in the vicinity of Paris and, according to Tetry,[8] the species are not very abundant in Lorraine. Only one specimen of *tyrtaeum* may have been available to Savigny and most if not all of it may have been discarded after the dissection that was needed to determine genital characters already thought to be systematically important. Certainly, none of Savigny's successors were able to find any material identified by him. Clearly then, there are no types. Michaelsen[9] seems to have believed that Savigny's taxon was an inadequately characterized species of *Lumbricus*. Later, Cernosvitov[10] recognized that Savigny's *E. tyrtaeum* is the same as Oerley's *O. lacteum*, but, without explanation or any formal action then or at any subsequent occasion, he merely mentioned the fact in a footnote.

Savigny's *tyrtaeum*, in his Tribe 2, has paired setae and six seminal vesicles but his *cyaneum*, in Tribe 4, was said to have widely paired setae and eight seminal vesicles. Neither of those differences now provides any real problem. Primarily, of course, because both taxa are known only from parthenogenetic morphs some of which are male sterile. In such circumstances, size of seminal vesicles in ix and/or x is liable to be much reduced, so markedly in x that the vestiges are unlikely to be recognized during routine dissections. *O. lacteum* has been said at various times to have setae closely paired, widely paired, or closely to widely paired within the same individual. Setae in a preclitellar region of many specimens identified as *lacteum* often do appear to be paired or even closely paired though wider pairing does characterize the postclitellar portion of the body. Finally, in Europe from the region of Paris north, all worms with a clitellum of six segments in xxx–xxxv have been referable to morphs of the taxon that now must be known as *tyrtaeum*.

Allolobophora tyrtaea Ribaucourt 1896, erected on a single specimen from Switzerland, was distinguished from *lacteum*, according to its author, by the absence of spermathecae and by other characters (such as the obvious nephropores) now known to be common to *tyrtaeum* and *cyaneum* or to be quite unimportant. Athecal morphs of *cyaneum* and *tyrtaeum* have not been recognized as yet but are to be expected if parthenogenesis has been established long enough. Spermathecae in male sterile octolasia are transparent and often are so small as to be easily overlooked.

Ribaucourt's taxon of course was inadequately characterized and the species has been referred to *Helodrilus*, to *Eophila*, and even to *Microeophila*. Good evidence for inclusion in any genus other than *Octolasion* is lacking.

Allolobophora tyrtaea, recorded from Austria by Wessely,[11] supposedly has dorsal pores present from 4/5 but the description, limited to external characteristics and without mention of spermathecae, provides no other reason for questioning identification as *O. tyrtaeum*.

(Genus?) eiseni

1884. *Lumbricus Eiseni* Levinsen, *Vidensk. Meddel. Naturhist. Forhandl. Copenhagen*, ser. 4, 5: pp. 311, 241. (Type locality, Deer Garden, Copenhagen. Types, 5, presumably in the Copenhagen Mus.)

1909. *Helodrilus (Bimastus) eiseni*, Michaelsen, *Mem. Indian Mus.* 1: p. 246.

1916. *Helodrilus (Bimastus) eiseni*, Stephenson, *Rec. Indian Mus.* 12: p. 352.

1923. *Allolobophora (Bimastus) eiseni*, Stephenson, (*The Fauna of British India*), *Oligochaeta*, p. 505.

1958. *Bimastos eiseni*, Gates, *Breviora*, No. 91: p. 5.

1968. "*Lumbricus eiseni*," Gates, 1968, *ibid.*, No. 299, p. 1.

---

[7] "Introuvable aux environs de Paris," *Bull. Sci. France Belgique* 35 (1901): p. 211.

[8] *Faune de l'est de la France (Lorraine)* (Nancy, 1938).

[9] *Das Tierreich* 10 (Berlin, 1900) p. 513.

[10] *Enterion tyrtaeum* "halte ich für identisch mit *Octolasium lacteum* Örley," *Zool. Anz.* 96 (1931): p. 203.

[11] *Jahresbericht Mus. Francisco Carolinum* 78 (1920): p. 18.

SYSTEMATICS

BACTERIA, in nephridia: Knop, 1926. *Zeitschr. Morph. Oekol. Tiere* 6: p. 601. CELLULASE: Tracey, 1951, *Nature* 167: p. 777. CHROMOSOMES: Muldal, 1949, *Proc. Linnean Soc. London* 161: p. 117; 1952, *Heredity* 6: pp. 59, 61. COCOONS: Svendsen, 1956, *Ann. Mag. Nat. Hist.*, ser. #12, 9: p. 732. ECOLOGY: Boyd, 1957, *Proc. Roy. Soc. Edinburgh*, B, 66: pp. 322–336. Svendsen, 1957, *Jour. Animal Ecol.* 26: p. 411, etc. Laverack, 1961, *Comp. Biochem. Physiol.* 2: p. 23. Lloyd, 1963, *Jour. Animal Ecol.* 32: p. 159. REPRODUCTION: Evans & Guild, 1948, *Ann. Mag. Nat. Hist.*, ser. 11, 14: pp. 655–658.

Athecal. Clitellum, saddle-shaped and then down to or nearly to *A*, or mostly annular but not as well developed ventrally, in (xxiii?) xxiv, xxv–xxxi, xxxii, (xxxiii, xxxiv?). Male pores, in xv, at or median to m*BC* and laterally in a small transverse cleft. Male tumescences, confined to median half of *BC* and to xv or xv–xvi(?). Female pores, in xiv, slightly lateral to *B*. Genital tumescences, *a,b*/xvi, xxiv, xxv. Setae, closely paired, behind the clitellum *BC* usually < *AA*, *DD* < ½*C*, back through the clitellum *a,b* follicles longer than the *c*, *d*, the GS follicles still more conspicuously protuberant into the coelom and with elongated and slenderized genital setae. Nephropores, obvious, irregularly alternating, and with asymmetry, usually between levels a little above *B* and almost half way from *D* to m*D*. First dorsal pore, at 5/6. Prostomium, tanylobous. Color, brown, slate, sometimes almost black. Segments, 75–113. Size, 30–48 by 2–4 mm.

Septa, none thickly muscular, 10/11 bulged anteriorly and 11/12 posteriorly by enlargement of the gut of xi. Pigment, in the circular muscle layer, apparently yellowish brown, brown or sometimes reddish brown. Longitudinal muscle layer, fasciculate. Calciferous sacs, in x, horizontal, reaching anteriorly or anterolaterally to or nearly to 9/10, opening into gut lumen at level of insertion of 10/11. Esophagus, much widened and bead-shaped in xi, half as thick in xii and of uniform width through xii–xiv, lumen vertically slitlike in xi, calciferous lamellae narrowing from xii. Gizzard, mostly in xvii (ending at ?). Typhlosole, beginning rather gradually in region of xx–xxiv, for some segments rather thickly lamelliform, ending abruptly (unamputated worms) in region of 88th–97th segments, leaving 9–15 metameres atyphlosolate. Extra-esophageal trunks, joining dorsal vessel in xii. Hearts, in (vii?) viii–xi. Nephridial bladders, *U*-shaped and with limbs of about equal length. Holandric. Seminal vesicles, in xi, xii. Epididymis, present and well developed or lacking or in various intermediate stages. Atrial glands, extending through xvi and reaching to *C*, in median half of *BC* only, or of decreasing size until almost confined to the parietes. Supraparietal GS glands, none. (Tubercula pubertatis and TP glands, lacking.)

Reproduction: Uniparental, according to Evans & Guild (1948), when juveniles are reared in isolation to maturity. Polyploidy was not recognized. The chromosome number, 32 (Muldal, 1949, 1952), is diploid. One of the questions that remains to be answered is whether reproduction ever is amphimictic.

Sperm are matured and often are aggregated on male funnels of normal size but so as to suggest that maturation is sparse. Sperm were not seen in the male duct nor in the spermatophores found on two individuals. The spermatophores suggest that copulation may have been attempted but without transfer of sperm. If sperm can be passed out through the male ducts, such gametes perhaps could be ejected into the cocoon as the latter is passed anteriorly over the male pores in which case pseudogamy or autogamy might be involved.

Pyriform to saccular testes, contained a few free sperm. Trophocytes,[12] only very slightly larger than their nucleus, each bore but several sperm. Maturation of sperm in the testes, rather than in the seminal vesicles, characterizes various parthenogenetic morphs and has not been recorded from species with obligatory biparental reproduction. The polymorphism also is of the sort associated with parthenogenesis.

Distribution. India, Naini Tal in the western Himalayas.

New Zealand, Stewart Island.

Denmark, Sweden, Germany, Poland, Czechoslovakia, Austria, Jugoslavia, Albania, Bulgaria, England, Ireland and Clare Island, Scotland and the outer Hebrides, Belgium, France, Switzerland, Italy, Portugal.

Azores, Madeira, Canary Islands, St. Helena.

South Africa.

The species was intercepted at least once by the U. S. Bureau of Plant Quarantine, and it is highly probable that individuals were introduced to America on numerous occasions before the quarantine became effective. This failure, of an obviously anthropochorous form, to colonize anywhere in the Americas seems so unusual as to need some sort of an explanation.

Habitats: Under bark of a fir stump, of a dead beech, of a fallen larch, in highly decayed beech logs, in fallen branches of beech litter, in moss, in litter layer over peat and moder sites in heather moor (with a mean *p*H of 4.4), peat cuttings, in dung pats, in bogs, swamps, marshes, springs, caves, mines, under peaty soil Among roots at distances 20 m. from shore and at depths of *ca.* 2 m. in sections of Lake Windermere that are always submerged.

Unspecified habitats at elevations of, 6,400 feet in the Himalayas, 300–1,900 m. in the Alps, 2,200–2,400 m. in the Dolomites, 200–700 m. in Switzerland, 1,000 m. and 1,649 m. in Jugoslavia, 1,900–2,800 m. in Italy.

Biology: The species appears to be a selective feeder, no soil particles were found in guts of specimens examined by the author and support is provided by the recorded habitats.

[12] Blastophores or cytophores.

The species is characterized as a surface or litter dweller (but nothing is known of how it spends unfavorable dry and cold periods), also as acid tolerant. On moorland soils of all major Islands of the Hebrides, except Canna, according to Boyd (1956), Levinsen's *eiseni* was dominant or in heather-moor shared dominance (Boyd, 1957) with *D. octaedra* and *L. rubellus*. The species aggregates in dung pats (Boyd) and in particular in sheep dung (Svendsen, 1957) as a result of movement at or near the surface. Individuals burrowed into soils with a *p*H of 4.1–4.3 (Laverack) within a few hours but also will burrow into soils with a *p*H of 4.4–5.0. Worms tend to go into fallen branches from beech litter (Lloyd) in the spring as the weather warms up.

Polymorphism: Material available to author provided a graded series of morphs, at one end showing well-developed atrial glands extending through xv–xvi and covering *BC*, terminating with one in which no trace of the glands was recognizable in the coelom (whether microtome sections would have showed intraparietal vestiges is unknown). Morphs also showed various stages of epididymal elimination. Supraparietal GS glands were lacking in all of the author's morphs and tubercula pubertatis never have been recorded so they and the associated TP glands may have been lost along with the spermathecae as in the genus *Bimastos*. Completely male sterile morphs were not seen but can be anticipated, at least in some future time.

A biparental population, from which the known *eiseni* morphs could have been isolated by their parthenogenesis, is unknown, perhaps because one never was sought. However, known *eiseni* morphs, like some *Bimastos* morphs in America, may be all that is left of an amphimictic species otherwise unable to survive the rigors of arctic or subarctic climates just below the limits of European glaciation. The distribution, providing no evidence of ability to colonize tropical lowlands, does not contra-indicate such an origin.

Parasites: Nematodes: Larvae of *Spiroptera turdi* Molin by Cori, C. J., 1898, in or associated with the ventral blood vessel, Scotland. *Dicelis filaria* Dujardin, 1845, coelomic cavities, Scotland (identification by Dr. S. Prudhoe, British Museum, who characterized it by "probably").

Systematics: Levinsen's *eiseni* as known today is not a species, according to criteria of the modern systematics, and admittedly is a group of morphs in which structure already has undergone some degradation as a result of parthenogenesis. An ancestral biparental population now seems likely to have had, in addition of course to spermathecae, tubercula pubertatis along with TP glands and supraparietal acinous GS glands associated with the enlarged follicles of the genital setae. The ancestral population may have had a longer clitellum of ten or so segments

or shorter clitella of variable location in the region of xxiii–xxxiv. Another possibility that may have to be considered is that not all morphs referred to *eiseni* have a common origin in a single biparental population. One bit of evidence hinting at something of the sort is provided, by the clitellum in xxii–xxxi and the setae that are not strictly paired, of the types of Friend's *merciensis* which was synonymized, perhaps erroneously, with *eiseni*. Specimens with larger somas, 64 by 5 mm., may have been relaxed or perhaps macerated but if with more than 113 segments may have belonged to a morph of different origin.

Remarks: Levinsen's *eiseni* has been characterized as "euro-americaine" in spite of the fact that it has never been found in the Americas, and by speleologists as a troglophile.

Oligochaete literature has only perfunctory characterizations of the *eiseni* anatomy and contains very little information about variation with respect to systematically important characters. The above account summarizes, at various places, unpublished data from the author's records.

Some of the recent records for "*Eisenia parva*," because of an erroneous synonymization of *eiseni* in *Bimastos parvus* mistakenly transferred to *Eisenia*, may be of Levinsen's species.

MEGASCOLECIDAE

1891. Megascolicidae (*part*[1]), Rosa, *Ann. Naturhist. Hofmus. Wien* **8**: p. 379.

1895. Megascolicidae (*part*), Beddard, *A Monogr. of the Order of Oligochaeta* (Oxford), p. 357.

1900. Megascolecinae (Megascolecidae), *part*, Michaelsen, *Das Tierreich* (Berlin) **10**: p. 161.

1907. Megascolecinae (Megascolecidae), *part*, Michaelsen, in Michaelsen & Hartmeyer, *Fauna Südwest-Australiens* **1**: p. 149.

1909. Megascolecinae (Megascolecidae), *part*, Michaelsen, *Mem. Indian Mus.* **1**: p. 118.

1910. Megascolecinae (Megascolecidae), *part*, Michaelsen, *Abhandl. Naturwiss. Ver. Hamburg* **19**, 5: p. 21.

1916. Megascolecinae (Megascolecidae), *part*, Michaelsen, *K. Vetensk. Akad. Handl. Stockholm* **52**, 13: p. 53.

1921. Megascolecinae (*part*), Michaelsen, *Arch. Naturgesch.* **86**, A: p. 141.

1923. Megascolecinae (Megascolecidae), *part*, Stephenson, (*Fauna of British India*), *Oligochaeta*, p. 165.

1928. Megascolecidae (Megascolecina), *part*, Michaelsen, *Handbuch der Zoologie* (Berlin) **2**, 2–8: p. 110.

1930. Megascolecinae (Megascolecidae), *part*, Stephenson, *The Oligochaeta* (Oxford), p. 828.

1937. Megascolecinae (Megascolecidae), *part*, Pickford, *A Monogr. of Acanthodriline Earthworms of South Africa* (Cambridge, England), p. 98.

1959. Megascolecidae, Gates, *Bull. Mus. Comp. Zool. Harvard College* **112**: p. 255. Acanthodrilinae (part)+Megascolecinae (part, Megascolecidae), Lee, 1959, *The Earthworm Fauna of New Zealand* (Wellington), pp. 32–33.

---

[1] Excluding all species without truly racemose prostates, as also in each instance for which "part" is indicated.

Digestive system, with intestinal origin behind ovarian segment. Vascular system, with a supra-esophageal trunk or trunks, extra-esophageals median to the hearts, hearts that are in part latero-esophageal and with the terminal pair behind the last testis segment. Setae, sigmoid, with simply pointed tip. Dorsal pores, present.

Male pores, behind female pores.[2] Spermathecae, in front of gonadal segments, with seminal chambers. Clitellum, multilayered, (female pore segment always included?). Ovaries, in xiii, fan-shaped and with numerous egg strings(?). (Ovisacs?) Ova, not yolky. Seminal vesicles, trabeculate. PROSTATES, RACEMOSE AND OF MESOBLASTIC ORIGIN.

Distribution: Ceylon, India, Burma, Malay Peninsula, Thailand, Indo-China, China, Korea, Japan, Malaysia, the Philippines, New Guinea, New Caledonia, Australia, New Zealand.

The proper range as thus indicated has been extended, presumably in large part, if not entirely, through transport by man of various species, to include many other parts of the world.

The proper megascolecid range is larger than that of the Moniligastridae by the area of Australia, New Guinea, and perhaps also New Caledonia and New Zealand. Nevertheless, it is noteworthy that no single megascolecid genus, not even the often overrated *Pheretima*, has a self-acquired range as great as that of the moniligastrid *Drawida*.

Systematics: Presence of a central and axial canal within the prostate glands now automatically excludes a species from the Megascolecidae.

In family definitions, male pores at first were said (Beddard, 1895) to be in xviii "usually" but subsequently (Michaelsen, 1900; Stephenson, 1923, 1930) the qualification was omitted. The pores usually are in xviii of most genera in all families of Michaelsen's Megascolecina but in occasional species or genera of those families may be in xvi, xvii, xix, xx or xxi, as also in certain genera of Michaelsen's Lumbricina. Location of the male pores obviously is not, by itself, definitive at supra-generic levels. The difficulty was shortly evaded (Michaelsen, 1900: p. 162) by specifying that megascolecin male pores were the external apertures of prostatic ducts—the vasa deferentia having joined those ducts outside of or even within the prostates. That single, unqualified characterization of the subfamily definition was, however, based on one of those extrapolations that are so common and so basic in the classical system inasmuch as termination of the vasa deferentia (often unrecognizable in dissections) rarely has been recorded. Subsequently, to meet requirements of a family tree, exceptions were made for species of a genus, *Plutellus*, in which the vasa deferentia may open, externally near to but independently of the prostatic pores, into

[2] Male gonoducts, except perhaps in *Exxus* Gates, 1959, opening into prostatic ducts entally.

the prostatic ducts within the parietes or at various levels of those ducts within the coelom. The evolutionary series provided by *Plutellus* is of considerable interest but the family tree that determined the grouping of genera had required elimination of the single supposed character-in-common of the classical Megascolecinae. However, exclusion of *Plutellus*, as well as other genera with tubular prostates, from the classical subfamily now recognized as a family does not enable definition of the group by location of the male pores. Though still usually in xviii, they are in xvii of the genera *Nelloscolex* and *Tonoscolex* and are in xix or xx in occasional species of *Pheretima*. Furthermore, the so-called megascolecin male terminalia are not even definitive as in *Exxus wyensis* Gates 1959 the male pores are not even in the same segment as prostatic pores. The male terminalia of that anomalous species actually are acanthodrilin in their locations though the prostates are racemose and of the pheretima sort. The single condition by which male pores throughout the family can be characterized is so common as to be hardly worth including in the definition.

Remarks: The morphological and phyletic heterogeneity of the classical Megascolecidae was somewhat simplified in the twenties by Michaelsen's separating off, as the Acanthodrilidae, all subfamilies characterized throughout by tubular prostates. Nevertheless, the classical megascolecinae—at that time raised to family status—still was based solely on esoteric phylogenetic speculations and remained morphologically undefinable by any single, significant "character-in-common."

The classification in Michaelsen's Tierreich monograph (1900), regarded as a triumph of arrangement which brought order into confusion and that constituted a remarkable advance in our understanding of the group, was retained with but minor modifications by Stephenson (1923, 1930) in both of his monographs. Although existence of convergence and polyphyly in the classical system was admitted and even emphasized by Stephenson, he may have been responsible for spread of a paralyzing belief that obvious systematic difficulties were incapable of resolution and that they were due to nature itself rather than to a phylogenetic and superficially circumscribed classification. Perhaps as one result, proposed neoclassical modifications in the system were not based on any substantial increase in knowledge of somatic anatomy and amounted to little more than reshuffling species or genera, sometimes so that morphological and phyletic heterogeneity of certain taxa was even greatly increased.

Meanwhile, a study of structure that had been neglected or even derogated in the classical system (Gates, 1937–1961) revealed necessity for increasingly more and more drastic revisions. Transfer of Indian species of one megascolecin genus to the megascolecid *Tonoscolex* as well as to the octochaetid genera, *Scolioscolides*, *Barogaster*, and *Travoscolides*, left one vast oceanic discontinuity instead of two in the distribution of *Megascolides* now common only to Australia, Oregon, and Washington. Indian species of another megascolecin genus were distributed among Indian megascolecid genera, *Nelloscolex*, *Nellogaster*, *Notoscolex*, and *Lennoscolex*, leaving *Woodwardia*, now *Woodwardiella*, purely Australian. Erection of a small south Indian octochaetid genus, *Celeriella*, returned *Spenceriella* to its native Australia. One bit of the polyphyly in the Indian section of another genus was removed by transfer of a species to *Tonoscolex* thus emphasizing a huge discontinuity in the distribution of the classical *Megascolex* now with a range comprising only a southern part of the Indian peninsula, Ceylon, Australia and perhaps New Zealand. Researches on a few south Indian and Ceylon forms, by Bahl and the author, indicate probable necessity of some splitting just within the Indian sections of *Megascolex* as well as of the classical *Notoscolex*.

Finally, in hope of dissipating the almost hypnotic spell that had handicapped logical thought on megadrile classification, a drastic revision of the old Megascolecidae that was begun with exclusion of the Ocnerodrilinae as an independent family (Gates, 1939) was at last (Gates, 1960) completed. Megascolecin genera with holoic nephridia and tubular prostates, *Diplotrema*, *Plutellus*, *Pontodrilus*, *Diporochaeta*, were transferred to the classical Acanthodrilinae (as variously added to by Michaelsen, Stephenson and Pickford) now regarded as a family and defined morphologically by three "characters-in-common." Megascolecin genera with meroic excretory systems and tubular prostates, *Spenceriella*, *Megascolides*, were transferred to the octochaetinae now recognized as an independent family and also defined morphologically by three characters-in-common.

Thus purged, the megascolecid family now is defined by a uniquely diagnostic character (capitalized in the family definition), the importance of which is suggested by the absence of such characters in the definitions of most other megadrile families. Nevertheless, other important megascolecid characters-in-common either are as yet unrecognized or are lacking. The glands that provide the diagnosis have disappeared completely and rapidly in various species after reproduction became parthenogenetic and such organs now seem to be less stable phylogenetically than is preferred for family and generic definition. However, spermathecae may well have been one of the earliest of new organs to appear in the evolution of the oligochaetes from their marine ancestors. Spermathecae still are almost universal throughout megadriles with biparental reproduction. Yet, those very organs, probably more ancient than the prostates, can and sometimes do disappear as rapidly and as

completely as the prostates after reproduction becomes parthenogenetic.

Monophyly for the family as herein defined was not previously suggested and certainly is not now maintained, though the range is compact and without large oceanic discontinuities. All that is claimed for the present is that we now have a family defined objectively by visible and easily recognized morphological characters-in-common instead of by reference to a family tree on which no two oligochaetologists could agree. Much more anatomical information probably will have to be accumulated to enable the insights that are prerequisite to a greater degree of systematic permanence.

KEY TO ORIENTAL GENERA OF MEGASCOLECIDAE

1. Spermathecal pores unpaired and at mV
    Comarodrilus[a]
    Spermathecal pores paired, lateral to mV..... 2
2. Nephridia stomate, and with preseptal funnels
    Perionyx
    Nephridia astomate, at least in some part of the
    body................................... 3
3. Gizzard in front of 7/8.................... 4
    Gizzard behind 7/8................*Pheretima*
4. Female pores in xiii, male pores in xvii....... 5
    Female pores in xiv, male pores in xviii....... 6
5. Calciferous glands stalked...........*Tonoscolex*[b]
    Calciferous glands unstalked and not constricted
    off from the esophagus..............*Nelloscolex*
6. One pair of large enteronephric nephridia per
    segment behind the clitellum............ 7
    Nephridia behind the clitellum not enteronephric or if so then more than two per segment.. 8
7. Gizzard in v, spermathecae bidiverticulate.*Lampito*
    Gizzard in vi, spermathecae unidiverticulate
    Nellogaster[c]
8. Setae eight per segment.................... 9
    Setae numerous per segment........*Megascolex*[d]
9. Excretory system behind the clitellum reduced
    to one pair per segment of clusters of 4–6
    small, exoic nephridia............*Lennoscolex*
    Excretory system not so reduced.....*Notoscolex*[e]

[a] *Comarodrilus*, known only from the original description of *C. gravelyi* Stephenson, 1915 is assumed for the present to have racemose prostates of mesoblastic origin and so to belong in the Megascolecidae. So little is known about somatic anatomy, that status of the genus and relationships of *C. gravelyi* are not determinable. The character by which the genus is keyed above now seems unlikely to be of more than specific value.

[b] Stephenson's *antrophyes* (*q.v.*p. 233) may be generically distinguishable from *Tonoscolex* by location of female and male pores in xiv and xviii respectively as well as by several other characters.

[c] *Vide* Gates, 1938, *Rec. Indian Mus.* 40: p. 427.

[d] The type species of *Megascolex* Templeton 1844 is Ceylonese. Whether any of the Australian species in that classical genus really belong there remains to be determined.

[e] The type species of *Notoscolex* 1886 is Australian and whether any Ceylon and South Indian species belong in the same genus with it remains to be determined.

Lampito

1866. *Lampito* Kinberg, *Ofvers. Akad. Förhandl.* **23**: p. 103. (Type species, *L. mauritii* Kinberg 1866.) *Megascolex*, (part), Perrier, *Compt. Rend. Acad. Sci. Paris* **102**: p. 877. (Excluding from the synonymy; *Amyntas, Nitocris, Pheretima, Rhodopis, Megascolex, Perichaeta* of Schmarda and Kinberg.)
1895. *Megascolex* (part), Beddard, *A Monogr. of the Order of Oligochaeta* (Oxford), p. 370.
1900. *Megascolex* (part), Michaelsen, *Das Tierreich* **10**: p. 212.
1907. *Megascolex* (part), Michaelsen, "Oligochaeta," in, *Die Fauna Südwest Australiens* (Jena) **1**: p. 163. *Lampito*, Michaelsen, *Mitt. Zool. Mus. Hamburg* **24**: p. 159.
1909. *Lampito*, Michaelsen, *Mem. Indian Mus.* **1**: p. 178.
1916. *Megascolex* (part), Michaelsen, *K. Svenska Vetensk. Handl.* **52, 13**: p. 57.
1923. *Megascolex* (part), Stephenson, (*The Fauna of British India*), *Oligochaeta*, p. 222.
1928. *Megascolex* (*Lampito*), Rao, *Half-Yearly Mysore Univ. Jour.* **2**: p. 31.
1930. *Megascolex* (part), Stephenson, *The Oligochaeta* (Oxford), p. 837.
1938. *Lampito*, Gates, *Rec. Indian Mus.* **40**: p. 404.
1960. *Lampito*, Gates, *Bull. Mus. Comp. Zool. Harvard College* **123**: p. 243.
1961. *Lampito*, Gates, *Ann. Mag. Nat. Hist.*, ser. 13, **3**: p. 653.

Digestive system, with a gizzard in v, longitudinal calciferous lamellae with free median margins in x–xiii, intestinal origin in region of xv–xvi, with a typhlosole but without caeca and supra-intestinal glands. Vascular system, with unpaired dorsal, ventral and supra-esophageal trunks but no subneural, paired extra-esophageals median to the hearts, paired lateroparietal trunks from the anal region connecting in xiii with extra-esophageal and supra-esophageal trunks. Hearts, of x–xiii lateroesophageal, of v–ix lateral. Excretory system, meroic, all nephridia avesiculate, paired clusters of astomate microic tubules on anterior faces of septa in v–xiii, xiv, ducts from some clusters opening into the pharynx, numerous astomate, v-shaped micronephridia per segment on parietes from (xiii or xiv?) xv posteriorly, one pair per segment of meganephridia with preseptal funnels from region of xx posteriorly, ducts passing to a median longitudinal supra-intestinal canal from which a pair of ductules open into gut in each segment. Septa, all present from 4/5. Dorsal pores, present.

Biprostatic, male pores, in xviii. Female pores, in xiv. Clitellum, annular. Sperm ducts, pass into ental ends of prostatic ducts(?). Spermathecae, with two small, digitiform diverticula from lateral and median faces of ducts. Metagynous, ovaries fan-shaped and with several egg strings.

Distribution: Palni and Cardomom Hills of South India, as indicated by the published records for endemic species, at elevations between 1,500 and 7,000 feet. The proper generic range, which perhaps should include lowlands surrounding the hills of the endemics, has been considerably extended in part at least by transport of the type species.

Remarks: Spermducts are assumed to pass into ental ends of the prostatic ducts though records are lacking for some of the species.

The genus presently comprises seven species. Six, *L. sylvicola* and *vilpattiensis* (Michaelsen, 1907), *marianae* and *palniensis* (Stephenson, 1924), *kumiliensis* (Aiyer, 1929) and *kempi* Gates, 1938, are known only from the original materials.

Lampito is close to *Nellogaster* Gates, 1938, presently known only from the Ceylonese type species. A common ancestor of the two genera probably was octovesiculate. Major changes in one or in both of the groups seemingly would have been as follows. Disappearance of seminal vesicles in ix–x or in x–xi, elimination of a third pair of vesicles associated with abortion of the first pair of testes, shifting of spermathecal pores—both pairs by the same amount or each pair differently, increase or reduction in number of spermathecae, duplication of the spermathecal diverticulum, elongation of the esophagus, increase in number of setae (some species of *Lampito*), enlargement of the intestinal typhlosole sometimes followed by its bifurcation ventrally.

KEY TO SPECIES OF *Lampito*

1. Holandric............................... 2
   Metandric............................... 3
2. Quadrithecal, setae 8 per segment...... *palniensis*
   Sexthecal, setae more than 8 per segment. *mauritii*
3. Male pores in a single field, typhlosole bifid ventrally........................... *marianae*
   Male pores not in a single field, typhlosole not bifid ventrally........................... 4
4. Spermathecal pores with same intrasegmental location in viii as in ix.................... 5
   Spermathecal pores postsetal in viii, presetal in ix................................ *kempi*
5. Spermathecal pores presetal......... *vilpattiensis*
   Spermathecal pores not presetal............ 6
6. Spermathecal pores equatorial....... *kumiliensis*
   Spermathecal pores postequatorial...... *sylvicola*

Lampito mauritii

1866. *Lampito mauritii* Kinberg, *Ofvers. K. Vetens.-Akad. Förhandl. Stockholm* **23**: p. 103. (Type locality, Mauritius. Types, in the Naturhistoriska Riksmuseet, Stockholm.)
1888. *Megascolex armatus*, Rosa, *Ann. Mus. Civ. Sto. Nat. Genova* **26**: p. 159; *Boll. Mus. Zool. Univ. Torino* **3**, **50**: p. 2.
1895. *Lampito mauritii* + *Megascolex armatus* + *M. madagascariensis* Beddard, *A Monogr. of the Order of Oligochaeta* (Oxford), pp. 369, 384, 385.
1900. *Megascolex mauritii* (part), Michaelsen, *Das Tierreich* **10**: p. 227. (Excluding, *P. Mauritiana* Beddard, 1892.)
1903. *Megascolex mauritii*, Michaelsen, *Die geographische Verbreitung der Oligochaeten* (Berlin), p. 92, etc.
1910. *Lampito "Mauritii,"* Michaelsen, *Abhandl. Naturwiss. Ver. Hamburg*, **19**: p. 10.
1916. *Lampito mauritii*, Stephenson, *Rec. Indian Mus.* **12**: p. 315.

1923. *Megascolex mauritii+M. trilobatus*, Stephenson, (*The Fauna of British India*), *Oligochaeta*, pp. 259, 279.
1925. *Megascolex mauritii*, Gates, *Rec. Indian Mus.* 27: p. 473.
1926. *Megascolex mauritii*, Gates, *Ann. Mag. Nat. Hist.*, ser. 9, 17: p. 440; *Jour. Bombay Nat. Hist. Soc.* 31: p. 183; *Jour. Burma Res. Soc.* 15: p. 207; *Rec. Indian Mus.* 28: p. 151.
1929. *Megascolex mauritii*, Gates, *Proc. U. S. Natl. Mus.* 75, 10: p. 5.
1930. *Megascolex mauritii*, Gates, *Rec. Indian Mus.* 32: p. 301.
1931. *Megascolex mauritii*, Gates, *ibid.*, 33: p. 361.
1932. *Megascolex mauritii*, Gates, *ibid.* 34: p. 374.
1933. *Megascolex mauritii*, Gates, *ibid.* 35: p. 491.
1936. *Megascolex mauritii*, Gates, *ibid.* 38: p. 388. Hla-Kyaw & Gates, *Jour. Roy. Asiatic Soc. Bengal* (*Sci.*) 2: p. 166.
1938. *Lampito mauritii*, Gates, *Rec. Indian Mus.* 40: p. 413.
1939. *Lampito mauritii*, Gates, *Jour. Thailand Res. Soc., Nat. Hist. Suppl.* 12: p. 78.
1955. *Lampito mauritii*, Gates, *Rec. Indian Mus.* 52: p. 80.
1960. *Lampito mauritii*, Gates, *Bull. Mus. Comp. Zool. Harvard College* 123: p. 243.
1961. *Lampito mauritii*, Gates, *American Midland Nat.* 66: p. 62; *Burma Res. Soc. 50th Anniv. Pub.* No. 1: p. 57.
1970. *Lampito mauritii*, Soota & Julka, *Proc. Zool. Soc. Calcutta* 23: p. 202.

ABNORMALITY: Gates, 1947, *Rec. Indian Mus.* 45: p. 79. ACCLIMATION TO LOW TEMPERATURE: Rao, 1962, *Science*, 137: p. 262. AXIAL GRADIENTS: Tandan, 1951, *Current Sci. Calcutta* 20: p. 214. CIRCULATORY SYSTEM: Vasudevan, 1939, *Rec. Indian Mus.* 41: p. 309. Gates, 1947, *ibid.* 45: p. 79. EXCRETORY SYSTEM: Bahl, 1924, *Quart. Jour. Micros. Sci.* 68: p. 72. GROWTH: Gates, 1948, *Growth* 12: p. 172. GROWTH RATES: Sitaramaiah, 1966, *Jour. Anima. Morph. Physiol.* 13, LUMINESCENCE: Gates, 1925, *Recl Indian Mus.* 27: p. 473. OXYGEN CONSUMPTION: Saroja, 1959, *Proc. Indian Acad. Sci.*, B, 49: p. 183; 1961, *Nature* 190: p. 930. PARASITES: (NEMATODA) Refuerzo & Reyes, 1959, *Philippine Jour. Animal Industry* 19: p. 55. PHYSIOLOGY: Saroja, 1959, *Proc. Indian Acad. Sci.*, B, 49; 1961, *Nature* 190: p. 930. *Proc. Indian Acad. Sci.*, B, 59; 1964, *Arch. Internatl. Physiol. Biochim.* 72. REGENERATION: Gates, 1959, *Wasmann Jour. Biol.* 17: p. 55. Santhakumari, 1963, *Jour. Animal Morph. Physiol.* 10. (Hyaluronic Acid in cuticle?) Rajalu & Krishnan, 1967, *Indian Jour. Exp. Biol.* 5. THERMAL ACCLIMATION: Rao, 1962, *Science*, 137; 1963, *Proc. Indian Acad. Sci.* 57, 58. Saroja & Rao, 1965, *Zeitschr. Vergleich. Physiol.* 50.

Sexthecal, pores larger than the female apertures, in region of *EG*, at 6/7–8/9. Male pores, at or lateral to *B*, in paired, circular, slightly raised porophores that extend from *A* into *CE*. (Genital markings, none.) Clitellum, setae retained, from 13/14 or a postsetal portion of xiii to 17/18. First dorsal pore, in region 10/11–12/13. Setae: 26–39/iii, 40–51/viii, 38–50/xii, 30–43/xx, vii/10–15, viii/11–16, xvii/4, xviii/0, xix/4, median setae on each side of xviii penial, circles interrupted midventrally, $AA = 1\frac{1}{2}$–4 AB, *a–d* enlarged and in an anterior part of the body ornamented. Color, grayish, yellowish, brownish. Segments, 157–201. Size, 95–155 by 3–6 mm.

Pigment, brown, sparse, associated with the circular muscle layer, mostly in wide transverse bands between segmental equators and the anterior intersegmental furrows. Intestinal origin, in xv. Typhlosole, quite insignificant. Holandric. Seminal vesi-

cles, in ix and xii. Prostates, confined to xviii, ducts straight, about 2 mm. long. Penial setae, with horseshoe-shaped to scoop-shaped tip, ornamented with closely crowded circles of large triangular teeth. Ovaries, fan-shaped, with several egg-strings. Ovisacs, small, acinous, in xiv. Spermathecae, duct barrel-shaped, about as wide as but much shorter than the ampulla.

Reproduction: Presumably biparental as sperm are exchanged during copulation. Cocoons may be deposited, in favorable circumstances, throughout the year. Sexual forms were found at Allahabad from July to January. After breeding is completed the clitellum disappears though its former existence is attested by a yellowish or brownish discoloration mostly in the dorsum. Male porophores may become almost unrecognizable though some faint hints of previous presence usually can be detected.

Distribution: Mergui (Mergui). Ross Island, Middle Point, Mount Harriet, Jinghighat, Minnie Bay, Haddo, Pahargaon, Aberdeen, Rajatgarh, Maya Bundar (Andaman Islands). Tavoy, Sinbyudaing, Tenasserim River along road to Thailand (Tavoy). Moulmein, Mupun (Amherst). Pyapon, Kyaiklat (Pyapon). Rangoon, Syriam (Hanthawaddy). Bassein, Coomzamu, Kochi (Bassein). Toungoo (Toungoo). Prome (Prome). Thayetmyo, Allanmyo, Thanbula (Thayetmyo). Sandoway, Ngapoli (Sandoway). Pyinmana, Pyigyaung (Yamethin). Magwe, Taungdwingyi, Natmauk (Magwe). Minbu (Minbu). Kyaukpyu (Kyaukpyu). Meiktila, Thazi, Mahlaing, Mondine, Magyidaung (Meiktila). Myingyan, Pagan, Taungtha, Kyaukpadaung (Myingyan). Akyab, Buthidaung, Maungdaw (Akyab). Kyaukse (Kyaukse). Pakokku (Pakokku). Mandalay, Tonbo, Kyaukkyone (Mandalay). Sagaing, Myotha, Tada-U, (Sagaing). Monywa, Mingin, Kalewa, Kindat (Lower Chindwin). Shwebo, Kyaukmyaung, Kin-U, Ye-U (Shwebo). Katha, Indaw Lake (Katha). Bhamo (Bhamo). Myitkyina (Myitkyina).

Thailand (Bawti and Not Theinko, probably will be found commonly in the lowlands). Malay Peninsula. Sumatra. Christmas Island. Nordwachter (Java Sea). Sumba. Kisser Island (north of Timor). Labuan and British North Borneo. Nias. Philippine Islands. Hainan. Hong Kong and on the mainland of China at Kowloon. New Caledonia.

Mauritius, Seychelles, Comoro Island, Madagascar, Zanzibar.

India, the peninsula and the Indogangetic plains, from Cape Comorin to Lahore, Kapurthala, and Siliguri. Maldive and Laccadive Islands including Minicoy. Ceylon.

Only in the lowlands of Burma, probably present in the six lowland districts for which there are as yet no records; Thaton, Myaungmya, Pegu, Insein, Tharrawaddy and Henzada. From sea level to 2,000 feet in Ceylon. To 2,500 feet (possibly 3,000?) in India but most records are from localities below a thousand feet.

The original home of the species must be assumed to be somewhere in the vicinity of the Palni and Cardamom Hills of South India. The distribution clearly resulted from transoceanic carriage and the agent probably was man during the last 2,000 years.

Habitats: Soil, manure heaps. Earth around roots of potted plants (Andaman Islands, Burma, India). The species seems to be especially successful in the light soils of the dry zone in central Burma, but hardly ever was found in jungle soils or well away from towns or former village sites.

Biology: Drought, in monsoon lands, usually poses a period of inactivity that may last from four to eight months. However, in soil around wells, hydrants, tanks, drainage ditches, irrigation canals, etc., that remains moist enough near end of monsoon rains, activity may continue through the dry season. A species as widely distributed as *L. mauritii* must have been introduced many times into areas such as the Shan Plateau, the summer resorts of peninsular India and of the Himalayan mountains. Absence in such places accordingly suggests that some climatic factor other than moisture restricts dispersal.

L. mauritii is geophagous and feeds underground where it presumably copulates. Copulation has not been observed.

Growth, at Allahabad, apparently involved or was associated (Gates, 1948) with increase in number of segments. (Also *cf.* Bhatti, 1962, *Proc. 14th Pakistan Sci. Conf. B*, p. 22.)

Castings: Intestinal ejecta of this species may be deposited on the surface of the ground as small piles of strings more or less completely constricted into nearly spheroidal pellets. Castings with a dry weight of seven ounces were secured (Hla Kyaw & Gates) in a plot ten feet square during twenty days but much more may have been washed away by heavy rains. Dry weight of castings deposited on the surface was estimated to be 0.425 ton per acre for a season of 100 days. The number of specimens of *L. mauritii* in Rangoon plot was found to be 97 (51-16-30). In that same plot there were also 71 earthworms of at least 5 other species.

Variation and Abnormality: Oviducts occasionally, e.g., in 28 of the 429 specimens in one series, unite mesially so as to open to the exterior by a single median pore.

Abnormal metamerism of various kinds (Gates, 1947) often had originated during embryonic development but the same sorts of metameric abnormality also arose during regeneration. Many abnormalities of organ location arose during head regeneration.

Regeneration: Head regeneration is possible (Gates, 1959) in an anterior direction at all levels back to region of 25/26. Complete replacement of lost seg-ments is possible back to 8/9 and eight segments may be regenerated at levels as far back as 25/26. Tail regeneration is possible in a posterior direction from 30/31 to the hind end. So much of the regenerative capacity of *L. mauritii* was learned without resort to experimentation. Heteromorphic regeneration, however, was not seen.

Luminescence: Fine masses that glow faintly for a short time only may be ejected, on appropriate stimulation (Gates, 1925), by a single individual. Material ejected more or less simultaneously by thirty to forty worms in a small receptacle may luminesce for several hours.

Parasites: SPOROZOA: *Apolocystis matthaii* (Bhatia & Setna, 1926), Bombay. Coccidium perichaetae[3] Beddard, 1888, coelom, Borneo. *Monocystis megascolexae* Kar, 1946, Calcutta.[4] NEMATODA: *Metastrongylus apri* larvae, Philippines (Refuerzo & Reyes, 1959). INSECTA: Larvae of *Sarophaga* sp., occasionally present in anterior ends of Indian worms, were not found in Burma. Predation internally by those larvae was believed (Gates, 1959) to be responsible for the amputations that resulted in head regeneration.

Remarks: Bands of color are quite unrecognizable externally but are obvious after the longitudinal muscle layer has been stripped off from the body wall. Pigment may be dense in the special longitudinal muscle band at mD.

The male pores i.e., the superficial openings through which sperm are extruded in this species presumably are the common apertures of the prostatic ducts and the penisetal follicles, though the literature seems to provide no definite supporting evidence.

Nephropore locations of the excretory tubules in x–xi were not recorded, and whether nephridia are present in ii–iv is unknown.

L. mauritii does not appear to be closely related to any of the known species of the genus.

Krishnan & Rajalu (1969: p. 1623), as a result of their histochemical studies of epidermal secretions by *L. mauritii*, suggest that the alarm pheromone of *L. terrestris* (Ressler *et al.*, 1968, *Science* 161: p. 597) is in the nature of a highly acidic sulphated mucopolysaccharide produced by the large mucus gland cells.

Lennoscolex

1960. *Lennoscolex* Gates, *Bull. Mus. Comp. Zool. Harvard College* 123: p. 241. (Type species, *Woodwardiella pumila* Stephenson, 1931.)

Digestive system, with one esophageal gizzard in v, intestinal origin behind xv, without caeca, typhlosoles, calciferous and supra-intestinal glands. Vascular sys-

---

[3] Absence of italics suggests that the name was not intended to be a technical specific name.

[4] Loubatières (1955, *Ann. Sci. Nat.* (Zool.), p. 17) lists *M. perichaetae* from *L. mauritii* but on what authority?

tem, with unpaired complete dorsal and ventral trunks, paired extra-esophageals median to the hearts, a supra-esophageal but no subneural (instead a pair of posterior lateroparietal trunks?), with latero-esophageal hearts in x–xii. Excretory system, meroic, (in anterior segments?), behind the clitellum with one pair per segment of clusters of four to six small, exoic and astomate, parietal and avesiculate ne-phridia. Septa, all present from 5/6. Dorsal pores, present. Setae, four pairs per segment, d gradually becoming more dorsal posteriorly, a and b of xviii penial (or lacking?). (Pigment, lacking.)

Biprostatic, male pores (through which penisetal follicles also open?) in xviii. Quadrithecal, pores minute and superficial, at 7/8–8/9, spermathecae di-verticulate. Female pore, median and slightly pre-equatorial, in xiv. Clitellum, annular, dorsal pores occluded, intersegmental furrows obliterated, setae retained, on xiv–xvii at least. Holandric. (Sperm ducts, pass into ental ends of prostatic ducts?) Seminal vesicles, in xi–xii. (Ovaries? Ovisacs?)

Distribution: Presumably South India (Travancore and Cochin only?). Two species now are domiciled in Burma, one of them also in Java, probably after recent accidental introduction by man.

Remarks: The definition is incomplete. Informa-tion as to nephridia of clitellar and preclitellar seg-ments is needed as Bahl's brief account (1942) appears to refer only to the intestinal region of one species.

Among possible candidates for inclusion in *Lenno-scolex* are *Notoscolex minimus, peermadensis, travan-corensis* Aiyer, 1929, *kayankulamensis* (Aiyer, 1929) and *tenmalai* (Michaelsen, 1910). All are in need of accurate characterization of vascular, excretory, diges-tive systems and, for taxa, also of the prostate glands.

Phylogenetically, *Lennoscolex* is of interest because of its association of (primitive?) simplicity in the digestive system with a highly specialized sort of meronephry.

KEY TO SPECIES OF *Lennoscolex*

Spermathecal pores close to *A*, last hearts in xii
javanica
Spermathecal pores close to *B*, last hearts in xiii
pumila

Lennoscolex javanica

1910. *Woodwardia javanica* Michaelsen, *Mitt. Naturhist. Mus. Hamburg* 27: p. 93. (Type locality, Buitenzorg, Java. Types, if still extant, in the Hamburg Mus.?)
1942. *Woodwardiella javanica*, Gates, *Bull. Mus. Comp. Zool. Harvard College* 89: p. 108.
1960. *Lennoscolex javanica*, Gates, ibid. 123: p. 243.
1961. *Notoscolex javanica*, Gates, *Burma Res. Soc. Anniv. Pub.* No. 1: p. 57.

Spermathecal pores, at or slightly lateral to *A*. Male pores, at centers of small indistinctly demarca ted porophores that reach to or slightly beyond *A* and *B*.

(Genital markings lacking?) Setae, *d* slightly above m*L* anteriorly, behind the clitellum more dorsal so that *DD* < *AA*. Clitellum, reaches into xiii, the anterior boundary indistinct. First dorsal pore, at 6/7. Prostomium, proepilobous. (Segments?) Size, 38–75 by 1–1¼ mm.

Septa, 5/6 membranous to slightly muscular, 6/7–11/12 muscular, 12/13 slightly muscular. Intestinal origin, in xviii. Last hearts, in xii. Prostates, in xviii, ducts 1½–2 mm. long, muscular, thickened ectally, coiled or in a *U*-shaped loop. Penial setae, 0.55–0.69 mm. long, 4–7 μ thick at base, 4–6 μ thick at midshaft, 1–3 μ at tip, sinuous in ectal half, sinuos-ities alternate; tip flattened and narrowed to a sharp point; ornamentation of rather triangular teeth in the sinuosities. Spermathecal duct much shorter than the ampulla from which it is not constricted; di-verticulum about half as long as main axis, slenderly club-shaped, from lateral face of duct just above parietes.

Reproduction: Presumably biparental, as sperm are exchanged in copulation.

Distribution: Mupun (Amherst). Boyagyi (Tha-ton). Java (Buitenzorg).

Remarks: The anatomy of this species is so like that of *L. pumila* as to warrant inclusion in the same genus at least until contraindication is provided by more precise characterization of its excretory system.

Lennoscolex pumila

1931. *Woodwardiella pumila* Stephenson, *Proc. Zool. Soc. London* 1931: p. 51. (Type locality, Bhamo. Types in the British Mus.)
1942. *Woodwardiella pumila*, Bahl, *Quart. Jour. Micros. Sci.* 84: p. 33. Gates, *Bull. Mus. Comp. Zool. Harvard College* 89: p. 112.
1945. *Notoscolex pumila*, Gates, *Proc. Natl. Acad. Sci. India* 15: p. 48.
1960. *Lennoscolex pumila*, Gates, *Bull. Mus. Comp. Zool. Harvard College* 123; p. 242.
1961. *Notoscolex pumila*, Gates, *Burma Res. Soc. 50th Anniv. Pub. No.* 1: p. 57.

Spermathecal pores, at or slightly median to *B*. Male pores, at or near centers of small transversely elliptical male porophores that reach to or slightly beyond *A* and *B*. Genital markings, often repre-sented by transversely elliptical unpaired areas of whitening in *AA* along 17/18 or 17/18–18/19. Setae, *d* at or slightly above m*L* anteriorly, behind clitellum more dorsal so that *DD* < *AA*. Clitellum, reaches into xiii, the anterior boundary indistinct. First dorsal pore, at 6/7–7/8. Prostomium, proepilobous, occasionally a furrow from posterior margin extend-ing along m*D* toward 1/2. Segments, 114 (± ?). Size, 30–90 by 1 mm.

Septa, 5/6 membranous or slightly muscular, 6/7–9/10 muscular, 10/11–11/12 slightly muscular. In-testinal origin, in xvi or xvii. Last hearts, in xiii. Prostates, in xviii or xviii–xix; ducts less than ½ mm.

long, straight, slender but muscular. Penial setae, 0.28–0.42 mm. long, 5–9 μ thick at base, 6–8 μ thick at midshaft, narrowing ectally to a short filament, shaft sinuous ectally; ornamentation sparse, variable, the spines or teeth usually recognizable only under oil immersion. Spermathecae, duct slender and shorter than the ampulla, diverticulum about as long as duct, digitiform to slenderly club-shaped, passing into lateral face of duct slightly ental to its middle and then upwards to open into widened portion of duct lumen.

Reproduction: Presumably biparental, as sperm are exchanged in copulation.

Habitat: Soil, in jungles as well as in and near towns and villages, but only in lowlands.

Distribution: Ye (Amherst). Pyapon, Thameintaw (Pyapon). Myaungmya, Wakema (Myaungmya). Sittang (Thaton). Rangoon, Kungyangon, Kyauktan (Hanthawaddy). Danubyu (Maubin). Bassein, Thinbawgyin, Padaukchaung (Bassein). Pegu (Pegu). Toungoo (Toungoo). Bhamo (Bhamo). (South India?)

Abnormality: Absence of one of the hearts of xiii and doubling of the dorsal trunk in certain anterior segments, presumably as a result of failure of embryonic anlage to unite in the usual manner, have been recorded (Gates, 1942: p. 117).

Regeneration: Tail regeneration is possible.

Parasites: Epizoic organisms sometimes noted are most numerous on and just behind the clitellum. The dominant species is colonial and supposedly vorticellid (a further identification has not been obtainable) on which there are various sorts of much smaller plants and animals. The dominant colonial organism from another lot of specimens was tentatively identified as a rotifer.

Remarks: Specimens that were longer than 70 mm. may have been relaxed.

L. pumila is presently distinguishable from *Notoscolex peermadensis* only by penisetal characters and from *N. kayankulamensis*, if excretory systems are alike, scarcely at all.

Nelloscolex

1939. *Nelloscolex* Gates, *Rec. Indian Mus.* **41**: p. 37. (Type species, *Woodwardia burkilli* Michaelsen, 1907.)

Digestive system, without intestinal caeca and supra-intestinal glands but with a massive gizzard in vi, rudimentary calciferous glands, an intestinal origin behind xiii, a lamelliform intestinal typhlosole. Calciferous tissues, in vertical lamellae within paired lateral pouches that are not constricted off from the gut in viii–xii, size and distinctness of the pouches increasing posteriorly. Vascular system, with unpaired dorsal, ventral and supra-esophageal trunks, paired extra-esophageal trunks median to the hearts,

without a subneural trunk (? but with lateroparietal trunks?), hearts of viii and anteriorly lateral, of ix–xii latero-esophageal (?). Excretory system, meroic. Setae, 8 per segment, *a–b*/xvii lacking. Dorsal pores, present from region of 8/9–9/10. Pigment, lacking. Septa, present at least from 6/7.

Biprostatic, male pores in seminal grooves confined to xvii. Female pores, anteromedian to *a*, in xiii. Quadrithecal, pores at 6/7–7/8. Clitellum, annular, intersegmental furrows obliterated, dorsal pores occluded, setae retained. Holandric, testes in ix–x. Seminal vesicles, rather small, acinous, in x–xi. Prostates, strap-shaped. Spermathecae, diverticulate. Ovaries, (fan-shaped? egg-strings?), in xii. (Ovisacs, present?).

Distribution: The western mountain wall of Burma from Akyab north into the Khasi Hills of Assam.

Remarks: Six specimens, of the *burkilli* size, that had to serve as types (possibly of three species) could not provide the desired information even if one had not been macerated and perhaps abnormal.

Loss of some or all of the ventral setae in the spermathecal segments may prove to be characteristic.

Relationships of *Nelloscolex* clearly are with *Tonoscolex* (*q. v.*) the only other megascolecid genus known to have the same regional anterior homoeosis.

Nelloscolex burkilli

1907. *Woodwardia burkilli* Michaelsen, *Mitt. Naturhist. Mus. Hamburg* **24**: p. 152. (Type locality, Buthidaung. Types, in the Indian and the Hamburg Museums.)
1909. *Woodwardia "burkillii,"* Michaelsen, *Mem. Indian Mus.* **1**: p. 162.
1923. *Woodwardia burkilli*, Stephenson, (*The Fauna of British India*), *Oligochaeta*, p. 185.
1932. *Woodwardiella burkilli*, Gates, *Rec. Indian Mus.* **34**: p. 364.
1939. *Nelloscolex burkilli*, Gates, *ibid.* **41**: p. 38.

Male pores, median to *B*(?) and close to anterior ends of nearly straight, longitudinal, seminal grooves. Spermathecal pores, in *AB*. Setae, *a* and *b* lacking (?) in vii–viii; *AB* < *CD* < *AA* < *BC*, *DD* ca. = ½*C*. First dorsal pore, at 9/10. Clitellum, 12/13–16/17. Prostomium, prolobous. Segments, 125. Size, 50 by 1 mm.

(Intestinal origin?). Typhlosole, high. Prostates, in xvii–xviii; duct looped. Spermathecae, large, duct shorter than ampulla; diverticulum from median face of duct near parietes, shorter than main axis, club-shaped, with very short stalk portion.

Reproduction: Presumably biparental as sperm are exchanged during copulation. The breeding season may extend into December.

Distribution: Buthidaung (Akyab).

Habitat: Damp spot in high forest.

Remarks: Known only from three or four specimens. The species apparently is able, in favorable circumstances, to remain active after the rainy season.

Nelloscolex strigosus

1939. *Nelloscolex strigosus* Gates, *Rec. Indian Mus.* **41**: p. 40. (Type locality, Dumpep. Type, in the Indian Mus.)

Male pores, in AB, each at posterior (?) end of a very small seminal groove within a region of parietal thickening on anterior wall of a single deep transverse depression. Spermathecal pores, at or very slightly lateral to A. Setae, a and b of vii–viii dehisced (?); AB ca. $= CD$. First dorsal pore, at 8/9(?). Clitellum, 12/13–16/17. Prostomium, pro-epilobous almost prolobous. (Segments?) Size, 97 by 2 mm. Intestinal origin, in xvi(?). (Typhlosole?) Prostates, in xviii–xx; duct in a U-loop, ectal limb muscular. Spermathecae, ampulla somewhat wider and shorter than the duct which may be slightly bulbous ectally; diverticulum from duct close to parietes, digitiform, longer than main axis.

Distribution: Dumpep (Khasi Hills, Assam).

Reproduction: A formed body in a spermathecal diverticulum entally and another in the ampulla of the same organ could have represented histolyzed aggregates of sperm that had been received during copulation. Information as to spermatozoal iridescence on the male funnels no longer is available. If sperm had been exchanged, biparental reproduction could be expected. However, the spermathecae are abnormal and possibly in more ways than is indicated by the shriveled terminal appendage of the diverticulum. The spermathecae, in comparison with those of a second specimen from Dumpep, appear to have a dwarfed main axis and a considerably hypertrophied secondary axis. Such a combination of conditions often has been found to characterize spermathecae of certain parthenogenetic morphs in the genus *Pheretima*.

Remarks: Known only from the macerated type. The species seemingly is distinguished by the length of the spermathecal diverticulum which would be even greater if the terminal appendage represents an aborted or collapsed portion of a seminal chamber.

Nelloscolex sp.

1939. *Nelloscolex* sp., Gates, *Rec. Indian Mus.* **41**: p. 43.

Male pores, at anterior ends of nearly straight seminal grooves along B. Spermathecal pores, in AB, just behind 6/7–7/8. Setae enlarged posteriorly, a and b of vii–viii dehisced(?), $AB > CD$, AA ca. or $> BC$, $DD < \frac{1}{2}C$. First dorsal pore, at 8/9. Clitellum, 12/13–16/17. Prostomium, prolobous. (Segments?) Size, 46 by $1\frac{1}{2}$ mm. Intestinal origin, in xiv. (Typhlosole?) Prostates in xviii–xxiv; duct in a U-shaped loop, ectal limb thicker and muscular. Spermathecae, duct almost confined to parietes, much shorter than ampulla, diverticulum from duct in or close to parietes, much shorter than main axis, slenderly club-shaped and in a_U-loop.

Distribution: Dumpep (Khasi Hills, Assam).

Remarks: The worm is fairly well preserved and the only one of the genus in which the intestinal origin was definitely determinable.

Because of the systematic importance of characters provided by the spermathecae and the male field in the closely related genus *Tonoscolex*, reference of the single specimen to *strigosus* seemed inadvisable. But a possibility that the individual just characterized might be only a normal specimen of *strigosus* contraindicated erection of another species until much more information about individual variation in *Nelloscolex* is available.

Perionyx

1872. *Perionyx* Perrier, *Nouv. Arch. Mus. Hist. Nat. Paris* **8**: p. 126. (Type species, *P. excavatus* Perrier, 1872.)
1889. *Megascolex* (part), Vaillant, *Hist. Nat. Anneles* **3**, 1: p. 62.
1893. *Perionyx*, Beddard, *Proc. Zool. Soc. London* 1892: p. 685.
1895. *Perionyx*, Beddard, *A Monogr. of the Order of Oligochaeta* (Oxford), p. 435.
1900. *Perionyx*, Michaelsen, *Das Tierreich* **10**: p. 207.
1907. *Perionyx* (part), Michaelsen, *Die Fauna Südwest Australiens* **1**, 1: 163.
1923. *Perionyx* (part), Stephenson, (*The Fauna of British India*), *Oligochaeta*, p. 318.
1930. *Perionyx* (part), Stephenson, *The Oligochaeta* (Oxford), p. 481.
1960. *Perionyx*, Gates, *Bull. Mus. Comp. Zool. Harvard College* **123**: p. 216.

REGENERATION: Gates: 1927, *Biol. Bull.* **53**: p. 351; 1941, *Jour. Exp. Zool.* **88**: p. 161; 1943, *Proc. Natl. Acad. Sci. India* **13**: p. 168; 1944, *Current Sci.* **13**: p. 16; 1947, *Rec. Indian Mus.* **45**: p. 75; 1951, *Proc. Acad. Sci. Bangalore*, B, **34**: p. 115; 1961, *Wasmann Jour. Biol.* **18**: p. 291 and *Bull. Mus. Comp. Zool. Harvard College* **123**: p. 221, etc.

Digestive system, without supra-intestinal and calciferous glands, intestinal caeca and typhlosoles, but with calciferous tissues in some portion (?) of the esophagus. Vascular system, with complete, unpaired, dorsal, ventral and supra-esophageal trunks, a subneural adherent to the parietes, paired extraesophageals median to the hearts, latero-esophageal hearts in x–xii. Nephridia, with preseptal funnels and postseptal loops that open in their own segment to the exterior through epidermal apertures. Setae, numerous, in a circle at equator of each segment from ii posteriorly. Dorsal pores, present.

Biprostatic, male pores (apertures of united sperm and prostatic ducts) in xviii. Female pore, intraclitellar and median. Clitellum, annular, setae retained. Ovaries, fan-shaped and with several egg-strings.

Distribution: Burma, Assam, the Himalayas west at least to Simla, peninsular India, possibly also Ceylon and Malaysia.

P. sansibaricus Michaelsen, 1891, was recorded from the island of Zanzibar some time before it was recognized in its Indian home territory. Possibly *P. violaceus* Horst, 1893, presently known only from the Malay Peninsula, Java and Sumatra, as well as certain other species from more distant localities, will eventually be found to be of Indian origin.

Transoceanic distributions, obviously a result of transportation of two species, are not included above in the generic range.

Systematics: *Perionyx*, in the classical system, became a polyphyletic congeries with species having lumbricin or perichaetin setae, one, two or no gizzards, calciferous glands of varied or unknown structure in different segments of a region including vii–xiv or no such glands, holomeganephridia or meromeganephridia or micronephridia alongside meganephridia, racemose or tubular prostates, hearts of x-xiii that were lateral, esophageal or latero-esophageal. Characters by which the genus actually was defined usually were qualified by, often, very frequently, always?, but when not so qualified should have been as none were without exceptions. Choosing between the phylogeny of Stephenson and one of the family trees of Michaelsen could solve none of the problems. Accordingly, the genus was recently redefined (Gates, 1960: p. 216) quite tentatively by characters of the generic type that could be assumed, with some degree of confidence, to be shared with other Asiatic species. The difficulties involved in classifying and identifying perionyxes were considered at length in the publication just cited and here attention perhaps should be directed only to the amazing plasticity that seems to characterize much of the genus as now defined. That somatic plasticity, responsible for many of the systematic difficulties, refers to an amazing regenerative capacity exemplified in part by the following: Epimorphic regeneration of a head in an anterior direction at any level of the body except in a short anal region. Epimorphic regeneration of a head in a posterior direction at most any level through the anterior half or even more of the body. Morphallactic regeneration of genital organs at levels way behind those of the normal locations. Conversion of heteromorphic tails, regenerated at levels where head formation appears to be impossible, into homomorphic regenerates. Regeneration, at levels well posterior to normal locations, of gonads that produce mature sperm and ova. Even after major portions of an individual are putrescent (1960: p. 211) a still living portion can be excised and will regenerate normally.

Characters that may prove to be uniform throughout much or all of the genus are: Pigment, red. Copulatory papillae and definite genital markings, lacking though slight epidermal modifications may be recognizable in immediate vicinity of male and/or spermathecal pores. Septal abortions, restricted to the region in front of v. Cephalization of the excretory system lacking or slight.

Seminal vesicles always are present in xii and those of either segment often are, or at least seem to be, united dorsally though arising independently. The gizzard, even when large, appears to be rather weak

and is often vestigial if not entirely lacking—a decision between those alternatives often possible only from microtome sections.

Although septa rarely seem to be thickly muscularized, many of those in the intestinal region often are thickened so as to considerably reduce coelomic space in the segments involved. Perhaps peritoneal cells on each side of the septum are markedly distended with fat or other reserve materials as happens in many species of various families to the peritoneal covering of nephridia. Individual variation already noticed may be due to seasonal cycles of activity and inactivity or to differences in metabolic states—starved worms for instance showed little or no evidence of such septal thickening. Some usefulness systematically is anticipated from modification of nephridial investment but data *re* modification of peritoneal linings of perionycil septa still are too fragmentary to provide any such indication.

Wording of the nephridial character in the generic definition was necessitated by a Nepalese giant species, unfortunately now without a certainly valid name, which has meromeganephridia. They supposedly arose by longitudinally splitting several times each of the two segmental nephridial anlage. Morphologically each of such meroic nephridia appears the same as a holomeganephridium. Meromeganephridia of that sort are presently unknown from any other species of any genus.

Burmese species, according to the information now available, seem to fall into two groups. Species of the larger group have the last hearts in xii and superficial male pores. Both characters may be primitive. Species of the other group have hearts in xiii and the primary male pores retracted, in various ways, into the parietes or into definite vestibula. Absence of penial setae in some species of each group may not be primitive but secondary and due to abortion of follicles that would, if present, produce modified shafts.

Parthenogenesis, which was responsible for so much systematic confusion in the classification of *Pheretima*, was not anticipated because of the genital and regenerative plasticity of *Perionyx*. However, the genital polymorphism of *P. bainii* Stephenson, 1915, for which H, R, and A morphs were recently (1960) recorded warns against repetition of mistakes made in work on the other genus.

For determination of intrageneric relationships, information from many species is needed about the following: Location of nephropores and whether those pores are obvious or inconspicuous. Levels at which nephridial ducts enter the parietes. Nephridial vesicles, whether present or absent and the shapes when present. Segmental locations of the hearts and whether they are lateral, esophageal or latero-esophageal. Chambers within the wall of the spermathecal ducts, their location, number, and use.

Also of interest and perhaps of some systematic importance will be data as to segment number and whether metameres are added after hatching, information about pigmentation pattern (possibly characteristic in certain species), also re location of junction of vasa deferentia with the prostatic ducts and the precise situation of male pores (not yet recognized in a number of species). Rudimentary pseudovesicles in xiii–xiv, like those of many pheretimas (in which they originally were thought to be ovisacs), have been recognized only in *P. macintoshi*.

Penial setae, to which so much systematic importance was attributed in the past now seem to be of much less significance in *Perionyx* as well as in certain other genera.

Remarks: The digestive system seems quite primitive though a gizzardless condition is believed to be a slowly attained secondary simplification. Although discrete calciferous glands, even calciferous lamellae, are lacking, calcium is freely excreted, so much as to drastically affect results of certain regeneration experiments. Cephalization is slight. None has been recorded for the nephridial system and dorsal pores are lacking at but few levels. Multiplication of the setae is the only universal specialization.

Intrageneric evolutionary modifications, often independently acquired in different parts of the range (1960), involve: Posterior translocation as well as elimination of the gizzard. Esophageal elongation. Appearance of hearts in xiii. Shifting of nephropores to provide an arrangement in four longitudinal ranks. Vesiculation of nephridial ducts. Multiplication of nephridia (once). Elimination of one, two, or three pairs of the primitive battery of eight seminal vesicles. Meroandric reduction (once). Modifying size of spermathecal battery somewhat as in *Pheretima*. Seemingly, considerable experimentation with development of seminal chambers in spermathecal ducts and eventually of extramural diverticula. Experimentation also with invagination of primary male pores in various ways and leading to development of penes for insertion into enlarged spermathecal pores.

That portion of the genus belonging to the Indian peninsula and Ceylon is not so well known (species fewer and less common?) and in various ways seems as if it might be more ancient. In the geologically younger Himalayas, species are more numerous, more common, and delimitation of species boundaries is more difficult. Endemics are lacking in the geologically still younger Indo-Gangetic plains.

The chief home of the genus, according to Stephenson (1923: p. 320), is the eastern Himalayas and Assam. Available distributional data seem to allow another explanation that is rather reluctantly recorded because of lack of time to investigate the geological implications.

Perionyx arose in peninsular India and from there reached Ceylon before the latter was separated off

and got into continental India across the Rajmahal-Garo gap (a recent physiographic feature) before the Himalayan foredeep was filled up. Absence of endemic perionyxes in the Gangetic Valley then might be due in part to lack of time for migration into the Gangetic alluvium and speciation there or because loss as well as vestigialization of the gizzard had unfitted various species for existence in the valley soils.

KEY TO BURMESE SPECIES OF *Perionyx*

1. Spermathecal pores, one pair, at 7/8........ 2
   Spermathecal pores, more than one pair...... 3
2. Penial setae, none..................... *ditheca*
   Penial setae, present.................... *viridis*
3. Spermathecal pores, two pairs............. 4
   Spermathecal pores, more than two pairs.... 13
4. Spermathecal pores, at 6/7–7/8............ 5
   Spermathecal pores, at 7/8–8/9............ 6
5. Spermathecal pores, at or median to *B. .montanus*
   Spermathecal pores, at or slightly median to
   mL............................... *nemoralis*
6. Last hearts, in xii........................ 7
   Last hearts, in xiii....................... 10
7. Holandric and with penial setae............ 8
   Metandric and without penial setae...... *rufulus*
8. Spermathecal seminal chambers (when recognizable) discrete........................... 9
   Spermathecal seminal chambers aggregated into
   two cauliflower-like clusters....... *shillongensis*
9. Intestinal origin, in region of xv–xvi.... *excavatus*
   Intestinal origin, behind xvi........... *turaensis*
10. Penial setae, present..................... *fossus*
    Penial setae, lacking..................... 11
11. Male pores, on penes..................... 12
    Male pores, not on penes........... *macintoshi*
12. Penes, shortly columnar, in paired
    invaginations...................... *modestus*
    Penes, distally bifid and in a median
    vestibulum............................ *horai*
13. Penial setae, present................... *tenuis*
    Penial setae, lacking............... *arboricola*

Perionyx intermedius, sp. dub., was erroneously listed for Burma by Michaelsen (1900, 1903) because of a belief that Sibpur was in that country instead of being in Bengal near Calcutta.

Because of the uncertainties as to characterization even of the common *P. excavatus*, the paucity of our knowledge re so many of the species, the probability that additional closely related forms are present in Burma, and also because of the unusual variability and somatic plasticity that now seems to prevail throughout much of the genus, the above key is unlikely on many occasions to be more than a slight aid to, certainly not a means of, identification.

Certain taxa, after careful consideration have been left out: *P. depressus* and *aborensis* Stephenson, 1914 from the Abor country, *himalayanus* Michaelsen, 1907, *inornatus* Stephenson, 1916, *kempi* Stephenson, 1914, *sikkimensis* (Michaelsen, 1907), from vicinity of Darjiling, all with spermathecal pores at 6/7–7/8, because information as to anatomy and individual variation is insufficient to distinguish them with certainty not only from each other but also from still other taxa. *P. foveatus* Stephenson, 1914, from the Abor country, because of uncertainty about systematically important genital as well as somatic anatomy. *P. koboensis* Stephenson, 1912, from the Abor country because it is not clearly distinguishable from *P. excavatus*. *P. annulatus* Stephenson, 1914, from the Abor country, placed in *Perionyx* because of its "general habitus" in spite of the presence of micronephridia (possibly only shreds of disintegrated peritoneum? GEG) because of uncertainty as to what genus the species may belong (*cf.* Gates, 1962: p. 218).

Perionyx arboricola

1890. *Perionyx arboricola* Rosa, *Ann. Civ. Mus. Sto. Nat. Genova* 30: p. 119. (Type locality, Cobapo, Leiktho Circle, Burma. Types, in the Genoa Mus.)
1895. *Perionyx arboricola*, Beddard, *A Monogr. of the Order of Oligochaeta* (Oxford), p. 438.
1900. *Perionyx arboricola*, Michaelsen, *Das Tierreich* 10: p. 209.
1903. *Perionyx arboricola*, Michaelsen, *Die geographische Verbreitung der Oligochaeten* (Berlin), p. 89.
1923. *Perionyx arboricola*, Stephenson, (*The Fauna of British India*), *Oligochaeta*, p. 326.
1936. *Perionyx arboricola*, Gates, *Rec. Indian Mus.* 38: p. 465.
1961. *Perionyx arboricola*, Gates, *Burma Res. Soc. 50th Anniv. Pub.* No. 1: p. 57.

Octothecal, pores minute, superficial, further from mV than usual, at 5/6–8/9. Male pores, in protuberances from flattened lateral tumescences in a median transverse depression reaching well into xvii and xix. Clitellum, xiv–xvi. Setae, 56–60 per segment, much further apart in the dorsum where there may be only 8–9. (Nephropores?) First dorsal pore, at 5/6. Prostomium, epilobous. Color, lacking ventrally. Segments, 110. Size, 70 by 5 mm.

Gizzard, in v. (Intestinal origin? Hearts? Nephridia?) Holandric, (testis sacs present?). Seminal vesicles, in xi, xii–xiv. Prostates, in xvi–xxiv, the duct elongate and in several segments, in a hairpin loop with ectal limb much thicker. (Penial setae, lacking.) Spermathecae, duct short, diverticulum club-shaped (and looped?).

Distribution: Cobapo, (Leiktho Circle, Toungoo), in the Karen hills at western margin of the Shan Plateau.

Habitat: Leaf axils of trees.

Remarks: The clitellum, prostates, prostatic ducts and testis sacs resemble the same organs in various species of *Pheretima*. However, further and appropriate deepening of the depression in xviii might well provide a vestibulum as in *P. horai.*

The species really is known only from a single mature specimen. Six other specimens in the type receptacle are juveniles possibly of a different species. Many hundreds, perhaps several thousands, of perionyxes from trees as well as other habitats in the region of the type locality, collected by Karen residents of the area, were examined in a laboriously wasteful effort to obtain further specimens. Nothing is known about relationships with other species.

Perionyx ditheca

1931. *Perionyx ditheca* Stephenson, *Rec. Indian Mus.* 33: p. 186. (Type locality, Thandaung. Types in the British Mus.)
1960. *Perionyx ditheca*, Gates, *Bull. Mus. Comp. Zool. Harvard College* 123: p. 223.

Bithecal, pores at 7/8. Male pores, at centers of small, round, closely paired, mesially directed protuberances in a depressed field. Clitellum, xiii–xviii. Setae, 38/v, 38/ix, 42/xii, 38/xix, 32/midbody (but approximate only). Nephropores, in a single rank on each side (?). First dorsal pore, at 5/6. Prostomium, epilobous, tongue open. (Pigmentation?) Segments, 106. Size, 55–70 by 1.5–2 mm.

Gizzard, in v. Intestinal origin, in (or behind?) xvii. Last hearts, in xii. (Nephridial vesicles?) Holandric. Seminal vesicles, in xi, xii. Prostates, in xviii, ducts slender and somewhat twisted. (Penial setae, none.) Spermathecae, duct as long as and half as thick as ampulla (seminal chambers?).

Distribution: Thandaung (Toungoo), in hills at western margin of Shan Plateau.

Remarks: Known only from the original account of four types. A sexthecal type, with pores at 6/7–8/9, is assumed to be hyperplasic, but if so other anatomy should give evidence of regeneration.

Relationships, if the bithecal condition is normal, presumably are more with *P. viridis* than any of the other species now known.

Perionyx excavatus

1872. *Perionyx excavatus* Perrier, *Nouv. Arch. Mus. Hist. Nat. Paris* 8: p. 126. (Type locality, Saigon. Types, in Paris Mus.)
1888. *Perionyx excavatus*, Rosa, *Ann. Civ. Mus. Sto. Nat. Genova* 26: p. 157.
1890. *Perionyx excavatus*, Rosa, *ibid.* 30: p. 121.
1891. *Perionyx excavatus*, Rosa, *Ann. Naturhist. Hofmus. Wien* 6: p. 404.
1895. *Perionyx excavatus*, Beddard, *A Monogr. of the Order of Oligochaeta* (Oxford), p. 436.
1900. *Perionyx excavatus*, Michaelsen, *Das Tierreich* 10: p. 208.
1903. *Perionyx excavatus*, Michaelsen, *Die geographische Verbreitung der Oligochaeten* (Berlin), p. 89.
1909. *Perionyx excavatus*, Michaelsen, *Mem. Indian Mus.* 1: p. 175.
1910. *Perionyx excavatus*, Michaelsen, *Abhandl. Naturwiss. Ver. Hamburg* 19, 5: p. 9.
1917. *Perionyx excavatus*, Stephenson, *Rec. Indian Mus.* 13: p. 375.
1918. *Perionyx fulvus*, Stephenson, *ibid.* 14: p. 16.
1921. *Perionyx excavatus*, Stephenson, *ibid.* 22: p. 760.
1923. *Perionyx excavatus+P. fulvus*, Stephenson, (*The Fauna of British India*), *Oligochaeta*, pp. 329, 333.

1924. *Perionyx excavatus*, Stephenson, *Rec. Indian Mus.* 26: p. 340.
1926. *Perionyx excavatus*+*P. fulvus*, Gates, *Ann. Mag. Nat. Hist.*, ser. 9, 17: p. 472; *Jour. Burma Res. Soc.* 15: p. 213; *Jour. Bombay Nat. Hist. Soc.* 31: p. 183. *Perionyx excavatus*. Gates, *Rec. Indian Mus.* 28: p. 162. Stephenson, *ibid.*, p. 258.
1927. *Perionyx excavatus*, Gates, *Biol. Bull.* 53: p. 351.
1929. *Perionyx excavatus*, Stephenson, *Rec. Indian Mus.* 31: p. 238.
1930. *Perionyx excavatus*, Gates, *ibid.* 32: p. 324.
1931. *Perionys excavatus*, Gates, *ibid.* 33: p. 417.
1932. *Perionyx excavatus*, Gates, *ibid.* 34: p. 546.
1933. *Perionyx excavatus*, Gates, *ibid.* 35: p. 549.
1936. *Perionyx excavatus*, Gates, *ibid.* 38: p. 466.
1939. *Perionyx excavatus*, Gates, *Jour. Thailand Res. Soc. Nat. Hist. Sup.* 12: p. 108.
1941. *Perionyx excavatus*, Gates, *Jour. Exp. Zool.* 88: p. 161.
1943. *Perionyx excavatus*, Gates, *Proc. Natl. Acad. Sci. India* 13: p. 168.
1955. *Perionyx excavatus*, Gates, *Rec. Indian Mus.* 52: p. 79.
1960. *Perionyx excavatus*, Gates, *Bull. Mus. Comp. Zool. Harvard College* 123: p. 224.
1961. *Perionyx excavatus*, Gates, *American Midland Nat.* 66: p. 62; *Burma Res. Soc. 50th Anniv. Pub.* No. 1: p. 57.
1970. *Perionyx excavatus*, Soota & Julka, *Proc. Zool. Soc. Calcutta*, 23: p. 204.

Quadrithecal, pores near mV, at 7/8–8/9. Male pores, anterolateral to tips of penial setae in small transverse protuberances within a single field distinctly demarcated only at anterior and posterior margins, each protuberance with a slightly irregular transverse groove containing apertures of 4–9 penisetal follicles. Clitellum, xiii–xvii. Setae, 41–48/v, 46–56/ix, 47–52/xii, 46–52/xx, viii/4–6. Nephropores, inconspicuous, in one rather irregular rank on each side of the body and near mL. First dorsal pore, at any of 2/3–5/6 but often at 4/5. Prostomium, epilobous, tongue open. Color, except in first few segments restricted to dorsum, often lacking in immediate vicinity of follicle apertures. Segments, 123–178. Size (clitellate specimens only), 30–180 by 3–7 mm.

(Gizzard, lacking or present but weak and then in v or vi?) Esophagus, widened, bead-shaped in xiii and there with calciferous ridges that extend into xiv and xi, valvular in xv. (Intestinal origin, in xv and/or xvi?) Last hearts, in xii. Nephridia, avesiculate. Holandric. Seminal vesicles, in xi, xii, last pair often continued in pockets of 12/13 back to level of 14/15. Prostates, in xviii, ducts short and straight. Penial setae, 0.60–0.69 mm. long, 15–25 μ thick, ornamented ectally with 6–16 circles of fairly large and elongately triangular spines, tip bluntly rounded or finely pointed or flattened and truncate. Spermathecae, large, duct short and stout, often with intramural seminal chambers of various sizes and locations near ental end of duct.

Reproduction: Presumably biparental as sperm are exchanged during copulation.

Breeding is possible, in favorable conditions, throughout the year.

Distribution: Mergui (Mergui). Little Andaman Island, Parnashala (Andaman Islands). Tavoy, Maungmagaun (Tavoy). Ye, Mupun, Kyaikmaraw, Moulmein, Martaban, Kya In, Meetan, Kawkareik (Amherst). Thaton, Bilin, Kyaikto, Sittang (Thaton). Rangoon, Twante, Kyauktan, Syriam, Kungyangon (Hanthawaddy). Wanetchaung, Hlegu, Hmawbi (Insein). Maubin, Yandoon, Danubyu, Pantanaw (Maubin). Pyapon, Thameintaw, Bogale, Thanchitaw, Dedaye, Kyaiklat (Pyapon). Myaungmya, Wakema (Myaungmya). Bassein, Kochi, Thinbawgyin (Bassein). Pauktawgwin (Salween). Pegu, Nyaunglebin, Thanatpin, Pazunmyaung (Pegu). Letpadan, Thonze (Tharrawaddy). Henzada, Ingabu, Zalun (Henzada). Sandoway, Ngapoli (Sandoway). Shwegyin, Toungoo, Thandaung, Tantabin, Daylo (Toungoo). Prome, Paukkaung, Laboo (Prome). Thayetmyo (Thayetmyo). Pyinmana, Pyigyaung (Yamethin). Taungdwingyi (Magwe). Kyaukpyu (Kyaukpyu). Kalaw, Yaungwhe, Tai-o, Inle Lake, Taungyi, Maymyo, Mogok, Namkham (Shan Plateau). Mount Popa (Myingyan). Akyab, Myohaung, Buthidaung, Maungdaw (Akyab). Bhamo, Teinzo (Bhamo). Myitkyina, Nyaungbin, Lonton, Mogaung, (Myitkyina). Masein, Paungbyin (Upper Chindwin). Falam (Chin Hills). Rangamati (Chittagong Hill Tracts, Bengal). Katlicherra (South Cachar). Langol Hills, Imphal, Loktak Lake, Dibrugarh, Sadiyah, Abor Country (Assam).

Darjiling west in the Himalayas to Kumaon district, Dehra Dun, Almora, Simla, Simla Hills. Calcutta, Sibpur, Rajshahi (Bengal). Talewadi (Bombay). Kodaikanal (South India). Kandy (Ceylon). Bangkok, Koulan (Thailand).

Malay Peninsula, Cochin China (now part of Vietnam). Malay Archipelago (Sumatra, Java, Bali, Madura, Mentawei, Borneo). Philippine Islands. Formosa. Hawaii. Dominica (West Indies). Reunion. Madagascar. Comoro.

From sea level to elevations of 9,000 feet.

P. excavatus seems to be equally at home in the lowlands of tropical Burma and in the Himalayas of the temperate zone at elevations up to 9,000 feet. No other species of earthworm is presently known to live in so many different kinds of climates. Although obviously with wide wandering habits (see Habitats) transportation presumably by man and accidentally must be responsible for most if not all of the extra-Asiatic range. The original home of the species is believed to be in the Himalayas.

Habitats: Under bark of standing and fallen trees. On epiphytic ferns in rain forest. Debris in axils of plantain leaves and in forks of trees. Submerged leaves in tree cavities. Leaves of aquatic plants. Under moss and ferns on spray-drenched rocks by waterfalls. Under logs, rocks, bricks at edge of pond. In rotten wood. Debris of eaves gutters, cracks, and other cavities in buildings of as many as four stories. Soil near manure piles or water courses. Manure heaps. Soil saturated with water from bathrooms, cookhouses. Dhobie compounds. Mud, at edge of lake and under 1½–3 feet of water. Black earth. Paddy field (but introduced with some sort of organic fertilizer?).

Considerable moisture and organic matter apparently are required.

Biology: Activity and also reproduction, in favorable circumstances, are year-round. As habitats dry out the worms disappear but where they go and how they hibernestivate is unknown. Worms, unless previously parasitized, thrive in the laboratory where reproduction, cocoon deposition, development, hatching and subsequent growth can easily be studied.

A marked propensity for wandering and especially for climbing is indicated.

Contents of the gut, always mostly organic matter, show the species to feed selectively.

Variation and Abnormality: Posterior amputation is so common that determination of normal range of variation in segment number has been greatly handicapped. A possibility that segments are formed after hatching introduces another complication. Frequency of anterior regeneration, also common, along with certain little understood changes that are not always associated with regeneration has hindered determination of gizzard location and segment of intestinal origin. Presence or absence of a gizzard as well as its location, if and when present, may have to be determined from microtome sections. Considerable variation in number and location of spaces in the wall of the spermathecal duct is anticipated.

Paired thickenings of postclitellar septa, that do not quite reach gut and parietes, or each other anteroposteriorly (the nephridia being between), may be obvious or even quite unrecognizable. Increase of muscular tissues seemingly is not involved (no muscular sheen). Peritoneal cells perhaps may be much distended by metabolic products (*cf.* fat storage in peritoneal cells investing the nephridia of species in various genera).

Divergence from normal in many specimens had arisen during cephalic regeneration. Some of such deviations from usual organization were recently recorded (1960: p. 207–210).

Regeneration: Amputation is common. Regenerates are nearly as common. The process is easily induced and rapidly completed. In every part of the body regeneration is possible in one direction and along most of the anteroposterior axis is equally possible in both directions. Homomorphic head regeneration is possible throughout a major anterior portion of the body, with equimery often attained at each level back to 17/18 but behind that level only in a short region and always as a result of morphallaxis. In a region between 17/18 and 4/5 hypermery often results. Most reproductive organs develop in head regenerates, often characterized by spermathecal and gonadal hyperplasia, but prostates are formed only by reorganization (morphallaxis) in the substrate.

Homomorphic tail regeneration is usual at all levels from 29/30 posteriorly and is possible at levels 20/21–28/29. Average daily rates, in tail regenerates, of segment formation, 3.81, and of increase in length, 1 + mm., for a period of two weeks in the laboratory at room temperature, were attained.

Heteromorphic head and tail regeneration is easily induced in appropriate regions of the body.

Individuals that had regenerated heads after loss of all sex organs except the prostates are known to have matured sperm and to have received sperm during copulation. Reproduction by such individuals seemingly can be assumed.

Parasites: SPOROZOA: *Monocystis longispora* Boisson, 1957, Indo-China. NEMATODA: *Scolecophilus mus* Timm, 1967, coelomic cavities, Cebu and Luzon Islands (Philippines), Bangkok, Rangoon. INSECTA: *Aphiochaeta scalaris* Loew 1866,[1] various parts of the body, Rangoon. Adult insects were bred in tightly closed jars from fragments excised at various locations along the major axis. However, entomologists who identified the parasites reported that the insect is a very general feeder and that the infestation "was accidental, for it seems unlikely that this fly would prove to be a true parasite."

Remarks: Copulation never was observed though many thousands of individuals were kept in the laboratory during a ten-year study of regeneration. The rather indefinite protuberances seen on the male field of preserved specimens may be capable of more marked elevation into rather penial-like bodies for insertion through fairly large spermathecal pores. Spermatophores never were found on the surface of the body and may be formed in the spermathecal duct after direct deposition therein of the sperm.

Uncertainty as to certain aspects of the specific characterization perhaps can be resolved only by study of microtome sections. Such information may have to be acquired before relationships with several less well-known species seemingly with much anatomical similarity to *excavatus* can be determined.

Perionyx fossus

1920. *Perionyx fossus* Stephenson, *Mem. Indian Mus.* 7: p. 214. (Type locality, Shillong, Assam. Type, in the Indian Museum.)
1923. *Perionyx fossus*, Stephenson, (*The Fauna of British India*), *Oligochaeta*, p. 331.

Quadrithecal, pores close to *I*, at 7/8–8/9. Male pores, at *D* or *E*, in a single transverse groove on roof of a deep, squarish depression. Clitellum, reaching into xiii and xvii. Setae, 52/v, 56/ix, 56/xii, 52/xix, 54/midbody. (Nephropores?) First dorsal pore, at 4/5. Prostomium, epilobous, tongue closed. Color,

---

[1] Now known as *Megaselia* (*Megaselia*) *scalaris* (Loew, 1866)?

restricted to dorsum (?). Segments, 136. Size, 86 by 3.5 mm.

Gizzard, large but soft, in vi. Intestinal origin, in xvii. Last hearts, in xiii. (Nephridia, avesiculate?) Holandric. Seminal vesicles, in xi, xii, in each segment united dorsally. Prostates, in xviii but bulging 17/18 and 18/19 to occupy a space equivalent to that of three or four segments, duct bent once or twice, thickened and with muscular sheen ectally. Penial setae, 0.45 mm. long, 18 μ thick, sigmoid but without nodulus, ornamentation ectally of a few small indentations, tip fairly sharply pointed. Spermathecae, large, duct half as long and a third as thick as ampulla, seminal chambers aggregated into two protuberant clusters at middle of duct.

Distribution: Shillong, Khasi Hills, Assam, at 4,500–5,000 feet.

Remarks: Known only from the original description of the holotype.

The statement (1923: p. 331) that the spermathecal pores are three-fourths of the circumference apart obviously is erroneous and may well be a *lapsus calami*.

Perionyx horai

1924. *Perionyx horai* Stephenson, *Rec. Indian Mus.* 26: p. 342. (Type locality, Cherrapunji, Assam. Types in the Indian Mus.)
1960. *Perionyx horai,* Gates, *Bull. Mus. Comp. Zool. Harvard College* 123: p. 229.

Quadrithecal, pores at *C*, at 7/8–8/9. Univestibulate and penile, penes 1 + mm. long, with digitiform distal bifurcations, male pores minute and presumably in anterior ends of deep grooves that pass from the vestibular roof down anterolaterally on the penes where they send a branch into each distal bifurcation. Clitellum, xiii–xvii(?). Setae, 54/v, 36/viii, 55/ix, 45–50/xii, 48/xix, 44/xx, 49/midbody, vii/3–6, viii/0, ix/3–6. (Nephropores?) First dorsal pore, at 4/5. Prostomium, epilobous, tongue open. Color, restricted to dorsum(?). Segments, 144. Size, 86 by 3–4 mm.

Gizzard, in v. Intestinal origin, in xix(?) but maximal width attained only in xx–xxi. Last hearts, in xiii. Holandric. Seminal vesicles, in xi, xii, in each segment united dorsally. Prostates, in xviii, ducts *ca.* 2 mm. long, entally slender and *U*-shaped, ectally straight and with muscular sheen, passing into lateral wall of vestibulum. (Penial setae, none.) Spermathecae, duct muscular, rather spindle shaped, with a number of small seminal chambers in a collar (interrupted once or twice) around ental neck portion.

Reproduction: Presumably biparental as sperm are exchanged during copulation.

Distribution: Cherrapunji, Khasi Hills, Assam, at 4,300 feet.

Habitat: Under stones.

Abnormality and Regeneration: The abnormalities that were recorded warrant anticipation of a regenerative capacity of the same order as in *P. modestus.* Attribution of the gizzard to vi (Stephenson, 1920) may have been due to abnormality known to have characterized at least one of the two types that were dissected.

Remarks: The species is known only from seven specimens none of which had a clitellum at maximal tumescence. The exceptional individual suggests that copulation takes place before clitellar tumescence is well developed if not before any evidence of clitellar differentiation is externally recognizable.

Closest relationships are with a west Himalayan form, *P. simlaensis* (Michaelsen, 1907) from which *P. horai* is distinguished at present only by two characters of unknown systematic importance. Little is known about either taxon and further information might require *P. horai* to be treated as an eastern subspecies of *P. simlaensis.*

Perionyx macintoshi

1883. *Perionyx "m'intoshii"* Beddard, *Ann. Mag. Nat. Hist.,* ser. 5, 12: p. 217. (Type locality, "Akhyab." Types, None.)
1886. *Perionyx "macintoshii,"* Beddard, *Proc. Zool. Soc. London* 1886: p. 209, footnote.
1889. *Megascolex "M'Intoshii,"* Vaillant, *Hist. Nat. Anneles, Paris* 3, 1: p. 86.
1895. *Perionyx "macintoshii,"* (part) Beddard, *A Monogr. of the Order of Oligochaeta* (Oxford), p. 438. (Excluding Indian specimens.)
1900. *Perionyx "m'intoshi"* (part), Michaelsen, *Das Tierreich* 10: p. 208. (Excluding Indian specimens.)
1903. *Perionyx "M'Intoshi"* (part), Michaelsen, *Die geographische verbreitung der Oligochaeten* (Berlin), p. 89. (Excluding Indian specimens. The Burma record is erroneously given as "Seebpore" which is in Bengal where the species would have been found only in shipments to the Botanical Garden.)
1910. *Perionyx "M'Intoshi"* (part), Michaelsen, *Abhandl. Naturwiss. Ver. Hamburg* 19, 5: p. 9. (Excluding Indian specimens.)
1923. *Perionyx "m'intoshi"* (part), Stephenson, (*The Fauna of British India*), *Oligochaeta,* p. 341. (Excluding Nepalese worms.)
1924. *Perionyx "m'intoshi,"* Stephenson, *Rec. Indian Mus.* 26: p. 341.
1952. *Perionyx "m'intoshi,"* Gates, *American Mus. Novitates,* No. 1555: p. 3.

?

1910. *Perionyx "Annandalei"* (part), Michaelsen, *Abhandl. Naturwiss. Ver. Hamburg,* 19, 5: p. 61. (Cherrapunji specimen only.)

Quadrithecal, pores at 7/8–8/9. Male pores, slightly larger than apertures of setal follicles, on posterior wall dorsally in deepened lateral portions of a transverse equatorial groove (which may be postsetal) in a slightly depressed field that may extend into xvii and xix. Clitellum, xii–xxiii(?). Nephropores, present from ii, dorsal, in two rather irregularly zigzagged ranks, *ca.* 14–15 intersetal intervals from mD. Setae, 33–51/ii, 49–53/iii, 53 ± /iv, 54–76/viii, 61–75/xii, 53–66/xx, viii/2–6, xix/2–4, none penial. First dorsal pore, at

5/6, 6/7. Prostomium, epilobous, tongue open. Color, reddish to slate. (Segments, 200–300?) Size, to 310 by 10–11 mm.

Septa, 6/7–19/20 thickly muscular. Pigment, red, denser in the dorsum. Gizzard, large, in vi. (Calciferous lamellae?) Intestinal origin, in region of 19/20. Last hearts, in xiii. Nephridia, avesiculate(?). Holandric. Seminal vesicles, in xi, xii, in each segment united dorsally. Prostates, in xviii (into xix?), ducts with muscular sheen ectally, *ca.* 2 mm. long, straight or sinuous. Spermathecae, duct muscular, slightly shorter than ampulla, slightly narrowed ectally, with one or two intramural seminal chambers entally on the median side.

Distribution: Akyab (Akyab), Gora, east of the Chindwin (Myitkyina). Dumpep, Cherrapunji (Khasi Hills, Assam, India). At elevations of 3,200–4,300 feet.

Habitats: Under stones in damp places.

Parasites: A nematode, *Scolecophilus lumbricicola* Baylis & Daubney, 1942, was reported from *P. macintoshi* in 1942 but the host was from Nepal and is now believed to belong to another and meronephric species for which the proper name presently cannot be selected.

Remarks: Seminal chambers in the spermathecae of Michaelsen's Cherrapunji specimen of *P. annandalei* (1910: p. 61) were as follows: "2 oder 3 an Zahl, bilden winzige, zu einige knotigen Papille verschmelzende Vorwölbungen."

The distribution cited above is based on assumptions that require proof. (1) That the holotype, supposedly from Akyab, is conspecific not only with Gora worms but also with certain others from Assam possibly including the Cherrapunji specimen referred to *P. annandalei.* (2) That giant forms from the Himalayas are meronephric and of a species? and genus? *cf.* Gates, 1952: pp. 7–10) lacking in the Indo-Burmese mountain wall and its extension westward into the Khasi Hills of Assam.

The unique type was characterized too briefly and then lost. Subsequent efforts to obtain specimens of a giant *Perionyx* in the immediate vicinity of Akyab as well as in nearby hills were unsuccessful. If the unique type was secured in Burma its source now would appear to have been away from Akyab town and fairly well up in the Arakan Yomas.

Few specimens (11 + ?) have been available for study and none of them have had a really tumescent clitellum.

Perionyx modestus

1922. *Perionyx modestus* Stephenson, *Rec. Indian Mus.* 24: p. 455. (Type locality, Cherrapunji, Assam. Types, in the Indian Mus.)
1923. *Perionyx modestus*, Stephenson, (*The Fauna of British India*), *Oligochaeta*, p. 344.
1960. *Perionyx modestus*, Gates, 1960, *Bull. Mus. Comp. Zool. Harvard College* 123: p. 231.

Quadrithecal, pores not as closely paired as usual, at 7/8–8/9. Male pores, on flattened ends of shortly columnar penes retractile into prostatic ducts which open laterally in an equatorial groove within a male field that reaches into FI and dislocates 17/18, 18/19. Clitellum, xiii–xvi(?). Setae, 31–63/ii, 43–68/iii, 48–84/viii, 40–60/xii, 44–54/xx, 42/midbody, viii/11–18, xviii/0, xix/6–11. Nephropores, just above ventral limit of pigmentation (always?). First dorsal pore, at one of 3/4–5/6. Prostomium, epilobous, tongue open. Color, faint in the ventrum. Segments, 174. Size, 85–206 by 3–6 mm.

Gizzard, in v. Intestinal origin, in xix or xx. Last hearts, in xiii. (Nephridia, vesiculate?) Holandric. Seminal vesicles, in xi–xii, in each segment usually united dorsally. Prostates, in xviii, duct 4–6 mm. long, in a U-shaped loop, ectal limb thicker and with muscular sheen. (Penial setae, none.) Spermathecae, large, duct shorter than ampulla, slightly muscular, seminal chambers small and in collar around neck of duct (but apparently opening in some specimens into ampulla).

Reproduction: Presumably biparental as sperm are exchanged during copulation.

Distribution: Cherrapunji to Dumpep, Khasi Hills, Assam.

Habitat: Under stones in muddy pools.

Abnormality and Regeneration: The abnormalities of several Dumpep specimens were attributable to regeneration and enabled characterization of regenerative capacity as follows: Anterior regeneration with equimery possible at all levels back to 15/16. Replacement of all reproductive organs except the male terminalia possible. A head probably can be regenerated anteriorly at any level in the first three quarters of the body and a tail probably can be regenerated posteriorly at any level behind 30/31. Heteromorphic regeneration as in *P. excavatus, millardi,* and *sansibaricus* is anticipated.

Parasites: Nematodes from the coelomic cavity of segment x have not as yet been identified.

Remarks: Types were aclitellate and may have been rather juvenile.

Perionyx montanus

1954. *Perionyx montanus* Gates, *Ark. Zool.* 6, 20: p. 433. (Type locality, Mount Kambaiti, Myitkyina district. Types, in the Naturhistoriska Riksmuseets, Stockholm.)

Quadrithecal, pores at 6/7–7/8, in AB or at B. Male field, reaching into DE, depressed, crossed by a single transverse ridge with a cluster of 3–4 apertures of setal follicles just lateral to mV on each side (male pores just lateral to the setal clusters?). Clitellum, xiii–xvii. Setae, 35/viii, 38/xii, viii/3–4. Nephopores, in a single irregular rank on each side of the body(?). First dorsal pore, at 4/5, 5/6. Prostomium, epilobous, tongue open. (Color?) Segments, 91–103. Size, 40–65 by 1–2 mm.

Gizzard, represented only by a slight thickening of the musculature (?) in v–vi. Intestinal origin, in xvi–xvii. Last hearts, in xii. Nephridia, avesiculate. Holandric. Seminal vesicles, acinous, in xi, xii. Spermathecae, small, duct shorter and narrower than ampulla, without external indication of presence of seminal chambers.

Prostates, in xviii, duct slender, nearly straight or with one or two slight quirks entally. Penial setae, 3–4 on each side (?), ectally with 6–8 circles of the same kind of teeth as in *excavatus*.

Distribution: Mount Kambaiti, (Myitkyina) at 2,000 meters.

Remarks: The species is known only from five poorly preserved specimens. Alcoholic preservation and maceration may not have been entirely responsible for inability to recognize pattern of pigment distribution though the pattern was obvious in specimens of *P. nemoralis* from the same container.

P. montanus belongs to an east Himalayan (Darjiling to Abor Country) group of little known species distinguishable from each other at present mainly by characteristics of uncertain systematic value.

Perionyx nemoralis

1954. *Perionyx nemoralis* Gates, *Ark. Zool.* 6, 20: p. 435. (Type locality, Mount Kambaiti, Myitkyina district. Types, in the Naturhistoriska Riksmuseets, Stockholm.)

Quadrithecal, pores minute, slightly median to mL, at 6/7–7/8. Male field, depressed, reaching to *HJ*, including two conical porophores with 5–7 penial setae protuberant from each apex, male pores about in *FG*, just lateral to penial setae. (Clitellum?) Setae, *ca.* 50/xx, vii–viii/14–18. Nephropores, in one irregular rank on each side of the body near mL. First dorsal pore, at 4/5(?). Prostomium, epilobous, tongue open. Color, restricted to dorsum and in each segment to an equatorial stripe. Segments, 100–110. Size, 60–65 by 1.5–2 mm.

Gizzard, represented only by a slight thickening of the musculature in vi(?). Intestinal origin, in xvi, xvii. Last hearts, in xii. (Nephridia, avesiculate?) Holandric. Seminal vesicles, in xi, xii. Prostates, in xviii, ducts straight except for an ental quirk. Penial setae, 5–7 on each side (?), ornamented ectally with 15–20 circles of the same sort of teeth as *in excavatus*. Spermathecae, duct longer than ampulla(?), widened entally (and there thicker than the ampulla?), seminal chambers represented by a broken circle of 5–7 small, beadlike protuberances from ental portion of duct.

Distribution: Mount Kambaiti (Myitkyina), at 2,000 meters.

Remarks: The species is known only from three aclitellate specimens in which alcohol had not yet obscured pattern of pigmentation.

P. nemoralis belongs to the same group of little known species as *P. montanus* from which it appears to be distinguished adequately by the reversal of a rather general generic trend to approximation of spermathecal pores at mV.

Differences from *P. depressus* (and its synonym *aborensis*?) are small and may prove to be unimportant when more is learned about distributions and individual variation. Originally, the distance of Kambaiti from the Arbor Country and the rugged terrain between, along with the morphological differences, was thought to warrant specific distinction for *P. nemoralis*.

Perionyx rufulus

1945. *Perionyx rufulus* Gates, *Jour. Roy. Asiatic Soc. Bengal,* (*Sci.*) 11: p. 74. (Type locality, Khezhabama, Naga Hills, Assam. Types, in the Indian Mus.)
1960. *Perionyx rufulus*, Gates, *Bull. Mus. Comp. Zool. Harvard College* 123: p. 219.

Quadrithecal, pores close to mV, at 7/8–8/9. Male pores, minute, presetal, each within a small transversely crescentic groove (concave posteriorly) which is in anterior part of a longitudinally elliptical area just lateral to *A*. Male field, transversely elliptical, reaching into *FG*. Clitellum, xiii–xviii. Setae, 24–28/ii, 27–37/iii, 37–43/viii, 45–57/xii, 53/xx, vii/4–6, viii/0, ix/0, xviii/0, xix/5. Nephropores, in two regular longitudinal ranks. First dorsal pores, at 5/6, 6/7. (Prostomium? Segments?) Size, 68–98 by 3 mm.

Gizzard, rudimentary, in v. Intestinal origin, in xix or xx. Last hearts, in xii. Metandric. Seminal vesicles, in xii, united dorsally and continued as a median band into xxiv. Prostates, in xviii, ducts 1½–2 mm. long, ental half slender and sigmoid, ectal half straight and with muscular sheen. (Penial setae, none.) Spermathecae, duct (with vertical ridges internally) as long as ampulla, lumen large (seminal chambers represented by one or two slight bulges of duct near ampulla?).

Distribution: Khezhabama, Naga Hills, Assam at 4,800 feet.

Remarks: *P. rufulus*, the only meroandric perionyx, is known only from the eleven types.

Perionyx shillongensis

1920. *Perionyx shillongensis* Stephenson, 1920, *Mem. Indian Mus.* 7: p. 213. (Type locality, Shillong, Assam. Type, in the Indian Museum.)
1923. *Perionyx shillongensis*, Stephenson, (*The Fauna of British India*), *Oligochaeta*, p. 357.

Quadrithecal, pores in *BC*, at 7/8–8/9. Male pores, "fairly conspicuous," in *CD*. Clitellum, xiii–xviii. Setae, 42/v, 46/ix, 49/xii, 48/xix, 41/midbody. (Nephoropores?) First dorsal pore, at 3/4. Prostomium, epilobous, tongue open. Color, restricted to dorsum. Segments, 120. Size, 66 by 3 mm.

Gizzard, fair-sized, in vi. Intestinal origin, in xvi. Last hearts, in xii. (Nephridia, vesiculate?) Holandric. Seminal vesicles, in xi, xii, in each segment united dorsally (?). Prostates, in xviii, duct short, straight, stout but without muscular sheen. Penial setae, 0.87 mm. long, 20 μ thick, ornamented ectally with *ca*. 8 rings of fine spines, tip slightly bowed and bluntly pointed. Spermathecae, large, duct stout and half as long as ampulla, seminal chambers aggregated into two cauliflower-like clusters sessile at ental of duct.

Distribution: Shillong, Khasi Hills, Assam, at 4,500–5,000 feet.

Remarks: Known only from two types.

The gizzard originally was said to be in vi and location in vii of a subsequent publication (1923) presumably is a misprint.

Relationships of the species are with *P. excavatus* from which *P. shillongensis* presently is rather dubiously distinguishable.

Perionyx sp.

1930. *Perionyx* sp., Gates, *Rec. Indian Mus.* **32**: p. 324. (Yebawgyi, Sandoway district.)
1932. *Perionyx* sp., Gates, *ibid.* **34**: p. 546. (Leiktho, Toungoo. Vicinity of Sindin, Zowai, Tharrabyin, Mergui. Northern part of Tavoy district.)
1933. *Perionyx* sp., Gates, *ibid.* **35**: p. 533. (Kamaungthwe River region of Tavoy district. Kyaukpyu. Akyab. Mount Popa, Myingyan. Taungyi and Maymyo, Shan Plateau. Falam, Tiddim, Haka, and other sites in the Chin Hills. Myitkyina. John Lawrence Island, Andaman Islands. *Perionyx excavatus*, Stephenson, *Proc. Zool. Soc. London* 1932: p. 930. (Fort Hertz, Myitkyina.)
1936. *Perionyx* sp., Gates, *Rec. Indian Mus.* **38**: p. 467. (Mogok. Myitkyina and Fort Hertz, Myitkyina. Thandaung.)
1955. *Perionyx* sp., Gates, *ibid.* **52**: p. 79. (Kyaiktiyo, Kumingyaung, Bilin, Thaton, Naunggala, Thaton district. Thanatpin, Pegu. Akyab, Myohaung, Akyab. Pathichaung, Daw Pakko, Shwenyaungbin, Daylo, Lalawata Ferry, Pelachi, Sah Der, Shwe-ta-dah, Myasawni Bridge, Toungoo.)

Remarks: Many thousands of juvenile and aclitellate specimens from various localities including those mentioned above were not identified as to species. All of the worms in which spermathecal pores or their sites could be recognized were quadrithecal. Many might have been referable to *P. excavatus* at maturity. Others differed from the latter as to some or all of the following. (1) Color, which was dark blue or slate, sometimes almost black, or a special shade of purple (at certain localities only) never seen in specimens definitely identifiable as *P. excavatus*. (2) Pattern of pigment distribution. (3) Wider separation of spermathecal and male pores. (4) Characteristics of the male field. (5) Penisetal characters, such as length, thickness, number of circles of spines ectally. (6) Muscularity of the gizzard. (7) Number and location of seminal chambers in spermathecal ducts.

Perionyx tenuis

1954. *Perionyx tenuis* Gates, *Ark. Zool.* 6, 20: p. 437. (Type locality, Mount Kambaiti, Myitkyina district. Type, in Naturhistoriska Riksmuseet, Stockholm.)

Octothecal, pores at 5/6–8/9. Male field, depressed, crossed by a transverse ridge with a cluster of penial setae just lateral to mV (male pores each on a nipplelike protuberance just lateral to the penial setae?). Clitellum, xiii–xvii(?). (Setae? Nephropores, in dorsum? First dorsal pore? Color?) Prostomium, epilobous, tongue open. Segments, 60 (+?). Size, 35 (+?) by *ca*. 1 mm.

Gizzard, represented only by a slight thickening of the musculature in v (?). Intestinal origin, in xvi(?). Last hearts, in xiii. Nephridia, avesiculate (?). Holandric. Seminal vesicles, in xi, xii. Prostates, in xviii(?), duct with a single loop entally. Penial setae, ectally with 10–11 circles of the same sort of teeth as in *excavatus*. Spermathecae, duct shorter than the ampulla and without external indication of presence of seminal chambers.

Distribution. Mount Kambaiti, (Myitkyina), at 2,000 meters.

Remarks: The species is known only from an anterior portion of a partly macerated holotype in which pigment was unrecognized (because of alcoholic preservation?). Though otherwise sexually mature, tumescence of clitellar epidermis was externally unrecognizable.

P. tenuis does not appear to be related to *arboricola*, the only other octothecal species of *Perionyx*. Relationships presumably are to be sought with sexthecal Himalayan species.

Perionyx turaensis

1920. *Perionyx turaensis* Stephenson, *Mem. Indian Mus.* **7**: p. 216. (Type locality, Tura, Garo Hills, Assam. Type, in the Indian Museum.)
1923. *Perionyx turaensis*, Stephenson (*The Fauna of British India*), Oligochaeta, p. 360.
1960. *Perionyx turaensis*, Gates, *Bull. Mus. Comp. Zool. Harvard College* **123**: p. 219.

Quadrithecal, pores at 7/8–8/9. Male pores, on small round papillae in a slight common depression. Clitellum, xiii–xvii Setae, 48/v, 56/xi, 54/xii, 44/xix, 55/midbody. (Nephropores?) First dorsal pore, at 4/5, 5/6. Prostomium, epilobous, tongue open or closed. Color, restricted to dorsum. Segments, 132. Size, 74 by 2 mm.

Gizzard, in vi. Intestinal origin, in xviii. Last hearts, in xii. (Nephridia, vesiculate?) Holandric, (a testis sac in x?). Seminal vesicles, in xi, xii, in xi united dorsally(?). Prostates, in xviii, ducts short and stout (straight?). Penial setae, 0.5 mm. long, 11 μ thick, shaft straight, ornamented ectally with six circles of fine spines, tip slightly curved, "cut off squarely and carries five or six fine spines." Spermathecae, duct thick and short but not distinctly demarcated from ampulla, seminal chambers represented by "a few small rounded knobs at the ental end of the duct, perhaps not always present."

Distribution: Tura (Garo Hills, Assam), at 3,500–3,900 feet.

Remarks: *P. turaensis* presently is distinguishable from *P. excavatus* only by the more posterior intestinal origin, in xviii instead of xv, and by the termination of penial setae ectally in five or six spines. Those characteristics, in case of a single worm, provide little support for specific distinctness in the genus *Perionyx*, for the following reasons. The intestine may appear to begin in various segments behind xv even in *P. excavatus* and for several reasons, compression by prostates, changes undergone during regeneration as a result of which or even for other reasons an anterior portion of the intestine may acquire a valvular or esophageal appearance. A penial seta of *P. excavatus*, after the tip has been broken off, may have a jagged distal margin similar in appearance to that figured by Stephenson (1923: p. 361).

Perionyx viridis

1933. *Perionyx viridis* Gates, *Rec. Indian Mus.* **35**: p. 551. (Type locality, Falam, Chin Hills district. Types, none.)

Bithecal, pores at 7/8. Male pores, in shallow slits(?), on oval tumescences convergent posteriorly or sometimes united. Clitellum, xiii-xvii. Setae, 42–47/xx, viii/6–8. (Nephropores?) First dorsal pore, at 3/4, 4/5. (Prostomium?) Color, restricted to the dorsum. Segments, 103–125. Size, 40–78 by 2–3 mm.

Gizzard, in v. Intestinal origin, in xv–xvi. Last hearts, in xii. (Nephridia, vesiculate?) Holandric. Seminal vesicles, in xi, xii–xiii, xvii. Prostates, in xviii, duct in two *J*-shaped loops with the short arm common to both loops, with muscular sheen ectally. Penial setae, 0.65–0.82 mm. long, 11–13 μ thick, shaft straight, ornamented ectally with 12–20 circles of long spines, tip slightly curved. Spermathecae, duct 1+ mm. long, nearly as thick as and longer than ampulla. Seminal chambers, unrecognizable externally, visible on anterior face of duct entally and then discrete (2–3) or in a ridge or a spheroidal cluster.

Distribution: Falam, Tiddim, Haka (Chin Hills).

Abnormality and Regeneration: A sexthecal individual with male pores in xvi and various other abnormalities almost certainly had regenerated anteriorly, probably in front of 17/18.

Remarks: A green color was characteristic of recently killed worms but disappeared shortly even though preserved in formalin. A pigment, possibly red and like that of many perionyxes, was not leached by the preservative.

Pheretima

1866. *Amyntas* + *Nitocris* + *Pheretima* + *Rhodopis* + *Perichaeta* Kinberg, *Öfvers. Vetenks. Akad. Förhandl. Stockholm* **23**: pp. 101, 102. (*Amyntas*, *Nitocris* and *Rhodopis*, with type species *aeruginosus*, *gracilis* and *javanica* respectively, were preoccupied. Type species of *Pheretima*, no designation.)

1867. "*Perichoeta*" (part), Vaillant, *Bull. Soc. Philom. Paris*, ser 6, **4**: p. 234.
1870. *Perichaeta*, Vaillant, *ibid.* **7**: p. 27.
1872. "*Perichoeta*" (part), Perrier, *Arch. Zool. Exp. Gen.* **1**: p. lxxiv.
1883. *Megascolex* (part) + *Perichaeta* (part), Beddard, *Ann. Mag. Nat. Hist.*, Ser. 5, **12**: p. 214.
1886. *Megascolex* (part), Perrier, *Compt. Rend. Acad. Sci. Paris* **102**: p. 877. (Excluding from the synonymy, *Lampito* and *Megascolex*.)
1888. *Perichaeta* (part) + *Megascolex* (part), Rosa, *Ann. Mus. Civ. Sto. Nat. Genova* **26**: p. 155.
1889. *Megascolex* (part), Vaillant, *Hist. Nat. Annelès* **3**, **1**: p. 62. (Excluding from the synonymy, *Megascolex*, *Lampito*, *Perionyx*, and *Pleurochaeta*.)
1890. *Perichaeta* (part), Benham, *Quart. Jour. Micros. Sci.* **31**: p. 233.
1892. *Perichaeta* (part), Horst, In: Weber, M., *Zool. Ergeb. Reise Niederländish Ost-Indien* **2**: p. 59.
1895. *Perichaeta*, Beddard, *A Monogr. of the Order of Oligochaeta* (Oxford), p. 388.
1899. *Amyntas*, Michaelsen, *Mitt. Naturhist. Mus. Hamburg* **16**: p. 3.
1900. *Amyntas*, Beddard, *Proc. Zool. Soc. London* **1900**: p. 612. *Pheretima* (part), Michaelsen, *Das Tierreich* **10**: p. 234. (Excluding, *P. burliarensis* and *lawsoni*.)
1907. *Pheretima*, Michaelsen, In: Michaelsen W. & Hartmeyer H., *Fauna Südwestaustraliens* **1**: p. 164. (*Pheretima montana* Kinberg 1866 designated as type of the genus.)
1914. *Pheretima*, Cognetti, *Nova Guinea* **5** (Zool.): p. 560.
1922. *Promegascolex* Cognetti, *Boll. Mus. Zool. Univ. Torino* **37**, 744: p. 3. (Type and only species, *P. mekongianus* n. sp.)
1923. *Pheretima* (part), Stephenson, (*The Fauna of British India*), *Oligochaeta*, p. 288. (Excluding *P. burliarensis*.)
1928. *Pheretima* Michaelsen, *Ark. Zool.* **20A**, 3: p. 4.
1930. *Pheretima* + *Megascolex* (part, including only *Promegascolex mekongianus*), Stephenson, *The Oligochaeta* (Oxford), p. 837.
1934. *Pheretima*, Michaelsen, *Quart. Jour. Micros. Sci.* **72**: p. 12. (A previous designation of *P. montana* Kinberg 1866 as type species is withdrawn and in its place *P. californica* Kinberg 1866 is designated as the type species.)
1936. *Pheretima*, Gates, *Rec. Indian Mus.* **38**: p. 389.
1960. *Pheretima*, Gates, *Bull. Mus. Comp. Zool. Harvard College* **123**: p. 246.

Digestive system, without supra-intestinal and calciferous glands (calciferous tissues in low, non-lamelliform ridges in region of xiii?) but with a gizzard that develops in viii. Vascular system, with unpaired dorsal, ventral, and supra-esophageal trunks, a complete subneural trunk adherent to the parietes, paired extra-esophageal trunks (median to the hearts) that are on the gut through x–xiii but then pass off to the subneural, lateroesophageal hearts in some of segments x–xiii. Excretory system, meroic—paired clusters of small astomate nephridia in iv–vi with ducts opening into the pharynx—astomate, exoic, very small, V-shaped parietal nephridia numerous in each segment back from iii—larger, stomate nephridia on both sides of the septa from 16/17 posteriorly, joining postseptal canals that pass to a longitudinal, supra-intestinal excretory duct opening at frequent intervals into the gut. All nephridia avesiculate. Setae, numerous, in a circle at equator of each segment from ii posteriorly. Dorsal pores, present.

Biprostatic, male pores (apertures of united sperm and prostatic ducts) postclitellar. Female pores, always minute, in xiv. Clitellum, annular, intersegmental furrows obliterated, dorsal pores occluded. Seminal vesicles, postseptal. Spermathecae, diverticulate and in front of x. (Ovisacs, none.) Metagynous, ovaries fan-shaped and with several egg strings.[1]

Distribution: The Andaman Islands, the mainland of Asia from the Chindwin-Irrawaddy axis of Burma east through Yunnan and Szechuan provinces of China to include Korea and Japan, thence south to include New Guinea, Java, and Sumatra.

The generic range, as just indicated, comprises the area where species are known to be endemic. Whether any species is native to the Marianas, Bismarck Archipelago, the Solomons, New Caledonia, and the islands to the east remains to be determined.

Several species have been carried almost everywhere by man. As one result of the many ensuing colonizations, boundaries of an autochthonous domain began to be discernible only in the thirties.

Systematics: *Pheretima* is "by far the largest genus of Oligochaeta," according to Stephenson (1930: p. 838), who thought there might be approximately 293 species with about 20 doubtful. Since 1930 many more have been added. Unfortunately, parthenogenetic polymorphism is common, and species now are known to have been erected needlessly on specimens of anatomically degraded morphs. Several species have been widely distributed by man since 1500 A.D. but before that was recognized, certain peregrine forms were given a new name almost every time they were found in a different island, country or continent.[2] Doubtless others of the earlier, inadequately characterized species, known only from the original descriptions, also will have to be synonymized. Nevertheless, there are good reasons for believing that many new species remain to be found in the large, unsurveyed sections of the pheretima domain.

So many species, especially in the Oligochaeta where only one other genus has been thought to be of even comparable size, constituted a temptation to split. One subgenus, *Parapheretima*, supposedly restricted to New Guinea, was erected by Cognetti[3] as early as

1914. Accepting that taxon, Michaelsen[4] distributed other species into three new subgenera, *Archipheretima*, *Pheretima*,[5] *Metapheretima*, each supposedly also geographically restricted. Finally, Michaelsen[6] redefined his *Archipheretima* and added two more subgenera, *Planapheretima*[7] and *Polypheretima*,[8] likewise believed to be geographically limited. Subgenera were distinguished mainly by genital characters and additionally only by presence or absence of a creeping sole and of intestinal caeca. Each of the supposedly distinguishing characters[7,8] or combination of characters, is suspected, or actually known, to have arisen independently in diverse parts of the generic range. Nor is information available, for many species, as to some of the subgeneric characters. Our knowledge of *Pheretima* is as yet too inadequate[9] to warrant recognition of subgenera.

Testis sacs, usually absent in *Megascolex* but usually present in *Pheretima*, provided one of the two distinctions between those classical genera. The primitive condition of those sacs in *Pheretima*, according to classical specialists, was paired. Reasons for such an assumption were not stated. The sacs seemingly were considered to be of systematic importance at species level also, and they usually are characterized in original descriptions. Self-contradictory statements such as, testis sacs paired but united,[10] probably refer to nothing more than a somewhat dumbbell shape[11] of an unpaired and transversely placed, subesophageal chamber containing a pair of testes and the associated pair of male funnels. Anteroposterior connections between sacs of a side (holandric species only), occasionally mentioned by some authors, never were found by the present writer and probably are lacking. Pairing or its absence often is not easily determined, and either character may be subject to some individual variation. Vertically U-shaped testis sacs of x filled with coagulum frequently were mistaken for seminal vesicles, and inclusion of the

---

[1] The generic definition still is regarded as tentative for reasons previously explained (Gates, 1960: p. 246–248). Geographic (known) variation in various endemics of the Shan Plateau cannot be characterized because of loss of specimens and records of worms collected after 1931.

[2] Specimens of *P. diffringens*, for instance, have been referred to as, or have been types of, *californica*, *campestris*, *cingulata*, *clerica*, *corticis*, *divergens*, *heterochaeta*, *heteropoda*, *indica*, *mirabilis*, *molokaiensis*, *nipponica*, *peregrina*, *perkinsi*, *sanctae-helenae*, *silvestrii*, *tajiroensis*, *torii*, possibly also *carnosa*, *directa*, *homoseta*, *kyamikia*, *morii*, *oyamai*, *pingi*. Some of those names still are in use.

[3] Cognetti di Martiis, L., *Nova Guinea* 5 (Zool.) (1914): p. 561.

[4] Michaelsen, W., *Ark. Zool.* 20, A, No. 3 (1928): p. 4. *Pheretima aberrans* Cognetti 1911 designated as type of *Parapheretima*.

[5] Michaelsen, 1928, designated *Pheretima montana* as type of the genus.

[6] Michaelsen later (*Quart. Jour. Micros. Sci.* 77 (1934)) divested *P. montana* of its dignity as a type species and designated *P. californica* Kinberg 1867 as the type of the genus and the subgenus of the same name.

[7] Supposedly distinguished by a creeping sole which has arisen independently in Borneo, Burma, and China.

[8] Supposedly distinguished by presence of more than two spermathecae in a segment. Spermathecae have been multiplied in caecal as well as acaecal species.

[9] The very important excretory system has been studied only in two or three anthropochorous species, all of which are of subg. *Pheretima*.

[10] In Michaelsen's descriptions, at least until 1932.

[11] The median constriction often varies considerably in depth from one worm to another and, when the sac is most distended by testicular coagulum, may be nearly or completely obliterated.

vesicles of xi in a similar sac[12] or in paired, vertical sacs usually was unrecognized. Characterizations of sacs and vesicles of many species accordingly require corroboration or correction.

Structures in xiii and/or xiv that are serially homologous with seminal vesicles sometimes were called ovisacs. In xiii, they cannot be such, as adults have no ovaries in xii. Ova never have been found in the organs of xiv. Function is unknown, and this author calls them pseudovesicles (an abbreviation for pseudoseminal vesicles). Their initial growth may be induced by early rudiments of gonads in xii and xiii. On abortion of those in xii and the beginning differentiation of those in xiii as ovaries, development of the pseudovesicles often ceases, in which case the vertically placed cords probably were overlooked in many dissections. Some thickening of the dorsal end of the rudiment often is recognizable, and sometimes a dorsal lobe is differentiated as in development of real seminal vesicles.

Spermathecal characters probably were grossly overrated in classical treatment of the genus. Two, number and location of spermathecal pores, erroneously assumed to be intraspecifically invariant as a general rule, always had to be determined as a first step in species identification. Some of the intraspecific variation now is known to have resulted from parthenogenetic degradation. Examination of long series has revealed occasional individual variation as to number of spermathecae even in species in which reproduction probably is obligatorily biparental. The number of species needlessly erected because of such variation is unknown.

Locations of the pores sometimes were imprecisely stated because of failure to define intersegmental furrow. As a result only major dislocations from intersegmental levels were recorded. Neglected entirely were pore and associated characters now known to be of considerable systematic usefulness. Among such characters are: size—ranging from minute (that of the female pore) to large, superficial or invaginate and in the latter case location of the minute, primarily spermathecal pore within the invagination. The latter can be small or large, intraparietal or coelomic. A large coelomic chamber may be readily recognizable or flattened out against the body wall so as to be easily overlooked. A small coelomic chamber may be superficially indistinguishable from the spermathecal duct. Genital markings sometimes are present in parietal as well as coelomic pore-chambers and, like the associated glands, are to be adequately characterized.

The male terminalia received from classical specialists a little more attention than did the ectal ends of the spermathecae. Presence of copulatory chambers usually was mentioned, but information as to contents,

tubular penes, penial bodies of several sorts, genital markings of various kinds, penial setae when present, glands either intramural or coelomic, is needed. Intraparietal chambers containing the minute, primary male pores were not described. Copulatory chambers as well as the parietal invaginations often (always?) are eversible and the markedly protuberant male porophores of various pheretimas probably are everted chambers. The kind of invagination cannot, of course, be determined from the description or sometimes even from the available specimens.

Other genital organs that also need more precise characterization are the glands often associated with superficial as well as invaginate genital markings. The glands may not be readily recognizable and sometimes are flattened against or even buried within the body wall.

Ovaries usually were not characterized in the past but now need no mention unless differing from the generic norm indicated in the definition above. Ovisacs probably are absent throughout the genus.

The second and only other character by which *Pheretima* was distinguished, in the classical system, from its supposedly parental genus, *Megascolex*, was somatic—location of the gizzard behind septum 7/8. The segmental situation was variously recorded as viii, ix, ix–x, even viii–x as often does seem at first glance to be the case. In every species or normal individual in which 8/9[13] is present, the author has found the gizzard obviously to be in viii. The hearts of viii, when complete, are in the same segment as the gizzard. If the ventral portions of those hearts are lysed during development, the unaborted dorsal portions pass to and ramify on the gizzard.

Location of intestinal caeca, at times with some description or illustration of shape, often was mentioned. Whether absence of caeca always should be assumed when the organs were not mentioned in systematic descriptions seems doubtful. Presence or absence, paired or unpaired, and segment of origin, are characters that now seem likely to be more useful in initial stages of species' identification than number and location of spermathecal pores. Additionally, the axes of secondary lobing now seem likely to be of interest. Complete retraction of caeca into the intestinal lumen can result in a record of "absence" unless the gut is opened. Occasional dorsal direction of the caeca within the segment of origin now appears not to be of systematic interest.

The excretory system is known only from Bahl's study, perhaps of three species. As for the vascular system, Bahl assumed that his description of the nephridia was universally characteristic throughout

---

[12] The condition sometimes was characterized as, testis sacs and seminal vesicles united which, of course, did not precisely state what was involved.

[13] Septum 8/9 of *P. posthuma* was believed by several authors to be 9/10 merely because of presence in front of it of two pairs of spermathacae. Significance of the known fact that the posterior pair of organs opened at 8/9 was completely unrecognized.

the entire genus. No good reasons for questioning that assumption are known at present, but the system has not been studied in other species. The excretory organs provide no evidence for relationship with *Megascolex* which is not present anywhere in the pheretima domain. Relationship with *Plionogaster* Michaelsen 1892, seemingly restricted to a small portion of the *pheretima* domain, is indicated by organization of the nephridial system.

Presence or absence of one or more of septa 8/9–10/11 usually has been deemed worthy of record. As already indicated, the septum or septa were not always correctly stated. Very delicate septa also may make correct determinations difficult. A complete but transparent and membranous septum 8/9 has been recognized in an occasional, unusually well-preserved specimen of a species in which that septum usually was thought to be lacking. Whether the partition, instead of being aborted (as usual?), was retained as in occasional individual variation is unknown. A ventral rudiment, sometimes recognizable, allows a suggestion that the membrane often may have been ruptured as the worm was pinned out for study of the internalia.

The circulatory system often was ignored in classical systematic descriptions and, if not, location of the last pair of hearts alone was deemed worthy of record. Our knowledge of the vascular system accordingly is based mainly on a study of *P. diffringens*[14] and *P. posthuma*[15] by morphologists. Both species are markedly anthropochorous. Bahl assumed the condition he found in a single species to be characteristic of the entire genus, but the present author shortly found that to be incorrect. More recently, connectives between the supra-esophageal and extra-esophageal trunks have been found at various levels and in segments with a pair of hearts. Bahl's non-contractile hearts of x–xi now seem likely to be only such connectives. In that case, the real hearts of x–xi are aborted in *posthuma* (and perhaps also in closely related species?) thereby providing another vascular character of systematic usefulness. Although subject perhaps to some occasional individual variation and in spite of the difficulties involved in making a correct determination, relationship of the hearts of x–xii or xiii to the two dorsal longitudinal trunks seems likely also to be of systematic importance. A complete subneural trunk, adherent to the parietes, and a pair of extra-esophageal trunks, have been found so often, when condition of the specimens permits recognition of the vessels, that inclusion of such characters in the generic definition now seems warranted.

Acinous masses among the pharyngeal nephridia dorsally in iv–vi, supposedly for production of blood corpuscles and plasma haemoglobin, are readily recognizable in *P. posthuma* but are less obvious in *P.*

diffringens and *hawayana* where they are thought to be vestigial. Similar glands of two other species were found to comprise a more or less distinct, complete or variously interrupted and lobed collar on the esophagus behind the gizzard. Glands at either site have been mentioned so rarely that any guess as to systematic usefulness now is unwarranted. Delicate, lobed organs on the dorsal blood vessel immediately in front of the septa through the intestinal region have been called lymph glands and phagocytic organs. Brown debris and other solid matter does accumulate there in some individuals of several species. Often unrecognizable in poorly preserved material and frequently unmentioned in systematic descriptions, the organs have been found in so many of the recently studied species as to justify a belief that they may be present in large sections of the genius if not in all of it.

Setal characters usually were and still are thought to be of systematic importance. Presence or absence in clitellar segments was frequently mentioned by Beddard in his revision of the genus (1900) then called *Amyntas*. Unfortunately, presence in that region is not always determinable by inspection externally or internally when clitellar tumescence is maximal. Furthermore, there may be, from one individual to another, considerable intersegmental as well as intrasegmental variation with regard to normal number and the percentage of dehisced shafts. Number of setae in several other anterior segments often is mentioned in descriptions of new species, but the character now is known to vary somewhat along the anteroposterior axis. Accordingly, for interspecific comparisons, counts should be of the same segments in each taxon. In the very few species, of which data for more than one or two worms are available,[16] considerable individual variation was found even when all setae were present. Counting often is handicapped by more or less frequent occurrence of local "bald" patches in which setae had been (temporarily?) dehisced.

Segment number of at least one specimen usually has been mentioned in systematic descriptions. The number does vary intraspecifically, within unknown limits usually, but no evidence is available as yet to indicate that the number increases in any pheretima after hatching or after diapause. Raw data from many more specimens than hitherto have been available are needed, but some editors require their elimination and substitution of averages. For definition and even more so for identification of species, an average, mean, median or mode is unlikely to be of any systematic use for some long time to come if indeed ever. Amputation, regeneration, are not uncommon in pheretimas. Extensive splitting of embryonic somites also is known. Now needed, then, is information as to the usual range

[14] Cecchini, C., *Arch. Zool. Napoli* 8 (1916).
[15] Bahl, K. N., *Quart. Jour. Micros. Sci.* 65 (1919).

[16] A fraction such as 50/xii in a setal formula indicates that setae of segment xii were counted on one worm, whereas 50-52/xii may mean only that a count was made on two specimens.

of individual variation in unamputated, unregenerate, and otherwise normal material.

Determination of interspecific and intrageneric relationships has been prevented by the inadequate characterization of so many taxa of an unusually large genus. Even for many Burmese endemics, little can be said about geographic variation because of loss, during the war, of collections as well as accumulated data. A dichotomous key to species of the genus still is impossible. The best that Michaelsen could do, in his 1900 monograph, was to distribute species into fourteen groups according to number and location of spermathecal pores. The species of each category then were listed, without any easily recognized order, along with other characters, in six columns of a table that occupies seventeen pages. Number and location of spermathecal pores still enable a first step in the tedious process of running down a species one has not seen before. However, number of the organs does vary, not only in parthenogenetic polymorphism but also in species that now appear likely to be characterized by obligatory biparental reproduction. Information in columns headed "Höchste bekannte Borstenzahl" and "Abstand der männliches Poren" is of little assistance for reasons already indicated.

Remarks: For intestinal caeca, maximal recorded extent only is recorded. Individual variation in number of segments penetrated is such as to warrant no further statement. (*Cf.* also comments on p. 150.) Caeca, is plural and in a pheretima précis indicates presence of a single pair.

Probably as important, if not more so, than the usual prostatic characters might be indication as to whether all or a part of either or both of the glands is retained in a juvenile state (as in some parthenogenetic morphs).

The prostomium so often is epilobous and with an open tongue that mention of the organ often has been omitted. Any specific divergence from that apparent norm probably should be noted.

A ventral portion of a delicate septum 8/9 often is recognizable in some species. If a rudiment is large enough to reach the gizzard, effort to determine if a dorsal portion was destroyed during handling of the specimen may be worth while. Recognition of a dorsal portion of the membrane in an occasional individual, perhaps better preserved or more carefully pinned open, could provide supporting evidence. However, as a rare aberration, a septum usually aborted in the species, may be retained *in toto* or, even more rarely, be muscularized. Some ventral rudiment of 9/10 must be present, in any holandric or pro-andric species, even if nothing more of it is recognizable than a funnel-bearing portion of the testis sac belonging to x. Abortion in a précis is not qualified as to completeness because the character appears to have no systematic usefulness.

Setae that can be or were called penial have been reported from several Burmese species: Between the

male porophore and an associated genital marking, *P. aculeata.* In a genital marking median to the secondary male pore, *P. andamanensis.* Just median to each male porophore, *P. harrietensis.* In a male porophore, between male pore and aperture of an associated composite gland, and just median to the gland pore, *P. osmastoni.* In copulatory chambers, campanulata morph of *P. houlleti.*

KEY TO INDO–BURMESE SPECIES
OF *Pheretima*

1. Intestine, without caeca................... 2
   Intestine, with caeca..................... 3
2. Male pores superficial, spermathecal pores at
   or behind 7/8.....................*taprobanae*
   Male pores invaginate, spermathecal pores
   (when present) at or in front of 6/7......*elongata*
3. Intestinal caeca arising in xxii, last hearts, in xii[a]
                                          bicincta
   Intestinal caeca arising in xxvii, last hearts in
   xiii................................... 4
4. Intestinal caeca, simple.................. 5
   Intestinal caeca, manicate................ 58
5. Septum 8/9, present and muscularized....... 6
   Septum 8/9, if present membranous, usually
   aborted.............................. 7
6. Genital markings, none[b]..............*fluvialis*
   Genital markings, paired, equatorial in xvii, xix
                                          posthuma
7. Male pores, superficial[c].................. 8
   Male pores, invaginate.................... 39
8. Genital markings, lacking................. 9
   Genital markings, present................. 22
9. Spermathecal pores, in dorsum..........*youngi*
   Spermathecal pores, not in dorsum......... 10
10. Bithecal................................ 11
    More than one pair of spermathecae......... 13
11. Spermathecal pores, at 5/6............... 12
    Spermathecal pores, between eq/vi and 6/7 *.glabra*
12. First dorsal pore at 5/6 male and spermathecal
    pores minute.......................*nugalis*
    First dorsal pore at 12/13, male and spermathe-
    cal pores not minute..   *papilio*
13. Quadrithecal............................ 14
    More than 2 pairs of spermathecae.......... 15
14. Male porophores small, prostatic ducts without
    marked thickening...................*faceta*
    Male porophores large, some portion of each
    prostatic duct markedly thickened......*doliaria*
15. Sexthecal............................... 16
    Octothecal............................. 20
16. First pair of spermathecal pores at 4/5..*pauxillula*
    First pair of spermathecal pores behind 4/5.... 17
17. Spermathecal pores, at 5/6–7/8............. 18
    Spermathecal pores, at 6/7–8/9........*carinensis*
18. Male pores, in a single porophore........*bellatula*
    Male pores, each in a discrete porophore....... 19

19. Pigment, lacking.....................*balteolata*
　　Pigment, red.........................*inclara*
20. Male pores, not in distinctly demarcated poro-
　　phores, in slight depressions ....*alexandri*
　　Male pores, not in depressions, in distinctly
　　demarcated porophores..................21
21. Spermathecal pores at 5/6–8/9, male poro-
　　phores transversely elliptical..............*feai*
　　Spermathecal pores behind 5/6–8/9, male pores
　　in larger porophores of variable shape ....*rimosa*
22. Spermathecal pores, in the dorsum...........23
　　Spermathecal pores, not in the dorsum........24
23. Sexthecal, pores intrasegmental, anteriorly in
　　vi–viii.............................*sulcata*
　　Octothecal, pores intersegmental, at 5/6–8/9
　　　　　　　　　　　　　　　　　rodericensis
24. GM glands, unstalked.....................25
　　GM glands, stalked, coelomic...............31
25. Sexthecal...............................26
　　Octothecal..............................30
26. Spermathecal pores, behind intersegmental
　　furrows...............................27
　　Spermathecal pores, intersegmental.........28
27. Spermathecal pores, just behind intersegmental
　　furrows............................*dolosa*
　　Spermathecal pores, midway between anterior
　　inter-segmental furrows and segmental
　　equators...........................*bournei*
28. Testis sac of xi, ventral, seminal vesicles and
　　hearts of the segment excluded..........*fucosa*
　　Testis sac of xi, vertically *U*-shaped, seminal
　　vesicles and hearts of the segment included ..29
29. Spermathecal diverticula, variously looped
　　　　　　　　　　　　　　　　　　terrigena
　　Spermathecal diverticula, not looped or so only
　　close to origin from the duct ....*carinensis*
30. Genital markings, one pair, extending through
　　the length of xviii..................*suctoria*
　　Genital markings, unpaired, but if paired in
　　xviii in front of or behind the setal equator
　　　　　　　　　　　　　　　　　　andersoni
31. Spermathecae, one pair...................32
　　Spermathecae, more than one pair...........33
32. Spermathecal pores, at 5/6 ....*minima*
　　Spermathecal pores, at 6/7..............*malaca*
33. Spermathecae, two pairs..................34
　　Spermathecae, more than two pairs..........35
34. Spermathecal pores, at 5/6–6/7..........*morrisi*
　　Spermathecal pores, at 7/8–8/9..........*robusta*
35. Spermathecae, three pairs................36
　　Spermathecae, four pairs.................37
36. Testis sacs, unpaired and ventral, hearts and
　　seminal vesicles of the segment excluded
　　　　　　　　　　　　　　　　　　hawayana
　　Testis sacs, unpaired, hearts and seminal
　　vesicles of the segments included .....*papuosa*
37. GM glands, with a central lumen........*aculeata*
　　GM glands, with no lumen.................38

38. Testis sacs, ventral, hearts and seminal vesicles
　　of the segments excluded...........*diffringens*
　　Testis sacs, extended dorsally, hearts and semi-
　　nal vesicles of the segments included......*exigua*
39. Spermathecal pores, at 5/6..............*papilio*
　　Spermathecal pores, not or not only at 5/6.....40
40. Genital markings, none...................41
　　Genital markings, present.................45
41. Male pore invaginations, confined to parietes...42
　　Male pore invaginations, coelomic...........44
42. Spermathecal pores, at 6/7–8/9.........*scitula*
　　Spermathecal pores, at 7/8–8/9.............43
43. Copulatory chamber apertures, in a median de-
　　pression with anteroposteriorly apposable
　　margins...........................*lorella*
　　Copulatory chamber apertures, superficial
　　　　　　　　　　　　　　　　　californica
44. Spermathecal pores, at 6/7–8/9......*umbraticola*
　　Spermathecal pores, at 7/8–8/9....*kengtungensis*
45. Male pore invaginations, confined to parietes..46
　　Male pore invaginations, coelomic..........51
46. GM glands, sessile.......................47
　　GM glands, stalked, coelomic...............49
47. Spermathecal pores, at 5/6.............*subtilis*
　　Spermathecal pores, behind 5/6............48
48. Genital markings, superficial...........*peguana*
　　Genital markings, invaginate.............*bahli*
49. Spermathecal pores, at 5/6–7/8........*anomala*
　　Spermathecal pores, at 6/7–8/9............50
50. Spermathecal pores, large, male porophores
　　without penial setae..............*harrietensis*
　　Spermathecal pores minute, each male poro-
　　phore containing a penial seta........*osmatoni*
51. Spermathecae, two pairs..................52
　　Spermathecae, more than two pairs..........54
52. Spermathecal pores, intrasegmental, in vii,
　　viii...............................*planata*
　　Spermathecal pores, intersegmental, at 7/8,
　　8/9.................................53
53. Genital markings, superficial, in xviii
　　　　　　　　　　　　　　　　　andamanensis
　　Genital markings, all invaginate........*insulanus*
54. Spermathecal pores, at 6/7–8/9............55
　　Spermathecal pores, at 5/6–8/9............57
55. Intestinal origin, in xv...................56
　　Intestinal origin, in xvi............*quadrigemina*
56. GM glands, associated with copulatory cham-
　　bers and with spermathecae...........*houlleti*
　　GM glands, associated only with copulatory
　　chambers.............................*virgo*
57. Spermathecal duct, seemingly much longer
　　than ampulla.......................*bipora*
　　Spermathecal duct, much shorter than ampulla
　　　　　　　　　　　　　　　　　malayana
58. Genital markings, lacking.................59
　　Genital markings, present.................60

59. Male pores, superficial..................*defecta*

   Male pores, in coelomic copulatory chambers
   birmanica

60. Spermathecal pores, 2 pairs behind 5/6–6/7
   canaliculata

   Spermathecal pores, 4 pairs, at or just behind
   5/6–8/9............................... 61

61. A creeping sole present in ventrum, ventralmost
   secondary intestinal caecum the longest
   arboricola

   A creeping sole lacking, dorsalmost secondary
   intestinal caecum the longest.........*manicata*

ᵃ Genital markings, rarely or occasionally, are lacking in species normally characterized by presence of such structures. Data that would have enabled construction of a much better key, because of war losses, have not been available. Accordingly, and also because of possibility of parthenogenetic organ deletions, at each dichotomy involving genital markings, trial of both alternatives might be profitable.

ᵇ *P. hupbonensis* drops out here because of lack of information about hearts and intestinal caeca. Hearts probably are present in xiii and the caeca probably arise in xxvii. However, uncertainty about male terminalia would drop the species out at No. 7. Available information contra-indicates synonymization.

ᶜ *P. immerita* drops out here because of lack of information about the male terminalia. Types were juvenile. Spermathecal pores were at 6/7–7/8 only, a location characterizing no other Burmese species.

KEY TO SOME OF THE ANATOMICALLY DEGRADED
PARTHENOGENETIC MORPHS OF BURMESE
SPECIES OF *Pheretima*

1. Spermathecae, absent...................... 2
   Spermathecae, present..................... 8

2. Male terminalia, present.................... 3
   Male terminalia, absent. AR and ARZ morphs
   of *alexandri, houlleti*. Published data *re*
   these species insufficient to enable identification of such morphs.

3. Male pores, superficial..................... 4
   Male pores, invaginate..................... 6

4. Male pores, in distinctly delimited porophores . 5
   Male pores, not in distinctly demarcated porophores, in slight depressions..........*alexandri*

5. Seminal grooves, lacking..................*illota*
   Seminal grooves, present in male porophores. *glabra*

6. Male pores, in parietal chambers............. 7
   Male pores, in copulatory chambers.......*houlleti*

7. Genital markings, superficial...........*elongata*
   Genital markings, invaginate..........*anomala*

8. Male terminalia, present. *Cf.* preceding key.
   Male terminalia, in part or all lacking........ 9

9. Prostates, one or both juvenile, rudimentary
   or lacking, male pores present.......*diffringens*
   Male terminalia, lacking.................... 10

10. Spermathecal pores, at 5/6–7/8..........*anomala*
    Spermathecal pores, at 6/7–8/9..........*houlleti*

Pheretima aculeata

1936. *Pheretima aculeata* Gates, *Rec. Indian Mus.* **38**: p. 390. (Type locality, Port Blair. Types, in the Indian Mus.)

Octothecal, pores minute, superficial, at 5/6–8/9. Male pores, in xviii, minute, superficial, each in a small circular disc. Female pore, median. Genital markings, small, one just median to each male porophore. Clitellum, xiv–xvi (but lacking ventrally in xv–xvi?). Setae, 30/iii, 36/viii, 58/xii, 8/xvi, 10/xv, 15/xvi, 61/xx, vi/10–11, vii/9–12, viii/9–13, xvii/11–15, xviii/3–4 (penial), xix/13–16. First dorsal pore, at 12/13(?). Prostomium, epilobous(?). Color, red in dorsum of preclitellar segments (pigment red and in circular muscle layer?). (Segments?) Size, 128 by 4 mm.

Septa, 8/9–9/10 aborted, 10/11 membranous, 12/13–13/14 muscular and thicker than 6/7–7/8. Intestinal origin, in xv. Caeca, simple, margins smooth or with slight septal constrictions, in xxvii (– ?). (Typhlosole? Hearts, of x–xiii latero-esophageal?) Holandric. Testis sacs, unpaired and ventral. Seminal vesicles, medium sized, in xi, xii, each with a long and digitiform primary ampulla. Prostates, in xvii–xx, ducts 2 mm. long, spindle-shaped. Penial setae, 0.58–0.6 mm. long, 85 μ thick entally, straight, ornamented ectally with short, transverse rows of fine spines, tip slightly concave on one side. Spermathecae, small, duct bulbous, diverticulum from duct in parietes, with stalk which may be longer than the main axis, a looped middle portion and an ovoidal to ellipsoidal seminal chamber. GM glands, in xvii, ovoidal, each with small central lumen, duct passing into parietes on median face of a penisetal follicle.

Reproduction: Presumably biparental as sperm are exchanged during copulation.

Distribution: Port Blair (Andaman Islands).

Parasites: Fifty fairly large cysts were present, in one specimen, on the gut dorsally in the first few intestinal segments.

Remarks: This species is known only from three types, two of which were aclitellate and one of which had not attained maximal clitellar tumescence. Further developments might have resulted in appearance of clitellar tumescence ventrally in xv–xvi, perhaps along with dehiscence of setae still remaining in xiv–xvi.

Genital markings, because of the depth of the circumferential bounding furrow perhaps should be characterized as within parietal invaginations. If so, the margins of an aperture into the invagination may be apposable so as to cover the marking completely. Support for that suggestion is provided by the glabrous surface of the marking. The latter may then be protrusible like a penis during copulation though without a male pore. If not retractile as is, the marking presumably will be flattened into a discoidal shape. Whether apertures of a penisetal follicle and of the GM gland are discrete or not was not determined. The seta in a follicle opening between the male porophore and the genital marking was assumed to be penial though its characters were not recorded. Shape may prove to be a sigmoid as usual and the shaft itself may be dehisced at full maturity.

A posterior diverticulum within the parietes from the duct of the GM gland may be the stalk of a posterior gland that had not yet developed or that had been inhibited from completing its growth.

Pheretima alexandri

1901. *Amyntas alexandri* Beddard, *Proc. Zool. Soc. London* 1900: p. 988. (Type locality, supposedly Calcutta. Type, in the British Mus.)
1914. *Pheretima lignicola* Stephenson 1914, *Rec. Indian Mus.* 8: p. 399. (Type locality, Dibrugarh. Types, in the Indian Mus.)
1916. *Pheretima lignicola*, Stephenson, *ibid.* 12: p. 335.
1923. *Pheretima alexandri* + *P. lignicola* + *P. suctoria* (part), Stephenson, (*The Fauna of British India*), *Oligochaeta*, pp. 291, 305, 311 (Excluding Andaman Island worms with discrete genital markings.)
1925. *Pheretima lignicola*, Gates, *Ann. Mag. Nat. Hist.*, Ser. 9, 16: p. 567. Stephenson, *Rec. Indian Mus.* 27: p. 61.
1926. *Pheretima lignicola*, Gates, *Jour. Bombay Nat. Hist. Soc.* 31: p. 182, etc.; *Jour. Burma Res. Soc.* 15: p. 211; *Ann. Mag. Nat. Hist.*, Ser. 9, 17: p. 463; *Rec. Indian Mus.* 28: p. 161.
1929. *Pheretima lignicola*, Stephenson, *ibid.* 31: p. 238.
1930. *Pheretima lignicola*, Gates, *ibid.* 32: p. 314.
1931. *Pheretima alexandri*, Gates, *ibid.* 33: p. 363.
1932. *Pheretima alexandri*, Gates, *ibid.* 34: p. 492.
1933. *Pheretima alexandri*, Gates, *ibid.* 35: p. 492.
1937. *Pheretima alexandri*, Gates, *ibid.* 39: p. 178.
1939. *Pheretima alexandri*, Gates, *Jour. Thailand Res. Soc. Nat. Hist. Suppl.* 12: p. 80.
1955. *Pheretima alexandri*, Gates, *Rec. Indian Mus.* 52: p. 80.
1956. *Pheretima alexandri*, Gates, *Evolution* 10: p. 218.
1961. *Pheretima alexandri*, Gates, *American Midland Nat.* 66: p. 62; *Burma Res. Soc. 50th Anniv. Pub. No. 1*: p. 57.

Octothecal, pores minute, superficial, more than ⅓C apart, at 5/6–8/9. Male pores, in xviii, minute, superficial, each in a rather circular area (often slightly depressed) between arms of a U-shaped ridge that is open mesially. Female pore, median. (Genital markings, none.) Clitellum, setae unrecognizable externally, xiv–xvi, occasionally reaching into xvii. Setae, enlarged in ii–ix, 59–78/xvii, 58–76/xx, vi/9–18, vii/10–20, viii/11–22, xvii/15–24, xviii/9–28, xix/16–24. First dorsal pore, at 12/13. Prostomium, rudimentary (in two small parts distinguished from adjacent marginal lobes of peristomium only by presence of pigment?). Color, in dorsum, pinkish to deeper red, yellowish to dark brown, slate, sometimes lacking in very narrow stripes at segmental equators. Segments 90–141. Size, 105–290 by 4–9 mm.

Septa, 8/9–9/10 aborted, 6/7–7/8 (and sometimes 5/6) much thickened, 10/11–11/12 less so. Pigment, in circular muscle layer, reddish brown. Intestinal origin, in xv (sometimes in xvi?). Caeca, simple, in xxvii–xx. Typhlosole, lamelliform, ending in region of xc. Hearts, in viii unaborted dorsal portions to gizzard, in ix (left or right) lateral, in x esophageal(?), in xi–xiii latero-esophageal. Blood glands, in v, also in a low and lobed espohageal collar. Holandric. Testis sacs, paired and vertical or unpaired and horseshoe-shaped, hearts of x–xi and vesicles of xi included. Seminal vesicles, large, especially the posterior pair, in xi, xii. Prostates, large, in xvi–xxii, duct muscular and variously looped or coiled. Spermathecae, rather small, duct markedly narrowed in the parietes, diverticulum from median face of duct at parietes, longer than main axis, with slender stalk and a variously looped wider portion entally.

Reproduction: Uniparental, in various morphs. Parthenogenesis in a number of morphs, because of male sterility and/or organ defects, is obligatory. Parthenogenesis now seems more likely to be optional than obligatory in certain morphs that do copulate and/or mature sperm. Whether reproduction is biparental (or obligatorily so) in any of the sperm maturing and/or copulating morphs, because of World War II destructions, cannot be suggested.

Distribution: Minnie Bay, Mount Harriet, Wimberleyganj (Andaman Islands). Mergui, Labaw (Mergui). Mayan Chaung, Posoe Chaung, Nyaungdonle Chaung, Pyinthadaw, Myittha, Kamaungthwe River, Tavoy, San Hlan, Maungmagaun, Heimza Basin (Tavoy). Ye, Moulmein, Martaban, Chaungson, Kyaikmaraw, Kya In, Kawkareik, Myawadi (Amherst). Kinmunsakhan, Kyaiktiyo, Kyaikto, Taungzun, Bilin, Thaton, Duyinzeik, Aungsaing (Thaton). Rangoon, Twante (Hanthawaddy). Bassein, Kochi (Bassein). Insein (Insein). Wakema (Myaungmya). Maubin, Yandoon, Danubyu (Maubin). Ler-muhtee, Paut-taw-gwin, Mewaing (Salween). Nyaunglebin (Pegu). Thonze (Tharrawaddy). Shwegyin, Tantabin, Toungoo, Thandaung, Blachi, Daylo,(Toungoo). Kyangin (Henzada). Prome, Laboo (Prome). Thayetmyo, Thanbula (Thayetmyo). Mala, Koopra, Loikaw (Karenni). Pygigyaung, Pyinmana (Yamethin). Magwe (Magwe). Mt. Popa (Myingyan). Mandalay (Mandalay). Sagaing (Sagaing). Kengtung (Kengtung). Na Kho Sheh, Pang Noi, Hpa Cha, Meung Nawng (Pang Long). Tan Yang (Mong Yai), also known as South Hsenwi). Nawnglon, Kat Pang, Man Peng (Mang Lun). Lashio, Taungyi, Kalaw, Maymyo (Shan States). Man Meh Hang Hsipaw). Kin-U, Kyaukmyaung (Shwebo). Mogok, Katha, Wuntho, Indaw Lake, Naba (Katha). Lawng Neu (Mong Lem, Yunnan). Selan, Namkham (North Hsenwi). Bhamo (Bhamo). Hopin, Nyaungbin, Myitkyina, N Sop Zup (Myitkyina). Mawkadaw, Mingin, Kalewa, Masein, Mawleik, Homalin (Upper Chindwin). From sea level to elevations of 4,000 feet or more.

Chiengrai, Chiengmai, Kosichang, Mu-ang Pong (Thailand).

Bombay, Jubbulpore, Calcutta (India).

Presence of colonies on the Andaman Islands, at Bombay and Jubbulpore, are a result of transportation and by man. The unique type came to Beddard from Kew Gardens in England but was believed to have come from Calcutta. Presence of the species in the Calcutta region has not been confirmed and it is possible that Calcutta merely happened to be the port from which plants were shipped. Absence of records for Calcutta is, however, of no significance. Since 1826 no real survey of the oligochaeta fauna has been made in the vicinity of territory immediately surrounding any museum.

How much of the Burma distribution also is due to transportation, perhaps in more than one manner, has not been determined. The most advanced parthenogenetic morphs yet recorded were secured near the Thailand border in Kengtung State on the Shan Plateau and at Kawkareik in Amherst district. The original home of the species now seems likely to have been east of the Salween.

Relationships of *P. alexandri* are unknown, but may prove to be with Thailand species as yet unstudied.

Habitats: Soil, in gardens, lawns, fields of ridges around paddy plots, open areas, bamboo and deciduous jungles, rain forests. In mud covered with water cress. In red lateritic soil of Shan Plateau. Under logs.

Biology: Activity usually is limited to five or six months of the year, at Rangoon from June into an early part of November. Breeding probably is from sometime in August and through September or perhaps an early portion of November.

Abnormality: Herniation of a spermatheca through the body wall, sometime before preservation, had been followed by deposition of a black (?) pigment in the exposed organ.

Also see section on polymorphism.

Polymorphism: A species must be defined primarily from its biparental population. Because of the present unfortunate lack of information about that population, characterization of some genital organs in the précis above is incomplete and may in part be incorrect. Especially needed is information regarding the spermathecal duct which was rather slender and without externally recognizable muscular sheen in morphs possibly still able to reproduce biparentally. The rather barrel-shaped ducts with their marked muscularity, of other morphs, are of so normal an appearance that aberration is not likely to be suspected. Indeed, the slender condition was thought for some time to be less normal. From an H morph that cannot yet be adequately characterized, at least one H_p morph has evolved. No A morph was seen but is anticipated unless the spermathecal battery and the male terminalia are equally susceptible to modification and deletion —a possibility for all other known morphs of *P. alexandri*, AR and I_4 intermediates, provide some support. A number of intermediate morphs could

have been listed or even named. Some idea of the number can be obtained by estimating the combinations that are possible of the characters listed below.

Spermathecal battery

Spermathecal diverticulum, normal, without differentiation into stalk and seminal chamber but of usual length or variously shortened, or reduced to a short rod, a mere knob or completely eliminated. Any one, two, or more or all of the spermathecae may be degraded in one way or another.

Spermathecal duct, normal (whatever that may be), or with muscle layer markedly thickened or markedly thinned.

Spermathecae, any one, two, or more, or all deleted without trace.

Male terminalia

Prostate glands, of normal size, variously reduced in size, or lacking.

Prostatic ducts, normal, lengthened and thinned, considerably shortened (when glands are lacking), represented only by a slight knob, or lacking.

Sperm ducts, passing into prostate or prostatic duct normally, or into ental end of a much shortened duct (in absence of prostate glands), or attenuating at various levels from xvii anteriorly. All combinations of those conditions possible in bilaterally asymmetrical individuals.

Male pore areas, normal, variously vestigialized, or deleted.

Oviducts, uniting normally to open by a single and median pore or failing to unite and opening separately. (The latter was recorded in 20 of 32 specimens in a Kentung series.) Oviducts originally opened separately. Hence doubling of the female pores in xiv is a reversion to an ancestral condition.

Seminal vesicles, normal, normal but with juvenile shape and texture, variously vestigialized, or deleted, in one or more sides of one or both segments.

Testis sacs, normal (but condition unknown), paired and then vertical or smaller and subesophageal or unpaired and horseshoe-shaped.

Testes, fertile, fertility much reduced or sterile.

Spermathecal deletions in parthenogenetic morphs indicate ways in which evolution may have proceeded, though more slowly, in the evolution of various species. Thus a sexthecal condition may have been derived from an octothecal by elimination of an anterior or of a posterior pair of spermathecae. A quadrithecal battery with pores at 6/7-7/8 could have arisen by consecutive deletion of the anterior and posterior pair (in either order) or even perhaps by the simultaneous deletion of both pairs. A quadrithecal battery with pores at 5/6-6/7 could have arisen by deletion twice (consecutively) of the last pair of spermathecae or perhaps by a simultaneous deletion of both pairs. Larger collections of parthenogenetic individuals can be expected to throw additional light on evolution of the spermathecal battery in the very large genus *Pheretima*.

Parasites: Sporozoa infested coelomic cavities (1933: pp. 493–494), in individuals of certain morphs, so much more frequently and in so vastly greater numbers as to have been thought to provide strong support for a belief in a parasitic origin of the abnormalities. Massive infestations in certain pheretimas sometimes were associated with dwarfism and what seemed to be an unhealthy condition. By far the largest specimens of *P. alexandri* that were found seemed healthy but had massive infestations of protozoa. Nematodes often were found but never in such large numbers. *Siconema siamense* Timm, 1966, at Rangoon.

Remarks: An olive green coloration of the *lignicola* type is believed to be an artifact and to have resulted from contact with some other kind of worm in the alcohol or from a staining action caused by some substance leached from another organism by the preservative.

P. alexandri is one of the two taxa that provided material for the first microscopical examinations of a more or less completely annular ridge on the esophagus just behind the gizzard. The collar, according to Stephenson (1916: p. 335), was composed of "follicles of blood glands like those. . . behind the pharynx in certain species" such as *P. posthuma* and *hawayana*.

Relationships of *P. alexandri* are unknown, but may prove to be with Thailand species as yet unstudied.

After the pheretima section of this opus had been finally typed a specimen of a related Thailand species was received. The species was so common in northern Thailand that the collector assumed one worm would enable the desired identification. Various efforts to secure additional material have been futile to date. As much of a précis as is possible is subjoined.

Pheretima sp.

Sexthecal, pores minute, superficial, at 6/7–8/9. Male pores, in xviii, minute, each in a slight depression surrounded, except mesially, by a slight ridge. Female pore, median. Genital markings, an elliptical area of grayish translucence, slightly diagonal and with anterior end more lateral, in each male pore depression (and behind that pore?). Clitellum, reaching into xiii and xvii. Prostomium, small, bilobed, distinguished from everted portion of buccal cavity by presence of pigment. Segments, 145.

Intestinal origin, in xvi. Caeca, simple, in xxvii–xxii. Typhlosole, ends in 94th segment, leaving 51 (?) atyphlosolate. Holandric. Seminal vesicles, large, especially in xii, filling coelomic cavities. Prostates, in xvi–xx. GM glands, soft, without definite, muscular ducts, 6–8 on each side in xviii–xix, narrowing to pass into parietes in xviii over site of epidermal transluence. Spermathecae, fairly large, reaching up to level of dorsal surface of gut.

Reproduction: Spermatozoal iridescence in the spermathecae showed that copulation had been completed, but on the male funnels only scattered flecks were iridescent. Reproduction is assumed to be biparental.

Parasites: NEMATODA: *Homungella siamense* and *Siconema siamense* Timm, 1966, Udorn, Thailand.

Remarks: Male pore areas are as in *P. alexandri* except possibly for presence of a genital marking which could have been so indistinct and so nearly coincident with the bottom of a male-pore depression in Indo-Burmese material as not to have been recognized. GM glands that are so conspicuous in the Thailand worms certainly were lacking in Burmese worms dissected by the author, perhaps because of parthenogenetic degradation. However, need for a more thorough sampling of Burmese populations now is indicated.

If the spermathecal battery has not undergone parthenogenetic degradation and is normal, the Thailand worm now seems likely to belong to a species distinct from but closely related to *alexandri*. The reasons on which that belief is based, are: (1) Absence of intraspecific variation as to number of spermathecae in amphimictic taxa. (2) Low probability of parthenogenesis adding a pair of spermathecae in front of 6/7.

Presence in Thailand of a biparental population of *alexandri* now can be expected with somewhat more confidence. Adequate information as to manner of reproduction will be required for determination of relationships.

Pheretima andamanensis

1907. *Pheretima andamanensis* Michaelsen, *Mitt. Naturhist. Mus. Hamburg* **24:** p. 164. (Type locality, N. Cinque Island. Type, in the Indian Mus. Two paratypes in the Hamburg Mus.)

1909. *Pheretima andamanensis*, Michaelsen, 1909, *Mem. Indian Mus.* 1: p. 194.

1923. *Pheretima andamanensis*, Stephenson, (*The Fauna of British India*), *Oligochaeta*, p. 292.

1932. *Pheretima andamanensis*, Gates, *Rec. Indian Mus.* **34:** p. 414.

1936. *Pheretima andamanensis*, Gates, *ibid.* **38:** p. 393.

Quadrithecal, pores (superficial?) small, transverse, ca. ¾C apart, at 7/8–8/9. Male pores, in xviii, minute, each (on a penis?) on lateral wall of a copulatory chamber with a transversely slit-like aperture. (Female pores closely paired?) Genital markings, thickly discoidal, one (with two pores) median to each secondary male pore. Clitellum, xiv–xvi. Setae, 52/xii, 58/xix, 54/xxvi. viii/11–12, xviii/10–15, one seta with bifid and ornamented tip in each genital marking, *a* of pre-clitellar segments enlarged. First dorsal pore, at 12/13. Prostomium, epilobous, tongue open. Color, in dorsum only, brownish (pigment red, in circular muscles?). Segments, 110. Size, 108–120 by 6–6½ mm.

Septa, 8/9–9/10 aborted, 12/13–13/14 most thickly muscularized. (Intestinal origin, in xv?) Caeca, simple, in xxvii–xxiii. Typhlosole, simple. (Hearts, of x or xi to xiii latero-esophageal?) Holandric. Testis sacs, unpaired and ventral. Seminal vesicles, large, in xi, xii, each with a long primary ampulla. Prostates, in xix–xxiii, ducts each in an S-shape. Spermathecae, duct thicker ectally and shorter than ampulla, diverticulum from median face of duct close to parietes, with "threadlike" stalk longer than main axis and looped, seminal chamber small, ovoidal. A lobed, annular gland on parietes around base of each spermathecal duct. GM gland, on anterior face of prostate, with a long stalk opening in the genital marking median to the aperture of penisetal follicle.

Distribution: North Cinque Island (Andaman Islands).

Remarks: The species is known only from three types on none of which had the clitellum attained more than slight tumescence.

GM discs, because of the thickness and the depth of the circumferential bounding furrow could be shortly columnar and within a parietal invagination. If so, the margin of the aperture may prove to be apposable so as to cover the disc completely. Because of its modifications the seta of the follicle opening in the genital marking presumably should be called penial—it is at least as close to the male pore as are the penial setae of the quadriprostatic species with acanthodrilin male terminalia—through here opening independently rather than close to or through a pore (*cf. P. houlleti*).

The collar around the spermathecal duct presumably is glandular but condition of the specimens did not permit recognition of external pores or markings in vicinity of the spermathecal pores. If such markings or glandular pores are present they should become recognizable as the softening of alcoholic specimens (inevitable unless the specimens become brittle) sets in.

Pheretima andersoni

1960. *Pheretima andersoni*, Gates, *Bull. Mus. Comp. Zool. Harvard College* 123: p. 250.

Several taxa, originally believed to be species, according to presently limited knowledge are so much alike as to cause doubt about their validity. The anatomical, reproductive and distributional data needed for solution of the problems have been unavailble ever since 1935 and are unlikely to be procurable in any near future. In the meanwhile, the taxa involved, considered to be either a species group or superspecies, are collectively characterized as below.

Octothecal, pores minute, superficial, *ca.* ⅔C or more apart, usually (and unless otherwise indicated to the contrary below) at 5/6–8/9. Male pores, in xviii, minute, superficial, each usually in a distinctly delimited, equatorial, discoidal porophore. Female pore, median. Genital markings, postclitellar, of fair size, never small tubercles, always present, in some part of the region of xvii–xxvii. Clitellum, setae unrecognizable externally. Setae, small, closely spaced,

usually a little further apart dorsally, without obvious differences in size, circles without uniform gaps. First dorsal pore, at 12/13. Prostomium, epilobous, tongue open. Color, usually only in dorsum, darker anterior to the clitellum than posteriorly, varying individually, light reddish, reddish, brownish red, brownish, grayish brown, chestnut, slate.

Septa, 8/9–9/10 aborted. Pigment, in circular muscle layer, reddish. Intestinal origin, in xv. Intestinal caeca, simple. Hearts, in viii unaborted dorsal portions to gizzard, in ix (left or right) lateral, in x esophageal(?), in xi–xiii latero-esophageal(?). Blood glands, in v and in an esophageal collar just behind the gizzard. Holandric. Testis sacs, unpaired, ventral. Seminal vesicles, in xi, xii. Prostates usually large, ducts usually each in a U-loop. Spermathecae, duct abruptly narrowed in parietes, diverticulum from anterior face of duct in body wall, longer than main axis, digitiform or slightly widened entally, variously coiled or looped sometimes in a more or less regular zigzag. GM glands, sessile, comprising a tough ectal portion and a softer internal portion sometimes obviously of vertical columnar strands.

Reproduction: Presumably biparental in all taxa. Sperm are known to be exchanged during copulation in most species and in the others, evidence provided by size of seminal vesicles, especially the pair in xii, and of the prostates, usually supports the same assumption.

Distribution: Peninsular Burma and north in the mainland east of the Irrawaddy-Sittang axis into Yunnan. The Irrawaddy-Sittang axis was crossed perhaps in the vicinity of Toungoo long enough ago for differentiation of races in the Pegu Yomas and also much further north perhaps in or near Myitkyina district. Two of the species are known to reach into Thailand.

Remarks: Studies of segment number, of typhlosoles, etc., if completed, might have enabled more adequate characterization of the taxa.

The order in which the species are considered is based primarily on geography, beginning with the southernmost which also happens to have been the one first erected.

KEY TO TAXA OF THE *andersoni* GROUP

1. Spermathecal pores at intersegmental furrows . 2
   Spermathecal pores slightly behind intersegmental furrows 14
2. Genital markings in three longitudinal ranks
   compta
   Genital markings not in three longitudinal ranks 3
3. Genital markings intrasegmental 4
   Genital markings across or crossing intersegmental furrows 5

4. Genital marking in xxii . . . . *labosa*
　　Genital markings in xviii . . . . *luxa*
5. Genital markings paired 6
　　Genital markings unpaired and median 7
6. Genital markings, one pair, across 18/19
　　　　　　　　　　　　　　　analecta rufula
　　Genital markings, more than one pair, usually
　　　behind 18/19 . . . . *longicauliculata*
7. Only one genital marking present 8
　　Two or more genital markings present, across
　　　consecutive intersegmental furrows
　　　　　　　　　　　　　　　andersoni (part)
8. Genital marking between equators of two con-
　　secutive segments 9
　　Genital marking not so restricted 12
9. Genital marking across 18/19
　　　　　　　　　　　　　analecta promota (part)
　　Genital marking not across 18/19 10
10. Genital marking across 19/20 11
　　Genital marking across 20/21 . . . . *andersoni* (part)
11. Soma size, 50–124 by 3–5 mm . . . . . *analecta analecta*
　　Soma size, larger . . . . *andersoni* (part)
12. Genital marking reaching into xviii
　　　　　　　　　　　　　analecta promota (part)
　　Genital marking behind xviii 13
13. Genital marking between equators of xix and xxi
　　　　　　　　　　　　　　　andersoni (part)
　　Genital marking between equators of xxii and
　　xxiv . . . . *sonella*
14. Spermathecal diverticulum, within a short,
　　opaque sac . . . . *velata*
　　Spermathecal diverticulum, not within an
　　opaque sac . . . . *choprai*

I. *Pheretima andersoni*

1907. *Pheretima andersoni* Michaelsen, *Mitt. Naturhist. Mus. Hamburg* **24**: p. 167. (Type locality, Amherst. Types in the Indian Mus.)
1909. *Pheretima andersoni*, Michaelsen, *Mem. Indian Mus.* **1**: p. 198.
1923. *Pheretima andersoni*, Stephenson, (*The Fauna of British India*), Oligochaeta, p. 293.
1930. *Pheretima andersoni*, Gates, *Rec. Indian Mus.* **32**: p. 305.
1931. *Pheretima andersoni*, Gates, *ibid.* **33**: p. 371.
1932. *Pheretima andersoni* + *P. velata* var. *alveata*, Gates, *ibid.* **34**: pp. 504, 539. (Type locality of *alveata*, Thaton. Types, none.)
1936. *Pheretima andersoni*, Gates, *ibid.* **38**: p. 395.
1955. *Pheretima andersoni*, Gates, *ibid.* **52**: p. 82.
1932. *Pheretima nemoralis* Gates, *ibid.* **34**: p. 531. (Type locality, Heimza Basin. Types, none.)
1933. *Pheretima* sp., Gates, *ibid.*, **35**: p. 548.

Male porophores, 3–6 intersetal intervals wide. Genital markings unpaired and median, transversely placed, across one or more (and then consecutively) of 18/19–26/27. Clitellum, rarely failing to reach 16/17, occasionally reaching slightly into xiii and/or xvii. Setae, 70–83/iii, 112–119/viii, 109–116/xii, 100–113/xx, vi/20–46, vii/25–44, viii/24–50, xvii/18–29, xviii/14–40, xix/20–47. Segments, 120–134. Size, 197–260 by 6–11 mm.

Septa, 6/7–7/8, 10/11–11/12 thickly muscular. Caeca, in xxvii–xviii. Blood gland collar, uninterrupted or in 2–10 lobes. Seminal vesicles, of xii much larger. Prostates, in xvi–xx, ducts 6–15 mm. long.

Distribution: Zowai, Sindin, Tharrabyin (Mergui). Maungmagaun, Nyinmaw, Migyaunglaung, Pyinthadaw, Siyigyan, Zinba, Heimza Basin (Tavoy). Ye, Amherst, Chaungson, Kyaikmaraw, Kya In, Kawkareik (Amherst). Thaton, Naung-gala, Bilin, Kyaiktiyo (Thaton). From sea level to elevations of 1,000 feet or more.

Habitats: Soils of undisturbed areas away from towns and villages. Never secured, for instance, in the vicinity of Moulmein where considerable collecting was done in each of several years.

Variation: A functional dorsal pore was present at 11/12 of two specimens. Occasionally the first functional pore was at 16/17 though in other worms the pore at that level had been occluded.

Genital markings of the types, were short, slenderly spindle-shaped, centered along six intersegmental furrows, 19/20–24/25. Worms with similar markings, at 5–8 of levels 18/19–26/27 but usually at 19/20–24/25, were found only in lowlands of Amherst district which is nearer the northern limit of the range. Elsewhere, and probably in most cases from uplands or nearby hills, markings were fewer, longer, more nearly elliptical and often reaching nearly to segmental equators as well as nearly to, to or even beyond male pore levels. Worms from the Maungmagaun hills near the seashore of a middle portion of Tavoy district had three markings across 20/21, 21/22, 22/23. Worms from a northern part of Tavoy district and others from near the southern limit of the range in Mergui district had two markings across 20/21 and 21/22. Only one marking was present, at 19/20 of some Thaton worms but at 20/21 of worms from Kawkareik near the Thailand border and from a southern Mergui site. A single genital marking on ten specimens from the vicinity of Thaton extended through three or four segments and could have resulted from union of two or three markings originally along 19/20, 20/21, and 21/22.

"Bald" areas had appeared in some individuals, in vii–viii or vi–viii. The areas seemingly were unpaired and reached well toward the intersegmental furrows as well as nearly to spermathecal pore levels. Setae had not been dehisced in two specimens from such areas.

Prostates occasionally were confined to xviii but were accordion-pleated and pushed 17/18 well anteriorly and 18/19 well posteriorly. Septum 5/6 occasionally seemed to be fairly thickly muscular.

Abnormality: No. 1. An extra spermatheca on the left side of ix was adiverticulate. No. 2. Right spermatheca of viii was lacking.

An individual from near the southern boundary of the range, otherwise indistinguishable from specimens previously referred to *P. andersoni* had a single marking, across 20/21, that was quite obviously dumbbell-shaped. The two types from which *P. nemoralis* alone is known had two discrete genital markings in the same area, separated from each other by a median space equal to *ca.* five intersetal intervals. A continuation of the trend already under way in the Mergui specimen of *P. andersoni* presumably would give a pair of longitudinally elliptical genital markings reaching to the equators of xix and xx. Septa 5/6–7/8 and 10/11–12/13 quite obviously were stronger than the others though characterized as membranous. A blood gland collar supposedly lacking may have been small and unrecognized. As other differences were lacking, the *nemoralis* types now are thought to be aberrant or mutant individuals of *P. andersoni*.

A similar modification or aberration may have been involved in a Kamaungthwe River pheretima (1933) that was not referable at the time to any known species. Each of two median genital markings, in that case across 19/20 and 20/21, then would have been separated into two markings but with retention of the original orientation, transverse to the long axis. However, soma size (110 by 5 mm.) was smaller than usual in *P. andersoni*, no septa were strongly thickened though 6/7, 10/11–11/12 were slightly muscular, and the prostatic ducts were only 3½ mm. long.

Parasites: Gregarine protozoa were found, occasionally in numbers, in the coelomic cavities, in the blood gland collar, on the parietes, and the septa. Large cysts were found in the parietes and the nerve cord. Ovaries of a few specimens contained gregarines along with pseudonavicellae cysts. Nematodes. *Adieronema magnum* Timm, 1967, occasionally were present in coelomic cavities of anterior segments.

Remarks: *P. andersoni* is not so easily obtained as some other species, perhaps because of the depths usually inhabited as well as the nature of the preferred sites. Vast numbers of the worms are easily obtainable at end of the breeding season(?) if one can only arrange to be present at the very day when they come to the surface, according to local inhabitants, in order to die.

The northernmost record is Kyaiktiyo. From there up into Karenni, through the sparsely inhabited and densely forested region between the Sittang and the Salween rivers, almost nothing is known about the earthworm fauna. A related species, *P. analecta*, does get into the Salween district and there resembles *P. andersoni* at least in setal numbers. Finding of intermediate forms in the unsurveyed region will not be surprising.

II. *Pheretima analecta*

1960. *Pheretima analecta*, Gates, *Bull. Mus. Comp. Zool. Harvard College* 123: p. 250.

Pheretima analecta promota

1933. *Pheretima analecta* var. *promota* Gates, *Rec. Indian Mus.* 35: p. 494. (Type locality, Pegu Yomas, Types, none.)
1936. *Pheretima promota*, Gates, *ibid.* 38: p. 451.
1960. *Pheretima analecta promota*, Gates, *Bull. Mus. Comp. Zool. Harvard College* 123: p. 250.
1936. *Pheretima pannosa* Gates, *Rec. Indian Mus.* 38: p. 441. (Type locality, Pegu Yomas near Toungoo. Types, none.)

Genital marking, between equators of xviii and xix or reaching to 17/18 and 19/20, transversely or longitudinally elliptical or circular. Setae, lacking dorsally in ii, 4–13/ii, 44–52/iii, 96–109/viii, 105–113/xii, 96–111/xx, vi/26–37, vii/28–36, viii/29–37, xvii/26–37, xviii/7–21, xix/12–27. Clitellum, not always reaching 13/14 and/or 16/17. Segments, 108. Size, 94–170 by 4–7 mm.

Septa, none thickly muscular. Intestinal caeca, in xxvii–xxiii. Typhlosole, low. Blood gland collar, low, interrupted. Prostates, in xvii–xxi, ducts 2–4 mm. long, rather slender but with muscular sheen, looped or coiled.

Distribution: Pegu Yomas of Tharrawaddy, Pegu and Toungoo districts, perhaps from latitude of Letpadan to that of Pyu or Toungoo.

Variation and Abnormality: The type, from which *P. pannosa* alone is known, is now believed to be a mutant, variant or abnormal individual in which the anterior pair of spermathecae did not develop.

Remarks: A genital marking, originally across 18/19 as still in some individuals, presumably has been anteroposteriorly extended in others.

Setal numbers, except in ii–iii, are closer to those of *P. andersoni* than of subspecies *analecta*.

Ancestors of the western subspecies of *P. analecta* may have crossed the Irrawaddy-Sittang axis in the region of Toungoo and then spread north and south in the Pegu Yomas. Those ridges provide the western boundary in lower Burma for pheretima endemicity.

Pheretima analecta rufula

1933. *Pheretima rufula* Gates, *Rec. Indian Mus.* 35: p. 543. (Type locality, Pegu Yomas in vicinity of Su Law and Pray Law. Types, none.)
1960. *Pheretima analecta rufula*, Gates, *Bull. Mus. Comp. Zool. Harvard College* 123: p. 250.

Genital markings, one pair, circular to longitudinally elliptical, 9–12 intersetal intervals wide, separated mesially by a space about equal to 8–10 intersetal intervals, just median to male pore levels, between equators of xviii and xix. Clitellum, xiv–xvi. Setae, lacking dorsally in ii, 1–11/ii, 108–119/xx, vi/33–39, vii/32–42, viii/35–45, xvii/33–45, xviii/27–31. Segments, 105–116. Size, to 150 by 6 mm.

Septa, 5/6–7/8 thickly muscular, 10/11–11/12 somewhat strengthened. Intestinal caeca, in xxvii–xxii. Blood gland collar, rudimentary, or ridgelike and then uninterrupted or lobed. Prostates, in xvii–

xix, ducts, 2–4 mm. long, muscular but rather slender.

Distribution: Northern portion of the Pegu Yomas, in vicinity of Su Law and Pray Law, west of Toungoo.

Variation and Abnormality: Of the type series of 98 specimens, 3 varied from the norm as to genital markings. Right marking, lacking (1 specimen). Right markings, across 17/18 rather than 18/19 (1). Left marking across 18/19 lacking, paired markings across 17/18 (1).

A left spermatheca doubled, in vii (1 specimen), in viii (1). Extra organs normal except for parietal union with another spermatheca so as to open by a single pore.

Parasites: Gregarine protozoa and cysts, in obvious numbers, were present on septa, parietes or in coelom of 17 of the 39 dissected specimens. Nematodes were seen in the coelom of only one worm.

Remarks: The range is shared with a subspecies of *P. papilio* the only other pheretima that is endemic west of the Irrawaddy-Sitting axis in lower Burma.

Pheretima analecta analecta

1932. *Pheretima analecta* Gates, *Rec. Indian Mus.* **34**: p. 501. (Type locality, Ko Haw Der. Types, none.)
1936. *Pheretima analecta*, Gates, *ibid.* **38**: p. 392.
1955. *Pheretima analecta*, Gates, *ibid.* **52**: p. 81.
1960. *Pheretima analecta analecta*, Gates, *Bull. Mus. Comp. Zool. Harvard College* **123**: p. 250.

Genital marking, transversely elliptical, reaching nearly to levels of male pores, between equators of xix and xx. Clitellum, occasionally extending slightly into xiii and/or xvii. Setae, lacking dorsally in ii, 5–12/ii, 39–63/iii, 69–89(132)/viii, 74–82(123)/xii, 61–81(118)/xx, vi/20–32(44), vii/20–37(50), viii/23–38(52), xvii/16–36, xviii/12–35. Segments, 82–102 but usually 92–102. Size, 50–124 by 3–5 mm.

Septa, none thickly muscular (?). Intestinal caeca, in xxvii–xxii. Typhlosole, present. Prostates, in xvii–xxi, ducts slender, in a *C* or *U*-shape, 2–2½ mm. long.

Distribution: Pauk-taw-gwin (Salween). Kyauk-kyi, Daylo Stream, Blachi, Shoko, Sah Der, Ko Haw Der (Toungoo) but on the Shan Plateau. The range is about from the western border of the Shan Plateau to the eastern boundary of Burma near the latitude of Toungoo town. Setal numbers of the single specimen from Salween District, are more like those of *P. andersoni* than of subspecies *analecta* elsewhere. The range is shared with *P. velata* (immediately following) but differences in size are sufficient to contra-indicate interbreeding.

Variation: Setal numbers (indicated above in parentheses) of a far eastern specimen were higher even than in *andersoni*.

Genital markings of 28 specimens in a lot (1932) of 207 diverged from the norm in various ways of which differences in number and location only were as follows: A pair of markings across 19/20 (9 speci-

mens). No marking on left (1) or right (1) side. An extra marking, across 18/19 (1), or 20/21 (15). Genital markings on 20/21 instead of 19/20 (1).

A half marking, as it were, on one side of 19/20 and another half marking on the opposite side of 18/19 or 20/21 were seen several times in later lots.

Parasites: Nematodes, *Adieronema mirabile* Timm, 1967, were present in coelomic cavities of 20 per cent of one series. Large cysts were present in the nerve cord, of five specimens, between ix–xxx. Nervous tissue in the cord of that region, in three of those worms, had disappeared entirely.

Remarks: The localities from which this subspecies was recorded, with one exception, are along or near to one of the two branches of a road that passes from Toungoo up onto the Shan Plateau. Away from those roads, the megadrile fauna south almost to the sea-coast and north to the Kalaw-Taungyi highway is unknown. The exceptional locality is in the Salween district that borders on Thailand.

III. *Pheretima velata*

1930. *Pheretima velata* Gates, *Rec. Indian Mus.* **32**: p. 321. (Type locality, Thandaung. Types, none.)
1932. *Pheretima velata* var. *typica* & *clavata*, Gates, *ibid.* **34**: pp. 538, 541. (Type locality of *clavata*, Shoko. Types, none.)
1933. *Pheretima velata*, Gates, *ibid.* **35**: p. 546.
1960. *Pheretima velata*, Gates, *Bull. Mus. Comp. Zool. Harvard College* **123**: p. 252.

Spermathecal pores, transverse slits, just behind 5/6–8/9. Genital markings, in front of eq/xviii, paired in xvii or median and crossing 16/17, 17/18. Clitellum, xiv–xvi. Setae, 90/iv, 105/v, 90/xix, vi/17–42, vii/18–39, viii/19–37, xviii/10–24, xix/17–32. Color, lacking in transverse bands with intersegmental furrows at their centers. Segments, 120–149. Size, 150–250 by 6–13 mm.

Septa, 5/6–7/8 thickened. Caeca, in xxvii–xx, sometimes with a few short ventral lobes. Blood glands, in a pair of lateral, earlike flaps on the esophagus. Testes, of x (usually?) in paired sacs. Seminal vesicles, large, especially those of xii. Prostates, in xvii–xxii, ducts 5–8½ mm. Spermathecal diverticulum, transparent but enclosed in an opaque sac at median side of longer duct.

Reproduction: Information as to maturation and exchange of sperm no longer is available but no reason has been recognized for suspecting parthenogenesis. Size of seminal vesicles suggests profuse maturation of sperm and the prostates appeared to be functional at maturity.

Breeding season includes October and part (at least) of November and perhaps a latter portion of September.

Distribution: Thandaung, Shoko, Leiktho Circle (Toungoo). Mala, Kwachi, Mawchi (Karenni). From the western edge of the Shan massif to the Salween in the region of 19° latitude. (*Cf.* Remarks, under subsp. *analecta.*)

Biology: Pigment was present only in the dorsum of younger specimens but in larger and older worms extended well into the ventrum sometimes so much so as to reach mV. The white bands may become smaller by later deposition of pigment in vicinity of the intersegmental furrows.

Parasites: Nematodes, as yet unidentified, were present in coelomic cavities of 48 of 52 Mala worms. At Shoko, *Siconemella burmensis* Timm, 1967, was found. Coelomic gregarines and large nerve cord cysts apparently were lacking.

Remarks: Such information as is now available about the species was provided by 87 specimens (27 still aclitellate) collected prior to 1933. Mala worms (52) though clitellate were small (stunted?), with pigment sparse anteriorly and absent posteriorly (an abnormal condition?).

Hearts of x were said in the original description to pass from the dorsal blood vessel to the ventral trunk. If that is correct (and not some sort of a *lapsus calami*) the hearts of x presumably were latero-esophageal, with blood present only in the posterior bifurcations— a condition that certainly has not been observed in any one of the many pheretimas examined during the last score of years. The anterior bifurcations presumably were not recognized because they were empty as the posterior bifurcations almost always are.

Genital markings are primarily intrasegmental and at first may have been paired. Median union, not geographically restricted, then presumably was followed by enlargement as well as invasion of xvi and xviii. A genital marking of somewhat similar appearance, but behind the male pore metamere, is thought to have resulted in *P. andersoni* by enlargement and anteroposterior union of two or three markings originally across consecutive intersegmental furrows. Intrasegmental locations are recorded for only one other species of the *andersoni* group. The slight posterior dislocation of the spermathecal pores also was recorded for but one other species of that group. The seemingly saccular spermathecal diverticulum certainly is unique in the group and may also be in the genus. The real diverticulum, as it grows in length, is forced to loop or coil within an opaque and fairly tough-walled sac of unknown structure (peritoneum + ?).

The range may prove to be much the same as that of subspecies *analecta* though the latter was reported to the east only from Salween District while *P. velata* was recorded only from Karenni just to the north. Interbreeding of the two species is unlikely, in the common eastern portion of the ranges, because of differences in body sizes. Such differences are less likely to prevent interbreeding of *P. velata* with *P. compta* or *P. longicauliculata.* Those two species,

next to be considered in that order, alone of the *andersoni* complex are known to reach into Thailand.

IV. *Pheretima compta*

1932. *Pheretima compta* Gates, *Rec. Indian Mus.* **34**: p. 511. (Type locality, Blachi. Types, none.)
1933. *Pheretima compta*, Gates, *ibid.* **35**: p. 524.
1939. *Pheretima compta*, Gates, *Jour. Thailand Res. Soc.* **12**: p. 84.
1960. *Pheretima compta*, Gates, *Bull. Mus. Comp. Zool. Harvard College* **123**: p. 252.

Genital markings, circular to shortly elliptical, in three longitudinal ranks, one at mV, the other two median to male pore levels, across some or all of 18/19–25/26. Clitellum, xiv–xvi. Setae, usually lacking dorsally in ii, 5–33/ii, 63–87/iii, 114–145/viii, 123–139/xii, 122–136/xx, vi/31–54, vii/31–56, viii/34–57, xvii/52–59, xviii/21–47, xix/44–61. (Segments?) Size, 86–235 by 6–8 mm.

Septa, 6/7–7/8, 10/11–11/12 muscular. Caeca, in xxvii–xxiii. Blood glands, in a pair of lateral, earlike flaps on the esophagus. Seminal vesicles, large, the posterior pair much the larger. Prostates, small, in xviii, ducts muscular but rather slender, 6–10 mm. long. Diagonal muscles, a pair of groups, in each of xvii, xix–xxviii.

Reproduction: Presumably biparental, at least in Thailand where sperm are exchanged during copulation (Burma records, lacking).

Breeding season seemingly includes October–November.

Distribution: Blachi (Toungoo). Mawchi (Karenni).

Ban Huai Rai (Phre Province), Thailand.

Parasites: *Nellocystis birmanica* Gates, 1933, attached to body wall, gut and septa in region of viii–xxv, at Blachi on the Shan Plateau.

Coelomic face of parietes from one end of body to another sometimes was covered with white or brown cysts. Larger, tougher cysts were imbedded in parietal musculature. Nematodes, *Mesonema burmense* Timm, 1967, also sometimes were present in coelomic cavities.

Remarks: The species is known only from 32 specimens. All secured in Burma to end of 1932 had been immature, posterior amputees, and/or more or less heavily parasitized. Whether massive infestation had delayed terminal development of genital organs is unknown and parthenogenesis may have been involved.

Also undetermined is whether the larger number of setae per segment and of genital markings per individual of the Thailand worms is geographically associated or merely normal in absence of large numbers of parasites.

The GM pattern, a combination of unpaired and median markings with paired markings at the same level, is unknown elsewhere in the genus and perhaps throughout the Megadrili. That pattern, at present, is about all that distinguishes the species from *P. andersoni* or *longicauliculata*. However, pattern, or perhaps more accurately certain aspects of it, are not subject to intra-specific variation. Differences in number of levels having the markings were recorded but are just such as now can be expected in a range large enough to extend from the western margin of the Shan Plateau into Thailand. Nor was any evidence found that suggested possibility of interbreeding with the seemingly closely related sympatric species. Perhaps information about histology of the GM glands, about typhlosoles along with data as to segment number and other characters will provide other evidence for specific distinctness.

V. *Pheretima longicauliculata*

1926. *Pheretima* sp., Gates, *Rec. Indian Mus.* **28**: p. 162.
1931. *Pheretima longicauliculata* Gates, *ibid.* **33**: p. 395. (Type locality, Tolo Senca. Types, none.)
1932. *Pheretima longicauliculata*, Gates, *ibid.* **34**: p. 525.
1933. *Pheretima longicauliculata* + *Pheretima* sp., Gates, *ibid.* **35**: pp. 533, 528.
1936. *Pheretima longicauliculata*, Gates, *ibid.* **38**: p. 423.
1939. *Pheretima longicauliculata*, Gates, *Jour. Thailand Res. Soc. Nat. Hist. Suppl.* **12**: p. 95.
1960. *Pheretima longicauliculata*, Gates, *Bull. Mus. Comp. Zool.* **123**: p. 252.
?
1939. *Pheretima* sp., Gates, *Jour. Thailand Res. Soc. Nat. Hist. Suppl.* **12**: p. 105.

Genital markings, paired, transversely elliptical, median to male pore levels, 5–10 intersetal intervals wide, 18–24 intervals apart mesially, across 17/18–29/30, perhaps usually 3–5 pairs across some of 18/19–23/24. Clitellum, usually reaching into xiii occasionally almost to the equator but rarely getting into xvii. Setae, none laterally and dorsally in ii, 2–9/ii, 61–69/iii, 92–105/viii, 94–113/xii, 89–110/xx, vi/34–49, vii/37–44, viii/38–44, xvii/35–38, xviii/24–32, xix/34–38. Segments, 137–140. Size, 140–244 by 7–10 mm.

Septa, 6/7–7/8, 10/11–11/12 muscular to thickly muscular. Caeca, in xxvii–xx. Typhlosole, small, short. Blood gland collar, uninterrupted or in four large flaps. Seminal vesicles, large, the posterior pair much the larger. Prostates, small, confined to xviii, ducts 6–8 mm. long, thick, muscular.

Reproduction : Presumably biparental as sperm are exchanged during copulation. Size of seminal vesicles suggests profuse sperm maturation but prostates are small and parthenogenesis may be optional.

Breeding season includes September–November.

Distribution : Kwachi, Mawchi, Loikaw (Karenni). Kalaw, Taungyi (Shan States). Kengtung, Nam Shi Pan, Tolo Senca (Kengtung). Mong Mong and Mong Mong Valley (Mong Mong, Yunnan). Pali, Lawng Neu (Mong Lem, Yunnan). At elevations of 3,000 feet to 4,500 and more.

Khun Tan Mountains, at 4,000 feet, Doi Sutep at 2,000 and 5,000 feet (Ban Huai Rai, Phre Province?), Thailand.

Habitats : Soil, including red (lateritic?) and dark, under grass, bamboo, and trees, in jungles and open spaces, in well-rotted leaves, heavily manured garden, manure.

Biology : Vestiges of follicle apertures in the epidermis show that setae in the dorsum of ii are dehisced during late juvenile growth. The number remaining in the ventrum may be reduced to one (2 specimens) but complete absence was not seen.

Variation : Transversely elliptical, paired areas of slight epidermal modification from which setae had disappeared, sometimes were present in vii–viii. Differences from surrounding epidermis are so slight as to scarcely warrant characterization as genital markings. The areas are symmetrically placed with reference to mV and as well as to the intersegmental furrows and appear to be like those recorded for the same axial region of certain Japanese species.

Genital markings also were present across 16/17 and 17/18 of a Doi Sutep (Thailand) worm. The Khun Tan specimen, also from Thailand, had a complete circle of setae in ii, the only time that condition was seen in an adult. Spermathecal diverticula of that same worm appeared to come off from the median face of the duct.

Two Phre Province specimens, not certainly referable to *longicauliculata*, differed as follows : Setal numbers, probably larger, vii/54, viii/51. Genital markings, a single pair, from eq/xix almost to 20/21, logitudinally elliptical. Male porophores, longitudinally elliptical instead of circular. Prostates, through xvii–xix, ducts 10 mm. long.

A seeming tendency toward depression of the male porophores so that they can perhaps be covered over may be associated with parthenogenesis.

Abnormality : The blind end of a slenderly digitiform outgrowth of the intestine, when it came in contact with the body wall, apparently had functioned so as to lyse a tunnel to the exterior. The epidermal aperture through which the growth protruded very slightly was perfectly circular. Lysis had been followed by healing from peritoneum to epidermis. Similar gut outgrowths in appropriate circumstances are believed to induce formation of lateral growths of cephalic or caudal nature.

Worms with juvenile spermathecae and seminal vesicles as well as rudimentary prostates but with fully developed clitella were once thought to have been affected by parasites. The abnormalities were associated with unusually heavy infestations but the latter may have been permitted by parthenogenesis.

Organ deletions such as are associated with the evolution of advanced standard morphs were not recognized but parthenogenesis now must be suspected when prostates are rudimentary and spermathecae as well as seminal vesicles remain juvenile through the breeding period. Spermathecal diverticula may be hypertrophied, as in certain parthenogenetic morphs, and certain peculiar conformations that seemed from the first rather abnormal in spite of the presence of sperm may also be attributable to parthenogenetic degradation.

Parasites: Sporozoan cysts, in one specimen, were so numerous as to fill testis sacs and cover the parietes. In one segment 60 cysts were counted. Other worms had parasites in the coelomic cavities of vii–xxv. Small cysts sometimes were present in or on the ventral vessel and larger cysts in the nerve cord. Parasitic bodies of an unusual reddish color were present on the gizzard and blood vessels of viii–ix. Each of a certain kind of organism was attached to the body wall by two fine filaments. Cestode intermediate stages were present. NEMATODA. *Homungella monodontium, Mesonema burmense, Siconema limonovatum* and *micronchium* Timm, 1966, 1967, Loikaw.

Remarks: Information now available indicates that size of the range is likely to rank only below that of *P. manicata* or of *feai* even if the western margin of the Shan massif is reached. Evidence for existence of externally distinguishable geographic races, as in case of *P. feai* but in distinction from *P. manicata*, is unrecorded for *P. longicauliculata*.

The 260-odd specimens that had been identified prior to 1936 suggest that *P. longicauliculata* may be more common or at least more easily obtainable than *P. velata* and *compta*. Indeed, in contrast with those two taxa, 15 specimens were secured during dry seasons. Four of them were found in a depression under three to four feet of moist rotten leaves but rock probably would have prevented going any deeper if the worms had so desired. Perhaps this species when normally active is nearer the surface than adults of species of similar size or even of the longer tonoscolexes with which *P. longicauliculata* otherwise would be competing west of the Salween.

To the north and northwest *P. longicauliculata* seemingly is replaced by four taxa next to be considered.

VI. *Pheretima labosa*

1932. *Pheretima labosa* Gates, *Rec. Indian Mus.* **34**: p. 543. (Type locality, Nam Hpen Noi. Types, none.)
1960. *Pheretima labosa*, Gates, *Bull. Mus. Comp. Zool. Harvard College* **123**: p. 252.

Genital marking, transversely oval, 9–12 intersetal intervals wide, in xxii. (Clitellum?) Setae, 79–90/ xx, vi/22–26, vii/23–27, viii/22–28, xvii/17–23, xviii/ 11–19, xix/15–21. Color, none (pigment lacking). (Segments?) Size, to 70 by 4 mm.

Septa, none thickly muscular (?). Blood glands,

in a low and lobed esophageal collar. Seminal vesicles, of xii much the larger.

Reproduction: All specimens of this species were too young (spermathecae still rudimentary and prostates confined to xviii) to provide information as to presence of mature sperm. The size of the seminal vesicles was such as to warrant an assumption that spermatogenesis is profuse in which case reproduction probably will be found to be biparental.

The breeding season, at least at Pang Wo, would seem to be late, including November at least.

Distribution: Pang Wo (Mang Lun), Burma and Nam Hpen Noi (Mong Mong), Yunnan, at elevations of 4,500–5,500 feet. (At those elevations, near the tropic of Cancer, breeding may be delayed until or into December.)

Habitats: Soil. Manure.

Variation and abnormality: Septum 8/9, membranous, was complete, at least in one specimen. Dorsal portions of a complete septum 8/9 could have been destroyed unwittingly, in the other 19 dissected worms, merely by pinning back the body wall for a mid-dorsal dissection.

Two half markings were present in one worm, the left in xxii, the right in xxiii, each ending abruptly at mV. This condition could have arisen by halving of one mesoblastic somite on the right side at either of the following levels, 19th, 20th or 21st. If so produced, the right intestinal caecum presumably should have arisen in xxviii rather than xxvii unless some sort of regularization is possible.

Remarks: *P. labosa* is known only from the original description of 22 juveniles with quite rudimentary spermathecae.

The genital marking appears to be primarily intrasegmental, as it "may" (1932: p. 543) extend slightly into xxiii and/or xxi. If the further growth, that now seems likely, does take place before development is completed, the marking could reach at maturity to the equators of xxi and xxiii. In that case, if the early stages had not been seen, the marking could be thought to result from a fusion of two *andersoni* markings that had been across 21/22 and 22/23.

Interbreeding with sympatric taxa of the *andersoni* group now seems unlikely as differences in body size alone contra-indicate any possibility of successful copulation.

VII. *Pheretima sonella*

1936. *Pheretima sonella* Gates, *Rec. Indian Mus.* **38**: p. 458. (Type locality, near Namkham. Types, in the Indian Mus.)
1960. *Pheretima sonella*, Gates, *Bull. Mus. Comp. Zool. Harvard College* **123**: p. 253.

Genital marking, almost reaching male pore lines, between equators of xxii and xxiv. Clitellum, nearly reaches eq/xiii and eq/xvii. Setae, so closely crowded in ventrum as to appear to be in a zigzagged arrangement, 90(−142)/iii, 56(−148)/viii, 116(−160)xii, 118(−141)/xx, vi/31–36(+?), vii/40–48, viii/31(+) −48, xvii/33–48, xviii/21–41, xix/32–44. (Segments? Color?) Size, 130–180 by 7–8 mm.

Septa, 5/6–7/8 muscular, 10/11–11/12 thickly muscular. Caeca, long and slender, with 7 or more shortly digitiform lobes on ventral side. Blood glands, in an esophageal collar with 5 large or numerous smaller lobes. Posterior seminal vesicles, larger. Prostates, in xvii–xx, ducts 15 mm. long, each in a *U*-loop, ental limb thicker.

Distribution. Between Namkham and Mao-Hsao (North Hsenwi), at an elevation of *ca.* 3,700 feet.

Abnormality: No. 1. Left spermathecae of vi–vii, lacking. Right spermatheca of viii, with an abnormal diverticulum.

Remarks: *P. sonella* is known only from the original description of 6 specimens (one damaged anteriorly) that belonged to a museum and so were not dissected as drastically as one or more otherwise would have been. The worms had been in alcohol since 1926 which prevented any record as to presence or absence of pigment. Adherence of cuticle to the epidermis (in spite of nine years in spirit) considerably hampered determination of setal numbers. The parenthetical figures in the precis above include counts of follicle apertures in which no setal tips could be seen (the shaft too deeply withdrawn into the parietes or perhaps dehisced). Each worm was clitellate and in absence of appropriately aged juveniles it was not possible to determine whether the genital marking was enlarged from an intrasegmental rudiment within xxiii or became single as a result of union of markings across 22/23 and 23/24. The marking itself provided no indications, such as marginal incisions at intersegmental levels, of being other than an enlargement of a single area.

Seminal vesicles of xi were small and may have been retained in a more or less juvenile state. If sperm had not been present in the spermathecae, parthenogenesis might have been suspected.

Successful copulation of individuals of same size as the types of *P. sonella* with individuals of sizes of types of *P. labosa* and *luxa* is believed to be unlikely. However, information as to variation of soma size in this as well as so many other instances is needed.

The Namkham area was vainly searched for endemics other than glyphidriles, on various occasions, mostly after 1926, by the author and/or trained collectors. Native jungle was mostly lacking. Cultivation may have exterminated haemerophobic species in much of the area and if so there may not have been sufficient time for such worms to get back into abandoned fields. Or, the worms may live at depths below those dug over by the author and his collectors. In that case, the types presumably were obtained fortuitously at end of the breeding season when worms had come to the surface "to die."

VIII. *Pheretima luxa*

1936. *Pheretima luxa* Gates, *Rec. Indian Mus.* 38: p. 427. (Type locality, near Namkham. Types, in the Indian Mus.)

Genital markings, paired, transversely oval, median to male pore levels, two presetal and two postsetal, in xviii. Clitellum, nearly reaches eq/xiii and eq/xvii. Setae, xvii/33, xviii/15, xix/36+. (Color? Segments?) Size, 190–280 by 10 mm.

Septa, 5/6 thickly muscular, 6/7–7/8, 10/11–11/12 very thickly muscular. Caeca, more or less deeply constricted by the septa. Blood glands, in an esophageal collar deeply incised into 10 lobes. Seminal vesicles, rather small(?). Prostates, in xvi–xx, ducts 14 mm. long, each in a *U*-loop, the ectal limb thicker.

Distribution: Between Namkham and Mao-Hsao (North Hsenwi), at an elevation of *ca.* 3,700 feet.

Variation and Abnormality: Male porophores of a distinctly delimited discoidal sort such as characterize all other taxa of the *andersoni* group were unrecognizable on both specimens. The minute and superficial male pores of one worm were said to be "each on a tiny conical swelling that is not distinctly demarcated basally." Genital markings were not recognized on one individual. Except for presence of copulatory sperm in the spermathecae, parthenogenesis (also note size of seminal vesicles) probably would have been suspected.

Remarks: *P. luxa* is known only from the original description of two specimens neither of which could be dissected as drastically as the author would have desired. Many of the remarks regarding types of *P. sonella* are equally applicable to the types of *P. luxa*.

IX. *Pheretima choprai*

1929. *Pheretima andersoni choprai* Stephenson, *Rec. Indian Mus.* 31: p. 233. (Type locality, Nyaungbin. Types, in the Indian Mus.)
1932. *Pheretima choprai*, Gates, *ibid.* 34: p. 532.
1960. *Pheretima choprai*, Gates, *Bull. Mus. Comp. Zool.* 123: p. 253.

Spermathecal pores, small transverse slits, nearly at mL, just behind 5/6–8/9. Genital markings, paired across 21/22, unpaired and median, across some or all of 17/18, 22/23–26/27. Clitellum, from 16/17 into xiii. Setae, 69/ix, 72/xii, 68–81/xx, (90 per segment at midbody, 100 per segment posteriorly?), vi/14–20, vii/18–27, viii/22–24, xvii/10–20, xviii/ 8–10, xix/19–22. Segments, 118–119. Size, 132–138 by 5 mm.

Septa, none thickly muscular (?). Caeca, from xxvii (into?). Blood glands, in a low esophageal collar. Prostates, in xv–xxi, ducts *ca.* 8 mm. long and most of each in a *U*-loop.

Distribution: Nyaungbin, at north end of Indawgyi Lake (Myitkyina).

Remarks: *P. choprai* is known only from the original two specimens. Color was said to be brownish purple but whether the brownish referred to alcoholic browning is unknown. Small dark dots at the segmental equators look much like setae except under high magnification and presumably were responsible for the disagreement as to number of setae—on xx originally said to be 81 but later (1932) thought to be 68–73. Number of setae per segment usually decreases posteriorly and because of the possibility that dark dots were counted as setae, the increase in number per segment posteriorly that was originally recorded requires confirmation.

Nyaungbin is above the 25th parallel and so constitutes the northern-most record for the *andersoni* group. The locality is perhaps fifty miles west of the Irrawaddy. *P. choprai* now seems to be the only species of that group to get across the Irrawaddy River though three subspecies are west of the Irrawaddy-Sittang axis in lower Burma. Presence of the species at Nyaungbin may, of course, be due to human transport just as was presence of a *papilio* subspecies at one site west of the Irrawaddy River in Prome district. Failure of various collectors to secure additional specimens in Myitkyina district does at first seem to favor transportation. That evidence, however, seems much less significant if one recalls collectors often were unable to secure specimens of other *andersoni* taxa in localities where the worms almost certainly were present.

Small tubercles such as were recorded (1932: p. 533) usually are associated with stalked and coelomic glands. The specimens, property of a museum, could not be dissected drastically and nothing was recorded even as to glands presumably associated with postclitellar genital markings. If stalked glands are present in preclitellar segments and sessile glands are not associated with postclitellar markings, relationships other than with the *andersoni* group probably will be involved.

The range of the *andersoni* group just about determines the area of certain endemicity for pheretimas in Burma. A north-south succession of species ranges, though here with some overlapping, is somewhat like that indicated below for species of the octochaetid genus *Eutyphoeus*.

Pheretima anomala

1907. *Pheretima anomala* Michaelsen, *Mitt. Naturhist. Mus. Hamburg* 24: p. 167. (Type locality, Botanical Gardens, Sibpur near Calcutta. Types, in the Indian and Hamburg Museums.)

1909. *Pheretima anomala*, Michaelsen, *Mem. Indian Mus.* 1: p. 189.

1923. *Pheretima anamola*, Stephenson, (*The Fauna of British India*), Oligochaeta, p. 294.

1925. *Pheretima anomala* + *P.* sp., Gates, *Ann. Mag. Nat. Hist.*, Ser. 9, 15, pp. 538, 543. *P. anomala* + *P. insolita*, Gates, *ibid.*, 16: pp. 567, 568. (Type locality of *insolita*, Rangoon. Types, in the U. S. Natl. Mus.)

1926. *Pheretima anomala* + *P. insolita*, Gates, *Jour. Bombay Nat. Hist. Soc.* 31: p. 183; *Jour. Burma Res. Soc.* 15:

pp. 207, 210; *Rec. Indian Mus.* 28: pp. 151, 161. Stephenson, *ibid.*, p. 236.

1929. *Pheretima anomala* f. *typica*, f. *insolita*, f. *centralis*, Stephenson, *ibid.*, 31: pp. 234, 236.

1930. *Pheretima anomala* + *P. insolita*, Gates, *ibid.*, 32: pp. 307, 312, 355.

1931. *Pheretima anomala*, Gates, *ibid.*, 33: p. 372.

1932. *Pheretima anomala* var. *typica, centralis, insolita*, Gates, *ibid.*, 34: pp. 387, 389, 390.

1933. *Pheretima anomala* var. *typica, centralis, insolita*, Gates, *ibid.*, 35: pp. 496, 502, 504.

1936. *Pheretima anomala*, *ibid.*, 38: p. 396.

1937. *Pheretima anomala*, Gates, *ibid.*, 39: pp. 179, 193.

1939. *Pheretima anomala*, Gates, *Jour. Thailand Res. Soc. Nat. Hist. Suppl.* 12: p. 82.

1951. *Pheretima anomala*, Gates, *Proc. Natl. Acad. Sci. India*, B, 21: p. 18.

1954. *Pheretima anomala* f. *typica, centralis, insolita*, Gates, *Breviora*, No. 37: p. 1.

1955. *Pheretima anomala*, Gates, *Rec. Indian Mus.* 52: p. 82.

1956. *Pheretima anomala*, Gates, *Evolution* 10, p. 224.

1960. *Pheretima anomala*, Gates, *Bull. Mus. Comp. Zool. Harvard College* 123: p. 254.

1961. *Pheretima anomala*, Gates, *American Midland Nat.* 66: p. 62. *Burma Res. Soc. 50th Anniv. Pub.* No. 1: p. 57.

Sexthecal, pores minute, each at center of a small disc in a small parietal invagination with a transversely slitlike aperture, secondary pores at 5/6–7/8. Male pores, in xx, minute, each at center of a small disc in a parietal invagination with transversely slitlike equatorial aperture, the invagination eversible to a columnar protuberance with male porophore at its distal end. Female pore, median. Genital markings, small, paired, disclike, each with obvious central pore and in a parietal invagination with transversely slitlike equatorial aperture, invaginations eversible to columnar protuberances with the discs at distal ends, in xvii–xix. Clitellum, xiv–xvi. Setae, small, closely spaced, 60–68/iii, 90–96/viii, 78–95/xii, 81–96/xiii, 6–26 (ventrally)/xvi, 61–70/xx, vi/17–25, vii–17–25, xvii/17–23, xix/16–18, xx/15–21. First dorsal pore, at 12–13. Prostomium, epilobous, tongue open (?). Color, only in dorsum, red, faint, or lacking behind the clitellum (pigment red? in circular muscles?). Segments, 119–130. Size, 80–200 by 3–7 mm.

Septa, 8/9–9/10 aborted, none very thickly muscular. Intestinal origin, in xv. Caeca, simple, margins with septal incisions, in xxvii–xxi. Typhlosole, simply lamelliform. Hearts, of x–xiii lateroesophageal (?) Holandric. Testis sacs, unpaired and ventral. Seminal vesicles, large, in xi,xii. Prostates, large, extending through some or all of xvi–xxiii, ducts 4–7 mm. long, muscular, looped. Spermathecae, duct slender and nearly as long as the indistinctly demarcated ampulla, diverticulum from anterior face of duct in parietes, longer than main axis, with short and slender stalk, looped or twisted middle region in which lumen is gradually widened, and a club-shaped seminal chamber. GM glands, mushroom-shaped, erect in coelom, with soft head and straight, spindle-shaped, muscular stalks.

Reproduction: Presumably biparental when circumstances permit as sperm are exchanged during copulation. However, circumstances often seem not to allow effective copulation to sperm maturing individuals and facultative parthenogenesis is anticipated. Parthenogenesis is obligatory because of male sterility in some morphs.

Distribution: Labaw, Tharrabyin, Sindin, Zowai (Mergui). Tavoy, Kyaukmedaung and north, Kameik, Pyin Tha Daing, Kataungni, Heinza Basin, Sitpye, Nyaungdon, Posoe Chaung, Kawlet Chaung, Siyigyan, Kamaungthwe River east (Tavoy). Ye, Mupun, Moulmein, Martaban, Chaungson, Kya In, Kyaikmaraw, Kawkareik (Amherst). Pyapon, Kyaiklat (Pyapon). Myaungmya, Myohaung (Myaungmya). Thaton, Kyaikto, Kumingyaung, Duyinzeik, Bilin (Thaton). Rangoon, Twante (Hanthawaddy). Insein, Damsite, Hmawbi, Wanetchaung (Insein). Maubin, Danubyu, Yandoon (Maubin), Pauktawgwin (Salween), Bassein, Coomzamu, Kochi (Bassein). Pegu, Paung, Thanatpyin, Nyaunglebin, Pazunmyaung (Pegu). Pegu Yomas from Tharrawaddy side, Zigon, Thonze, Myagyaung (Tharrawaddy). Henzada, Zalun, Ingabu, Letpadan, Kyangin (Henzada). Loikaw, Lerbako, Mala, Koopra (Karenni), S'nite (Mong Pai). Toungoo, Kyaukkyi, Maw Pah Der, Daylo Stream, Blachi, Ko Haw Der, Shoko, Leiktho Circle, Thandaung, Tantabin (Toungoo). Ngapoli, Taungup (Sandoway). Prome (Prome). Thayetmyo, Thanbula (Thayetmyo), Pyinmana, Pyigyaung (Yamethin). Akyab, Buthidaung, Maungdaw (Akyab). Kengtung, Taungyi, Kalaw, Maymyo, Lashio (Shan States), Loi Se (Mang Lun). Mong Ko (Kengtung). Pa Mung, Tan Yang (South Hsenwi). Man Neh Hang (Hsipaw). Man Hung, Nam Hpen, Na Kho Sheh, Ha Hta, Meung Nawng (Pang Long). Namkham (North Hsenwi). Mong Mong (Yunnan). Tiddim, Falam (Chin Hills.) Naba (Katha). Namsamkyin Island, Pantha, Mawleik, Kalewa (Chindwin districts). Bhamo (Bhamo). Myitkyina, Mohnyin, Mogaung, Hopin, Nyaungbin, Sumprabum (Myitkyina). In the dry zone districts of central Burma was found only in Yamethin and may have been lacking in Magwe, Minbu, Meiktila, Myingyan, Pakokku, Mandalay, Kyaukse, Sagaing, and Shwebo. Perhaps to be expected in Kyaukpyu district and Arakan Hill Tracts.

Chiengrai Thailand).

Sibpur (Bengal). Pashok (Darjiling District). Dehra Dun, Lachhiwala (United Provinces). Saffraha (Deccan).

From sea level to elevations of more than 4,000 feet.

The range extends from the southern border of peninsular Burma to the north of Myitkyina and into Thailand as well as the Chinese province of Yunnan. Presence of a colony near the center of the Indian Peninsula, of another in the Dun at the foothills of the western Himalayas proves that the species is anthropochorous. Accordingly *P. anomala* now seems unlikely to be autochthonous in any area lacking other endemic pheretimas. If that is so, presence west of the Chindwin-Irrawaddy axis has involved transportation, which in turn warrants a question as to how much of the range east of the axis may also have been acquired by accident.

A region including Karenni and Kengtung was thought (Gates, 1956: p. 215), because of the high incidence of the H morph in collections therefrom, to be nearer the original home of the species than middle Burma.

Habitats: Soil, of open spaces, under lawns, under trees, in ditches, in and near towns as well as in jungles. Rotting leaves. Manure.

Biology: In Rangoon, *P. anomala* was usually the last of the species to appear after the rainy season had begun. Sexual worms usually were not found until well into August. Breeding probably occurs from late in August through September and perhaps part of October. Most individuals disappear from strata of activity during October though an occasional specimen can be secured during November and early December. The period of inactivity, which may comprise five to eight months presumably is spent in deep diapause about which nothing is known nor of the level to which the worms retire hibernestivation.

Abnormality: One of the hearts of xiii was lacking in several specimens. The intestine seemingly began in xiv (of specimens) or in xvi (of four) but these were about the only strictly somatic aberrations that were recognized during dissections of hundreds of worms. Divergence of septa 8/9–10/11 from the H norms was associated with the genital organization.

The "testicular chambers" were of various sorts but always seemingly unlike the H and the R testis sacs.

Abnormal in the reproductive system often is normal or at least usual in parthenogenetic morphs.

Polymorphism: Evolution of parthenogenetic polymorphism is well advanced in *P. anomala* with three major kinds of morphs common and numerous minor, intermediate or other morphs rarer.

H morph

This morph has a full complement of reproductive organs each of which is capable of functioning in a normal manner. Furthermore, sperm are matured and, when opportunity permits, exchanged in copulation or at least passed on to a copulatory partner even though none are received in return. The H accordingly designates the biparental population from which other morphs have evolved mostly by organ deletions of the kinds that parthenogenesis makes possible. Accordingly, the species is defined by and from the biparental H morph rather than from any of its degraded derivatives.

Reproduction: Although believed to be capable of biparental reproduction after effective copulation and real fertilization, opportunities for that sort of reproduction frequently have seemed so limited as to require assuming facultative parthenogenesis.

Distribution: Rarely obtained even in most localities where it is known to be present. Absence, except in India and perhaps in the Chin Hills district, should not be assumed till after careful search.

Unusually high incidence (to about 20 per cent) of this morph in collections from eastern mainland Burma was thought to indicate that there the species was nearer its original home than in middle Burma.

A morphs

Athecal. Male pores, in xx. Genital markings, present in some or all of xvii–xix, xxi–xxiv.

Testes and male funnels, in some or all of v–xii, in x–xi usually in small chambers unlike the H testis sacs, in other segments free. Seminal vesicles, lacking or rudimentary.

Reproduction: Sperm cannot be received during copulation because of absence of spermathecae. Externally adhesive spermatophores are lacking. Testes remain juvenile through maturity (248 of 250 specimens in one series). Maturation of sperm by the ovaries is unknown in this species. Reproduction then presumably must be parthenogenetic, especially in the Indian localities where sperm maturing morphs are absent.

Parasites: Sporozoan (?) trophozoites often were present in or on the blood vessels of vii–xiii, sometimes so numerous as to conceal from view the gizzard, the ventral face of the esophagus or the hearts.

Distribution: Chiengrai (Thailand). Sibpur, (Bengal), Pashok (Darjiling district), Dehra Dun and Lachhiwala (United Provinces), Saffraha on the Deccan (Central Provinces), India. Other *anomala* morphs are unknown at each of those places and probably are absent from most of the Indian localities. From various Burmese localities (e.g., Tiddim, Falam) only A morphs were recorded but until much more thorough search has been made, absence of others probably should not be assumed.

Remarks: Male funnels of v–ix may end blindly or the sperm duct from a male funnel of v, joined by all other funnel ducts of the same side, may unite with a prostatic duct of xx just as in the H morph. One or both testes may be fragmented so that there are several of smaller size in a segment instead of the normal two.

Worms of A morphs once (Stephenson, 1929) were thought to be of a secondarily evolved male sex. However, seminal vesicles in which sperm mature are lacking and testes remain juvenile throughout the breeding season (248 of 250 specimens in one series). In spite then of all the extra testes, the worms morphologically are only "pseudomasculinized" and physiolog-

ically are female in a parthenogenetic sense. The hyperandry is now regarded as a reversion, enabled by parthenogenesis, to or toward a polygonadal condition of some ancient ancestor.

Precise characterization of septal relationships and testicular chambers along with number of testes, male funnels and their relationships with functional sperm ducts, and including rare somatic abnormalities, would enable a long list of distinct and presumably genetically determined morphs.

So many of the evolutionary changes resulting in parthenogenetic polymorphism are of a negative sort that it is worth while to emphasize two major changes in the evolution of the *anomala* A that may be called positive as they involve increase rather than decrease in number of organs, namely of the gonads and of the mushroom-shaped GM glands (added in xxi–xxiv).

R morphs

Anarsenosomphic. Genital markings and GM glands, none. Male deferent ducts attenuate, anywhere in the region of xv–xxx, without acquiring an external aperture. (Otherwise as in the H morph.)

Reproduction: Sperm are matured and probably profusely but are within a closed system comprising testis sacs sealed off from the coelomic cavities, seminal vesicles that open only into the testis sacs, and sperm ducts which attenuate distally to a filament through which the sperm cannot pass. Maturation of sperm in ovaries of xiii never was recognized. Sperm occasionally are received during copulation with some fertile individual of a morph with functional male terminalia and in such circumstances reproduction may be biparental. However, many individuals of the R morphs are not so situated as to be able to copulate with an individual of the rare H morph. Parthenogenesis accordingly is assumed to be the usual method of reproduction even in a morph that profusely matures seemingly normal sperm.

An individual of the H morph that copulated with an R partner would, unfortunately, receive no sperm and even after copulation was completed could reproduce only by parthenogenesis.

Distribution: More than a score of administrative districts of Burma. Absence of records for any district may be of little or no significance though it should be noted that these morphs and the H were not found in the Chin Hills.

Parasites: Cysts with pseudonavicellae spores were found frequently in coelomic cavities of ix–xv and similar cysts as well as trophozoites were seen even in quite small juveniles. Blood vessel parasites such as those found in the A morphs were not seen in any individuals of the R morphs that were dissected.

Remarks: Little evidence was found to indicate evolution of an AR morph was under way and the R morphs may be even more stable than the A morphs.

Other Morphs

I. First order intermediate morphs. These differ from the A only in having a spermatheca, sometimes normally developed, otherwise more or less rudimentary, at one of the normal H sites.

I₂. Second order intermediate morphs differ from the H in absence of some genital markings and associated glands, on one or both sides of the body, or in absence of male terminalia only on one or both sides of the body, or in various combinations of the two organ deletions.

I₃. Third order intermediates differ from the R morph in absence of one or more of the spermathecae. Records of only one specimen are available, of a quadrithecal individual lacking the anterior pair of spermathecae.

Although it seems possible at present to read the evolutionary changes involved in either direction, it has been assumed that intermediate morphs represented less advanced stages of evolution to more advanced morphs such as the completely athecal or entirely anarsenosomphic rather than a reversion to a stage of less advanced degradation.

Pseudo-intermediate morphs, are of various sorts. They are characterized by presence of male terminalia on one or both sides in xviii, xix, or xxi, presence of a spermatheca in ix, presence of a genital marking and its associated gland on one or both sides of xx where the male terminalia should be.

Presence of the male terminalia in xix or in xxi might now be explainable as a result respectively of deletion of mesoblastic somites at any one of the 17th to 19th levels or of halving of mesoblastic somites at any one of those levels. Such a causation presumably will be confirmed if the intestinal caeca are displaced anteriorly or posteriorly by a single segment.

Some of the changes involved are of such a nature as to suggest reversion to some more remote ancestral stage than that of the biparental population.

Yet other morphs result from gynic changes in A morphs. (1) Ovaries and gonoducal funnels in xii and xiii but with only one female pore, that of xiv, ducts from the funnels of xii joining the sperm ducts. (2) Testes of xii partly feminized so that the gonad appears to be hermaphroditic. (3) Ovaries of xii mature but those of xiii juvenile. (4) Presence of one or more small but softish seminal vesicles.

Presence in A morphs of one or more small but not rudimentary vesicles sometimes is associated with maturation of a limited amount of a sperm. An individual of such a morph presumably could transfer sperm to a copulatory partner of an H or of an R morph in order to further confuse the genetics of an already complicated situation. Perhaps some cytologist will care to make a few predictions about what is involved in the *anomala* polymorphism.

The single prostate of one curious morph had to open to the exterior in xx at its usual site by its duct passing down through the mushroom gland and its stalk. The GM gland presumably developed as usual from an epidermal ingrowth and subsequently was penetrated by the developing prostatic duct which according to Stephenson is of mesodermal origin.

Remarks: Nothing is known about the relationships of this truly anomalous species. Closest relatives presumably are to be sought east of Burma.

Pheretima arboricola

1936. *Pheretima arboricola* Gates, *Rec. Indian Mus.* **38**: p. 399. (Type locality, Karen Hills of Toungoo district.)

Octothecal, pores minute, superficial, at 5/6–8/9. Male pores, in xviii, minute, superficial, each in a circular discoidal porophore. Female pore, median. Genital marking, transversely elliptical, posteriorly in xviii and reaching almost to male porophores. Clitellum, xiv–xvi. Setae, 51/viii, 46/xii, 44/xx, vi/21, vii/23, viii/22, xvii/19, xviii/0, xix/20. First dorsal pore, at (10/11?), 11/12. Body wall of ventrum behind clitellum thickened as a sort of "creeping sole" that decreases in width posteriorly, and in which setae are smaller and more closely crowded than elsewhere. (Prostomium?) Color, only in dorsum, reddish, (pigment in circular muscle layer?). Segments 66 (+?). Size, 49(+?) by 3 mm.

Septa, 8/9–9/10 aborted, none especially thickened. Intestinal origin, in xv. Caeca, manicate, ventralmost secondary caecum the longest, in xxvii(–?). (Typhlosole?) Hearts, of ix on one side only and lateral, of x–xiii (latero-esophageal?). Holandric. Testis sacs, unpaired and ventral. Seminal vesicles, large, in xi, xii. Prostates, in xvii–xxi, duct short and slender. Spermathecae, duct shorter than ampulla, abruptly narrowed in body wall, diverticulum from anterior face of duct at parietes, elongately digitiform, in part shortly looped in a more or less regular zigzag, lumen narrow, arborescent, lateral branches ending in very small chambers. GM gland, sessile on the parietes.

Distribution: Karen Hills on Shan Plateau (Toungoo).

Habitat: Under bark of trees, in rain forest.

Remarks: *P. arboricola* is known only from the type which may have been a posterior amputee. The worm was obtained during a futile, several years search for another tree-living form, *Perionyx arboricola*, in which many hundreds of specimens were collected from trees, their epiphytic ferns and any other niches that were detected.

A "creeping sole" is associated in *arboricola* with compound intestinal caeca, in *lacertina* Chen, 1946, with simple caeca so short as to reach only into xxvi, but in *bambophila* Chen, 1946, also from Szechuan, with absence of intestinal caeca as in a small group of Borneo species for which the subgenus *Planapheretima* Michaelsen, 1934, was erected.

The sole appears to have evolved independently in different parts of the *Pheretima* range and therefore cannot define, regardless of absence or presence and shape of intestinal caeca, a monophyletic subgenus.

Pheretima balteolata

1932. *Pheretima balteolata* Gates, *Rec. Indian Mus.* **34**: p. 425. (Type locality, Pang Wo. Types, none.)
1936. *Pheretima balteolata*, Gates, *ibid.*, **38**: p. 402.

Sexthecal, pores minute, superficial, at 5/6–7/8. Male pores, in xviii, minute, superficial, each nearer lateral margin of a transversely or longitudinally elliptical area of epidermal modification, *ca.* 12 intersetal intervals wide that may reach nearly to 17/18 and 18/19. Female pore, median. (Genital markings, aside from the male porophores, none.) Clitellum, setae unrecognizable externally, xiv into xvii, sometimes to the equator. Setae, 2–8/ii, 75/iii, 108/viii, 110/xii, 96–99/xx, vi/22–25, vii/23–28, xvii/18–24, xviii/10–16, xix/24–26. First dorsal pore, at 12/13. Prostomium, epilobous(?). Color, none, (pigment, none). Segments, 110. Size, 60–89 by 3 mm.

Septa, 8/9–9/10 aborted, 5/6–7/8 and 10/11 muscular. Intestinal origin, in xv. Caeca, simple, in xxvii–xxii. (Typhlosole?) Hearts, of ix lateral but usually lacking on one side, of x–xiii (latero?)-esophageal. Blood glands, in v. Holandric. Testis sacs, unpaired, median. Seminal vesicles, large, in xi, xii. Prostates, in xvii–xx, ducts slender, 2–6 mm. long, each in a hairpin loop. Spermathecae, duct slender, shorter than ampulla, abruptly narrowed in body wall, diverticulum from anterior face of duct near parietes, about as long as main axis, slender, slightly widened at ental end.

Reproduction: Presumably biparental but the only evidence now available is the large size of the seminal vesicles.

Distribution: Peng Sai, Pang Wo (Mang Lun). Teung Cong (Mong Lem, Yunnan). At elevations of 3,500–4,500 feet.

Habitats: Rotting leaves, soil in wet ravines.

Remarks: Known only from six specimens.

Pheretima bellatula

1932. *Pheretima bellatula* Gates, *Rec. Indian Mus.* **34**: p. 427. (Type locality, Teung Cong. Types, none.)
1936. *Pheretima bellatula*, Gates, *ibid.*, **38**: p. 403.

Sexthecal, pores minute, superficial, at 5/6–7/8. Male pores, in xviii, minute, superficial, both in a transversely elliptical area of epidermal thickening extending slightly into xvii and xix. Female pore, median. (Genital markings, other than the male porophore, none.) Clitellum, xiv–xvi. Setae, present ventrally in xv–xvi, 30/iii, 41/viii, 44/xii, 38–42/xx, vi/8–12, vii/11–14, xvii/10–12, xviii/0, xix/10–12. First dorsal pore, at 11/12. Prostomium, epilobous(?). Color, brownish (pigment red and in cir-

cular muscle layer?). (Segments?). Size, 37–72 by 2–3½ mm.

Septa, 8/9 present but membranous, 9/10 aborted. Intestinal origin, in xv. Caeca, simple, in xxvii–xxiv. (Typhlosole?) Hearts, of ix lateral but usually lacking on one side, of x–xiii (latero?)-esophageal. Holandric. Testis sacs, unpaired, ventral. Seminal vesicles, large, in xi, xii. Prostates, in xvii–xx, ducts slender, *ca.* 2 mm. long, looped. Spermathecae, duct slender, shorter than ampulla, narrowed in body wall(?), diverticulum from anterior face of duct near parietes, shorter than main axis, slightly widened entally.

Reproduction: Presumably biparental as sperm are exchanged during copulation.

Distribution: Teung Cong (Mong Lem, Yunnan). Nam Mang (Mang Lun). At elevation of *ca.* 4,000 feet.

Habitats: Rotting leaves. Sandy soil.

Remarks: Presence of sperm in individuals lacking any evidence of past or present clitellar tumescence still is noteworthy.

P. bellatula is known only from four specimens.

Pheretima bicincta

1875. *Perichaeta bicincta* Perrier, *Compt. Rend. Acad. Sci. Paris* **81**: p. 1044. (Type locality, supposedly in Mindonoro or in Luzon, Philippine Islands. Types, in the Paris Mus.).
1942. *Pheretima bicincta*, Gates, *Bull. Mus. Comp. Zool. Harvard College* **89**: p. 119.
1961. *Pheretima bicincta*, Gates, *Burma Res. Soc. 50th Anniv. Pub. No. 1:* p. 57.

Decathecal, pores minute and superficial, well towards mL, at 4/5–8/9. Male pores, in xviii, minute, superficial, each within a porophore of variable shape. Female pores, anteromedian to *a*. Genital markings when present, paired, in xviii or on 18/19. Clitellum, some setae retained in xiv–xv, reaching from 13/14 nearly to setae of xvi. Setae, somewhat enlarged in preclitellar segments except x, 31–38/iii, 42–50/viii, 43–56/xii, 46–48/xx, v/9–11, vi/10–12, vii/10–14, viii/11–15, xvii/10–13, xviii/4–9(one often included in each male porophore), xix/9–13. First dorsal pore, at (11/12), 12/13. Prostomium, epilobous, tongue open. Color, in dorsum only, reddish. Segments, 77–101. Size, 40–80 by 2–3 mm.

Septa, 5/6–7/8 muscular, 8/9 aborted, 9/10–12/13 muscular to thickly muscular. Pigment, reddish brown, associated with circular muscle layer. Intestinal origin, in xv. Caeca, simple, small, in xxii or reaching into xxi. Typhlosole, present from xxii, simply lamelliform, ending in region between 50th and 73d segments. Ventral and (single) dorsal trunks, complete (subneural?). Hearts, of viii aborted, of ix lateral, of x–xii latero-esophageal. Holandric. Testis sacs, unpaired, horseshoe-shaped to annular, hearts included. Seminal vesicles, in xi, xii, small to medium-sized, of xi included in testis sac. Prostates, in xvi–xix, xx, ducts J- to C-shaped, muscular, 2+ mm. long. Spermathecae, small, subesophageal, duct shorter than ampulla. Diverticulum, from anterior face of duct at parietes, usually shorter than main axis, comprising a slender stalk and an ovoidal to ellipsoidal seminal chamber. GM glands, sessile on parietes.

Reproduction: Parthenogenesis is suspected, because of constant absence of sperm in mature individuals examined by the author. Moreover, unilateral absence of male terminalia (*cf.* Abnormality, below) and lack of genital markings when GM glands are present may be instances of parthenogenetic degradation.

Distribution: Rangoon (Hanthawaddy). Found there only in flower pots on one veranda. Taiwan (Taipei), Java (Malang). Christmas Island. Penang. Philippine Islands (locality unknown). Werua Island, Kapingamarangi Atoll, southern Caroline Islands. Hawaii. Mexico. Maine (greenhouses) Florida (new record). St. Thomas. Grenada. Trinidad. India (Trivandrum in Travancore and Hyderabad in the Deccan).

The original home of *P. bicincta* is unknown.

Abnormality: An extra spermatheca mesially in ix, the parietal portion of a prostatic duct along with a typical male porophore mesially in xviii, a typical porophore and prostate mesially in xix, in a Mexican specimen were associated with doubling of the nerve cord. The worm had 125 segments.

Male terminalia of the left side, in a specimen from a Maine greenhouse, were lacking.

Remarks: Pigment has been seen by the author only in specimens that had been quite recently preserved. GM glands, lacking in Mexican and Formosan worms, sometimes are present and presumably with external pores even though the epidermis is not sufficiently modified for genital markings to be recognizable.

Pheretima bipora

1901. *Amyntas biporus* Beddard, *Proc. Zool. Soc. London* **1900**: p. 908. (Type locality, Malay Peninsula. Types, in British Mus., having been dried at some time, now are of little value.)

1903. *Pheretima biporus*, Michaelsen, *Die geographische Verbreitung der Oligochaeten*, (Berlin), p. 908.

1917. *Pheretima annandalei* Stephenson, 1917, *Rec. Indian Mus.* **13**: p. 386. (Type locality, Singgora. Type, possibly a posterior amputee, in the Indian Mus.)

1931. *Pheretima gemella* Gates, *Rec. Indian Mus.* **33**: p. 379. (Type locality, Ye. Types, none.)

1932. *Pheretima gemella*, Gates, *Ibid.*, 34: 519. *Pheretima bipora*, Stephenson, *Ann. Mag. Nat. Hist.*, Ser. 10, 9: p. 213.

1933. *Pheretima gemella*, Gates, *Rec. Indian Mus.* 35: p. 528.

1934. *Pheretima bipora* + *P. annandalei*, Gates, *ibid.*, 36: pp. 257, 256.

1939. *Pheretima bipora* + *P. annandalei*, Gates, *Jour. Thailand Res. Soc. Nat. Hist. Suppl.* 12: pp. 83, 81.

1949. *Pheretima annandalei* + *P. balingensis* + *P. flocellana* Gates, *Bull. Raffles Mus. Singapore*, No. 19: pp. 6, 10,

15. (Type locality of *balingensis and flocellana*, Baling. Types, in the Raffles Mus.)

1955. *Pheretima gemella*, Gates, *Rec. Indian Mus.* 52: p. 87.

1960. *Pheretima bipora*, Gates, *Bull. Mus. Comp. Zool. Harvard College* 123: p. 254. (The paragraph beginning, "Coelomic copulatory chambers," on p. 254, belongs under *P. birmanica* on p. 255.)

Octothecal, pores minute, invaginated into coelomic chambers opening through transverse slits ½C or more apart at 5/6–8/9. Male pores, in xviii, minute, each at ventral end of a penis pendent from roof of a copulatory chamber with a transversely slitlike equatorial aperture. Female pore, median. Genital markings, small, usually circular, when external presetal and/or postsetal in some of vii–ix, xvi–xix, singly in vicinity of a spermathecal pore, in pairs of transverse rows or groups or in unpaired and median groups, when invaginate in xix–xx (rarely in xxi or xvii) each on roof of a coelomic (pseudocopulatory) chamber with a transversely slitlike, equatorial aperture in line with and like that of the copulatory chambers, or each on ventral end of a rather penislike but columnar protuberance from roof of the chamber. Clitellum, setae usually unrecognizable externally, nearly to or to 13/14 and 16/17. Setae, 24–36/ii, 30–42/iii, 37–47/iv, 43–64/viii, 44–64/xii, 43–70/xx, vi/19–43, vii/21–44, viii/24–40 (+ 20–24 dorsally), xvii/12–19, xviii/4–15, xix/9–19, xx/9–12 (when pseudocopulatory chambers present). First dorsal pore, at 11/12, 12/13. Prostomium, epilobous, tongue open. Color, only in dorsum except in several anterior segments, red to bluish gray, brownish red or brownish, (pigment, in circular muscle layer? red?). Segments, 79–105. Size, 32–170 by 2–7 mm.

Septa, none thickly muscular, 8/9–9/10 aborted. Intestinal origin, in xv. Caeca, simple, xxvii–xxii. Typhlosole, simply lamelliform, ending in region of 56th–62nd segments. Hearts, unaborted dorsal portions in viii passing to gizzard, in ix (left or right) lateral, in x (latero-?)-esophageal, in xi–xiii lateroesophageal. Connectives between extra-esophageals and the supra-esophageal, one pair, in x. (Blood glands, in v?) Lymph glands, in intestinal segments. Holandric. Testis sacs, unpaired, ventral or in x annular (hearts included), in xi cylindrical or U-shaped (vesicles included). Seminal vesicles, small to large, in xi, xii. Prostates, in xvii–xx, ducts 1¼–5 mm. long, straight or curved. Spermathecae, small, diverticulum with a slender, sinuous or looped stalk and a shortly sausage-shaped or ellipsoidal to ovoidal seminal chamber. GM glands, composite (always?), stalked and coelomic.

Reproduction: Presumably biparental in some of those individuals that copulate but data as to presence and absence of sperm on male funnels or in spermathecae no longer are available. Parthenogenesis is anticipated in other strains though organ absence, as in A, R, and AR morphs, was not found.

Distribution: Ye (Amherst). Heimza Basin, Ky-

TABLE 3

VARIATION IN NUMBER AND LOCATION OF
PSEUDOCOPULATORY CHAMBERS

Number of chambers in segment				Number of specimens	Side with asymmetrical organ
xvii	xix	xx	xxi		
1	2			1	
	1			3	
	2			87	
	2	1		9	Left, 4. Right, 5.
	2	2		26	
	1	2	1	2	Right
	2	1	1	1	Left

aukmedaung, Kameik, Pyinthadaing, Kataungni, Nyinmaw, Nyaungdonle Chaung, Nyaungdon, Pyinthadaw, Mayan Chaung (Tavoy). Zowai (Mergui). Singgora, Tale Sap, Thailand.

Kaki Bukit (Perlis), Baling (Kedah State), Malay Peninsula.

The range, as now known, extends through about ten degrees of longitude, reaching up into mainland Burma (and also in Thailand?) and in the peninsula down into the range of *P. malayana*.

Biology: *P. bipora* is geophagous but along with the earth (often red and then of lateritic origin?) in its gut, organic materials sometimes were found (1949), including knobby and woody twigs 5–6 mm. long and *ca.* ½ mm. thick.

Variation: Testis sacs were subesophageal in specimens referred to *P. annandalei.* That condition is more likely to be associated with biparental reproduction as also is the large size of seminal vesicles that characterized each of those seven worms. Other kinds of testis sacs (as in *P. gemella, balingensis*), as well as the small size of seminal vesicles, are more likely to be associated with parthenogenesis.

Pseudocopulatory chambers were lacking in one mature Burmese worm (*gemella*) and in the 24 peninsular specimens that were referred to *P. annandalei* (7) or *balingensis* (17). The Burmese worm lacked postclitellar genital markings which were present and superficial on each of the peninsular specimens.

The spermathecal-pore chamber is cylindrical and of about the same caliber as the spermathecal duct from which it is almost indistinguishable externally. The coelomic portion is considerably elongated and is looped once or twice. So much elongation with reference to the smallness of the spermatheca at its ental end looks like an abnormality but any correlation of length of chamber with parthenogenesis or with time since establishment of that method of reproduction remains to be determined.

Penes are ½–3 mm. long. Some of that variation may be correlated with differences in size of the soma which are greater than usual (*cf.* measurements in the précis above). Although size of pseudocopulatory

chambers seemed to be about the same relative to soma size, the genital marking therein was sometimes (Baling) on the roof of the chamber or (Burma) at end of a millimeter long, columnar protuberance.

Abnormality: No. 1. Main axis of one spermatheca bifid entally, so that there are two distinct ampullae. No. 2. An extra spermatheca present in v, with its pore at 4/5. No. 3. Trithecal, spermathecal pores at 8/9 and on left side of 7/8. Size of soma, 63 by 3 mm.

Absence of glandular material at ental end of bundles of stalks passing to genital markings, frequent in Burma, presumably is an abnormality. Possibly some of the differences in testis sacs should be similarly designated.

Parasites: Protozoa were numerous (1931: p. 381 and 1932: p. 521) in ovaries that reached to level of the dorsal blood vessel but which contained few or no ova. Masses of coelomic sporozoa were associated with enlargement of ix–xi, loss of some ventral setae in x and a locally transparent body wall. Masses of coelomic cysts frequently were present in ix–xi. Some of those cysts were brittle, breaking under pressure with a "distinct crackling noise." Nematodes. *Filiponema burmense* Timm, 1967, were present in coelomic cavities anteriorly of most specimens that were dissected.

Some of the small specimens may have been dwarfed as a result of the presence of numerous parasites during the period of postnatal growth. However, parthenogenesis may have enabled fairly heavy parasitism without marked effect on soma size.

Remarks: Penes can be protruded to the exterior through the secondary male pores. Thereafter, copulatory chambers may themselves be everted so as to add to effective length of the intromittent organs. Pseudocopulatory chambers also are eversible.

Relationships are with *P. malayana* and appear to be especially close through southern forms without invaginate genital markings that were collected in a northwestern corner(?) of the *malayana* range.

Pheretima birmanica

1888. *Perichaeta birmanica* Rosa, *Boll. Mus. Zool. Univ. Torino* 3, 50: p. 2. *Ann. Mus. Civ. Sto. Nat. Genova* 26: p. 164. (Type locality, Bhamo. Types, in the Genoa Mus.)

1895. *Perichaeta birmanica*, Beddard, *A Monogr., of the Order of Oligochaeta*, (Oxford), p. 405.

1900. *Amyntas birmanicus*, Beddard, *Proc. Zool. Soc. London* 1900: p. 637. *Pheretima birmanica*, Michaelsen, *Das Tierreich* 10: p. 255.

1903. *Pheretima birmanica*, Michaelsen, *Die geographische Verbreitung der Oligochaeten* (Berlin), p. 94.

1909. *Pheretima birmanica*, Michaelsen, *Mem. Indian Mus.* 1: p. 110.

1923. *Pheretima birmanica*, Stephenson, (*The Fauna of British India*), *Oligochaeta*, p. 295.

1926. *Pheretima birmanica*, Gates, *Rec. Indian Mus.* 28: p. 152.

1930. *Pheretima birmanica*, Gates, *ibid.*, 32: p. 307.

1931. *Pheretima birmanica*, Gates, *ibid.*, 33: p. 372.

1932. *Pheretima birmanica*, Gates, *ibid.*, 34: p. 428.

1933. *Pheretima birmanica*, Gates, *ibid.*, 35: p. 510.

1945. *Pheretima birmanica*, Gates, *Sci. & Culture, Calcutta* 10: p. 403.

1955. *Pheretima birmanica*, Gates, *Rec. Indian Mus.* **52**: p. 83.
1960. *Pheretima birmanica*, Gates, *Bull. Mus. Comp. Zool. Harvard College* **123**: p. 255. (The paragraph beginning, "Coelomic copulatory chambers," on p. 254, belongs under *P. birmanica*.)
1961. *Pheretima birmanica*, Gates, *Burma Res. Soc. 50th Anniv. Pub.* No. 1: p. 57.

Sexthecal, pores minute, transverse slits, each on a tubercle on the roof of a small parietal invagination, secondary apertures nearly ⅓C apart(?), at 5/6–7/8. Male pores, in xviii, minute, each in a longitudinal groove on a conical penial body in a small copulatory chamber with an equatorial aperture. Female pore, median(?). (Genital markings, none.) Clitellum, without externally recognizable setae(?), xiv–xvi. Setae, "*ca.* 70 per segment," vi/24–30, vii/25–33, xviii/12–18. First dorsal pore, at 12/13. Prostomium, epilobous, tongue open. Color, in dorsum, brownish to slate (pigment reddish and in circular muscle layer?). Segments, 112. Size, 100–160 by 4–7 mm.

Septa, 8/9–9/10 aborted, 5/6–6/7 "thickened." Intestinal origin, in xv. Caeca, manicate, dorsalmost of 3–6 secondary caeca the longest, in xxvii–xxiii. Typhlosole, simply lamelliform, small, ending in region of 81'st–85'th segments. Hearts, in viii unaborted dorsal portions to gizzard, in ix (left or right) lateral, in x (latero?)-esophageal, in xi–xiii latero-esophageal(?). Holandric. Testis sacs, (unpaired?) and ventral. Seminal vesicles, small, in xi, xii. Prostates, in xvi–xx, ducts each in a *U*-shaped loop. Spermathecae, duct seemingly widened and thickened within the parietes but not distinguishable externally from the parietal invagination, with narrow lumen ental to diverticular junction, diverticulum from duct in parietes, longer than the main axis, looped in a more or less regularly zigzag manner.

Reproduction: Records of sperm maturation and transfer no longer are available but the seminal vesicles, usually retained at maturity in an early juvenile state, indicate that maturation at most would be sparse. Parthenogenesis is anticipated but it should be noted that none of the aberrations or deletions so often associated with that method of reproduction in *Pheretima* have been recognized in *birmanica*.

Distribution: Toungoo, Maw Pah Der (Toungoo). Taungyi, Maymyo, Lashio (Shan States). Tan Yang (South Hsenwi). Pa Maung, A Kin De, Na Ko Sheh, Pang Noi, Hpa Cha, Meung Nawng (Pang Long). Kengtung (Kengtung). Nawng Kham (Mang Lun). Bana (Mong Lem, Yunnan). Kutkai, Namkham (North Hsenwi). Mogok, Katha, Naba (Katha). Bhamo (Bhamo). Myitkyina (Myitkyina).

Dehra Dun (western Himalayas, India).

Probably fairly common at Myitkyina, Bhamo, Kutkai, Lashio and vicinity, Kengtung. Only a few specimens were obtained at other localities.

Found in lowland plains only at Myitkyina and Bhamo (and at foot of Shan massif near Toungoo?)

which are north of the tropic. All other localities were in jungle-covered hills or on the Shan Plateau.

Presence on the Dun at the foot of the western Himalayas very probably is due to direct transportation from Burma. How much of the known range in Burma also is attributable to human carriage cannot now be estimated. Each of the towns at which the species was secured in numbers had or had had in the past exotic potted plants.

Habitats: Soils, red, dark, rich, gravelly, in fields, manured gardens, ridges of paddy fields, of open areas under grass, bamboo, bush and hill jungles. Mud. Manure.

Abnormality: Aberrant organ locations of a specimen (1933: p. 510) probably could have been explained by abortion on the left side of the mesoblastic somite at the 8th level.

Regeneration: Tail regenerates, at levels from 52/53 posteriorly, were of 2–4 segments only. Frequency of amputation was high in Dehra Dun worms, 16 of 18 being posterior amputees (with or without regeneration). Conditions observed in those worms (1945) enabled this postulate, after posterior amputation at certain levels a posterior portion of the typhlosole is resorbed until number of atyphlosolate segments (in unregenerate amputees) approximates that of normal and complete individuals.

Remarks: Relationships now appear to be with *P. defecta* as *birmanica* can be derived from a form much like that species by the following changes: Differentiation of a discoidal porophore around each spermathecal pore. Retraction of each porophore into a small parietal invagination. Invagination of the male porophore into the coelomic cavity as an eversible copulatory chamber and elevation from the chamber roof of the small porophore into a conical penial body.

Pheretima bournei

1890. *Perichaeta bournei* Rosa, *Mus. Civ. Sto. Nat. Genova* **30**: p. 110. (Type locality, Cobapo. Type, in the Genoa Mus.)
1895. *Perichaeta bournei*, Beddard, *A Monogr. of the Order of Oligochaeta* (Oxford), pp. 403, 395, 398.
1900. *Amyntas bournei*, Beddard, 1900, *Proc. Zool. Soc. London* 1900: p. 635; *Pheretima bournei*, Michaelsen, *Das Tierreich* **10**: p. 257.
1903. *Pheretima bournei*, Michaelsen, *Die geographische Verbreitung der Oligochaeten* (Berlin), p. 94.
1909. *Pheretima bournei*, Michaelsen, *Mem. Indian Mus.* **1**: p. 110.
1923. *Pheretima bournei*, Stephenson (*The Fauna of British India*), *Oligochaeta*, p. 296.
1932. *Pheretima bournei*, Gates, *Rec. Indian Mus.* **34**: p. 441.
1933. *Pheretima bournei*, Gates, *ibid.*, **35**: p. 510.

Sexthecal, pores minute, superficial, each in a very small tubercle in a slight depression and about midway between the anterior intersegmental furrow and the segmental equator, in vi–viii. Male pores, in xviii, minute, superficial, each in a small disc. Female pore, median. Genital markings, small, circular to transversely elliptical, one just posteromedian to each spermathecal pore, two or three in a longitudinal row in xviii median to each male porophore, one equatorial, one presetal, one postsetal. Clitellum (setae?), xiv–xvi (not always reaching 16/17?). Setae, lacking in ii, 55–63/xx, vi/17–20, vii/17–22, xvii/19–25, xviii/14–16, xix/23. First dorsal pore, at 11/12, 12/13. Prostomium, epilobous. Color, in dorsum only, reddish to slate (pigment red and in circular muscle layer?). Segments, 90–130. Size, 117–150 by 5–6½ mm.

Septa, 8/9–9/10 aborted, 5/6–7/8 thickly muscular. Intestinal origin, in xv. Caeca, simple, xxvii(–?). (Typhlosole?) Hearts, of x–xiii latero(?)-esophageal. Blood glands, in v. Holandric. Testis sacs, paired, ventral. Seminal vesicles, large, in xi, xii. Prostates, in xvii–xx, duct in a long hair-pin loop, ectal limb thicker. Spermathecae, duct slender, shorter than the ampulla, diverticulum from anterior face of duct at parietes, longer than the main axis, with a single u-shaped loop near the duct and an ellipsoidal seminal chamber that is shorter than the stalk. GM glands, sessile on the parietes (stalks confined to body wall?).

Reproduction: Presumably biparental though the only supporting evidence that now can be proffered is the large size of seminal vesicles and prostates (not juvenile).

Distribution: Cobapo, Blachi, Leiktho Circle, on the Shan Plateau, (Toungoo). Kwachi (Karenni).

Parasites: Gregarine trophozoites and pseudonavicellae cysts in one specimen were present in the ovaries.

Remarks: Aside from the type (or types) five specimens have been recorded.

A mesially facing, crescent-shaped area just lateral to each male pore perhaps should be called a genital marking.

Beddard (1900) was disposed to think that *P. bournei* really was *P. hawayana*. The species was allowed by Michaelsen (1900) and is distinguished by the intrasegmental location of the spermathecal pores as well as by the presence of genital markings in a presetal portion of xviii.

Pheretima californica

1866. *Pheretima californica* (part) Kinberg, *Ofvers. K. Vetenks-Akad. Förhandl. Stockholm* 23: p. 102. (Excluding octothecal specimens. Type locality, Sausolita Bay, California. Type, and paratypes from San Francisco, in the Stockholm, Mus.)
1912. *Pheretima browni* (part) Stephenson, *Rec. Indian Mus.* 7: p. 274. (Excluding sexthecal specimens of the type series. Type locality, Tengyueh. Types, in the Indian and British Mus.)
1927. *Pheretima modesta* Michaelsen, *Boll. Lab. Zool. Portici* 21: p. 88. (Type locality, Yi Leang. Types?)
1931. *Pheretima browni*, Gates, *Rec. Indian Mus.* 33: p. 372.
1932. *Pheretima molesta* Gates, *ibid.* 34: p. 420. (Type locality, Kutkai. Types, none.)
1936. *Pheretima californica*, Gates, *ibid.*, 38: p. 404.
1955. *Pheretima "osliformica,"* Gates, *ibid.*, 52: p. 83.
1961. *Pheretima californica*, Gates, *Burma Res. Soc. 50th Anniv. Pub.* No. 1: p. 57.

CASTINGS, organic carbon and moisture: Hassan *et al.*, 1956, *Ann. Agric. Sci. Cairo* 1: pp. 328, 329, 333, 334. PARASITES: Puytorac & Tourret, 1963, *Ann. Parasitol. Humaine Comp.* 38.

Quadrithecal, pores minute, superficial, *ca.* ⅓C or more apart, at 7/8–8/9. Male pores, in xviii, minute, each on a very small tubercle often at end of a columnar protuberance from the roof of an eversible, deep, transversely slitlike parietal invagination. Female pore, median. (Genital markings, none.) Clitellum, setae unrecognizable externally, xiv–xvi, occasionally reaching into xiii. Setae, enlarged in some or all of iii–ix and more so in v–vi (sometimes smaller in x than posteriorly?), 20–30/iii, 30–48/viii, 40–60/xii, 40–61/xx, viii/12–20, xvii/16–24, xviii/9–18, xix/15–23. First dorsal pore, at 11/12. Prostomium, epilobous, tongue open. Color, in dorsum, reddish, brownish, slate in front of the clitellum, posteriorly usually brownish. Segments, 85–115 but usually 103–112. Size, 50–132 by 3–5 mm.

Septa, 8/9–9/10 aborted, 5/6 or 6/7–7/8 and 10/11–13/14 muscular. Pigment, reddish brown, in circular muscle layer. Intestinal origin, in xv (sometimes in xvi?). Caeca, simple, without or with short lobes of dorsal and/or ventral sides, in xxvii–xxii. Typhlosole, low but lamelliform, usually irregular and sometimes translucent posteriorly, ending in region of 64th–74th segments. Hearts, of vi–vii lateral (when complete), in viii aborted dorsal portions to the gizzard, of ix (left or right) lateral, of x (latero?)-esophageal, in xi–xiii latero-esophageal. Blood glands, in an esophageal (and sometimes unusually thick) collar behind the gizzard. Holandric. Testis sacs, ventral, usually unpaired (sometimes paired?). Seminal vesicles, in xi, xii, medium-sized to large. Pseudovesicles, in xiii and/or xiv, often unrecognized (because so rudimentary?). Prostates, in xvi–xxii, ducts muscular, 1½–3 mm. long, much narrowed ectally and looped just before passing into a male invagination the roof of which at most is only slightly above the parietes. Spermathecae, small to medium-sized, the anterior pair usually in vii, duct shorter than ampulla, much narrowed in parietes but less so near ampulla, diverticulum from anterior face of duct near parietes, longer than the main axis, with a short, slender stalk and a long as well as wider seminal chamber that is variously looped.

Reproduction: As sperm are exchanged during copulation, reproduction is assumed to be biparental. Organ deletions, such as characterize so many parthenogenetic morphs, are almost unknown.

Distribution: Nampakka, Kutkkai (North Hsenwi), Mogok (Katha). Meung Nawng (Pang Long). Khamkho, Lewanga, Nawangkai, Putao, Kankiu, (Myitkyina).

Tengyueh, Yi Leang (Yunnan).

Malay Peninsula.

China (Yunnan, Szechuan, Kiangsu, Kiangsi, Hupei, Hunan, Anhwei, Chekiang Provinces). Hong Kong.

Taiwan. Japan (but not previously listed from there under its own name). Okinawa. Bali. Australia. Marquesas, Hawaiian and Easter Islands. United States (California, Louisiana, New York). Mexico, Costa Rica, Panama, Barbados, Brazil, Peru, Argentina. Greece. Azores. Madeira. Egypt, Transvaal. St. Helena and Ascension Islands. Lebanon.

Much of the distribution as now known resulted from transportation by man. The original home of *californica* is thought to be somewhere in China but how much of the range in that country (only partly known?) was self-acquired remains to be learned. Also unknown is whether the species made its own way into Burma from the north or whether it was introduced.

P. californica was introduced to Hawaii and California before 1852. Eisen was unable to find the species while collecting in the latter state during the last years of the 19th century and concluded that Kinberg must have mixed his labels. It is quite possible that colonization was not permanent in the state from which the species derived its name.

Habitats: Soil. Recorded once (under another name) from inner part of a Japanese cave.

Biology: *P. californica* is geophagous. However, the gut of one worm, from lxxi to the hind end, was filled with plant debris in which there was very little earth.

Variation and Abnormality: Two variations that may have been correctly recorded in the past are included in the précis. Sexthecal individuals with normal spermathecae in vii and pores at 6/7 have been seen (1 from Mexico, 1 from Hawaii, the latter with sperm in all spermathecae).

Septa 5/6–7/8 and 10/11–12/13 have been said at various times to be slightly muscular, muscular, fairly thickly muscular.

Abnormal No. 1. Right sperm duct ends with a bulbous swelling in xv. Male terminalia of right side, lacking (St. Helena).

No. 2. Left heart of xiii, lacking. Two hearts present in xiii on the right side, the anterior one seemingly without connection to the dorsal trunk (Malay Peninsula).

No. 3. Spermathecae, adiverticulate. Ampulla of one spermatheca completely bifid down to the duct.

Distention, according to a previous record, was so great that one spermathecal ampulla was constricted by each of septa 4/5–6/7.

Regeneration: Tail regenerates, of 2–5 segments only, have been seen. They were at various levels behind 66/67. At 62/63 only a pygomere was regenerated.

Parasites: Macroscopically recognizable coelomic parasites were found infrequently. Parietal cysts (with a diameter of 1 + mm.) were present in xiv–1 of 3 Hawaiian worms and smaller cysts were present on the parietes of a Hawaiian juvenile. PARASITES:

Microsporidia, *Coccospora gatesi* Puytorac & Tourret 1963 (Ascension Island).

Remarks: The male invagination rarely was noticed or properly characterized, in part because of partial eversion, in part because of failure to look into the equatorial slits that were dismissed as male pores without further characterization. Rarely the roof of the invagination is raised above the general peritoneal level sufficiently to allow, with some care, its removal with the male porophore attached. Whether that condition represents a state of complete retraction or a variation from a more usual condition of restriction to the parietes has not been decided. Connective tissue-binding loops of the slender ectal portion of the prostatic duct to the parietes over site of the invagination sometimes creates an appearance of the chamber reaching slightly into the coelom.

"Nephridialblasen" were mentioned by two authors and were thought to be two pairs or at least one pair per segment. *P. californica* was not studied by Bahl who assumed, from his study of two or perhaps three species, that vesicles were lacking throughout the genus.

Pheretima canaliculata

1932. *Pheretima canaliculata* Gates, *Rec. Indian Mus.* **34**: p. 408.
　　(Type locality, Blachi. Types, none.)
1936. *Pheretima canaliculata*, Gates, *ibid.*, **38**: p. 410.

Quadrithecal, pores small, superficial, transverse slits, each in a tubercle, at anterior margins of vi–vii. Male pores, in xviii, minute, superficial (each in a small discoidal porophore?). Female pore, median. Genital markings, one pair, between equators of xvii and xviii, slightly median to male pore levels, a seminal groove from each marking passing postero-laterally to nearer male pore. Clitellum, without externally recognizable setae, xiv–xvi. Setae, small, not so close together dorsally, 80–82/viii, 85–88/xii, 80–87/xx, vi/22–30, xvii/27, xviii/16–22, xix/25. First dorsal pore, at 7/8 or posteriorly. Prostomium, epilobous, tongue open(?). Color, in dorsum, slate anterior to clitellum, posteriorly reddish (pigment in circular muscle layer and reddish?). Segments, 111. Size, 147 by 5–7 mm.

Septa, 9/10 aborted, others membranous, 8/9 (usually present?) and 10/11 very delicate. Intestinal origin, in xv. Caeca, manicate, dorsalmost secondary caecum shortest, some secondary caeca occasionally with tertiary lobing, in xxvii–xxv. (Typhlosole?) Hearts, of ix (left or right) lateral, of x (latero?)-esophageal, of xi–xiii latero-esophageal(?). Blood glands, in v–vi. Holandric. Testis sacs, unpaired, ventral. Seminal vesicles, large, in xi, xii. Prostates, in xvii–xx, ducts short, stout, straight or *L*-shaped. Spermathecae, duct shorter than the ampulla, almost confined to the parietes and not narrowed therein, diverticulum from anterior face of duct in parietes, longer than main axis, with a short, straight stalk, a slightly thicker and looped middle portion as well as a terminal ovoidal to ellipsoidal seminal chamber. GM glands, one mass without stalks for each marking, under the longitudinal musculature.

Reproduction: Presumably biparental as sperm are exchanged during copulation.

Breeding season includes September.

Distribution: Blachi, Thandaung (Toungoo) on the Shan Plateau.

Remarks: Microscopic anatomy of the GM glands is expected to provide systematically useful characters that may help in determination of interspecific relationships.

Seminal grooves are rare in the Megascolecidae and in Burma are not presently thought to be indicative of relationships to other species such as *P. papilio*.

P. canaliculata is known only from 8 specimens, 7 of which were posterior amputees. Perhaps the species provides choice tidbits for some predator.

Pheretima carinensis

1890. *Perichaeta carinensis* Rosa, *Ann. Mus. Civ. Sto. Nat. Genova* **30**: p. 107. (Type locality, Meteleo. Types, in the Genoa Mus.)

1895. *Perichaeta carinensis*, Beddard, *A Monogr. of the Order of Oligochaeta* (Oxford), pp. 404, 395, 398.

1900. *Amyntas carinensis* Beddard, *Proc. Zool. Soc. London* **1900**: p. 625. *Pheretima carinensis*, Michaelsen, *Das Tierreich* **10**: p. 260.

1903. *Pheretima carinensis*, Michaelsen, *Die geographische Verbreitung der Oligochaeten* (Berlin), p. 95.

1909. *Pheretima carinensis*, Michaelsen, 1909, *Mem. Indian Mus.* **1**: p. 110.

1923. *Pheretima carinensis*, Stephenson, (*The Fauna of British India*), *Oligochaeta*, p. 297.

1930. *Pheretima pinguis* Gates, *Rec. Indian Mus.* **32**: p. 319. (Type locality, Thandaung. Types, none.)

1932. *Pheretima carinensis*, Gates, *ibid.* **34**: p. 446.

1933. *Pheretima carinensis*, Gates, *ibid.* **35**: p. 521.

Sexthecal, pores minute, superficial, less than ½C apart, at 6/7–8/9. Male pores, in xviii, minute, superficial, each in a discoidal porophore which may be united with a genital marking. Female pore, median. Genital markings, lacking or when present one or two pairs behind the clitellum, two in xviii, or across 16/17, or 17/18, or across each of 16/17 and 19/20. Clitellum, some setae usually retained (?), xiv–xvi/n or to 16/17. Setae, 66/xii, vii/15–25, viii/16–24, xvii/19–34, xviii/0–22. First dorsal pore, at 12/13. Prostomium, epilobous, tongue open. Color, only in dorsum, reddish, reddish brown, brownish, slate, (pigment in circular muscle layer?). Segments, 146–155. Size, 120–205 mm. by 6½–7 mm.

Septa, 8/9–9/10 aborted. Intestinal origin, close to 15/16. Caeca, simple, in xxvii–xvii, with several short ventral pockets. Typhlosole, small, simply lamelliform. Hearts, in ix lateral and usually present on both sides, in x (latero?)-esophageal, in xi–xiii latero-esophageal. Blood glands, in v. Holandric. Testis sacs, unpaired, *U*-shaped, hearts and anterior

vesicles included. Seminal vesicles, large, xi, xii–xiii. Prostates, in xvii–xxii, duct 8–13 mm. long, in one hairpin or several smaller loops. Spermathecae, large, duct shorter than ampulla and gradually narrowed in parietes, diverticulum from anterior face of duct in body wall, usually longer than main axis, slenderly club-shaped and slightly wider entally, looping or coiling restricted to ectalmost portion. GM glands, stalkless, sessile on parietes, each flat mass usually larger than its marking.

Reproduction: Presumably biparental as sperm are exchanged during copulation.

Distribution: Su Law on the western side of the Pegu Yomas (Toungoo), Metelindaung, Ko Haw Der, Shoko, Leiktho Circle, Blachi, on the Shan Plateau (Toungoo). Mawchi, Koopra (Karenni). *P. carinensis* is one of the very few endemic pheretimas that got across the Sittang without human intervention.

Parasites: NEMATODA: *Siconema saccaturum* Timm, 1966, Loikaw. *Adungella major* Timm, 1967, Koopra. *Adungella gatesi* Timm, 1967. Large cysts (protozoan?) frequently were present in the nerve cord and sometimes in considerable numbers. Thus, nervous tissue was lacking in one worm inside the nerve cord sheath from viii–lxvi.

Remarks: Fewer than 175 specimens had been available to end of 1932 and records of later lots, of course along with the worms, were destroyed during the war.

Male porophores vary considerably in size and sometimes are associated with glandular (?) material like that above discrete genital markings. Porophores occasionally are postequatorial.

Genital markings of xviii usually are median to the male porophores and then are circular to transversely elliptical, rarely smaller than the porophores. More lateral markings in the same segment are equatorial, of about the same size as the male porophores with which they may be united into dumbbell-shaped areas. Markings that are across intersegmental furrows may be transversely or longitudinally elliptical. Markings were lacking on 54 worms.

Pheretima defecta

1930. *Pheretima defecta* Gates, *Rec. Indian Mus.* **32**: p. 308. (Type locality, Labaw. Types, none.)

1931. *Pheretima jacita*, Gates, *ibid.*, **33**: p. 391. (Type locality, Ye. Types, none.)

1932. *Pheretima defecta* + *P. jacita*, Gates, *ibid.*, **34**: pp. 391, 440.

1933. *Pheretima jacita*, Gates, *ibid.*, **35**: p. 530.

1960. *Pheretima defecta*, Gates, *Bull. Mus. Comp. Zool. Harvard College* **123**: p. 255.

Sexthecal, pores minute, *ca.* ½C apart, at or just behind 5/6–7/8. Male pores, in xviii, minute, superficial, each in a small deeply demarcated tubercle within a distinctly delimited, longitudinally elliptical, glabrous area that does not cross 17/18 or 18/19. Female pore, median. (Genital markings, except for the glabrous areas, lacking.) Clitellum, without externally recognizable setae, xiv–xvi. Setae, 80–88/xx, vi/27–35, vii/27–35, xvii/21–27, xviii/13–21, xix/22–27. First dorsal pore, occasionally at 11/12, usually at 12/13. Prostomium, epilobous, tongue open. Color, in dorsum, slate anterior to the clitellum, posteriorly brownish (pigment reddish and in circular muscle layer?). Segments, 58–127. Size, 68–170 by 5–7 mm.

Septa, 8/9, 9/10 (usually?) aborted, none especially thickened. Intestinal origin, in xv (occasionally in xvi?). Caeca, manicate, the dorsalmost of 5–12 secondary lobes the longest, one or more usually with tertiary lobes, in xxvii–xxiv. (Typhlosole?) Hearts, of ix (left or right) lateral, of x esophageal(?), of xi–xiii latero-esophageal. Blood glands, in v–vi. Testis sacs, paired, ventral. Seminal vesicles, large, in xi, xii. Prostates, large, in xvii–xxii, ducts muscular, in a crescent or a U-shape, 4–6 mm. long. Spermathecae, small, duct shorter than ampulla and abruptly narrowed in the parietes, diverticulum from anterior face of duct at body wall, longer than main axis, with a short stalk and a slightly wider portion looped in a more or less regular zigzag.

Reproduction: Presumably biparental, in worms that exchange sperm during copulation. Support for that assumption is provided by the seminal vesicles which at first appear to be in contact with gizzard as well as prostates. The large size is suggestive of massive spermatogenesis. Prostates also are large and at maturity, seemingly in a functional state. Whether such forms have option of parthenogenesis remains to be determined.

Breeding season includes September and perhaps some part of October.

Distribution: Ye (Amherst). Kamaungthwe River region (intermediate morphs, only), Kyaukmedaung, Kataungni, Kameik, Pyin Tha Daing (Tavoy), Labaw (Mergui).

The range as now known includes peninsular Burma north to the Ye River.

Habitats: Mainly, if not wholly, jungles away from larger towns.

Polymorphism: None of the standard major morphs were found but an AR is, perhaps, to be expected though distinction from AR morphs of other Burmese species will be difficult if not impossible until more information is available about the H morph of *defecta*. Intermediates were of fourth and/or fifth sort, being aprostatic (but still with prostatic ducts) and athecal. Rudimentary prostates, confined to xviii, were present in one worm and another worm lacked the left posterior spermatheca. Testis sacs, in some specimens, were lacking.

Seminal vesicles of intermediate morphs were juvenile or rudimentary and for individuals of such morphs, reproduction, involving copulation with each other, must be parthenogenetic. Perhaps eggs of such intermediates can be fertilized by sperm from an individual

of the H morph which presumably is the one characterized in the précis above.

Parasites: Nematodes, *Siconemoides disarmatus* Timm, 1967, were present in coelomic cavities of anterior segments of individuals of the H as well as of other morphs. Sporozoan cysts were present in large numbers within coelomic cavities of posterior segments of intermediates only.

Remarks: Records of only 80-odd specimens have been available. Some of the worms were poorly preserved, some were immature, many lacked fifty or fewer segments of their tails.

Dislocation of the spermathecal pores behind the intersegmental furrows (some specimens) was about the smallest that could be detected.

How glabrous areas containing the male pores should be characterized is uncertain. Tissue associated with the male pore tubercle may be glandular but intraparietal tissue associated with the glabrous areas (perhaps thickenings of the epidermis) seemed less likely (some specimens at least) to be glandular.

Pheretima diffringens

1869. *Megascolex diffringens* Baird, *Proc. Zool. Soc. London* 1869: p. 40. (Type locality, Plas Machynlleth, North Wales. Types, in the British Mus.)
1909. *Pheretima heterochaeta*, Michaelsen, *Mem. Indian Mus.* 1: p. 189.
1910. *Pheretima heterochaeta*, Michaelsen, *Abhandl. Naturhist. Ver. Hamburg*, 19: pp. 11, 83.
1912. *Pheretima divergens* var. *yunnanensis* Stephenson, *Rec. Indian Mus.* 7: p. 274. (Type locality, Tengyueh. Type in the Indian Mus.)
1914. *Pheretima heterochaeta*, Stephenson, *Rec. Indian Mus.* 8: p. 399.
1917. *Pheretima heterochaeta*, Stephenson, *ibid.*, 13: p. 385.
1921. *Pheretima heterochaeta*, Stephenson, *ibid.*, 22: p. 760.
1922. *Pheretima heterochaeta*, Stephenson, *ibid.*, 24: p. 433.
1923. *Pheretima heterochaeta*, Stephenson (*The Fauna of British India*), *Oligochaeta*, p. 302.
1924. *Pheretima heterochaeta*, Stephenson, *Rec. Indian Mus.* 26: p. 339.
1926. *Pheretima heterochaeta*, Gates, *ibid.*, 28: p. 156.
1930. *Pheretima heterochaeta*, Gates, *ibid.*, 32: p. 310.
1931. *Pheretima heterochaeta*, Gates, *ibid.*, 33: pp. 387, 437.
1932. *Pheretima heterochaeta*, Gates, *ibid.*, 34: p. 524.
1933. *Pheretima heterochaeta*, Gates, *ibid.*, 35: p. 529.
1934. *Pheretima mirabilis*, Gates, *ibid.*, 36: p. 260.
1936. *Pheretima diffringens*, Gates, *ibid.*, 38: p. 412.
1937. *Pheretima diffringens*, Gates, *ibid.*, 39: pp. 183, 198.
1955. *Pheretima diffringens*, Gates, 1955, *ibid.*, 52: p. 86.
1961. *Pheretima diffringens*, Gates, *Burma Res. Soc. 50th Anniv. Pub. No. 1*: p. 57.

BLOOD GLANDS: Stephenson, 1924, *Proc. Roy. Soc. London*, B, 97: p. 188. CIRCULATORY SYSTEM: Cecchini, 1914, *Arch. Zool. Italiano* 8: p. 7. Hertling, 1921, *Zool. Anz.* 52: p. 182. EXCRETORY SYSTEM: Bourne, 1894, *Quart. Jour. Micros. Sci.* 36: p. 33. Bahl, 1919, *Quart. Jour. Micros. Sci.* 64: p. 104. LEUCOCYTES AND LEUCOCYTOPOIETIC ORGANS: Kindred, 1929, *Jour. Morph.* 47: p. 435. LYMPH GLANDS: Thapar, 1918, *Rec. Indian Mus.* 15: p. 71. LYMPHOCYTES: Cognetti, 1923, *Boll. Mus. Zool. Univ. Torino* 38: 5, p. 3. MOVEMENTS: Bourne, 1891, *Quart. Jour. Micros. Sci.* 32: p. 52. NERVE CELLS: Leonard, 1933, *Jour. Alabama Acad. Sci.* 4: pp. 24–25. PARASITES (PROTOZOA): Hesse, 1909, *Arch.*

Zool. Paris, Ser. **5, 3.** Tolosani, 1916, *Monit. Zool. Italiano* **27**: p. 217. Bhatia & Chatterjee, 1925, *Arch. Protistenk.* **52.** Cognetti, 1925, *Monit. Zool. Italiano* **36.** Setna, *Parasitology* **19.** (NEMATODA): Timm, 1959, *Pakistan Jour. Sci. Res.* **11.** PHARYNGEAL GLANDS: Stephenson, 1917, *Quart. Jour. Micros. Sci.* **62**: p. 265. POLYMORPHISM: Gates, 1956 *Evolution* **10**: p. 224. REGENERATION: Liebmann, 1942 *Jour. Exp. Zool.* **91**: p. 379. REPRODUCTION: Omodeo, 1952 *Caryologia* **4**: p. 259. SPERMATOGENESIS: Cognetti, 1925 *Boll. Zool. Mus. Univ. Torino* **49**, 33: p. 2. SEPTAL SPHINCTERS: Bahl, 1919, *Quart. Jour. Micros. Sci.* **64**: pp. 104, 111. TYPHLOSOLE: Hertling, 1923, *Zeitschr. Wissensch. Zool.* **120**: p. 168.

Octothecal, pores minute, superficial, *ca.* ¼C apart or a little more, at 5/6–8/9. Male pores, in xviii, minute, superficial, each in a small, circular to transversely elliptical disc. Female pore, median. Genital markings small, circular to shortly elliptical discs with central translucence, presetal, in some or all of vi–ix, just median to spermathecal pore levels and/or about in *BC*, postsetal, in some or all of v–viii and just in front of spermathecal porophores, occasionally one or more near each male porophore in xviii. Clitellum, occasionally with a few setae externally recognizable ventrally in xvi, in xiv–xvi, occasionally failing to reach 13/14 and/or 16/17, occasionally extending slightly into xvii and/or xiii in which case 13/14 and/or 16/17 are obliterated, sites of dorsal pores at those levels still recognizable though pores are not functional. Setae, size decreases laterally from *a* to *d* in iii–xiii, and in iii–vi *a, b,* or *c* follicles protrude conspicuously into the coelom, more widely separated in ventrum of iv–vii, enlarged setae ornamented ectally, sometimes smaller in x than in xi and posteriorly, 21–28/iii, 26–46/viii, 39–46/xii, 39/54/xx, vi/6–11, vii/8–14, viii/10–16, xvii/13–18, xviii/8–16, xix/12–17. First dorsal pore, occasionally at 10/11 or 12/13, usually at 11/12. Prostomium, epilobous, tongue open. Color, only in dorsum except in first few segments, reddish, almost purple, reddish brown, yellowish brown, light brown, chocolate, slate, grayish, often lacking in narrow equatorial strips. Segments, 79–121 but usually 105–118. Size, 45–170 by 3–6 mm.

Septa, 8/9–9/10 aborted, 5/6–7/8 and 10/11–11/12 or 12/13 slightly muscular to muscular (5/6–7/8 occasionally thickly muscular?). Pigment, in circular muscle layer, reddish brown. Intestinal origin, occasionally in xv (?), usually in xvi. Caeca, simple, in xxvii–xxii. Typhlosole, simply lamelliform from xxvii and ending in region of 71st–86th segments. Hearts, of v–vii (when complete) lateral, in viii unaborted dorsal portions to gizzard, in ix (left or right) lateral, in x usually aborted (?), in xi–xiii lateroesophageal. Blood glands, in iv–vi (vestigial?), and in an iridescent collar that sometimes is unusually well developed, behind the gizzard. Lymph glands, in intestinal segments. Holandric. Testis sacs, unpaired and ventral. Seminal vesicles, xi, xii. Pseudovesicles, in xiii, xiv. Prostates, in xvi-xxii,

ducts muscular, to 6 mm. long, looped. Spermathecae, duct slender, usually shorter than but often not distinctly demarcated from the ampulla, diverticulum from anterior face of duct at parietes, slender, with long stalk and small, usually spheroidal to ellipsoidal seminal chamber. GM glands, stalked, coelomic, bound down to parietes or retained within body wall, composite.

Reproduction : Parthenogenetic throughout most of the range, in part at least because of male sterility. Spermatogenesis was found to be aberrant and seemingly is prematurely terminated. Almost no records of presence of mature and presumably normal sperm on the male funnels or in the spermathecae now are available.

Breeding may be possible at any time as the species seems to remain active throughout the year whenever conditions are favorable.

Distribution : Thandaung, Kalaw, Htamsang, Taungyi, Mogok, Manchio, Namkham (Shan Plateau). Bhamo, Namkhai (Bhamo). Putao, Kankiu, Sumprabum, Hpungin Hka, Hpunchan Hka, Tutuga, Lawanga (Myitkyina). Tengyueh (Yunnan). Unknown in tropical lowlands.

Rangamati (Chittagong Hill Tracts). Imphal, Cherrapunji (4,300 feet), Dumpep (Khasi Hills), Sadiya, Kobo (Abor country), Myntaung Valley, Shillong, Assam.

Yunnan, Szechuan, Fukien, Kiangsi, Kiangsu, Chekiang, Anhwei Provinces, China. Hong Kong, Hainan, Taiwan, Korea, Japan, Okinawa[17]. In Szechuan, at elevations of 1,100–12,000 feet.

Sumatra, Java (elevations unknown). Philippines, (Mt. Province, only). Australia, New Caledonia, New Zealand. Fiji (elevations unknown) and Hawaiian Islands.

United States (Oregon, California, Nebraska, Arkansas, Texas, Louisiana, Tennessee, Mississippi, Alabama, Florida, Georgia, South Carolina,[17] North Carolina, Virginia, Pennsylvania, New York, Connecticut, Maine. Mexico, Guatemala,[18] Costa Rica (at 250–2,000 m.), Salvador (at 700–1,965 m.) Panama,[17] San Domingo, Trinidad,[18] Colombia, Peru, Brazil, Argentina.[17]

Wales, England, Scotland, Holland,[18] Sweden,[18] Germany, Poland, Russia, Italy, Sardinia, France, Portugal, Azores.

Egypt, Rhodesia, Natal, Transvaal, Cape Province (South Africa), Anjouan. Cape Verde Islands. Madagascar. St. Thomas. St. Helena.

Adam's Peak, Peradeniya (in the hills, Ceylon).

Bababudan Hills (Mysore), Naduvatam, Kotagiri, Benhope, Fern Hill, Coonoor (Nilgiri Hills), Kodaikanal, Tiger Shola (Palni Hills), South India.

---

[17] First record for the country.
[18] Based on previously unpublished records of interceptions by the U. S. Bureau of Plant Quarantine and identifications by the author.

Punjab and Northwestern Frontier Province, Bang la Desh. Simla, Naini Tal (Western Himalayas), Nepal, Darjiling and vicinity, Gangtok (Sikkim), to 7,000+ feet, Eastern Himalayas. Elevations, in South India, 4,500–7,500 feet. All lowland records for the Asiatic mainland, 500 feet and up are for places north of the tropic.

The species was found in earth brought directly to Holland from Java (1878), to a Berlin botanical garden from San Domingo and Peru (1892), to Honolulu from China (1900), to German botanical gardens from the West Indies (1901), such dates indicating only time of publication rather than of interception which may have been earlier. During the last fifteen years, twenty specimens, intercepted by the U. S. Bureau of Plant Quarantine and identified by the author, were from Japan, Australia, New Zealand, Hawaii, Trinidad,[18] Guatemala, Colombia, Azores, England, Portugal, Holland,[18] Sweden.[18] Many of the associated plants were such as to have been from greenhouses or conservatories.

A study of records of transport and introduction of live plants presumably can provide useful estimates as to dates of introduction to various regions for this as well as other earthworms. Meanwhile the oligochaete literature provides for *P. diffringens*, publication dates prior to which, and perhaps by a period of twenty years or more, introduction could have been made: Mauritius–1834, England–1849, California and Hawaii–1852, Wales and St. Helena–1869, France–1870, Java–1878, Australia and hill stations of South India–1886, Scotland–1890. By 1896, the species already had been recorded from each of the 5 major islands of Hawaii. The first record for Australia, 1886, was of presence in earth of plant pots from a Sydney nursery that was believed to have received the worms from Mauritius.

Such evidence together with that provided by the habitat records and by the overseas gaps proves that distribution as presently known is largely a result of human activity and also suggests that *P. diffringens* may have been transported more widely and more often than the better known lumbricids. Restriction of importation only to those areas from which the species now is known, in view of what has been recorded about plant transport, is unlikely. Escapes and expulsions from greenhouses and conservatories must have been very numerous during the last century and more during which period exhausted earth must have been discarded at intervals of several years. Failure to colonize outdoors in New England, certain European countries, the tropical lowlands of India, etc., accordingly is significant.

The original home of the species is to be sought in the northern portion of the *Pheretima* domain and perhaps in China, but how much of the range in that country, Korea, Japan and neighboring islands was self-acquired remains to be determined.

Habitats: Soil, of open places, jungles, potato fields, earth with large admixture of bark in grounds of a mill. Moist river bottom forest (Louisiana). Deep woods (Virginia). Muddy pool and stream banks. Under stones and logs, in rotten wood, compost, manure and manured gardens, pit with compost and cow dung, decaying rubbish. Axils of plantain leaves. Caves (Alabama, North Carolina, Japan). Earth with potted plants. Greenhouses, and there only: Oregon, Illinois (Chicago, Urbana), Maine (6 greenhouses, in 3 towns, of a small central part of the state), Wales, England, Scotland (Denmark and Sweden?) Germany, Poland, Russia (Holland?), France. Botanical gardens, Italy, Portugal, Connecticut (New Haven), but such records certainly are imprecise as to habitat and the worms may have been secured in associated forcing houses. Earth in vicinity of or near a greenhouse and there only, Nebraska. Indoor culture bed of an earthworm farm (Maine). Outdoor culture beds of earthworm farms, Arkansas, Florida.

Biology: *P. diffringens* was said to be absent in Scottish greenhouses with a temperature below 60°F but to be present in those at or above that degree. The warmer the house, the livelier the worms and the more readily autotomized.

An ability, by a sudden contraction, to throw itself considerable distances relative to its size, attracted attention in various parts of the world and resulted in the winning of vernacular names such as: snakeworm (St. Helena), crazy-worm (Brazil), black wriggler (Alabama).

Individuals of this species were believed to suffocate comparatively quickly in water (but see Habitats) and to die after a few days in wet clay or marly clay soil. The same author (Vaillant, 1872, *Ann. Mag. Nat. Hist.* 9: p. 372) said he observed contraction on proper stimulation by extremities (still with normal appearance) though a middle portion of the worm already was decayed. Such a death gradient, if normal, seems to be the reverse of one recorded (Gates, 1960, *Bull. Mus. Comp. Zool. Havard College* 123: p. 211) for *Perionyx sansibaricus*.

Variation and Abnormality: Divergence from usual as to number of female pores in xiv has been recorded for parthenogenetic morphs of *P. diffringens* and also for biparental species but without significant information as to frequency. More variation as to intestinal origin is indicated by the literature than the author encountered in his own studies. Determination of the character is handicapped by the fact that the superficially apparent condition often is not real. Wide separation of paired testis sacs (recorded for some *diffringens* morphs) has not been noted in biparental species with unpaired, subesophageal sacs and is believed to be associated with the uniparental reproduction. Comment here on those organ modifications and deletions that are standard in parthenogenetic evolution is unnecessary.

Abnormalities characterizing two Teesta Valley specimens (Gates, 1937: p. 200) now probably could be shown to result from splitting of mesoblastic somites.

Polymorphism: A biparental morph (H) characterized as in the précis above never has been recognized. All individuals of *P. diffringens* that were studied now are believed to have been of parthenogenetic morphs. An athecal morph is unknown. Two in a series of 60 South Indian specimens from Kodaikanal lacked the posterior pair of spermathecae. No other record of spermathecal deletions has been found. Reports of spermathecal abnormality also are rare. The spermathecal battery accordingly seems to be unusually resistant to parthenogenetic modification. The male terminalia, on the contrary, are susceptible and second order intermediate morphs are numerous. Although individuals of such morphs are common, absence of male porophores is unrecorded. If an R stage has been reached in *P. diffringens*, specimens of it have not yet been recognized. H_p morphs also are common.

Regeneration: Tail regenerates were seen at various levels behind 57/58. Maximal number of segments in a regenerate was 12 at 58/59. Minimal number was 2 at 111/112. Tail regeneration began simultaneously (Liebman, 1942) in all individuals operated on at various times during December–February.

Economic importance: Presence of *P. diffringens* in greenhouses and in the earth with potted plants has been considered disadvantageous. The worms were said to wander widely, presumably at night, to multiply rapidly until all plant beds were infested and finally to choke drainage pits.

Parasites: SPOROZOA: *Apolocystis michaelseni* (Hesse, 1909), Italy. *Choanocystoides costaricensis* Cognetti, 1925 and testis sacs, Costa Rica. *Dirhycocystis globosa* (Bhatia & Chatterjee, 1925), Bombay. *Grayallia quadrispina* Setna, 1927, coelomic cavities, Bombay. *Monocystis michaelseni* Hesse, 1909, coelom, Italy. *Nematocystis hessei* Bhatia & Chatterjee, 1925, Lahore. *N. lumbricoides* Hesse, 1909, Lahore.[19] NEMATODA: *Synoecnema anseriforme* and *S. hirsutum* Timm, 1959, coelomic cavities, Bangla Desh.

Pseudovesicles were unusually large in a worm with a heavy infestation of nematodes. Prostates were of normal size and appearance.

Remarks: Two Formosan worms were suspected (Gates, 1959, *American Mus. Novitates*, No. 1941: p. 7) of being "close to, if not identical with, the ancestral hermaphroditic morph from which the parthenogenetic previously-known strains of *P. diffringens* diverged since acquisition of ability to reproduce asexually." Male porophores differed from those of *P. diffringens* in the same ways as those of subse-

[19] *Rhynchocystis awatii* Bhatia & Seta, 1926, India. (Listed by Loubatières, 1955: p. 194.)

quently studied male sterile morphs from Hawaii. While the Formosan specimens had copulated, the male porophores now do not seem to be quite normal. Whether a different species or an early stage of uniparental modification (perhaps in association with facultative parthenogenesis) was involved is unknown. The only other differences recorded from *P. diffringens* morphs were in the setal numbers e.g., 32(?)–42/iii, 64–66/viii, 60–78/xii, 64–75/xx, 81/1xxiv, 72/xc, vi/17, vii/20–22, viii/20, xviii/12–14, which also were as much greater than in the Hawaiian worms with similar male pore regions. Several "species" differ from *P. diffringens*, according to our small present knowledge, almost only by the greater number of metameres and of setae per segment.

Pheretima doliaria

1931. *Pheretima doliaria* + *P. referta* Gates, *Rec. Indian Mus.* **33**: pp. 374, 405. (Type localities, Mong Ko. Types, none.)
1932. *Pheretima doliaria* + *P. referta*, Gates, *ibid.*, **34**: pp. 415, 406.
1933. *Pheretima doliaria*, Gates, *ibid.*, **35**: p. 524.

Quadrithecal, pores minute, superficial, widely separated, at or just behind 7/8–8/9. Male pores, in xviii, minute, superficial, each in a fairly large porophore of variable size and shape. Female pore, median. Genital markings, except for male porophores and glabrous areas sometimes in vicinity of male and spermathecal pores, lacking. Clitellum, setae unrecognizable externally, xiv–xvi. Setae, 50–68/xx, circles without regular median gaps, viii/18–26, xvii/10–24, xviii/3–17, xix/11–20. First dorsal pore, at 12/13. Prostomium, epilobous, tongue open. Color, reddish to reddish grey, slate anteriorly, reddish brown to brownish posteriorly (pigment red and in circular muscle layer?). Segments, 109–135. Size, 81–129 by 4–6 mm.

Septa, 8/9–9/10 nearly always, 10/11 occasionally, aborted, none especially muscular, or 5/6–7/8 muscular to thickly muscular. Intestinal origin (in xv?), gut very gradually widening until region of xx–xxiii where full intestinal width obviously is attained. Caeca, simple, in xxvii–xix. Typhlosole, large, from xxvii. Hearts, of ix (left or right) lateral, of x–xiii (latero?)–esophageal. Holandric. Testis sacs, unpaired (occasionally paired?), ventral. Seminal vesicles, large, in xi, xii. Prostates, large, in xvi–xxiii, ducts 4–7 mm. long, a middle or more ectal portion very much thickened and muscular. Spermathecae, often in vii–viii, duct small, markedly narrowed in parietes, coelomic portions bulbous with muscular wall but lumen fairly large, diverticulum from median face of duct at parietes, with slender but muscular stalk straight or looped, seminal chamber pear-shaped to ellipsoidal or elongate and then looped.

Reproduction: Presumably biparental as sperm are exchanged during copulation. Sperm maturation appears to be profuse. Abortion of 10/11 enables anterior vesicles to surround the gizzard. Seminal vesicles and prostates, on first opening the worm, appear to fill coelomic cavities all the way from viii through xxiii.

The breeding season includes October or a later portion of it and November but probably is earlier in manure.

Distribution: Loikaw (Karenni). Taungyi (Shan States)? Mong Ko (Kengtung). Pang Wo (Mang Lun). Nam Shi Pan, Nam Hpen Noi, (Mong Mong, Yunnan). At elevations of 3,500–5,500 feet. Known to be present in many localities on the Shan Plateau that cannot now be listed.

An uncertain Taungyi record, for immatures, remains unconfirmed.

Habitats: Soil. Manure.

Variation: Even in the 23 specimens secured before 1932 considerable variation had been seen though epidermal erosion and frequency of amputation obviated recording some of it. Male pores, for instance were near posterior ends of longitudinally elliptical and anteromesially convergent porophores, or at centers of smaller circular porophores, in lateral parts of mesially facing C-shaped areas. Additional markings were lacking or were perhaps represented by glabrous areas near spermathecal pores, or in xviii and xix and there without setae. Spermathecal pores were exactly at intersegmental levels or behind though only slightly. Septum 10/11 was present or aborted. Hearts of x were thought to be lacking in some specimens but may only have been unrecognized because of lack of blood therein. Pairing of testis sacs supposedly was seen but requires confirmation from better material. Spermathecal diverticula were looped or lacked the looping. If *P. referta* now is correctly synonymized, even the number of spermathecae varied.

None of the variation recorded, and much never was published, is now believed to have been associated with parthenogenesis. *P. doliaria*, accordingly, varies with respect to nearly every character by which pheretimas have been defined.

As a pair of spermathecae now is known to be lacking in an occasional individual of species with biparental reproduction, without indications of abnormality, somite abortion or sterility, *P. referta*, only known from the holotype, seemingly should be synonymized. Further support was provided by the variation in spermathecal diverticula of the quadrithecal individuals.

Parasites: NEMATODA: *Siconema microrum* Timm, 1966, Nam Hpen Noi. *Succamphida robustum* Timm, 1966, Nam Hpen Noi. Gregarines and their cysts, so frequently massed in coelomic cavities of other species (especially perhaps in parthenogenetic individuals) had not been found up to 1933.

Remarks: Dehiscence of setae, sometimes with

abortion of follicles, was noted in the ventrum of viii, in circles of xvii and xix just in front of and just behind male porophores, but not in ii (even in the dorsum) of any specimen.

If the narrowing of the gut in xv–xxiii results from compression by the large prostates as individuals become mature, the real segment of intestinal origin presumably can be determined from juveniles as soon as they are identifiable.

Whether the glabrous areas should be considered genital markings and if so what systematic importance they might have was not determined. Soft and nonmuscular tissue, in vicinity of parietal portions of prostatic ducts, may have been glandular but nothing is known about the nature of genital-marking-like areas that included the male porophores.

The much thickened and muscular portion of the prostatic duct may function as an ejaculatory bulb and perhaps much more efficiently than sperm-duct bulbs of the octochaetid *Eutyphoeus*.

Pheretima dolosa

1932. *Pheretima dolosa* Gates, Rec. Indian Mus. 34: p. 443. (Type locality, Teung Cong. Types, none.)

Sexthecal, pores minute, superficial, just behind intersegmental furrows, in vi–viii. Male pores, in xviii, minute, superficial, each in a transversely elliptical disc. Female pore, median. Genital markings, paired, transversely elliptical, 2 intersetal intervals wide, 4–6 intersetal intervals median to levels of spermathecal pores, and 5–6 intervals apart mesially, presetal in vii–viii, circular and almost in contact at mV, presetal in xviii, transversely elliptical, slightly smaller, more widely separated, postsetal in xviii. Clitellum, setae unrecognizable externally, xiv–xvi. Setae, lacking dorsally in ii, 7/ii, 42/xx, vi/19, vii/18, xvii/15, xviii/6, xix/15. First dorsal pore, at 11/12(?). Prostomium, epilobous(?). Color, only in dorsum, brownish (pigment red and in circular muscle layer?) (Segments?) Size, 84(+?) by 4 mm.

Septa, 8/9–9/10 aborted, 5/6–7/8 and 10/11–11/12 muscular. Intestinal origin, in xv. Caeca, simple, in xxvii–xxiv. (Typhlosole?) Hearts, of ix lateral on both sides, of x–xiii (latero?)–esophageal. Holandric. Testis sacs paired, separated, ventral. Seminal vesicles, large, in xi, xii. Prostates, in xvii–xx, ducts 2½ mm. long, in a U- or a C-shape, thicker ectally. Spermathecae, duct longer than ampulla, slender, narrowed in body wall, diverticulum from anterior face of duct at parietes, with short seminal chamber and longer stalk. GM glands, sessile on parietes.

Reproduction: Presumably biparental though the only evidence that now can be cited in support is the large size of the seminal vesicles.

Distribution: Teung Cong (Mong Mong, Yunnan).

Remarks: Known only from the type which was a posterior amputee.

Pheretima elongata

1872. *Perichaeta elongata* Perrier, *Nouv. Arch. Mus. Hist. Nat. Paris* **8**: p. 124. (Type locality, Peru. Types, in the Paris Mus.)

1926. *Pheretim aelongata*, Gates, *Ann. Mag. Nat. Hist.*, Ser. 9, **17**: p. 444. *Jour. Bombay Nat. Hist. Soc.* **36**: p. 183; *Jour. Burma Res. Soc.* **15**: p. 208; *Rec. Indian Mus.* **28**: p. 153.

1929. *Pheretima elongata*, Stephenson, *ibid.*, **31**: p. 237.

1930. *Pheretima elongata*, Gates, *ibid.*, **32**: p. 309.

1931. *Pheretima elongata*, Gates, *ibid.*, **33**: p. 378.

1932. *Pheretima elongata*, Gates, *ibid.*, **34**: p. 391.

1933. *Pheretima elongata*, Gates, *ibid.*, **35**: p. 525.

1936. *Pheretima elongata*, Gates, *ibid.*, **38**: p. 413.

1939. *Pheretima elongata*, Gates, *Jour. Thailand Res. Soc. Nat. Hist. Suppl.* **12**: p. 87.

1955. *Pheretima elongata*, Gates, *Rec. Indian Mus.* **52**: p. 86.

1960. *Pheretima elongata*, Gates, *Bull. Mus. Comp. Zool. Harvard College* **123**: p. 256.

1961. *Pheretima elongata*, Gates, *American Midland Nat.* **66**: p. 62. *Burma Res. Soc. 50th Anniv. Pub.* No. 1: p. 57.

1970. *Pheretima elongata*, Soota & Julka, *Proc. Zool. Soc. Calcutta* **23**: p. 202.

CYTOLOGY: Golgi elements and mitochondria of intestinal epithelium and in eggs. Kamat, 1956, *Proc. Indian Acad. Sci.* **44**: p. 95. HEPARIN: See metachromasia. INTESTINAL PROTEINASE: Kamat, 1955, *Jour. Animal Morph. Physiol.* **2**: p. 79; 1957, *ibid.*, **4**: p. 60. METACHROMASIA: Of ovarian ova. Joshi & Kamat, 1961, *Current Sci.* **30**: p. 422. PARTHENOGENIC POLYMORPHISM: Gates, 1956, *Evolution* **10**: p. 224. ECONOMIC IMPORTANCE: Puttarudriah & Sastry, 1961, *Mysore Agric. Jour.* **36**: p. 4. VARIATION AND ABNORMALITY: Gates, in MS.

Polythecal, pores minute, superficial, in paired groups of 2–5 at 5/6–6/7. Male pores, in xviii, minute, each in a small disc on median wall entally of a deep, eversible parietal invagination with a longitudinally crescentic aperture. Female pore, median. Genital markings, transversely elliptical, 3–4 intersetal intervals wide, presetal, widely paired, in some or all of xix–xxiv. Clitellum, setae may be retained ventrally in xiv and xvi, xiv–xvi. Setae, circles with a regular midventral gap, *a* and *b* enlarged in iv–xiii, *a* enlarged posteriorly, 50/ii, 50–66/iii, 81/v, 67–104/viii, 54–80/xii, 55–75/xx, vi/13–17, xvii/12–17, xviii/7–15 (continued into male pore invagination on median wall), xix/12–15. First dorsal pore, at 12/13. Prostomium, rudimentary or lacking, Color, none. Segments, 136–297. Size, 75–300 by 3–6 mm.

Septa, 8/9–9/10 aborted, 5/6–7/8 thickly muscular. Intestinal origin, in xv. (Caeca, none.) Typhlosole, simply lamelliform, ending in region of 116th to 137th segments. Ventral blood vessel, incomplete, usually aborted between hearts of ix and 7/8. Hearts, of viii usually completely aborted, of ix lateral and usually present on both sides, of x esophageal often receiving a vessel passing vertically up to it from the extra-esophageal, of xi–xiii latero-esophageal but often lacking or abnormal in xiii. Holandric. Testis sacs unpaired and annular, seminal vesicles of xi, hearts of x–xi and usually the dorsal blood vessel included. Seminal vesicles, small, in xi, xii. Pseudovesicles, of xiii larger than the seminal vesicles, in xiv

often lacking or too rudimentary to be recognized. Prostates, in xvi–xxi, ducts 2–6 mm. long, each in a hairpin loop with the ectal limb much thicker. Spermathecae, small, duct shorter than the ampulla, diverticulum from duct at or near parietes with ovoidal to ellipsoidal seminal chamber and a longer stalk. GM glands, sessile on the parietes.

Reproduction: Probably parthenogenetic because of male sterility in athecal and various intermediate morphs, in many if not most of the areas to which the species has been introduced.

Distribution: Mount Harriet, Minnie Bay, Wrightmyo, Rajatgarh (Andaman Islands). Pyapon (Pyapon). Moulmein (Amherst). Maubin (Maubin). Nyaunglebin, Pazunmyaung (Pegu). Henzada (Henzada). Toungoo (Toungoo). Nyaungnyo (Prome). Sandoway, Ngapoli (Sandoway). Akyab (Akyab). Mount Popa (Myingyan). Mandalay, Tonbo, (Mandalay). Sagaing (Sagaing). Nawng Khaw (Mang Lun) and Mongnai (Southern Shan States). Hsipaw, Maymyo (Northern Shan States). Mogok (Katha). Mawlaik (Upper Chindwin). Bhamo (Bhamo). Myitkyina, Lonton (Myitkyina). From sea level to elevations of *ca.* 3,500 feet.

Bangkok, Chiengmai (Thailand). Malay Peninsula. Formosa.

Sumatra, Java, Celebes, Philippines, New Caledonia, Sumba, Salibaboe, Kei, Flores, Truk, Yap, Coloni, Palao, Bonin, Okinawa, Riu Kyu, and Hawaiian Islands. Puerto Rico, Haiti, Cuba, Panama, Venezuela, Peru, French Guiana, Guyana, and Surinam.

Egypt. Madagascar. Comores.

India (lowlands of tropical portions). Pakistan (Karachi). Ceylon.

Possibly as yet unrecognized in its presumed home somewhere in a region including Borneo, Celebes, and the Philippines.

Specimens from Kobana-jima, near Iriomote, Riu Kyu Islands, were referred (Ohfuchi, 1956, *Jour. Agric. Sci. Tokyo* **3**: p. 148) to *P. elongata*. Genital markings, male porophores, muscularization of 8/9, intestinal origin, as well as the intestinal caeca, disprove the identification. Specimens referred to *P. biserialis* (*ibid.*, p. 151) probably were of *P. elongata*.

Habitats: Black mud. Soils: red, black cotton soil (India, Burma), black taro soil (Hawaii), taro fields, in yard with animal feces of a dairy farm, gardens (Hawaii, several times). Under rubbish and compost, sometimes down to a depth of 12–18 inches below the surface. Hog lots of several breeding stations, centers or stock farms (Philippines). Botanic gardens.

Biology: Year round activity in favorable situations seems probable, perhaps along with year round breeding.

Number of segments in juveniles 16–54 by 1½–3 mm., the smallest available, was 136–281, the latter number in a worm 50 by $2\frac{1}{2}$ mm. No evidence indicative of post-hatching production of new metameres has been recognized.

Castings: When found on the surface, are cords, about 2 mm. thick and 20–40 mm. long, or in irregular piles 10–20 mm. high and 20–35 mm. wide.

Variation, Abnormality and Polymorphism: Distinguishing individual variations from induced ontogenetic abnormality and such in turn from changes made during evolution of parthenogenetic polymorphism is a matter of some difficulty in *P. elongata*. Organ and regional homoeoses, including one specimen from Guyana comparable to the type of *Allolobophora* "*relictus*," could have resulted from suppression of or halving of embryonic somites. A complete ventral vessel though already slightly narrowed between hearts of ix and 7/8, in the three smallest juveniles that were obtainable, enables attribution of absence of that part of the ventral trunk in older worms to a post-hatching abortion. Whether that abortion is usual in a biparental population or has appeared since reproduction became parthenogenetic has not yet been determined. Conditions of the vascular system in xiii were for a time thought to show that a pair of hearts in xiii was being eliminated. Perhaps equally good reasons could be adduced for thinking that absence in xiii is primitive and that a pair is now being added in various strains. However, the direction of evolution is to be read, the "blueberry" organs certainly seem to be abnormal and of a sort that might be induced during development. How much of the variation in relationship of the hearts of x with a commissure between the extra- and the supraesophageals is real or due to inability to recognize vessels when empty also remains to be determined.

The most common morph, in the author's experience, is the A. Much less common are the numerous kinds of first order intermediate morphs with one to three spermatheca in one or two, occasionally even three, of the four ancestral locations. Very rare indeed is a symmetrical reduction of each of the four spermathecal groups to a single organ. The classical characterization of the spermathecal battery as usually of "two pairs" (*cf.* Stephenson, 1923: p. 299) presumably was based on assumption that a complete battery ought to be quadrithecal. If along with the spermathecal battery, conditions of the vascular system in xiii and in x, are taken into consideration literally hundreds of morphs could be listed.

The maximal number of organs in a group of spermathecae, as *elongata* now is known, is five, but if, as the author long has suspected, *stelleri* (Michaelsen, 1891) is a synonym, the maximal number in a group can be as many as 17. In that case, the morph characterized in the précis above may be an H_p in which each group of spermathecae already has undergone considerable reduction in number.

Regeneration: Few tail regenerates have been recognized and amputation in spite of the length may be rare. However, one tail regenerate has been recorded, and at 98/99. The typhlosole in that worm ended in the 70th segment as a result of being lysed through 18 segments at least.

Economic importance: *P. elongata*, along with *Pontoscolex corethrurus* (Pattarudriah & Sastry, 1961, *Mysore Agric. Jour.* **36**: p. 4), was found to have rendered a South Indian soil compact, hard, cloddy, thus hampering digging, preparation of the soil, rooting of vegetable crops, and normal percolation. Severe seepage losses from flooded taro patches on Kauai (Hawaiian Islands), from rice paddies on Formosa from mountain terraces of Ifugao (Philippine Islands), are attributed to worms that were found (Gates, MS) to belong to various parthenogenetic morphs of *P. elongata*. That species constituted only 15.6 per cent of the population in the compacted, Mysore, field soils.

Parasites: SPOROZOA: *Rhynchocystis awatii, mamillata, Stomatophora bulbifera* Bhatia & Setna 1926, seminal vesicles, Bombay. NEMATODA: *Metastrongylus apri* (Gmelin 1790) Vostokov 1905, Philippines. *Synoecnema fragile* Magalhaes 1905, coelom, Rio de Janeiro, Brazil.

Remarks: Relationships clearly are with the classical *P. stelleri* in which Michaelsen (1900) recognized seven subspecies.

Pheretima exigua

Octothecal, pores minute, superficial, at or just behind 5/6–8/9. Male pores, in xviii, minute, superficial, each in a discoidal porophore. Female pore, median. Genital markings, rather small tubercles or discs, usually paired, in vii–ix and xvii–xx. Clitellum, with or without externally recognizable setae, xiv–xvi. First dorsal pore, usually at 12/13. Prostomium, epilobous, tongue open. Color, none (? some sparse pigment in dorsal portion of circular muscle layer?). Segments, 73–103. Size, 24–75 by 2–3 mm.

Septa, 8/9–9/10 aborted, 6/7–7/8 slightly, 10/11–14/15 very slightly strengthened. Intestinal origin, in xvi. Caeca, simple, in xxvii–xxiv. Typhlosole, rather high, simply lamelliform. Hearts, of ix (left or right) lateral, of x esophageal(?), of xi–xiii lateroesophageal(?). (Blood and lymph glands?) Holandric. Testis sacs, unpaired, of x annular and including hearts, of xi cyclindrical (or annular?) including seminal vesicles as well as hearts. Seminal vesicles, small, in xi, xii. Prostates, in xvii–xx, ducts $1\frac{1}{2}$ mm. long, looped or nearly straight. Spermathecae, duct shorter than ampulla, narrowed in parietes, diverticulum from duct at or in body wall, about as long as or longer than main axis, with long and unlooped stalk, shorter ovoidal to ellipsoidal seminal chamber. GM glands, stalked, coelomic (composite?).

Reproduction: Parthenogenesis, in absence of records re sperm maturation and copulation, is anticipated. Seminal vesicles seemingly are retained in a juvenile stage through maturity so that any spermatogenesis would be sparse. Rudimentary prostates and absence of prostates often are associated with male sterility.

Distribution: Eastern Burma.

Remarks: Finding at Bhamo and Myitkyina of four worms with characters of southern forms was thought (1936: p. 401) to contra-indicate recognition of geographical races. Probability of parthenogenesis and a possibility that southern forms had been taken to the north by man just as was the nominate race of *D. longatria* now are believed to warrant recognition of the two subspecies.

P. exigua is close to *P. diffringens*, perhaps also to *P. sigillata* Gates, 1936, from the Malay Peninsula.

Pheretima exigua exigua

1929. *Pheretima minuta* (non Beddard 1901) Gates, *Proc. U. S. Natl. Mus.* 75, 10: p. 18. (Type locality, Lashio. Types, none.)

1930. *Pheretima exigua* (part), Gates, *Rec. Indian Mus.* 32: p. 310. (Excluding Nyaungbinkwin specimens.)

1932. *Pheretima exigua* var. *typica*, Gates, *ibid.*, 34: p. 512.

1933. *Pheretima exigua* var. *typica*, Gates, *ibid.*, 35: p. 525.

1936. *Pheretima exigua*, Gates, *ibid.*, 38: p. 415.

1939. *Pheretima exigua*, Gates, *Jour. Thailand Res. Soc. Nat. Hist. Suppl.* 12: p. 89.

1955. *Pheretima exigua*, Gates, *Rec. Indian Mus.* 52: p. 86.

1960. *Pheretima exigua*, Gates, *Bull. Mus. Comp. Zool. Harvard College* 123: p. 260.

Spermathecal pores, at intersegmental furrows 5/6–8/9. Genital markings, smaller than male porophores, presetal, occasionally unpaired and median in vii, xviii or xix, closely paired in vii, viii, xix, widely paired in one or more of xvii–xx. Setae, occasionally present in ventrum of xvi, 23–29/iii, 32–40/viii, 33–40/xii, 0–8/xvi, 35–46/xx, vi/11–14, vii/11–15, viii/13–16, xvii/10–18, xviii/9–15, xix/11–17. Segments, 86–91. Size, 24–48 by 2–3 mm.

GM glands, coelomic, with long stalks.

Distribution: Mala, Loikaw (Karenni). Kengtung (Kengtung). Kalaw, Taungyi, Maymyo, Lashio and vicinity (Shan States), Nawng Kham, Kat Pang (Mang Lun). Tan Yang (Mong Yai). E Nai (North Hsenwi). Bhamo (Bhamo). Myitkyina (Myitkyina). At elevations of 4,000 feet or more. At Bhamo and Myitkyina probably from lowlands.

Mu'ang Pong on the upper Me Yom (Ban Muang), Thailand.

Habitats: Soils, including red (lateritic?), gravelly, in gardens (some richly manured), or open areas, under bamboo, banyans, in jungles, leaf mould.

Variation: Each spermathecal pore often is in a translucent area or a small tubercle with convex surface. Anteriorly and/or posteriorly in such a porophore there may be a still smaller marking presumably indicating location of opening of a gland. The gland sometimes is stalked and coelomic and the delicate stalk occasionally (but probably incorrectly) was thought to pass into the spermathecal duct. A collar of soft tissue, possibly glandular, occasionally was recognized around an ectal portion of the spermathecal duct.

Records indicative of spermathecal deletions or aberrations are lacking.

Pheretima exigua austrina

1930. *Pheretima exigua* (part) Gates, *Rec. Indian Mus.* 32: p. 310. (Including only Nyaungbinkwin worms.)

1931. *Pheretima exigua*, Gates, *ibid.*, 33: p. 378.

1932. *Pheretima exigua* var. *austrina* Gates, *ibid.*, 34: p. 514. (Type locality, Leiktho Circle. Types, none.)

1933. *Pheretima exigua* var. *austrina*, Gates, *ibid.*, 35: p. 525.

1936. *Pheretima austrina*, Gates, *ibid.*, 38: p. 400.

1939. *Pheretima austrina*, Gates, *Jour. Thailand Res. Soc. Nat. Hist. Suppl.* 12: p. 82.

1955. *Pheretima austrina*, Gates, *Rec. Indian Mus.* 52, p. 83.

Spermathecal pores, just behind 5/6–8/9. Genital markings, larger than male porophores, paired, across one or more of 17/18, 18/19, 19/20, reaching well toward segmental equators, occasionally a pair of (smaller?) presetal markings in ix or one median. Setae, present (usually?) in clitellar segments, 15–21/iii, 24–33/viii, 28–37/xii, –14/xvi, 24–38/xx, vi/9–15, vii/10–15, viii/10–16, xvii/8–13, xviii/6–10, xix/10–12. Segments, 73–103 but perhaps usually 86–103. Size, 33–75 by 2–3 mm.

GM glands, on parietes, stalks confined to body wall(?).

Distribution: Nyaungbinkwin (Mergui). Myittha, Siyigyan, Migyaunglaung, Kyaukmedaung to Kameik, Heimza Basin (Tavoy). Ye, Kya In, Kawkareik (Amherst). Kyaiktiyo, Kumingyaung (Thaton). Ler-mu-htee, Pauk-taw-gwin (Salween). Leiktho Circle (Toungoo). Bhamo (Bhamo). Myitkyina (Myitkyina).

Base of Doi Sutep, at *ca.* 1,100 feet, Thailand.

Habitats: *P. austrina* was obtained in vicinity of larger towns only in the north, at Bhamo and Myitkyina, to which the taxon presumably was taken by man. Although precise records now are lacking, most of the specimens certainly were secured from jungles away from all except small villages.

Parasites: Small, spheroidal, transparent cysts, in four Kamaungthwe River worms, distended testis sacs and seminal vesicles.

Remarks: All or nearly all setae of x occasionally were lacking, a condition that has been reported from other species.

Absence or vestigialization of prostates were not recorded for *P. austrina* but shortness of gland stalks (perhaps usually confined to the parietes) may also be parthenogenetic degradation or elimination.

Pheretima faceta

1932. *Pheretima faceta* Gates, *Rec. Indian Mus.* **34**: p. 422. (Type locality, John Lawrence Island. Types, in the Indian Mus.)
1960. *Pheretima faceta*, Gates, *Bull. Mus. Comp. Zool. Harvard College* **123**: p. 260.

Quadrithecal, pores minute, superficial, anteriorly in viii–ix. Male pores, in xviii, minute, superficial, each at center of a small, transversely elliptical disc. Female pores, closely paired (?). (Genital markings, none.) Clitellum, setae unrecognizable externally, xiv–xvi. Setae, 26/iii, 48/xii, 52–56/xx, viii/16–22, xvii/14–17, xviii/10–15, xix/14–18. First dorsal pore, at 12/13. Prostomium, epilobous, tongue open. Color, only in dorsum, reddish to brownish or slate (pigment red, in circular muscle layer?). Segments, 93–114. Size, 75–113 by 4 mm.

Septa, 8/9 aborted, 9/10 membranous and attached peripherally to 10/11. Intestinal origin, in xv. Caeca, simple, in xxvii–xxiii. Typhlosole, simply lamelliform. Hearts, in ix (on one side only) lateral, in x (latero?)–esophageal, in xi–xiii latero-esophageal. Blood glands, in v–vi. Lymph glands, present in intestinal segments. Holandric. Testis sacs, unpaired and ventral. Seminal vesicles, in xi, xii, fairly large and bulging septa, each with a fairly large primary ampulla of a distinctly different texture. Prostates, small, in xvii–xix, ducts 1 mm. long and nearly straight. Spermathecae, duct shorter than ampulla and not noticeably narrowed in body wall, diverticulum from anterior face of duct at parietes, longer than the main axis, stalk sinuous or zigzag-looped, seminal chamber slightly thicker and of variable length.

Reproduction: Presumably biparental, as sperm are exchanged during copulation.

The breeding season, judging from condition of the material that has been available, includes October–December.

Distribution: John Lawrence Island and vicinity of Port Bonington, North Andaman Island (Andaman Islands).

Worms from Truk Islands were referred (Ohfuchi, 1941, *Studies Palao Tropical Biol. Sta.* **2**: p. 279) to this species but spelled as *"faseta."* Spermathecal pores, male porophores, genital markings, etc., contradict the identification. Absence of diverticula on some spermathecae of the Caroline worms warrants a suspicion that parthenogenesis was involved.

Remarks: Known only from five specimens.

Spermathecal pores of the four types (John Lawrence Island) are nearly midway between intersegmental furrows and eq/viii–ix but on the North Andaman specimen are just behind 7/8–8/9.

Whether two very closely paired female pores or a single median pore characterized this species could not be determined at the time the specimens were examined.

Pheretima feae

1888. *Perichaeta feae* Rosa, *Boll. Mus. Zool. Univ. Torino* **3**, 50: p. 2. *Ann. Mus. Civ. Sto. Nat. Genova* **26**: p. 161. (Type locality Kawkareik. Types, in the Genoa Mus.)
1895. *Perichaeta feae*, Beddard, *A Monogr. of the Order of Oligochaeta* (Oxford), pp. 388, 396, 398, 404.
1900. *Amyntas feae*, Beddard, *Proc. Zool. Soc. London* **1900**: p. 643. *Pheretima feae*, Michaelsen, *Das Tierreich* **10**: p. 266.
1903. *Pheretima "Feae,"* Michaelsen, *Die geographische Verbreitung der Oligochaeten* (Berlin), p. 95.
1909. *Pheretima feae*, Michaelsen, *Mem. Indian Mus.* **1**: p. 110.
1916. *Pheretima feae*, Stephenson, *Rec. Indian Mus.* **12**: p. 335.
1923. *Pheretima feae*, Stephenson (*The Fauna of British India*), *Oligochaeta*, p. 299.
1926. *Pheretima feae*, Gates, *Rec. Indian Mus.* **28**: p. 154.
1932. *Pheretima feae*, Gates, *ibid.* **34**: p. 516.
1945. *Pheretima feae*, Gates, *Ark. Zool.*, Ser. 2, **9**: p. 438.
1960. *Pheretima feai*, Gates, *Bull. Mus. Comp. Zool. Harvard College* **123**: p. 262.
1961. *Pheretima feae*, Gates, *Burma Res. Soc. 50th Anniv. Publ. No. 1*: p. 57.

Octothecal, pores minute, superficial, at 5/6–8/9. Male pores, in xviii, minute, superficial, each in a thick, transversely elliptical disc 1½ intersetal intervals wide. Female pore, median. (Genital markings, none.) Clitellum, setae unrecognizable externally, extending into xiii and xvii, sometimes almost reaching equator of one or both of those segments. Setae, small and closely spaced both dorsally and ventrally, 103/xx, vi/18–25, vii/17–28, viii/18–26, xvii/24–30, xviii/16–33, xix/26–34. First dorsal pore, at 12/13. Prostomium, epilobous, tongue open. Color, in dorsum, almost black but with a metallic sheen, (pigment red and in circular muscle layer?). Segments, 90–160, perhaps usually 130–160. Size, 180–380 by 7–12 mm.

Septa, 8/9–9/10 aborted, 5/6–7/8 muscular, 10/11–11/12 much thickened. Intestinal origin, in xv. Caeca, simple, in xxvii–xxiii. (Typhlosole?) Hearts, in viii unaborted dorsal portions to gizzard, of ix (left or right) lateral, in x (latero?)–esophageal, in xi–xiii latero-esophageal(?). Blood glands, in v and just behind the gizzard in an esophageal collar of 6–8 lobes. Holandric. Testis sacs, unpaired, ventral. Seminal vesicles, large, in xi, xii, the second pair much larger and in posterior pockets of 12/13. Spermathecae, duct shorter than ampulla, narrowed in the parietes, diverticulum from anterior face of duct in body wall, longer than main axis, digitiform, often in part looped.

Reproduction: Probably biparental though records *re* exchange of sperm no longer are available. Large prostates appeared to be functional at maturity and size of the seminal vesicles, especially of the posterior pair, suggests profuse maturation of sperm.

Distribution: Kawkareik (Amherst). Mount Kambaiti (Myitkyina). At elevations of *ca.* 1,000 feet (at Kawkareik) to 2,300 meters (Mount Kambaiti).

Transportation by man presently seems not to have been involved in extending the range. Origin through local establishment of an inornate mutant race of *P. andersoni* in the Dawna Hills, seemingly a possibility when *P. feae* was known only from vicinity of its type locality, now is contra-indicated by a presence so much further north and indeed where *P. andersoni* is unknown. Acquisition of such a range, because of the rugged nature of the country involved, would seem to have required, unless migration originally from the north was aided by river transport, considerable time. Even if a southward migration was so assisted, crossing the Salween and climbing the various intervening ranges to reach Kambaiti, would have required much time, as the available evidence provides little reason for believing that advance normally is other than through rather than over the earth.

The major portion of the *feae* range presumably is in Thailand, Indo-China, and Yunnan.

Biology: No specimens were secured by digging and the species is believed to live at depths greater than trained collectors were willing or able to dig up with their primitive tools. At such depths the species would be geophagous.

Some morning, or perhaps mornings, in October–November and doubtless at end of the breeding season, but exact date variable and seemingly unpredictable, worms are found after sunrise crawling around on the surface. So snakelike in color, appearance, and activity are the animals that coolies refused to gather them even for wages much higher than usual. Worms, according to the local mythology, come to the surface to die and such evidence as could be obtained does seem to indicate that all wander around on the surface, without attempting to get back inside the roil, until killed by heat or the ultraviolet in the sun's says.

At Kambaiti worms probably were sexually mature in May.

Variation: The very dark color with metallic sheen seems to be characteristic. One exception had a light brownish color that usually is associated with a reddish pigment in the circular musculature.

Absence of macroscopically recognizable variation internally as well as externally, in a range that now seems rather large for an endemic pheretima, also is suggestive of considerable age.

Parasites: Gregarines and masses of pseudo-navicellae spores were present in ovaries of some Kawkareik worms but not in enough of them to indicate that parasites were responsible for surface wanderings.

Remarks: The single specimen of *P. feae* available to Stephenson provided much of such information as we have about the nature of the esophageal collar just behind the gizzard.

Caecal origin has been attributed, perhaps mistakenly, to three different segments.

Relationships now appear to be closest with *P. andersoni* from which *P. feae* presently is distinguished only by the absence of genital markings and associated glands as well, of course, as by the characteristic darker color. *P. feae* presumably reached lower Burma so much later than *P. andersoni* that time has been insufficient to get across the Salween to say nothing of the Sittang.

Pheretima fluvialis

1939. *Pheretima fluvialis* Gates, *Jour. Thailand Res. Soc. Nat. Hist. Suppl.* **12:** p. 89. (Type locality, Chiengsen Kao, at bank of the Mekong River. Types, in the U. S. Natl. Mus.)

Octothecal, pores tiny slits, each at center of a transversely elliptical, grayish translucent area, at 5/6–8/9. Male pores, in xviii, tiny transverse slits each at center of a discoidal porophore (each retractile into a slight parietal invagination so as to be covered over by a lateral flap?). Female pore, median. (Genital markings, none.) Clitellum, setae unrecognizable externally, xiv–xvi or slightly into xvii. Setae, 99/ii, 103/iii, 106–118/viii, 106/xii, 101/xx, vi/12–19, vii/14–21, viii/15–22, xvii/15–21, xviii/4–13, xix/10–20. First dorsal pore, at 12/13. (Prostomium?) Color, grayish to brownish gray ventrally as well as dorsally, pigment of postclitellar segments lacking or sparse and in narrow equatorial bands. (Segments?) Size, 365–555 by 6–8 mm.

Septa, 4/5–8/9 thickly muscular, 9/10 lacking, 10/11 and several consecutive septa slightly muscular. Intestinal origin, in xv. Caeca, simple (from xxvii?). Typhlosole, simply lamelliform, from xxvii(?) into region of cl. Hearts, unaborted dorsal portions in viii passing to gizzard, in ix (on right or left side) lateral, in x (?), in xi–xiii latero–esophageal, supra–esophageal bifurcates in x the branches passing to the extra–esophageals (?). Blood glands, in iv–vi. Lymph glands, present. Holandric. Testis sacs unpaired and ventral. Seminal vesicles, small, vertically placed on posterior faces of 10/11, 11/12. Prostates, small, thin, confined to xviii, ducts large, muscular, 10 mm. long, in a *U*-shaped loop. Spermathecae, duct much shorter than the ampulla, diverticulum from anterior face of duct close to parietes, as long as or longer than main axis, with a straight muscular stalk shorter than the duct and a zigzag-looped seminal chamber enclosed in a thick sheath.

Reproduction: Dissected clitellate specimens were said not to be sexual. Prostates were thought to be juvenile, the spermathecae perhaps not fully developed. Spermatozoal iridescence was lacking both on male funnels and in the spermathecae. Testes were said to be undischarged by which was meant retained in a juvenile condition. Such a combination of conditions is now known to be sufficient justification for expecting parthenogenesis.

Breeding season, if clitella were at maximal tumescence, would include January.

Distribution: Chiengsen Kao, Thailand.

Habitats: Mud or sand of banks of the Mekong River.

Remarks: Known only from the original description of the types.

Relationships are with *P. juliani* (Perrier, 1875), known only from the types secured at Saigon, and perhaps less closely with *P. posthuma*.

Pheretima fucosa

1933. *Pheretima fucosa* Gates, *Rec. Indian Mus.* 35: p. 526. (Type locality, in region between Kyaukmedaung and Kameik. Types, none.)

Sexthecal, pores minute, superficial, at 6/7–8/9. Male pores, in xviii, minute, each in a discoidal porophore 3–4 intersetal intervals wide. Female pore, median. Genital markings, paired, transversely elliptical, 9–11 intersetal intervals wide, separated mesially by a space of equal width, just median to male porophores, across 17/18 and 18/19, almost reaching segmental equators. Clitellum, without externally recognizable setae, xiv–xvi. Setae, small, closely spaced, circles usually unbroken except sometimes at mD, 92–110/xx, vii/39–45, viii/38–46, xvii/29–39, xviii/27–33, xix/35–39. First dorsal pore, usually at 12/13. (Prostomium?) Color, in dorsum, red anterior to clitellum, posteriorly lighter red to reddish brown or brownish (pigment reddish and in circular muscle layer?). Segments, 114–115. Size, 98–120 by 5–6 mm.

Septa, 8/9–9/10 aborted, 5/6–7/8, 10/11–12/13, slightly muscular. Intestinal origin, in xv. Caeca, simple, in xxvii–xvii. (Typhlosole?) Hearts, of ix (left or right) lateral, of x (latero?)–esophageal, of xi–xiii latero-esophageal(?). Blood glands, in v, and in a small esophageal collar behind the gizzard. Holandric. Testis sacs, unpaired, ventral. Seminal vesicles, large, in xi, xii, pushing 12/13–13/14 back to 14/15. Prostates, in xvii–xx, ducts 4½–6½ mm. long, each in a U-loop with ectal limb thicker. Spermathecae, duct about as long as ampulla, narrowed in parietes, diverticulum from anterior face of duct at parietes, elongately digitiform, in a flat mass of loops. GM glands, without discrete stalks, protuberant into coelom.

Reproduction: Heavy parasitic infestation, such as characterized the types, is much more likely to be found in parthenogenetic individuals and provides the only justification for suspecting parthenogenesis. However, the condition of the genital organs, so far as the account shows, is more like that to be expected if reproduction is biparental.

Distribution: Northeast of Tavoy, in the vicinity of the Kamaungthwe River, between Kyaukmedaung and Kameik (Tavoy).

Parasites: Small ovoid cysts were numerous, in coelomic cavities of the type, behind xviii. Larger cysts were present (4 paratypes), in a posterior half of the body, laterally on (or in?) the nerve cord. A large nematode 30 mm. long, in coelomic cavities of vii–xiv alongside gut, a small nematode in dorsal blood vessel above the gizzard, two kinds of stalked protozoa on coelomic face of body wall in region of iv–vi were associated, in one earthworm, with six 1½ mm. thick, reddish cysts on ventral parietes of iii–v. NEMATODA: *Siconema ovicostatum* Timm, 1966. Kamaungthwe River region.

Remarks: *P. fucosa* is known only from the original description of six, more or less massively parasitized specimens. Whether the parasitism had resulted in any modification of systematically important structure is unknown as also whether the massive infestation was associated with male sterility and/or parthenogenesis. Relationships with *terrigena*, about which almost as little is known, presumably are close. Material from the fairly high, east-west belt between the ranges (as now known) may enable recognition of another example of north-south speciation.

Pheretima glabra

1932. *Pheretima glabra* + *P. tenellula* Gates, *Rec. Indian Mus.* 34: p. 395, 398. (Type locality of *glabra*, Nam Hpen Noi, of *tenellula*, Kwang Yeh.)

1936. *Pheretima vieta* Gates, *ibid.*, 38: p. 462. (Type locality, Peng Sai. Types, none.)

1960. *Pheretima glabra*, Gates, *Bull. Mus. Comp. Zool. Harvard College* 123: p. 262.

Bithecal, pores very small transverse slits midway between eq/vi and 6/7. Male pores, in xviii, minute and superficial(?), each at posterior end of a longitudinal seminal groove within a distinctly demarcated, elongately elliptical porophore. Female pore, median. (Genital markings, aside from the male porophores none.) Clitellum, setae unrecognizable externally, xiv–xvi. Setae, 44–49/iii, 58–63/viii, 48–54/xii, 39–66/xx, vi/0–5, xvii/0–8, xviii/0–5, xix/4–11. First dorsal pore, at 12/13. Prostomium, epilobous. Color, none (pigment, none). Segments, 114–122. Size, 30–110 by 1½–5 mm.

Septa, 8/9–9/10 aborted, 10/11 membranous or aborted, 5/6–7/8 muscular. Intestinal origin, in xix(?). Caeca, simple, in xxvii–xviii. (Typhlosole?) Hearts, in viii unaborted dorsal portions to gizzard, in ix lateral but usually aborted on one side, of x aborted(?), of xi–xiii latero-esophageal(?). Holandric. Seminal vesicles, in xi, xii. Prostates, in xvii–xxiii, ducts 4½–8 mm. long, variously looped or twisted. Spermathecae, duct slender and shorter than the ampulla but not narrowed in the body wall, diverticulum from duct at parietes and longer than main axis.

Reproduction: Probably parthenogenetic in all specimens for which records still are available.

Polymorphism: The evolution of parthenogenetic polymorphism appears to be well advanced and morphs other than those hitherto recorded are expected.

Thecal morphs

Male porophores, between 16/17 and 18/19, setae lacking between porophores in xviii. Size, 42–65 by 2 mm. Septum 10/11, present. Intestinal caeca, small, extending only through 1–3 segments. Hearts of x (latero?)–esophageal. Testis sacs, unpaired, U-shaped and vertical, including hearts of x, xi and vesicles of xi. Seminal vesicles, small (juvenile?). Prostates, rather small, in xvii–xix.

Reproduction: Presumably parthenogenetic, as no evidence of maturation and reception of sperm was recognized in specimens obtained toward end of breeding season.

Distribution: Peng Sai (Mang Lun), at an elevation of ca. 3,500 feet.

Abnormality: Two diverticula, on opposite sides of one spermathecal duct.

Remarks: Known only from 16 specimens in which an ental portion of each spermathecal diverticulum had been retained in a juvenile condition or else was abnormal. The morphs, if H$_p$, may differ but slightly from the unknown H except perhaps with regard to spermathecal diverticula.

A morphs

1

Male porophores, between eq/xvii and 18/19, setae lacking between the porophores in xvii–xviii as well as behind the porophores in xix. Size, 75–110 by 4–5 mm. Septum 10/11, aborted. Intestinal caeca, long enough to reach forward into xviii–xx. Hearts of x, aborted. Testis sacs, paired, widely separated, ventral. Seminal vesicles, large. Prostates, medium-sized, ducts 4½ mm. long.

Reproduction: Parthenogenesis is anticipated even if sperm are matured (for which records no longer available but note size of seminal vesicles).

Distribution: Nam Hpen Noi (Mong Lem State, Yunnan). At an elevation of ca. 5,500 feet.

Habitat: Manure.

Remarks: Known only from 24 specimens. Epidermis of a midventral region from 16/17 to eq/xix is smooth and glistening but lacks the clitellar coloration and tumescence that is present in individuals of the next morph.

2

Male porophores, between 15/16 and 18/19, setae occasionally present between porophores in xvii but

usually lacking there in xviii as well as behind the porophores in xix. Clitellum, continued ventrally between porophores into xix. Size, 30–77 by 1½–3 mm. Septum 10/11, aborted. Intestinal caeca, long enough to reach forward into xviii. Hearts of x, aborted(?). Testis sacs, paired and ventral. Seminal vesicles, large. Prostates large, in xvi–xxiv, ducts 6–8 mm. long.

Reproduction: Parthenogenesis is anticipated even if sperm are matured (concerning which records no longer are available but note size of seminal vesicles).

Distribution: Kwang Yeh, Pang Wo (Mang Lun). Ang Lawng Mountain, Bana (Mong Lem State, Yunnan). At elevations of 4,500–6,000+ feet.

Habitats: Hard earth in a pine grove. Soil of wet and shady ravine as well as of a dry and sunny hilltop. Wet leaves.

Variation: Paired female pores were present in three specimens.

Remarks: Known only from 49 specimens. Anterior migration of spermathecal pores from 6/7 would seem to be required in an economical phylogenetic derivation of the thecal morphs. However, pores at the equator of vi in *P. youngi* appear to have arrived there by migration back from 5/6 and further dislocation in the same direction would then give the *glabra* position on the anteroposterior but not on the dorso-ventral axis. Location of the pores in the dorsum of *P. sulcata* and *youngi*, so far from the *glabra* site, now seem to contraindicate close relationships. Location of spermathecal pores in parthenogenetic morphs, according to information now available, is the same as in the ancestral biparental morphs.

Pheretima harrietensis

1925. *Pheretima harrietensis* Stephenson, *Rec. Indian Mus.* 27: p. 59. (Type locality, Mount Harriet. Type, in the Indian Mus.)
1932. *Pheretima harrietensis*, Gates, *ibid.*, 34: p. 463.

Sexthecal, pores fairly large(?), transverse slits, ca. ¼C apart, at 6/7–8/9. Male pores, in xviii, minute, each in a small conical papilla protrusible from a slight invagination. Female pore, median (?). Genital markings, small circular discs, in two median patches of ca. 9 irregular transverse rows each, pre- and post-setal in xviii but reaching slightly into xvii and xix. Clitellum, setae present ventrally, in xiv–xvi. Setae, 38/v, 75/ix, 90/xii, 91/xix, vii/24, viii/25, xvii/31, xviii/30. First dorsal pore, at 12/13. Prostomium, epilobous, tongue open. Color, only in dorsum, reddish to slate (pigment, red? in circular muscle layer?). Segments, 123. Size, 200 by 11.5 mm.

Septa, 8/9 aborted, 9/10 membranous and adherent peripherally to 10/11(?). Intestinal origin, in xv. Caeca, simple, in xxvii(– ?). (Typhlosole?). Hearts, of x esophageal (?), of xi–xiii latero–esophageal(?). Lymph glands, in the intestinal segments. Holandric. Testis sacs, unpaired and ventral(?). Seminal vesicles, large, in xi, xii, each with a long primary ampulla. Pseudovesicles, well developed in xiv. Prostates, in xvii–xxi, ducts short, curved, thicker ectally. Penial setae, *ca.* 1 mm. long, straight, ornamented ectally with transverse rows of fine spines, in a follicle just median to each prostatic duct. Spermathecae, duct shorter than ampulla and abruptly narrowed in parietes, diverticulum from anterior face of duct at parietes, much longer than main axis, only slightly dilated at ental end. Glandular collar on parietes around each spermathecal duct. GM glands, stalked, coelomic.

Distribution: Mount Harriet, at 900 feet, South Andaman Island (Andaman Islands).

Remarks: Known only from the type which was in two fragments. The worm was found "under a very large casting on the jungle floor." The alcohol in which the worm was preserved became blackish.

The circumferential furrows bounding the genital markings are deep and the markings sometimes are slightly depressed below the general epidermal level. A similar deep furrow around a discoidal male porophore capable of protrusion in a rather conical shape might be interpreted as an invagination. Whether the external aperture is a primary or second spermathecal pore remains to be determined.

Material accumulated in the saccular dorsal portion of a pseudovesicle of xiv—corpuscles, monocystid spores, nematode ova, setae—presumably would have passed eventually into the coelomic cavity aggregated into a brown body.

Pheretima hawayana

1891. *Perichaeta hawayana* Rosa, *Ann. Hofmus. Wein* 6: p. 396. (Type locality, Hawaii. Type, in the Vienna Mus.)
1912. *Pheretima hawayana*, Stephenson, *Rec. Indian Mus.* 7: p. 276.
1921. *Pheretima hawayana*, Stephenson, *ibid.* 22: p. 760.
1931. *Pheretima hawayana* var. *typica*(?), Gates, *ibid*, 33: p. 382.
1932. *Pheretima hawayana* var. *typica*(?), Gates, *ibid.*, 34: p. 433.
1936. *Pheretima hawayana*, Gates, *ibid.*, 38: p. 417.
1939. *Pheretima hawayana*, Gates, *Jour. Thailand Res. Soc. Nat. Hist. Suppl.* 12: p. 93.
1961. *Pheretima hawayana*, Gates, *Burma Res. Soc. Anniv. Pub. No.* 1: p. 57.

BLOOD GLANDS: Stephenson, 1924, *Proc. Roy. Soc. London,* B, 97: p. 192. CALCIFEROUS GLAND: Stephenson & Prashad, 1919, *Trans. Roy. Soc. Edinburgh* 52: pp. 439, 460. DEVELOPMENT, of excretory system: Bahl, 1922, *Quart. Jour. Micros. Sci.* 66. EXCRETORY SYSTEM: Beddard, 1888, *ibid.,* 28: p. 401. Bahl, 1919, *ibid.,* 64: pp. 103, 111. LIFE HISTORY: Ball & Curry, 1956, *Circ. Bull. Michigan State Univ. Agric. Exp. Sta.* 222. LYMPH GLANDS: Thapar, 1918, *Rec. Indian Mus.* 15: p. 71. PARASITES: Puytorac & Tourret 1963, *Ann. Parasitol. Humaine Corp.* 38. PHARYNGEAL GLANDS: Stephenson, 1917, *Quart. Jour. Micros. Sci.* 62: p. 267. PROSTATES: Stephenson & Ram, 1919, *Trans. Roy. Soc. Edinburgh* 52: p. 439. REACTIONS: Harper, 1905, *Biol. Bull.* 10: 1909, *Jour. Comp. Neurol.* 19. RESPIRATION: Mendes & Valente, 1953, *Bol. Fac. Fil. Ci. Univ. S. Paulo (Zool.).* 18:

p. 92. Mendes & Nonata, 1957, *ibid.,* 21: pp. 153–166. Mendes & Almeida, 1963, *ibid.,* 24: pp. 43–65. VARIATION: Gates, 1965, *Proc. Biol. Soc. Washington* 78.

Sexthecal, pores minute, superficial, *ca.* $\frac{1}{4}$–$\frac{1}{3}$C apart, at 5/6–7/8. Female pore, median. Male pores, in xviii, minute, superficial, *ca.* $\frac{1}{4}$C apart, each in a small transversely elliptical disc. Female pore, usually single and median. Genital markings, usually present, small discs, postsetal, paired and slightly median to levels of spermathecal pores in some of vi–ix, in xviii median to male porophores and in clusters of 1–11. Clitellum, occasionally reaching 13/14, rarely reaching 16/17, dorsal pores occluded and intersegmental furrows obliterated but setae usually present ventrally in xvi, occasionally also in one or both of xiv–xv. Setae, size as well as intersetal intervals increasing through ii–vi especially ventrally, much smaller in vii though somewhat larger ventrally than in viii and posteriorly, 17–21/iii, 36–40/viii, 44–49/xii, 0–16/xvi, 48–56/xx, 47–56/xxx–xci, vi/4–8, vii/10–15, xvii/15–20, xviii/10–16, xix/16–20. First dorsal pore, usually at 10/11. Prostomium, epilobous, tongue wide and open. Color, in dorsum, brownish, reddish brown, yellowish, reddish, almost purple, bright red, slate, grayish. Segments, 70–101, usually 91–98. Size, 56–156 by 3–6 mm.

Septa, 5/6–7/8 and 10/11–13/14 somewhat thickened, 8/9–9/10 aborted. Pigment, reddish brown, in circular muscle layer. Intestinal origin, in xv. Caeca, simple but with several short lobes of ventral and/or dorsal side, in xxvii–xxiv. Typhlosole, rudimentary or even lacking anteriorly, slightly widened but interrupted and with diagonal ridges in some part of xli–lxiv. Hearts, unaborted dorsal portions in viii passing to gizzard, in ix (left or right only) lateral, in x (latero?)-esophageal, in xi–xiii latero-esophageal. Blood glands, in some or all of iv–vi. Lymph glands, present in intestinal segments. Holandric. Testis sacs, usually unpaired, ventral. Seminal vesicles, medium-sized to large, in xi, xii. Pseudovesicles, cordlike or with some dorsal widening and then often with brown debris, in xiii, xiv. Prostates, through some or all of xvi–xxiv, ducts 3–5 mm. long, often with a considerable portion rather thickly muscular. Spermathecae, medium-sized to large, duct usually shorter than ampulla and quite slender but narrowed in parietes, diverticulum from anterior face of duct at parietes, shorter than main axis, slender, slightly and asymmetrically widened entally. GM glands, stalked, coelomic, composite.

Reproduction: Presumably biparental as sperm are exchanged during copulation. Before appearance of the clitellum, according to evidence provided by a number of specimens, copulation may take place.

Distribution: Kawngmu, Namkham, Kutkai (North Hsenwi). Lashio (Shan States). Mogok, (Katha, but on the Shan Plateau). At elevations down to 2,500 feet. (Rangamati, Bangla Desh?)

Imphal, Shillong (Khasi Hills, at 4,500–5,000 feet), Assam.
Fakirapul, Dacca, Bangla Desh. Kurseong, Ramnee, at elevations of 3,000–6,000 feet, India. Lahore, Pakistan. Ceylon, in the hills.
Dor Kiu Ko Ma, at 1,450 m., Thailand.
Singapore (almost at sea level), Kuala Lampur, Malay Peninsula.
China (Yunnan, Szechuan, Kwangtung, Fukien, Chekiang Provinces). Hong Kong, Taiwan, Borneo, Java, Samoa, Tahiti, Fiji and Hawaiian Islands. United States. Mexico (common but no data as to elevations), Guatemala, Salvador. Bermuda. Barbados. French Guiana, Brazil, Uruguay (greenhouses), Argentina, Chile. England, France, Poland, and Russia (greenhouses only). Madeira. Egypt. St. Helena.

Much of the distribution, as known today, clearly is a result of transportation and the greenhouse habitat proves that man has been involved. Introduction to Hawaii and California was before 1852.

In São Paulo, Brazil, *P. hawayana* is (Mendes & Valente, 1953) "the commonest species" in gardens.

The original home of the species is believed to be in China but how much of the distribution (only partly known?) in that country is due to man remains to be determined.

Relationships may be closer to Chinese species with similar genital markings and associated glands than to species with the same number of spermathecae.

Habitats: Under stones. Gardens. Soil of open areas. Greenhouses (Oregon, Illinois, Tennessee, New Jersey, New York, Massachusetts, Maine, Uruguay, England, France, Poland, Russia). Drainage area for sewage—where it is abundant in Dacca, Bangla Desh. Premises of a bait dealer, Jackson, Michigan (bait obtained from Florida).

P. hawayana may have become established in Illinois greenhouses before 1888.

Biology: Although *P. hawayana* gets down into the plains of India and Burma only north of the tropic, and though it appears to have evolved in the temperate zone, the species sometimes is common in tropical lowlands as at Singapore. Colonization outside of greenhouses, in northeastern United States and in Europe, is unknown though introductions must have been numerous. Ability to survive at 6,000 feet in the eastern Himalayas might have been thought to guarantee success in settling places such as New Jersey, Massachusetts, and England. Yet, only in Alabama, Florida, Louisiana, and California has outdoor domicile been achieved. (To which list of states now (1972) can be added, Texas, Mississippi, Georgia, South Carolina.)

Autotomy may be fairly easily induced. The anteriormost level at which self-amputation has been noted is 64/65.

P. hawayana is geophagous.

Variation and Abnormality: For data on and some consideration of both subjects, a recent article (Gates, 1965) may be consulted. Of interest, especially for a species believed to have obligatory biparental reproduction, is the variation in the spermathecal battery.

Regeneration: Four segments can be replaced at the anterior end. Tail regeneration, probably always hypomeric, is possible at levels from 39/40 posteriorly. Maximum number of segments recorded is 22 at 58/59. Only 3–8 segments were differentiated at levels behind 67/68 but as many as 5 were regenerated at 89/90.

Parasites: CILIATA: *Maupasella nova* Cepede, 1901, Lahore. *M. vacuolata* Lom, 1959, Madagascar. GREGARINA: *Apolocystis michaelseni* (Hesse, 1909). coelom, France, Italy. *Monocystis macrospora* Hesse, 1909, coelom, France. *Nematocystis anguillula* Hesse, 1909, France. *Stomatophora coronata* Hesse, 1904, France, Pakistan (Lahore). *S. diadema* Hesse, 1909, France, Pakistan (Lahore), COCCIDIA: *Orcheobius cruzi* Carino & Pinto, 1927, Brazil. MYXOSPORIDIA. *Cariniella carinii* Pinto, 1927, Brazil. MICROSPORIDIA: *Coccospora gatesi* Puytorac & Tourret, 1964, longitudinal muscles, Hawaii.

Nematocystis vermicularis, Stomatophora borealis, coronata, and *diadema* were reported from *P. barbadensis.* Whether the hosts were referable to *hawayana* or to *morrisi* is not ascertainable but *hawayana* is more likely to have been examined.

P. hawayana may have been the species in which Magalhaes (*Arch. Parasitol.* **9** (1905): pp. 306, 310) found numerous cysticercoids, and in the posterior end of which an insect, *Notochaeta aldrichi* Lopes, 1942, completes its larval period in *ca.* 70 hours.

Cysts on body wall similar to those of *Coccospora gatesi* and *Zygocystis pheretimae* were noted commonly in various pheretimas of Burma but never were identified. Cysts were especially numerous in individuals of parthenogenetic morphs.

Remarks: Little can be said as yet about relationships. They may prove to be closest with species having similar stalked GM glands, such as *P. diffringens, morrisi, robusta, loveridgei.*

Cocoons, according to Stinauer (*in lit.* 1951) are round and 3.5 mm. thick, with an average of 1.0 embryos per capsule.

Pheretima houlleti

Sexthecal, primary pores minute and within parietal invaginations opening through transverse slits somewhat less than ⅓C apart at 6/7–8/9. Male pores, in xviii, minute, each on a penial body within a copulatory chamber opening to exterior through an equatorial aperture. Female pore, median. Genital markings, very small, distinctly delimited, circular, translucent areas. Clitellum, xiv–xvi. Setae, often with modified tips, retained at maturity in clitellar segments. Prostomium, epilobous, tongue open. Color, only in dorsum, often lacking or sparse in a narrow equatorial strip, reddish, reddish brown, brownish, slate in preclitellar segments, (pigment reddish brown and in circular muscle layer?).

Septa, 8/9–9/10 aborted, 5/6–7/8 thickened. Intestinal origin, in xv. Caeca, simple, in xxvii–xxii. Typhlosole, present. Hearts, of viii with unaborted dorsal portions to gizzard, of ix lateral but usually aborted on one side, in x(latero?)-esophageal, in xi–xiii latero-esophageal. Blood glands, in v–vi. Lymph glands, in intestinal segments. Holandric. Testis sacs, unpaired, ventral. Seminal vesicles, in xi, xii. Pseudovesicles, in xiii, xiv, quite rudimentary. Spermathecae, duct ectal to diverticular junction with narrow lumen that opens into a parietal invagination without externally recognizable demarcation from the duct itself, diverticulum with short, slender stalk and wider, elongate, seminal chamber that is variously looped. GM glands, long stalked, usually coelomic but sometimes buried in parietes.

Three forms, hitherto regarded as species, are treated herein as morphs of a single species. Two of the taxa were distinguishable by somatic characters but those of the reproductive system, being more easily determinable, had more use. However, some genital characters, such as conformation of penial bodies were not determined by classical specialists. The failure was responsible, at least in part, for the uncertainty involved in some identifications of the *houlleti* complex as well as for present inability to refer various records (including the first for Africa) to any of the currently recognized morphs. All of the genital differences between the morphs, as is now clear from study of other species, may be attributable to presence or absence of parthenogenetically allowed or induced, anatomical modifications. Whether distinguishing somatic characters of one taxon, unfortunately that with the oldest name, are similarly involved in the method of reproduction is unknown. Material that might have provided useful information on the matter was destroyed during World War II and has not been available during the last twenty or more years. Information as to individual variation of somatic characters within the H and the smaller H_p morphs is needed and more especially so from the original homeland of the complex wherever that may be.

Conceivably an amphimictic population ancestral to all the *houlleti* morphs already could have been differentiated into two (or more?) geographical races prior to intervention of parthenogenesis.

In view of the need for confirmation of the relationships herein proposed, pertinent bibliographic references for the morphs are retained separately in hope of avoiding difficulties such as arose from the erroneous synonymization of *B. parvus* and Levinsen's *eiseni* (*q.v.* on previous pages).

The original home of the complex is believed to be somewhere in southeast Asia and may have included an eastern portion of Burma. Frequent transportation by man handicaps recognition of that home.

H morph

1890. *Perichaeta campanulata* Rosa, *Ann. Mus. Civ. Sto. Nat.* **30**: p. 115. (Type locality, Palon. Types, in the Genoa Mus.)

1893. *Perichaeta "Houlleti"* (part), Horst, In: Weber, M., *Zool. Ergeb. einer Reise in Niederländisch Ost-Indien* (Leiden), **3**: p. 64.

1895. *Perichaeta houlleti* (part), Beddard, *A Monogr. of the Order of Oligochaeta* (Oxford), p. 424. (Excluding all except worms with trilobed penial bodies and with penial setae.)

1897. *Perichaeta houlleti* (part), Michaelsen, *Abhandl. Senckenbergischen Naturforsch. Ges.* **21**: p. 234. (Exclusions as above.)

1900. *Pheretima houlleti* (part), Michaelsen, *Das Tierreich* **10**: p. 273. *Amyntas houlleti* (part), Beddard, *Proc. Zool. Soc. London* 1900: p. 613. (Exclusions as above.)

1903. *Pheretima "Houlleti"* (part), Michaelsen, *Die geographische Verbreitung der Oligochaeten* (Berlin), p. 97. (Exclusions as above.)

1909. *Pheretima houlleti* (part), Michaelsen, *Mem. Indian Mus.* **1**: p. 110.

1910. *Pheretima houlleti* (part), Michaelsen, *Abhandl. Naturhist. Ver. Hamburg* **19**, 5: p. 11. (Exclusions as above.)

1923. *Pheretima houlleti* (part), Stephenson, (*The Fauna of British India*), *Oligochaeta*, p. 304. (Exclusions as above.)

1925. *Pheretima wimberleyana* Stephenson, *Rec. Indian Mus.* **27**; p. 62. (Type locality, Wimberleyganj, Andamans. Types, in the Indian Mus.)

1926. *Pheretima houlleti* var. *tortuosa* Gates, *Ann. Mag. Nat. Hist.*, Ser. 9, **17**, p. 454. (Type locality, Rangoon. Types, none.) *Pheretima houlleti tortuosa*, Gates, *Rec. Indian Mus.* **28**: p. 157. *Pheretima "Houlleti"* var. *tortuosa*, Gates, *Jour. Burma Res. Soc.* **15**: p. 209.

1927. *Pheretima campanulata*, Gates, *Ann. Mus. Civ. Sto. Nat. Genova* **52**: p. 230.

1930. *Pheretima campanulata*, Gates, *Rec. Indian Mus.* **32**: pp. 307, 311.

1931. *Pheretima campanulata*, Gates, *ibid.* **33**: p. 373.

1932. *Pheretima campanulata*, var. *typica* + var. *penetralis*, Gates, *ibid.*, **34**: pp. 452, 460.

1933. *Pheretima campanulata*, var. *typica* + var. *penetralis*, Gates, *ibid.*, **35**: p. 511.

1936. *Pheretima campanulata*, Gates, *ibid.* **38**: p. 406.

1937. *Pheretima campanulata*, Gates, *ibid.* **39**: pp. 180, 197.

1955. *Pheretima campanulata* (part), Gates, *ibid.*, **52**: p. 84. (Excluding athecal and partially thecal morphs.)

1956. *Pheretima campanulata*, (H morph), Gates, *Evolution* **10**: p. 220.

1961. *Pheretima campanulata*, (part), Gates, *American Midland Nat.* **66**: p. 62. *Burma Res. Soc. 50th Anniv. Pub. No.* 1: p. 57. (Excluding athecal and partially thecal morphs.)

Genital markings, when present externally, in vicinity of spermathecal pores and near 6/7–8/9. First dorsal pore, at 11/12. Setae, enlarged and more or less irregularly spaced in ventrum of iii–ix, 20–26/iii, 33–50/viii, 44–65/xii, 48–62/xx, vii/11–15, viii/14–19, xvii/11–22, xviii/4–7, xix/10–21, 1–3 penial setae in follicles opening into copulatory chambers. Segments, 92–140 (usually between 110–125). Diameter, 4–7 mm. Length 92–200 mm.

Seminal vesicles, large. Prostates, in xvi–xxi, ducts 5–10 mm. long, each in a C- or U-shaped loop. Copulatory chambers, nearly spheroidal, with GM glands on anterior and posterior faces. Penial body, columnar, with trilobed tip, male pore on middle lobe. Internal genital markings, one on each of two lateral lobes at tip of penial body, others at or near base of penial body or elsewhere on wall of copulatory chamber, one on anterior and one on posterior wall of each spermathecal invagination. Penial setae, slightly sigmoid or with nearly straight shaft, 0.47–0.9 mm. long, tip bifid, with single terminal spine or rounded, sparsely ornamented ectally with small teeth.

Reproduction: Sperm, matured profusely, are exchanged during copulation. Accordingly, reproduction is assumed to be biparental, at least ordinarily. Facultative parthenogenesis, however, is anticipated.

Distribution: Wimberleyganj, Minnie Bay, Mount Harriet, Port Blair, Viper Island (Andaman Islands). Victoria Point, Sindin, Zowai, Tharrabyin, Mergui, Dubyinchaung, Labaw, Wuzinok, Kala Island, Nyaungbinkwin (Mergui). Tavoy, Myittha, Maung-magaun, Mindat, Kamaungthwe River, Kanyindaung, Nyinmaw, Nyaundonle Chaung, Pyinthadaw, Posoe Chaung, Mayan Chaung, Kawlet Chaung, Migyaung-laung, Zinba, Siyigyan, Heimza Basin (Tavoy). Ye, Moulmein, Mupun, Chaungson, Kyaikmaraw, Kya In, Kawkareik (Amherst). Thaton, Boyagyi, Kyai-kto, Kyaiktiyo, Taungzun, Kumingyaung, Duyinzeik, Aungsaing, Naung-gala, Bilin, Sittang (Thaton). Thameintaw, Kyaiklat (Pyapon). Kungyangon, Kyauktan, Syriam, Rangoon, Twante (Hantha-waddy). Insein, Hlawga, Hlegu, Dam-site, Tauk-kyan (Insein). Myaungmya, Myohaung (Myaung-mya). Maubin, Danubyu, Pantanaw (Maubin). Bassein, Coomzamu, Kochi, Padaukchaung (Bassein). Pegu, Paung, Nyaunglebin, Pazunmyaung, Pegu Yomas (Pegu). Ngapugale, Tharrawaddy, Thonze, Pegu Yomas (Tharrawaddy). Henzada, Ingabu (Henzada). Ler-mu-htee, Pauk-taw-gwin, Mewaing (Salween). Loikaw, Koopra, Mala (Karenni). Pathichaung, Daw Pakko, Daylo stream, Sah Der. Myasawni Bridge, Kyaukkyi, Shwegyin, Maw Pah Der, Thandaung, Palon, Leiktho, Blachi, Pa Taw Lo, Toungoo, Tantabin, Pegu Yomas (Toungoo). Prome, Laboo, Arakan Yomas (Prome). Sandoway, Ngapoli, Andrew Bay, Taungup, Yebawgyi, Myaya (Sando-way). Ramree Island (Kyaukpyu). Akyab, Padali, Kyauktaw, Myohaung, Buthidaung, Maungdaw (Akyab). Paletwa (Arakan Hill Tracts). Pyinmana, Pyigyaung (Yamethin). Ang Lawng Mt., Mong Ma (Yunnan). Kat Pang (Mang Lun). Pa Mung, Loi Pang Pra, Tan Yang (Mong Yai). Na Shai (Pang Long). Kengtung (Kengtung). Kalaw, Taungyi, Maymyo (Shan States). Selan, Namkham (North Hsenwi). Katha, Wuntho, Indaw Lake (Katha). Anidaung (Lower Chindwin). Masein, Paungbyin,

Kalewa, Homalin, Mawleik, Pantha (Upper Chind-win). Falam (Chin Hills). Bhamo (Bhamo). Myitkyina, Hpunchan Hka, Sumprabum, Khamko, Lawange, Masum Zup, Putao, Kankiu, Kadrangyang, Kawapang, Tiang Zup, Weshi, N Sop Zup, Chinkram Hka, Supkaga (Myitkyina). From sea level to elevations of 6,000 + feet.

Kalimpong and Pashok (Darjiling district, India). Kuala Lumpur (Malay Peninsula).

Sumatra (Singkarah).

Some of the records of *P. houlleti* for China, Sumatra, Java, may prove to be of this morph. Var. *bidenryoana* Ohfuchi, 1956 (*Jour. Agric. Sci. Tokyo* 3: p. 169) from the Riu Kyu Islands, has a thickly muscularized septum 8/9. That and other characters contradict inclusion in *P. houlleti*.

Habitats: In soil of gardens, lawns, open areas, bamboo and deciduous jungles, rain forests. Gravelly soil. Under bark of trees and under logs. In rotting logs and among wet leaves. On tree ten feet from ground (while searching for *Perionyx arboricola*.)

Biology: Activity is limited to the rainy season except in occasional niches, presumably blessed with more than the usual amount of moisture. Clitellate specimens were secured in Rangoon from late in August through December and once one in January. Breeding probably is mostly during August–September. The worms are geophagous. Deposition of castings on the surface (in towerlike forms?) was suspected but did not take place when worms were kept in the laboratory in large pots.

Abnormality: Some of the homoeosis in a specimen of this morph (1926: p. 160) may have resulted from splitting of embryonic somites. Markedly protuberant male porophores of Kadranyang and Kawapang worms (with large numbers of gregarines in coelomic cavities of postprostatic segments) were said (1955: p. 85) to be abnormal but without further specification.

Parasites: Except as just noted in the paragraph immediately above, no records regarding parasitism of this morph survived the war. About all that now should be said is that heavy parasitic infestations of protozoa were much less frequent than in certain other morphs.

A morph

1926. *Pheretima houlleti* var. *rugosa* Gates, *Ann. Mag. Nat. Hist.*, Ser. 9, 17: p. 459. (Type locality, Rangoon. Types, none.) *Pheretima "Houlleti"* var. *rugosa*, Gates, *Jour. Burma Res. Soc.* 15: p. 210. *Pheretima houlleti rugosa*, Gates, *Rec. Indian Mus.* 28: p. 157.

1930. *Pheretima houlleti* var. *rugosa*, Gates, *ibid.*, 32: p. 311.

1932. *Pheretima rugosa*, Gates, *ibid.*, 34: p. 398.

1933. *Pheretima campanulata* var. *rugosa*, Gates, *ibid.*, 35: p. 512.

1936. *Pheretima campanulata* f. *rugosa*, Gates, *ibid.*, 38: p. 409.

1939. *Pheretima campanulata* f. *rugosa*, Gates, *Jour. Thailand Res. Soc. Nat. Hist. Suppl.* 12: p. 83.

1955. *Pheretima campanulata* f. *rugosa*, Gates, *Rec. Indian Mus.* 52: p. 85.

1956. *Pheretima campanulata*, A morph, Gates, *Evolution* 10: p. 220.

Genital markings, none externally. Setae, of clitellar segments with irregular transverse constrictions, and with or without bifurcation at tip, none in copulatory chambers (no penial setae). Testes, juvenile. Seminal vesicles, juvenile. Copulatory chambers, with 2–3 GM glands on anterior faces and 3–5 on posterior faces. Penial body, not trilobed, with 4–6 genital markings near tip. (Otherwise as in H morph.)

Reproduction: All individuals of this athecal morph for which information still is available were male sterile. Reproduction obviously must be parthenogenetic.

Distribution: Victoria Point, Mergui, Labaw (Mergui). Ye, Moulmein, Kyaikmaraw, Kawkareik (Amherst). Aungsaing, Duyinzeik, Thaton, Bilin, Kyaikto, Taungzun, Kumingyaung, Sittang (Thaton). Kungyangon, Twante, Rangoon (Hanthawaddy). Taukkyan, Dam Site (Insein). Danubyu, Maubin, Pantanaw (Maubin). Myaungmya (Myaungmya). Bassein (Bassein). Pegu, Pegu Yomas, Paung (Pegu). Ngapugale, Pegu Yomas (Tharrawaddy). Ingabu (Henzada). Prome, Laboo, Arakan Yomas (Prome). Pauk-taw-gwin, Mewaing (Salween). Shwegyin, Kyaukkyi, Toungoo, Blachi, Maw Pah Der, Daylo Stream, Thandaung, Pa Taw Lo, Leiktho Circle, Pegu Yomas (Toungoo). Pyigyaung, Pyinmana (Yamethin). Kengtung (Kengtung). Kalaw, Taungyi, Maymyo, Lashio (Shan States). Namkham (North Hsenwi). Katha, Wuntho (Katha). Bhamo (Bhamo). Anidaung, Masein, Paungbyin, Homalin, Mawleik, Kalewa (Upper Chindwin). Falam (Chin Hills).

Mu'ang Pong, Ban Muang, (Thailand).

Parasites: Sporozoa, possibly of several sorts, usually are present in coelomic cavities, often in especially large masses in v–vi (Gates, 1933: p. 520).

Intermediate morphs

Numerous intermediate morphs, first, fourth, and fifth orders were found. Many of these morphs are male sterile. Nevertheless, some individuals had sperm in a more or less normal diverticulum on one or more of the remaining spermathecae as a result of copulation with an individual of some male fertile morph.

Distribution: Labaw (Mergui). Myohaung (Myaungmya). Rangoon (Hanthawaddy). Bassein (Bassein). Prome, Laboo, Arakan Yomas (Prome). Yamethin (Yamethin). Wuntho (Katha). Kengtung (Kengtung). Taungyi (Shan States). Bhamo (Bhamo). Recent introduction to many of those localities seems probable.

Kyaikto, Duyinzeik, Aungsaing (Thaton), Pauktaw-gwin, Mewaing (Salween). Kyaukyi, Daylo Stream, Thandaung, Pa Taw Lo, Shwegyin, several localities in the Pegu Yomas (Toungoo), Koopra, Mala (Karenni).

AR morph

Few more than a dozen records of specimens of this morph survived the war but they may well have constituted the majority of those that had been secured. Most of the individuals were secured at Myohaung and Laboo (recent introduction?), Toungoo, Duyinzeik and Thaton.

ARZ morph

This rare terminal morph, as yet unrecorded for any other species of earthworm, is known only from two specimens secured at Duyinzeik, Thaton District.

Larger H_p morph

1929. *Pheretima houlleti* (part), Stephenson, *Rec. Indian Mus.* **31**: p. 237. (Excluding all specimens except those with tip of penial body characteristically bilobed.)
1932. *Pheretima campanulata* var. *meridiana* Gates, *ibid.*, **34**: p. 457. (Type locality, Myittha. Types, none.)
1933. *Pheretima campanulata* var. *meridiana*, Gates, *ibid.*, **35**: p. 511.
1936. *Pheretima meridiana*, Gates, *ibid.*, **38**: p. 434.
1955. *Pheretima meridiana*, Gates, *ibid.*, **52**: p. 88.
1961. *Pheretima meridiana*, Gates, *Burma Res. Soc. 50th Anniv. Pub. No. 1*: p. 57.

- - - - - -

1887. *Perichaeta houlleti*, Beddard, *Proc. Zool. Soc. London* **1887**: p. 389.
1937. *Pheretima meridiana*, Gates, *Bull. Mus. Comp. Zool. Harvard College* **80**: p. 358.

Genital markings, usually lacking externally. First dorsal pore, at 11/12. Setae, none in copulatory chambers (no penial setae), 20–26/iii, 30–40/viii, 44–54/xii, 48–60/xx, vii/11–17, viii/15–26, xvii/15–21, xviii/10–16, xix/15–20. Segments, 119–131. Diameter, 3–8 mm. Length, 110–200 mm.

Seminal vesicles, juvenile. Prostates, in xvi–xxi, ducts each in a C-shaped loop. Copulatory chambers, with smooth dorsal surface, with 1–3 GM glands on anterior faces. Penial body, with bilobed tip (lacking one of the markings of the H morph). Internal genital markings, one at tip of penial body, one on median wall of copulatory chamber, one on anterior and one on posterior wall of each spermathecal invagination.

Reproduction: Believed to be parthenogenetic though records as to sperm maturation, presence of sperm in spermathecae, and the condition of the testes no longer are available. Seminal vesicles, at full maturity, are juvenile as in male sterile individuals of other morphs.

Distribution: Victoria Point, Sindin, Zowai, Tharrabyin, Dubyinchaung (Mergui). Myittha, Kyaukmedaung, Kameik, Pyin Tha Daing, Kataungni, Zinba, Siyigyan, Tanbin Hills, Pyinthadaw, Posoe Chaung, Mayan Chaung, Kawlet Chaung, Heimza Basin (Tavoy). Ye (Amherst). Thaton, Duyinzeik, Naunggala (Thaton). Pegu Yomas (Pegu). Su Law, Pray Law and other localities in the Pegu

Yomas, Thandaung, Blachi, Shwegyin, Sah Der, Maw Pah Der, Kyaukkyi (Toungoo). Ler-mu-htee, Pauk-taw-gwin, Mewaing (Salween). Myohaung (Akyab). Paletwa (Arakan Hill Tracts). Tan Yang (South Hsenwi). Man Peng, Nawng Lon (Mang Lun). Lashio, Taungyi, Maymyo (Shan States). Nyaung-bin, Kadranyang, Supkaga (Myitkyina). From sea level to elevations of 4,000 feet.

Bahamas. (Nepal, at 6,500 feet?)

Habitats: Soil, under grass, bamboo and rain forest.

Abnormality: An extra and normal copulatory chamber in xvii (1 specimen).

Related morphs: An A morph with penial bodies having the characteristically bilobed tip of the penial body is known only from a single individual secured at Na Ko She on the Shan Plateau. First order intermediate morphs are unknown.

Remarks: External genital markings, in H morph locations, were recorded for five specimens.

A conspicuous columnar protuberance from the wall of a copulatory chamber with a genital marking at its distal end should not be mistaken for a penial body when the latter is concealed within folds of the chamber wall.

The Nepal worm, recorded as *campanulata*, had no penial setae, and penial bodies were not characterized. Various intermediate morphs also have no penial setae.

Smaller H$_p$ morph

1872. *Perichaeta houlleti* Perrier, *Nouv. Arch. Mus. Hist. Nat. Paris* 8: p. 99. (Type locality, Calcutta. Types, in the Paris Museum.)

1893. *Perichaeta "Houlleti"* (part), Horst, In: Weber, M., *Zool. Ergeb. einer Reise in Niederländisch Ost-Indien* (Leiden) 3: p. 64. (Excluding here and below, all worms with bilobed or trilobed tips of penial bodies.)

1895. *Perichaeta houlleti*, (part), Beddard, *A Monogr. of the Order of Oligochaeta* (Oxford), p. 424.

1897. *Perichaeta houlleti* (part), Michaelsen, *Abhandl. Senckenbergischen Naturforsch. Ges.* 21: p. 234.

1900. *Amyntas houlleti* (part), Beddard, *Proc. Zool. Soc. London* 1900: p. 613. *Pheretima houlleti* (part), Michaelsen, *Das Tierreich* 10: p. 273.

1903. *Pheretima "Houlleti"* (part), Michaelsen, *Die geographische Verbreitung der Oligochaeten* (Berlin), p. 97.

1909. *Pheretima houlleti* (part), Michaelsen, *Mem. Indian Mus.* 1: p. 110.

1910. *Pheretima houlleti* (part), Michaelsen, *Abhandl. Naturhist. Ver. Hamburg* 19, 5: p. 11.

1922. *Pheretima houlleti*, Stephenson, *Rec. Indian Mus.* 24: p. 434.

1923. *Pheretima houlleti* (part), Stephenson, (*The Fauna of British India*), *Oligochaeta*, p. 304.

1926. *Pheretima houlleti* var. *typica*(?), Gates, *Ann. Mag. Nat. Hist.* Ser. 9, 17: p. 450 and *Jour. Burma Res. Soc.* 15: p. 208. *Pheretima houlleti* (part), Gates, *Jour. Bombay Nat. Hist. Soc.* 31: pp. 182, 183. *Pheretima houlleti typica*, Gates, *Rec. Indian Mus.* 28: p. 157.

1927. *Pheretima houlleti*, Gates, *Ann. Mus. Civ. Sto. Nat. Genova* 52: p. 229.

1929. *Pheretima houlleti* (part), Stephenson, *Rec. Indian Mus.* 31: p. 237.

1930. *Pheretima houlleti* var. *typica*. Gates, *ibid.*, 32: p. 311.

1931. *Pheretima houlleti*, Gates, *ibid.*, 33: p. 389.

1932. *Pheretima houlleti*, Gates, *ibid.*, 34: p. 464.

1933. *Pheretima houlleti*, Gates, *ibid.*, 35: p. 529.

1936. *Pheretima houlleti*, Gates, *ibid.*, 38: p. 419.

1937. *Pheretima houlleti*, Gates, *ibid.*, 39: pp. 181, 203.

1939. *Pheretima houlleti* (part), Gates, *Jour. Thailand Res. Soc. Nat. Hist. Suppl.* 12: p. 94.

1955. *Pheretima houlleti*, Gates, *Rec. Indian Mus.* 52: p. 87.

1956. *Pheretima houlleti*, Gates, *Evolution* 10: p. 222.

1945. *Pheretima houlleti*, Gates, *Proc. Natl. Acad. Sci. India* 15: pp. 48, 54.

1960. *Pheretima houlleti*, Gates, *Bull. Mus. Comp. Zool. Harvard College* 123: p. 263.

1961. *Pheretima houlleti*, Gates, *American Midland Nat.* 66: p. 62; *Burma Res. Soc. 50th Anniv. Pub.* No. 1: p. 57.

CULTURE: Tembre & Dubash, 1959, *Jour. Bombay Nat. Hist. Soc.* 56: p. 645. PARASITES: Sporozoa: Puytorac & Tourret, 1963, *Ann. Parasitol. Humaine Comp.* 38.

Genital markings, none externally. First dorsal pore, in region of 7/8–11/12. Setae, *a–c* enlarged but size decreasing laterally, 24–31/iii, 37–55/viii, 46–61/xii, 51–61/xx, vii/11–17, viii/16–25, xvii/12–16, xviii/5–12, xix/12–15, lacking in copulatory chambers (no penial setae). Segments, 90–120. Diameter, 3–5 mm. (but usually nearer lower than upper limits). Length, 40–130 mm.

Typhlosole, lamelliform but with low vertical ridges laterally, ending in region of 63nd to 72nd segments. Seminal vesicles, juvenile. Prostates, in xvi–xxiii, ducts variously looped. Copulatory chambers, with 1–3 GM glands on anterior faces, often with a gland on posterior face. Penial body, slenderly conical, male pore at distal end. Spermathecae, large reaching to dorsal parietes, diverticular junction ental to parietes. Internal genital markings, one on median wall of each copulatory chamber, 1–3 others on or near base of penial body, one on anterior wall of each spermathecal invagination.

Reproduction: Many specimens with maximal clitellar tumescence show no evidence of having matured sperm or of having copulated. Such worms, with juvenile testes and small male funnels, presumably are male sterile, in which case reproduction, at least usually, must be parthenogenetic. Sperm obviously had been matured and also were present in spermathecal diverticula (1945: p. 54) of some Allahabad individuals that seemingly would have been referable anatomically to the smaller H$_p$ morph. However, exchange of sperm during copulation does not guarantee that reproduction is biparental nor even that penetration of ova by sperm is necessary for initiation of parthenogenetic development.

Distribution: Mount Harriet, Minnie Bay (Andaman Islands).

Nyaungbinkwin, Mergui, Kala Island, Wuzinok, Labaw (Mergui). Kanyindaung, Mindat, Tavoy, Myittha, Sinbyudaing, Maungmagaun, Kyaukmedaung, Kameik, Pyin Tha Daing, Kataungni, Nyinmaw, Nyaungdonle Chaung, Tanbin Hills, Kawletchaung, Zinba (Tavoy). Ye, Moulmein, Mupun, Chaungson, Martaban, Kyaikmaraw, Kya In Seik

Kale, Kya In, Kawkareik (Amherst). Thaton, Aungsaing, Naung-gala, Taungzun, Kumyingyaung, Duyinzeik, Kyaikto, Mokpalin, Bilin, Boyagyi, Sittang (Thaton). Rangoon, Thongwa, Kungyangon, Twante, Kyauktan (Hanthawaddy). Insein, Hmawbi, Taikkyi, Wanetchaung, Hlawga, Taukkyan (Insein). Pyapon, Bogale, Thameintaw, Thanchitaw, Kyaiklat (Pyapon). Myaungmya, Wakema, Myohaung (Myaungmya). Maubin, Yandoon, Danubyu, Pantanaw (Maubin). Bassein, Thinbawgyin, Coomzamu, Kochi. Henzada, Ingabu, Kyangin (Henzada). Pegu, Thanatpin, Pegu Yomas, Paung, Nyaunglebin, Pazunmyaung (Pegu). Tharrawaddy, Letpadan, Thonze (Tharrawaddy). Prome, Laboo, Paukkaung (Prome). Sandoway, Doedaung Hill, Ngapoli, Andrew Bay, Taungup, Yebawgyi, Myaya (Sandoway). Pauk-taw-gwin (Salween). Maw Pah Der, Shwenyaungbin, Thandaung, Pathichaung, Leiktho, Blachi, Shwegyin, Kyaukkyi, Toungoo, Tantabin, Pegu Yomas (Toungoo). Thayetmyo, Thanbula (Thayetmyo). Kyaukpyu (Kyaukpyu). Loikaw, Koopra, Mala (Karenni). S'nite (Mong Pai). Pyigyaung, Pyinmana (Yamethin). Magwe (Magwe). Minbu (Minbu). Akyab, Myohaung, Buthidaung, Maungdaw (Akyab). Paletwa (Arakan Hill Tracts). Kyaukpadaung, Mt. Popa (Myingyan). Mandalay (Mandalay). Sagaing (Sagaing). Nam Hpen Noi, Lawng Meu (Mong Lem, Yunnan). Kengtung, Nam Shi Pan (Kengtung). Loi Se, Man Peng, Nawng Kham, Nawng Lon, Kwang Yeh (Mang Lun). Tan Yang, Ta Pung (South Hsenwi = Mong Yai). Nam Hung, Meung Nawng (Pang Long). Kutkai, Namkham (North Hsenwi). Man Meh Hang (Hsipaw). Lashio, Taungyi, Kalaw, Maymyo (Shan States). Anidaung (Lower Chindwin). Kin-U (Shwebo). Katha, Wuntho, Naba, Indaw Lake (Katha), Bhamo (Bhamo). Lonton, Mogaung, Kamaing, Hopin, Myitkyina, Chinkram Hka, N Sop Zup, Putao (Myitkyina). Laungbyin, Masein, Paungbyin, Mawleik, Homalin (Upper Chindwin). Falam, Tiddim (Chin Hills). From sea level to elevations of 5,500 feet.

Cherrapunji (Assam). Bawti, Not Theinko, Bangkok, Mu'ang Pong, Chiengmai, from sea level to 1,000 feet (Thailand).

India (tropical lowlands, Indo-Gangetic plains to Dehra Dun, Bhim Tal and Rawalpindi). Ceylon (lowlands). Malay Peninsula (Kuala Lumpur, Aring, Singapore).

Java. Philippine Islands (Manila). Fiji Islands. (Yap, Caroline Islands?)

Salvador (at elevations of 300–975 m.). Georgia, Florida (new records). Cuba. Sierra Leone.

P. houlleti has been recorded from Indo-China, Sumatra, Madagascar, Comoro, France (Nice, Paris— in greenhouses) and French Guiana, but information that would enable recognition of the morph was not provided. Chinese records for Kiangsu, Hopei, and

Hupeh provinces are erroneous—other species were involved. The questioned Yap record is based on P. yapensis Ohfuchi, 1941.

P. houlleti now appears to be primarily a lowland tropical species. It was not found in large collections from Darjiling at 6,000 feet although it probably has been taken there many times in soil with potted plants. However, Sims (Jour. Bombay Nat. Hist. Soc. 60: 87) recently recorded finding of two specimens at 6,500 feet in Nepal during November.

Habitats: Soil (red or black) of gardens, lawns, fields, open areas, under bamboo, various types of jungle and rain forests. Mud of marshy areas, around wells, at banks of tanks, ponds, streams, rivers, and lakes. Under bark of trees (rain forests). Under stones (Shan Plateau). On epiphytic ferns. In earth around roots of potted plants (Burma— common, India).

Biology: Fully mature worms were collected, both at Rangoon and Allahabad, during July–January. Breeding presumably continues throughout such portions of the year as conditions favor activity. Hatching is after 25–30 days in laboratory cultures (Tembe & Dubash, 1959) but if temperature is below 28°C and relative humidity is below 50 per cent 45–50 days may be required.

The worms are geophagous. Undigested residues, in nature, probably are deposited below the surface as did happen when a few worms were kept in very large laboratory pots. When in earth around roots in a small plant pot, deposition often was on the surface. Castings, in such circumstances, were cordlike, scattered or in low piles, never in towers.

Abnormality: The homoeosis of one specimen could have resulted, according to the data still available (1936: p. 421), from halving, on the left side, of eleven mesoblastic somites, two of those at the 7th, 8th, and 9th levels, one of those at the 10th and 11th levels, the one at the 18th level and seven of those at the 19th to 26th levels.

Polymorphism: Four specimens, of intermediate morphs in which evolution toward an athecal stage is under way, were found. Each of those specimens had characteristic copulatory chambers, penes, and spermathecae.

Parasites: Zygocystis pheretimae Puytorac & Tourret 1963, parietal peritoneum, Malay Peninsula.

Remarks: Characteristics of a penial body, especially if "buried" as it were between folds of the chamber wall, are unlikely to have been determinable in the usual sort of microtome sections secured from field-preserved material. When a penial body is concealed in the same way it can be easily overlooked even in a dissection in which case a columnar protuberance with a genital marking at its distal end could be mistaken for the male porophore.

Pheretima hupbonensis

1931. *Pheretima hupbonensis* Stephenson, *Proc. Zool. Soc. London*
   1931: p. 61. (Type locality, Hup Bon, Thailand. Type,
   in the British Mus.)
1939. *Pheretima hupbonensis*, Gates, *Jour. Thailand Res. Soc.
   Nat. Hist. Suppl.* 12: p. 95.

Quadrithecal, pores transverse slits *ca.* ¼C apart, at
7/8–8/9. Male pores, in xviii, towards lateral margins of suckerlike areas that nearly reach eq/xvii and
eq/xix, each pore overhung by a small papilla.
Female pore, median (?). (Other genital markings,
none.) Clitellum, xiv–xvi. Setae, enlarged in ii–ix,
32/v, 45/ix, 54/xii, 68/xix, ix/9, xviii/8. First dorsal
pore, at 11/12. (Prostomium?) Segments, 142.
Size, 225- by 99 mm.

Septa, 8/9–9/10 aborted. Intestinal origin, in xv.
Caeca, not manicate, in xxvii (?–?), vertical lobes
from ventral side decreasing in size anteriorly, some
of the larger ones with tertiary lobing. (Typhlosole?
Hearts?) Holandric. Testis sacs, ventral (paired?).
Seminal vesicles, in xi, xii. Prostates, in xvii–xix,
ducts soft, short but muscular. Spermathecae, duct
almost confined to parietes, diverticulum shorter than
the main axis, with short and muscular stalk, with a
digitiform seminal chamber in a crescentic curve or
with one or two *U*-shaped loops of different size.

Distribution: Hup Bon, in southeastern Thailand.

Remarks: This species is known only from the
original description of the holotype. Attribution of
the intestinal caeca to xvii presumably is an uncorrected printer's error. Testis sacs of a side were
said to be continuous which seems unlikely. Stephenson seems to suggest that the small papilla overhanging the male pore is an everted copulatory pouch.

Pheretima illota

1932. *Pheretima illota* Gates, *Rec. Indian Mus.* 34: p. 397. (Type
   locality, To Noi. Types, none.)
1960. *Pheretima illota*, Gates, *Bull. Mus. Comp. Zool. Harvard
   College* 123: p. 266.

Athecal. Male pores, in xviii, minute and superficial, each in a small transversely oval area within a
circular discoidal porophore that extends slightly into
xvii and xix. Female pore, median. (Genital markings, aside from the male porophores, none.) Clitellum, setae unrecognizable externally, xiv–xvi. Setae,
lacking ventrally in ii, small and closely crowded,
16–17/ii, 94–103/xx, xvii/18–20, xviii/6–9, xix/18–19.
First dorsal pore, at 12/13. Prostomium, epilobous.
Color, none (pigment none?). Segments, 120. Size,
149–160 by 5–6 mm.

Septa, 8/9–9/10 aborted, 5/6–7/8 and 10/11 muscular. Intestinal origin, in xv. Caeca, simple, in xxvii–
xxii. (Typhlosole?) Hearts, of ix lateral but usually
lacking on one side, of x–xiii (latero?)-esophageal.
Blood glands, in v. Holandric. Testis sacs, ventral
and unpaired. Seminal vesicles, small, in xi, xii, each

with large primary ampulla. Prostates, in xvii–xxi,
ducts 3–4 mm. long.

Reproduction: Presumably uniparental. (*Note*, absence of spermathecae. Male sterility is anticipated.)

Distribution: To Noi (Mong Lem State, Yunnan).
At an elevation of *ca.* 6,000 feet. Specimens from
four localities on Ishigaki Island, Riu Kyu Islands
were referred (Ohfuchi, 1941, *Studies Palao Tropical
Biol. Sta.* 2: p. 137) to *P. illota.* Muscularization of
8/9 as well as of other septa, pigmentation, male
porophores, etc., contradict the identification.

Habitat: Village manure heap.

Polymorphism: Athecal and possibly male sterile
individuals suggest that evolution of parthenogenetic
polymorphism is fairly well advanced. If intermediate morphs that could have provided clues to
characterization of an ancestral H ever were available,
records of them did not survive the war. One candidate for the H morph might be *P. youngi* but at
present too little is known about modifications of
specific norms arising during evolution of parthenogenetic polymorphism to warrant further consideration.

Remarks: Known only from the original description
of two clitellate individuals.

Pheretima immerita

1931. *Pheretima immerita* Gates, *Rec. Indian Mus.* 33: p. 389.
   (Type locality, Mong Ko. Types, none.)
1933. *Pheretima immerita*, Gates, *ibid.*, 34: p. 410.

Quadrithecal, pores minute, superficial(?), *ca.* ¼C
apart, at 6/7–7/8. Male pores, in xviii, (invaginate?).
Female pore, median(?). Genital markings, postsetal, paired, and median to the spermathecal pores
or unpaired and median in vii, paired across 17/18 or
postsetal in xviii, in vii each with a central pore.
(Clitellum?) Setae, 50–55/xx, vii/26–30, xviii/17–27.
First dorsal pore, at 12/13. Prostomium, zygolobous.
Color, only in dorsum and there lacking in narrow
but obvious equatorial stripes, red (pigment in circular muscle layer?). Segments, 92–114. Size, 85–110
by 4–5 mm.

Septa, 8/9–9/10 membranous, the latter inserted on
parietes along with 10/11, none thickly muscular.
Intestinal origin, in region of insertion of 15/16.
Caeca, simple but with shortly digitiform lobes protuberant from the dorsal side, in xxvii–xxiv. Typhlosole, rudimentary, short. Hearts, of ix lateral, of
x (latero?)-esophageal, of xi–xiii latero-esophageal.
Holandric. Testis sacs, paired (?) in x, unpaired in
xi. Seminal vesicles, in xi, xii. (Spermathecae and
GM glands?)

Distribution: Mong Ko (Kengtung). Pang Wo
(Mang Lun). At elevations of *ca.* 3,000 to 4,500 feet
on the Shan Plateau.

Remarks: Known only from 9 immature worms in
which genital organs were juvenile. The longitudinal
musculature seems to be unusually thick and strong.
P. immerita cannot be properly defined until various

genital characters are known but in the meanwhile is distinguished from all of its congeners in southeast Asia by the location of the four spermathecal pores at 6/7–7/8.

Pheretima inclara

1932. *Pheretima inclara* + *P. porrecta* Gates, *Rec. Indian Mus.* **34**: pp. 439, 444. (Type locality, Peng Sai. Types, none.)
1933. *Pheretima porrecta*, Gates, *ibid.*, **35**; p. 542.
1936. *Pheretima inclara*, Gates, *ibid.*, **38**: p. 422.

Sexthecal, pores minute, superficial, at or just behind 5/6–7/8. Male pores, in xviii, minute, superficial, each in a very small conical protuberance from a small disc or within a larger porophore much like a genital marking and reaching towards, to or into xvii and xix. Female pore, median. (Genital markings, other than the areas containing male pores, none.) Clitellum, setae unrecognizable externally, xiv–xvi. Setae, present or lacking in ii, *ca.* 125/xx, vi/25–33, vii/26–33, xvii/27–34, xviii/11–24, xix/27–36. First dorsal pore, at 12/13. Prostomium, epilobous(?). Color, only in dorsum, reddish to slate, (pigment red and in circular muscle layer?). Segments, 123–127. Size, 120–254 by 6–10 mm.

Septa, 8/9–9/10 aborted, 5/6–7/8 and 10/11–11/12 muscular to very thick. Intestinal origin, in xv. Caeca, simple, in xvii–xix. (Typhlosole?) Hearts, of ix lateral but usually aborted on one side, of x–xiii (latero?)-esophageal. Blood glands, in v. Holandric. Testis sacs, ventral, unpaired. Seminal vesicles, large, in xi, xii, displacing septa. Prostates, large, in xvii-xx but displacing septa, ducts 2–13 mm. long, straight, each in a *J* or a *U*-shape or otherwise looped. Spermathecae, duct slender, usually shorter than ampulla, abruptly narrowed in body wall, diverticulum from anterior face of duct at or near the parietes, longer than the main axis, digitiform, only slightly widened entally, sinuous, or very shortly looped.

Reproduction: Presumably biparental as sperm are exchanged during copulation.

Distribution: Peng Sai (Mang Lun). Loikaw (Karenni). At elevations of *ca.* 3,500 feet.

Habitat: Soil, in dense jungles.

Parasites: Parasites, presumably sporozoa, were numerous in the post-gizzard (glandular?) collar, a niche very rarely occupied by parasites that are macroscopically recognizable.

Biology: The breeding season includes October.

Remarks: Known only from 11 specimens, 3 of which had died before preservation and 6 of which (from Loikaw) appeared to be unhealthy though parasites were not recognized.

The male porophores of the *porrecta* type were in beautifully differentiated, large, areas of epidermal thickening each of which was characterized by a markedly glabrous marginal band distinctly delimited peripherally as well as centrally from the major portion with a finely roughened (punctate?) surface.

Absence of anything of the sort was the main reason for erection of the appropriately named *P. inclara*.

Pheretima insulanus

1930. *Pheretima insulanus* Gates, *Rec. Indian Mus.* **32**: p. 312. (Type locality, Kala Island. Types, none.)
1932. *Pheretima insulanus*, Gates, *ibid.*, **34**: p. 419.

Quadrithecal, external pores small transverse slits at edge of pigmented region and slightly more than $\frac{1}{3}$C apart, at 7/8–8/9. Male pores, in xviii, minute, each on a slightly flattened conical penis in a columnar and erect copulatory chamber with a small equatorial aperture. Female pore, median. Genital markings, all internal, very small, circular and translucent areas, one on the anterior wall of each spermathecal invagination (?), others on roofs of copulatory chambers. Clitellum, setae unrecognizable externally, xiv–xvi. Setae, small, circles without conspicuous breaks at median lines, 55/xix, 60–65 per segment behind xx, viii/32, xvii/18, xviii/16, xix/17. First dorsal pore, at 12/13. Prostomium, epilobous, tongue open. Color, in the dorsum, slate anterior to the clitellum, reddish brown posteriorly (pigment red and in circular muscle layer?). Segments, 67(+ ?). Size, 79(+ ?) by 5 mm.

Septa, 8/9–9/10 aborted. Intestinal origin, in xv. Caeca, simple, with slightly incised margins, in xxvii–xxiv. (Typhlosole?) Hearts, in viii unaborted dorsal portions to the gizzard (?), in ix (left or right) lateral, in x (latero?)-esophageal, in xi–xiii latero-esophageal (?). Lymph glands, present in intestinal region. Holandric. Testis sacs, paired and ventral. Seminal vesicles, medium-sized, in xi, xii, each with a distinct primary ampulla. Prostates, in xv–xxi, ducts looped or curved. Spermathecae, duct shorter than ampulla, diverticulum from anterior face of duct at parietes, longer than main axis, in a ball of coils, with a short stalk, the remainder widened, thin-walled, with a constriction marking off a shortly pear-shaped wider terminal portion. GM glands, stalked, coelomic, composite, one passing into parietes on anterior face of each spermathecal duct, others on the copulatory chambers.

Reproduction: Possibly biparental though in absence of records *re* spermatozoal iridescence only seminal vesicles and the prostates support such an assumption. Though medium-sized, seminal vesicles perhaps should not be expected to be any larger when breeding has ended.

Distribution: Kala Island (Mergui).

Habitat: Presumably soil. The worm was found wandering on a shaded sandy beach just above high-tide mark at ten o'clock of an October morning. Such worms come to the surface to die, according to the local mythology.

Remarks: Spermathecae were in vii, viii as so often in *P. californica* and *schmardae*.

A minute and primary spermathecal pore is believed

to be within a spermathecal invagination containing on its anterior wall a genital marking but the invagination was not seen. Size is expected to be limited by the thickness of the parietes in absence of coelomic penetration.

P. insulanus is known only from the holotype, a posterior amputee without regeneration.

Relationships seem to be closest with *P. houlleti*. Differences however were originally believed to contra-indicate the type being a quadrithecal mutant of that species. Classical specialists, because of number and location of spermathecal pores to which so much importance was attached, presumably would have thought relationships of *P. insulanus* and *lorella* to be closer.

Pheretima kengtungensis

1931. *Pheretima kengtungensis* Gates, *Rec. Indian Mus.* 33 : p. 394. (Type locality, Mong Ko. Types, none.)
1932. *Pheretima kengtungensis*, Gates, *ibid.*, 34 : p. 420.

Quadrithecal, external pores medium-sized slits, at 7/8–8/9. Male pores, in xviii, minute, each on a short, rather stout penis within a longitudinally ellipsoidal copulatory chamber. Female pore, median. (Genital markings, none.) Clitellum, setae unrecognizable externally, xiv–xvi. Setae, regularly spaced, circles unbroken at median lines, 68/xx, viii/16, xviii/0(6?). First dorsal pore, at 12/13. (Prostomium?) Color, in dorsum, reddish brown to slate in front of the clitellum, posteriorly reddish brown (pigment red and in circular muscle layer?). Segments, 109. Size, 108 by 4 mm.

Septa, 8/9–9/10 aborted. Intestinal origin, in xv. Caeca, simple, in xxvii–xxii. (Typhlosole?) Hearts, in ix (left or right) lateral, in x (latero?)-esophageal, in xi–xiii latero-esophageal(?). Holandric. Testis sacs, unpaired, ventral. Seminal vesicles, in xi, xii, large, posterior pair pushing 12/13 back to 13/14. Prostates, in xvi–xxi, ducts 7 mm. long, thicker ectally, variously wound, into dorsal face of copulatory chambers. The latter, with thick anterior and posterior walls but with a transversely slitlike lumen. Spermathecae, duct rather stout, slightly shorter than ampulla, slightly narrowed at body wall, diverticulum from median face of duct near parietes, about as long as main axis but looped and bound to duct.

Reproduction: Possibly biparental though in absence of records *re* spermatozoal iridescence only the seminal vesicles and the prostates support such an assumption.

Distribution: Mong Ko (Kengtung).

Remarks: Known only from the holotype with prostomium and i not normal because of some past injury.

How to characterize the spermathecal pores is unknown. The primary pores may be minute and then invaginate or (probably?) if external then relatively large.

The copulatory chambers may have a large gland in each anterior and posterior wall, perhaps somewhat as in *P. schmardae*. If so a single aperture of each gland may have been within a more or less distinctly demarcated genital marking.

All ventral setae between the male pores probably had been dehisced but 6 follicle apertures or their rudiments were recognized. Loss of those setae may have been an individual peculiarity.

Relationships of the species, because of number and location of the spermathecae, presumably would have been thought by classical specialists to be with *P. insulanus* and *lorella* (*q.v.*).

Pheretima lorella

1936. *Pheretima lorella* Gates, *Rec. Indian Mus.* 38 : p. 425. (Type locality, Loi Pang Pra. Types, none.)

Quadrithecal, pores minute, in very slight invaginations with small transversely slitlike apertures at 7/8–8/9. Male pores, in xviii, minute, each at ventral end of a slenderly conical penis, 1 mm. long, on roof of a thick-walled, longitudinally ovoidal, parietal chamber opening dorsolaterally into an unpaired transversely rectangular depression 8–11 intersetal intervals wide with apposable anterior and posterior margins. Female pore, median. (Genital markings, none.) Clitellum, setae unrecognizable externally, xiv–xvi. Setae, small, closely and regularly spaced, slightly larger ventrally, circles only with a slight gap at mD, 91/xii, viii/12–15, xviii/0. First dorsal pore, at 12/13. Color, in dorsum, reddish to slate in preclitellar region, reddish posteriorly (pigment red and in circular muscle layer?). (Segments?) Size, 100–140 by 5–6 mm.

Septa, 8/9–9/10 aborted, none thickly muscular. Intestinal origin, in xv. Caeca, simple, in xxvii–xix, with 7–10 small ventral lobes posteriorly. (Typhlosole?) Hearts, of ix (left or right) lateral, in x esophageal, in xi–xiii latero-esophageal(?). Holandric. Testis sacs, unpaired, ventral. Seminal vesicles, fairly large, xi, xii, the posterior pair pushing 12/13 and other septa back to contact with 15/16. Prostates, in xvi–xxi, ducts 5–6 mm. long, C-shaped or spirally coiled. Spermathecae, duct about as long as ampulla, bulbous ectally, abruptly narrowed in parietes, diverticulum longer than main axis, gradually widened entally, spirally wound or variously bent but not zigzag looped, (with little or no externally recognizable indication(?) of demarcation into stalk and chamber but with muscular sheen ectally).

Reproduction: Presumably biparental as sperm are exchanged during copulation. Condition of the seminal vesicles and prostates also supports that assumption.

The breeding season includes September. Dorsal pores or their sites may have been visible at 14/15–15/16 because clitellar tumescence was not yet maximal.

Distribution: Loi Pang Pra (Mong Yai). At an elevation of 4,000 + feet.

Habitat: Leaf covered, sandy soil.

Abnormality: A second slightly shorter diverticulum on one spermatheca, lacked spermatozoal iridescence.

Parasites: Nematodes were numerous in the coelomic cavities of anterior segments.

Remarks: The species is known only from the four types.

The male terminalia are similar to those of a Singapore form, *Pheretima arcuata*, except that in the latter penes are within chambers that extend into the coelomic cavity rather than being confined to the parietes. The Singapore worms may belong to a parthenogenetic morph but there presently is little reason for suspecting relationship to the Burmese species.

Pheretima malaca

1933. *Pheretima maculosa* (non *P. "maculosus"* Hatai 1930) Gates, *Rec. Indian Mus.* 35: p. 534. (Type locality, between Kyaukmedaung and Kameik. Types, none.)

1936. *Pheretima malaca* Gates, *ibid.*, 38: p. 429. (Nom. nov. pro *maculosa* Gates 1933.)

1960. *Pheretima malaca*, Gates, *Bull. Mus. Comp. Zool. Harvard College* 123: p. 267.

Bithecal, pores minute, superficial, at 6/7. Male pores, in xviii, minute, superficial, each in a large, circular disc nearly reaching 17/18–18/19. Female pore, median(?). Genital markings, very small, circular discs, in transverse rows across or close to 17/18, 18/19, occasionally one marking postsetal in vii or viii median to spermathecal pore levels. Clitellum, xiv–xvi. Setae, present ventrally in all clitellar segments, 56/iii, 77/viii, 61/xii, 4–9/xiv, 3–10/xv, 6–11/xvi, 50–57/xx, vii/26–31, xvii/10–12, xviii/4–10, xix/12–14. First dorsal pore, at 10/11, 11/12, 12/13. Prostomium, epilobous(?). Color, only in dorsum, red (pigment red and in circular muscle layer?). Segments, 109–119. Size, 46–82 by 2–4 mm.

Septa, 8/9–9/10 aborted, none especially thickened. Intestinal origin, in xv. Caeca, simple, in xxvii–xxii. (Typhlosole?) Hearts, of ix lateral but usually present only on one side, of x–xi aborted, of xii–xiii (latero?)-esophageal. Holandric. Testis sacs paired and vertical, unpaired and *U*-shaped or annular, anterior vesicles included. Seminal vesicles, juvenile(?), in xi, xii. Prostates, small, in xviii, duct 1 mm. long, muscular, straight or in a hairpin loop. Spermathecae, duct shorter than ampulla, abruptly narrowed in parietes, diverticulum from anterior face of duct at parietes, shorter than main axis, with muscular(?) stalk, looped middle portion and a shortly ovoidal seminal chamber. GM glands, stalked, coelomic.

Reproduction: Parthenogenesis is expected, no indications of maturation or exchange of sperm having been found.

Distribution: Kyaukmedaung to Kameik (Tavoy).

Lashio (Shan Plateau). From sea level to elevations of *ca.* 3,500 feet.

Habitat: Soil, dense jungles.

Abnormality: Spermathecae were abnormal in three, rudimentary in three, of eight dissected types. Diverticula may not have been quite normal in the other two types.

Parasites: Types were heavily parasitized. Cysts were present in the nerve cord of the anterior half of the body (type). Coelomic gregarines were present in the type and paratypes.

Remarks: The species is known only from eighteen specimens.

A postsetal median genital marking was present (one worm) in each of x–xii. Markings may also be present within the male porophore.

Prostates, as well as the seminal vesicles, may have been juvenile. All worms may have been fourth order intermediate morphs evolving in direction of an AR morph.

P. malaca does not now appear to be closely related to any of the other five bithecal species with pores at or near 6/7. Derivation from some form with a larger number of spermathecae, because of the probability of parthenogenesis, requires consideration. A sexthecal hermaphroditic morph much like *P. papulosa* presumably could have given rise to the *malaca* morph by abortion of the anterior and posterior pair of spermatheca and leaving only a middle pair, a change that has been made in mutant individuals of several species.

Pheretima malayana

1901. *Amyntas malayanus* + *A. bosschae* + *A. pulauensis* + *A. evansi* Beddard, *Proc. Zool. Soc. London* 1900: pp. 893, 892, 904, 907. (Type localities, of *malayanus*, Aring; of *paulauensis*, Pulau Bidang; of *evansi*, Biserat. Types, in the British Mus.)

1903. *Pheretima malayana* + *P. "Evansi"* + *P. pulauensis* + *P. "Bosschae"* (part), Michaelsen, *Die geographische Verbreitung der Oligochaeten* (Berlin), pp. 97, 95, 99, 94.

1932. *Pheretima malayana* + *P. baruana* + *P. evansi* + *P. pulauensis*, Stephenson, *Ann. Mag. Nat. Hist.*, Ser. 10, 9: pp. 220, 209, 213, 233. (Type locality of *baruana*, Khota Baru. Type, in the British Mus.)

1939. *Pheretima evansi* + *P. baruana*, Gates, *Jour. Thailand Res. Soc. Nat. Hist. Suppl.* 12: pp. 87, 88.

1949. *Pheretima malayana* + *P. fovella* + *P. strellana*, Gates, *Bull. Raffles Mus. Singapore* 19: pp. 27, 21, 34. (Type localities: of *fovella*, Kaki Bukit; of *strellana*, Baling. Types, in the Raffles Mus.)

Octothecal, pores (secondary?) transverse slits somewhat median to mL, at 5/6–8/9. Male pores, in xviii, minute, each at ventral end of a penis pendent from the roof of a small copulatory chamber with a transversely slitlike equatorial aperture. Female pore, median. Genital markings, small and circular, paired, in transverse rows or in median and unpaired clusters, presetal and/or postsetal in some of vi–ix, xvii–xxiv. Clitellum, setae usually unrecognizable externally, nearly to or to 13/14 and 16/17 or reaching slightly into xiii and/or xvii. Setae, 23–25/ii, 20–31/iii, 26–37/iv, 33–48/viii, 35–56/xii, 38–51/xx, vi/18–23, vii/19–24, viii/19–23, xvii/10–17, xviii/9–18, xix/11–17. First dorsal pore, at 11/12, 12/13. (Prostomium?) Color, red, in dorsum only except sometimes in first few segments (pigment red and in circular muscle layer?). Segments, 77–120. Size, 42–165 by 2–6 mm.

Septa, none thickly muscular, 8/9–9/10 aborted. Intestinal origin, in xv. Caeca, simple, in xxvi-xxiii. Typhlosole, lamelliform, with vertical ridges on the sides anteriorly, from caecal segment into region of 53rd–65th segments. Hearts, unaborted dorsal portions in viii passing to gizzard, in ix (left or right) lateral, in x (latero-?)-esophageal, in xi-xiii latero-esophageal. Blood glands, in v. Lymph glands, in intestinal segments. Holandric. Testis sacs, of x annular (hearts included), of xi cylindrical (vesicles included). Seminal vesicles, (large?), in xi, xii. Prostates, in xvi-xxiv, ducts, 1½-5 mm., straight or looped. Spermathecae, small, duct much shorter than ampulla, diverticulum with an elongately ellipsoidal seminal chamber and a muscular but slender and looped stalk to median face of duct at parietes. GM glands, composite, stalked and coelomic.

Reproduction: Presumably biparental as sperm are exchanged during copulation.

Distribution: Biserat (Jalor State), Thailand.

Kaki Bukit (Perlis), Aring, Khota Baru (Kelantan), Pulau Bidan, Baling (Kedah), Malay Peninsula.

The little information now available suggests that the range though overlapping that of *bipora* may be mainly to the south.

Biology: *P. malayana* is geophagous. Earth in the intestines has been black (1949) but with red grains (of laterite?) or red (and then of lateritic origin?).

Remarks: Setal counts in the precis were obtained from the author's specimens with soma sizes of 42–110 by 2–4 mm. In the larger worms, 120–165 by 4½–6 mm., that were seen by previous specialists setal numbers may have been higher.

The primary spermathecal pore now is believed to be invaginate but if so the pore invagination must be confined or nearly so to the parietes. If the pore is not invaginate and the primary aperture is a large, superficial slit, another distinction from *P. bipora* is provided.

Male pores of types of *P. malayana*, *evansi*, and *baruana* were not adequately characterized. If synonymy above is correct, copulatory chambers were not recognized, perhaps because of eversion or, in a retracted state, because of small size and coverage by other tissues. Each copulatory chamber in *baruana* presumably contains a genital marking (lacking in chambers of other forms) in addition to a penis.

Relationships now appear to be closest with *P.*

bipora but much more information is needed about several characters in the *malayana* range.

Pheretima manicata

Octothecal, pores minute, superficial, at 5/6–8/9. Male pores, in xviii, minute, superficial, each in a distinct porophore. Female pore, median. Genital markings, one pair, in xviii. Clitellum, setae present at least ventrally in xvi, sometimes present also in xiv and/or xv, sometimes reaching 13/14 but very rarely 16/17. Setae, circles without ventral or regular dorsal gap, a little more widely spaced in the dorsum, 46–58/iii, 60–72/viii, 60–77/xii, 9–32/xvi, 46–73/xx, vi/17–30, vii/18–32, viii/20–32, xvii/18–29, xviii/0–20, xix/18–30. First dorsal pore, occasionally at 11/12, usually at 12/13. Prostomium, epilobous, tongue open. Color, in dorsum, reddish brown to slate anteriorly, posterior to clitellum, reddish brown, brownish to light brown (pigment reddish and in circular muscle layer?). Segments, to 95. Size, to 120 by 6½ mm.

Septa, 8/9–9/10 aborted, none thickly muscular. Intestinal origin, in xv (occasionally in xvi?). Caeca, manicate, dorsalmost of 4–11 secondary caeca the shortest, tertiary lobes sometimes present, in xxvii-xxii. Rudimentary manicate caeca, in some or all of xxviii, xxix, xxx. Typhlosole, lacking or rudimentary (?). Hearts, of ix (left or right) lateral, of x esophageal(?), of xi-xiii latero-esophageal(?). Blood glands, in v-vi. Holandric. Testis sacs, unpaired, ventral. Seminal vesicles, large, in xi, xii-xiv. Prostates, large, in xv-xxiii, ducts 1½-7 mm. long, thicker ectally. Spermathecae, duct shorter than ampulla, abruptly narrowed in parietes, diverticulum from anterior (or median?) face of duct at parietes, longer than main axis, with short and straight muscular stalk having a narrow lumen, an elongate and variously looped seminal chamber, the ectal portion about as thick as the stalk but with wider lumen, a more ental portion wider and with much thinner wall. GM glands, one mass for each marking, sessile on or in the parietes.

Reproduction: Records of sperm maturation and of copulatory transfer are lacking. However, there is no reason to suspect that reproduction is other than biparental. Size of seminal vesicles suggests massive spermatogenesis and the prostates of adults appeared to be functional.

Polymorphism: Numerous collections of *P. manicata* secured between 1932 and 1942 were destroyed and additional material never became available. Such data as still were available in 1960 were thought to warrant recognition of two subspecies. Characterization of either with respect to segment number, size, as well as some anatomy is inadequate because of the high percentage of posterior amputees and immatures as well as condition of the material in earlier collections. Divergence from conditions prevailing in topotypical material probably was greater than can now be indicated.

Distribution: Burma, from the Irrawaddy-Sittang axis east into Yunnan and Thailand, south through the two northern districts of Tennasserim to the Ye River.

Pheretima manicata manicata

1931. *Pheretima suctoria* var. *manicata* Gates, *Rec. Indian Mus.* **33**: p. 414. (Type locality, Moulmein. Types, none.)
1932. *Pheretima manicata* var. *typica*, Gates, *ibid.*, **34**: p. 526.
1936. *Pheretima manicata*, Gates, *ibid.*, **38**: p. 432.
1960. *Pheretima manicata manicata*, Gates, *Bull. Mus. Comp. Zool. Harvard College* **123**: p. 268.

Male porpophores, small, circular discs or tubercles. Genital markings, slightly median to male porpophores, longitudinally elliptical, reaching or crossing 17/18 and/or 18/19.

Distribution: Ye, Moulmein, Chaungson, Kyaikmaraw (Amherst). Thaton (Thaton). All specimens were obtained from lowland sites.

The Ye River marks the region where the *manicata* range meets the *defecta* range. From Thaton north to the vicinity of Thandaung (east of Toungoo) on the Shan Plateau there are presently no records for *P. manicata*. However, there is no reason at present for suspecting that the discontinuity is real.

Remarks: Origin of the intestine in xvi instead of xv, and pairing of testis sacs, were recorded but confirmation from better material is desirable.

Pheretima manicata decorosa

1932. *Pheretima manicata* var. *decorosa* + *P. m.* var. *supina* Gates, *Rec. Indian Mus.* **34**: pp. 528, 529. (Type localities, Teung Cong and Nam Hpen Noi. Types, none.)
1933. *Pheretima manicata* var. *decorosa*, Gates, *ibid.*, **35**: p. 537.
1939. *Pheretima manicata*, Gates, *Jour. Thailand Res. Soc. Nat. Hist. Suppl.* **12**: p. 97.
1960. *Pheretima manicata decorosa*, Gates, *Bull. Mus. Comp. Zool. Harvard Coll.* **123**: p. 268.

Male porpophores, varying in size and shape, circular (and then often small), crescentic and concave mesially, longitudinally elliptical, and then sometimes crossing 17/18 and/or 18/19. Genital markings, transversely elliptical, postsetal, in xviii (?), usually paired.

Distribution: Thandaung (Toungoo). Mala, Koopra, Loikaw, Mawchi (Karenni). Teung Cong, Nam Hpen Noi (Mong Mong, Yunnan). At elevations of 3,500–5,500 feet.

Doi Sutep, at 5,200–5,500 feet, Doi Angka (Doi Intanon) at 4,600 feet, Thailand.

Possibly to be expected from the western edge of Shan massif between latitude of Toungoo and that of Mong Mong well into the mountains of Thailand.

Habitats: Soil, in jungles. Wet leaves. Under rotting logs. Manure.

Variation: The clitellum of two Thailand worms (Doi Sutep) reached to 16/17. Three Thailand worms (Doi Angka) had unpaired and median genital markings, one apparently postsetal in xvii but displacing 17/18, the other perhaps primarily postsetal in xix but reaching into xx.

Abnormality: No. 1. Male pores, in xix, nearer lateral margins of elliptical porophores. Genital markings, transversely elliptical, postsetal in xix or reaching into xx (? 19/20 unrecognizable ventrally). The clitellum reached 16/17, the only record of so much posterior extent for a Burmese worm. Halving of both mesoblastic somites at the seventeenth level might have explained location of the male pores. Unfortunately the segment in which the intestinal caeca came off from the gut was not determinable or if so was not recorded.

Transfer of capacity for developing prostates from xviii to xx if not also or previously to xix was involved in the ancestry of *P. anomala* and a supposedly similar mutational shift was originally thought to warrant recognition of a variety (*supina*).

No. 2. Posterior pair of spermathecae, lacking. Setae, of this Koopra worm, were fewer per segment than in others that were counted.

Parasites: *NEMATODA: Adungella manicatae* Timm, 1967, Loikaw. Parietal cysts, 2 mm. long, presumably of protozoan origin(?) in one Doi Angka worm, instead of being white as in all previous cases, were reddish to yellowish.

Remarks: Presence of sperm in spermathecae of Thailand worms supports assumptions as to method of reproduction in Burmese individuals.

Hearts of x may have been unrecognizable because empty in specimens where they were thought to have been aborted.

Pheretima minima

1893. *Perichaeta minima* Horst, in: Weber, M., *Zool. Ergeb. Reise Niederländisch Ost.Indien* **3**: p. 66. (Type locality, Tjibodas, Java. Holotype, in the Leiden Mus.)
1942. *Pheretima humilis* Gates, *Bull. Mus. Comp. Zool. Harvard College* **89**: p. 120. (Type locality, soil in pots with plants, Judson College, Rangoon, Burma. Type locality and types destroyed during World War II.)
1960. *Pheretima minima*, Gates, *ibid.*, **123**: p. 278.
1961. *Pheretima minima*, Gates, *Zool. Meded. Leiden* **37**: p. 298. (After examination of the type.) (*Pheretima humilis*, Gates, *Burma Res. Soc. Anniv. Pub.* No. 1: p. 57.)

Bithecal, pores minute, superficial, widely separated, at 5/6. Male pores, in xviii, minute, superficial, each in a circular disc. Female pore, median. Genital markings, small, circular tubercles with central translucence, presetal, paired and then at or near spermathecal pore levels or much more mesially in some of v-viii, unpaired and median in some of xvii, xix-xxi, rarely postsetal and then at or near spermathecal pore levels in v. Clitellum, setae unrecognizable externally, xiv–xvi. Setae, 40–55/ii, 40–47/iii, 75/vi, 60/vii, 54–58/viii, 46–52/xii, 3–12/xvi (ventrally), 38–44/xx, v/26–29, vi/27–33, xvii/10–12, xviii/6–10, xix/10–12. First dorsal pore, at 11/12,

12/13. Prostomium, epilobous, tongue open. (Unpigmented.) Segments, 85–97. Size, 20–56 by 1.5–2.0 mm.

Septa, 8/9 complete (usually or always?) but membranous, 9/10 aborted. Intestinal origin, in xv. Caeca, simple, in xxvii–xxiii. Typhlosole, simply lamelliform, from xxvii to region of xlviii–lxiv. Hearts, unaborted dorsal portions in viii passing to gizzard, in ix lateral, in x–xi (latero?)-esophageal, in xii–xiii lateroesophageal. Holandric. Testis sacs unpaired, U-shaped to annular, hearts of x–xi and anterior vesicles included. Seminal vesicles, small (juvenile?), in xi, xii. Prostates, in some or all of xvi–xxi, ducts 1–2 mm. long, twisted or in a hairpin loop. Spermathecae, large, diverticulum from median face of duct near parietes. GM glands, stalked and coelomic.

Reproduction: Presumably parthenogenetic. Male sterility is suspected in all morphs as yet referred to P. minima.

Distribution: Rangoon.

Buitenzorg and Tjibodas, Java. North Queensland, Australia. Oahu Island, Hawaii. (China, Hainan and Japan?) Louisiana, Mississippi, Alabama, Georgia, Florida, South Carolina, Virginia (new records).

The original home of the species is unknown.

Habitats: Earth, in flower pots (Rangoon). Never found outside of plant pots in Burma. Possibly imported shortly before it was discovered there.

Banana grove, clayey soil, porous and crumbly soil under trash (Hawaii).

Variation: Lengths greater than 30 mm. had been attained only by worms secured at elevations of 1,600–2,500 feet on an Oahu mountain.

Abnormality: Most spermathecae of specimens studied by the author certainly were abnormal. Normal shape of the organs is unknown.

Polymorphism: Athecal, monothecal, bithecal, and monoprostatic morphs are known. Both prostatic ducts are about equally developed in the monoprostatic morphs in which one prostatic duct sometimes appears to be a direct continuation of the sperm duct. An H morph is unknown or has not been recognized as such but is assumed to be bithecal. Several parthenogenetic morphs of China and Japan that have been given species names (cf. Gates, 1961: p. 306–308) may have had a common origin with the morphs presently referred to P. minima.

Remarks: Margins of v and vi in vicinity of spermathecal pores may be markedly tumescent but without the translucent areas usually indicative of closely crowded pores of composite glands. Characteristic genital markings were present in one specimen in which there were no corresponding glands at least in the coelomic cavities. Occasionally genital markings are lacking. Male porophores usually appear to be asymmetrically placed with reference to the equator of xviii and often seem to be postsetal. When symmetrically placed the porophores nearly reach both 17/18 and 18/19. Glands (in or on the parietes?) may be associated with the male porophores.

Adequate and correct characterization of minima genitalia awaits recognition of an H morph. Meanwhile, relationships would seem to be closest to the parthenogenetic P. zoysiae Chen, 1933, from Chekiang, P. fungina and muta Chen, 1938, from Hainan, each of which is distinguished from P. minima by quantitative differences, such as in size of soma and in number of setae. P. ishikawai Ohfuchi, 1941, known only from the original description of three types secured in a cave of a Japanese mountain, is not distinguishable from P. minima.

Pheretima morrisi

1892. *Perichaeta morrisi* Beddard, *Proc. Zool. Soc. London* **1892**: p. 166. (Type locality, Penang. Type, in the British Mus.)

1912. *Pheretima browni* (part) Stephenson, *Rec. Indian Mus.* **7**: p. 273.

1926. *Pheretima hawayana* var. *lineata* Gates, *ibid.*, **28**: p. 154. (Type locality, Taungyi. Types, none.)

1931. *Pheretima hawayana* var. *lineata* Gates, *ibid.*, **33**: p. 385.

1932. *Pheretima hawayana* var. *lineata* Gates, *ibid.*, **34**: p. 434.

1933. *Pheretima hawayana*, Gates, *ibid.*, **35**: p. 529.

1936. *Pheretima morrisi*, Gates, *ibid.*, **38**: p. 437.

1939. *Pheretima morrisi*, Gates, *Jour. Thailand Res. Soc. Nat. Hist. Suppl.* **12**: p. 100.

1955. *Pheretima morrisi*, Gates, *Rec. Indian Mus.* **52**: p.89.

1961. *Pheretima morrisi*, Gates, *Burma Res. Soc. 50th Anniv. Pub.* No. 1: p. 75.

1968. *Pheretima morrisi*, Gates, *Jour. Nat. Hist. London* **2**: p. 253. (Variation and distinctions from P. *loveridgei* n. sp., which may have been mistaken in the past for P. *morrisi*.)

Quadrithecal, pores minute, superficial, nearly $\frac{1}{2}$C apart, at 5/6–6/7. Male pores, in xviii, minute, superficial, each in a small, discoidal porophore. Female pore, median but occasionally paired. Genital markings, very rarely lacking, small discs, presetal and usually nearer segmental equators than the intersegmental furrows, unpaired and median in some or all of v–viii or vi–ix, xviii, paired and just median to spermathecal pore levels or more mesially, in some of vi–ix, occasionally two or more in xviii or xix and there nearer intersegmental furrows, almost constantly two just median to each male porophore and close to eq/xviii but with one pre- and the other post-setal. Clitellum, usually with setae ventrally in xvi, occasionally also in xiv and/or xv, often not reaching 16/17 and/or 13/14. Setae, 26–39/iii, 45–51/viii, 42–56/xii, 4–22/xvi, 46–59/xx, vi/16–28, xvii/16–23, xviii/10–17, xix/16–23. First dorsal pore, at 10/11. Prostomium, epilobous, tongue open. Color, in dorsum only except sometimes in i–iii or iv, grayish, yellow, brownish red, brownish, slate. Segments, 75–102, but usually 92–98. Size, 40–150 by 2½–6 mm.

Septa, 8/9–9/10 aborted, none thickly muscular. Pigment, reddish, in circular muscle layer. Intestinal origin, in xv. Caeca, simple, with several short lobes on ventral side, in xxvii–xxiv. Typhlosole, quite rudimentary or lacking. Hearts, of vi–vii (when complete) lateral, in viii unaborted dorsal portions to the gizzard, in ix (left or right) lateral, in x (latero?)-esophageal, in xi–xiii latero-esophageal. Blood glands, in v–vi, and in a small esophageal collar behind the gizzard. Lymph glands, in intestinal segments. Holandric. Testis sacs, paired, ventral. Seminal vesicles, of medium size, in xi, xii. Pseudovesicles, rudimentary, vertical cords, at most with slight thickening dorsally. Prostates, in xvii–xxiii, ducts 2–3 mm. long. Spermathecae, large enough to reach to or nearly to dorsal parietes, duct slender, shorter than ampulla, narrowed in the parietes, diverticulum from anterior face of duct at parietes, shorter or longer than main axis, slender, slightly and gradually widened entally. GM glands, stalked, coelomic, composite.

Reproduction: Presumably biparental as sperm are exchanged during copulation.

P. morrisi, like other species that are exotic in Burma, is able to remain active, in certain sites (presumably those with adequate moisture), throughout the dry season. Breeding, accordingly, in such favorable situations, is likely to be year round.

Distribution: Myitkyina (Myitkyina), Bhamo (Bhamo). Mogok (Katha). Namkham, Kutkai (North Hsenwi). Lashio, Maymyo, Taungyi, Htamsang, Kalaw (Shan States). Hsenwi (North Hsenwi). Tan Yang (Mong Yai). Nawnglon, Nawng Kham (Mang Lun). Meung Nawng (Pang Long). Kengtung, Mong Ko (Kentung). Lawng Neu (Mong Lem), Tengyueh, (Yunnan). Imphal (Manipur). Mount Popa (Myingyan). Usually at elevations of 2,500–4,800 + feet. Found, in the lowlands only at Myitkyina and Bhamo, in tropical Burma only on the Shan Plateau and Mount Popa. In April, at Taungyi and Htamsang, *P. morrisi* was the commonest species of earthworm.

Chiengmai (Thailand), at 1,000 feet.

Yunnan, Szechuan, Chekiang, Fukien Provinces, Hainan (China). Hong Kong. To elevations of 8,000 + feet in Szechuan. Taiwan.

Taiping, Batu Barendam, Malacca, Kuala Lumpur, Penang, Singapore (Malay Peninsula), Sumatra, Hawaii (elevations unknown).

United States: Texas, Louisiana, Mississippi, Alabama, Georgia, Florida, New York, Maine (greenhouses). Mexico,[20] Guatemala,[20] Barbados, Peru, Brazil, Argentina, Chile. England, Spain,[20] Italy,[20] Madeira,[20] Cape Verde Island, St. Helena, Ascension. Pakistan. India (Bombay, Brindaban, Udaipur, Dehra Dun, Kalimpong). From sea level (presumably) at Bombay to 6,500 feet in the Himalayas.

Types of *P. morrisi* were at first said to be from Penang but that origin was questioned by Michaelsen and much later the types (in the British Museum)

were found to be labeled as from Hong Kong. Recently the species has been recorded from both places.

Much of the distribution, as known today, clearly is a result of transportation and the greenhouse habitat and the interceptions prove that man has been involved. The original home of *P. morrisi* is believed to be in China but how much of the distribution in that country (probably but partly known) is due to man remains to be determined.

Habitats: Soil (sometimes red and then presumably of lateritic origin) of gardens, lawns, open areas, under bamboo, under banyan, of jungles. Greenhouses (New York and Maine, only records for those states), England (only record for that country). To be expected in greenhouses of other countries. Habitats, in the Malay Peninsula, are unknown and may have been artificial.

Variation and Abnormality: Absence of the anterior pair of spermathecae was recorded for two specimens. Six are known to have had one or two spermathecae in viii, the pore or pores at 7/8, the single organ of viii in one of those worms being adiverticulate.

Regeneration: Tail regeneration has been noted at various levels behind 66/67. Number of segments in regenerates, 7–13, the maximal number at 67/68, the minimal obtained as far back as 82/83.

Parasites: Cysts containing pseudonavicella spores, nematode eggs, and "multicellular bodies of a greenish color" were found in coagulum discharged through the more anterior dorsal pores.

Some records of parasites infesting *P. hawayana* quite possibly should have been attributed to *P. morrisi* which for long was not distinguished from the other species.

Remarks: Relationships of *P. morrisi* presumably are with Chinese species having similar genital markings and GM glands, species such as *P. aspergillum, diffringens, hawayana* and *robusta*, but probably are closer with *hawayana*.

Slenderness of spermathecal duct and the diverticulum, as in *hawayana*, is characteristic of the species. The diverticulum may be nearly straight, sinuous, with some looping of an ectal portion. A distally terminal portion may be pear-shaped, spheroidal, or ellipsoidal but spermatozoa are not confined to that portion which looks like a typical seminal chamber of various other species. Here, however, sperm usually extend through all of ental half of the diverticulum or occasionally even more.

Surface of a dorsal portion of pseudovesicles once was roughened so as to have a regularly granular appearance.

Pheretima nugalis

1931. *Pheretima nugalis*, Gates, *Rec. Indian Mus.* **33**: p. 402. (Type locality, Kyaikmaraw. Types, none.)
1932. *Pheretima nugalis*, Gates, *ibid.*, **34**: p. 401.
1936. *Pheretima nugalis*, Gates, *ibid.*, **38**: p. 439.
1961. *Pheretima nugalis*, Gates, *Zool. Meded. Leiden* **37**: p. 305.

---

[20] Based on unpublished records of interceptions by the U. S. Bureau of Plant Quarantine (specimens identified by the author).

Bithecal, pores minute and superficial, *ca.* ¼C apart, in very small tubercles, at 5/6. Male pores, in xviii, minute and superficial, each in a transversely elliptical, discoidal porophore reaching nearly to 17/18 and 18/19. Female pore, median. (Genital markings, none.) Clitellum, setae unrecognizable externally, xiv-xvi. Setae, v/11–15, vi/13–16, xvii/12–15, xviii/6–10, xix/12–14. First dorsal pore, at 5/6. Prostomium, epilobous, (tongue open?). Color, only in dorsum, brownish, (pigment red? In circular muscle layer?). Segments, 70–87. Size, 25–30 by 2–2½ mm.

Septa, 8/9–9/10 aborted, none strongly thickened. Intestinal origin, in xv. Caeca, simple, in xxvii–xxv. (Typhlosole?) Hearts, of x-xiii latero-esophageal(?). Holandric. Testis sacs, unpaired and ventral. Seminal vesicles, large, in xi, xii. Prostates, in xviii–xxiv, ducts slender, straight or in a *U*-loop, *ca.* 1½ mm. long, Spermathecae, large, duct shorter than ampulla, diverticulum from anterior face of duct near parietes, about as long as main axis, with a long stalk and a short seminal chamber.

Reproduction: Presumably biparental as sperm are exchanged during copulation.

Distribution: Kyaikmaraw (Amherst).

Remarks: The widening and irregularity of a middle portion of the spermathecal diverticulum in the types (1931: fig. 34, p. 403 and *cf.* fig. 4, p. 440, 1936) may have been an aberration.

This species is known only from published records of five specimens obtained at the type locality. The collector was able to get additional material whenever the spot was revisited but each time was able only to find a couple or so specimens.

Pheretima osmastoni

1907. *Pheretima osmastoni* Michaelsen, *Mitt. Naturhist. Mus. Hamburg* 24: p. 163. (Type locality, Port Blair. Types, in the Indian and Hamburg Museums.)
1909. *Pheretima osmastoni*, Michaelsen, *Mem. Indian Mus.* 1: p. 191.
1923. *Pheretima osmastoni*, Hertling, *Zeitschr. Wiss. Zool.* 120: p. 180. Stephenson, (*The Fauna of British India*), *Oligochaeta*, p. 307.
1925. *Pheretima osmastoni*, Stephenson, *Rec. Indian Mus.* 27: p. 61.
1932. *Pheretima osmastoni*, Gates, *ibid.*, 34: p. 476.
1960. *Pheretima "osmatoni,"* Gates, *Bull. Comp. Zool. Harvard College* 123: p. 268.

Sexthecal, pores minute, *ca.* 2/7C apart, each at center of a small porophore slightly retractile(?) into parietes, at 6/7–8/9. Male pores, in xviii, minute, towards lateral margin of a transversely elliptical thick disc which bears centrally the opening of a penisetal follicle and mesially that of a gland, the disc protrusible from a slight invagination. Female pores, closely paired (?). Genital markings, small, circular discs, each with a central pore, closely crowded in irregular transverse rows, postsetal, median or paired patches in viii, x, xii, xiii (but often lacking).

Clitellum, xiv-xvi. Setae, 28/v, 50/ix, 58/xiii, 65–84/xx, vii/8–14, viii/10–17, xviii/10–20, (present ventrally in xvi ?). First dorsal pore, at 12/13. Prostomium, epilobous, tongue open. Color, only in dorsum, red, lacking in longitudinal strips each of which does not reach an intersegmental furrow but does include aperture of a setal follicle. Segments, 126–153. Size, 185–320 by 8–11 mm.

Septa, 8/9 membranous, 9/10 membranous and attached peripherally to 10/11. Intestinal origin, in xv. Caeca, simple, in xxvii–xxiv. Typhlosole, simply lamelliform, 1+ mm. high, ending in civ or anteriorly. Hearts, in ix (on both sides) lateral, of x (latero-?)-esophageal, of xi–xiii lateroesophageal. Lymph glands, large, in intestinal segments. Holandric. Testis sacs, unpaired and ventral. Seminal vesicles, large, in xi, xii–xiii, each with a long primary ampulla. Pseudovesicles, in xiii and xiv, with saccular dorsal ampullae. Prostates, in xvii–xxii, ducts long, looped, thicker ectally. Penial setae, 1.13–1.25 (+?) mm. long, 0.20–0.28 mm. thick, straight, ornamented ectally with circles of very fine teeth, tip rounded. Spermathecae, duct shorter than ampulla, diverticulum from anterior face of duct in parietes, longer than main axis, seminal chamber very short and small. Collar (glandular?) usually on parietes around spermathecal duct, better developed in an anterior portion. Gland to male porophore, in xv–xxv, bilobed, composed of numerous glands each with a muscular stalk passing to a common *T*-shaped duct with thick and straight horizontal limbs and a slender, vertical leg especially narrowed in the parietes. GM glands, coelomic, with long stalks.

Reproduction: Presumably biparental as sperm are exchanged during copulation.

Breeding season may include December though some of the specimens secured in that month were not fully mature.

Distribution: Wimberleyganj-Port Blair, Mount Harriet, Minnie Bay, Port Bonington (Andaman Islands).

From sea level to 900 feet.

Habitat: Soil in dense jungle.

Remarks: Depression of the deeply delimited spermathecal porophore presumably was responsible for producing an appearance of a large secondary spermathecal pore. Male porophores also are so deeply delimited as to seem to be in invaginations but if so margins of the invagination may be apposable so as to cover the porophore and close off the invagination from the exterior.

A second penisetal follicle was present in the most recently examined specimen and was median to the duct of the bilobed gland but a discrete follicle aperture was not recognized. The large, bipartite gland is an aggregation of glands just like those of the discrete genital markings but with ducts united (?) to form one massive stalk.

The prostatic gland develops much more slowly or later than the duct.

Pheretima papilio

1961. *Pheretima papilio*, Gates, *Zool. Meded. Leiden* **37**: pp. 309, 311, 312. *Burma Res. Soc. 50th Anniv. Pub.* No. 1: p. 57.

Bithecal, pores nearly ½C apart, near 5/6 in anterior margin of vi. Male pores, in xviii, small slits, each within a large, tough, longitudinally elliptical porophore in region between eq/xvii and eq/xix. Female pore, median. (Genital markings, except for the porophores, none.) Clitellum, setae unrecognizable externally, xiv-xvi. Setae, 60–80/xx, 67–71/li, v/23-36, xvii/18–22, xviii/3–14, xix/16–24. First dorsal pore, at 12/13. Prostomium, epilobous, tongue open. Color, only in dorsum, red, to slate in front of clitellum, to brownish posteriorly, (pigment red and in circular muscle layer?). Segments, 87–114. Size, 60–138 by 3–6 mm.

Septa, 8/9–9/10 aborted, none markedly muscular. Intestinal origin, in xv. Caeca, simple but with shortly digitiform protuberances of dorsal and/or ventral margins, in xxvii–xxiii. Typhlosole, none(?). Hearts, of ix lateral but usually present only on one side, of x-xiii (latero?)-esophageal. Blood glands, in v. Holandric. Testis sacs, unpaired and ventral. Seminal vesicles, large, each with a primary ampulla within a cleft of dorsal end, in xi, xii. Prostates, large, in xvi-xxi, ducts 2½–8 mm. long, each in a loop with the ectal limb thickened. Spermathecae, duct shorter than ampulla, diverticulum from anterior face of duct at or within parietes, comprising a short stalk, a more or less regularly zigzag-looped, thicker middle portion and an ellipsoidal terminal chamber.

Reproduction: Presumably biparental as sperm are exchanged during copulation.

Distribution: Mergui, Tavoy, Amherst, Thaton, Toungoo Districts, in the latter across the Sittang Valley in the Pegu Yomas and in the same district up on the Karen Hills of the Shan Plateau. In Tenasserim Division probably from the sea to the Thailand border.

Remarks: Gaps in the longitudinal musculature at mD under intersegmental furrows may be present as far forward as 5/6 but sphinctered dorsal pores were recognized in front of 12/13 only at 11/12 and then but rarely.

The invagination that is so obvious when protuberant into v of *P. hiulca* may be present, though much smaller, in other forms and there responsible for an appearance of thickening of the spermathecal duct within the parietes. If so the real duct must be characterized in this species as much narrowed within the body wall.

Four geographical races originally were thought to be present but intermediates were found in Thaton where much collecting must be repeated as well as at a trans-Irrawaddy site in Prome District and through the more difficult Pegu Yomas.

P. papilio does not appear to be closely related to any of the other bithecal Burmese species with spermathecal pores at or near 5/6. Relationships may be with taxa having sulcate male porophores. An octothecal representative of an ancestral type has not been found and may no longer be in existence. A subsequent stage in the sequence may be represented by the sexthecal *P. sulcata* presumably belonging in the mountains of the peninsular region and in the vicinity of the Thai border. A quadrithecal stage in the sequence may then be represented today by *P. youngi* in which the pores of 5/6 recently had migrated back to eq/vi. From such a pre-*youngi* stage, *P. papilio* then would be derived by elimination of the posterior pair of spermathecae and a slight dislocation posteriorly of the pores of the anterior pair.

The subspecies of *P. papilio*, like those of the *Drawida caerulea-decourcyi* group, evidence nice evolutionary series of increasing specializations in a south to north direction.

Pheretima papilio papilio

1930. *Pheretima papilio* Gates, *Rec. Indian Mus.* **32**: p. 316. (Type locality, San Hlan. Types, none.)
1932. *Pheretima papilio* var. *typica*, Gates, *ibid.*, **34**: p. 401.
1933. *Pheretima papilio* var. *typica*, Gates, *ibid.*, **35**: p. 539.
1960. *Pheretima papilio papilio*, Gates, *Bull. Mus. Comp. Zool. Harvard College* **123**: p. 270.

Spermathecal pores, superficial (?). Male porophores, thickenings (glandular?) of an outer portion of the parietes and beneath the longitudinal musculature in which seminal grooves are lacking. Prostatic ducts, 2½–4 mm. long.

Distribution: Sindin, Zowai, Tharrabyin (Mergui). San Hlan, Kamaungthwe River region (Tavoy). Ye, Kawkareik (Amherst).

Specimens from Iriomote Island, Riu Kyu Islands, were referred (Ohfuchi, 1956, *Jour. Agric. Sci. Tokyo* **3**: p. 140) to *P. papilio*. Color, location of spermathecal pores and of intestinal origin, all contradict the identification.

Remarks: If an invagination like that of *hiulca* but much smaller is present the external spermathecal aperture may be secondary in which case the primary pore presumably is minute and invaginate.

An evolutionary derivation of *P. papilio*, on an assumption that it belongs to a sulcate lineage, was suggested above though subspecies *papilio*, presumably the most primitive, lacks the furrows.

Pheretima papilio insignis

1932. *Pheretima papilio* var. *insignis* Gates, *Rec. Indian Mus.* **34**: p. 404. (Type locality, Thaton. Types, none.)
1960. *Pheretima papilio insignis*, Gates, *Bull. Mus. Comp. Zool.* **123**: p. 270.

Male porophores, thicker than in *papilio* but only slightly protuberant into coelomic cavity of xviii, and with a circuitous seminal groove (1932: fig. 5) from the male pore.

Distribution: Thaton (Thaton). Near west bank of Irrawaddy River somewhat south of Prome (Prome), at a former village site.

Probably fairly common throughout much of Thaton district from which the species was carried, presumably by man, to the trans-Irrawaddy site. The Telegu collector who secured the Prome worms, when subsequently questioned, was able almost instantly to mention a particular locality in Thaton from which the worms could have been taken.

Records of 15 specimens from the type locality are all that are now available.

Remarks: Adequate characterization of the sub-species is impossible because of loss of most of the material. More variation in characters of systematic importance is to be expected than in the range of subspecies *papilio*.

Pheretima papilio hiulca

1932. *Pheretima papilio* var. *hiulca* Gates, *Rec. Indian Mus.* **34**: p. 402. (Type locality, Blachi. Types, none.)

1933. *Pheretima papilio* vars. *hiulca* and *fracta* Gates, *ibid.*, **35**: p. 539. (Type locality of *fracta*, Su Law or Pray Law. Types, none.)

1960. *Pheretima papilio hiulca*, Gates, *Bull. Mus. Comp. Zool. Harvard College* **123**: p. 270.

Spermathecal pores, each at posterior margins of an opening into an invagination or at center of a discoidal porophore within such an invagination that reaches into coelomic cavity of v though not at once so recognizable because of concealment within a seemingly much thickened spermathecal duct. Male porophores, much thickened and protuberant into coelomic cavity of xviii. Prostatic ducts, 5–8 mm. long. The posterior wall of the spermathecal invagination is much thickened and mainly composed of a tough porophore.

Distribution: Blachi, Karen Hills of Toungoo District on the Shan Plateau. Su Law and Pray Law, on the eastern side of the Pegu Yomas (Toungoo).

Remarks: The protuberance into coelom of xviii in the Yomas specimens (var. *fracta*) was smaller and *hemispheroidal* but in the Blachi worms was in the shape of cone 1½ mm. high containing a small lumen with a pointed protuberance possibly bearing the male pore.

Records now available are of 17 specimens only.

Pheretima papulosa

1896. *Perichaeta papulosa* Rosa, *Ann. Civ. Mus. Sto. Nat. Genova* **36**: p. 525. (Type locality, Balighe. Types, in the Genoa Mus.)

1900. *Amyntas papulosus*, Beddard, *Proc. Zool. Soc. London* 1900: p. 644. *Pheretima papulosa*, Michaelsen, *Das Tierreich* **10**: p. 291.

1901. *Amyntas papulosus*, Beddard, *Proc. Zool. Soc. London* **1900**: p. 892.

1903. *Pheretima papulosa*, Michaelsen, *Die geographische Verbreitung der Oligochaeten* (Berlin), p. 98.

1922. *Pheretima papulosa* var. *sauteri* Michaelsen, *Capita Zool.* **1, 3**: p. 26. (Type locality, Dorf Kosempo. Type, in the Leyden Mus.)

1932. *Pheretima composita* Gates, *Rec. Indian Mus.* **34**: p. 430. (Type locality, Kengtung, Types, none.) *Pheretima papulosa*, Stephenson, *Ann. Mag. Nat. Hist.*, Ser. 10, **9**: p. 224.

1939. *Pheretima papulosa*, Gates, *Jour. Thailand Res. Soc Nat. Hist. Suppl.* **12**: p. 101.

1959. *Pheretima papulosa*, Gates, *American Mus. Nat. Hist. Novitates*, No. 1941: p. 18.

?

1933. *Pheretima rockefelleri* Chen, *Contrib. Biol. Lab. Sci. Soc. China*, (*Zool.*) **9**: p. 238. (Type locality, Linhai, Chekiang. Types, one in the U. S. Natl. Mus. Others?)

1935. *Pheretima rockefelleri*, Gates, *Lingnan Sci. Jour.* **14**: p. 454.

Sexthecal, pores minute, superficial, ¼C or more apart, at 5/6–7/8. Male pores, in xviii, minute, superficial, *ca.* ¼C apart, each in a small area or disc. Female pore, median. Genital markings, small circular discs, in transverse rows, in vii–ix (pre- and post-equatorial), xvii (in pre- and post-equatorial patches of 2–3 rows), in xviii pre- and post-equatorial, in xix (pre-equatorial). Clitellum, xiv–xvi. Setae, usually present ventrally in xiv–xv and in xvi always, sigmoid or straight, with short transverse rows of very fine spines ectally; 54–60/v, 61/ix, 60/xii, 62–66/xiii, 4–6/xvi, 56–62/xix, 57–67/xx, vi/13–16, vii/14–17, xvii/14–19, xviii/11–15, xix/16–22. First dorsal pore, at 12/13. Prostomium, epilobous. Color, none(?). (Pigment, lacking?). Segments, 96–119. Size, 45–78 by 3–5 mm.

Septa, 8/9–9/10 aborted, 5/6–7/8 muscular, 10/11–12/13 slightly muscular. Intestinal origin, in xvi (usually?). Caeca, simple, in xxvii–xxii. (Typhlosole?) Hearts, of x-xiii (latero?)-esophageal. Holandric. Testis sacs unpaired, of x horseshoe-shaped, of xi cylindrical, hearts of x-xi and vesicles of xi included. Seminal vesicles, small (juvenile?), in xi, xii. Prostates, in xvi–xxi, ducts 3–4 mm. long, looped, thicker ectally. Spermathecae, duct slender(?), as long as ampulla(?), abruptly narrowed in parietes, diverticulum from anterior face of duct at body wall, with short, slender stalk and longer, looped seminal chamber. GM glands, coelomic, stalked.

Reproduction: Data as to maturation and exchange of sperm in Burmese worms no longer are available but parthenogenesis is suspected, primarily because of the small size of the seminal vesicles.

Distribution: Kengtung (Kengtung). Mong Mong Valley (Yunnan, China).

Biserat (Jalor State, Thailand). Sumatra (Balighe). Formosa (Dorf Kosempo). (Chekiang Province, China? Hong Kong? Taiwan?)

Remarks: Some 30-odd specimens have been referred to *P. papulosa* and 65 to *P. rockefelleri*.

Male porophores seem to be rather variable, fairly large circular discs (Formosa), without definite demarcation mesially and laterally (Sumatra), grayish crescentic areas facing mesially (Burma), slightly protuberant and rather conical but surrounded by a circumferential furrow so deep as to suggest a shallow invagination (*P. rockefelleri*). Genital marking rows may reach to level of spermathecal pores or, behind the clitellum, to beyond the male pore levels.

Var. *sauteri* was distinguished by location of intestinal caeca in xxix–xxvi of the single specimen which may have been abnormal. *P. rockefelleri* is distinguished by quantitative differences; larger soma size, more segments, more setae, genital markings fewer and larger, first dorsal pore at 11/12 instead of 12/13. Spermathecae are rudimentary or abnormal in one type (U. S. N. M.) and prostates were lacking on one or both sides of others. Concurrent trends toward an AR morph (also *cf. P. malaca*) in a population presumably already parthenogenetic seemingly are indicated.

Specimens from the Riu Kyu Islands, were referred (Ohfuchi, 1956, *Jour. Agric. Sci. Tokyo* 3: p. 164) to *P. papulosa* var. "*sauteria.*" The muscularization of 8/9, as well as other characters, contradict the identification.

Pheretima pauxillula

1936. *Pheretima pauxillula* Gates, *Rec. Indian Mus.* 38: p. 442. (Type locality, Kutkai. Types, in the Indian Mus.)
1960. *Pheretima pauxillula*, Gates, *Bull. Mus. Comp. Zool. Harvard College* 123: p. 270.
?
1946. *Pheretima digna* Chen, *Jour. West China Border Res. Soc.*, B, 16: p. 132. (Type locality, Sha-P'ing Pa, Chungking, Szechuan Province, China. Types, none?)

Sexthecal, pores minute, superficial, at 4/5–6/7. Male pores, in xviii, minute and superficial, each at center of a circular discoidal porophore almost reaching 17/18 and 18/19. Female pore, median. (Genital markings, none.) Clitellum, setae usually unrecognizable externally, xiv-xvi. Setae, v/15, vi/19, xvii/12–14, xviii/7–11, xix/13–14. First dorsal pore, 12/13. Prostomium, epilobous(?). Color, none (unpigmented). Segments, 84–91. Size, 25–47 by 1–2 mm.

Septa, 8/9–9/10 aborted, none thickly muscular. Intestinal origin, in xvi. Caeca, simple, in xxvii-xxv. (Typhlosole?) Hearts, of ix lateral but usually lacking on one side, of x (latero?)-esophageal, of xi–xiii lateroesophageal(?). Holandric. Testis sacs horseshoe-shaped or annular (?), above the dorsal blood vessel, vesicles of xi included. Seminal vesicles, small, in xi, xii. Prostates, in xvi-xxi, ducts *ca.* 1¼ mm. long, muscular, spindle-shaped, with a quirk in entalmost portion. Spermathecae, duct slender, as long as to longer than the ampulla, diverticulum from duct at or within the parietes, shorter than the main axis, seminal chamber ovoidal, spheroidal to club-shaped, slightly thicker and shorter than the stalk.

Reproduction: Absence of spermatozoal iridescence in spermathecae and on male funnels, together with the juvenile state of the seminal vesicles, especially in worms collected during the probable breeding season, now are sufficient reasons for suspecting male sterility and parthenogenesis.

Distribution: Kutkai (North Hsenwi). Peng Sai (Mang Lun). At elevations of *ca.* 3,500–4,500 feet.

Habitats: Soil in compound of Dak bungalow (Kutkai). Rotten leaves in dense jungle (Peng Sai).

Remarks: Known only from 18 specimens. The types were obtained from a site to which they could have been introduced with potted plants from some other part of the country. The species never was represented in rather large collections subsequently secured in and from the immediate vicinity of Kutkai.

P. digna Chen, 1946, known only from the original description of five specimens secured at Chungking, Szechuan, has a middle portion of its spermathecal diverticula looped. Other differences from *P. pauxillula* appear unimportant. Some of the types (number not stated) have no spermathecae in v, a loss, if the quadrithecal worms are of the same species, that is to be expected after reproduction becomes parthenogenetic. No instances are presently known in which evolution of parthenogenetic polymorphism has resulted in, as it were, unlooping a middle portion of a spermathecal diverticulum.

Pheretima peguana

1890. *Perichaeta peguana* Rosa, *Ann. Mus. Civ. Sto. Nat. Genova* 30: p. 113. (Type locality, Rangoon. Type, in the Genoa Mus.)
1895. *Perichaeta peguana*, Beddard, *A Monogr. of the Order of Oligochaeta* (Oxford), p. 403.
1898. *Perichaeta peguana*, Rosa, *Ann. Mag. Nat. Hist.*, Ser. 7, 2: p. 289.
1899. *Amyntas peguanus*, Michaelsen, *Mitt. Naturhist. Mus. Hamburg* 16: p. 7.
1900. *Amyntas peguanus*, Beddard, *Proc. Zool. Soc. London* 1900: p. 628. *Pheretima peguana*, Michaelsen, *Das Tierreich* 10: p. 292.
1903. *Pheretima peguana*, Michaelsen, *Die geographische Verbreitung der Oligochaeten* (Berlin), p. 98.
1909. *Pheretima peguana*, Michaelsen, *Mem. Indian Mus.* 1: p. 110.
1923. *Pheretima peguana*, Stephenson, (*The Fauna of British India*), *Oligochaeta*, p. 308.
1925. *Pheretima peguana*, Gates, *Ann. Mag. Nat. Hist.*, Ser. 9, 16: p. 561.
1926. *Pheretima peguana*, Gates, *ibid.*, 17: p. 463; *Jour Bombay Nat. Hist. Soc.* 31: pp. 182, 183; *Jour. Burma Research Soc.* 15: p. 211; *Rec. Indian Mus.* 28: p. 161.
1929. *Pheretima peguana*, Stephenson, *ibid.* 31: p. 238.
1930. *Pheretima peguana*, Gates, *ibid.*, 32: p. 318.
1931. *Pheretima peguana*, Gates, *ibid.*, 33: p. 404.
1932. *Pheretima peguana*, Gates, *ibid.*, 34: p. 481. Stephenson, *Bull. Raffles Mus., Singapore*, No. 7: p. 49.
1933. *Pheretima peguana*, Gates, *Rec. Indian Mus.* 35: p. 540.
1936. *Pheretima peguana*, Gates, *ibid.*, 38: p. 444.
1937. *Pheretima peguana*, Gates, *Bull. Mus. Comp. Zool. Harvard College* 80: p. 326.
1939. *Pheretima peguana*, Gates, *Jour. Thailand Res. Soc. Nat. Hist. Suppl.* 12: p. 102.

1955. *Pheretima peguana*, Gates, *Rec. Indian Mus.* **52**: p. 89.
1957. *Pheretima peguana* + *P. saigonensis* Omodeo, *Mem. Mus. Sto. Nat. Verona* **5**: pp. 325, 327. (Type locality of *saigonensis*, Saigon. Types, 3, in Verona Mus.?)
1961. *Pheretima peguana*, Gates, *American Midland Nat.* **66**: p. 62; *Burma Res. Soc. Anniv. Pub.* No. **1**: p. 57.

PARASITES (PROTOZOA): Dissanaike, 1953, *Ceylon Jour. Sci.* **25**. Boisson, 1957, *Ann. Sci. Nat. Paris* **19**. (NEMATODA): Timm, 1962. *Biologia, Lahore* **8**.

Sexthecal, pores minute, superficial, *ca.* 2/7C apart, at 6/7–8/9. Male pores, in xviii, minute, each in a circular porophore on posterior wall near roof of a deep vertical parietal invagination with transversely slitlike lumen and aperture, the latter at the segmental equator. Female pore, median. Genital markings, transversely elliptical, with firm glistening surface and obvious central aperture, two pairs, across 17/18, 18/19. Clitellum, setae unrecognizable externally, xiv–xvi. Setae, *a, b* follicles of postclitellar segments enlarged and more protuberant into coelomic cavities, 30–40/iii, 45–55/viii, 55–66/xii, 55–65/xx, vii/15–24, viii/17–25, xvii/11–16, xviii/6–14, xix/11–17. First dorsal pore, at 12/13. Prostomium, epilobous, tongue open. Color, only in dorsum, reddish. Segments, 97–124, but usually 110–124. Size, 140–240 by 5–8 mm.

Septa, 8/9–10/11 aborted, 11/12–12/13 muscular. Pigment, reddish, in circular muscle layer. Intestinal origin, in xv. Caeca, simple, in xxvii–xxii. Typhlosole, rather small, to rudimentary, ending in region of 59th–86th segments. Hearts, in viii unaborted dorsal portions to gizzard, in ix (usually on one side only) lateral, in x (latero?)-esophageal, in xi–xiii latero-esophageal. Blood glands, in v. Lymph glands, in intestinal segments. Holandric. Testis sacs, paired and ventral. Seminal vesicles, large of xi reaching forward alongside gizzard, of xii smaller, each with a large primary ampulla distinctly demarcated though not protuberant from the main mass. Pseudovesicles, rudimentary, in xiii, (in xiv too small to be detected in dissections?). Prostates, in xvi–xxi, ducts 3–5 mm. long. Spermathecae, small to medium-sized, duct shorter than ampulla, with thick wall and narrow lumen, much narrowed in the parietes, diverticulum from anterior face of duct at parietes, longer than main axis, with slender stalk, longer and thicker mid-portion irregularly looped, terminal seminal chamber spheroidal to ovoidal. GM glands, nearly spheroidal, with thick muscular wall and small lumen, slightly protuberant into coelom.

Reproduction: Presumably biparental as sperm are exchanged during copulation. Evidence indicative of parthenogenesis has not been seen.

Distribution: Victoria Point, Nyaungbinkwin, Mergui, Kala Island, Labaw (Mergui). Tavoy, Maungmagaun, Tanbin Hills, Migyaunglaung, Nyaungdon, Pyinthadaw, Kawlet Chaung, Siyigyan, Zinba, Heimza Basin (Tavoy). Ye, Moulmein,

Chaungson, Mupun, Martaban, Kyaikmaraw, Kya In, Kya In Seik Kale, Kawkareik, (Amherst). Thaton, Boyagyi, Bilin, Sittang, Kyaikto, Taungzun, Naunggala, Duyinzeik, Aungsaing (Thaton). Pyapon, Thanchitaw, Thameintaw, Bogale, Dedaye (Pyapon). Myaungmya (Myaungmya). Rangoon, Syriam, Kyauktan, Twante, Kungyangon, Thongwa, Kayan (Hanthawaddy). Insein, Hmawbi, Taikkyi, Wanetchaung, Taukkyan, Hlegu, Hlawga (Insein). Maubin, Yandoon, Pantanaw, Danubyu (Maubin). Bassein, Coomzamu, Kochi, Thinbawgyin, Padaukchaung (Bassein). Ler-mu-htee (Salween). Pegu, Paung, Thanatpin (Pegu). Tharrawaddy, Ngapugale, Letpadan, Thonze (Tharrawaddy). Henzada, Ingabu, Zalun (Henzada). Prome, Paukkaung, Laboo (Prome). Sandoway, Ngapoli, Taungup, Kywegale (Sandoway). Toungoo, Shwegyin, Tantabin, Kyaukkyi, Maw Pah Der (Toungoo). Thayetmyo, Thanbula, Allanmyo (Thayetmyo). Kyaukpyu, Ramree Island, Myebon (Kyaukpyu). Pyinmana, Pyigyaung (Yamethin). Taungdwingyi (Magwe), Akyab, Padali, Myohaung (Akyab). Meiktila, Mahlaing (Meiktila). Mt. Popa (Myingyan). Mandalay, Myohaung (Mandalay). Sagaing, Tada-U (Sagaing). Kin-U, Kyaukmyaung (Shwebo). Mogok, Naba (Katha). Bhamo (Bhamo). Lonton, Myitkyina, Alam (Myitkyina). Mingin (Upper Chindwin).

From sea level to 3,800 feet. Not found in the Andaman Islands, the Arakan Hill Tracts and the Chin Hills as well as in the Pegu Yomas and most of the Shan Plateau, nor within central Burma in Minbu, Pakkoku, Kyaukse, and Lower Chindwin districts. Presence at Mingin in Upper Chindwin district, at Maw Pah Der in the Toungoo Karen Hills, at Mogok in Katha district, may be due to recent introductions. The species is anthropochorous and much or even all of the Burma distribution may have resulted from transportation. If, however, *P. peguana* is autochthonous anywhere in Burma it is most likely to be so in the peninsular section from Victoria Point north to the Salween district and presumably also in adjacent portions of Thailand and the Malay Peninsula.

Chantaboon, Bangkok, Chiengmai, Chiengrai, Doi Sutep, Ban Sa-iep (in the basin of the Me Yom), Mu'ang Pong (Ban Muang), Thailand. From sea level to 1,600 feet (at Doi Sutep). Saigon, Vietnam.

Penang, Taiping, Singapore, Malay Peninsula.

Java (Tandjong Priok, Djakarta). Borneo (Lombok, Bay, Labuan).

One specimen recently was recorded (Halder & Julka, 1967, *Current Sci.* **36**: p. 467) from St Paul's Cathedral, Calcutta. Accidental transfer by man from southeast Asia presumably was involved. Although that city is the seat of the Zoological Survey of India, little is known of the megadrile fauna of Calcutta and vicinity.

Specimens from Ishigaki I., Riu Kyu Is., were referred (Ohfuchi, 1956, *Jour. Agric. Sci. Tokyo* **3**:

p. 171) to *P. peguana*. Muscularity of septum 8/9, the genital markings, and the intestinal origin contradict the identification.

Habitats: Soil of gardens, lawns, banana groves, and numerous other sites in cities and towns as well as in various kinds of adjacent jungles.

Biology: Individuals of this species appear early in the rainy season and they may remain active, in appropriate conditions at Rangoon, well into the dry season. Clitellate specimens were secured in the period July–February during a considerable portion of which breeding may be possible.

Variation and Abnormality: *P. peguana* has a pattern of GM location so rigidly determined that the only divergence therefrom is absence of one or more of the markings. In a series of 834 specimens, genital markings were absent as follows: On left side at 17/18 (4 specimens), on right side (2), on both sides (1), on both sides at 18/19 (1). Additional records previously or subsequently, all absent except one on left side at 17/18 (1), except the anterior pair (1).

Instances just cited are thought to be of individuals in which the missing organs never developed. In an aberrant specimen the posterior genital markings may have completely disappeared without leaving any recognizable trace though associated glands were represented at the proper sites by nondescript vesicles with white or brown debris. Male funnels, sperm ducts and male terminalia of the right side failed to develop in one worm in which the right genital markings were wholly in xviii. That worm as well as several other abnormals suggest that the genital markings and their glands, in spite of appearances, really originate in xviii, but perhaps get pushed anteriorly or posteriorly as the male pore invaginations develop. Other abnormals now can be explained as a result of abortion of or halving of mesoblastic somites during embryonic development. Thus #3 and #4 (1955) resulted respectively from abortion of the right or the left somite (also #3, 1926) at the ninth level. Two specimens (#6, 7, 1925), resulted from halving of the right somite at the 18th level. Two specimens (#3, 1925, #2, 1926), may have aborted the right somite at the 16th level. Abortion of the right somite at 17th level (#5, 1925) presumably was associated with a later halving of the right somite at the 18th level. Conditions of another (#4, 1925) presumably could have arisen by abortion of the left somite at the 9th level and halving of the left somite at the 18th level.

Parasites: SPOROZOA: *Monocystis minor, Dirhynchocystis sacciformis, Mastocystis denticulata* and *tuzeti* Boisson, 1957, seminal vesicles, Vietnam. NEMATODA: *Synoecnema gatesi* Timm, 1962, coelomic cavities, Singapore. Cestoda. *Amoebotaenia cuneata* (Linstow, 1872) Burma.

Remarks: The male-pore invagination is eversible to a one mm. high, columnar, penial body with the male porophore on its ventral end. Invaginations of relaxed specimens usually were more or less completely everted but in strongly contracted individuals always were fully retracted. Invaginations for reception of intromittent organs during copulation seemingly have not yet evolved.

Septum 8/9, according to Beddard (1900: p. 628) "was distinctly present in specimens examined by myself" presumably from Penang, but the record never has been confirmed. Although it is possible for 8/9 to be retained and perhaps even thickened, such a development now seems likely to be found only very rarely and not in more than a single individual at any one time or place.

Hearts of x were found only in one specimen which was believed to have retained an ancestral condition that usually is lost during development. Admittedly, empty hearts of x often are unrecognizable but so large a number of failures to see hearts that should be present seems unusual.

The differences by which *P. saigonensis* was distinguished are quantitative, trivial, and of characters liable to more or less individual variation in the genus. Furthermore, septum 10/11 always is absent in *P. peguana* as well as in the types of *saigonensis*.

P. peguana illustrates one of the difficulties encountered when attempting to monograph a poorly known region without boundaries of a nature to be respected by earthworms. In the original manuscript, the species was said to be related to another known only from a few individuals that had been carried by man to the remote type locality. To permit discussion of that relationship, the related form, though not known to be present in any region adjoining Burma, was included during revision of the manuscript to which also had to be added some comment on *saigonensis*. After the revised copy had been typed in supposedly final state, a single specimen of yet another related form was received from Thailand. Before passing on to statements, as in case of *P. andersoni*, of known differences distinguishing the related forms, attention is called to the following comment.

Genital markings, of relaxed as well as of strongly contracted material, of many thousands of specimens of *P. peguana* (the species often was used in the intermediate zoology course of Rangoon University to illustrate annelid anatomy), and also of all individuals known to be in museum collections, showed no slightest trace of being withdrawn into the parietes.

Pheretima bahli

1939. *Pheretima peguana*, Kirtisinghe, *Spolia Zeylanica* 21: p. 89. (Histology of gut of specimens from Colombo, Ceylon.)

1945. *Pheretima bahli* Gates, *Spolia Zeylanica* 24: p. 85. (Type locality, Colombo, Ceylon. Types, supplied by Kirtisinghe, originally in Prof. Bahl's collection, probably have been lost.)

1953. *Pheretima peguana*, Dissanaike, *Ceylon Jour. Sci.* 25: p. 161. (Protozoan parasites of Colombo specimens.)
1959. *Pheretima bahli*, Refuerzo & Reyes, *Philippine Jour. Animal Industry* 19: pp. 56, 57, 58. (Parasitized by larvae of *Metastrongylus apri*.)

Spermathecal pores, fairly widely separated but less than $\frac{1}{2}$C apart. Male pores, each at tip of a small, conical, penial body protuberant mesially from lateral wall of deep parietal invagination the lumen of which is about twice as wide as the external aperture. Genital markings, invaginate, each on posterior or anterior face of a parietal invagination with a transversely slitlike aperture, two pairs, at 17/18 and 18/19. Setae, 30/iii, 49/viii, 63/xii, 63–68/xx, vii/6–9, viii/(4), 10–13, xvii/0–7, xviii/2–6, xix/0–8. Segments, 92–119. Size, 110–150 by $4\frac{1}{2}$–5 mm.

Typhosole, ends in region of 77th–99th segments. Hearts, of x lacking (?), of xi esophageal (?), of xii–xiii latero-esophageal. Seminal vesicles, large, especially those of xi that may reach forward to 7/8. Pseudovesicles, of xiii just above the ovaries and each with a free dorsal lobe, in xiv smaller. Prostates, in xvii–xxi, ducts 3–4 mm. long, each in a U-shaped loop with the ectal limb thicker. GM glands, sausage-shaped or ellipsoidal. Spermathecae, small, seminal chamber shortly ovoidal to ellipsoidal.

Reproduction: Presumably biparental as sperm are exchanged during copulation.

Distribution: Ceylon (Colombo). Manila and Clark Field, Sual (Pangasinan), Luzon, Philippine Islands.

The species is exotic in Ceylon and almost certainly also in the Philippines where it has no close relatives. Presence of *P. bahli* in both those regions presumably resulted from transportation by man. Known relatives are in Burma and Thailand. In the latter of those countries or in some adjacent region the homeland of *P. bahli* should be sought.

P. bahli and *P. peguana* as yet have not been found in the same locality.

Variation: Only thirty specimens were available for study, but all were much alike though from places as widely separated as Ceylon and the Philippines.

Parasites: SPOROZOA: *Zeylanocystis burti* and *fernandoi* Dissanaike 1953, Colombo. (Hosts were misidentified as *P. peguana*.) CESTODA: *Amoebotaenia* sp., gut wall, Philippines.

Remarks: Externally, male areas at first appear to be quite different from those of *P. peguana*, the only similarity being in the transversely slitlike apertures of the male-pore invaginations. The male field of *P. bahli* appears as a very slightly concave, transversely elliptical area having laterally on each side three fairly large, transverse slits, two each at 17/18, eq/xviii and at 18/19. But on opening the parietal invaginations, the *peguana* genital markings were found.

If *P. bahli* is specifically distinct from *P. peguana*, which of course remains to be demonstrated, the former seemingly can be derived directly from the latter almost only by invaginating the superficial genital markings of *P. peguana* and at the same time elongating the associated GM glands.

Although unknown in the region with which this publication deals, *P. bahli* is included with a prediction that the home of the taxon is somewhere in that area or just outside it to the east.

Pheretima sp.

Sexthecal, pores minute, superficial, less than $\frac{1}{2}$C apart, at 6/7–8/9. Male pores, minute, invaginate, each on a small, shortly conical and glabrous protuberance from lateral wall of a parietal chamber that opens to the exterior just lateral to a rather thick but low, median ridge between 17/18 and 18/19. Genital markings, two pairs, each marking with an obvious central aperture and within a slight parietal chamber that opens at site of an intersegmental furrow (17/18 or 18/19) and just lateral to the median ridge. Apertures of the six parietal chambers, in a deep transversely rectangular depression of the ventrum surrounded by a ridge with a distinctly octagonal shape, anterior and posterior sides long, the six lateral sides equally short. Clitellum, setae unrecognizable externally, between 13/14 and 16/17. Setae, small, deeply retracted. Prostomium, epilobous, tongue open. Color, in dorsum, light red to brownish. Segments, 114. Size, 95 by *ca.* 6 mm.

Septa, 8/9–9/10 aborted, 5/6–6/7, 11/12–12/13 slightly muscular. Pigment, reddish, in circular muscle layer. Intestinal origin, in xv. Intestine, deeply sacculated in xix–xxvi. Caeca, simple, ventral margins slightly incised, in (xxvii?)–xx. Typhlosole, lamelliform from xxvii, ending abruptly in the 88th segment, leaving 26 atyphlosolate. Hearts, of ix lateral (right side), lacking (?) in x, of xi esophageal (?), of xii–xiii latero-esophageal. Blood glands, obvious, in v–vi. Lymph glands, present, at least behind xxvi. Holandric. Testis sacs, paired and ventral, rather widely separated from each other mesially. Seminal vesicles, fairly large. Prostates, in xvii–xix, ducts 3+ mm. long, each in a U-loop, the ectal limb thicker and with muscular sheen though rather slender. Spermathecae, rather small to medium-sized, duct slender and shorter than ampulla, diverticulum from anterior face of duct (or anteromesially) near parietes, much coiled or looped, with a stalk, a slightly thicker middle region and a terminal seminal chamber that is small and almost spheroidal. GM glands, as in *peguana* and *bahli*.

Reproduction: Iridescence was lacking on plicate male funnels, only a few flecks visible within testis sacs. Discrete egg-strings were lacking on the ovaries although the clitellum seemed to be at maximal tumescence. No organ degradation suggestive of parthenogenesis was seen, but evidence for biparental reproduction also was unrecognized.

Remarks: The only specimen, of a supposedly common species, was received for identification after final typing of the sections dealing with *P. peguana* and *bahli*. Efforts to secure further material were futile.

The male field, of the present worm, is still more depressed than in *P. bahli*. Further evolution might lead to a state in which contraction of muscles in the circumferential ridge could close off an opening into the depression except at time of copulation, if indeed that state already has not been reached.

The worm is believed to be from Thailand.

Pheretima planata

1926. *Pheretima planata* Gates, *Ann. Mag. Nat. Hist.*, Ser. 9, 17: p. 411. (Type locality, Rangoon. Topotypes in the U. S. Natl. Mus.) *Jour. Burma Res. Soc.* 15: p. 122; *Jour. Bombay Nat. Hist. Soc.* 30: pp. 182,183; *Rec. Indian Mus.* 28: p. 162. Stephenson, *ibid.*, p. 256.
1929. *Pheretima planata*, Stephenson, *ibid.*, 31: p. 238.
1930. *Pheretima planata*, Gates, *ibid.*, 32: p. 320.
1931. *Pheretima planata*, Gates, *ibid.*, 33: p. 405.
1932. *Pheretima planata*, Gates, *ibid.*, 34: p. 411.
1933. *Pheretima planata*, Gates, *ibid.*, 35: p. 541.
1936. *Pheretima planata*, Gates, *ibid.*, 38: p. 446.
1939. *Pheretima planata*, Gates., *Jour. Thailand Res. Soc. Nat. Hist. Bull.* 12: p. 103.
1955. *Pheretima planata*, Gates, *Rec. Indian Mus.* 52: p. 90.
1960. *Pheretima planata*, Gates, *Bull. Mus. Compl Zool. Harvard College* 123: p. 270.
1961. *Pheretima planata*, Gates, *American Midland Nat.* 66: p. 62; *Burma Res. Soc. 50th Anniv. Pub. No. 1*: p. 57.
1970. *Pheretima planata*, Soota & Julka, *Proc. Zool. Soc. Calcutta* 23; p. 204.

Quadrithecal, pores minute, superficial, each in a very small circular area of slight epidermal modification close to 6/7–7/8, in vii–viii. Male pores, in xviii, minute, each in a circular area on roof of an eversible copulatory chamber with a transversely slitlike aperture at the segmental equator. Female pore, median. Genital markings, very small circular areas of translucence, 1–4 just median to each spermathecal pore, in vii–viii, 8–13 on roof and walls of each copulatory chamber. Clitellum, setae present, xiv–xvi. Setae, 60–69/iii, 75–87/viii, 63–78/xii, 56–65/xx, vii/35–42, xvii/17–22, xviii/8–14, xix/16–20. First dorsal pore, occasionally at 10/11, usually at 11/12. Prostomium, lacking (or rudimentary?). Color, in dorsum, reddish brown to slate. Segments, 115–142. Size, 64–176 by 4–7 mm.

Septa, 6/7–7/8 muscular, 8/9–9/10 aborted, 10/11–12/13 slightly muscular. Pigment, brownish red, in circular muscle layer. Intestinal origin, in xv. Caeca, simple, xxvii–xx. Typhlosole, small, at first with diagonal ridges on each side, ending in region of 121st segment. Hearts, in viii unaborted dorsal portions to gizzard, in ix lateral but usually lacking on one side, of x (latero?)-esophageal, of xi–xiii latero-esophageal. Lymph glands, present in intestinal segments. Blood glands, in iv–vi and in a low, blade-like ridge or an irregularly interrupted band just

behind the gizzard. Holandric. Testis sacs, paired, of x ventral, of xi vertical and including seminal vesicles. Seminal vesicles, in xi, xii, each with a digitiform primary ampulla. Prostates, in xvi–xxi, duct in a *U*-shaped loop. Spermathecae, large enough to reach dorsal parietes, duct elongate, slender, lumen abruptly narrowed in region of diverticular junction, diverticulum longer than main axis, with short stalk, thicker middle portion in wall of which there are numerous microscopic chambers, and an elongately ellipsoidal, terminal seminal chamber. GM glands, composite, stalked, coelomic.

Distribution: Garai-berana, Corbyn's Cove, Navy Bay, Port Blair (Andaman Islands). Dubyinchaung, Palaw (Mergui). Tavoy, Maungmagaun, Myittha (Tavoy). Ye, Moulmein, Mupun, Kawkareik (Amherst). Thaton, Sittang, Kyaikto, Bilin (Thaton). Pyapon, Bogale, Dedaye, Kyaiklat, Thameintaw (Pyapon). Myaungmya, Wakema, Myohaung (Myaungmya). Maubin, Yandoon, Danubyu (Maubin). Bassein, Coomzamu, Thinbawgyin, Padaukchaung (Bassein). Rangoon, Kayan, Thongwa, Kungyangon, Twante, Kyauktan, Syriam (Hanthwaddy). Insein, Hmawbi, Taikkyi, Wanetchaung, Hlegu, Hlawga (Insein). Henzada, Zalun, Ingabu (Henzada). Pegu, Pegu Yomas, Madauk, Pazunmyaung, Nyaunglebin, Thanatpin (Pegu). Tharrawaddy, Ngapugale, Letpadan, Myagyaung, Thonze (Tharrawaddy). Paukkaung (Prome). Sandoway, Ngapoli, Arakan Yomas (Sandoway). Toungoo, Shwegyin, Tantabin, Pegu Yomas, Kyaukkyi, Pathichaung, Pa Taw Lo (Toungoo). Thanbula (Thayetmyo). Kyaukpyu (Kyaukpyu). Pyigyaung, Pyinmana (Yamethin). Akyab, Padali, Buthidaung, Maungdaw (Akyab). Paletwa (Arakan Hill Tracts). Namkham (North Hsenwi). Lashio (Shan States). Indaw Lake, Wuntho (Katha). Bhamo (Bhamo). Kamaing, Loinon, Nyaungbin, Lonton, Myitkyina, Mogaung, Mohnyin, Alam, Weshi, N Sop Zup (Myitkyina). Mawleik, Masein (Upper Chindwin). From sea level to elevations of more than 3,000 feet.

Katlicherra (South Cachar, Assam). Siliguri (Bengal). Dacca (Bangla Desh). Chiengmai, Ko Chang (Thailand). Kuala Lumpur (Selangor, Malay Peninsula).

The species was not obtained, except at Pyinmana and Pyigyaung, in the dry zone of central Burma, nor in sections of Tavoy district and the Shan Plateau where considerable collecting was done. The only specimen from Lashio was from grounds of the Dak bungalow. Such rest houses often have in their compounds exotic cultivated plants with which worms could have been introduced. *P. planata* usually was obtained only in short series and certainly is not very common in Rangoon but large numbers were obtained on the Andaman Islands, at Padali (Akyab), Kyaukpyu, Sandoway, and at Palaw Township in Mergui.

The oceanic gap in the range as now known requires transportation in one direction or another. The original home was thought for a time to be perhaps in far eastern Burma but the seeming absence of the species in most of the Shan Plateau provides little support for a Burma origin. Introduction to Burma from the Anadamans now seems unlikely. The species may prove to be markedly anthropochorous but in absence of any information as to relationships with other species any suggestion as to a possible original home could be only a guess.

Abnormality : Testis sacs of x, lacking (2 specimens). Seminal vesicles, of xi lacking (2) or rudimentary and obviously functionless (others), of xii lacking (6) or rudimentary (4). Prostates, rudimentary and confined to xviii (7), lacking (1). GM glands of copulatory chambers, represented only by stalks without glandular heads (1). Those aberrations were found in Garaiberana worms of which at least 10 of the 41 specimens were opened. The worms were "sickly looking" and each of those that was dissected had parasitic cysts on the body wall, gut, and septa 7/8–13/14. Such aberrations are common in the parthenogenetic morphs of certain species. Heavy infestation by coelomic parasites often has been noted in parthenogenetic individuals of other species.

Similar aberrations and parasitism were not recorded for mainland *planata* and athecal as well as anarsenosomphic individuals probably never were recognized. Unfortunately, data as to presence or absence of sperm on male funnels and in spermathecae no longer are available. The only justification for suspecting parthenogenesis on the mainland is the small size of the seminal vesicles.

Parasites : CESTODA: *Amoebotaenia cuneata* (Linstow, 1872), cysticercoids, Burma.

Pheretima posthuma

1868. *Perichaeta posthuma* Vaillant, *Ann. Sci. Nat.*, Ser. 5, 10: p. 228. (Type locality, Java. Types in the Paris Mus.)
1912. *Pheretima posthuma*, Stephenson, *Rec. Indian Mus.* 7: p. 278.
1917. *Pheretima posthuma*, Stephenson, *ibid.*, 13: p. 385.
1923. *Pheretima posthuma*, Stephenson, (*The Fauna of British India*), *Oligochaeta*, p. 309.
1925. *Pheretima posthuma*, Gates, *Rec. Indian Mus.* 27: p. 237. *Ann. Mag. Nat. Hist.*, Ser. 9, 16: p. 567.
1926. *Pheretima posthuma*, Gates, *ibid.* 17: p. 464; *Rec. Indian Mus.* 28: p. 162; *Jour. Bombay Nat. Hist. Soc.* 31: p. 182; *Jour. Burma Res. Soc.* 15: p. 212.
1930. *Pheretima posthuma*, Gates, *Rec. Indian Mus.* 32: p. 321.
1931. *Pheretima posthuma*, Gates, *ibid.*, 33: p. 405.
1932. *Pheretima posthuma*, Gates, *ibid.*, 34: p. 487.
1933. *Pheretima posthuma*, Gates, *ibid.*, 35: p. 543.
1936. *Pheretima posthuma*, Gates, *ibid.*, 38: p. 448.
1937. *Pheretima posthuma*, Gates & Hla Kyaw, *Jour. Roy. Asiatic Soc. Bengal*, (*Sci.*) 2: p. 123.
1939. *Pheretima posthuma*, Gates, *Jour. Thailand Res. Soc.*, *Nat. Hist. Suppl.* 12: p. 104.
1955. *Pheretima posthuma*, Gates, *Rec. Indian Mus.* 52: p. 91.
1961. *Pheretima posthuma*, Gates, *American Midland Nat.* 66: p. 62; *Burma Res. Soc. 50th Anniv. Pub. No.* 1: p. 57.

1970. *Pheretima posthuma*, Soota & Julka, *Proc. Zool. Soc. Calcutta* 23: p. 204.

ABNORMALITY: Beddard, 1886, *Ann. Mag. Nat. Hist.*, Ser. 5, 17: p. 93. Beddard, 1900, *Proc. Zool. Soc. London* 1900: p. 642. Stephenson, 1917, *Rec. Indian Mus.* 13: p. 385. Malaviya & Verma, 1961, *Proc. Natl. Acad. Sci. India*, B 31: p. 94. ANATOMY: Bahl, 1950, *Indian Zool. Mem., Lucknow* 1 (4th ed.). AXIAL GRADIENTS: Tandan, 1951, *Current Sci.* 20: p. 214; 1952, *ibid.*, 21: p. 51. BLOOD GLANDS: Stephenson, 1924, *Proc. Roy. Soc. London* B, 97; p. 180. CASTINGS. Bahl, 1934, *Quart. Jour. Micros. Sci.* 76: p. 567. Roy, 1957, *Proc. Zool. Soc. Calcutta* 10: p. 84. Dubash & Ganti, 1963, *Proc. Natl. Acad. Sci. India* 23: pp. 193–197. CIRCULATORY SYSTEM: Stephenson 1913, *Trans. Roy. Soc. Edinburgh* 49: p. 764. Bahl, 1921, *Quart. Jour. Micros. Sci.* 65: p. 354. COCOONS: Bahl, 1922, *Quart. Jour. Micros. Sci.* 66: p. 56. COELOMIC POUCHES: Beddard & Fedarb, 1902, *Proc. Zool. Soc. London* 1902: p.164. CULTURE: Tembe & Dubash, 1959, *Jour. Bombay, Nat. Hist. Soc.* 56: p. 645. CYTOPLASMIC INCLUSIONS: Sharga, 1928, *Allahabad Univ. Studies* 4: p. 177. Nath & Bhatia, 1944, *Proc. Natl. Inst.Sci. India* 10: p. 232. Malhotra, 1957, *Res. Bull. Panjab Univ. Zool. No.* 118: p. 367. DISTRIBUTION: Gates, 1937, *Rec. Indian Mus.* 39: pp. 182, 207. Yeolekar & Rao, 1934, *Proc. Indian Sci. Congr.* 21: p. xx. ECOLOGY: Gates, 1945, *Proc. Natl. Acad. Sci. India* 15: p. 48. EMBRYOLOGY: Bahl, 1922, *Quart. Jour. Micros. Sci.* 66: p. 56. EXCRETION: Bahl, 1954, *Quart. Jour. Micros. Sci.* 85: p. 347; 1946, *ibid.*, 87: p. 359; 1947, *Biol. Rev.* 22: p. 130. EXCRETORY SYSTEM: Bahl, 1919, *Quart. Jour. Micros. Sci.* 64: p. 69. Matthew, 1950, *Jour. Zool. Soc. India* 2: p. 144. GOLGI APPARATUS: Nath, 1930, *Quart. Jour. Micros. Sci.* 73: p. 477; 1933, *ibid.*, 76: p. 138. HISTOCHEMICAL STUDIES OF LIPIDS IN OOGENESIS: Nath, 1958, *ibid.*, 94: p. 475. LOMBRICINE: Dubey, 1963, *Indian Jour. Chem.* 1. LYMPH GLANDS: Thapar, 1918, *Rec. Indian Mus.* 15: pp. 71, 74. MALE GONODUCTS AND PORES: Lloyd & Powell, 1911, *Jour. Bombay. Nat. Hist. Soc.* 21: p. 289. Michaelsen, 1934, *Arch. Neerlandaises Zool.* 1: p. 111. NEUROSECRETION: Dogra, 1967, *Gen. Comp. Endocrinol.* 9 and 1968, *Jour. Zool. London*, 156. OOGENESIS: Sharga, 1932, *Nature* 130: p. xxx. PARASITES: (PROTOZOA): Ghosh, 1918, *Rec. Indian Mus.* 15: p. 129; 1921, *Bull. Carmichael Med. College* 2: p. 12; 1923, *Jour. Roy. Micros. Soc.* 1923: pp. 423, 425. Bhatia & Chatterjee, 1925, *Arch. Protistenk.* 52: p. 199. Bhatia & Gulati, 1927, *ibid.*, 57: p. 100. Ray & Chatterjee, 1936, *Proc. Indian Sci. Congr.* 23: p. 345. Boisson, 1957, *Ann. Sci. Nat. Paris*, 19: pp. 72, 74. Raychaudhuri et al., 1969, *Arch. Protistenk.* 111. (NEMATODA): Timm, 1959, *Pakistan Jour. Biol. Agric. Sci.* 2: p. 42; 1959, *Pakistan Jour. Sci. Res.* 11: p. 58; 1960, *Proc. Helminthol. Soc. Washington* 27: p. 77. PARIETO-ENTERIC MUSCLES: Saksena, 1943, *Current Sci.* 12:, p. 120. PHARYNGEAL GLAND CELLS: Stephenson, 1917, *Quart. Jour. Micros. Sci.* 62: p. 261. POPULATIONS & CAST PRODUCTION: Khan & Ashraf, 1967. *Science* (*India*) 5. VARIATION: Malaviya & Verma, 1959, *Proc. Natl. Acad. Sci. India* 29: p. 39. Gates, 1960, *Bull. Mus. Comp. Zool. Harvard College* 123: p. 271. WATER RELATIONSHIPS: Bahl, 1934, *Quart. Jour. Micros. Sci.* 76.

Octothecal, pores minute, superficial, ¼C or more apart, in small translucent areas just in front of 5/6–/8/9. Male pores, in xviii, minute, each in a small disc on median wall near roof of a slight, eversible invagination with longitudinally crescentic aperture. Female pore, median. Genital markings, paired, small, circular or nearly so, equatorial, slightly median to male pore levels, restricted to region between xiv and xxx but usually present only on xvii and xix. Clitellum, reaching to or nearly to 13/14 and 16/17. Setae, small, closely spaced in unbroken circles; retained in ventrum of clitellum, 106–129/viii, 63–75/xii, 60–95/xx, vi/36–43, vii/38–44, viii/36–43, xvii/15–20, xviii/16–22, xix/16–20. First dorsal pore, at 12/13. Prostomium, small, epilobous, tongue short and usually open. Color, light to dark gray. Segments, 91–124, usually 111–121. Size, 60–140 by 3–8 mm.

Septa, 5/6–7/8 thickly muscular, 8/9 muscular and complete, 9/10 aborted. Pigment, brown, in circular muscle layer. Intestinal origin, in xv. Caeca, simple, in xxvii–xxiv. Typhlosole, simply lamelliform, from xxvii, ending in region of 85th to 95th segments, leaving 20–27 atyphlosolate. Subneural, ventral and (single) dorsal trunks, complete. Hearts, of vii and ix lateral (one in ix usually in part or wholly aborted or vestigial), unaborted dorsal portions in viii passing to gizzard, in x–xi lacking, in xii–xiii latero-esophageal. Paired loops with nonmuscular walls between supra- and extra-esophageal trunks in x–xi. Blood glands, in iv–vi. Lymph glands present posteriorly from xxv (at least). Holandric. Testis sacs, unpaired, ventral in x, vertically U-shaped in xi. Seminal vesicles, of xi rather small and included in testis sac, of xii larger. Pseudovesicles of xiii, small. Prostates, in xv–xxi, ducts 2½–3½ mm. long, each in a U-shaped loop. Spermathecae, rather small, duct shorter than ampulla, diverticulum with short stalk from median face of duct near ampulla and longer ellipsoidal seminal chamber. GM glands, sessile on parietes.

Reproduction: Presumably biparental, as sperm are matured and exchanged. However, seminal vesicles are rather small and the anterior pair is included in the testis sac of xi, conditions often associated with parthenogenesis. Uniparental reproduction will be expected in the Philippines if the aberrant worms referred to *P. posthuma* by Beddard (1886) were not misidentified (but see aberrations below).

Distribution: Jinghighat, Minnie Bay, Port Blair (Andaman Islands). Mergui (Mergui). Tavoy (Tavoy). Moulmein, Mupun, Chaungson, Kya In, Kawkareik (Amherst). Sittang (Thaton). Pyapon, Thanchitaw, Thameintaw, Bogale, Dedaye (Pyapon). Myaungmya (Myaungmya). Maubin, Yandoon, Danubyu (Maubin). Bassein (Bassein). Rangoon, Thongwa, Kyauktan, Twante (Hanthawaddy). Insein, Hlegu, Taikkyi, Wanetchaung (Insein). Henzada, Kyangin (Henzada). Pegu, Pazunmyaung, Thanatpin (Pegu). Tharrawaddy, Thonze, Ngapugale, Keinbingyi, Letpadan (Tharrawaddy). Prome, Sanbot, Nyaungyi, Okshitbin, Paukkaung, Laboo (Prome). Sandoway, Nagapoli, Taungup (Sandoway). Kyaukpyu (Kyaukpyu). Akyab (Akyab). Toungoo, Kyaungnagwa (Toungoo). Pyigyaung, Pyinmana (Yamethin). Thayetmyo, Allanmyo, Thanbula (Thayetmyo). Magwe, Taungdwingyi, Yenangyaung (Magwe). Minbu (Minbu). Pakokku (Pakokku). Kyaukpadaung, Myingyan (Myingyan).

Meiktila, Mahlaing (Meiktila). Kyaukse (Kyaukse). Mandalay (Mandalay). Sagaing (Sagaing). Monywa, Ingyindaung, Powindaung, Anidaung, Okma (Lower Chindwin). Ye-U, Kin-U, Kyaukmyaung (Shwebo). Katha, Naba, Indaw Lake (Katha). Bhamo (Bhamo). Myitkyina (Myitkyina). Laungbyin, Mingin, Pantha, Masein, Homalin, Kalewa (Upper Chindwin).

India, Rangamati (Chittagong) and the Indo-Gangetic Valley from Dacca through Bengal, Bihar, United Provinces, Punjab and Rajputana, also at Gwalior and Bombay. Thailand, Chiengmai, Ban Sa-iep, Mu-ang Pong, Ban Huai Rai. Malay Peninsula (one record, without further specification as to locality, has not been confirmed). Vietnam.

Formosa, Sumatra, Java, Christmas Island, Celebes, Groot Bastaard, Ternate, Amboine, Nias, Sebesi, Sumba (Soemba), Flores, Philippine Islands, Santa Cruz, New Hebrides, Tinian. Florida (new record). (Records for Egypt, Bahamas, Argentina, France, require confirmation as a Chinese species, *P. hupeiensis*, has been mistaken for *P. posthuma*.)

The original home of *P. posthuma* presumably is in a southeast portion of Asia comprising Thailand and the states formerly in French Indo-China. Overseas distribution is attributable to human transportation.

In India and Burma *P. posthuma* has been found only from sea level to elevation of *ca.* 2,500 feet and the species now appears to be restricted to tropical lowlands.

Habitats: Sands of river banks (Burma, India), light sandy soils away from river banks (Burma, India) of lawns, gardens, other open spaces and in groves. Mud, manure pile (once, one worm). Individuals of this species are "innumerable" in wet and unshaded banks of the Chindwin, Irrawaddy, Sittang and Salween rivers. The sandy surface, where moisture and other conditions are favorable, may be covered with castings for miles.

Biology: Activity is uninterrupted at Burmese and Indian sites where moisture is adequate during the dry season. Elsewhere, quiescence, at unknown depths in the soil, is imposed during the dry season, even at most favorable Allahabad sites from early May into July. Physiological characteristics of the inactive state are unknown. Breeding apparently is possible, in favorable conditions, throughout the year. Hatching, after 8+ weeks (Bahl, 1922) or after 30–37 days in laboratory cultures (Tembe & Dubash, 1959) but, if temperature is below 28°C and relative humidity is below 50 per cent, 45–50 days may be required. Cocoons may be deposited after a period of four weeks from attainment of maturity, possibly after 8+ weeks from hatching. Usually only one worm hatches from a cocoon. After breeding, clitellar tumescence disappears leaving clitellar segments with a slight yellow or brown coloration, intersegmental furrows reappear and setae become obvious except in the dorsum. Clitellar regression may be completed before sperm in spermathecal seminal chambers are lysed.

P. posthuma is geophagous and feeds underground where it presumably copulates. Copulation has not been observed. Burrows supposedly (Bahl, 1934) have but a single opening at the surface.

Castings: Low pyramidal piles of small spheroidal to ovoidal pellets (Gates, 1930. Bahl, 1934: fig 3 and 1950: fig. 40). Pellet piles of *L. mauritii*, as well as the heaps of more or less deeply and irregularly constricted faecal cords of other earthworms, superficially may seem to be like the castings of *posthuma*. Attribution to the latter species of all "small castings" (Roy, 1958) is likely to have been erroneous. A statistically significant increase of total nitrogen, organic matter, and calcium carbonate content, over percentages in parent soil, was found (Dubash & Ganti) in castings.

Variation and Abnormality: The GM pattern appears to be invariable as to intrasegmental situation (equatorial and slightly median to male pore levels) but subject to more variation as to number and anteroposterior location (*cf.* Gates, 1926, 1932, 1960; Malaviya, 1959) than in *P. peguana*.

Variation in abortion of postgizzard septa is indicated by the literature but has not been confirmed. Statements as to absence of 8/9 and presence of 9/10 probably are erroneous insofar as normal worms are concerned. Early mistakes appear to have been due to an assumption that presence of four spermathecae between two consecutive septa proved that an intervening septum was absent.

Male deferent ducts, according to Michaelsen (1934), continue through the muscular layer of the prostatic duct to discrete openings on the male tubercle but separate male and prostatic pores were not recognized in Burmese specimens.

Sexthecal and athecal forms, as well as other aberrations in the spermathecal battery, were recorded (Beddard, 1886) from the Philippines but no evidence for evolution of genital polymorphism was recognized in Burma or in India where the species is widely used as a laboratory type in college courses. As in case of supposed absence of intestinal caeca (Beddard, 1900) confirmation is lacking and misidentifications are suspected.

One to three extra prostates (Stephenson, 1917. Gates, 1926), in xvii, xix, xx, have been seen. In some of such instances, splitting of mesoblastic somites may have been involved. Such individuals are be of interest except for the extra prostates may be of interest in connection with origin of the quadriprostatic condition characterizing *Exxus wyensis* Gates 1959.

Some of the deviations from normal in an unusually aberrant individual (Gates, 1925) perhaps could have resulted from hypomeric anterior regeneration.

Regeneration: Anterior regeneration was not obtained in preliminary experiments and no records of regenerates, at either end of the body, have been found in the literature.

Parasites: CILIATA: *Anoplophrya lloydi* Ghosh, 1918, seminal vesicles, Calcutta (= *A. lumbrici*?). *A. pheretimii* Raychaudhuri *et al.*, 1969, gut lumen of xv–xl, Calcutta. *Parabursaria "pheretima"* Ghosh, 1921, seminal vesicles, Calcutta. *Maupasella nova* Cépède, 1910, gut, Lahore and Indo-Cnina. SPOROZOA: *Monocystis bengalensis* and *lloydi* Ghosh, 1923, seminal vesicles, Calcutta. *M. "pheretimi"* Bhatia & Chatterjee, 1925, seminal vesicles, Lahore, Bombay. *M. minima* Boisson 1957, testis sacs, Indo-China. *Stomatophora diadema* Hesse, 1909, seminal vesicles, Calcutta. *Dirhynchocystis globosa* (Bhatia & Chatterjee 1925), seminal vesicles, Lahore and Bombay. *Craterocystis myonemata* Boisson, 1957, testis sacs, Indo-China.[21] NEMATODA: *Perodira pheretimae* Timm, 1960, testis sacs and seminal vesicles. *Synoecnema anseriforme* Timm, 1959, *S. hirsutum* Timm, 1959. *Pharyngonema "pheretimai"* Timm, 1959, coelomic cavities, East Pakistan.

Masses of coelomic cysts, such as are frequently found in parthenogenetic forms, never were observed in Burmese and Allahabad individuals.

Remarks: A species is mentioned just once in the description of the vascular system (Bahl, 1921) and then only in the title of figure 1 but figure 5 is of *P. posthuma* even though the septum behind the gizzard is erroneously labeled 9/10 and the intestinal origin is placed too far back. The absence of hearts in x–xi of *P. posthuma* is no more characteristic of the genus *Pheretima* than are the castings. Studies of nephridial development, supposedly of *P. posthuma*, in part were made in England where the species is unknown. Specimens of *P. rodericensis, hawayana*, or *diffringens* are most likely to have been studied.

Partly because of activity in the dry months of April–June, partly because of preference for sandy habitats, Bahl (1943) concluded that *P. posthuma* "is able to live in drier conditions" than species of *Eutyphoeus*. Disregarded was the fact that dry season activity was allowed by heavy watering of gardens or near saturation of soils close to "way-side water-taps." Activity in river-bank sands continues only as long as they are obviously moist.

The distribution of small castings in experimental plots at Calcutta was believed (Roy, 1958) to indicate that *P. posthuma* is, in contrast with *Eutypheous waltoni*, a shade-loving or photosensitive species.

Blood glands of *P. posthuma* were believed (Stephenson, 1924) to be functional rather than vestigial as in *P. diffringens* and *hawayana*.

---

[21] *Echinocystis globosa* Bhatia & Chatterjee 1925, was recorded from *P. heterochaeta*, a synonym of *P. diffringens*, but Bhatia (1938, *Sporozoa, Fauna of British India*) lists the host as *P. posthuma*.

Relationships of *P. posthuma* are uncertain but now appear to be with *P. fluvialis* Gates, 1939, known only from the original description of several specimens from Thailand, and *P. juliani* (Perrier, 1875) known only from types collected in Indo-China.

Amino acids of the soil are not depleted (Dubash & Ganti, 1964, *Current Sci., Bangalore*, **33**. p. 219) by passage through the gut of *P. posthuma*. On the contrary arginine is increased.

Pheretima quadrigemina

1932. *Pheretima quadrigemina* Gates, *Rec. Indian Mus.* **34**: p. 486. (Type locality, Victoria Point. Types, none.)

Sexthecal, external pores small transverse slits just behind 6/7–8/9. Male pores, in xviii, minute, each on a short and thickly columnar protuberance from roof of a copulatory chamber with a transversely slitlike equatorial aperture. Female pore, median. Genital markings, internal only, each on end of a shortly and thickly columnar protuberance from walls of copulatory chambers. Clitellum, with some setae (ventrally?), xiv–xvi. Setae, circles with gaps near mD of variable width, 52/xx, vii/21–25, viii/19–25, xvii/16–18, xviii/6, xix/16. First dorsal pore, at 12/13. (Prostomium?) Color, in dorsum, red anterior to clitellum, pinkish posteriorly (pigment red and in circular muscle layer?). Segments, 115. Size, 64–72 by 3–4 mm.

Septa, 8/9 (usually?) and 9/10 aborted, none thickly muscular. Intestinal origin, in xvi. Caeca, simple, in xxvii–xx. (Typhlosole?) Hearts of ix (left or right) lateral, of x (latero?)-esophageal, of xi–xiii latero-esophageal(?). Blood glands, in v. Lymph glands, in intestinal region. Holandric. Testis sacs, paired(?), ventral. Seminal vesicles, large, in xi, xii, pushing 12/13 back to 13/14. (Pseudo-vesicles?) Prostates, in xvi–xx, ducts slender, looped. Spermathecae, small, duct shorter than ampulla and with thickly muscular wall(?), diverticulum from anterior face of duct in the parietes, longer than main axis, with a short stalk, zigzag looped ental portion gradually widened distally and ending with a short, pear-shaped chamber. GM glands, stalked (composite?), on anterior and posterior faces of the copulatory chambers.

Reproduction: Lacking records regarding copulation and sperm maturation, in absence of contra-indications, and because condition of seminal vesicles and prostates is consonant therewith, reproduction is assumed to be biparental.

The breeding season includes at least some part of October.

Distribution: Victoria Point (Mergui).

Remarks: Whether primary spermathecal pores are minute and within a very small parietal invagination or are larger and superficial unfortunately is not known. Dislocation of the pores, back from the intersegmental furrows, at the time the species was erected, seemed to be of considerable importance. Dislocation in either direction from those furrows is rare as a species character and has not been recognized as an individual variation or mutation. Origin through cephalic regeneration, in the genus *Pheretima*, is highly improbable.

Failure of septum 8/9 to abort has been recognized as an occasional variation in other pheretimas, and presumably was also in one type of *P. quadrigemina*.

Pheretima rimosa

1931. *Pheretima rimosa* Gates, *Rec. Indian Mus.* **33**: p. 408. (Type locality, Mong Ko. Types, none.)
1932. *Pheretima rimosa*, Gates, *ibid.*, **34**: p. 534.
1936. *Pheretima rimosa*, Gates, *ibid.*, **38**: p. 453.
1960. *Pheretima rimosa*, Gates, *Bull. Mus. Comp. Zool. Harvard College* **123**: p. 274.

Octothecal, pores minute, superficial, widely separated, close to anterior margins of vi–ix. Male pores, in xviii, minute, superficial, each toward lateral margin of a large porophore that varies in shape and size and that may cross into xvii and/or xix. Female pore, median. (Genital markings, except for the male porophores, none.) Clitellum, without externally recognizable setae, xiv–xvi. Setae, lacking dorsally in ii or ii–iii but in ventrum nearly straight and ornamented ectally, somewhat further apart in preclitellar segments, circles often with irregular gaps dorsally in preclitellar segments but posteriorly without regular gaps, 1–7/ii, 4–16/iii, 28–38/viii, 35–43/xii, 42–56/xx, vi/7–18, vii/8–18, viii/7–19, xvii/14–19, xviii/5–11, xix/15–20. First dorsal pore, at 11/12 or 12/13. Prostomium, epilobous, tongue open. Color, in dorsum, reddish, brownish (pigment red and in circular muscle layer?). Segments, 104–119. Size, 60–122 by 3–5 mm.

Septa, 8/9 present, 9/10 aborted, none markedly muscular. Intestinal origin, in xv. Caeca, simple, ventral side sometimes shortly lobed, in xxvii–xxiii. Typhlosole, very small (rudimentary?). Hearts, in viii unaborted dorsal portions to gizzard, in ix (left or right) lateral, in x (latero?)-esophageal, in xi–xiii latero-esophageal(?). Holandric. Testis sacs, unpaired, ventral. Seminal vesicles, large, in xi, xii. Prostates, large, in xv–xxi, ducts to 7 mm. long, each in a *U*-shaped loop. Spermathecae, duct shorter than ampulla, narrowed in the parietes, lumen very narrow ectal to diverticular junction, diverticulum from anterior face of duct near body wall, shorter than main axis, with muscular stalk and longer seminal chamber.

Reproduction: Presumably biparental in those worms that exchange sperm during copulation. Size and condition of seminal vesicles, occupying a space equivalent to lengths of viii–xiv or xv (and the anterior pair sometimes passing through 10/11), indicate profuse maturation of sperm. Prostates also are large and seemingly in a functioning state at maturity.

However, male porophores often seem indistinct and that, together with the parthenogenesis of male sterile morphs justifies a suspicion that even profusely sperm-maturing individuals may have option of parthenogenesis.

Breeding season, for worms in manure, includes August. Elsewhere breeding seemingly would be late in October and in November.

Distribution: Peng Sai, Mong Ko, Man Peng, Nam Mang, Pang Wo (Mang Lun). Mong Ko (Kengtung). Kutkai (North Hsenwi). Ang Lawng Mountain, To Noi (Mong Mong), Nam Hpen Noi, Bana (Mong Lem), Yunnan. At elevations of 3,500 to more than 6,000 feet.

Kutkai specimens were obtained in grounds of the Dak bungalow where there were exotic flowering plants of kinds often carried around the country in pots or kerosene tins. The species never was procurable elsewhere in or near Kutkai.

Habitats: Soil, including sandy, under trees including pines. Wet leaves, rotting leaves. Manure heap.

Variation: Last two pairs of spermathecae often were in viii.

Polymorphism: The only major morph that was recorded is the R but there were H$_p$ morphs. Second order intermediates (evolving towards the R) may have been more common. *P. rimosa* is, however, one of those very variable species about which little was published but which was represented by numerous specimens in many collections.

Regeneration: A tail regenerate at 89/90, with metameric differentiation apparently complete, had only four segments. Tail regeneration was common.

Remarks: Erosion resulting from rubbing against each other during mule-back transportation up and down hill damaged the male areas of many specimens. Those areas do sometimes have much of the appearance of genital markings and are associated with intraparietal tissue that may be glandular. However, no discrete male porophore is demarcated within the area. Depressions figured on C-shaped male porophores may have been sites of seminal grooves.

Frequency of posterior amputation was responsible for inadequate characterization of size and of segment number.

Pheretima robusta

1872. *Perichaeta robusta* Perrier, *Nouv. Arch. Mus. Hist. Nat. Paris*, **8**: p. 112. (Type locality, Mauritius or Manila. Types, in the Paris Mus.)
1929. *Pheretima ornata* Gates, *Proc. U. S. Natl. Mus.* **75**, 10: p. 20. (Type locality, Lashio. Types, in the Indian and the U. S. Natl. Mus.)
1931. *Pheretima ornata*, Gates, *Rec. Indian Mus.* **33**: p. 404.
1932. *Pheretima ornata*, Gates, *ibid.*, **34**: p. 421.
1933. *Pheretima ornata*, Gates, *ibid.*, **35**: p. 538.
1936. *Pheretima robusta*, Gates, *ibid.*, **38**: p. 454.
1955. *Pheretima robusta*, Gates, *ibid.*, **52**: p. 92.
1954. *Pheretima robusta*, Gates, *Ark. Zool.* Ser. 2, **6**: p. 438.
1955. *Pheretima robusta*, Gates, *Rec. Indian Mus.* **52**: p. 92.

1961. *Pheretima robusta*, Gates, *Burma Res. Soc. Anniv. Pub.* No. 1: p. 57.

Quadrithecal, pores minute, superficial, *ca.* $\frac{1}{3}$C apart, often in circular, translucent areas, at 7/8–8/9. Male pores, in xviii, minute, superficial, each in a thick disc. Female pore, median. Genital markings, always(?) present, small, *ca.* 1 intersetal interval wide, circular to elliptical, often one in each spermathecal porophore and just behind the spermathecal pore, less often one just in front, more frequently one somewhat median to a pore and presetal, often two just median to each male porophore, one pre- and other post-setal, others paired (apparently never median and unpaired), usually presetal, in some or all of vii–viii, xviii. Clitellum, without externally recognizable setae, xiv–xvi. Setae, circles with slight gaps at mV and mD, enlarged at least ventrally in some of ii–ix and then ornamented ectally with transverse rows of fine spines, 30–40/iii, 40–54/viii, 44–68/xii, 46–72/xx, to 80/xxv, (to 100/xxviii?) viii/19–31, xvii/15–31, xviii/10–25, xix/14–31. First dorsal pore, at 11/12. Prostomium, epilobous, tongue open. Color, in dorsum, often lacking or unrecognizable at setal equators, slate, dark brown, dark chestnut, reddish brown (pigment reddish and in circular muscle layer?). Segments, 79–136. Size, 33–180 by 2–9 mm. but usually from 85 by 4 mm. upwards.

Septa, 8/9–9/10 aborted, 5/6–7/8 muscular, 10/11–12/13 muscular to thickly muscular. Intestinal origin, in xv (occasionally in xvi?). Caeca, simple, with very short lobings of the ventral side, in xxvii–xxii. Typhlosole, simply lamelliform, from xxvii (to region of?) xxii. Hearts, in vi–vii (when complete) lateral, in viii unaborted dorsal portions to gizzard, in ix (left or right) lateral, in x (latero?)-esophageal, in xi–xiii latero-esophageal(?). Blood glands, in an esophageal collar that is rudimentary and ridge-like to larger and lobed, behind the gizzard. Lymph glands, present in intestinal segments. Holandric. Testis sacs, usually unpaired and ventral. Seminal vesicles, large, in xi, xii. Pseudovesicles, in xiii–xiv. Prostates, in xvi–xxi, ducts to 6 mm. long, looped or coiled. Spermathecae, duct shorter than ampulla, narrowed in parietes, diverticulum from anterior face of duct near body wall, with slender but muscular stalk and shorter, wider, spheroidal to ellipsoidal seminal chamber. GM glands, stalked, coelomic, composite.

Reproduction: Parthenogenetic, often because of male sterility. An occasional individual with mature sperm on its male funnels has been seen but only once was spermatozoal iridescence recognized in spermathecal seminal chambers.

The breeding season, in Burma, includes May–October.

Distribution: Kawngmu, Tengyueh (Yunnan). Mount Kambaiti (Myitkyina). Namkham, Kutkai (North Hsenwi). Lashio and vicinity (Shan States).

SYSTEMATICS

Pang Noi, Hpa Cha, Meung Nawng (Pang Long). At elevations of 2,000 feet to 2,300 meters.

India (Darjiling, at *ca.* 6,000 feet).

China, (Yunnan, Szechuan, Chekiang, Fukien, Kiangsu, Kiangsi, Hupei Provinces), Hong Kong, Taiwan, Quelpart Island, Korea, Japan, Okinawa. In Szechuan, to elevations of 12,000 feet.

(Manila, West Indies, Madagascar, Mauritius?)

The American record is based on a British Museum specimen that was labeled *Pheretima mandhorensis* 1904, 10.5.1401. West Indies. coll. Beddard. *P. mandhorensis* is a synonym of *P. hawayana.* The label probably was erroneous and as likely with respect to locale as to the taxon name. The Philippine record was based on a specimen supposedly collected by Barrot at Manila in 1852. The species never again was found in the Philippines from which a number of collections now have been identified. The first Mauritius record is based on 3 individuals collected prior to 1872 by Desjardins. Live worms obtained by Beddard from Kew (in or before 1892) but supposedly from Mauritius were inadequately characterized as types of *Perichaeta mauritiana,* a name that may be a synonym of *robusta* or as Beddard later (1900) thought, of *hawayana.* Whether Michaelsen had other specimens from the island is uncertain. Types of *P. zavatarii* Cognetti, 1909, secured at an unspecified Madagascarene locality almost certainly are referable to the Chinese species. *P. robusta* doubtless has been transported by man, and individuals may actually have been obtained at Manila, Mauritius, and Madagascar from earth with transported plants or from some locality reached by such plants. However, no single specimen of the species was among the hundreds of earthworms intercepted by the U. S. Bureau of Plant Quarantine during the last fifteen years. Such evidence as is available, accordingly indicates that the species has not been successful in spreading throughout any considerable section of the world outside of the *Pheretima* domain.

The species is now thought to have originated somewhere in China but much more collecting will be required to reveal the original home and to show how much of the range as presently known was self acquired.

Habitats: Soil; fertile, rocky, in jungles and in manured gardens.

Biology: *P. robusta* is believed to be a temperate zone species that now seems unlikely to be able to colonize tropical lowlands.

Variation and Abnormality: Most, if not all, of the markedly variant and aberrant conditions that were seen are now known to be normal, that is to the extent of being usual in certain stages of the evolution of parthenogenetic polymorphism. Further comment or characterization of modifications that now appear to be standard, not only in different species of the

same genus but also in other genera and families, seems unnecessary.

Polymorphism: Whether any individual of a normal H morph has been seen is unknown. Many specimens certainly are of male sterile H_p morphs. Athecal individuals on which *P. sheni* Chen, 1935 was erected are likely to be referable to *P. diffringens* or to *P. robusta* than to *P. hawayana* or *P. morrisi.* However, A, R, and AR morphs, if seen were never recognized as belonging to *P. robusta.* Many of the specimens studied are referable to one or another of numerous intermediate morphs of several orders. Thus, some, with various spermathecal deformities seemingly are evolving toward an athecal state. Others, already aprostatic, apparently are evolving toward an R or toward an AR stage.

A Szechuan morph with "invaginate" male pores could be placed in *robusta* because asymmetrical individuals showed the slightly invaginate condition on one side of xviii along with the normal superficial location, presumably characteristic of the species, on the other side of the same segment.

Characterization of the H morph (as in the précis above) was secured by combining those characters, of the numerous available morphs, that seemed most likely to be normal.

Parasites: None of the observations on parasites of the author's specimens were published and none of the parasites, at date of writing, have been identified.

Remarks: Each spermathecal pore often is within an area of grayish translucence which may be almost flat or quite convex. The porophore obliterates the intersegmental furrow, equally indenting apparently the adjacent metameres. The pores, as they first become distinguishable, always seem to be exactly in line with the intersegmental furrows. General translucence of the porophore is not to be confused with that of a genital marking that may be contained therein.

A circumferential furrow delimiting any kind of a porophore usually is slight but in many specimens of *robusta* that furrow has become such a very deep groove that the male pores and sometimes even the postclitellar genital markings seemed to be at ventral ends of thick columns.

Male porophores sometimes have, in addition to the male pore, a genital marking or perhaps more accurately the grayish translucence of a central portion of such a marking. Presence of openings of a GM gland on male porophore in several species long had been suspected because number of glands was greater than the number of markings. The rotten condition (politely called macerated) of certain specimens, for which identifications were requested, not only enabled recognition of the GM pores on the male porophores but also the fact that much if not all of the central translucence of a marking was due to close crowding of minute pores of the composite glands.

Relationships of *P. robusta* presumably are with species having the same sort of GM glands but probably are closest with *P. aspergillum* (Perrier, 1872) a Chinese species also in need of characterization from a certainly biparental population.

The history of our knowledge of *P. robusta*, in several respects, is rather typical. The species was erected, at least in part, on a worm collected in 1852 but was not recognized, either in China or as Chinese, until 1935. Meanwhile, in a period from 1892 to 1938, specimens of *P. robusta* had been made types of ten other species. One of the synonyms probably is still in use. Since World War II, nothing has been added to our knowledge of this species, indeed to our understanding of many others native to Burma and China, countries in which systematic surveys of megadrile faunas had been under way prior to that war.

Pheretima rodericensis

1879. *Perichaeta rodericensis* Grube, *Philosoph. Trans. Roy. Soc. London* **168**: p. 554. (Type locality, Rodriguez. Type, in the Breslau Mus.? Other types, in the British Mus.)

1937. *Pheretima rodericensis*, Gates, *Bull. Mus. Comp. Zool. Harvard College* **801**: p. 365. (Synonymy.)

CIRCULATORY SYSTEM: Bahl, 1921, *Quart. Jour. Micros. Sci.* **65**. Hertling, 1921. *Zool, Anz.* **52**: p. 185. COCOONS: Bahl, 1922, *Quart. Jour. Micros. Sci.* **66**: p. 56. EXCRETORY SYSTEM: Bahl, 1924, *ibid.* **68**. HISTOLOGY, blood vessels: Sterling, 1909, *Jenaische Zeitschr. Naturwiss.* **44**: pp. 330–332. NEPHRIDIA, development: Bahl, 1922. PARASITES: (PROTOZOA): Hesse, 1904, *Bull. Assoc. Français Avanc. Sci. Paris* **268**. Drzewecki, 1907, *Arch. Protistenk.* **10**. Hesse, 1909, *Arch. Zool. Paris*, Ser. 5, **3**. Bhatia, 1924, *Quart. Jour. Micros. Sci.* **68**. Lom, 1959, *Mem. Soc. Zool. Tchecoslowakei* **23**. SAUERSTOFFVERBRAUCH: Lang & Wenig, 1936, *Vest. Cesk Zool. Spol. Praze* **3**: pp. 34–35. SPERMATOGENESIS: Hesse, 1909, *Arch. Zool. Exp. Gen.* Ser. 4, **10**: pp. 427–441. TYPHLOSULE: Hertling, 1923, *Zool. Anz.* **120**: p. 175. VARIATION: Nemec & Zizala, 1917, *Sborn. Zool. Prague* **1**: pp. 69–82.

Octothecal, pores minute, superficial, well within dorsum and there *ca.* $\frac{1}{5}$–$\frac{1}{8}$C apart, at 5/6–8/9. Male pores, in xviii, minute, superficial, each in a small porophore. Female pore, median. Genital markings, paired, *ca.* 4–6 intersetal intervals wide, just median to male porophores, across(?) 17/18 and/or 18/19 (rarely 19/20). Clitellum, without externally recognizable setae, xiv–xvi but usually not reaching 13/14 and/or 16/17. Setae, in spermathecal segments and especially in vii enlarged and modified (occasionally or usually?), 20–30/ii, 30–40/iii, 23–38/viii, 40–49/xii, 42–48/xx, 45–51/xxx, 45–50/xl, 43–50/l, 40–47/lx, 40–47/lxx, 41–51/lxxx, vi/(±) 20 + 7, vii/(±) 22 + 7, viii/(±) 27 + 4–8, xvii/20, xviii/16–20, xix/21. First dorsal pore, at 11/12 or 12/13. Prostomium, epilobous, tongue open. Color, in dorsum, reddish, reddish brown, rich brown, grayish, slate. Segments, 80–100, but usually 91–100. Size, 55–150 by 3–10 mm.

Septa, 8/9–9/10 lacking, 5/6–7/8 slightly muscular, 10/11–13/14 muscular. Pigment, reddish, in circular muscle layer, dense in front of clitellum, spare posteriorly. Intestinal origin, in xv. Caeca, simple, in xxvii–xxiv. Typhlosole, simply lamelliform, 1–1½ mm. high, ending in region of 57th to 72d segments. Hearts, in viii unaborted dorsal portions to gizzard, in ix (left or right) lateral, of x (latero?)-esophageal, of xi–xiii latero-esophageal. Lymph glands, present in intestinal segments. Holandric. Testis sacs, unpaired and ventral. Seminal vesicles, in xi, xii, medium-sized to large, filling the coelomic cavities. Pseudovesicles, in xiii, xiv, (rarely recognized probably because often quite rudimentary). Prostates, in xvi–xxii, ducts 4–8 mm. long, muscular, thicker ectally, coiled or looped. Spermathecae, rather small, duct shorter than ampulla and narrowed in parietes, diverticulum from median face of duct at parietes, often longer than main axis, with slender stalk, wider and longer seminal chamber often markedly moniliform and with 4–7 constrictions. GM glands, shortly stalked, coelomic, closely crowded, a number associated with each marking.

Reproduction: Presumably biparental as sperm are exchanged during copulation. Further support for the assumption is provided by the condition of seminal vesicles and prostates, also by absence of organ deletions that are so characteristic of parthenogenetic polymorphism.

Distribution: China (?). Indo-China (?). Christmas Island. Australia. New Caledonia. United States (Alabama, Georgia, Florida, Maine). Bermuda, Jamaica, Puerto Rico, Dominica, Martinique, St. Kitts, Grenada, Trinidad, Venezuela, Guyana, Surinam, and French Guiana. England, France, Italy, Switzerland, Germany, Poland, Bohemia, Russia. Lagos(?), South Africa. Rodriguez, Madagascar, Mauritius, Moheli, Nossi Be.

Listing for China and Lagos (Nigeria), is questioned because the records are based on worms from Kew Gardens without confirmation from collections made subsequently at Foochow and Lagos or anywhere else in the same continents or surrounding regions. Inclusion of the other Asiatic area is questioned for the same reason, as the specimen involved was intercepted by the U. S. Bureau of Plant Quarantine from soil with a camelia supposedly brought from Indo-China (not previously recorded).

New Caledonia is nearest to, though still outside of, the pheretima domain. There is accordingly no valid record of the occurrence of this anthropochorous species anywhere in the region where it must have originated! The distribution and its apparent restrictions suggest that the home of the species should be sought in tropical lowlands.

Habitats: Soil. Earth with potted plants (Poland). Greenhouses, and there only (?) in the following: Maine, England, France, Italy, Switzerland, Germany, Poland, Czechoslovakia, Russia. Several times locality was listed only as "Botanical Garden" but greenhouses in the gardens seem more probable especially as several other sites were recorded as greenhouse, forcing house, or some equivalent, in a botanical garden. Cave (Puerto Rico).

Regeneration: Tail regeneration is possible at levels behind 54/55 but only 2–5 segments were differentiated in those specimens for which records are available.

Parasites: CILIATA: *Maupasella vacuolata* Lom, 1959, Prag, (Czchoslovakia), Madagascar. *Lomiella bergeri* Puytorac & Rakotoarivelo, 1964, Madagascar. SPOROZOA. *Monocystis macrospora* Hesse, 1909, coelom, France. *Nematocystis anguillula* Hesse, 1909, France. *Stomatophora coronata* Hesse, 1904, France. *Stomatophora simplex* Bhatia, 1924, Kew.

Remarks: Bahl's English studies of excretory and vascular systems and their development presumably were made on *rodericensis* still common at Kew Gardens or with less probability on *P. diffringens* as *P. posthuma* is unlikely to have been obtainable in Britain. Instead of the two, closely paired, longitudinal excretory ducts on the gut dorsally that Bahl originally thought all pheretimas have, *rodericensis* (Bahl, 1924) is provided with but one and that median. Other divergence, from a pattern of nephridial anatomy (assumed to be universally characteristic throughout the genus), is then perhaps to be anticipated. Species of Michaelsen's *Archipheretima* now seem most likely to provide such differences.

Development of the spermatheca may be associated with lysis of pigment and with abortion of some setal follicles in the median region between spermathecal pore levels.

Genital markings may appear first in xviii and then in subsequent growth obliterate 17/18 and 18/19 so as to appear to cross the levels of those two furrows.

Pheretima scitula

1936. *Pheretima scitula* Gates, *Rec. Indian Mus.* **38**: p. 457. (Type locality, Port Blair. Types, in the Indian Mus.)

Sexthecal, pores minute, superficial, *ca.* ⅓C apart, each at center of a very small circular porophore on ectal end of spermathecal duct, at 6/7–8/9. Male pores, in xviii, minute, invaginate, each on a thick circular disc (or short column?) on roof of a deep parietal invagination with a transversely slitlike aperture. Female pore, median. (Genital markings, none.) Clitellum, xiv–xvi. Setae, 29/iii, 37–44/viii, 44–46/xii, 44–46/xx, vii/12–25, viii–19–25, xvii/14–19, xviii/12–14, xix/13–17 (present ventrally in xvi?). First dorsal pore, at 12/13. Prostomium, epilobous, tongue open. Color, only in dorsum, red, almost to blueish or slate anteriorly (pigment not, in circular muscle layer?). (Segments?) Size, 100–200 by 5 mm.

Septa, 8/9–9/10 aborted, none thickly muscular. Intestinal origin, in xvi. Caeca, simple, in xxvii (– ?). (Typhlosole? Hearts, of x–xiii latero-esophageal?)

Holandric. Testis sacs, unpaired and ventral. Seminal vesicles, medium-sized, in xi, xii, each with a fairly large primary ampulla. Prostates, in xvii–xix, ducts 2–3 mm. long, muscular, each in a short loop. Spermathecae, small, duct almost confined to but not narrowed in the parietes, diverticulum from anterior face of duct in the parietes, much longer than main axis, with muscular stalk, thinner walled middle portion looped in part in a zigzag manner, and a short spheroidal to ellipsoidal seminal chamber.

Reproduction: Presumably biparental as sperm are exchanged during copulation.

Sperm already were present in spermathecal seminal chambers of two aclitellate worms but were lacking in the same chambers of the clitellate individual.

Distribution: Port Blair, South Andaman (Andaman Islands).

Remarks: This species is known only from seven specimens. The clitellum, distinguishable only on one worm, probably was not at maximal tumescence.

Record of this species from the Caroline Islands (Ohfuchi, 1956) is mistaken as the description clearly indicates.

Pheretima subtilis

1943. *Pheretima subtilis* Gates, *Ohio Jour. Sci.* **43**: p. 104. (Type locality, Meung Nawng. Types, none.)
1961. *Pheretima subtilis*, Gates, *Zool. Meded. Leiden* **37**: p. 308.

Bithecal, pores minute, superficial, widely separated, at 5/6. Male pores, in xviii, minute, each on a small tubercle on median wall of a slight parietal invagination with thin lateral wall. Female pore, median. Genital markings, unpaired, median, transversely elliptical, almost reaching segmental equators, across 22/23, 23/24. Clitellum, xiv–xvi. Setae, 42/ii, 69/iii, 86/viii, 76/xii, 16/xiv, 18/xv, 20/xvi, 76/xx, vi/39, xvii/23, xviii–16, xix/24. First dorsal pore, at 11/12(?). Prostomium, epilobous, tongue open(?). Color, none (pigment lacking). (Segments?) Size, 60 by 4 mm.

Septa, 8/9–9/10 aborted, none very thickly muscular. Intestinal origin, in xv. Caeca, simple, with smooth dorsal and ventral margins. (Typhlosole?) Hearts, of ix (both sides) lateral, (of x–xi aborted?), of xii–xiii latero-esophageal(?). Holandric. Testis sac of x ventral, of xi U-shaped and including vesicles. Seminal vesicles, of xi small, of xii medium-sized. Prostates, in xvii–xix, ducts *ca.* 3½ mm. long, slender, each in a U-shaped loop. Spermathecae, duct slender, shorter than ampulla, narrowed in parietes, diverticulum from median face of duct near body wall, with stalk as well the slightly longer seminal chamber looped in a more or less regular zigzag. GM glands sessile on parietes.

Reproduction: Presumably biparental as sperm are exchanged during copulation.

Distribution: Meung Nawng (Pang Long State) on the Shan Plateau.

Habitat: Manured ground in a low and damp garden.

Remarks: This species is known only from the original description of the type which was softened and slightly eroded.

Pheretima suctoria

1907. *Pheretima suctoria*, Michaelsen, *Mitt. Naturhist. Mus. Hamburg* **24**: p. 165. (Type locality, Andaman Islands. Types, in the Indian and the Hamburg Museums.)
1909. *Pheretima suctoria*, Michaelsen, *Mem. Indian Mus.* **1**: p. 196.
1923. *Pheretima suctoria* (part), Stephenson (*The Fauna of British India*), *Oligochaeta*, p. 311. (Excluding, *P. suctoria*, Stephenson, 1922.)
1931. *Pheretima suctoria* var. *typica*, Gates, *Rec. Indian Mus.* **33**; p. 412.
1936. *Pheretima suctoria*, Gates, *ibid.* **38**: p. 461.

Octothecal, pores minute, superficial, *ca.* ¼C apart, at 5/6–8/9. Male pores, in xviii, minute, superficial, each in a small disc. Female pores, closely paired. Genital markings, 2, circular to transversely elliptical, just median to the male porophores, in xviii. Clitellum, setae unrecognizable externally, xiv–xvi. Setae, 34/iii, 25–38/v, 35–38/x, 58/xii, 60–70/xiii, 66/xx, 80/xxvi, vi/10, vii/10–14, viii/10–16, xvii/20, xviii/ 4–8, xix/22. First dorsal pore, at 12/13. Prostomium, epilobous, tongue closed. Color, only in dorsum, purple, in a pre- and in a post-setal band in each metamere (pigment, red? in circular muscle layer?) Segments, 103–123. Size, 70–140 by 4–7 mm.

Septa, 8/9–9/10 aborted, none especially muscular but 11/12–13/14 thicker than the others. Intestinal origin, in xv (?). Caeca, simple, in xxvii–xxii. (Typhlosole, none.) Hearts, of x (latero?)-esophageal, of xi–xiii latero-esophageal (?). Lymph glands, present in intestinal segments. Holandric. Testis sacs, unpaired and ventral. Seminal vesicles, in xi, xii, large. Prostates, small, in xvii–xix, ducts short, straight, or looped. Spermathecae, duct shorter than ampulla, diverticulum from anterior face of duct at parietes, longer than main axis, with a short stalk and a more or less regularly zigzagged seminal chamber. GM glands, sessile on the parietes.

Distribution: Andaman Islands. Camorta (Nicobar Islands).

Habitat: Jungle, where it was found "creeping over the ground" (Camorta), in March.

Remarks: The species is known only from nine specimens.

Setae of ii–ix are larger, straighter and more widely separated than in subsequent segments. The spermathecal duct may be slender to muscular.

Pheretima sulcata

1932. *Pheretima sulcata* Gates, *Rec. Indian Mus.* **34**: p. 424. (Type locality, Kawkareik. Types, none.)
1960. *Pheretima sulcata*, Gates, *Bull. Mus. Comp. Zool. Harvard College* **123**: p. 278.

Sexthecal, pores minute, transverse slits, each in a white (unpigmented?) area of the dorsum, anteriorly in vi–viii. Male pores, in xviii, minute, superficial, each in a small circular disc connected by a narrow strip of modified epidermis to a genital marking and at posterior end of a seminal groove passing back from the marking along that strip. Female pore, median. Genital markings, two, circular, separated by a median space wider than the markings, between eq/ xvii and eq/xviii. Clitellum, setae unrecognizable externally, xiv–xvi. Setae, vii/38, xviii/12. First dorsal pore, at 11/12. Prostomium, epilobous(?). Color, only in dorsum, reddish to slate, (pigment red and in circular muscle?). (Segments?) Size, 96 by 3 mm.

Septa, 8/9–9/10 aborted. Intestinal origin, in xv. Caeca, simple. (Typhlosole?) Hearts, last pair in xiii. Holandric. Testis sacs, paired(?) and ventral(?). Seminal vesicles, rather small, in xi, xii. Prostates, fairly large, duct short, much widened just before passing into body wall. Spermathecae, small, duct slender, slightly shorter than ampulla, diverticulum from anterior face of duct near parietes, with short stalk, irregularly looped and thicker middle portion, a shortly ovoidal terminal chamber. GM glands (really glandular?) represented by a thickening of the outer portion of the body wall without protuberance into coelomic cavities.

Reproduction: Possibly parthenogenetic. The smallness of the seminal vesicles often is associated with male sterility. Absence of the left spermatheca of vi and the small size of the others may be an indication of an early trend in evolution of an athecal morph.

Distribution: Kawkareik (Amherst).

Remarks: Known only from a clitellate type that was soft and lacked the left anterior spermatheca. The looped middle portion of the spermathecal diverticulum may be part of the seminal chamber.

Pheretima taprobanae

1892. *Perichaeta taprobanae* Beddard, *Proc. Zool. Soc. London* **1892**: p. 162. (Type locality, Ceylon. Types, in the University Mus., Oxford.)
1937. *Pheretima taprobanae*, Gates, *Bull. Mus. Comp. Zool. Harvard College* **80**: p. 371.

Bithecal, pores minute, superficial, slightly behind 7/8. Male pores, in xviii, minute, superficial, slightly postsetal, each in an indistinctly demarcated circular area between 18/19 and presetal furrow of xviii. Female pore, median. Genital markings, small, circular, presetal, paired, somewhat median to spermathecal pore levels in vi–xi, in xviii–xxii. Clitellum, xiv–xvi. Setae, present in each of the clitellar segments at least ventrally, 70–81/v, 80/viii, 77/x, 74/xii, 26/xvi (ventrally), 70/xix, 75/xx, 67/xxv, 54/xxvi, viii/34–41, xvii/20–25, xviii/14–19, xix/19–25. First dorsal pore, at 12/13. Prostomium, epilobous, tongue open and with a longitudinal groove at mD. Color, white. Segments, 95–136. Size, 80–145 by 4–7 mm.

Septa, 8/9 present but membranous, 9/10 aborted, 10/11–11/12 membranous, 12/13–13/14 thickly muscular, 14/15 muscular. Intestinal origin, in the region of insertion of 15/16. (Intestinal caeca, none.) Typhlosole, present and lamelliform from xxiii, ending in region of the 71st segment. Hearts, (of x?), of xi–xii (latero-?)esophageal. Holandric. Testis sacs unpaired and annular, hearts of x–xi and vesicles of xi included. Seminal vesicles, in xi, xii, (juvenile?). Prostates, small, in xviii, ducts 5–6 mm. long, in a single hairpin loop or in 2–3 shortly U-shaped loops, muscular, ental half slenderer, narrowing again in the parietes. (GM glands, none.) Spermathecae, small, long enough to reach only to 9/10, duct abruptly narrowed in parietes, diverticulum from anterior face of duct at parietes, about as long as main axis, with slender stalk and spheroidal to ovoidal seminal chamber.

Reproduction: Male sterility is suspected because of absence of spermatozoal iridescence on male funnels and in spermathecae. Further evidence indicative of parthenogenesis is provided by the juvenile condition of seminal vesicles, the rudimentary male funnels and by the slenderness of the sperm ducts of specimens in which the clitellum was at maximal tumescence.

Distribution: India (a botanical garden in Trivandrum, Travancore). Ceylon (Peradeniya, botanical garden?). Brazil (Rio de Janeiro, botanical garden). Hawaii (a botanical garden, Oahu). Madagascar (St. Marie).

Remarks: Although the genital markings might well be expected to be associated with stalked, coelomic glands of the kind that is present in *P. diffringens* and *hawayana*, no traces of glands were found. Possibly a parthenogenetic degradation is involved.

The species never has been recognized in the *Pheretima* domain. Nothing is known about its relationships. The original home presumably is in the tropics and perhaps in New Guinea.

Pheretima terrigena

1932. *Pheretima terrigena* Gates, *Rec. Indian Mus.* **34**: p. 482. (Type locality, Karen Hills, Toungoo. Types, none.)
1933. *Pheretima terrigena*, Gates, *ibid.*, **35**: p. 546.

Sexthecal, pores minute, superficial, each in a very small tubercle, at 6/7–8/9. Male pores, in xviii, minute, superficial, each in a small disc. Female pore, median. Genital markings, paired, median to male pore levels, transversely and shortly elliptical, fairly widely separated mesially, across some or all of 17/18–21/22. Clitellum, 13/14–xvi/eq. Setae, lacking dorsally in ii(?), present in clitellar segments(?), vii/17–20, viii/18, xvii/21–28, xviii/15–20, xix/19–29. First dorsal pore, at 12/13. Prostomium, epilobous(?). Color, only in dorsum, reddish

to brownish red or brownish (pigment red and in circular muscle layer?). Segments, 112. Size to 89 by 4 mm.

Septa, 8/9–9/10 aborted, none thickly muscular(?). Intestinal origin, in xvi(?). Caeca, simple, in xxvii–xix. (Typhlosole?) Hearts, of x–xiii (latero?)-esophageal. Blood glands, in v. Holandric. Testis sacs, unpaired, ventral(?) in x, U-shaped in xi and including the vesicles. Seminal vesicles, in xi, xii, the latter large, displacing septa and with protuberant primary ampulla. Prostates, in xvii–xix, duct slender. to 9 mm. long, coiled. Spermathecae, duct shorter than ampulla, abruptly narrowed in body wall, diverticulum from anterior face of duct at parietes, longer than main axis, variously coiled or looped. GM glands, sessile on parietes.

Reproduction: Presumably biparental but the only support that can be cited from publications is the large size of the posterior seminal vesicles. Prostates also were not juvenile.

Distribution: Karen Hills of the Shan Plateau (Toungoo). Loikaw (Karenni).

Parasites: Nematodes, *Adungella major* Timm, 1967, in a Loikaw worm, were present in coelomic cavities of ix–xiv.

Remarks: Known only from four specimens, one an anterior amputee, two posterior amputees.

The beheaded fragment from Loikaw, if correctly identified, warrants an anticipation that the range extends across the Salween and into Thailand. The unknown locality at which the types were secured is assumed to be in the Karen Hills near the western margin of the Shan Plateau, as presence in the Toungoo plains or across the valley and on the Pegu Yomas now seems unlikely.

Relationships presumably are closest with *P. fucosa* known only from a few specimens secured in jungles along the Kamaungthwe River northeast of Tavoy town. Unfortunately, nothing is known as to what happens to either taxon in the almost unexplored region between the two ranges as now known.

Pheretima umbraticola

1932. *Pheretima umbraticola* Gates, *Rec. Indian Mus.* **34**: p. 484. (Type locality, Kawkareik. Types, none.)

Sexthecal, pores minute, superficial(?), in very small circular tubercles at 6/7–8/9. Male pores, in xviii, minute, each on wall of a copulatory chamber with large, transversely, slitlike equatorial aperture. Female pore, median. (Genital markings, none.) Clitellum, without externally recognizable setae, xiv–xvi. Setae, 45–60/xx, vii/14–20, viii/16–20, xvii/8–14, xviii-8–10. First dorsal pore, at 12/13. (Prostomium?) Color, in dorsum, slate anterior to the clitellum, brownish posteriorly (pigment reddish and in the circular muscle layer?). Segments, 125–135. Size, 115–122 by 6–7 mm.

Septa, 8/9–9/10 aborted, 4/5 to 7/8 increasingly thickened. Intestinal origin, in xv. Caeca, simple, in xxvii–xxiii. (Typhlosole? Hearts, of xi–xiii latero-esophageal?) Blood glands, in v. Holandric. Testis sacs, paired and then vertical, unpaired (and then *U*- or horeseshoe-shaped or annular?). Seminal vesicles, small, each with distinct primary ampulla (and retaining juvenile shape?), in xi, xii, the anterior pair within testis sacs of xi. Pseudovesicles, in xiii (also xiv?). Prostates, in xvi–xx, ducts 2½–4½ mm. long, in a hairpin loop with ectal limb thicker. Spermathecae (size?), duct shorter than ampulla and gradually narrowed in the parietes, diverticulum from duct ectally, longer than the (small?) main axis, with a long, thick-walled stalk and a shorter, slightly wider and ellipsoidal seminal chamber.

Reproduction : Parthenogenesis, in absence of records *re* sperm maturation and copulation, is suspected. Seminal vesicles seemingly are retained in a juvenile state and in xi nearly fill testis sacs which should be distended at maturity by the coagulum associated with profuse spermatogenesis.

Distribution : Kawkareik, near the Dawna Hills (Amherst).

With the possible exception of several specimens picked up near Kawkareik on the road to Thailand, nothing is known of the megadrile fauna of Dawna Hills and their southern continuations.

Remarks : Male pores were not seen and no penes were found. If GM glands had been on the coelomic face of the copulatory chamber they probably would have been recognized but because the original description of the copulatory chambers was not more thorough no suggestion now is possible as to presence of glands in the chamber wall. The large aperture presumably was associated with a large chamber which probably could be completely everted as a large protuberance perhaps more or less thickly columnar. But, there is no spermathecal invagination into which so large a copulatory organ could be inserted.

In absence of any one of the various kinds of penes or penial bodies that are associated with coelomic copulatory chambers, perhaps the latter organs in *P. umbraticola* can be protruded in such a way as to form suckerlike structures for adhering to the epidermis in the vicinity of spermathecal pores. The primary spermathecal pore clearly is minute. Whether it is to be regarded as superficial is not perfectly clear. If the pore-bearing tubercle can be drawn into the body wall in such a way as to be completely covered over ordinarily there presumably would be involved a primitive spermathecal invagination. On the contrary, it is suspected that a circumferential groove bounding the porophore can be deepened in such a way as to produce an appearance of the slight retraction indicated in the original description.

If reproduction really is parthenogenetic, some decision as to the kind of morph involved is prerequisite to any systematic judgment. Accordingly, all specimens (the four types from which the taxon alone is known) are assumed to have been of an H_P morph in which seminal vesicles (and testes?) were retained in a juvenile condition. That assumption enables recognition of *P. umbraticola* at species level and obviates search for an H morph from which the Kawkareik worms could be derived by some of the many organ modifications or deletions that are permitted by parthenogenesis.

Pheretima virgo

1901. *Amyntas virgo* Beddard, *Proc. Zool. Soc. London* 1900: p. 895. (Type locality, Tale, Siam. Types, in the British Mus.)

1903. *Pheretima virgo*, Michaelsen, *Die geographische Verbreitung der Oligochaeten* (Berlin), p. 101.

1931. *Pheretima mamillana* Gates, *Rec. Indian Mus.* 33: p. 400. (Type locality, Ye. Types, none.)

1932. *Pheretima mamillana* + *P. mendosa* Gates, *ibid.*, 34: pp. 470, 473. (Type locality of *mendosa*, Zowai, Types, none.) *Pheretima virgo*, Stephenson, *Ann. Mag. Nat. Hist.*, Ser. 10, 9: p. 236.

1933. *Pheretima mamillana* + *P. mendosa*, Gates, *Rec. Indian Mus.* 35: pp. 537, 538.

1934. *Pheretima virgo*, Gates, *Rec. Indian Mus.* 36: p. 264.

1936. *Pheretima mamillana*, Gates, *ibid.*, 38: p. 430.

1939. *Pheretima virgo*, Gates, *Jour. Thailand Res. Soc. Nat. Hist.* Sup. 12: p. 104.

1949. *Pheretima mamillana*, Gates, *Bull. Raffles Mus. Singapore*, no. 19: p. 32.

1955. *Pheretima mamillana*, Gates, *Rec. Indian Mus.* 52: p. 88.

1960. *Pheretima virgo*, Gates, *Bull. Mus. Comp. Zool. Harvard College* 123: p. 275.

?

1901. *Amyntas perichaeta* Beddard, *Proc. Zool. Soc. London* 1900: p. 896. (Type locality, Patalung State? Type, in the British Mus.)

1932. *Pheretima perichaeta*, Stephenson, 1932, *Ann. Mag. Nat. Hist.*, Ser. 10, 9: p. 227.

1939. *Pheretima perichaeta*, Gates, *Jour. Thailand Res. Soc. Nat. Hist. Suppl.* 12: p. 103.

Sexthecal, pores large, close to mL, at 6/7–8/9. Male pores, in xviii, each minute and on a porophore within a copulatory chamber opening to the exterior through a transversely slitlike equatorial aperture. Female pore, median. Genital markings, small, circular, present only within the copulatory chambers, sometimes at distal ends of shortly columnar porophores. Clitellum, without externally recognizable setae, reaching nearly to, to, or slightly beyond 13/14 and 16/17. Setae, (slightly enlarged in some of the preclitellar segments?), 19–25/ii, 30–40/iii, 38–51/viii, 43–56/xii, 44–56/xiii, 47–56/xx, vii/21–28, viii/21–30, xvii/13–18, xviii/5–13, xix/14–19. First dorsal pore, at 12/13. Prostomium, epilobous, tongue open. Color, in dorsum, red, reddish, reddish brown, brownish, slate. Segments, 97–157 but usually 141–157. Size, 51–210 by 2½–7 mm.

Septa, 6/7–7/8, 10/11–12/13 slightly strengthened to muscular, 8/9–9/10 aborted. Intestinal origin, in xv. Caeca, simple, in xxvii–xvii. Typhlosole, fairly high, lamelliform, ending in region of 80th segment. Hearts, of viii with unaborted dorsal portions passing to gizzard, of ix (left or right) lateral, of x esophageal(?), of xi–xiii latero-esophageal. Blood glands, in v and in discrete lobes around esophagus just behind the gizzard. Lymph glands, present at least from xxvii posteriorly. Holandric. Testis sacs, ventral, usually paired, occasionally unpaired. Seminal vesicles, small to medium sized, in xi, xii. Pseudovesicles, in xiii, xiv, each with a free dorsal lobe as in early juvenile stages of functional vesicles. Prostates, small, confined, to xvi–xviii, xvii–xviii, or xviii, ducts 2–4 mm. long, slightly curved or in a single U-loop. Copulatory chambers, large, hemispheroidal(?), prostatic ducts passing into middle of dorsal face. GM glands, one finely acinous, 2–4 others usually not, with short stalks passing to markings within the chambers. Male porophores, on roofs of the copulatory chambers, each usually completely retracted and appearing as a thick, firm annulus but with the roof of the deep central concavity protrusible to provide with the annulus a rather conical intromittent organ perhaps having at its tip the male pore. Spermathecae, duct shorter than the ampulla, rather thick, gradually widened within the parietes, diverticulum usually from lateral face of the duct, usually looped, often partly in a more or less regularly zigzagged manner, with a muscular stalk of variable length, a thicker middle portion with thinner wall and also of variable length, a terminal ellipsoidal seminal chamber.

Reproduction: Information as to spermatozoal iridescence as well as condition of testes, seminal vesicles, and prostates, no longer is available, but parthenogenesis is confidently anticipated in many morphs.

Copulatory chambers when fully everted and with a fully protruded penial body distally, presumably could deposit sperm directly at or near aperture of diverticulum into spermathecal duct.

Distribution: Kinmunsakhan (Thaton). Kawkareik, Ye (Amherst). Eindawaza, Myigyaunglaung, Shintapi, Kale-aung in or near the Forest Reserve of the Heimza Basin area, Nyinmaw, Nyaungdon, Nyaungdonle Chaung, Tanbin Hills, Pyinthadaw, Posoe Chaung, Mayan Chaung, Zinba, Siyigyan, Kamaungthwe river region to the east (Tavoy). Sindin, Zowai, Tharrabyin, Victoria Point (Mergui).

Tale, peninsular Thailand.

Kaki Bukit (Perlis), Baling (Kedah), in the northernmost part of the Federation of Malaysia.

The range now seems to run through ten degrees of latitude, about from 16° to 6°. In spite of considerable collecting in Thaton and Amherst districts the species never was found south of Kinmunsakhan and west of Kawkareik until at the very Tavoy border of Amherst district. Perhaps man was responsible for getting *P. virgo* across the Salween.

Variation and Abnormality: Specimens from Malaysia were smaller than Burmese worms, 52–85 by 2½–4 mm. as compared with 132–210 by 4–7 mm. Associated with the smaller size were fewer setae, 21–30/iii, 24–33/iv, 31–42/viii, 36–49/xii, 40–54/xx, vii/15–20, viii/15–21, xvii/12–16, xviii/6–9, xix/14–18 (those figures omitted from the précis). Pigment, in postclitellar portions of Perlis worms varied from dense deposition to absence.

The single heart of ix was on the left side in 49 Burmese specimens, on the right side in 60. Testis sacs in a certainly unpaired state were found only in Baling worms. The spermathecal diverticula of worms from the Malay Peninsula arose from the duct anteriorly, anteromesially, or mesially but not laterally. Seminal vesicles of xiii, recorded by Stephenson (1932), probably are unusually large pseudovesicles. Absence of the right heart of xiii in 1 of 109 specimens presumably is to be considered an abnormality as also the presence of an extra copulatory chamber along with a prostate and duct as well as GM glands, on the left side in xvii of one individual. Other abnormalities, at least, now seem likely to have arisen during the evolution of parthenogenetic polymorphism.

Polymorphism: Information that might have enabled designation of a biparental population no longer is available and the species accordingly had to be defined with reference to Burmese worms (*mamillana*) free of obvious anomalies or aberrations. The prostates of those worms seem small and the ducts short as compared with the same organs in many biparental forms of the same size in other species.

All specimens originally referred to *mendosa* and to *virgo* were of morphs in which genital organs already had undergone parthenogenetic degradation leading toward elimination of spermathecae as well as prostates. Two specimens were thought to be of an A morph but may already have shown some evidence of a trend toward an AR. AR morphs of *virgo* may have been secured but would have been very difficult to distinguish, at least with the reduced information now available, from AR morphs of *houlleti* and *alexandri*. Evolution of parthenogenetic polymorphism, as previously mentioned (1960), seems to be most advanced near the center of the range as now known.

Parasites: Vermiform gregarines were present in coelomic cavities of Baling worms. Nematodes were present in anterior coelomic cavities, in 14 of 24 Mergui district worms in which nematode eggs were loose in the fluid, piled up in heaps on the ventral parietes, or aggregated into discs. Gregarines and pseudonavicella cysts were present in much hypertrophied ovaries of 8 of the 14 worms. In 3 of the 8, not a single ovum was recognizable in host ovaries.

If maturation of the female gametes had not been prevented by the parasites, all ova must have been shed during the breeding season. NEMATODA: *Siconema turgidum* Timm, 1966, Burma.

Remarks: Relationships may be closest with the H morph of *P. houlleti*. Both can be derived from a common ancestor by slight differences in the manner of invaginating spermathecae and male pores.

P. perichaeta, known only from the holotype, is distinguishable only by absence of hearts in xiii (an aberration?), and by slenderness of the spermathecal duct and diverticulum, the latter long and in a ball of closely compacted loops.

Pheretima youngi

1932. *Pheretima youngi* Gates, *Rec. Indian Mus.* **34**: p. 406. (Type locality, Pang Wo. Types, none.)

Quadrithecal, pores very small transverse slits, each in a white area (without pigment?) in the dorsum, at eq/vi and at 6/7. Male pores, in xviii, small, longitudinal slits, each in a longitudinally elliptical tough porophore reaching slightly into xvii and xix. Female pore, median. (Genital markings, aside from the porophores, none.) Clitellum, setae unrecognizable externally, xiv–xvi. Setae, 29–33/v, 40–48/xx, vi/20–25 + 5–9, xvii/13–18, xviii/4–7, xix/15–18. First dorsal pore, at 10/11. Prostomium, epilobous (?). Color, only in dorsum, reddish brown to yellowish brown (pigment red and in circular muscle layer?). Segments, 89–97. Size, 68–90 by 4 mm.

Septa, 8/9–9/10 present, none especially muscular. Intestinal origin, in xv. Caeca, simple, in xxvii–xxiv. (Typhlosole?) Hearts, of ix lateral but usually present only on one side, of x–xiii (latero?)-esophageal. Blood glands, in iv–v. Holandric. Testis sacs, unpaired, ventral. Seminal vesicles, large, displacing the septa, in xi, xii. Prostates, in xvii–xx, duct 2–3 mm. long, each in a U-shaped loop. Spermathecae, duct shorter than ampulla and narrowed abruptly in body wall, diverticulum from anterior face of duct in parietes, longer than main axis, with long and looped stalk, shortly ellipsoidal seminal chamber. Tough (non-glandular?) tissue present in body wall above male porophores but not protuberant into coelom.

Reproduction: Biparental reproduction is anticipated although the only reason that now can be cited in support is the large size of the seminal vesicles. Large prostates also appeared to be functional.

Distribution: Pang Wo (Mang Lun). Teung Cong (Mong Mong, Yunnan). At elevations of 4,000–4,500 feet.

Habitats: Soil. Wet leaves.

Remarks: Known only from five clitellate specimens, two of which were posterior amputees.

Male pores were nearer the median or the lateral margins of the porophores. A portion immediately around each male pore was protruded in a rather conical fashion but presumably can be flattened out in which case the male porophores may be slightly concave.

The anterior spermathecal pores are slightly further from mD than the others.

The intestine seemed to be narrowed in xv–xxi.

Tonoscolex

1933. *Tonoscolex* Gates, *Rec. Indian Mus.* **35**: p. 484. (Type species, *Notoscolex birmanicus* Gates, 1927).
1936. *Tonoscolex*, Gates, *ibid.* **38**: p. 379.

Digestive system, without intestinal caeca and supra-intestinal glands but with a massive gizzard in vi, paired calciferous glands, an intestinal origin in xiv, and a lamelliform typhlosole. Calciferous glands, with short stalks passing to esophagus laterally just in front of a septum.

Vascular system, with unpaired dorsal, ventral and supra-esophageal trunks, paired extra-esophageals median to the hearts, paired lateroparietal trunks (from the posterior end?) but no subneural, hearts of viii and anteriorly lateral, of ix and posteriorly latero-esophageal. Excretory system, meroic—massed astomate micronephridia around gut in vi (v?) with numerous ductules that open into portions of the pharynx belonging to iii–iv—astomate, V-shaped, exoic micronephridia on parietes from iv posteriorly and especially numerous in clitellar segments—beginning at or behind lvi[1] 8–30 stomate V-shaped meganephidia per segment with preseptal funnels and terminal canals joining a transverse duct on each side and in front of each septum, the two ducts of a segment uniting just before opening into the intestine at apex of the typhlosole. Setae, ventral couples of xvii lacking. Dorsal pores, present. Segments, more than 150(?). Pigment, lacking. Septa, present at least from 5/6.

Biprostatic, male pores in xvii and in seminal grooves that extend into another segment. Female pores, anteromedian to *a*, in xiii. Quadrithecal, pores at 6/7–7/8. Clitellum, annular, intersegmental furrows obliterated, dorsal pores occluded, setae retained. Holandric. Testes in ix–x. Prostates, strap-shaped, extending through several segments, ducts short, soft (little muscularity?) and joined entally by the sperm ducts. Ovaries, (fan-shaped? with numerous egg-strings?), in xii. Ovisacs, present(?). Spermathecae, each with a diverticulum shorter than the main axis.

Distribution: On the Shan Plateau from the Karen Hills of Toungoo district and from Karenni on the eastern border of Burma, in middle Burma from Mawleik in the Upper Chindwin District and from Mansum in Myitkyina district, north into the Himalayas. In India, west in the Himalayas to the vicinity of Darjiling and south in Assam to the Khasi Hills.

A Pyinmana (Yamethin) record is assumed to be of worms obtained in the foothills of the Shan Plateau

---

[1] In last 56–91 segments.

to the east of the town as no species of *Tonoscolex* has yet been reported from the Pegu Yomas. Although *Tonoscolex* is known from the Khasi section of the Garo-Jaintia spur into Assam, the genus was not found anywhere on the western mountain wall of Burma and may not have crossed the Sittang.

Biology: Tonoscolexes, at least after removal from their native habitats, seem to be the most sluggish of all Burmese earthworms and in that respect are like some of the larger Australian endemics. The oriental forms when dropped into strong spirit only quiver slightly once or twice.

Burmese species are geophagous and macroscopically recognizable bits of organic matter were not found in their egesta. Leaf litter in forest reserves where tonoscolexes are numerous was (before World War II) burned off every year during the dry season. Whatever the food may be, in the tropical soils supposedly poor in organic matter, it is sufficient to enable growth to a size that is truly gigantic for megadrilous oligochaetes.

Activity, as indicated by deposition of intestinal ejecta on the surface, is restricted to the rainy season which, in the eastern hills, includes May–October. Breeding probably is restricted to a short terminal portion of the period of activity. Embryonic development seems likely to require all of the six months of drought until the next rainy season is well under way. Postnatal development, at least for the giant forms, must require more than one rainy season, perhaps (see *T. conversus*) even three or four. No data were published that would indicate that postnatal growth involves formation of new segments though such growth does seem highly probable for the larger forms.

The season of activity is spent in eating the way through the soil and in depositing the fecal earth on the surface in heaps called castings. Burning the litter in forest reserves at Kalaw showed (Gates, 1927) that "the bare ground between the bases of the trees and shrubs in such burnt-over areas is almost literally covered with the tower-like castings." The towers were those last formed at the end of the rains and so had not been washed away. Castings of *birmanicus*, reaching heights of 200–250 mm., were deposited in 3–4 days. A casting 150 mm. high, part of which already had been washed away during a heavy shower, had a dry weight of $3\frac{1}{4}$ pounds. Estimates of the tonnage of finely divided soil brought to the surface in an acre during six months, based on observations during May and during October, i.e., at the beginning and at the end of the rainy season, seemed too huge. However, it was stated that the amounts so deposited "must be very great." Whether deposition continues at the same rate throughout the entire six months cannot now be stated.

Phylogeny: *Tonoscolex* and *Nelloscolex* clearly are much more closely related to each other than to any megascolecids now known. Origin and main evolu-

tion within an area centered in Burma now seems possible. A common ancestor, prior to appearance of the chief distinguishing character of *Tonoscolex-Nelloscolex*, may already have reached a fairly advanced stage of evolution, with the gizzard in vii (two segments behind the location in the ancestral megadrile), with the esophagus elongated so that the intestinal origin was in xv, with hearts in xii–xiii, and possibly with calciferous tissues already concentrated within esophageal lamellae of segments ix–xiii. However, the ancestral form seemingly must have had a primitive battery of seminal vesicles—four pairs, two preseptal and two postseptal—just as today still characterizes the lumbricid genus *Allolobophora*. Some sort of primitive meronephry may have characterized the ancestral stage but our ignorance of the anatomy of the excretory organs in so many of the extant forms requires limitation of this discussion to other systems.

The only uniquely diagnostic character of the group among megascolecids, the anterior regional homoeosis, presumably appeared before separation of *Nelloscolex* and *Tonoscolex* as simultaneous development in two or more diverging stocks seems less likely. The homoeosis, involving all organs from the gizzard to the prostates, perhaps could have arisen in several ways but hitherto evidence has been found only for one of them, excalation or elision of one segment in front of the gizzard. The process presumably involved disappearance of the intersegmental furrow demarcating segments i and ii from each other, disappearance of the setae of ii (nephridia of that segment may already have been lost), reduction in size of the united metamere but retention of the prostomium seemingly contraindicates attaining the homoeosis by "swallowing the anterior end," in the glossoscolecid manner.

An early stage in evolution of the calciferous glands is shown by *Nelloscolex* in which calciferous lamellae of viii–xii are within lateral swellings of the esophagus not as yet constricted off from the gut. Size of the pouches increases posteriorly, those of viii still quite rudimentary. In *Tonoscolex* discrete calciferous glands are constricted off from the esophagus so completely as to open into it only through short and slender stalks. Elimination of the seminal vesicles of ix–x in a line leading to *T. horai* was associated sooner or later with disappearance of calciferous glands in viii–ix as well as of the hearts of xii. In a second line to which most species belong, vesicles of ix–x were retained but those of viii–ix were eliminated. In that lineage all calciferous glands were retained only by *T. monorchis*, the glands of viii having disappeared in the ancestry of the remaining species. Perichaetin setae have appeared twice, once in the *horai* lineage and in the northern part of the generic range, once in the *ferinus* lineage and then near the southern boundary of that range.

The discussion above has proceeded on certain assumptions that may well be erroneous and knowledge of the internal organization of the calciferous glands along with information as to excretory and vascular systems of various species may require considerable modifications not only in the phylogeny but also in the systematics.

Systematics: Little has been added to our knowledge of *Tonoscolex* since 1935 primarily because of inability to replace or add to collections of subsequent years that were destroyed during World War II along with the older material and all of the records. The genus appears to be definitely distinguished from other megascolecids even though characterizations of the vascular system and of internal structure of calciferous glands leave much to be desired. The excretory system of three species was studied by Bahl who assumed, as usual, that the structure he found was uniform throughout the genus. Prostates, though obviously racemose, do differ from those of *Pheretima* by their shape—in *Tonoscolex* being more narrowly elongate and almost bandlike but thicker and with smoother margins mesially, thinner laterally and there more apt to be incised by the septa. Nothing is known of the microscopic structure of the prostates or of any other organ except the nephridia.

Known interspecific differences as to location of seminal vesicles as well as of the last hearts, perhaps also as to number of calciferous glands and of setae (*cf.* key to the species of *Tonoscolex*), now seem likely to be associated with differences in the excretory system (see *T. horai*) of so much greater importance than the common holandry as to require recognition at some hierarchical level in the classification. Individual variation in number of calciferous glands was almost nil and absence of the glands in segment x (instead of at either end of the series) of one specimen was regarded as an abnormality. Calciferous glands were restricted to x–xii in each of the thirteen specimens of *T. horai* that were dissected. Only one specimen of the misnamed *T. monorchis* was examined but presence of glands in viii–xii reasonably can be expected to characterize the species if not also a distinct genus.

Species hitherto recorded from within the political boundaries of Burma appear, with but one exception, to be quite uniform with regard to structure that often does provide systematically useful characters. (Compare, for instance, in the species précis, prostomium, clitellum, location of first dorsal pore, and note constant absence of color, the slight differences in location of spermathecal pores—always in *BB* which is not very large.) Too little is known about some other structure. Except for a record as to one specimen in each of several species next to nothing is known about segment number. Setal characters were of very little use in dealing with recorded material because of variation from one region of the body to another, from one individual to another in the same segment, because of small size and deep retraction so that follicle apertures were not certainly distinguishable thus preventing even estimations of relative widths of intersetal intervals, and, strangely, at times because setae were completely lacking. Thus, in an August collection from Taungyi that contained only 209 juveniles, 172 did not have a single seta, 4 had a few scattered setae and only 33 were provided with eight complete chaetal ranks. Adults obtained in September from the same place were all setigerous. Although empty setal follicles were not recognized in the juveniles, faint markings in the longitudinal musculature probably did indicate locations of aborted follicles. Whether setal follicles would have reappeared during the next month if the juveniles had not been preserved and whether adults secured during September had gone through a period of setal dehiscence in their earlier development cannot now be stated.

The seminal grooves, which appear long before the clitellum and in a variety of shapes and locations—even with reference to situation of the male pores, enabled segregation of juveniles when no other evidence for specific distinctness was recognized. In spite of the importance allowed to those grooves (*cf.* key to species of *Tonoscolex*) the characters they provided did not seem helpful in estimating interspecific relationships. Shapes occasionally appeared to be somewhat different in fully clitellate individuals perhaps as a result of the development of parietal depressions or even of local expansions within the male field. Other and perhaps fortuitous furrows that appear in fully mature and strongly contracted specimens can cause confusion as did happen in connection with Stephenson's species. Finally, in an occasional specimen, the grooves are abnormally shortened for reasons unknown though interference with development by large numbers of parasites was suspected as also was parthenogenesis.

Copulatory organs of the male fields, special protuberances, tumescences or "tags," first noted by Stephenson, now seem to be helpful in defining and distinguishing species though perhaps also without value as indicators of interspecific relationships. Nothing is known about the development or functioning of these organs. Information as to existence of tumescent and detumescent, protruded and retracted, states is needed for maximum systematic usefulness. The organs of *T. birmanicus* always were present and tumescent in every adult specimen for which records now are available. Shortly u-shaped and laterally directed tumescences seem to be similarly characteristic of adults of *T. depressus* and *lunatus*. Paired slits that were mistaken for male pores in *T. oneili* open into deep invaginations from which the "tags" are protruded. *T. triquetrus*, as well as *T. ferinus*, has a single median invagination in the region occupied by the *oneili* pair but protrusions were not recorded. Whether *T. montanus* for which no organs were recorded has invaginations cannot now be stated.

KEY TO SPECIES OF *Tonoscolex*

1. Setal arrangement lumbricin................ 2
   Setal arrangement perichaetin.............. 12
2. Calciferous glands in viii–xii..........*monorchis*
   Calciferous glands in ix–xii................. 3
3. Seminal grooves in xvi–xvii............... 4
   Seminal grooves in xvii–xviii.............. 6
4. Seminal grooves divergent anteriorly, posterior ends at or near *A*, anterior ends in lateral part of *BC*.............................*parvus*
   Seminal grooves not divergent anteriorly..... 5
5. Seminal groove transversely *V*-shaped, anterior and posterior ends equidistant from m*V*..............................*quartus*
   Seminal grooves not *V*-shaped, longitudinal
                                   conversus
6. Spermathecal duct straight................. 7
   Spermathecal duct looped...............*oneili*
7. Seminal grooves united anteriorly into a *U*-shape.........................*birmanicus*
   Seminal grooves discrete.................. 8
8. Copulatory organs two pairs of "tags" in the angles formed by lateral bends of anterior and posterior ends of the seminal grooves
                                   striatus
   Copulatory organs lacking or if present not two pairs....... 9
9. Spermathecal pores close to m*V*, spermathecal diverticulum zigzag looped.........*montanus*
   Spermathecal pores not close to m*V*, spermathecal diverticulum not looped......... 10
10. Anterior ends of seminal grooves enclosed by *U*-shaped tumescences................. 11
    Anterior ends of seminal grooves not so enclosed.........................*triquetrus*
11. Seminal grooves with a single curvature and ending posteriorly at or near *C*......*depressus*
    Seminal grooves with a double curvature and ending posteriorly near *B*............*lunatus*
12. Seminal vesicles in x–xi, calciferous glands in ix–xii.............................*ferinus*
    Seminal vesicles in viii and xi, calciferous glands in x–xii......................*horai*

Note: Blood glands, according to Bahl (1941: p. 461), are lacking.

Tonoscolex birmanicus

1926. *Notoscolex sp.*, Gates, *Rec. Indian Mus.* **28**: p. 151.
1927. *Notoscolex birmanicus* Gates, *Ann. Mag. Nat. Hist.*, Ser. 9, **19**: p. 609. (Type locality, Maymyo. Types, None?)
1929. *Notoscolex birmanicus*, Gates, *Proc. United States Natl. Mus.* **75**, 10: p. 14.
1931. *Notoscolex birmanicus*, Gates, *Rec. Indian Mus.* **33**: p. 360.
1932. *Notoscolex birmanicus*, Gates, *ibid.* **34**: p. 367.
1933. *Tonoscolex birmanicus*, Gates, *ibid.* **35**: p. 484.
1941. *Tonoscolex birmanicus*, Bahl, *Quart. Jour. Micros. Sci.* **82**: p. 445.

Male pores, in a single *U*-shaped seminal groove with each limb ending posteriorly at or near m*BC* in

xviii. Copulatory organs, one pair, transversely ovoidal, attached by median ends or by very short stalks in *AA* anterior to the seminal groove. Spermathecal pores, large, longitudinal slits, closely paired in *AA*. Setae, lumbricin, $AB < CD < BC < AA, DD > \frac{1}{2}C$. Clitellum, eq/xii, xiii–xvi/eq ventrally or 16/17 dorsally. First dorsal pore, at 10/11. Prostomium, prolobous. (Segments?) Size, to 600 (+?) by 10–15 mm.

Calciferous glands, in ix–xii. (Typhlosoles?) Last hearts, in xii. Seminal vesicles, in x, xi. Prostates, in xvii–xxxv, duct 3–8 mm. long, straight or with one or two quirks near parietes. Sperm ducts, discrete, opening separately into the prostatic duct. Spermathecae, duct short, stout, almost confined to parietes, diverticulum small, wart-shaped, from duct near ampulla.

Distribution: Maymyo and vicinity. The western boundary of the range probably is at the edge of the Shan plateau. Other boundaries cannot now be suggested.

Habitats: Red (lateritic) soil of lawns, gardens, tennis and volley ball courts as well as at roadsides in the town. Open areas around the town. Nearby forest reserves.

Seasonal Occurrence: May to October.

Habits: Intestinal ejecta are deposited (Gates, 1927) throughout the rainy season (May–September) in towerlike to irregularly conical or pyramidal piles that may be 250 mm. tall, 400 mm. in circumference, and with a dry weight of over $3\frac{1}{4}$ pounds. All passages through the pile, except the last used, are filled. Castings are built up to a height of 200–250 mm. in three or four days. In order to cast on the surface, some worms worked their way up through a 3-inch layer of "consolidated road metal."

Remarks: Length measurements, or estimates, of 3 to 7 feet have been reported by trustworthy residents of the town who had been in a position to see an occasional worm alive on the surface after unusually heavy rain. Lengths of 10 feet were reported to the writer on several occasions. One worm, crossing the road to Mandalay, was long enough to impel a conscientious Buddhist driver to bring his passenger bus to a sudden stop to avoid the sin of "taking life."

Copulatory organs always were turgescent but whether there is any possibility of retraction into the parietes like the copulatory "tags" of certain other species cannot now be stated.

T. birmanicus seemed to be just as common throughout the town during the twenties and thirties as in surrounding area where there must have been much less human interference with the environment.

Tonoscolex conversus

1930. *Notoscolex conversus* Gates, *Rec. Indian Mus.* **32**: p. 301. (Type locality, at or near Pantha, Paungbyin or Mawlaik in the Chindwin Valley. Types, none.)

Male pores, near posterior ends of seminal grooves that pass forward into xvi, anterior portion of each groove nearly straight, posterior portion slightly curved and concave mesially. Setae, lumbricin, $AB < CD < BC < AA$. First dorsal pore, at 10/11. (Prostomium? Segments?) Size, to 500 (+?) by 8 (+?) mm.

Calciferous glands, in ix–xii. (Typhlosole?) Last hearts, in xii. Seminal vesicles, in x, xi. Sperm ducts, discrete throughout, often widely separated.

Distribution: Chindwin Valley, near the river, between Mawlaik, Pantha and Paungbyin (Upper Chindwin District).

Parasites: NEMATODA: *Tonoscolecinema setosum* Timm, 1967, coelomic cavities.

Remarks: Known only from quite immature specimens secured in the latter part of July.

The worms were obtained by digging in the vicinity of the large, characteristic castings. Individuals in the upper 2–3 inches of soil were about 5 inches long (alive). Others, 10–12 inches long (alive), were 3–6 inches further from the surface, the largest individuals deeper still. If at each of those depths a different generation was present, maturity would be attained at earliest only toward end of the third period of activity.

Tonoscolex depressus

1929. *Notoscolex depressus* Gates, *Proc. United States Natl. Mus.* 75, 10: p. 14. (Type locality, Maymyo. Holotype, destroyed. Paratypes in the U. S. Natl. Mus. and the British Mus.) *Notoscolex choprai* Stephenson, *Rec. Indian Mus.* 31: p. 230. (Type locality, Nyaungbin. Types, in the Indian Mus.)
1932. *Notoscolex depressus* + *N.* sp. f. *prima*, Gates, *Rec. Indian Mus.* 34: pp. 368, 373.
1933. *Tonoscolex depressus* (part), Gates, *ibid.* 35: p. 484. (Excluding, var. *scutatus*.)
1936. *Tonoscolex depressus*, Gates, *ibid.* 38: p. 379.
1941. *Tonoscolex depressus*, Bahl, *Quart. Jour. Micros. Sci.* 82: p. 445, 456.
1952. *Tonoscolex depressus*, Gates, *American Mus. Nat. Hist. Novitates* 1555: p. 10.
1955. *Tonoscolex depressus*, Gates, *Rec. Indian Mus.* 52: p. 79.
?
1930. *Notoscolex* sp. (B), Gates, *ibid.* 32: p. 300.

Male pores, at *B*, anteriorly in *f*-shaped seminal grooves ending in xviii at or just median to *C*, the anterior end of each groove at *A* between limbs of a shortly *U*-shaped tumescence in xvii. Spermathecal pores, small, at or slightly median to *A*. Setae, lumbricin, $AB < CD < BC < AA$, DD ca. = $\frac{1}{2}C$. First dorsal pore, at 9/10, 10/11. Clitellum, eq/xii, xiii–xvi/eq or 16/17 dorsally. Prostomium, prolobous. Segments, 304–334. Size, 75–500 (+?) by 3–7 (+?) mm.

Calciferous glands, in ix–xii. Typhlosole, simply lamelliform, present from xiv–xv (ending 85–90 segments from posterior end?) Last hearts, in xii. Seminal vesicles, in x, xi. Prostates, in xvii–xxiv, ducts 3–8 mm. long and with one or two small quirks

ectally. Spermathecae, duct stout, shorter than ampulla, narrowed in parietes, diverticulum wartlike to longer than duct and then club-shaped.

Reproduction: Presumably biparental ordinarily in worms that mature and exchange sperm during copulation. Parthenogenesis (possibly male sterility?), is suspected in some individuals, especially those with small seminal vesicles, without seminal grooves or with rudimentary (short) furrows. Breeding season, in September.

Distribution: Pyinmana (Yamethin). Maymyo, Taungyi (Shan States). Loi Se Mountain, Nam Mang (Mang Lun). Nyaungbin, Lonton, Mansum, Hopin (Myitkyina). (Chindwin valley, in region between Mawlaik and Pantha?) At elevations up to 5,000 feet.

Yamethin district worms are assumed to have been secured in nearby hills of the Shan plateau rather than from those to the west.

Habitats: Red (lateritic) soil. Soil under tall grass, rocky or sandy soil under leaves, "dense rain forest of bamboo."

Abnormality: Seminal grooves, short, occasionally even lacking (Pyinmana). Calciferous glands of x, lacking (1, Taungyi).

Parasites: Gregarine cysts filled seminal vesicles of some worms and occasionally were sparsely scattered in coelomic cavities throughout much of the body. Cysts in one worm nearly covered parietes between prostatic and anal regions. Nematodes forcing their way from the coelom to the exterior may have been responsible for the unusually wide lumen and abnormally thin walls of some sperm ducts. NEMATODA: *Tonoscolecinema burmense* Timm, 1967, Pyinmana, coelomic cavities.

Remarks: Setae were lacking in 130 of 140 specimens in an August collection from Taungyi. Vestiges of a few follicles, or a few setae, were present in three individuals. Seven were fully setigerous. All worms subsequently obtained from Taungyi (September), as well as those from other localities, were setigerous.

Many individuals referred to this species become clitellate and fully mature when little more than 75 mm. long. Chindwin Valley giants, with a size of 500 by 9 mm., were not otherwise distinguishable from juveniles with seminal grooves secured at known *depressus* localities. Would the giants have become mature late in August or in September, or is the gigantism caused by a kind of parasitic castration, are questions that cannot now be answered.

In the south, *T. depressus* may be confined to the region between the Sittang and the Salween rivers but to the north where the Salween also may have been reached, the Chindwin as well as the Irrawaddy were crossed in one direction or another. How far toward or into Manipur from the west bank of the Irrawaddy the range may extend into wholly unexplored territory remains to be determined.

Tonoscolex ferinus

1933. *Tonoscolex ferinus* Gates, *Rec. Indian Mus.* 35: p. 487. (Type locality, Koopra. Types, none?)

Male pores, near anterior ends of seminal grooves passing from near mV lateroposteriorly into region of *EG* in xviii. A single deep transverse slit, (from which a copulatory tag or tags can be everted or protruded?), in *BB* just in front of anterior ends of the grooves. Spermathecal pores, minute(?), slightly median to *A*. Setae, perichaetin, *AA* wide throughout but *ZZ* wide only anteriorly, 23–29/ii, 21–29/vii, 31–39/xii, 37–41/xx, 34–41/posteriorly. First dorsal pore, at 9/10. (Clitellum? Prostomium? Segments?) Size, to 125 by 5 mm.

Calciferous glands, in ix–xii. (Typhlosole?) Last hearts, in xii. Seminal vesicles, in x, xi. Prostates, in xvii–xxii, ducts 1–1½ mm. long. Spermathecae, ducts stout, shorter than ampullae and narrowed in the parietes, diverticulum from duct at parietes, shorter than main axis, club-shaped and wider entally.

Distribution: Koopra (Karenni).

Abnormality: Two spermathecal pores only, spermathecae of each segment being united mesially within the parietes.

Remarks: A U-shaped thickening of the ventral parietes that contains the seminal grooves is more or less markedly protuberant into coelomic cavities.

The species is known only from five types, four of which were aclitellate. The fifth showed only slight indications, in xiii–xvi, of initial development of clitellar glandularity. Other reproductive organs appeared to be mature.

Tonoscolex horai

1922. *Megascolex horai* Stephenson, *Rec. Indian Mus.* 34: p. 432. (Type locality, Cherrapunji. Type, immature, in the Indian Mus.)
1923. *Megascolex horai*, Stephenson (*The Fauna of British India*), *Oligochaeta*, p. 247.
1934. *Tonoscolex horai*, Gates, *Rec. Indian Mus.* 36: p. 256.
1945. *Tonoscolex horai*, Gates, *Jour. Roy. Asiatic Soc. Bengal*, (*Sci.*) 11: p. 89.

Male pores, pre- or post-equatorial and near anterior or posterior ends of short seminal grooves at *B*. Spermathecal pores small transverse slits, at or close to *B*. Setae, perichaetin, *AA* and *ZZ* wide but usually *ZZ > AA* 11–16/ii, 16–19/iii, 18–22/viii, 20–32/xii, 21–32/xx. Clitellum, 12/13–16/17. First dorsal pore, at 10/11 (11/12). Prostomium, epilobous, tongue open. Segments, 116–188. Size, 50–110 by 2–2½ mm.

Calciferous glands, in x–xii. Typhlosole, simply lamelliform, in xiv–lxxxiv. Last hearts, in xi. Seminal vesicles, in viii and xi. Prostates, in xvii–xviii, duct short, slender but muscular, looped. Spermathecae, duct shorter than ampulla, diverticulum from duct slightly ental to parietes, nearly as long as main axis, slenderly club-shaped.

Reproduction: Probably biparental as sperm are exchanged during copulation.

Distribution: Cherrapunji, Dumpep (Khasi Hills, Assam).

Abnormality: No. 1. Intersegmental furrow 1/2, lacking, (Whether maceration or regeneration was involved was not decided.) Setae, lacking on one side of iii and in viii but the number in xii and in xx is greater than in normal worms.

Remarks: Only 46 specimens have been available but fortunately many were clitellate.

Seminal grooves, straight, comma- or crescent-shaped, may be restricted to xvii. Each is within an elliptical area of parietal thickening that is longitudinally placed or slightly diagonal.

The excretory system has been characterized (Gates, 1945) quite tentatively as follows. Pharyngeal micronephridia, in one pair of clusters, in v. Parietal micronephridia, in two transverse rows per segment in xiii–xvi, in one transverse row of 12 per segment behind xix. Enteromeganephridia, with preseptal funnels, one or two pairs per segment, on the parietes, in a posterior portion of the body.

Cylindrical testis sacs may be present.

The seminal vesicles show that *T. horai* is in a different line of descent from other presently known species of the genus. The common ancestor of the two lines is assumed to have had four pairs of seminal vesicles in the homoeotic equivalent of ix–xii as in many lumbricids today. Whether absence of calciferous glands in ix and of hearts in xii is primitive or whether a pair of each of those organs was eliminated in evolution of the species remains to be determined. The excretory system does seem to be more primitive than in those Burmese forms that Bahl studied. The perichaetin arrangement of the setae is of course specialized as in *T. ferinus*. Intermediate stages between the full perichaetin and the lumbricin arrangements, such as are recorded for *Celeriella*, are unknown in *Tonoscolex*.

Tonoscolex lunatus

1929. *Notoscolex lunatus* Gates, *Proc. United States Natl. Mus.* 75, 10: p. 16. (Type locality, Maymyo. Holotype, lost. Paratypes, in the U. S. Natl. Mus. and the British Mus.)
1930. *Notoscolex lunatus*?, Gates, *Rec. Indian Mus.* 32: p. 299.
1931. *Notoscolex* sp., Gates, *ibid.* 33: p. 361.
1932. *Notoscolex lunatus*, Gates, *ibid.* 34: p. 369.
1933. *Tonoscolex lunatus*, Gates, *ibid.* 35: p. 490.
1936. *Tonoscolex lunatus*, Gates, *ibid.* 38: p. 80.
                                                                    ?
1930. *Notoscolex* sp., (C), Gates, *ibid.* 32: p. 301.

Male pores, anteriorly in doubly curved seminal grooves (often with posterior limb straighter and then groove shaped like a question mark), the anterior end between arms of a laterally directed, shortly U-shaped tumescence near mV, the posterior end in xviii at or slightly lateral to *B*. Spermathecal pores, small, at or slightly median to A or close to mV. Clitellum, eq/xii, xiii–xvi/eq ventrally or 16/17 dorsally. Setae, lumbricin, $AB < $ or $= CD$, $BC < $ or $= $ or $> AA$. First dorsal pore, at 9/10 (10/11–12/13?). Prostomium, epilobous, tongue open. Segments, 314. Size, 85–215 (+?) by 3–7 (+?) mm.

Calciferous glands, in ix–xii. (Typhlosole?) Last hearts, in xii. Seminal vesicles, in x, xi. Prostates, in xvii–xxiv, ducts 1–2 mm. long. Spermathecae, duct usually shorter than ampulla, stout, narrowed in parietes, diverticulum slightly longer than duct to longer than main axis, club-shaped or elongately digitiform.

Reproduction: Presumably biparental ordinarily in worms that exchange sperm during copulation. Parthenogenesis (possibly male sterility?) is suspected in some individuals, especially those with abnormally small seminal vesicles and spermathecae. The breeding season includes September.

Distribution: Maymyo (Shan States). Peng Sai, Nam Mang (Mang Lun). Namkham (North Hsenwi). Homalin (Upper Chindwin). (Vicinity of Mawlaik, Pantha and Paungbyin in the Chindwin valley?) At elevations up to 3,500 feet.

Habitats: Red (lateritic) soil. Damp soil in groves. Leaf-covered, sandy soil on wooded hillside.

Abnormality: Male deferent ducts (Peng Sai), more especially toward the posterior ends, filamentous. Spermathecae and seminal vesicles of the same worms appeared to be abnormally small relative to body size (85–121 by 3 mm.). Setal ranks, irregular but without change in number of setae (3 Namkham and others). Shortness of the spermathecal diverticulum may also be an abnormality in some specimens as well as shortness of the ampulla relative to the duct. Paucity of ova in ovaries of several worms may have been a normal condition for end of the reproductive season.

Sperm ducts, if not wholly aborted, always attenuate in parthenogenetic R morphs, but here ducts attenuate in spite of presence of prostates.

Parasites: Gregarine cysts were massed in the coelom of ix–xi and similar cysts were scattered more sparsely in coelomic cavities posteriorly.

Nematode ova, accumulated in an ectal portion of the prostatic duct of Chindwin giants (all three forms), produced an appearance as of a muscular bulbus ejaculatorius. A gravid nematode also was found in the distended portion of one prostatic duct. Presumably that parasite could have continued on through the parietal portion of the duct to the exterior. Other nematodes were found in the coelomic cavities of ix–x. *Tonoscolecinema parvum* Timm, 1967, Namkham.

Remarks: Although sexual maturity is not to be expected much before September, two Namkham worms, secured in May after several weeks of rain, already had normally developed seminal grooves.

Chindwin Valley giants, still aclitellate though reaching a size of 533 by 9 mm., were otherwise distinguishable from *T. lunatus* only by location of first dorsal pore at 11/12. As in case of the *depressus*-like giants, sperm ducts of a side were nowhere in contact with each other. Gigantism presumably has the same cause as in *T. depresssus* (q.v.).

T. lunatus presently is distinguishable from *T. depressus* only by the double curvature of the seminal grooves and/or by the more median location of the posterior portion of those grooves.

The range of *T. lunatus* may not equal that of *T. depressus* as there are no records of its presence in lower Burma between the Sittang and the Salween rivers where the latter species was found. However, *T. lunatus* was collected at Namkham and can be expected further north along the Shweli River.

Tonoscolex monorchis

1916. *Megascolides "oneilli"* var. *monorchis* Stephenson, *Rec. Indian Mus.* **12**: p. 313. (Type locality, "Darjiling to Soom." Type, in the Indian Mus.)
1923. *Notoscolex oneili* var. *monorchis*, Stephenson, (*The Fauna of British India*), Oligochaeta, p. 214.

Male pores, minute, in AB. (Seminal grooves.) Spermathecal pores, at A. Setae, lumbricin, $a–b$/xviii also lacking, $AB < CD < BC < AA$, DD ca. $= \frac{2}{3}C$. (Clitellum?) First dorsal pore, at 9/10. Prostomium, prolobous. Segments, 188. Size, 115 by 5 mm.

Calciferous glands, kidney-shaped, in viii–xii. Typhlosole, with numerous closely set transverse folds on each side. Last hearts, in xii. Seminal vesicles, in x, xi. Prostates, in xix, ducts considerably coiled. Spermathecae, duct almost confined to the parietes, diverticulum digitiform and shorter than the ampulla.

Distribution: Between Darjiling and Soom, at 5–7,000 feet.

Remarks: Known only from an immature specimen. Abnormality is suspected because of presence of a seminal vesicle on one side of xii and supposed absence of testes and male funnels in x though vesicles are present in xi. If any abnormalities are due to halving of mesoblastic somites at the ninth level, presence of calciferous glands in viii might be without systematic or phylogenetic importance. However, the reported locations of organs behind viii do not allow mesoblastic splitting.

Shape of calciferous glands hints at an internal structure differing from that of other species in the genus.

Tonoscolex montanus

1932. *Notoscolex* sp., f. *secunda*, Gates, *Rec. Indian Mus.* **34**: p. 373.
1936. *Tonoscolex montanus* Gates, *ibid.* **38**: p. 381. (Type locality, Taungyi. Types, none?)
1941. *Tonoscolex montanus*, Bahl, *Quart. Jour. Micros. Sci.* **82**: p. 445.

Male pores, slightly lateral to mV and at or near anterior ends of seminal grooves that bend sharply at level of postsetal secondary furrow of xvii to pass back parallel to mV in longitudinally elliptical areas of parietal thickening across adjacent secondary annuli of xvii–xviii. Male field, containing grooves and tumescences, of equal width through xvii–xviii. Spermathecal pores, closely paired transverse slits in AA. Setae, lumbricin, $AB < CD < AA < BC$. Clitellum, xiii–xvi. First dorsal pore, at 9/10. (Prostomium? Segments?) Size, 140–265 by 6–8 mm.

Calciferous glands, in ix–xii. (Typhlosole?) Last hearts, in xii. Seminal vesicles, in x, xi. Prostates, in xvi–xxi, ducts 2–2½ mm. long and with 2–3 very small quirks close to parietes. Spermathecae, duct shorter than ampulla and narrowed in parietes, diverticulum elongately digitiform, looped entally in a rather regular zigzag within an opaque sac so as to seem club-shaped and shorter than the main axis.

Reproduction: Presumably biparental as sperm are exchanged during copulation.

Breeding season probably in September.

Distribution: Taungyi (Shan States).

Remarks: Records now are available only for 40 specimens. Setae were lacking in 26 of 28 immature August specimens. One August worm was fully setigerous but in the other specimens vestiges of follicle apertures were recognizable only in a few post-clitellar segments.

No invagination from which copulatory tags could be protruded was recognized in the first lot of mature specimens.

Tonoscolex oneili

1914. *Megascolides "oneilli"* Stephenson, *Rec. Indian Mus.* 8: p. 377. (Type locality, Janakmukh, Abor Country. Type, in the Indian Mus.)
1916. *Megascolides "oneilli"* f. *typica,* Stephenson, *ibid.* 12: p. 314.
1923. *Notoscolex oneili* (part) Stephenson, (*The Fauna of British India*), Oligochaeta, p. 212. (Excluding, var. *monorchis.*)
1934. *Tonoscolex "oneilli,"* Gates, *Rec. Indian Mus.* 36: p. 254.
1936. *Tonoscolex oneili,* Gates, *ibid.* 38: p. 383.
1952. *Tonoscolex oneili,* Gates, *American Mus. Nat. Hist. Novitates* 1555: p. 11.

Male pores, about at B, at angles of L-shaped seminal grooves extending back into xviii, the longer posterior limb along or close to C or directed towards c/xviii, the transverse limb at eq/xvii. Copulatory organs, transversely ovoidal, just in front of anterior ends of seminal grooves, retractile into deep transversely slitlike invaginations. Spermathecal pores, very small, at or median to A. Setae, lumbricin, $AB < CD < BC =$ or $< AA, DD > ½C$. Clitellum, xiii–xvi/eq. First dorsal pore, at 10/11. Prostomium, prolobous or pro-epilobous. Segments, 230–244. Size, 105–185 by 5–6 mm.

Calciferous glands, in ix–xii. (Typhlosole?) Last hearts, in xii. Seminal vesicles, in x, xi. Prostates, in xvii–xx, ducts spindle-shaped entally and zigzag-looped ectally. Spermathecae, duct large, much

larger than ampulla, rather spindle-shaped but looped, looping concealed by connective tissue, diverticulum with short stalk and thin-walled ental portion looped within an opaque sac and seemingly shortly club-shaped.

Reproduction: Presumably biparental as sperm are exchanged during copulation.

The breeding season in damp forest of Myitkyina district may be rather late.

Distribution: Mayan, Mansum (Myitkyina). Janakmukh, Abor country of the eastern Himalayas, Assam.

Habitats: The only record now available, "dense rain forest of bamboo."

Parasites: Nematodes, as usual, were present in coelomic cavities of ix–x and also in parietal cysts posteriorly.

Remarks: Information now available was derived from four specimens. Indian worms were said to have a light olive-green color.

Spermathecal pores sometimes seem to be longitudinal slits crossing the intersegmental furrow into two segments.

The posterior limb of the seminal groove may be slightly concave laterally. Supposed male pores of the types are the apertures of invaginations from which copulatory tags can be protruded.

The only information now available as to internal structure of *Tonoscolex* calciferous glands is provided by Stephenson's characterization (1914) of the organs of *T. oneili* as "lamellar."

Enteromeganephridia are present from about lxx posteriorly.

A ventral portion of septa 8/9 and 9/10 bearing the testes of the Mayan worm is pouched way forward. The septum is strongly muscular and closure of the pouch would provide a sort of testis sac from which the male funnels would be excluded.

"A minute folded structure in xiii was examined macroscopically, but was found not to be ovarian in structure" (Stephenson, 1914: p. 379) but whether ovisacs were involved remains to be decided.

Tonoscolex parvus

1932. *Notoscolex* sp., f. *tertia,* Gates, *Rec. Indian Mus.* 34: p. 373.
1936. *Tonoscolex parvus* Gates, *ibid.* 38: p. 386. (Type locality, Taungyi. Types, none?)

Male pores, at or near hind ends of crescentic seminal grooves concave anteromesially, the posterior end of each groove at or close to A, the front end in lateral part of BC, Grooves, in xvii–xvi, with an anterior portion of each in a diagonally placed elliptical area of parietal thickening. Genital field, wider in xvi than in xvii. Spermathecal pores, minute, in AA, slightly nearer to A than to each other. Setae, lumbricin, fine, $AB < CD < AA =$ or $<$ or $> BC$. Clitellum, xiii–xvi/eq. First dorsal pore, at (8/9?)

9/10. (Prostomium? Segments?) Size, 65–130 by 3–4½ mm.

Calciferous glands, in ix–xii. (Typhlosole?) Last hearts, in xii. Seminal vesicles, in x, xi. Prostates, in xvi–xx, xxi, ducts short, bifurcated entally, with 1 or 2 very small quirks ectally. Spermathecae, duct shorter than ampulla, abruptly narrowed in parietes, diverticulum shorter than the main axis, sausage-shaped or pear-shaped to almost spheroidal.

Reproduction: Presumably biparental as sperm are exchanged during copulation.

Breeding season probably is in September.

Distribution: Taungyi (Shan States).

Remarks: Information is now available only for 53 specimens. Setae were entirely lacking in 15 of 36 immature August worms. All setae were present in 21 immature August worms and in each of the worms secured in September.

T. parvus is distinguished from the Chindwin Valley giant, *T. conversus*, by its smallness as well as by the anterior divergence of the seminal grooves.

Tonoscolex quartus

1932. *Notoscolex* sp., f. *quarta*, Gates, *Rec. Indian Mus.* **34**: p. 374. (Type locality, Taungyi. Types, none?)

Male pores, at posterior ends of *V*-shaped seminal grooves placed so that the *V* points straight laterally with the open end facing mesially in a male field that is equally wide through xvi–xvii.

Calciferous glands, in ix–xii. (Typhlosole?) Last hearts, in xii. Seminal vesicles, in x, xi.

Distribution: Taungyi (Shan States).

Remarks: Information is now available only for five August juveniles, of which one had no setae at all though the other four were fully setigerous.

The seminal grooves are characteristic and quite different from those of all other species.

Tonoscolex striatus

1914. *Notoscolex striatus* + *N. stewarti* Stephenson, *Rec. Indian Mus.* **8**: pp. 380, 382. (Type locality of both species, Rotung. Types, in the Indian Mus.)
1923. *Notoscolex striatus* + *N. stewarti*, Stephenson, (*The Fauna of British India*), *Oligochaeta*, pp. 218, 216.
1934. *Tonoscolex striatus*, Gates, *Rec. Indian Mus.* **36**: p. 255.

Male pores, in seminal grooves which are at *A* in xvii but are sharply bent laterally at levels of 16/17 and 17/18. Copulatory organs, "tags," two pairs, from anterior margins of xvii and xviii, in the angles of the seminal grooves. Spermathecal pores, at or median to *A*. Setae, lumbricin, (*a*, *b*/xvi, xviii also lacking?). $AB = CD < AA < BC$, $DD > \frac{1}{2}C$. Clitellum, xiii–xvi. First dorsal pore, at 9/10, 10/11. Prostomium, prolobous, tongue open. Segments, 216–297. Size, 85–210 by 3½–6 mm.

Calciferous glands, in ix–xii. (Typhlosole?) Last hearts, in xii. Seminal vesicles, in x, xi. Prostates, in xvii–xviii, ducts short and with a U-shaped bend.

Spermathecae, duct almost confined to the parietes (?), diverticulum from duct in parietes, somewhat shorter than the main axis, club-shaped.

Reproduction: Presumably biparental as sperm, according to the iridescence on male funnels, had been matured.

Distribution: Rotung, Renging (Abor country in the eastern Himalayas of Assam), at elevations of *ca.* 1,300 feet.

Remarks: Only a few individuals have been available. The clitellum was fully developed on two specimens.

Absence of calciferous glands in ix of a type of *stewarti*, in view of the similarity in all other structure, was regarded as an abnormality.

The green color supposedly characterizing the types of *stewarti* was lacking or unrecognized in individuals of *striatus*. For that as well as other reasons the suggestion was advanced that the green color was foreign and obtained from specimens of *Eutyphoeus* in the same container.

Tonoscolex sp.

1930. *Notoscolex* sp., Gates, *Rec. Indian Mus.* **32**: p. 301.

Juveniles, 62 by 3 mm., obtained along with *Bimastos parvus* from black soil at swampy sites in Kalaw, cannot now be placed in any species. The worms were distinguished from all other tonoscolexes by location of the first dorsal pore at 12/13.

Tonoscolex triquetrus

1930. *Notoscolex depressus*? Gates, *Rec. Indian Mus.* **32**: p. 299.
1932. *Notoscolex triquetrus* Gates, *ibid.* **34**: p. 370. (Type locality, between Thandaung and Leiktho. Types, none?)
1933. *Tonoscolex triquetrus*? + *T. depressus* var. *scutatus*, Gates, *ibid.* **35**: pp. 491, 485.

Male pores, near anterior ends of seminal grooves that terminate slightly lateral to mV in xvii, the grooves straight and diagonally placed or with an angle, ending in xviii at or lateral to mBC, within a U-shaped thickening of the ventrum. Male field, wider in xviii, with a deep transversely slit-like depression in *BB* just in front of anterior ends of the grooves. Spermathecal pores, minute, at or slightly median to *A*. Setae, lumbricin. Clitellum, xiii–xvi First dorsal pore, at 9/10. Segments, 312. Size, 148–450 by 5–10 mm.

Calciferous glands, in ix–xii. (Typhlosole?) Last hearts, in xii. Seminal vesicles, in x, xi. Prostates, in xvii–xxvii, ducts 3–7 mm. long and with two small U-shaped loops ectally. Spermathecae, duct bulbous, narrowed in parietes, diverticulum small, digitiform, shorter than the duct.

Distribution: Thandaung to Leiktho (Karen hills, Toungoo). Loikaw, Koopra (Karenni).

Parasites: Nematodes, *Tonoscolecinema triquetrum* Timm, 1967, were massed in the coelomic cavities of ix–x in the immature Koopra giants. Smallish cysts with pseudonavicellae spores were present in the seminal vesicles. Larger cysts, also with pseudonavicellae spores, were flat against the parietes from ix posteriorly. Mature and clitellate Koopra specimens with a maximum size of 160 by 6½ mm. had no readily recognizable parasites of any kind.

Remarks: The slitlike depression or invagination anteriorly in xvii may be a median equivalent to the paired invaginations from which in *T. oneili* the copulatory organs are protruded but no tags were recorded for *triquetrus*.

Length of seven strongly contracted and well-preserved Koopra giants varied from 340–450 mm. but three were incomplete posteriorly. Gonads were juvenile and spermathecae as well as seminal vesicles were small though the worms were collected in September. Clitellate specimens from the same locality at most were 160 by 6½ mm.

Some or all of the ventral setae in ii–x were lacking in one series. Setal ranks, in another series, were more or less irregular especially behind the clitellum. Even when ranks were regular there was considerable variation in relative sizes of the intersetal intervals, not only from one region of the body to another but in the same segment of different individuals.

Oviducts of the Koopra giants unite mesially in the parietes to open by a single median female pore, a condition unrecorded elsewhere in the genus.

(Genus?) antrophyes

1924. *Megascolides antrophyes*, Stephenson, *Rec. Indian Mus.* 26: p. 130. (Type, locality, Siju Cave. Type, in the Indian Mus.)
1940. *Megascolides antrophyes*, Gates, *ibid.* 42: p. 128.

Quadrithecal, pores slightly nearer *A* than to each other, in viii–ix slightly behind 7/8–8/9. Male pores, about at *A*, pre-equatorial in xviii, each at anterior end of a crescentic furrow, concave laterally, that passes back into next segment. Female pore (at mV?). (Genital markings?) Clitellum, annular, eq/xiii–xvii. Setae, enlarged posteriorly where $AB < CD < AA < BC$, $DD < \frac{1}{2}C$. First dorsal pore, at 10/11. (Unpigmented?) Segments, 112. Size, 35 by 1 mm.

Gizzard, in v (?). Calciferous glands, two pairs, in xii, xiii (xi?), extra-mural (slenderly stalked?). Intestinal origin, in xiv. (Typhlosole, caeca, supra-intestinal glands?) Last hearts, in xiii. Excretory system, meroic. Holandric. (Seminal vesicles?) Prostates, racemose, reaching back into xxix. (Junction of male and prostatic ducts?) Spermathecal duct shorter than ampulla and with a sinuous diverticulum reaching from middle or ental portion to ental end of ampulla. (Glands opening at hind end of seminal furrows?)

Reproduction: Presumably biparental as sperm are

matured—the male funnels having been recognized because of their spermatozoal iridescence.

Distribution: Siju Cave, 2,000 feet from entrance, Garo Hills, Assam.

Remarks: Known only from the type specimen now probably of little value because of alcoholic deterioration. Location of each organ requires confirmation from better material because of possible relationships with genera characterized by a regional anterior homoeosis.

Prostates, according to Stephenson (1924), were tubular which allowed inclusion of his species in the classical congeries called *Megascolides*. If prostates are indeed tubular, *M. antrophyes* would now go into the Octochaetidae but other structure and the distribution suggested relationships with *Nelloscolex* and *Tonoscolex*. Examination of the type (Gates, 1940) indicated that the prostates, though long and rather slender, are more likely to be racemose and like those of *Tonoscolex*.

MICROCHAETIDAE

1895. *Microchaetinae* (part, Geoscolecidae), Beddard, *A Monogr. of the Order of Oligochaeta* (Oxford), p. 664. (Excluding *Criodrilus* and *Hormogaster*.)
1900. Microchaetinae (Glossoscolecidae) + *Alma* + *Criodrilus* (part, South American species only), Michaelsen, *Das Tierreich* 10: pp. 447–462, 465–467, 468–469.
1902. Alminae (Glossoscolecidae), Duboscq, *Arch. Zool. Exp. Gen.*, Ser. 3, 10, N & R: pp. xcvii–cvi.
1915 Microchaetinae + Criodrilinae (part, South American species only) + Alminae (Glossoscolecidae), Michaelsen, *Ergeb. 2 Deutsch. Zentral-Afrika Exped. 1910–1911, 1,* (*Zool. 1*): pp. 276–278.
1918. Microchaetinae (Lumbricidae), Michaelsen, *Zool. Jahrb. Syst.* 41: p. 305.
1921. Microchaetidae, Michaelsen, *Arch. Naturgesch.* 86, A: p. 141.
1923. Microchaetinae (Lumbricidae), Stephenson, (*The Fauna of British India*), Oligochaeta, p. 490.
1928. Microchaetidae (Lumbricina), Michaelsen, *Handbuch Zool.* 2, 2–8: p. 107.
1930. Microchaetinae (Glossoscolecidae), Stephenson, *The Oligochaeta* (Oxford), p. 899.
1959. Microchaetidae, Gates, *Bull. Mus. Comp. Zool. Harvard College* 121: p. 255.

Gizzard, lacking or if present esophageal and in front of testis segments. Calciferous glands, lacking or if present (*Microchaetus* and *Kynotus*) restricted to region of ix–x. Intestine, with or without some initial muscular strengthening but without a definite gizzard. (Vascular system?) Excretory system, of holoic nephridia except in *Tritogenia*. Setae, sigmoid, in eight longitudinal rows, none penial, sexual setae when present not of the grooved lumbricid kind. Dorsal pores, lacking.

Male pores, not postclitellar. Spermathecae, (behind testis segments?),[1] adiverticulate. Clitellum, multilayered. Ovaries, in xiii, (fan-shaped?). (Ovi-

[1] Except the rudimentary organs of *Glyphidrilus stuhlmanni* Michaelsen, 1897?

sacs?) Ova, not yolky. Seminal vesicles, trabeculate. Prostates with muscular ducts, none.

Distribution: Costa Rica, Colombia, Ecuador, eastern Brazil, Africa from Egypt to Cape Province, Madagascar, India, Ceylon, Burma, Malay Peninsula, Malay Archipelago.

Remarks: The definition above essentially is that of the classical system. Stephenson and Michaelsen agreed on the content of the taxon which was regarded as a subfamily by Stephenson as well as by Michaelsen at one time or another. Michaelsen finally allowed family status. One character in common, throughout all of the group, is absence of functional spermathecae in front of the gonadal segments. Whether that character is uniquely diagnostic of a group with a monophyletic origin remains to be determined. The wide oceanic gaps in the distribution have not been explained.

Little is known of any of the genera all of which doubtless require adequate characterization of somatic anatomy now known to be systematically important. Only one genus is Asiatic.

Glyphidrilus

1889. *Glyphidrilus* Horst, *Tijdschr. Nederlandsche Dierk.* Ver. 2, 2: p. lxxvi. (Type species, *G. weberi* n. sp.)
1890. *Bilimba* Rosa, *Ann. Mus. Civ. Sto. Nat. Genova* 29: p. 386. (Type species, *B. papillata* n. sp.)
1893. *Glyphidrilus + Annadrilus* Horst, in Weber, *Zool. Ergeb. Reise Niederländisch Ost-Indien, Leiden* 3: pp. 37, 44. (Type species of *Annadrilus, A. quadrangulus* n. sp.)
1895. *Glyphidrilus + Bilimba + Annadrilus*, Beddard, *A Monogr. of the Order of Oligochaeta* (Oxford), pp. 679, 686, 680.
1900. *Glyphidrilus*, Michaelsen, *Das Tierreich* 10: p. 459.
1918. *Glyphidrilus*, Michaelsen, *Zool. Jahrb. Syst.* 41: p. 343.
1923. *Glyphidrilus*, Stephenson, (*The Fauna of British India*), *Oligochaeta*, p. 490.
1930. *Glyphidrilus*, Stephenson, *The Oligochaeta* (Oxford), p. 901.
1958. *Glyphidrilus*, Gates, *Rec. Indian Mus.* 53: p. 53.

Digestive system, with an esophageal gizzard in region of vii–viii, an intestinal origin behind xiv, and an intestinal typhlosole but without caeca, calciferous and supra-intestinal glands. Vascular system, with dorsal (single), ventral (complete) and subneural trunks, the latter adherent to parietes(?), a supra-esophageal trunk in v–xiv, a sub-esophageal complex on the gut in viii–xiii, paired anterior and posterior (from anal region?) lateroparietal trunks passing to subesophageal complex in region of xii–xiv but no extra-esophageal trunks, hearts in part latero-esophageal. Excretory system, of holoic (avesiculate?) nephridia, aborted in many preclitellar segments. Septa, all present at least from 5/6. Prostomium, zygolobous. Anus, dorsal. Color, white (unpigmented). Body, nearly circular in cross section anteriorly, then becoming elliptical, posteriorly quadrangular (after preservation), with setae at the four corners and *DD* only slightly larger than *AA*.

A longitudinal lamellar ridge protuberant at maturity from body wall on each side, in *BC*, through

several of the clitellar segments. Clitellum, annular (?), intersegmental furrows not obliterated, setae retained, (including all segments with reproductive apertures?). Genital apertures, all minute and superficial. Holandric. Seminal vesicles, in ix–xii. Polythecal, pores at three or more intersegmental furrows. (Ovaries fan-shaped and with several egg-strings? Ovisacs?)

Distribution: Tanzania (one species). Ceylon, India, Burma, Malay Peninsula, Sumatra, Java, Flores, Borneo, Celebes. Hainan. Certainly to be expected in Thailand and states of Indo-China and perhaps also in a southern portion of China.

No species of *Glyphidrilus* is known to be anthropochorous and there now seems to be no good reason for questioning endemicity of each species in the area from which it is known. An African origin of the genus always has been assumed and is required by its supposed direct descent in the classical manner from *Microchaetus* and, of course, also merely by its inclusion in the Microchaetidae otherwise unrepresented to the east of that continent. Oligochaetologists have given little or no consideration to the puzzles posed by the genus. How did *Glyphidrilus* get to India and Ceylon from Africa? If overland, how is absence in the intervening region explained? How has a genus apparently confined to media saturated with fresh water crossed so many oceanic straits such as that between Ceylon and India?

Glyphidriles probably are present, in suitable habitats throughout most of Burma, India, Ceylon, and the Malay Peninsula. Whether absence of species in the region from the Sind desert west through to Palestine is real or not remains perhaps to be determined. Certainly more information about the distribution of African species is needed.

Systematics: Species usually were erected, in the past, on few specimens and were characterized mainly or wholly by reference to certain external characters. The latest monograph dealing with the group (Michaelsen, 1918) has no key to species but substitutes a table comparing five genital characters and two somatic (length and segment number, a single measurement and count in each case). The single specimen, or even a short series, as in *Eutyphoeus*, may not permit proper characterization of the species, either because of belonging to a strain that has lagged behind most of the population with respect to an important anatomical modification or because of a presumably ontogenetic reversion to some former ancestral condition.

Externally recognizable genital characters are subject to considerable individual variation. Genital markings seem to be much alike in various species and so variable as to number and location that even after examination (Gates, 1958) of numerous specimens doubt as to systematic usefulness of the structures is unresolved.

Genital apertures are minute and superficial. Spermathecal pores are not easily recognized. Female pores are unlikely to provide characters of systematic usefulness. Location of male pores may be determinable only in microtome sections. Sites have been recorded for three species, at 21/22 in *G. quadrangulus* (Horst, 1893), at 27/28 in *G. weberi* Horst, 1889, at 21/22, 27/28, 29/30 in *G. annandalei* Michaelsen, 1910. Determination of species norms is needed for several reasons, but the character, regardless of presence or absence of individual variation, seems unlikely to be of use in specific identifications.

Uniformity of the internal genitalia along with the individual variation in the external reproductive structures such as the wings and genital markings, necessitates even more than usual attention to somatic anatomy.

The digestive and vascular systems (excretory system not yet carefully studied) provide, beyond exceptions of the sort mentioned in the first paragraph of this section, characters that not only are uniform throughout a species but that differ from one species to another. Some of those characters are mentioned in the next section.

The sub-esophageal vascular complex and its relationships with other major trunks may seem to be rather variable but some if not all of the supposed differences may prove to be due only to inability to recognize, in a dissection (possibly even in sections?), empty blood vessels.

Exact characterizations of gizzard location may have to be determined from microtome sections. A glyphidrile gizzard often seems to have some small part in a second segment, as in the Lumbricidae. The appearance in some cases may be due to adherence of a septum to the gizzard in such a manner as to seem to be inserted thereon. Such a septum frequently is so delicate that determination of relationships in a dissection is impossible. The determinations are not expected to be of especial systematic importance but should throw light on the question of phylogenetic transfer of the organ posteriorly, whether by a macromutational transfer at a single step from completely in front of one septum to completely behind it or whether by a much slower and more gradual method. Whatever the answer, the condition has seemed to be capable of characterization as mainly in one particular segment according to a species norm.

Phylogeny: An ancestral proto-glyphidrilus from which the better known species could be derived was envisioned (Gates, 1958) as follows. Digestive system, with a single esophageal gizzard in vi (perhaps already one segment behind a primitive megadrile location), an intestinal origin in xv (perhaps already two or three segments behind that of an ancestral megadrile). A circulatory system with a single, complete dorsal blood vessel and in each segment a pair of connectives between that vessel and the ventral

TABLE 4

Species of Glyphidrilus	Gizzard in segment	Intestinal origin in segment	Dorsal vessel aborted in front of hearts of	Last hearts in segment	Latero-esophageal hearts in segment
annandalei Michaelsen, 1910	viii	xv	vii	xi	x,xi
ceylonensis Gates, 1945	vii	xvi	0	xi	ix–xi
gangeticus Gates, 1958	viii	xvi	ix	xi	xi
kuekenthali Michaelsen, 1896	viii	xv	vii	xi	
malayanus Michaelsen, 1902[a]	viii	xvi	ix	xi	x,xi
papillatus	viii	xv	vii	xi	x,xi
tuberosus Stephenson, 1916	vii	xv		xi	

[a] This species lives under the water by the million but is known only from two types one of which is incomplete! Michaelsen commented: "Es muss bei dieser Sachlage bedauerlich erscheinen, dass der Sammler dieser Art von den Millionen Thieren, die ihm zur Verfügung standen, nur zwei mitnahm; aber auch für diese zwei Exemplare sei ihm Dank gesagt." Specimens washed out from a cubic foot (as nearly as possible) of mud from the shore of Lake Victoria, Rangoon, were counted to obtain rough estimates of the population in the lake. By the million, would have been equally applicable there.

trunk. Excretory system, holoic. The vascular system became typical with appearance of supra- and sub-esophageal trunks along with paired latero-parietals. The generically characteristic wings presumably were an early innovation and they may have been of greater length then than in any extant species. Too little is known about somatic anatomy of too many species to warrant more than mention of changes that appeared in certain species.

None of the changes indicated by the table are characteristic of the genus or of the microchaetid family and, probably as in other families, have appeared independently at various times and in diverse lineages.

Sperm ducts of *G. weberi* Horst, 1889, open to the exterior in common with anomalous glands called prostates that may have been retained from some remote ancestor with glands resembling those of *Sparganophilus*. The *weberi* terminalia would seem to be a glyphidrile equivalent of the microscolecin condition so important in classical systematics.

Folded organs in ix and cell masses opposite are now believed to have been male funnels and testes respectively. Such extra organs could have been developed during head replacement if regenerative capacity is similar to that of *Criodrilus* and *Perionyx*. Another possibility is that the sex-testiculate condition is a persistence of or a reversion to an ancestral condition. The rudimentary vesicles, regularly present in xiii–xiv and occasionally also found in xv–xvi, that are serially homologous with the seminal vesicles of xi–xii, suggest existence of gonads during early embryology in xii–xvi.

The glyphidrile wings may prove to be unique among megadriles if not also among oligochaetes. They presumably always have been associated with two microchaetid spermathecal characters: Location of the spermathecae behind the testiculate segments (=x–xi) and, less uniformly, presence of more than one pair of spermathecae per metamere. Transfer of ancestral pregonadal spermathecae through the testis segments and still further posteriorly, according to the classical systematics, was evidenced by the persistence in *G. stuhlmanni*, of rudimentary spermathecae opening at 10/11. The condition of the spermathecal battery in *Plutellus pandus* suggests that new organs were added one segment at a time behind the end of the series. Eventually or perhaps even simultaneously spermathecae were aborted at the front end of the series in one segment at a time. Intrasegmental multiplication of the spermathecae may have resulted from fragmentation of early paired embryonic anlage.

Remarks: The glyphidrile wings are richly vascularized which suggests a respiratory function for those organs which are, however, developed only just before sexual maturity.

Several factors presumably are responsible for our slight knowledge of well-distributed species that are present, in appropriate environments, by the "millions." One, perhaps of some importance in certain cases, is indicated by M. Mathur's comment on the beautiful Saharanpur series of *gangeticus* that he collected—"Found only at a depth of about 18 inches, along banks of a river." Glyphidriles, believed to be obligatorily limicolous (and presumably intolerant of salt water), perhaps can have interesting zoogeographical value, like that claimed for salt-intolerant fish throughout the world.

Glyphidrilus is the only genus that now requires consideration of an Afro-Asiatic relationship involving either Gondwana or some Wegenerian approximation.

The region of Tethys, the ancestral Mediterranean, from the Mesozoic into the Tertiary (Darlington, 1957, Zoogeography) was one of continuous modification, with changing shorelines, changing barriers and bridges, changing archipelagos, as well presumably as deviations in watersheds perhaps sufficient to enable glyphidriles to get from Africa into the oriental region. Or, may the single African species have been referred to the wrong genus?

KEY TO CEYLONESE AND INDO-BURMESE
SPECIES OF *Glyphidrilus*

1. Dorsal blood vessel complete . . . . *ceylonensis*
   Dorsal blood vessel aborted anteriorly[a] 2
2. Gizzard in vii or mainly so . . . . *tuberosus*
   Gizzard in viii or mainly so 3

3. Intestinal origin in xv 4
   Intestinal origin in xvi . . . . *gangeticus*
4. Genital markings present in *A A* 5
   Genital markings lacking in *A A* . . . . . . *birmanicus*
5. Wings end in front of xxx . . . . *papillatus*
   Wings end behind xxx . . . . *annandalei*

[a] *G. spelaeotes* drops out here as the dorsal trunk was not characterized. Hearts of viii, mentioned by Stephenson, may be the first pair. If so the trunk might then be aborted in front of the hearts of viii which would distinguish the species from *papillatus*.

Glyphidrilus birmanicus

1958. *Glyphidrilus birmanicus*, Gates, *Rec. Indian Mus.* **53**: p. 54. (Type locality, in Burma but otherwise unknown. Types, none.)
?

1931. *Glyphidrilus papillatus* (part?), Gates, *ibid.* **33**: p. 431. (Worms from Mongkung, Namkham, Maymyo?)
1933. *Glyphidrilus papillatus* (part?) + *G.* sp., (part), Gates, *ibid.* **35**: pp. 603, 604. (Worms from Bana, Mong Mong, and Na Hang?)
1958. *Glyphidrilus* sp., Gates, *ibid.* **53**: p. 62. (Worms from Rangoon and Toungoo.)

Wings, begin in one of xxi–xxiii and end in one of xxix–xxx. Genital markings, primarily postsetal, numerous, lateral to *B*, in xii–xxvi, xxix–xxxiv. Spermathecal pores, at 13/14–17/18. (Female and male pores?) Clitellum, xii, xiii–xlii, lxvi. (Segments?) Size, 103 by 5–6 mm.

Gizzard, in viii (or also partly in vii?). Intestinal origin, in xv. Typhlosole, thickly lamelliform, begins in xviii. Dorsal blood vessel, aborted in front of hearts of vii, hearts of x, xi (last two pairs) lateroesophageal.

Reproduction: Presumably biparental as sperm are exchanged during copulation.

Distribution: Unknown but probably including some if not all of the following: Bana, Mongmong, Na Hang, Mongkung, Lashio, A Kin De, Namkham, Maymyo, Toungoo, Rangoon.

All except the last two localities are on the Shan Plateau.

Habitats: As for *G. papillatus*.

Glyphidrilus papillatus

1890. *Bilimba papillata*, Rosa, *Ann. Civ. Mus. Sto. Nat. Genova* **29**: p. 386. (Type locality, Cobapo. Type, aclitellate, "mal conservato," in the Genoa Museum.)
1895. *Bilimba papillata*, Beddard, *A Monogr. of the Order of Oligochaeta* (Oxford), p. 687.
1896. *Glyphidrilus papillatus*, Michaelsen, *Abhandl. Senckenbergischen Naturforsch. Ges.* **23**: p. 195.
1900. *Glyphidrilus papillatus*, Michaelsen, *Das Tierreich* **10**: p. 459.
1903. *Glyphidrilus papillatus*, Michaelsen, *Die geographische Verbreitung der Oligochaeten* (Berlin), p. 134.
1918. *Glyphidrilus papillatus*, Michaelsen, *Zool. Jahrb. Syst.* **41**: p. 344.
1923. *Glyphidrilus papillatus*, (part), Stephenson, (*The Fauna of British India*), *Oligochaeta*, p. 493. (Excluding from the synonymy, *G. papillatus*, Stephenson, 1920 = *G. gangeticus*, Gates, 1954.)

1926. *Glyphidrilus papillatus*, Gates, *Ann. Mag. Nat. Hist.*, Ser. 9, 17: p. 472; *Jour. Burma Res. Soc.* 15: p. 218; *Jour. Bombay Nat. Hist. Soc.* 31: p. 183; *Rec. Indian Mus.* 28: p. 169.
1930. *Glyphidrilus papillatus*, (part?), Gates, *ibid.* 32: p. 352.
1931. *Glyphidrilus papillatus*, (part?), Gates, *ibid.* 33: p. 431. (Some if not all of the Shan Plateau material perhaps of another species?)
1933. *Glyphidrilus papillatus* (part?), Gates, *ibid.* 35: p. 603. (Excluding all but Bassein and Tharrawaddy worms?)
1958. *Glyphidrilus papillatus*, Gates, *ibid.* 53: p. 60.

?

1926. *Glyphidrilus* sp., Gates, *ibid.* 28: p. 169.
1933. *Glyphidrilus* sp., Gates, *ibid.* 35: p. 604.
1958. *Glyphidrilus* sp., Gates, *ibid.* 52: p. 93.

Wings, begin in one of xvii–xix and end in one of xxiii–xxvii. Genital markings, primarily postsetal, numerous (8–40), unpaired and median in xi–xxi, xxiii–xxxiii, paired and lateral to *A* in xii–xxx. Spermathecal pores, at 13/14–16/17. (Female and male pores.) Clitellum, xiii, xiv–xl (±?). Segments, 130–330. Size, 74–120 by 3–6 mm.

Gizzard, in viii (or also partly in vii?). Intestinal origin, in xv. Typhlosole, thickly lamelliform, begins in xviii. Dorsal blood vessel, aborted in front of hearts of vii, hearts of vii, ix lateral, of x, xi latero-esophageal.

Reproduction: Presumably biparental as sperm are exchanged during copulation.

Breeding, in favorable conditions, may be year-round.

Distribution: Moulmein, Kawkareik (Amherst). Pyapon, Thanchitaw Kyaiklat (Pyapon). Sittang (Thaton). Rangoon, Kayan, Kyauktan (Hanthawaddy). Bassein, Thinbawgyin (Bassein). Nyaunglebin, Thanatpin (Pegu). Tharrawaddy, Letpadan, Thonze (Tharrawaddy). Henzada, Ingabu, Zalun (Henzada). Cobapo (Toungoo), Prome (Prome). Taungdwingyi (Magwe). Minbu (Minbu). Sagaing (Sagaing). Monywa, Okma, Anidaung, Powindaung, Ingyindaung, Laungbyin, (Lower Chindwin). Bhamo, Woreabone (Bhamo). Tiangzup (Mitkyina).

Records of the species for Lucknow almost certainly are of *G. gangeticus*. A record of *G. papillatus* for Hainan (Chen, 1938, *Cont. Biol. Lab. Sci. Soc. China, Zool.* 12: p. 426), without information as to internal anatomy, probably also is incorrect.

Habitats: Banks of lakes, ponds, tanks, rivers, in irrigation ditches, swamps, buffalo wallows, in almost any soil with organic matter that is saturated throughout the year. Elsewhere, in soil that is submerged only during the rainy season, as in paddy fields. Never found in the sandy river shores where *P. posthuma* is so common.

Biology: Quiescence during a considerable portion of the year is imposed at sites where soils dry out after the rains are over. In the ooze and mud that is covered with water throughout year, activity appears to be continuous. Freshly deposited cocoons of some

glyphidrile though not certainly of *papillatus* were obtained from one of the Rangoon lakes late in the hot season.

Remarks: Prior to 1932 all adult individuals were referred to *papillatus* because of lack of marked discontinuity (Gates, 1931: p. 432) in wing locations and, in accordance with classical practice, after very cursory inspection of internal anatomy. A revision of that material and study of the adults collected during 1932–1941 revealed existence of several taxa (at least three) possibly worthy of species rank. The little now known about internal anatomy of Burmese species was derived mostly from several specimens that had been sent to Stephenson in the late twenties or early thirties and that came back to the writer only some time after World War II.

Glyphidrilus sp.

1931. *Glyphidrilus papillatus* (part), Gates, *Rec. Indian Mus.* 33: p. 431.
1958. *Glyphidrilus* sp., Gates, *ibid.* 53: p. 62.

Wings, begin in one of xxi–xxiii and end in one of xxv–xxvi. Genital markings, paired in *AA*(?).

Remarks: As the discontinuity in segmental position of the anterior ends of the wings was only one segment, worms characterized as above were referred to *papillatus*. Whether subsequent material confirmed the discontinuity along with restriction of the wings to four to six segments and whether those characters were correlated with divergence in internal anatomy from that of *papillatus* cannot now be stated.

Glyphidrilus spelaeotes

1924. *Glyphidrilus spelaeotes*, Stephenson, *Rec. Indian Mus.* 26: p. 133. (Type locality, Siju Cave, Garo Hills, Assam. Types, in the Indian Museum.)

Wings, in xviii–xxiv, xxv/2. Genital markings, primarily postsetal, unpaired and median in xi, xvii–xviii. paired in *AA* of xix, xxvii–xxviii, lateral to *B* in xiv–xvii and xxv. Spermathecal pores, at 13/14–15/16, at *A*, *B*, *C*, *D*, and in *BC*. (Female and male pores?) Clitellum, (xiv–xv?)–xxx. Setae, ornamented ectally. Color, none. Segments, 310. Size, 175 by 2–3 mm.

Gizzard, in viii. Intestinal origin, in xv. (Typhlosole? Dorsal blood vessel aborted in front of hearts of viii? Latero-esophageal hearts?)

Remarks: The species is known only from the original account. Of the nine (+?) specimens that were available only one was thought to be mature (clitellum probably not fully tumescent) and only two had genital markings. Hearts of viii were mentioned by Stephenson. Any abortion of the dorsal trunk then could be only in front of 8/9.

The information available is insufficient to warrant a guess as to specific distinctness from *G. papillatus*. The record is of geographical importance and warrants a belief that glyphidriles will be found in the Arakan Yomas, the western mountain wall between India and Burma.

MONILIGASTRIDAE

1880. Moniligastridae Claus, *Grundzüge der Zoologie* (4th ed.) 1: p. 480.
1884. Moniligastridae, Vejdovsky, *System und Morphologie der Oligochaeten* (Prag), p. 63.
1888. Moniligastridae, Rosa, *Boll. Mus. Zool. Univ. Torino* 3, 41: p. 9.
1890. Moniligastridae, Benham, *Quart. Jour. Micros. Sci.* 31: p. 221.
1891. Moniligastridae, Rosa, *Ann. Naturhist. Hofmus. Wien* 6: p. 379.
1895. Moniligastridae, Beddard, *A Monogr. of the Order of Oligochaeta* (Oxford), p. 192.
1900. Moniligastridae, Michaelsen, *Das Tierreich* 10: p. 109.
1909. Moniligastridae, Michaelsen, *Mem. Indian Mus.* 1: pp. 117, 136.
1910. Moniligastridae, Michaelsen, *Abhandl. Naturwiss. Ver. Hamburg* 19, (5): p. 20.
1919. Moniligastrinae (Moniligastridae), Smith & Green, *Proc. U. S. Natl. Mus.* 55: p. 45.
1921. Moniligastridae, Michaelsen, *Arch. Naturgesch.* 86, A: p. 141.
1923. Moniligastrinae (Moniligastridae), Stephenson, (*The Fauna of British India*), *Oligochaeta*, p. 117.
1928. Moniligastridae (Phreoryctina), Michaelsen, *Handbuch der Zoologie, Berlin* 2, 2–8: p. 106.
1930. Moniligastrinae (Moniligastridae), Gates, *Rec. Indian Mus.* 32: p. 264. Stephenson, *The Oligochaeta* (Oxford), p. 813.
1939. Moniligastridae, Gates, *Jour. Thailand Res. Soc. Nat. Hist.* Sup. 12, 1: p. 72.
1959. Moniligastridae, Gates, *Bull. Mus. Comp. Zool. Harvard College* 121: p. 255.
1962. Moniligastridae, Gates, *idem.* 127: p. 299.

Digestive system, with gizzards behind ovarian segment, an intestinal origin behind xvii, *with paired "enterosegmental" organs on the intestine dorsally,* but without typhlosoles, calciferous and supra-intestinal glands. Vascular system, with dorsal, ventral and subneural trunks, the latter adherent to the parietes, paired extra-esophageal trunks lateral to the hearts and with connectives to the dorsal trunk (directly or indirectly) as well as to the subneural (asymmetrically), but without a supra-esophageal, hearts lateral, last two (*uniting mesially to open into the dorsal trunk through a short vertical vessel?*) two segments in front of ovarian metamere. Excretory system, of holoic and vesiculate nephridia, lacking in ii, (the bladder caecal and from tubule just prior to entry into parietes?). Prostomium, prolobous but separated from i, protuberant from roof of buccal cavity behind level of 1/2. Setae, sigmoid and single pointed, (penial and copulatory setae lacking), four pairs per segment. (Dorsal pores, none.)

Clitellum, unilayered(?), annular, intersegmental furrows not obliterated, setae retained, including male and female pore segments. Male pores, behind spermathecal pores and in front of female pores, the latter near B. *Testes and male funnels, intraseptal, in paired dorsal protuberances*[1] *of the septum.* (Seminal vesicles; none.) Sperm ducts, each opens to exterior through a "prostate" and at or close to the intersegmental furrow next behind that of the septum bearing the male funnels. Ovaries, vertically elongated and bandlike(?), *both in a chamber closed off mesially from small peri-esophageal and neural spaces.* Oviduct funnels, vertically elongated. Ova, large, yolky. Ovisacs, dorsal, elongate and backwardly directed, simple pockets of the posterior septum of the ovarian metamere. Spermathecae, attached to posterior face of a septum with ampullae dorsal to the gut.

Reproduction: Iridescence commonly seen in sperm ducts of many species has seemed indicative of presence of mature male gametes. A similar iridescence almost never was seen on male funnels in the very many testis sacs of Burmese species opened during a score of years. Nevertheless, microtome sections of almost every adult inspected by the author showed testis sacs distended by coagulum and sperm aggregated on male funnels. Just as in those specimens, at maturity coagulum always distended testis sacs of moniligastrids except in very rare abnormal individuals. Supposed smallness of the sacs (actually undistended) in certain species is not significant as the types were juvenile and in them coagulum had not yet had a chance to accumulate.

Degradation of genital anatomy, such as characterizes taxa of other families after establishment of parthenogenesis, never was found in the Moniligastridae. Male sterility was suspected only in very rare individuals with aberrations presumably induced by external interference with normal development.

Accordingly, reproduction of all moniligastrids, in absence of any good evidence indicative of the contrary, is assumed to be biparental. Inability to recognize spermatozoal iridescence in the spermathecal and testicular coagulum, suggesting marked physiological difference from other megadriles, may be significantly associated with the unique anatomy.

Further comment on reproduction, at generic and specific level, is unnecessary.

Distribution: Southeast and eastern Asia, from (Ceylon?), South India to Manchuria, Korea, also Japan, the Philippines, Borneo, Sumatra.

The range, as thus indicated by autochthonous forms has been extended as a result of transportation of species of *Drawida*, presumably by man and recently.

The family range is large and in size may be exceeded only by that of the Lumbricidae. One genus, *Drawida*, has a self-acquired distribution almost if not exactly equal to that of its family. The area involved is much greater than that of the proper *Pheretima* domain. Careful surveys, when made in most of Asia–Malaysia, are expected to reveal new species equalling if not excelling the number of

---

[1] Usually called "testis sacs."

pheretimas. Possibly *Drawida* alone may prove to have the greater number of species.

Habitats: In cultivated soils in certain sections of China, one *Desmogaster* supposedly is abundant. A somewhat similar habitat may be anticipated for anthropochorous drawidas. Some human interference with the environment probably is tolerated by several non-colonizing species of *Drawida* but hastirogasters and desmogasters were found, in Burma, only in undisturbed jungle. In addition, then, to casualness of most collecting, clearing of forests, cultivation, and other human activities may have been responsible for a lack of records that allowed characterization of a considerable eastern part of the moniligastrid range as a region where (Stephenson, 1923: p. 30, and 1930: p. 673) "the mighty genus *Pheretima* has crushed all competitors."

A supposed predilection for water or saturated habitats has been questioned (Gates, 1962: p. 304), and is belied by the gut which not only is always giceriate but throughout almost all of the family shows a more or less marked increase in number of gizzards. Many species certainly are terricolous and none are known to be aquatic or limicolous.

Systematics and Phylogeny: Certain characters of the classical system have been of little systematic usefulness primarily perhaps because of imprecision in description or because of failure to secure data regarding individual variation. Other characters mentioned in classical writings now are known to be useless. Among them are: shape of the testis sacs, shape of the prostates when the organs are glandular, perhaps also the number of gizzards.

Moniligastrid prostates, like those of eudrilids, were believed by classical authorities to be modified thickenings of the male ducts. Presence, occasionally in both families, of extra prostates not joined by the vasa deferentia, shows that the organs are ingrowths from the body wall with which posterior ends of the sperm ducts unite during development. Moniligastrid prostates usually develop so as to have two glandular layers, an internal one which is continuous with the epidermis, and the outer on the coelomic face of the organ. Shape of the latter layer often varies widely and fortuitously. The middle layer, originally muscular and continuous with the parietal musculature usually has a specifically characteristic shape. The shape-determining layer, herein called capsule, though muscular in some species and in earlier growth stages of other taxa, in adult and postreproductive stages (after preservation) may be yellow or red, brittle, translucent to transparent and without recognizable structure though retaining the typical shape. Whether all, or any, of those modifications take place *in vitro* or *in vivo* is unknown. The coelomic glandular layer may be variously reduced or even eliminated, and in the latter instances such a prostate is distinguished from the glandular kind as muscular.

Systematically important characters in two genera are provided by a portion of the spermatheca called by classical writers an atrium. The term is unfortunate partly because some species now are known to have a pre-atrial chamber as well as because the term sometimes means, in classical writing, prostate. Furthermore, in many species the atrium is a diverticulum, a lateral outgrowth from the main axis and is so characterized in the key below and elsewhere hereinafter. However, a secondary outgrowth from the main spermathecal axis admittedly does not have, in the Moniligastridae, the same function as the diverticulum in many megascolecid, acanthodrilid, and octochaetid species, to wit, reception and storage of sperm after copulation. The diverticulum (or atrium) of moniligastrids usually is empty at maturity, and its function is unknown. Sperm must then be stored in the ampulla.

A recent discussion[2] of moniligastrid anatomy, systematics, and phylogeny may be consulted for further information. Especially noteworthy is the demonstration therein that the classical characterizations of andry and gyny are not applicable to conditions in the Moniligastridae. Although the author has found previously unrecorded characters to be of systematic interest, others are likely to be needed, especially if the number of undescribed species approximates the number now anticipated. Of some importance may prove to be characters of the nephridia (especially of the caecum or bladder, also level at which ducts enter the parietes) and of the body wall (whether musculature is pinnate or fasciculate, also lateral thickenings of the body wall).

Derivation of moniligastrids from a postulated protomegadrilid involved: Separation of prostomium completely from peristomium. Extension posteriorly of the esophagus so that the gizzard was in or behind xii, perhaps at first with an intestinal origin not much behind xv. Development of enterosegmental organs, of extra-esophageal trunks lateral to the hearts and of a parietal subneural trunk. Abortion of nephridia of ii, of all but two or three pairs of gonads of segments included in the clitellum. Development of solid testis sacs as a result of reversal in direction of growth of the testis anlage (Gates, 1962: p. 301). Modification of B or X glands into capsular organs and acquisition by each of one or two pairs of an ability to attract into their anlage the growing male deferent ducts. Closure of a peripheral portion of the ovarian segment from ventral and central portions containing nerve cord and gut respectively, along with reduction in size of each of those coelomic spaces by approximation of the two bounding septa. Invagination of the entalmost portion of the spermatheca, its ampulla, very deeply into the coelom and so as to be connected with an external pore by an almost filamentous duct.

---

[2] Gates, 1962, *Bull. Mus. Comp. Zool. Harvard College* 127.

Primitive characters, such as location of male and spermathecal pores at or near *B*, restriction of gizzards (two or three) to xii–xiv or xv, an anterior location of esophageal valve, etc., have been found only in *Drawida* formerly regarded (except for Moniligaster) as terminal and of most recent origin. In *Desmogaster*, hitherto regarded as most primitive, as well as in *Hastirogaster* and *Eupolygaster*, number of gizzards is further increased, gizzards are further back and consequently the esophagus is still longer.

Certain evolutionary developments already recognized in three or more genera are: Setae dehisced and follicles aborted in some anterior segments or, in two genera, throughout the body, presumably early in ontogenetic development as gaps in the parietal musculature at probable sites of setal meridians still are recognizable in adults. Alternation, more or less irregularly, of nephropores due to lateral or mesial growth of nephridial ducts within the parietes to open to the exterior in region *AB*, or near mD, rather than in the vicinity of *CD*. Lateral dislocation of male and spermathecal pores to or almost to *CD*. Local lateral thickening of the circular muscle layer. Doubling of the dorsal blood vessel anteriorly (about which little is known). Abortion of nephridia in the segment behind the testis-sac septum. Elongation of male deferent ducts, penetration of those ducts into the parietes prior to junction with the prostates. Elimination of the glandular investment on prostates (and/or of the capsule also?). Invagination into the coelom of copulatory chambers and development therein of protrusible penes. Completion of closing off the ovarian chamber from the parietes.

Much of recent evolution in the moniligastrid family may well have been concerned with elaboration of genital structure, in part at least to make copulatory exchange of sperm more efficient. Increase in number of and posterior dislocation of the gizzards, the major macroscopically recognizable modification of somatic organization, perhaps increases effectiveness of the digestive system in dealing with a poor source of nutrition.

The number of uniquely diagnostic characters (italicized in the definition), unparalleled in megadriles, suggests achievement of considerable evolutionary stability, so much so in fact that classification of the family must be based very largely on more easily modified genital structure. However, a short postgonadal portion of the gut seemingly is involved in a mutational explosion. The gut always is giceriate, but variation in number and in location of the gizzards (*cf.* Gates, 1962, for tabulated data), usually seems rampant as in no other megadrile group.

Remarks: One or more species, in each genus, has attained "giant" size, along with an increase in number of segments.

A typical clitellum rarely has been recognized on live worms. A marked tumescence of clitellar epidermis, recorded several times, may have developed in the preservative. The site of the organ often is indicated, at least in preserved material, by a distinctive coloration, sometimes yellow but usually red. A yellow appearance of the clitellar region occasionally (or always?) may be due to the color of the ovisacs and ovarian chamber as seen through a transparent or translucent body wall. The red color, definitely is in or associated with the epidermis and in many instances is known to have been "developed" slowly by a postmortem chemical reaction usually involving formalin. Pigment, when present *in vivo*, according to all records now available, is located beneath the epidermis.

The male pores of moniligastrids are almost as close to the corresponding male-funnel septum as in any oligochaete. Yet, those very ducts may be hundreds of times longer than in any other megadrile. Perhaps an explanation will be found to be associated in some significant way with absence of seminal vesicles.

Gizzards so frequently are in consecutive segments as to warrant an assumption that such is the case unless the contrary is indicated in a précis. Although discrete glands are lacking, throughout the entire family, calcium-secreting tissues may be present though the only recorded evidence is that of granules seen in gut lumen behind gizzards of *Drawida japonica* (Michaelsen, 1892).

Presence of a supra-esophageal trunk was reported for at least one Indian species, but whatever was involved almost certainly is not equivalent to the vessel of that name in other megadriles.

Information about one or more systematically important characters is lacking for each species and, as a result, determination of relationships has been hindered and construction of keys has been handicapped. Actually, almost any character that has been found to be useful in defining apparently good species can crop up elsewhere as an individual or a racial variation.

Probably less than a hundred specimens, of the three genera *Desmogaster*, *Eupolygaster*, and *Hastirogaster*, have been studied. Further work perhaps will enable recognition of intergeneric differences in somatic anatomy.

KEY TO GENERA OF MONILIGASTRIDAE

1. Female pores at or just behind 13/14, ovaries
   in xiii................................. 2
   Female pores and ovaries anteriorly........... 3
2. Male pores at 11/12 and 12/13......*Desmogaster*
   Male pores at 11/12 only..........*Hastirogaster*
3. Female pores at or just behind 12/13, ovaries
   in xii.........................*Eupolygaster*
   Female pores at or just behind 11/12, ovaries
   in xi.................................... 4

4. Dichotomously branched glands in compact masses on Y-shaped spermathecal atria
Moniligaster
Dichotomously branched glands lacking..*Drawida*

Moniligaster Perrier, 1872, as restricted by Michaelsen (1900), is found at the southern tip of peninsular India but the northern limit of the range is unknown.
Eupolygaster Michaelsen, 1900, presently comprises two species, *E. modiglianii* (Rosa, 1896), from Sumatra, and *E. caerulea* (Horst, 1894), from west Borneo.

Desmogaster

1890. *Desmogaster*, Rosa, *Ann. Civ. Mus. Sto. Nat. Genova* 29: p. 369. (Type species, *D. doriae* Rosa, 1890.)
1895. *Desmogaster*, Beddard, *A Monogr. of the Order of Oligochaeta* (Oxford), p. 205.
1900. *Desmogaster*, Michaelsen, *Das Tierreich* 10: p. 110.
1922. *Desmogaster*, Stephenson, *Proc. Zool. Soc. London* 1922: pp. 136, 138, 144.
1923. *Desmogaster*, Stephenson (*The Fauna of British India*), *Oligochaeta*, p. 119.
1930. *Desmogaster*, Gates, *Rec. Indian Mus.* 32: p. 264. Stephenson, *The Oligochaeta* (Oxford), p. 813.

Male pores, at 11/12 and 12/13. Female pores, at or just lateral to B and at or slightly behind 13/14. Testes, in 10/11 and 11/12. Ovaries, in xiii. Spermathecae, adiverticulate.

Distribution: Burma, China, Borneo, Sumatra.

The Irrawaddy-Sittang axis, except possibly north of Myitkyina, now appears to constitute the western boundary of the *Desmogaster* range. Beyond Burma, in mainland Asia, the genus is known only from the immediate vicinity of Soochow, Kiangsu province, China. The abundant presence there of *D. sinensis* Gates, 1930, and the failure to find it elsewhere in the lower Yangtze Valley during a search "with utmost care" by Chen, warrants a thesis of successful colonization after introduction by man, perhaps rather recently. The Siamese border in peninsular Burma and the Salween River further north are not expected to constitute the eastern boundary of a range that includes Borneo and Sumatra.

Each species, with the possible exception of *D. sinensis*, is assumed to be endemic in the area where it is now known.

Systematics: Few specimens of any one species have been available for study. Species were distinguished, in the classical system, mainly by size of soma and by number of gizzards. In 39 specimens of *D. sinensis*, the longest series of a desmogaster that has been available, gizzards, according to Chen (1933, *Contrib. Biol. Lab. Sci. Soc. China, Zool.* 9: p. 188), were "normally in xiv–xvii." Such little information as is available for Burmese species of this as well as other genera indicates that number and location of gizzards is likely to have quite limited systematic usefulness.

Information as to the musculature and with regard to nephridia, especially as to the caeca and the level of entry into the parietes of the nephridial ducts may prove to be systematically useful at one level or another.

KEY TO SPECIES OF *Desmogaster*

1. Bithecal................................ 2
   Quadrithecal[a]............................ 4
2. Spermathecal pores at 7/8..............*doriae*
   Spermathecal pores at 8/9................ 3
3. Size, 260–290 by 9–10 mm..............*schildi*
   Size, 137–171 by 4–5 mm...............*ferina*
4. Prostates without glandular investment[b]...... 5
   Prostates with glandular investment.......... 6
5. Size, 35–57 by 3–4 mm., gizzards in xv–xvii
   büttikoferi
   Size, 260 by 8 mm., gizzards 7 in xiv or xv posteriorly.....................*stephensoni*
6. Prostates sessile on parietes................ 7
   Prostates not sessile...................*planata*
7. Size, 86–115 by 4–5 mm., vas deferens passes into parietes and back into coelom before entering prostate...................*albalabia*
   Size, 290–540 by 8–12 mm., vas deferens passes directly into prostate.................*sinensis*

[a] *D. horsti* Beddard 1895, from Berg Singalang, Sumatra, supposedly distinguished from *doriae* by its eight gizzards, drops out here as number of spermathecae is unknown. Gizzards are in xvii–xxiv. Prostates are elongate but the vas was thought not to pass into the ental end. If spermathecal pores are at 7/8 the species can be presently distinguished from *doriae* only by the smaller somatic size and the fewer gizzards.
[b] *D. giardi* Horst, 1899, from Nanga Raoen, Borneo, drops out here. If prostates are muscular, as they may be, there is little to distinguish it from *stephensoni* Michaelsen, 1934, from the same place in Borneo.
D. büttikoferi Michaelsen, 1922, is known only from Dutch central Borneo.

Desmogaster albalabia

1930. *Desmogaster albalabia*, Gates, *Rec. Indian Mus.* 32: p. 265. (Type locality, Thandaung. Types, none.)
1955. *Desmogaster albalabia*, Gates, 1955, *ibid.* 52: p. 56.

Quadrithecal, pores on ventral faces of small tubercles on posterior margins of vii–viii and at or slightly median to C. Male pores, small transverse slits in or just median to CD and in line with intersegmental furrows, each in a transverse groove on a nearly circular, white, discoidal porophore. Setae, lacking in ii–iii, small very closely paired, $AA = 2$–$4\frac{1}{2}$ BC, $DD > \frac{1}{2}C$. Nephropores, of iii–x in DD, (xi–xii and posteriorly?). Clitellum, rose-colored, x–xviii. Color, white (unpigmented). Segments, 236. Size, 86–115 by 4–5 mm.

Gizzards, 3–5, in xiv–xix. Intestinal origin, in xxvi, xxvii (±1?). Commissures from extra-esophageals, as in *doriae*. (Dorsal blood vessel doubled anteriorly? Hearts? Nephridia, of xi–xii, and posteriorly?) Sperm ducts, shortly looped, passing into parietes (but not deeply) in the AB gap and back again into coelom to pass into anterior face of a sessile prostate with a spheroidal capsule. Ovarian chamber, closed off from parietes.

Distribution: Thandaung (Karen Hills, Toungoo), on the Shan Plateau.

Remarks: Known only from the 9 types most of which were clitellate.

The species now appears to be close to *sinensis* from which it is distinguished as in the key. The dorsal blood vessel in *sinensis*, according to Chen (1933), is double from xi or x anteriorly.

Desmogaster doriae

1890. *Desmogaster doriae*, Rosa, *Ann. Civ. Mus. Sto. Nat. Genova* 29: p. 369. (Type locality, Metelindaung, Leiktho Circle. Types, in the Genoa Museum.)
1895. *Desmogaster doriae*, Beddard, *A Monogr. of the Order of Oligochaeta* (Oxford), p. 205.
1897. *Desmogaster doriae*, Rosa, *Ann. Civ. Mus. Sto. Nat. Genova* 37: p. 340.
1900. *Desmogaster doriae*, Michaelsen, *Das Tierreich* 10: p. 111.
1923. *Desmogaster doriae*, Stephenson (*The Fauna of British India*), *Oligochaeta*, p. 119.
1930. *Desmogaster doriae*, Gates, *Rec. Indian Mus.* 32: p. 268.
1933. *Desmogaster doriae*, Gates, *ibid.* 35: p. 417.
1934. *Desmogaster doriae*, Gates, *ibid.* 36: p. 264.
1955. *Desmogaster doriae*, Gates, *ibid.* 52: p. 56.

Bithecal, pores in CD, on anteroposteriorly flattened porophores (possibly cut off from posterior margins of vii but seemingly) at 7/8. Male pores, minute, in or close to CD, each within an area of epidermal whitening. Setae, lacking in ii–v or vi, $AA = 1\frac{1}{2}$-$4BC$, $DD > \frac{1}{2}C$. Nephropores, of iii or iv to x in DD, lacking in xi–xii, posteriorly in median or rarely lateral half of BC and variously located in DD but not regularly alternated. Clitellum, dark crimson or red (after preservation), x–xviii. Circular muscle layer of body wall segmentally thickened in region of mL on each side of body behind gizzard region. Color, brown (pigmented). Segments, 240–330. Size, 350 (−500?) by 12–16 mm.

Gizzards, 9–12, in xvi–xxix. Intestinal origin, in xxxvi (±1). Dorsal blood vessel, double anterior to 5/6. Hearts, of x–ix lateral, in x and xi continuous at mD but below dorsal trunk and joined by connectives from extra-esophageals that run dorsally on posterior faces of 10/11–11/12. Nephridia, of xi–xii lacking (in adults), posteriorly with ducts passing into parietes near AB. Sperm ducts, 35–40 mm. long, each with a hairpin loop on posterior face of its septum, passing deep into parietes in AB and, after emerging into coelom at CD, passing to ental end of an elongate (11–14 mm.) prostate, circular to shortly elliptical in cross section, glandular investment thicker than the slenderly and elongately digitiform capsule. Ovarian chamber, not closed off from the parietes (?).

Distribution: Metelindaung, Leiktho, Blachi, Thandaung (Toungoo). Kwachi, Mawchi (Karenni).

All localities are on the Shan Plateau and at elevations of 3,000 to 4,500 feet.

Habitats: Soil, in undisturbed jungle away from villages, and below depths usually dug over by collectors.

Biology: After the rainy season, late in October or early in November, large numbers of these worms come to the surface during the night and wander aimlessly around until killed (by ultraviolet rays of the sun?) the next morning. Only once was the author able to arrange for a person with equipment for preservation to be on hand at time of the mortal migration and the collector then was a small child who took the worms home from an early morning walk in her sun helmet.

Regeneration: Most of the available specimens had been amputated posteriorly at one time or another. Regeneration of a considerable number of tail segments is possible.

Remarks: Of the 22 specimens that were available, 2 were clitellate. One mislabeled specimen (of unknown history) has been for some years on exhibit in the Agassiz Museum at Harvard University.

Desmogaster ferina

1943. *Desmogaster ferina*, Gates, *Ohio Jour. Sci.* 43: p. 91. (Type locality, Tingpai. Types, none.)
1955. *Desmogaster ferina*, Gates, *Rec. Indian Mus.* 52: p. 58.

Bithecal, pores at or near C at 8/9. Male pores, (minute?), slightly median to C. Setae, lacking in ii to viii–x, small closely paired, AA much $> BC$, $DD > \frac{1}{2}C$. Nephropores, of iii–x in DD, (in xi–xii?), posteriorly in BC and DD but not regularly alternated. (Clitellum? Segments?) Body, slightly flattened dorsoventrally. Color, none (unpigmented). Size, 137–171 by 4–5 mm.

Gizzards, 6–7, in xx–xxvi. Intestinal origin, in xxxv (?). Connectives between extra-esophageals and dorsal trunk, on posterior faces of 10/11 and 11/12. (Dorsal blood vessel doubled anteriorly? Nephridia, of xi–xiii, and posteriorly?) Nephridial ducts, into parietes near AB. Sperm ducts, short, passing deep into parietes at AB gap and into prostates without emerging again into the coelom. Prostates, 2 mm. long, capsule elongate and elliptical in cross section. Ovarian chamber, closed off from the parietes.

Distribution: Tingpai (Myitkyina).

Habitat: Soil beneath logs beside a brook.

Remarks: Known only from two specimens, one juvenile, the other possibly of mature size (glandular investment of prostatic capsule not that of sexual maturity?). Junction of the sperm duct and the prostate may be entally rather than within the parietes, but with the deferent duct concealed within the glandular investment.

This species now seems to be most like *D. schildi* Rosa, 1897, from Pahang, Sumatra, which is inadequately characterized and presently distinguishable from the Burmese species only by the larger somatic size and by location of the male pores lateral to CD. Gizzards were said to be nine, in xxiii–xxxi, but variation in number and in location now must be expected until invariability is established for any particular species.

Desmogaster planata

1931. *Desmogaster planata*, Gates, *Rec. Indian Mus.* **33**: p. 331. (Type locality, Ye. Types, none.)
1933. *Desmogaster planata*, Gates, *ibid.* **35**: p. 417.
1955. *Desmogaster planata*, Gates, *ibid.* **52**: p. 58.

Quadrithecal, pores at or slightly median to CD, in slight tubercles at posterior margins of vii, viii(?) or at 7/8, 8/9. Male pores, in or just median to CD. Setae, lacking in ii–iii, small, closely paired, $AA = 1\frac{1}{2}$–2 BC ($DD > \frac{1}{2}C$?) Nephropores, of iii–ix in DD (of xi–xii?), posteriorly in BC and DD but not regularly alternated. (Clitellum?) Color, white, except for scattered black flecks of pigment anteriorly. Segments, 211. Size, 110–123 by 5–6 mm.

Gizzards, 4–6, in xix–xxiv. Intestinal origin, in xxix. Connectives between extra-esophageals and dorsal trunk, on posterior faces of 10/11–11/12. (Dorsal blood vessel doubled anteriorly? Nephridia of xi–xii and posteriorly?) Nephridial ducts, into parietes near AB. Sperm ducts, passing deep into parietes in AB and into prostates without again emerging into the coelom. Prostates, erect, strap-shaped (capsule?). Ovarian chamber, closed off from the parietes(?).

Distribution: Ye (Amherst). Mawchi (Karenni).

Habitat: Soil in jungles away from villages and cultivation.

Remarks: This species is known only from four specimens. All were immature. Gizzards, in each of the three types, were four, in xx–xxiii, but in the Karenni worm were six in xix–xxiv. A glandular investment on the prostatic capsule has been assumed but confirmation is required from fully sexual specimens.

Hastirogaster

1930. *Hastirogaster*, Gates, *Rec. Indian Mus.* **32**: p. 275. (Type species, *H. livida* Gates, 1930.)

Male pores, at 11/12. Female pores, at or just lateral to B and at or slightly behind 13/14. Spermathecal pores, at 8/9. Testes, in 10/11. Ovaries, in xiii. Spermathecae, adiverticulate.

Distribution: Burma, Sumatra.

The genus was found in Burma only well above the latitude of Mandalay, between the Salween axis and the Chindwin-Irrawaddy axis.

Each species, as also in *Eupolygaster* and *Moniligaster*, is assumed to be endemic in the area where it is now known.

KEY TO SPECIES OF *Hastirogaster*

1. Size 1100–1500 by 20–24 mm., gizzards four
   .... *houteni*
   Size of body much smaller, gizzards more numerous 2
2. Spermathecal pores at or near mBC, last gizzard in xxv .... *browni*
   Spermathecal pores in or close to CD, last gizzard in xxix or xxx .... *livida*

H. houteni (Horst, 1897) is based on a specimen secured at Tapanuli, Sumatra. Subsequently var. *rookmaakeri* Michaelsen, 1931, also from Sumatra, was named.

Hastirogaster browni

1907. *Eupolygaster "Browni"* Michaelsen, *Mitt. Naturhist. Mus. Hamburg* **24**: p. 143. (Type locality, Lashio. Type, in the Indian Museum.)
1909. *Eupolygaster browni*, Michaelsen, 1909. *Mem. Indian Mus.* **1**: p. 139.
1923. *Eupolygaster browni*, Stephenson (*The Fauna of British India*), *Oligochaeta*, p. 120.
1930. *Hastirogaster browni*, Gates, *Rec. Indian Mus.* **32**: p. 275.
1933. *Hastirogaster browni*, Gates, *ibid.* **35**: p. 418.

Male pores, just median to C (?). Spermathecal pores, on small anteroposteriorly flattened flaps at mBC (?). Setae, lacking in ii–viii (?), small and closely paired posteriorly, $AA > BC$, $DD > \frac{1}{2}C$. (Nephropores? Clitellum?) Color, brownish (pigmented or merely browned by alcohol?). Segments, 293. Size, 150 by 6 mm.

Gizzards, 6 (or 7?), in (xviii?) xix–xxiv. (Intestinal origin?) Sperm ducts, pass into ental ends of prostates after emerging into coelom from previously entered parietes. Prostates, flattened (rather strap-shaped?), each in an erect hairpin loop bound together more strongly towards closed end of loop, (a terminal ental portion with less muscularity?). (Ovarian chamber?).

Distribution: Lashio (Northern Shan States).

Habitats: No information available. Considerable collecting in and around Lashio during the thirties provided no specimens. The type, undoubtedly a chance find, probably was secured from a site that has undergone considerable modification since arrival of the railroad and introduction of potato cultivation.

Remarks: The species is known only from the immature type on which (when last examined) genital apertures still were not recognizable.

Hastirogaster livida

1930. *Hastirogaster livida* Gates, *Rec. Indian Mus.* **32**: p. 276. (Type locality, Pantha. Types, none.)
1955. *Hastirogaster livida*, Gates, *ibid.* **52**: p. 58.

Male pores, in CD. Spermethacal pores, minute, on ventral faces of protuberances seemingly cut off from viii, in CD. Genital markings, whitened area (of epidermal tumescence?) around each male pore. Setae, lacking in first 9–23 segments, small and very closely paired, $AA = 2$–$2\frac{1}{2}$ BC, $DD > \frac{1}{2}C$. Nephropores, of iii–x in DD, lacking in xi–xii(?), posteriorly alternating irregularly between levels in BC and DD. Clitellum, very dark blue (almost black), x–xx. (Segments?) Size, 144 (+?) by 10 mm. Gizzards, 7 (−9?), in (xxii?) xxiii–xxx. Intestinal origin, in xlii. Nephridia, lacking in x–xi, xiii–xiv (?). Sperm ducts, long end looped, passing deeply into parietes in AB and not emerging again into coelom. Prostates, elongate (13–14 mm.), ca. 1 mm. thick and nearly circular in cross section. Ovarian chamber, closed off from parietes.

Distribution: Pantha (Upper Chindwin).

Habitat: Undisturbed jungle soil.

Remarks: The species is known only from two anterior fragments, one of which was clitellate. A distinct "golden brownish" appearance mentioned in the field notes may have referred only to the clitellum the color of which certainly underwent a postmortem change. After preservation the color was brownish posteriorly but in front of the clitellum blueish (more accurately slate-colored?).

Drawida

1900. *Drawida* Michaelsen, *Das Tierreich, Berlin,* 10: p. 114. (Type species, *Moniligaster barwelli* Beddard 1886, an unfortunate designation, cf. Gates, 1937, *Bull. Mus. Comp. Zool. Harvard College* 80: p. 306.)
1909. *Drawida,* Michaelsen, *Mem. Indian Mus.* 1: p. 117.
1910. *Drawida,* Michaelsen, *Mitt. Naturhist. Mus. Hamburg* 27: p. 48.
1922. *Drawida,* Stephenson, *Proc. Zool. Soc. London,* 1922: p. 141.
1923. *Drawida,* Stephenson (*The Fauna of British India*), *Oligochaeta,* p. 124.
1930. *Drawida,* Gates, *Rec. Indian Mus.* 32: p. 279. Stephenson, *The Oligochaeta* (Oxford), p. 814.
1931. *Drawida,* Gates, *Rec. Indian Mus.* 33; 334. *Moniligaster* (part), Michaelsen, *Zool. Jahrb. Syst.* 61: p. 530.
1933. *Drawida,* Gates, *Rec. Indian Mus.* 35: p. 418.
1940. *Drawida,* Kobayashi, *Sci. Rept. Tohoku Univ.* (4, Biol.) 15: p. 261.
1945. *Drawida,* Gates, *Jour. Roy. Asiatic Soc. Bengal,* (Sci.) 11: p. 55.
1962. *Drawida,* Gates, *Bull. Mus. Comp. Zool. Harvard College* 127: p. 305.

CALCIFEROUS GLANDS: Michaelsen, 1892, *Arch. Naturgesch.* 58, 1: p. 232. ENTEROSEGMENTAL ORGANS: Stephenson, 1926, *Rec. Indian Mus.* 28: p. 252. EXCRETORY SYSTEM: Stephenson, 1922, *Proc. Zool. Soc. London,* 1922. TYPE SPECIES: Gates, 1937, *Bull. Mus. Comp. Zool. Harvard College* 80: p. 306. VASCULAR SYSTEM: Bourne, 1894, *Quart. Jour. Micros. Sci.* 36. Chapman, 1939, *Rec. Indian Mus.* 41.

Gizzards, in region of xii–xxvii. Last connectives between extra-esophageal and dorsal trunks on posterior face of 9/10, another pair associated with 8/9. Hearts, in each of viii–ix, after joining connectives from extra-esophageal trunks unite mesially above gut and then communicate with dorsal trunk through a short vertical vessel in median plane. Septa,

5/6–9/10 strengthened (usually thickly muscular), parietal insertion of 9/10 dislocated posteriorly, 10/11–11/12 approximated. Nephropores, present from iii.

Male pores, at or near 10/11. Female pores, at or just behind 11/12. Spermathecal pores, at 7/8. Clitellum, including x–xiii at least. Testes, in 9/10. Prostates, in x. Ovaries, in xi.

Distribution: India, Burma, Malay Peninsula, Thailand, Indo-China, China, Korea, Manchuria, Siberia, Japan, Philippine Islands, Borneo. Ceylon (?). (Possibly also Sumatra and Java?)

The natural generic range obviously is greater than that of the much larger genus, *Pheretima,* long supposed to be "all conquering." Introduction may have been responsible for Ceylon records of two species of dubious status. The northern limit of the *Drawida* range, above that of the *Pheretima* range according to Kobayashi (1940), is indicated by a line drawn from Chihfen through Hsinking to Tumen at ca. 43° 55′ north latitude.

Several species are established in areas to which they must have been transported. *D. nepalensis,* perhaps originally from the Himalayas, in the Andaman Islands and Java (Buitenzorg). *D. longatria longatria,* originally from the deltas region of lower Burma, at Palembang, Sumatra. *D. bahamensis* (Beddard, 1892), perhaps originally from China, in Bahamas, Puerto Rico, Juan Fernandez Islands, the Philippines. *D. japonica* (Michaelsen, 1892), perhaps originally from the Indian Himalayas, Yunnan and Szechuan, in Japan. *D. barwelli* (Beddard, 1886) sp. dub., erected on a series from Manila, recorded from India, Caroline Islands and Soemba. *D. parva* (Bourne, 1894), sp. dub., erected on South Indian specimens, recorded from Aru Island. *D. burchardi* Michaelsen, 1902, sp. dub., erected on a specimen from Indragiri, Sumatra, may have been introduced.

Remarks: Attention is invited to the remarks regarding atrium *versus* diverticulum on p. 239.

Spermathecal pores sometimes, in *bullata, flexa, gracilis, molesta, montana, victoriana,* seem to be very slightly in front of 7/8. Actual positions in younger material were not determined. Appearances in such cases may be of little or no importance.

KEY TO BURMESE SPECIES OF *Drawida*

1. Prostates glandular 2
   Prostates muscular 25
2. Prostate-like GM glands present . . . . *limella*
   Prostate-like GM glands absent 3
3. Spermathecae adiverticulate 4
   Spermathecae diverticulate 9
4. Male pores on penes in spheroidal muscular
   chambers . . . . *beddardi*
   Male pores not on tubular penes 5

5. Prostates long, to 15 mm . . . . *spissata*
   Prostates much shorter 6
6. Vas deferens rather short 7
   Vas deferens long . . . . *delicata*
7. Prostatic capsule small and mostly within
   parietes . . . . *rara*
   Prostatic capsule larger, in coelom 8
8. Prostatic capsule club-shaped . . . . *constricta*
   Prostatic capsule spheroidal . . . . *kempi*
9. Genital markings closely crowded into tumes-
   cent patches . . . . *montana*
   Genital markings not so aggregated 10
10. Genital markings lacking or not associated
    with solid glands 11
    Genital markings at least in part associated
    with solid glands (*cf.* note below) 17
11. Spermathecal diverticulum digitiform 12
    Spermathecal diverticulum saccular 16
12. Prostates erect in coelom 13
    Prostates sessile on parietes 15
13. Male pores on penislike porophores . . . . . *tenellula*
    Male pores not on such porophores 14
14. Male pores in porophores apparently belonging
    to xi . . . . *bullata*
    Male pores in porophores apparently belonging
    to x . . . . *vulgaris*
15. Ventral setae of v–vii markedly enlarged . . *gracilis*
    Ventral setae not so enlarged . . . . *lacertosa*
16. Male pores superficial . . . . *rangoonensis*
    Male pores invaginated and on penes . . . . . *molesta*
17. Spermathecal diverticulum digitiform 18
    Spermathecal diverticulum saccular 22
18. GM glands with translucent walls 19
    GM glands with opaque walls 20
19. Gizzards, 4–5, in xii–xvii . . . . *assamensis*
    Gizzards, 3–5, in xvi–xxiv . . . . *flexa*
20. GM glands with soft walls . . . . *nana*
    GM glands with firm muscular walls 21
21. Spermathecal diverticulum 2–3 mm. long . . *tumida*
    Spermathecal diverticulum (at least in *tumida*
    range) much longer . . . . *longatria*
22. Male pores superficial, no markedly protu-
    berant porophores 23
    Male pores on some sort of protuberant
    porophores 24
23. Vas deferens short . . . . *papillifer*
    Vas deferens long . . . . *nagana*
24. Male porophores apparently restricted to x
    . . . . *victoriana*
    Male porophores belong to x–xi but on pro-
    trusion appear independent of both . . *nepalensis*
25. Gizzards, 1–4 . . . . *caerulea*
    Gizzards, 7–9 . . . . *decourcy*

Two species, *nagana* and *victoriana*, are entered as if with solid glands. If such glands are lacking the species will run down in the key to No. 16. There, *nagana* is distinguished from *molesta* by absence of penes, from *rangoonensis* by location of male pores at level of 10/11, absence of an annular lip around male pores (as well as by presence of seminal grooves?). Similarly, *victoriana* is distinguished from *molesta* by absence of penes and from *rangoonensis* by the protuberant porophores.

Male porophores of some species are known only in a state of maximal protuberance. Characterization of the primary male pores then is uncertain, i.e., whether in a superficial and possibly discoidal porophore or whether invaginate and then in a chamber confined to the parietes or reaching into the coelom. Spermathecal pores likewise may be invaginated into chambers that have not always been described. In addition to data about nephropore locations, information now may be useful as to levels at which nephridial ducts enter the parietes, presence or absence of a nephridial vesicle (caecal?) and shape of it when present, and also as to abortion in certain anterior segments of the nephridia themselves. Passage of the male ducts into the prostates directly or indirectly and then sometimes without or after emerging again into the coelom, also are characters that now appear to be systematically useful.

Pigment, rarely present, is associated, or mostly so, with the circular muscle layer of the body wall. A red color of clitellar segments, lacking in immature and postsexual forms preserved in the same way, as well as in live adults, is due to fine granules in the outermost portion of the epidermis and perhaps precipitated as a result of action of the preservative.

Drawida assamensis

1945. *Drawida assamensis* Gates, *Jour. Roy. Asiatic Soc. Bengal,* (*Sci.*) 11: p. 59. (Type locality, Dumpep, Khasi Hills, Assam. Types, in the Indian Mus.)

Male pores, minute, in line with 10/11 and in median portion of *BC,* each at tip of a rather conical protuberance with base reaching postsetal and presetal secondary furrows, respectively, of x and xi. Spermathecal pores, minute, just median to *C.* Genital markings, very small, circular, in vicinity of spermathecal pores, one in each male porophore. Setae, AA ca. $= BC,$ DD ca. $= \frac{1}{2}C.$ Nephropores, at or close to D but slightly more dorsal in viii (in x?). (Clitellum? Pigmentation? Segments?) Size, 50–79 by $1\frac{1}{2}$–2 mm.

Gizzards, 4–5 in xii–xvii. (Intestinal origin? Connectives from extra-esophageals? Nephridia of x?) Sperm ducts, slender, very short, no hairpin loops, each passing to ental end of a prostate directly. Prostatic capsule, slenderly digitiform. Spermathecae, each duct, ca. 2 mm. long and passing into posterior face near parietes of a slenderly digitiform diverticulum that is erect in vii and there slightly more than 1 mm. long. Ovarian chamber, closed off from parietes(?). GM glands, spheroidal, protuberant into coelom.

Distribution: Dumpep, (Khasi Hills, Assam).

Remarks: Known only from four aclitellate but probably otherwise mature types. Male porophores, on retraction, presumably will be characterized as discoidal.

If the GM glands are trustworthy evidence, relationships are with a *longatria* group of species though closer relatives probably still are unknown.

Drawida beddardi

1890. *Moniligaster beddardii* Rosa, *Ann. Mus. Civ. Sto. Nat. Genova* 29: p. 379. (Type locality, unknown. No type.)

1894. *Moniligaster beddardi*, Bourne, *Quart. Jour. Micros. Sci.* 36: p. 374.

1895. *Drawida barwelli* (part), Beddard, *A Monogr. of the Order of Oligochaeta*, p. 200.

1900. *Drawida barwelli* (part), Michaelsen, *Das Tierreich* 10: p. 116.

1910. *Drawida barwelli* (part), Michaelsen, *Abhandl. Naturwiss. Ver. Hamburg* 19: p. 8.

1923. *Drawida barwelli* (part), Stephenson (*The Fauna of British India*), *Oligochaeta*, p. 133.

1924. *Drawida barwelli* var. *hehoensis* + *D.* "*fluvaitilis*" Stephenson, *Rec. Indian Mus.* 26: pp. 324 and 325. (Type localities, Heho plain and White Crow stream, both near Yaungwhe. Types, in the Indian Mus.)

1926. *Drawida tecta* Gates, *idem* 28: p. 148. (Type locality, Yaungwhe. Types, in the U. S. Natl. Mus.)

1931. *Drawida hehoensis*, Gates, 1931, *ibid.* 33: p. 340.

1933. *Drawida hehoensis*, Gates, *ibid.* 35: p. 443.

1962. *Drawida beddardi*, Gates, *Bull. Mus. Comp. Zool. Harvard College* 127: p. 312.

Male pores (secondary), transverse slits at 10/11 in median portion of BC, primary pores minute and on ventral ends of short, tubular penes pendent from roofs of eversible, spheroidal, muscular chambers protuberant more or less conspicuously into coelom. Spermathecal pores (external), rather large, at or just median to CD. (Genital markings, lacking but one or both margins of male apertures may be whitened and with thickened epidermis.) Setae, longer in posterior segments, $AA =$ or $< BC$, $DD =$ or $> \frac{1}{2}C$. Clitellum, red, in ix–xiv. Nephropores, at or close to D (in x?). Color, white, (unpigmented). Segments, 145–188. Size, 24–120 by 2–5 mm.

Gizzards, 3–4, in xii–xix. (Intestinal origin?) Connectives from extra-esophageals, usually(?) on anterior face of 8/9. (Nephridia of x?) Sperm ducts, rather short and each passing into a prostate slightly below ental end and directly. Prostates, not unusually long, capsule digitiform, erect or bent, glandular investment restricted to ental end or continued to or nearly to penial chamber. Spermathecae, adiverticulate, duct 6–7 mm. long, an ectal portion (usually or mostly confined to parietes), thickened and conical or pine-cone-shaped. Ovarian chamber, closed off from parietes.

Distribution: Yaungwhe and vicinity, Kalaw, Taungyi, Maymyo (Shan States), at elevations of 3–4,000 feet. (Chiengmai, Thailand?)

Remarks: Anterior connectives between extra-esophageal and dorsal vessels at least once were within septum 8/9. Some variation in prostatic glandularity (possibly geographical) also was found, the investment limited in Maymyo worms to a short ental portion of the capsule. Possibly correlated with that reduction is location of gizzards nearer the ovarian chamber.

Whitened and rather conical protuberances from adjacent margins of x–xi in vicinity of the external male pores are not porophores, nor genital markings, seemingly mere tumescences. The penes are completely protrusible and at copulation presumably are inserted into spermathecal ducts. Whether the large and superficial spermathecal aperture is to regarded as primary or secondary remains to be determined.

Relationships of *beddardi* presumably are closest with a group of species characterized by the supposedly primitive adiverticulate condition of the spermathecae. A *beddardi* group of Burmese species comprises *D. constricta, delicata, spissata* and *rara*.

Drawida bullata

1930. *Drawida constricta* (part), Gates, *Rec. Indian Mus.* 32: p. 282. (Excluding adiverticulate worms.)

1933. *Drawida bullata* + *D. fucosa* Gates, *ibid.* 35: pp. 424 and 439. (Type localities, Henzada and Kalewa. Types, none.)

1962. *Drawida bullata*, Gates, *Bull. Mus. Comp. Zool. Harvard College* 127: p. 313.

Male pores, transverse slits, median to mBC, at or behind 10/11. Spermathecal pores, small, transverse slits, at or just median to C. Genital markings, areas of epidermal thickening in which clitellar coloration is lacking, paired and in median portion of AC or median but reaching well into BC, in vii–xiii, usually present in the middle or the last two annuli of x–xi. Setae, $AA < BC$, ($DD = ?$). Nephropores, at or close to D (in x?). Clitellum, red, ix/n–xiv/n. Color, white (unpigmented). (Segments?) Size, 20–180 by $1\frac{1}{2}$–7 mm.

Gizzards, 1–5, in xiii–xviii. (Intestinal origin?) Connectives from extra-esophageals, on anterior face of 8/9. (Nephridia of x?) Sperm ducts, each with a terminal portion in x seemingly somewhat thickened and 4–70 mm. long, passing to ental end of a prostate directly. Prostates, short, erect or bent mesially, capsule to 2 mm. long, digitiform or slightly widened entally. Spermathecae, each duct passing in vii into a digitiform diverticulum that is erect or looped and *ca.* 5 to more than 70 mm. long. Ovarian chamber, closed off from parietes. (GM glands, none.)

Distribution: Kawkareik (Amherst), Henzada (Henzada), Toungoo, Tantabin (Toungoo), Prome, Laboo (Prome), Thanbula (Thayetmyo), Pyigyaung (Yamethin), Taungdwingyi (Magwe), Minbu (Minbu), Mandalay (Mandalay), Ye-U (Shwebo), Kalewa (Upper Chindwin). Possibly also including some or all of the following; Monywa (Lower

Chindwin), Mingin, Laungbyin, Homalin (Upper Chindwin).

Abnormality: Ental portion of main spermathecal axis bifurcated and with two ampullae.

Remarks: Size of Lower Burma worms ranges from 20 by $1\frac{1}{2}$ to 95 by 5 mm. Kalewa worms are much longer and slightly thicker.

An annular tumescence sometimes surrounds each male pore. Male porophores were anteroposteriorly flattened, rather pointed protuberances apparently mostly or entirely belonging to xi. Complete retraction has not been seen.

The spermathecal diverticulum is shortest in Henzada worms where it is slightly widened at the parietes. It is somewhat longer in Tantabin worms where it may be slightly sinuous and also in Minbu worms where it may be irregularly constricted or even very shortly zigzagged.

Relationships, as indicated by the anatomy, are with members of the *gracilis* group.

Intraspecific changes in sperm ducts and in spermathecal atria are the same as in *longatria*.

Drawida caerulea

Male pores, invaginate, external apertures large transverse slits at 10/11 in median half of *BC*. Spermathecal pores, each on a marking within a parietal invagination opening to the exterior through a transverse slit in the median half of *BC*. (Genital markings, lacking externally.) Setae, $AA < BC$ anteriorly but may $= BC$ posteriorly, $DD > \frac{1}{4}C$. Nephropores, at or close to D, more dorsal in viii and xii, present in x of adults. Clitellum, red, x–xiii (only?). Color, bluish (pigment in circular muscle layer?) Segments, 178. Size, 50–80 by 3–4 mm.

Gizzards, 1–4, in xiii–xviii. (Intestinal origin?) Connectives from extra-esophageals, on posterior faces of 8/9–9/10. Nephridia of x, present in adults (?). Sperm ducts, long. Prostates, muscular. Ovarian chamber, closed off from parietes.

Distribution: Central Burma, between Arakan Yomas and the Shan Plateau, at elevations below 1,000 feet. Never found in Rangoon and the deltas region to the south. Worms from Monywa, Masein, Mingin, in the Chindwin Valley, now are referable only to species.

Regeneration: Tail regenerates were seen.

Remarks: A thickened parietal portion of the spermathecal duct, somewhat like that present in *D. rara*, *constricta*, and *delicata*, and characteristic of the southern subspecies has evolved in the north into a much larger organ presumably not comparable with the digitiform and saccular structures that are called atria in other species.

A northern and a southern subspecies, linked by intermediates in the Mandalay region, are recognized.

Relationships clearly are with the Assamese *D. decourcyi*.

Drawida caerulea caerulea

1926. *Drawida caerulea* Gates, *Rec. Indian Mus.* **28**: p. 143. (Type locality, Nyaunglebin. Paratypes, in the British and U. S. Natl. Mus.)

1930. *Drawida caerulea* (part), Gates, *ibid.* **32**: p. 279. (Excluding some Chindwin Valley specimens.)

1931. *Drawida caerulea*, Gates, *ibid.* **33**: p. 338.

1933. *Drawida caerulea* var. *typica*, Gates, *ibid.* **35**; p. 429.

1939. *Drawida caerulea*, Chapman, *ibid.* **41**: p. 119. (Circulatory system.)

1962. *Drawida caerulea caerulea*, Gates, *Bull. Mus. Comp. Zool. Harvard College* **127**: pp. 319, 361.

Sperm ducts, slender, less than 70 mm. long (?), in a cluster of loops that is much smaller than the testis sac. Spermathecae, adiverticulate, duct widened ectally and there subspheroidal to shortly conical and usually or mostly buried within the parietes but opening to the exterior indirectly through an aperture in a circular marking on roof of a parietal invagination.

Distribution: Kayan, Thongwa (Hanthawaddy), Thinbawgyin (Bassein). Pegu, Thanatpin, Madauk, Pazunmyaung, Nyaunglebin (Pegu). Thonze (Tharrawaddy). Toungoo (Toungoo). Pyinmana (Yamethin). Minbu (Minbu). Myingyan, Kyaukpadaung (Myingyan). Mandalay, Tonbo, Kyaukkyone (Mandalay). Sagaing, Ava, Myotha (Sagaing). Ye-U (Shwebo). (Monywa? Lower Chindwin).

Abnormality: (No. 1) An extra spermatheca, smaller than the others, in ix on right side, with pore at 8/9. This could perhaps have been considered in the past as a unilateral reversion to the quadrithecal condition presumably characteristic of an ancestral moniligastrid.

(No. 2) Left spermatheca, lacking. Other reproductive organs of that side one segment anterior to usual location (ovarian chamber in two discrete portions). The anterior homoeosis is believed to have resulted from abortion of the left mesoblastic somite at the 8th level during early embryonic development.

Remarks: Male terminalia are eversible as high porophores presumably for insertion into spermathecal pores during copulation. The somewhat smaller distal portion bears a nipplelike protuberance on which the male pore is located and contains a spheroidal muscular bulb that is retractile into a larger and also thick-walled chamber of longitudinally ellipsoidal shape.

Drawida caerulea rasilis

1930. *Drawida caerulea* (part), Gates, *Rec. Indian Mus.* **32**: p. 279. (Including only some of the Chindwin Valley worms.)

1933. *Drawida caerulea* var. *rasilis* Gates, *ibid.* **35**: p. 433. (Type locality, Chindwin Valley, north of Monywa. Types, none.)

1962. *Drawida caerulea rasilis*, Gates, *Bull. Mus. Comp. Zool. Harvard College* **127**: p. 319.

Sperm ducts, much longer than 70 mm., thickened, each in a cluster of loops that is as large as or larger than the testis sacs. Spermathecae, each duct passing posteroventrally into median face of a large diverticulum touching 6/7 and 7/8, reaching nearly to the dorsal parietes, with a conical, ovoidal or elongate dome shape. Diverticular roof thicker than the walls and strongly ridged.

Distribution: Chindwin Valley north of Monywa.

Remarks: Male porophores were not seen at maximal protrusion.

Just what should be considered the primary spermathecal pore is unknown and presumably will remain so until development has been studied. The transverse external aperture certainly seems to be secondary or tertiary according to whether the longitudinal slit in the genital marking on the roof of the parietal invagination is an enlarged primary pore or not. In the latter case, the primary pore would be the opening of the spermathecal duct into an atrial diverticulum and the external aperture would be tertiary. Then, what is the primary pore in subsp. *caerulea* and how is the terminal portion of its spermathecae to be characterized?

Drawida constricta

1929. *Drawida constricta* Gates, *Proc. U. S. Natl. Mus.* **75**, 10: p. 8. (Type locality, Mandalay. Paratypes, in the British Mus. and the U. S. Natl. Mus.)

1930. *Drawida constricta* (part), Gates, *Rec. Indian Mus.* **32**: p. 282. (Excluding worms with diverticulate spermathecae.)

1933. *Drawida constricta*, Gates, *ibid.* **35**: p. 434.

Male pores, on rather conical porophores in median half of *BC*. Spermathecal pores, just median to *C*. Genital markings, whitened areas of epidermal thickening in last two annuli of a segment, usually paired in x–xi and extending from m*BC* to *B*, *A*, or even mesially, occasionally paired in vii–viii, xii, median but extending lateral to *B* in ix (x–xi), xii. Setae, $AA < BC$, DD ca. $= \frac{1}{2}C$. Nephropores, at or close to *D*. Clitellum, red, ix/n–xiv/n. Color, white (unpigmented). Segments, 148. Size, 73–95 by 3–$4\frac{1}{2}$ mm.

Gizzards, 2–3, in xiv–xviii. (Intestinal origin?) Connectives from extra-esophageals, on posterior face of **8**/9. (Nephridia of x?) Sperm ducts, rather short, slender, each passing into ental end of a prostate directly. Prostates, short, erect in coelom, capsule club-shaped, narrowed ectally, bent over towards nerve cord. Spermathecae, adiverticulate, duct slightly thickened ectally in 7/8 and within the parietes. Ovarian chamber, closed off from parietes. (GM glands, none.)

Distribution: Prome (Prome), Thayetmyo (Thayetmyo), Pakokku (Pakokku), Mandalay (Mandalay), Kalewa (Upper Chindwin). Possibly also at some or all of the following, Monywa (Lower Chindwin), Laungbyin, Mingin, Homalin (Upper Chindwin).

Remarks: Male pores probably develop at 10/11. Male porophores on complete retraction may be distinctly demarcated circular areas extending well (and equally?) into x–xi.

Relationships, at least in so far as indicated by the spermathecae, are with a *beddardi* group, and in that group presumably are closest with *spissata* and *rara*.

Drawida decourcyi

1914. *Drawida decourcyi* Stephenson, *Rec. Indian Mus.* **8**: p. 373. (Type locality, Upper Rotung, Abor country, Assam, India. Types, in Indian Mus.)

1923. *Drawida decourcyi*, Stephenson (*The Fauna of British India*), *Oligochaeta*, p. 136.

1934. *Drawida decourcyi*, Gates, 1934, *Rec. Indian Mus.* **36**: p. 235.

1961. *Drawida decourcyi*, Gates, *Bull. Mus. Comp. Zool. Harvard College* **127**: p. 319.

Male pores (secondary), large crescentic slits at 10/11 in median half of *BC*, primary pores each on a club-shaped penial body within a central depression of a papilla on roof of prostate. Spermathecal pores, each in a circular area on anterior wall of a deep invagination with transverse aperture median to *C*. (No genital markings.) Setae, AA ca. $= BC$, $DD > \frac{1}{3}C$. (Nephropores? Clitellum?) Color, bluish (pigment in circular muscle layer?). Segments, 226. Size, 100 by 8 mm.

Gizzards, 7–9, in xiv–xxvii. (Intestinal origin? Connectives from extra-esophageals? Nephridia of x?) Sperm ducts, very long, 640 mm., each with loops in a cluster larger than the testis sac, passing into ental end of prostate directly. Prostates, muscular, 5 mm. long, bent posteriorly, rather conical but with ental end bluntly rounded. Spermathecae, duct 20 mm. long, muscular, slightly widened ectally, passing into the diverticulum within parietes of vii. Diverticulum, strap–shaped, lumen slitlike in cross section, emerging into coelom anteriorly in vii, the ental portion erect just behind 6/7. Ovarian chamber, closed off from parietes(?).

Distribution: Upper Rotung, Renging (Abor country, Assam), at altitudes of ca. 2,000 feet.

Remarks: Known only from three types, none mature. One is a posterior amputee.

The prostate appears to be of about the same structure as copulatory chambers of various pheretimas and like those invaginations presumably is eversible during copulation.

Relationships of *D. decourcyi* obviously are with *D. caerulea* from which it is distinguished by the strap shape of spermathecal diverticula, the larger number of gizzards and the much greater length of male gonoducts. As differences are almost entirely quantitative, discovery of intermediates linking the two taxa will not be unexpected.

Drawida delicata

1930. *Drawida* sp., Gates, *Rec. Indian Mus.* **32**: p. 298.
1962. *Drawida delicata* Gates, *Bull. Mus. Comp. Zool. Harvard College* **127**: p. 319. (Type locality, Mergui. Types, none.)

Male pores, on anteromesially directed porophores in median part of *BC*. Spermathecal pores, close to *C* (?). Genital markings, lacking, but epidermis whitened in median portion of *BC* between equators of x–xi and 10/11. Setae, $AA < BC$ $(DD = ?)$. (Nephropores? Clitellum? Color, bluish? Segments?) Size, 23–27 by 2 mm.

Gizzards, 4, in xii–xv. (Intestinal origin? Connectives from extra-esophageals? Nephridia of x?) Sperm ducts, long, each in a cluster of loops that is as large as the testis sac, passing into prostate directly. Prostates, erect in coelom, capsule digitiform or club-shaped and with slender stalk. Spermathecae, adiverticulate, duct short (only one loop), thickness increased three to four times as it passes into parietes on posterior face of 7/8. (Ovarian chamber, closed off from parietes? GM glands, none?)

Distribution: Mergui (Mergui).

Remarks: Known only from the original description of four softened and immature specimens. Male poropores may normally be discoidal but capable of protrusion so as to show no indication of origin from x and xi or they may result from eversion of small invaginations.

Relationships of the species presumably are with forms not as yet collected in the little-known territory east and south of Mergui.

Drawida flexa

1929. *Drawida flexa* Gates, *Proc. U. S. Natl. Mus.* **75**, 10: p. 10. (Type locality, Kawkareik. Paratype, in U. S. Natl. Mus.)
1931. *Drawida flexa*, Gates, *Rec. Indian Mus.* **33**: p. 338.
1933. *Drawida flexa*, Gates, *ibid.* **35**: p. 436.
1962. *Drawida flexa*, Gates, *Bull. Mus. Comp. Zool. Harvard College* **127**: pp. 320, 360.

Male pores, small, superficial, at or slightly median to m*BC* and at or in line with 10/11, each in a nearly circular area between posterior tertiary furrow of x and presetal secondary furrow of xi. Spermathecal pores, small transverse slits, at or just median to *C*. Genital markings, 2–20, each with a circular translucent area, usually transverse, unpaired and median, presetal in viii, x–xii, postsetal in vii–x, in setal annulus of vii and x, paired, in *BC*, presetal in viii–xii, postsetal in vii and x, in setal annuli of viii–xi. Setae, $AA >$ or $< BC$, DD slightly $> \frac{1}{2}C$. Nephropores, at or close to *D*, more dorsal in viii, occasionally also in ix (x?). Clitellum, red, in x–xiii or reaching to eq/ix and eq/xiv. Color, none (unpigmented). Segments, 238. Size, 50–112 by 3–5½ mm.

Gizzards, 3–5, in xvi–xxiv. Intestinal origin, in xxvii (±1?). Connectives from extra-esophageals,

on anterior face of 8/9. (Nephridia of x?) Sperm ducts, short, thickened portion in x only 1–2 mm. long, passing into ental ends of prostates directly. Prostates, 2–3 mm. long, erect or recumbent, usually in a *J*- or *U*-shape, capsule digitiform. Spermathecae, diverticulum digitiform, 1–3 mm. long, in vii, erect, bent at right angles or coiled on parietes. Ovarian chamber, closed off from parietes. GM glands, small, spheroidal, (translucent?), beneath longitudinal musculature (lacking in male porophores).

Distribution: Kamaungthwe River region to Kameik, Kawletchaung, Pyinthadaw, Nyaungdon, Siyigyan, Zinba, Migyaunglaung (Tavoy), Ye, Kyundo, Kawkareik (Amherst), Thaton, Duyinzeik, Naunggala, Bilin, Taungzun, Kinmunsakhan, Boyagyi, Kyaikto, Sittang (Thaton).

From a northern portion of Tavoy district to the Sittang River and from the coast to the Thailand border (at least?).

Regeneration: Tail regenerates were found.

Remarks: Male pores of this species seemingly can be protruded but slightly.

Relationships clearly are with *D. longatria* from which *D. flexa* is distinguished, within its own range, by quantitative differences such as length of spermathecal diverticula and of male gonoducts, size of GM glands, absence of a solid gland in male porophores. Characterization of retracted states of the male porophores in *D. longatria* may provide further differences.

Drawida gracilis

1925. *Drawida gracilis* Gates, *Ann. Mag. Nat. Hist.* Ser. 9, **16**: p. 660. (Type locality, Rangoon. Types, none.)
1926. *Drawida gracilis*, Gates, *Rec. Indian Mus.* **28**: p. 144. *Jour. Burma Res. Soc.* **15**: p. 204.
1931. *Drawida gracilis*, Gates, *Rec. Indian Mus.* **33**: p. 339.
1933. *Drawida gracilis*, Gates, *ibid.* **35**: p. 441.
1961. *Drawida gracilis*, Gates, *American Midland Nat.* **66**: p. 62.
1962. *Drawida gracilis*, Gates, *Bull. Mus. Comp. Zool. Harvard College* **127**: pp. 324, 361.

Male pores, rather small transverse slits in median half of *BC*, each in a disc extending into x–xi that is protrusible as a rather more conical porophore. Spermathecal pores, at or slightly median to *C*. Genital markings, whitened areas of epidermal thickening, paired, in x–xi, one in front of and one behind each porophore, sometimes reaching into *AA* and rarely even to m*V*, unpaired and median, in *AA*, *BB* or even *CC* and presetal in viii–ix. Setae, $AA < BC$, DD ca. $= \frac{1}{2}C$, *a* and *b* enlarged in ii–xiii, especially so in v–vii, and with ornamentation more extensive than on *c* and *d*. Nephropores, at or near *D* (in x?). Clitellum, red, in x–xiii (+?). Color, white (unpigmented). Segments, 140–340. Size, to 180 by 3 mm.

Gizzards, 1–4, in xiv–xxi. Intestinal origin, in xxii (±?). Connectives from extra-esophageals, on anterior face of 8/9. (Nephridia of x?) Sperm ducts, rather long but slender throughout, each looped in a small cluster on each side of 9/10, both clusters together smaller than testis sac, passing into ental end of prostate directly. Prostates, sessile, flat, and circular or more protuberant and rather conical, capsule small, shortly ovoidal (or spindle-shaped?) and partly within parietes (pointed end entally or ectally?), 0.5–1.25 mm. long. Spermathecae, duct *ca.* 5 mm. long, diverticulum stoutly digitiform, 1.0–1.5 mm. long, with thick wall and narrow lumen, erect in vii. Ovarian chamber, not closed off from parietes(?). (GM glands, none.)

Distribution: Thaton, Kyaikto (Thaton), Thinbawgyin (Bassein), Rangoon, Thongwa, Twante (Hanthawaddy), Hlawga, Wanetchaung, Taukkyan (Insein), Pegu, Thanatpin (Pegu), Tharrawaddy, Pegu Yomas (Tharrawaddy), Letpadan (Henzada), Prome, Paukkaung (Prome), Sandoway (Sandoway), Thanbula (Thayetmyo), Ramree (Kyaukpyu), Taungdwingyi (Magwe), Tiddim (Chin Hills). From sea level to 5,000 feet.

Regeneration: Tail regenerates were found.

Abnormality: (No. 1) Testis sac, male deferent duct and ovisac lacking on one side of body. (No. 2) Two pairs of male porophores, prostatic capsules (glandular investments in a single mass on each side of body) and male deferent ducts. Nature of gonads in xi could not be determined and a second pair of testis sacs was not found. Ovisacs were present as usual.

Remarks: One specimen was clitellate but only an anterior portion was secured. Adults may have been at depths (under water?) below those dug over by the collectors. Juveniles at least may remain active during the dry season and even into April.

Relationships presumably are with those Burmese species that also lack GM glands but that do have digitiform spermathecal atria. The *gracilis* group comprises *D. bullata, gracilis, lacertosa, tenellula,* and *vulgaris.*

Drawida kempi

1914. *Drawida kempi + D. rotungana + D. pellucida,* Stephenson, *Rec. Indian Mus.* **8:** pp. 376, 372, 368. (Type localities, vicinity of Rotung, Assam. Types, in the Indian Mus.)
1923. *Drawida kempi + D. rotungana + D. pellucida* f. *typica* (part), Stephenson, (*The Fauna of British India*), *Oligochaeta,* pp. 144, 155, 150.
1934. *Drawida kempi,* Gates, *Rec. Indian Mus.* **36:** p. 238.

Male pores, minute, at or slightly lateral to mBC, each on a slight, soft protuberance. Spermathecal pores at C. Genital markings, translucent areas of epidermal thinness (?). Setae, $AA = BC, DD > \frac{1}{2}C$, (*a* and *b* of x lacking?). (Nephropores? Clitellum?) Color, greenish. Segments, 125 (165–187?). Size, 75 by 5 mm.

(Pigmentation?) Gizzards, 4, in xv–xix. (Intestinal origin? Anterior connectives from extra-

esophageals? Nephridia of x?) Sperm ducts, short, each with several loops, passing into prostate directly (?). Prostatic capsule, spheroidal, ventral face imbedded in parietes. Spermathecae, adiverticulate. Ovarian chamber, closed off from parietes. (GM glands, none?)

Distribution: Egar Stream between Rotung and Renging (and in vicinity of Rotung?), Abor country, Assam.

Remarks: The species was erected on a sexual but aclitellate specimen that may be a posterior amputee. The green coloration may have been an artifact. Types of *rotungana* are immature and the specimen of *pellucida* from the same region is juvenile and too immature to warrant more than a guess at identification. GM glands were not seen and are assumed to be lacking.

Drawida lacertosa

Male pores, in median half of BC, each at end of firm distal portion of a rather conical porophore the softer proximal portion of which extends into x and xi. Spermathecal pores, fairly large transverse slits, about at C. Genital markings, a small area of epidermal whitening in front of and behind each male porophore, median areas of epidermal thickening reaching laterally to A, C or any level in between, presetal in viii–x (rarely xii) or postsetal in viii–ix, occasionally occuping entire length of a segment. Setae, $AA < BC, DD = $ or slightly $> \frac{1}{2}C$. Nephropores, at or close to D (in x?). Clitellum, red, ix/n–xiii/n. Color, white (unpigmented). Segments, 212. Size, 80–144 by 5–6$\frac{1}{2}$ mm.

Gizzards, 3–5, in xiii–xx. (Intestinal origin?) Connectives from extra-esophageals, on anterior face of 8/9. Sperm ducts, long, thickened in x, each passing into a depression at or near middle of dorsal face of prostatic capsule directly. Prostates, sessile on parietes, the thick intraparietal stalk continued into tougher distal portion of the male porophore. Spermathecae, with muscular duct and short (muscular?), thickly digitiform diverticulum erect in vii. Ovarian chamber, closed off from parietes. (GM glands, none.)

Remarks: Protrusion of male porophores probably was maximal. The retracted state was not seen but there is some reason for believing that the softer, annular proximal portion of the porophore is an everted parietal invagination. The distal portion certainly has the glabrous appearance of tissue that is not exposed ordinarily to the external environment.

The species seems to be much less tolerant of human interference with the environment than *D. papillifer* and *longatria.* Few, if any, of the specimens were secured within town limits. Human activities of one sort or another but perhaps more especially agriculture may then be responsible for a seeming discontinuity in distribution.

Relationships, as indicated by the anatomy, are with species of the *gracilis* group.

Two subspecies, for reasons indicated below, were recognized.

Drawida lacertosa lacertosa

1930. *Drawida lacertosa* Gates, *Rec. Indian Mus.* **32**: p. 284. (Type locality, Ngapoli. Types, none.)
1933. *Drawida lacertosa* var. *typica*, Gates, *ibid.* **35**: p. 445.

Gizzards, 4–5, in xvi–xx. Prostates, larger and reaching up higher in coelom. Prostatic capsule, *T*-shaped, the longitudinally ellipsoidal ental portion more distinctly bilobed (1933: p. 466, fig. 8a). Spermathecal duct, elongate, looped entally, an ectal portion *ca.* 5 mm. long especially thickened and straight on posterior face of 7/8.

Distribution: Sandoway, Ngapoli (Sandoway).

Remarks: Differences from *sepulta* are quantitative and such as are to be expected in an uninterrupted species range. Whether the gap between western and eastern populations is real and if so whether it resulted from extinction in the intermediate area or from transportation, presumably by man, of an apparently haemerophobic species remains to be determined.

Drawida lacertosa sepulta

1931. *Drawida sepulta* Gates, *Rec. Indian Mus.* **33**: p. 352. (Type locality, Kya In. Types, none.)
1933. *Drawida lacertosa* var. *sepulta*, Gates, *ibid.* **35**: p. 447.
1962. *Drawida lacertosa sepulta*, Gates, 1962, *Bull. Mus. Comp. Zool. Harvard College* **127**: p. 361.

Gizzards, 3–5, in xiii–xix. Prostates, low circular patches on the parietes. Prostatic capsule, club-shaped, ental portion indistinctly bilobed (1933: p. 446, fig. 8b–c). Spermathecal duct, 10–14 mm. long, gradually thickened ectally.

Distribution: Chaungson, Martaban, Kya In, Kawkareik (Amherst), Thaton (Thaton), Shwegyin, Blachi, Su Law and Pray Law in the Pegu Yomas (Toungoo).

Abnormality: (No. 1) Left side only. Testis sac vesiculate, collapsed. Male funnel, transparent. Sperm duct, rather short. Prostate, rudimentary. The left testis probably had aborted without maturing sperm but not until the male funnel had reached usual size.

Remarks: Spermathecal and male deferent ducts probably are shorter and male porophores slightly smaller than in *lacertosa*.

The subspecies, according to information now available, would seem to have crossed Sittang River in the vicinity of Toungoo.

Drawida limella

1934. *Drawida limella* Gates, *Rec. Indian Mus.* **36**: p. 241. (Type locality, Amingaon, Assam. Type, in the Indian Mus.)

Male pores, in *AB* or at *B* (?), at 10/11 (?), on rather conical porophores belonging equally to x–xi. Spermathecal pores, at m*BC*. Genital markings, postsetal in *BC* of vii, transversely oval and with central pores. Setae, *AA* < *BC*. (Nephropores?) Clitellum, whiteish, ix–xiv. Color, bluish. (Segments?) Size, 58 by 2 mm.

Gizzards, 3, in xiii–xv (?). (Intestinal origin? Connectives from extra-esophageals? Nephridia of x?) Sperm ducts, short, no loops, passing to prostates directly. Prostatic capsule, shortly digitiform. Spermathecae, adiverticulate, duct rather thick. Ovarian chamber, not closed off from parietes (?). GM glands, prostatelike, capsule slenderly digitiform.

Distribution: Amingaon (Kamrup district, Assam).

Remarks: Known only from the original description of a sexual specimen in but fair condition. Male porophores probably were at maximal protrusion. On retraction, primary male pores may be invaginate.

Relationships are with the *willsi* species group of peninsular India from the northern border of the Deccan almost down to Cape Comorin. The prostatelike GM glands (*cf.* Gates, 1962: p. 358) are present only in that group.

Drawida longatria

Male pores, after preservation usually on markedly protuberant porophores (but when porophores are retracted, within small parietal invaginations?). Spermathecal pores, small transverse slits, at or just median to *C*. Genital markings, unpaired and median or paired, circular to elliptical and then transverse or longitudinal, each usually with a wide opaque peripheral portion distinctly demarcated from a translucent central portion, occasionally translucent areas alone recognizable (more often in vicinity of spermathecal pores). Setae, *AA* < or *ca.* = *BC*, *DD* > ½*C*. Nephropores, at or close to *CD*, often more dorsal in viii, lacking in x. Clitellum, red (occasionally yellowish), ix/n, x–xiii, xiv/n. Color, white (unpigmented). Segments, 183–240. Size, 50–200 by 3–9 mm.

Gizzards, 1–5, in xiii–xxiii. Intestinal origin, in xxii or xxiii (± ?). Connectives from extra-esophageals, on anterior face of 8/9. Nephridia, of x aborted in adults, ducts into parietes near *D*. Sperm ducts, long, a thickened postseptal portion of each in a cluster of loops that may be much longer than the testis sac, passing into ental end of prostate directly. Prostates, elongate, spirally coiled, variously twisted or curved, capsule digitiform. Spermathecae, with digitiform diverticula in vii, usually irregularly coiled, 2–180 mm. long. Ovarian chamber, closed off from parietes. GM glands, with muscular wall, spheroidal, ellipsoidal, flask-shaped, usually interrupting longitudinal musculature.

Distribution: Burma, in the following districts; Mergui, Tavoy, Amherst, Thaton, Pyapon, Myaungmya, Hanthawaddy, Maubin, Bassein, Insein, Henzada, Pegu, Tharrawaddy, Prome, Sandoway, Toungoo, Yamethin, Thayetmyo, Magwe, Meiktila, Pakokku, Akyab, Mandalay, Sagaing, Lower Chindwin, Shan States, Upper Chindwin, Bhamo, Myitkyina. Sumatra (recorded only from Palembang).

Presence of the species in Sumatra is due to transportation, presumably by man.

The proper range of the species may prove to be central Burma from the latitude of Mandalay-Sagaing southward to the Bay of Bengal, possibly with some natural extension into the Shan Plateau in the vicinity of Toungoo.

Abnormality: (No. 1) Testis sacs, represented only by opaque areas in 10/11 at ental ends of male gonoducts. Testes presumably aborted during development.

Remarks: Much of the structure that provides characters by which moniligastrids are distinguished from each other is subject, in *D. longatria*, to considerable individual as well as geographical variation. An usual number of local races for size of the species range was indicated by massive material provided by intensive collecting within that area. Discontinuities in distribution of some races complicated the situation. Some of the gaps, at least, resulted from transportation, presumably by man. Only those races that now can be characterized (though not uniformly or satisfactorily) from data secured prior to 1933 are allowed.

Relationships clearly are closest with *D. tumida* and *flexa*.

Drawida longatria longatria

1925. *Drawida longatria*, Gates, *Ann. Mag. Nat. Hist.*, Ser. 9, 16: p. 50. (Type locality, Rangoon. Topotypes, in U. S. Natl. Mus.)

1926. *Drawida longatria* (part), Gates, *Rec. Indian Mus.* 28: p. 144; *Jour. Burma Res. Soc.* 15: p. 205; *Jour. Bombay Nat. Hist. Soc.* 31: p. 183. (Excluding all except Rangoon worms.)

1930. *Drawida longatria* var. *typica* (part only?), Gates, *Rec. Indian Mus.* 32: p. 286. *Moniligaster straeleni* Michaelsen, *Meded. K. Nat. Mus. Belgium* 6, 2: p. 1, *Mem. Mus. Hist. Nat. Belgique* (*Hors. Series*) 2, 5: p. 4. (Type locality, Palembang, Sumatra. Types, in Hamburg and Brussels Mus.?)

1931. *Drawida longatria* var. *typica* (part only?), Gates, *ibid.* 33: p. 343.

1933. *Drawida longatria* var. *typica* (part only?), Gates, *ibid.* 35: p. 448.

1937. *Drawida longatria*, Gates, *Bull. Mus. Comp. Zool. Harvard College* 80: p. 309.

1939. *Drawida longatria*, Chapman, *Rec. Indian Mus.* 41: p. 117.

1961. *Drawida longatria*, Gates, *American Midland Nat.* 66: p. 62; *Burma Res. Soc. 50th Anniv. Pub.* No. 1, p. 57.

1962. *Drawida longatria longatria*, Gates, *Bull. Mus. Comp. Zool. Harvard College* 127: p. 328.

?

1929. *Drawida longatria*, Stephenson, *Rec. Indian Mus.* 31: p. 228.

Male pores, each at pointed posterior end of a longitudinally ovoidal body (1930: fig. 5) containing a large, horizontal, flask-shaped gland and on a markedly protuberant longitudinal ridge in *BC* between equators of x–xi. Genital markings, indistinct or even unrecognizable in region of spermathecal pores though glands are present, elsewhere distinctly demarcated and located at one or more of the following sites; unpaired and in *AA*, presetal in xi–xiii, paired, in *BC* or *AC*, presetal in x, xii. Nephropores, in viii usually dorsal to *D*. Segments, to 220. Size, to 180 by 8 mm.

Gizzards, 2–5, in xv–xx. Sperm ducts, each with a thickened portion in a cluster of loops as large as or larger than testis sac. Prostates, 5–6 mm. long. Spermathecal diverticula, 140–180 mm. long.

Distribution: Rangoon, Thongwa (Hanthawaddy), Insein (Insein), Myaungmya (Myaungmya), Pyapon, Thanchitaw (Pyapon), Maubin, Yandoon (Maubin), Henzada, Letpadan (Henzada), Pegu, Tantabin (Pegu), Bhamo (Bhamo), Myitkyina (Myitkyina). Palembang (Sumatra).

Worms with male porophores and longitudinal ridges like those of this race were secured at the following localities; Nyaungbinkwin (Mergui), Bassein, Coomzamu (Bassein), Madauk, Nyaunglebin, Pazunmyaung (Pegu), Toungoo, Shwegyin (Toungoo), Prome (Prome), Sandoway, Andrew Bay, Ngapoli (Sandoway), Akyab (Akyab), Mandalay (Mandalay), Pakokku (Pakokku), Monywa (Lower Chindwin), Homalin (Upper Chindwin). Possibly also at Kamaing (Myitkyina). Information as to genital markings and internal organization no longer is available.

Common in the deltas region south of Tharrawaddy where other races of the species are lacking. Transportation, presumably by man, is responsible for presence of this race in Sumatra, Mergui as well as Bhamo and Myitkyina sites, presumably also various sites in the Arakan division and other localities isolated from the nominate range.

Biology: Breeding probably is toward end of the rainy season, in September–October (possibly also November), at a time when ovarian chamber and ovisacs are distended (?). Later on, only a milky fluid is present in chamber and sacs. As the fluid is resorbed sacs shrink and the chamber collapses. Eventually, only a little granular brownish debris, occasionally aggregated into small balls like the brown discs of other kinds of worms, is left in the sacs. As those changes are taking place clitellar segments acquire a faint brownish (sometimes almost black) coloration. Genital markings, late in the dry season, may become quite indistinct or even unrecognizable though sites sometimes are distinguishable by a fine wrinkling of the epidermis.

After the rainy season, individuals in sites with adequate moisture may remain active even into April.

Remarks: Although the ridges and the porophores they bear are so prominent, retraction now seems unlikely to involve more than flattening to level of rest of the external surface by withdrawing solid glands deeper into the body. Complete retraction may leave the male pore within a slight invagination.

Drawida longatria deminuta

1930. *Drawida longatria* var. *deminuta* Gates, *Rec. Indian Mus.* **32**: p. 287. (Type locality, Pyinmana. Types, none.)
1931. *Drawida longatria* var. *deminuta*, Gates, *ibid.* **33**: p. 343.
1933. *Drawida longatria* var. *deminuta* (part), Gates, *ibid.* **35**: p. 454.

Male pores, each at pointed posterior end of a small ovoidal body (*ca.* one-quarter the size of that in subsp. *longatria*, fig. 6, 1930) containing a small, horizontal, flask-shaped gland and on a markedly protuberant longitudinal ridge in *BC* between equators of x–xi. Genital markings, unpaired and in *AA*, presetal in viii and xii, postsetal in viii and ix, paired, presetal in *AB* of x–xii but reaching into *AA* except in xii, presetal in *BC* of viii, postsetal in *BC* of viii and ix, 1 or 2 anteriorly in each porophore ridge. Size, 110–170 by 4–6 mm.

Gizzards, 2–4, in xv–xx. Sperm ducts, each with a thickened portion in a cluster of loops two to three times size of testis sac. Spermathecal diverticula, long.

Distribution: Pyinmana, Pyigyaung, Ywadaw (Yamethin), Magwe (Magwe).

Remarks: The first lot of Pyinmana specimens was found on the surface after a heavy rain (*cf. D. doriae*).

Length of male gonoducts, in absence of measurements, can be estimated only from size of the cluster of its loops.

Drawida longatria tortuosa

1931. *Drawida longatria* var. *tortuosa* Gates, *Rec. Indian Mus.* **33**: p. 345. (Type locality, Chaungson. Types, none.)

Male pores, each at tip of a very small teatlike protuberance from posterior margin of a genital marking in x that reaches, displaces posteriorly, or seems to cross 10/11. Genital markings, one or two of x in longitudinal sequence with that bearing the male pore, in vicinity of spermathecal pores, in *AA* of viii–xii and at mV or paired and close to mV or to *A*, presetal and in *AB* of x–xi, one or two close to m*BC* of vii and ix, usually small and confined to a secondary annulus. Size, 60–130 by 3–5 mm.

Gizzards, 3–5, in xvi–xxi. Sperm ducts, very long, thickened portion in a cluster of loops two to four

times size of testis sac. Spermathecal diverticula, 140 mm. long. GM glands, rather small, often buried in the parietes.

Distribution: Chaungson. Presence there may be due to human introduction from the Toungoo Karen Hills.

Remarks: Male porophores capable of marked protrusion apparently are lacking in this form. The teatlike protuberance presumably can be flattened so that only the male opening is recognizable.

Drawida longatria ordinata

1930. *Drawida longatria* var. *ordinata* Gates, *Rec. Indian Mus.* **32**: p. 288. (Type locality, Toungoo. Types, none.)
1933. *Drawida longatria* var. *ordinata*, Gates, *ibid.* **35**: p. 453.

Male pores, each at pointed end (fig. 7, 1930) of a posteroventrally directed and rather conical protuberance belonging to x–xi and containing a large, vertical, flask-shaped gland. Genital markings, in vicinity of spermathecal pores and variously located in *BC* of viii–ix, paired, in setal annulus of x and in line with male porophores, presetal in *BB* of ix–xii, with large translucent area and narrow margin. Size, 175–200 by 8–9 mm.

Gizzards, 4, in xvi–xix. Spermathecal diverticula, 56 mm. long. GM glands, large.

Distribution: Toungoo and vicinity.

Remarks: A median genital marking, in ix, was present only in 3 specimens of the original lot of 44.

Drawida longatria planata

1931. *Drawida longatria* var. *planata* Gates, *Rec. Indian, Mus.* **33**: p. 344. (Type locality, Shwegyin. Types none.)
1933. *Drawida longatria* var. *planata*, Gates, *ibid.* **35**: p. 452.

Male pores, each at pointed median end of a rather flattened protuberance (fig. 2, 1931), gradually widened laterally, containing a large flask-shaped gland. Genital markings, (unrecognizable on male porophores), small, indistinctly delimited, restricted to a secondary annulus and more often the middle one, in ix–xi. Size, 85–195 by 4–6 mm.

Gizzards, 3–5, in xviii–xxiii. Sperm ducts and spermathecal diverticula, long(?). GM glands, small (the solid sort lacking in vicinity of spermathecal pores?).

Distribution: Shwegyin (Toungoo), Paung (Pegu). Possibly also in vicinity of Nyaunglebin.

Remarks: The male pore protuberances do not here rise above general body level but are within a median depression. Retraction presumably involves withdrawing the solid gland dorsolaterally deeper into the body so as to bring the male pore, apparently in *AA* during protrusion, into a median portion of *BC*.

Drawida longatria verrucosa

1931. *Drawida longatria* var. *verrucosa* Gates, *Rec. Indian Mus.* **33**: p. 347. (Type locality, Tharrawaddy. Types, none.)

1933. *Drawida longatria* var. *verrucosa*, Gates, *ibid.* **35**: p. 458. (Including only specimens listed under "A.")
1962. *Drawida longatria verrucosa*, Gates, *Bull. Mus. Comp. Zool. Harvard College* **127**: p. 330.

Male pores, each at tip of a very small, anteriorly directed teatlike protuberance from an elongately elliptical distinctly demarcated disc (fig. 3, 1931) in *BC* between equators of x–xi. Genital markings, distinctly demarcated except in vicinity of spermathecal pores, often numerous (to 17) and then in part closely crowded, 1–3 in each male disc, unpaired and median in any annulus of xiii, postsetal in x–xii, paired, in *AB*, presetal in viii–xi, postsetal in x, larger and in some portion of *AD*, postsetal in vii–viii and presetal in xii. Segments, 207–240. Size, to 160 by 6–9 mm. Gizzards, 2–4, in xv–xxi. Sperm ducts and spermathecal diverticula, long.

Distribution: Tharrawaddy, Ngapugale, Keinbyingyi, Myagyaung (Tharrawaddy), Letpadan (Henzada).

Addenda

Specimens from localities mentioned below cannot now be referred to subspecies.

TAVOY DISTRICT

1933. *Drawida longatria* var. *deminuta* (part), Gates, *Rec. Indian Mus.* **35**: p. 454.

Male porophores, similar to those of *deminuta* but directed posterolaterally. Genital markings, lacking except on ridges bearing male porophores and in vicinity of spermathecal pores. Size, to 180 by 8 mm. Gizzards, 2–6, in xv–xxii. Spermathecal diverticula, 75–80 mm. long.

BASSEIN DISTRICT

1933. *Drawida longatria* var. *deminuta* (part), Gates, *Rec. Indian Mus.* **35**: p. 454. (Bassein, Kokya, Kochi.)

Male porophores, similar to those of *deminuta* but sometimes directed posteromesially. Gizzards, 2–4, in xv–xx.

BLACHI-LEIKTHO SECTION OF TOUNGOO KAREN HILLS

1933. *Drawida longatria* var. *tortuosa* (part), Gates, *Rec. Indian Mus.* **35**: p. 457.

Male porophores, similar to those of *tortuosa*. Genital markings, except in male pore ridge and in vicinity of spermathecal pores, lacking. Size, 90–140 by 3–4 mm. Gizzards, 3–5, in xvi–xx. Spermathecal diverticula, longer than 10 mm.

THAZI AND VICINITY

1929. *Drawida longatria* (part), Gates, *Proc. U. S. Natl. Mus.* **75**, 10: p. 12.

1933. *Drawida longatria* var. *deminuta* (part), Gates, *Rec. Indian Mus.* **35**: p. 454.

Male porophores, similar to those of *deminuta*. Genital markings, one in front of each male porophore, others in vicinity of spermathecal pores, additionally presetal and median in *AA* of xi–xiii. Gizzards, 1–3, in xiv–xviii.

MANDALAY DISTRICT

1929. *Drawida longatria* (part), Gates, *Proc. U. S. Natl. Mus.* **75**, 10: p. 12.
1933. *Drawida longatria* var. *typica*, Gates, *Rec. Indian Mus.* **35**: p. 448. (Mandalay-Tonbo worms.)

Male porophores and ridges somewhat like those of subspecies *longatria* (differences due to partial retraction?). Genital markings, lacking (or unrecognized?) except in vicinity of spermathecal pore or one or two others, median in *AA* of x–xi or paired and in *BC* of xii. Size, to 65 by 3½ mm. Gizzards, 2–3, in xiv–xviii. Spermathecal atria, 20–35 mm. long.

1933. *Drawida longatria* var. *verrucosa* (part), Gates, *Rec. Indian Mus.* **35**: p. 458. (Tonbo-Kyaukyone worms.)

Male areas, somewhat like those of *verrucosa* and discoidal or slightly more protuberant, with one or two spheroidal glands though markings are unrecognizable, only a posterior portion of each porophore protuberant, rather conical and anteriorly directed, containing a very small flask-shaped gland. Genital markings, in vicinity of spermathecal pores, median and in various annuli of ix–xii, paired and laterally in *BC* of viii–x. Gizzards, 2–3, in xiii–xviii. Sperm ducts, rather short. Spermathecal diverticula, 2–4 mm. long.

SAGAING DISTRICT

1933. *Drawida longatria* var. *tortuosa* (part), Gates, *Rec. Indian Mus.* **35**: p. 457. (Sagaing.)

Genital markings, in vicinity of spermathecal pores, occupying most of ventrum in viii or ix–xiii. Size, 60 by 3 mm. Gizzards, 2–3, in xiii–xvii. Spermathecal diverticula, *ca.* 10 mm. long.

1933. *Drawida longatria* var. *deminuta* (part), Gates, *Rec. Indian Mus.* **35**: p. 454. (Kaungmudaw.)

Male porophores and ridges somewhat like those of *deminuta*. Genital markings, in vicinity of spermathecal pores, longitudinally elliptical and postsetal in *AB* of x, circular and in *AA* or *BC* of viii–xiii. Size, to 70 by 3.5 mm. Gizzards, 2–4, in xiv–xviii. Spermathecal diverticula, *ca.* 20 mm. long.

SHAN STATES

1926. *Drawida longatria*, Gates, *Rec. Indian Mus.* **28**: p. 144.

The species was obtained but rarely in the Shan plateau outside of the Blachi-Leiktho area, once each and then but few specimens at Mongnai, Youngwhe, and Taungyi. Male porophores of Mongnai worms were irregularly columnar protuberances and genital markings were median in middle annuli of viii–x.

Drawida molesta

1933. *Drawida molesta* Gates, *Rec. Indian Mus.* **35**: p. 463. (Type locality, Victoria Point. Types, none.)

Male pores, transverse slits at ventral ends of tubular penes protruding from parietal invaginations near m*BC*. Spermathecal pores, transverse slits, just median to *C*. Genital marking, a median area of epidermal thickening reaching m*BC* in anterior two thirds of x. A longitudinally elliptical, whitened area reaching into x–xi, centered around each male pore. Setae, $AA < BC$, $(DD = ?C)$. Nephropores, at or close to *D*, more dorsal in viii, (in x of adults?). Clitellum, red, x–xiii $(+?)$. (Pigmentation? Segments?) Size, to 146 by 5 mm.

Gizzards, 3–5, in xiii–xxi, not or only partly in consecutive segments. Intestinal origin, in xxix (± 1). Connectives from extra-esophageals, on anterior face of 8/9. Sperm ducts, long, (the thickened postseptal portion?) of each in a cluster of loops as large as the testis sac, passing into ental end of prostate directly. Prostates, 8–9 mm. long, digitiform, duct(?) apparently widened in parietes. Spermathecae, with slender duct, 8 mm. long, and saccular diverticulum 8 mm. long, in vii. Ovarian chamber, not closed off from parietes (?). (GM glands, none.)

Distribution: Victoria Point (Mergui).

Remarks: Seeming location of the spermathecal pore in vii (rather than on 7/8) may be fictitious because of slight eversion of the spermathecal duct.

Penes probably were partly protruded in each of the 28 types. Complete retraction is expected to bring penes into the parietes leaving a transverse, large, secondary male aperture at 10/11.

GM glands probably are lacking though absence is not mentioned in the original description.

The supra-esophageal vessel previously recorded now seems likely to have been one of a pair of unimportant vessels, the second one having been recognizable because it was empty.

Similarity in male terminalia suggests relationships with *beddardi* but the saccular diverticula as large as is known, favor diverticulate species like *nepalensis* and *rangoonensis*. Nearer relatives presumably are to be sought to the east and south.

Drawida montana

1945. *Drawida montana* Gates, *Jour. Roy. Asiatic Soc. Bengal,* (*Sci.*) **11**: p. 68. (Type locality, Dumpep, Khasi Hills, Assam. Types, in the Indian Mus.)

Male pores, minute, superficial (no porophores?), at 10/11 and slightly lateral to *B*. Spermathecal

pores, very small transverse slits seemingly in vii and with annular lips reaching 7/8 (?), just median to *C*. Genital markings, small, circular, translucent areas closely crowded into tumescent patches of various shapes and sizes, in vii–viii, x–xi. (Clitellum?) Setae, $AA < BC$, DD ca. $= \frac{1}{2}C$. Nephropores, at or close to *D* behind x, progressively but quite gradually further above *D* in iv–viii or ix. (Pigmentation? Segments?) Size, 104–184 by 4–5 mm.

Gizzards, 5, in xiii–xviii. (Intestinal origin? Connectives from extra-esophageals?) Nephridia of x, present in adults. Sperm ducts, long, a postseptal portion of each in a cluster of loops that is larger than the testis sac, passing into a prostate slightly below ental end and directly. Prostates, digitiform, looped, 6 mm. long, glandular investment lacking ectally, duct narrowed deep in parietes. Spermathecae, with saccular diverticula, 5 mm. long, erect in viii. (Ovarian chamber closed off from parietes?) (GM glands, none or concealed within the parietes?)

Distribution: Dumpep (Khasi Hills, Assam).

Remarks: Known only from the original description of two specimens, one juvenile, the other more mature but not sexual. If slight eversion of the entalmost portion of a spermathecal duct produces the annular tumescence, the pores may really be at 7/8.

Long-stalked, coelomic glands are associated with closely crowded genital markings in several species of *Pheretima* but that kind of GM gland is unknown in *Drawida*. The longitudinal musculature over GM sites is uninterrupted in the types of *D. montana*. The larger specimen, though not mature, was believed to be old enough for glands to have become visible in the coelom though retention and concealment within the body wall is not impossible.

Relationships are believed to be with species presently unknown.

Drawida nagana

1945. *Drawida nagana* Gates, *Jour. Roy. Asiatic Soc. Bengal,* (*Sci.*) **11**: p. 69. (Type locality, Khezabana, Assam. Types, in the Indian Mus.)

Male pores, minute (no porophores), in median half of *BC*, at level of 10/11 but each in a slight longitudinal groove (seminal furrow?) reaching into x–xi. Spermathecal pores, minute, just median to *C*. Genital markings, small, transverse, with swollen margins and depressed translucent centers, in AA of vii, AB of viii, median in ix or in AB or BC, in AA of x, in AA or AB of xi, median in xii. Setae, small, dark, closely paired, $AA < BC$, DD ca. $= \frac{1}{2}C$. Nephropores, at or close to *D*, in iii–viii slightly more dorsal and especially so in viii, (in x of adults?). (Pigmentation? Segments?) Size, 66–84 by 3 mm.

Gizzards, 4–6, in xiv–xx. (Intestinal origin? Connectives from extra-esophageals? Nephridia of x?) Sperm ducts, very long, each with a postseptal portion in a cluster of loops as large as or larger than testis sac, passing into prostate directly. Prostatic capsule, ca. 4 mm. long, slenderly club-shaped and wider entally, spirally twisted, glandular investment lacking near parietes, abruptly narrowed in body wall. Spermathecal diverticula, saccular, 4–5 mm. long, in vii. (Ovarian chamber not closed off from parietes? GM glands?)

Distribution: Khezabana (Naga Hills, Assam), at an elevation of 4,800 feet.

Remarks: Known only from three immature, alcoholic types. Pigment, as in *papillifer.* is expected. To permit entry in the key, presence of solid clear glands is assumed (*cf.* notes under key). Relationships presumably are closest with *D. papillifer.*

Drawida nana

1933. *Drawida longatria* var. *nana* Gates, *Rec. Indian Mus.* **35:** p. 461. (Type locality, Rangoon. Types, none.)
1962. *Drawida nana*, Gates, *Bull. Mus. Comp. Zool. Harvard College* **127:** p. 330.

Male pores, presumably on anteroposteriorly flattened, rather conical porophores about at mBC on 10/11 and seemingly independent of both x and xi. Spermathecal pores, very small transverse slits at or just median to C. Genital markings, paired, small translucent areas at or near A, one in front of each male porophore on x, others presetal in vii–xii. (Setae? Nephropores? Clitellum? Pigmentation? Segments?) Size, 40–66 by 3 mm.

Gizzards, 2–3, in xiii–xv. (Intestinal origin?) Connectives from extra-esophageals, on anterior face of 8/9. (Nephridia of x?) Sperm ducts, long, each with a thickened postseptal section in a cluster of loops as large as testis sac, passing into ental end of prostate directly. Prostates, long, spirally coiled or zigzag looped, capsule digitiform (?). Spermathecal diverticula, digitiform, 12–15 mm. long, in vii. Ovarian chamber, not closed off from parietes (?). GM glands, spheroidal, with soft but opaque walls.

Distribution: Rangoon.

Remarks: Known only from four aclitellate July–January specimens, possibly secured from soil with potted plants. Search throughout the municipal area for additional material, continued for several years, was futile. Failure to secure additional specimens anywhere in Burma also supports an exotic origin. However, much probably remains to be learned about species as small as *D. nana* even in the best known sections of Burma.

Pores previously recorded for the *longatria* sort of GM gland are now believed to have been artefacts if not observational errors. If definite pores are present, as was thought originally, in the *nana* markings, the associated gland then will be of a quite different sort than in *D. longatria* and will provide another character for distinguishing the two species.

Drawida nepalensis

1907. *Drawida nepalensis* Michaelsen, *Mitt. Naturhist. Mus. Hamburg*, **24:** p. 146. (Type locality, Gowchar, near Katmandu, Nepal. Type, in the Indian Mus.)
1909. *Drawida nepalensis* + *D. burchardi*, Michaelsen, *Mem. Indian Mus.* **1:** pp. 147, 149.
1910. *Drawida nepalensis*, Michaelsen, *Abhandl. Naturwiss. Ver. Hamburg*, **19:** p. 8.
1917. *Drawida nepalensis*, Stephenson, *Rec. Indian Mus.* **13:** p. 372. (Rangamati.)
1923. *Drawida nepalensis* + *D. burchardi*, Stephenson (*The Fauna of British India*), *Oligochaeta*, pp. 146, 134.
1924. *Drawida troglodytes* Stephenson, *Rec. Indian Mus.* **26:** p. 129. (Type locality, Siju Cave, Garo Hills, Assam. Type, in the Indian Mus.)
1925. *Drawida burchardi* + *D. hodgarti* + *D. papillifer* (part), Stephenson, *ibid.* **27:** pp. 50 and 51.
1926. *Drawida cacharensis* Stephenson, *ibid.* **28:** p. 251. (Type locality, Katlichera, South Cachar, Assam. Types, in Indian Mus.)
1929. *Drawida nepalensis*, Stephenson, *ibid.* **31:** p. 229.
1930. *Drawida nepalensis*, Gates, *ibid.* **32:** p. 290.
1931. *Drawida nepalensis*, Gates, *ibid.* **33:** p. 348.
1933. *Drawida burchardi*, Gates, *ibid.* **35:** p. 426.
1934. *Drawida nepalensis* + *D. troglodytes*, Gates, *ibid.* **36:** pp. 242 and 253.
1961. *Drawida nepalensis*, Gates, *Burma Res. Soc. 50th Anniv. Pub.* No. 1, p. 57.
1962. *Drawida nepalensis*, Gates, *Bull. Mus. Comp. Zool. Harvard College* **127:** p. 331.
1970. *Drawida nepalensis*, Soota & Julka, *Proc. Zool. Soc. Calcutta* **23:** p. 201.

Male pores, obvious, at or median to mBC, each usually on or near end of a protuberant ventrally directed porophore apparently independent of both x and xi. Spermathecal pores, small, transverse slits, just median to C. Genital markings, one small, circular translucent area on lateral or anterior face of each male porophore, a similar area in vii just anterior to each spermathecal pore, also greyish or whitened areas in which epidermis may or may not be thickened, one in front of and one behind each male porophore, others (when present) paired or unpaired and variously located in vii–xi. Setae, AA = or only slight > or < BC, DD ca. = or slightly > $\frac{1}{2}C$. Nephropores, at or near D, somewhat more dorsal in viii or vii–viii, lacking in x (and xii?) of adults. Clitellum, ix/n–xiv/n. Color, (unpigmented). Segments, 129–180. Size, 78–130 by 4–5 mm.

Gizzards, 2–4, in xii–xx (xxiii?), often some or all not in consecutive segments. Intestinal origin, in xxvii (±1). Connectives from extra-esophageals, on on anterior face of 8/9. Nephridia, of x lacking in adults, ducts into parietes at D. Sperm ducts, a postseptal portion of each thickened and in a cluster of loops that may be larger than the testis sac, passing into ental end of prostate directly. Prostatic capsule, 2–4 mm. long, slenderly club-shaped, slightly and gradually widened entally, glandular investment continued to parietes. Spermathecal diverticula, saccular, 3–5 mm. long, an ectal portion of variable length stalklike, in vii. Ovarian chamber, closed off from parietes.

GM glands of male porophores and in front of spermathecal pores solid, spheroidal, with translucent walls.

Distribution: Mount Harriet, Port Blair, and other stations (Andaman Islands). Mandalay (Mandalay). Naba (Katha). Mong Mong Valley (Mong Mong, Yunnan). Homang (South Hsenwi). Kyaukme (Hispaw). Lashio (Northern Shan States). E Nai, Namkham (North Hsenwi), Kalewa (Upper Chindwin). Bhamo (Bhamo). Nyaungbin, Namma, Myitkyina, Weshi (Myitkyina). Sandoway (Sandoway). Akyab, Myohaung (Akyab). Rangamati (Chittagong Hill Tracts). Siju Cave (Garo Hills).

Katlicherra (South Cachar), Amingaon, Balipara Frontier Tract (Assam). Nagrota (Kulu district, Punjab). Teesta Bridge (Darjiling district) and Jalpaiguri (Bengal). Kierpur (Purneah district, Bihar). Gowchar (Nepal). Dehra Dun (United Provinces). Lahore (Pakistan).

Java (Buitenzorg).

From sea level to over 4,000 feet.

Overseas distributions presently are attributable to transportation by man. The original home is unknown, but may be in the Himalayas. Much, if not all of the mainland distribution also may have resulted from transportation.

Habitats: Soil. Earth at base of bamboo clumps. Mud under water cress.

Regeneration: Head regenerate of six segments at 6/7 (once). This is the only record of cephalic regeneration in the genus. Equimeric head regeneration presumably is possible at each level back to and including 6/7. Tail regenerates were seen.

Remarks: Our knowledge of the structure of the enterosegmental organs (Stephenson, 1926: pp. 252–253) was derived from this species.

Relationships, as indicated by the GM glands and the spermathecal diverticula, seem to be with *papillifer*.

Appendix to *D. nepalensis*

1926. *Drawida rangoonensis* (part), Gates, *Rec. Indian Mus.* 28: p. 146. (Bassein worms with definite male porophores.)
1930. *Drawida rangoonensis* (part), Gates, *ibid.* 32: p. 291. (Sandoway worms with definite male porophores.)
1931. *Drawida abscisa* Gates, *ibid.* 33: p. 336. (Type locality, Sandoway. Types, none.)
1933. *Drawida abscisa*, Gates, 1933, *ibid.* 35: p. 420.
1961. *Drawida abscisa*, Gates, *American Midland Nat.* 66: p. 62.
1962. *Drawida nepalensis* f. *abscisa*, Gates, *Bull. Mus. Comp. Zool. Harvard College* 127: p. 331.

Male pores, in median half of *BC*, each in a concave depression on ventral face of a protuberant male porophore. Genital markings, one or more small, circular translucent areas near each spermathecal pore, often also in vii–xii, but slightly larger and with opaque rim indistinctly demarcated peripherally, usually unpaired though rarely median, without a solid gland. (Nephropores in x of adults?) Color, none (unpig-

mented). Segments, 116–137. Size, 40–80 by 2½–4 mm.

Gizzards, 3–5, in xiii–xix, in consecutive segments and always present in xv–xvi. (Intestinal origin?) Connectives from extra-esophageals, within septum 8/9. (Nephridia of x, in adults?) Prostates, 3–5 mm. long, glandular investment usually not reaching parietes. Spermathecal diverticula, 2–3 mm. long. GM glands, present only in vicinity of spermathecal pores, small, solid, translucent.

Distribution: Mong Mong (Mong Mong, Yunnan), Lashio (Shan States). Kutkai, Namkham (North Hsenwi). Myitkyina, Tingpai, Kawapang, Kadranyang, Nawangkai (Myitkyina). Bassein (Bassein). Sandoway (Sandoway). From sea level to 4,500 feet. Probably introduced accidently at Rangoon, Bassein, and Sandoway. The original source apparently is to be sought in the region north of Myitkyina.

Colonization at Rangoon apparently was unsuccessful as worms were not found there after 1933.

Remarks: Male porophores were protuberant in all specimens referred to this taxon. Although the *abscisa* condition always was easily recognizable, little difference in appearance from *D. nepalensis* porophores is anticipated in a completely retracted condition.

Spermathecal ampullae or diverticula, ovisacs and ovarian chambers usually seemed to be juvenile even in clitellate worms. Parthenogenesis was suspected but the wrong organs, i.e., ovisacs and ovarian chambers, were juvenile at maturity(?).

Drawida papillifer

Male pores, very small, superficial (no protrusible porophores). Spermathecal pores, very small, at or slightly median to *C*. Genital markings, small, circular to shortly elliptical and transverse areas of translucence, in vii–viii and x–xi near spermathecal and male pores, occasionally in other positions on vii–xii. Setae, $AA < BC$, $DD =$ or $> \frac{1}{2}C$. Nephropores, about at *D*, somewhat more dorsal in viii, occasionally also in ix or iii–vii, present in x. Clitellum, red, occasionally whitish, ix/n–xiv/n. Color, bluish or purplish (pigment in the circular muscle layer).

Gizzards, 2–4, in xii–xx. Connectives from extra-esophageals, on anterior face of 8/9. Nephridia of x, present in adults. Sperm ducts, passing into prostates near ental ends and directly. Prostatic capsule, slenderly club-shaped, widened entally. Spermathecal diverticula, saccular, with a short stalk, 2–3 mm. long, in vii. Ovarian chamber, not closed off from parietes. GM glands, small (with thin, transparent walls?), spheroidal to ellipsoidal, usually beneath longitudinal musculature.

Distribution: From Mergui north nearly to Mandalay and through the western mountain wall into India.

Remarks: The two subspecies now recognized are distinguished by location of male pores.

Drawida papillifer papillifer

1917. *Drawida papillifer + D. affinis + D. hodgarti* Stephenson, 1917, *Rec. Indian Mus.* **13**: pp. 370, 368, 366. (Type locality of each species, Rangamati, Chittagong Hill Tracts, Bengal. Types, in the Indian Mus.)

1922. *Drawida rosea* Stephenson, 1922, *ibid.* **24**: p. 430. (Type locality, Cherrapunji, Assam. Type, in the Indian Mus.)

1923. *Drawida papillifer + affinis + D. hodgarti + D. rosea*, Stephenson (*The Fauna of British India*), *Oligochaeta*, pp. 148, 132, 140, 155.

1933. *Drawida ancisa* Gates, *Rec. Indian Mus.* **35**: p. 421. (Type locality, Sandoway.)

1934. *Drawida papillifer + D. affinis + D. hodgarti*, Gates, *ibid.* **36**: pp. 245, 234, 237.

1962. *Drawida papillifer papillifer*, Gates, *Bull. Mus. Comp. Zool. Harvard College* **127**: pp. 340, 361.

Male pores, in x, at or just lateral to *B* or (rarely) nearer to m*BC*, each in a whitened semicircular area with base at 10/11 or in a shorter portion of the same area but demarcated anteriorly by a furrow. Segments, 115–164. Size, 70–121 by 3–5 mm.

Gizzards, 2–4, in xiii–xx. Intestinal origin, in xxiii (±1?). Sperm ducts, short, 5–10 mm. long. Prostates, 2–3 mm. long, glandular investment often lacking near parietes.

Distribution: Sandoway (Sandoway), Ramree (Kyaukpyu), Akyab, Kyauktaw, Buthidaung, Maungdaw (Akyab), Paletwa (Arakan Hill Tracts). Rangamati (Chittagong Hill Tracts). Cherrapunji (Khasi Hills, Assam). From sea level to 4,300 feet.

Habitats: Soil. Under stones. Mud at bottom of pools.

Abnormality: (No. 1) Spermathecal ampulla in ix but ectal portion of duct associated normally with 7/8. (No. 2) Left spermathecal pore, at 8/9, the spermatheca in ix. Male organs of left side, including male pore, lacking.

Drawida papillifer peguana

1925. *Drawida peguana* Gates, *Ann. Mag. Nat. Hist.*, ser. 9, **15**: p. 316. (Type locality, Rangoon. Paratypes, in the British Mus. and the U. S. Natl. Mus.)

1926. *Drawida peguana*, Gates, *Rec. Indian Mus.* **28**: p. 146; *Jour. Burma Res. Soc.* **15**: p. 205; *Jour. Bombay Nat. Hist. Soc.* **31**: p. 182.

1930. *Drawida peguana*, Gates, *Rec. Indian Mus.* **32**: p. 291.

1931. *Drawida peguana*, Gates, *ibid.* **33**: pp. 348, 351.

1933. *Drawida peguana*, Gates, *ibid.* **35**: p. 466.

1961. *Drawida papillifer*, Gates, *American Midland Nat.* **66**: p. 62.

1962. *Drawida papillifer peguana*, Gates, *Bull. Mus. Comp. Zool. Harvard College* **127**: pp. 342, 361.

?

1890. *Moniligaster* sp., Rosa, *Ann. Mus. Civ. Sto. Nat. Genova* **29**: p. 380.

Male pores, at 10/11, longitudinal or diagonal slits, in median half of *BC*. Segements, 168. Size, 50–130 by 3–6 mm.

Gizzards, 2–5, in xii–xix. Intestinal origin, in xxii (±1?). Sperm ducts, very short, almost straight in

x. Prostates, 3–5 mm. long, glandular investment lacking near parietes.

Distribution: Mergui, Zowai (Mergui). Migyaunglaung, Pyinthadaw, Kawlet Chaung, Siyigyan (Tavoy). Ye, Moulmein, Martaban, Kyaikmaraw, Kya In, Kawkareik (Amherst). Thaton, Duyinzeik, Naunggala, Bilin, Aungsein, Sittang, Boyagyi, Kyaikto, Kyaiktiyo, Taungzun, Kinmunsakhan (Thaton). Dedaye, Thanchitaw (Pyapon). Rangoon, Syriam, Kyauktan, Twante, Thongwa, Kungyangone (Hanthawaddy), Maubin, Danubyu (Maubin). Pegu, Pegu Yomas (Pegu). Insein, Damsite, Wanetchaung, Taikkyi, (Insein). Kokya (Bassein). Pegu Yomas, Myagyaung (Tharrawaddy). Henzada, Kyangin, Ingabu (Henzada). Toungoo, Leiktho, Sah Der, Kyaukkyi, Pegu Yomas (Toungoo). Prome, Laboo, Paukkaung, Arakan Yomas along road through Taungup Pass (Prome). Arakan Yomas along road through Taungup Pass, Ngapoli (Sandoway). Thayetmyo, Thanbula (Thayetmyo). Ywadaw (Yamethin). Magwe (Magwe). Meitkila, Mahlaing (Meiktila). Kyaukpadaung (Myingyan). From sea level to elevations of *ca.* 4,000 feet.

Habitats: Soil, in jungles, gardens, lawns where the species was found more often near bases of trees.

Regeneration: Tail regenerates were found. One, at 65/66, comprised 40+ segments.

Abnormality: (No. 3) Diverticulum of one spermatheca, lacking.

Remarks: Rosa's specimens from Palon (1890), with "colore carneo livido" is more likely to have belonged here than to *caerulea*. The internal anatomy was not studied because of poor preservation.

Addenda

I

1917. *Drawida rangamatiana* Stephenson, *Rec. Indian Mus.* **13**: p. 369. (Type locality, Rangamati, Chittagong Hill Tracts, Bengal. Type, in the Indian Mus.)

1923. *Drawida rangamatiana*, Stephenson (*The Fauna of British India*), *Oligochaeta*, p. 153.

1934. *Drawida rangamatiana*, Gates, *Rec. Indian Mus.* **36**: p. 252.

Male pores, just beind 10/11 (?), in median half of *BC*. Spermathecal pores, at posterior margin of vii (?). Segments, 237. Size, 137 by 7.5 mm.

Gizzards, 4, in xvi–xix.

Remarks: This dubious species was erected on an aclitellate worm that was not sexual. Examination of the type (1934) enabled some corrections of the original characterization but showed (perhaps only because of immaturity) no differences from *papillifer* except as indicated above. An area bearing each male pore is slightly more elevated than in *papillifer* but much less so than protruded porophores of species such as *nepalensis* or *longatria*. Although the male pores certainly seemed to be in xi, confirmation is needed. Location of spermathecal pores immediately in front of 7/8 may be more apparent than real. Nothing is known as to pigmentation, intestinal origin, nephridia of x. GM glands were not recognizable from the coelom and were assumed to be absent, further dissection of a unique and already damaged specimen being deemed inadvisable. Removal of the longitudinal musculature might have revealed presence of glands like those of *papillifer*.

II

1925. *Drawida papillifer* (part), Stephenson, *Rec. Indian Mus.* **27**: p. 51.
1934. *Drawida pomella* Gates, *ibid.* **36**: p. 250. (Type locality Amingaon, Assam. Types, in the Indian Mus.)

Male pores, slightly behind 10/11 and lateral to *B*. Gizzards, 3–4, in xiii–xviii. GM glands, small but recognizable from the coelom.

Remarks: This equally dubious species was erected on 21 specimens originally (1925) referred to *papillifer*. Ovisacs are rudimentary and spermathecal ampullae small. Location of male pores in xi, previously unrecorded in the genus, was thought (1934) to warrant distinction from *papillifer*.

Drawida rangoonensis

1925. *Drawida rangoonensis* Gates, *Ann. Mag. Nat. Hist.*, Ser. 9, **15**: p. 230. (Type locality, Rangoon. Types, none.)
1926. *Drawida rangoonensis*, Gates, *Jour. Bombay Nat. Hist. Soc.* **31**: p. 182. *Jour. Burma Res. Soc.* **15**: p. 206. *Rec. Indian Mus.* **28**: p. 146 (excluding worms with male porophores).
1930. *Drawida rangoonensis* (part), Gates, *ibid.* **32**: p. 291. (Excluding worms with male porophores.)
1931. *Drawida rangoonensis*, Gates, *ibid.* **33**: p. 349.
1933. *Drawida rangoonensis*, Gates, *ibid.* **35**: p. 468.
1939. *Drawida rangoonensis*, Chapman, *ibid.* **41**: p. 119.
1961. *Drawida rangoonensis*, Gates, *American Midland Nat.* **66**: p. 62.
1962. *Drawida rangoonensis*, Gates, *Bull. Mus. Comp. Zool. Harvard College* **127**: pp. 346, 361.

Male pores, very small transverse slits in median half of *BC*, rims usually tumescent in a more or less completely annular lip, each located in a whitened area between 10/11 and eq/x, in x. Spermathecal pores, very small transverse slits, at or just median to *C*. Genital markings, an area of translucence in median half of *BC* in setal annulus of x, whitened areas of epidermal thickening, paired in *BC* of viii–xi, median and reaching *A*, *B*, or m*BC*, presetal in x, postsetal in viii–ix, xii–xiii. Setae, *AA* < *BC*, *DD* ca. = ½*C*, slightly enlarged in preclitellar segments. Nephropores, at or close to *D*, more dorsal in viii–ix, lacking in x of adults. Clitellum, red, ix/n–xiv/n. Color, white (unpigmented). Segments, 160–167. Size, 40–92 by 2½–4 mm.

Gizzards, 2–4, in xii–xvii. Intestinal origin, in xxiv (±1). Connectives from extra-esophageals, on anterior face of 8/9. Nephridia of x, lacking in adults.

Sperm ducts, long, postseptal portion of each in a cluster of loops that may be as large as testis sac, passing into ental end of prostate directly. Prostates, erect, 1–1½ mm. long, capsule almost digitiform but slightly and gradually widened entally. Spermathecal diverticulum, saccular, 3–5 mm. long, in vii. Ovarian chamber, closed off from parietes. (GM glands, none.)

Distribution: Moulmein (Amherst), Sittang (Thaton), Pyapon, Kyaiklat (Pyapon), Rangoon, Twante, Kyauktan, (Hanthawaddy), Maubin, Yandoon (Maubin), Bassein (Bassein), Pegu (Pegu), Sandoway, Ngapoli, Arakan Yomas near road through Taungup Pass (Sandoway), Myohaung (Mandalay), Bhamo (Bhamo), Myitkyina (Myitkyina). Transported, by man, from Lower Burma to Bhamo and Myitkyina?

Life history: The breeding season probably is from late in August through September (and into October?). Most individuals of this species become inactive in some unknown state and at unknown depths as the rainy season ends but an occasional specimen does sometimes remain active, presumably in some cool and moisture-blessed niche, into December or even January. The species would now seem to be less common than *rara*.

Abnormality: (No. 1) Left spermathecal diverticulum in viii.

Remarks: The annular lip around the male pore, possibly formed by everting ectalmost portion of the prostatic duct, seems to be all that this species has for a male porophore.

Relationships presumably are with the *gracilis* group from all of which *D. rangoonensis* now seems to differ mainly in the vesiculation of an ental portion of the spermathecal diverticulum.

Drawida rara

1925. *Drawida rara* Gates, *Ann. Mag. Nat. Hist.*, Ser. 9, **15**: p. 321. (Type locality, Rangoon. Topotype in U. S. Natl. Mus.)
1926. *Drawida rara*, Gates, *Rec. Indian Mus.* **28**: p. 147. *Jour. Burma Res. Soc.* **15**: p. 206. *Jour. Bombay Nat. Hist. Soc.* **31**: p. 182.
1930. *Drawida rara*, Gates, *Rec. Indian Mus.* **32**: p. 291.
1931. *Drawida rara*, Gates, *ibid.* **33**: p. 351.
1933. *Drawida rara*, Gates, *ibid.* **35**: p. 469.
1936. *Drawida rara*, Hla Kyaw & Gates, *Jour. Roy. Asiatic Soc. Bengal*, (*Sci.*) **2**: p. 166.
1961. *Drawida rara*, Gates, *American Midland Nat.* **66**: p. 62.
1962. *Drawida rara*, Gates, *Bull. Mus. Comp. Zool. Harvard College* **127**: pp. 348, 361.

Male pores, usually on slight teatlike protuberances from circular areas across 10/11 in median half of *BC*. Spermathecal pores, just median to *C*. Genital markings, usually whitened areas of epidermal thickening, in *BC* and paired, postsetal in vii, presetal and/or postsetal in viii and x, in setal annuli of x–xi (often translucent and depressed), unpaired, presetal and/or postsetal in *BB* of viii, x–xi. Setae, *AA* < *BC*,

$DD = \frac{1}{2}C$. Nephropores, at or close to D, lacking in x of adults. Clitellum, red, ix/n–xiii/n. Color, white (unpigmented). Segments, 154. Size, 35–80 by 3–4 mm.

Gizzards, 3–5, in xii–xviii. Intestinal origin, in xxiii ($\pm$1). Connectives from extra-esophageals, on anterior face of 8/9. Nephridia of x, lacking in adults. Sperm ducts, rather short, slender, each passing into a prostate directly. Prostates, sessile, of circular outline, discoidal to dome-shaped, capsule small, spindle-shaped and in the parietes. Spermathecae, adiverticulate, duct slightly thickened ectally in 7/8 and the parietes. Ovarian chamber, closed off from parietes. (GM glands, none.)

Distribution: Kamaungthwe River (Tavoy), Ye, Mupun, Moulmein, Chaungson, Kya In (Amherst), Thanchitaw, Dedaye (Pyapon), Bassein, Kokya, Kochi, Coomzamu (Bassein), Thaton, Duyinzeik, Naunggala, Bilin, Aungsaing, Sittang (Thaton), Syriam, Kungyangone, Kyauktan, Rangoon (Hanthawaddy), Insein, Dam site, Taikkyi, Hlawga, Taukkyan, Hmawbi, Wanetchaung (Insein), Danubyu (Maubin), Pegu, Thanatpin (Pegu), Tharrawaddy, Letpadan, Thonze, Myagyaung (Tharrawaddy), Henzada (Henzada), Prome, Arakan Yomas along road through Taungup Pass (Prome), Thayetmyo, Allanmyo (Thayetmyo), Pyinmana, Pyigyaung (Yamethin), Magwe, Taungdwingyi (Magwe), Minbu (Minbu), Myohaung (Mandalay), Falam (Chin Hills), Wuntho (Katha), Myitkyina (Myitkyina).

Habitats: Soil. In lawns, Rangoon.

Biology: The breeding season probably is from late in August till into October. Hibernestivation, in some unknown state and at unknown depths, usually begins as the rainy season ends but some individuals remain active into November and December presumably in specially favorable niches.

Regeneration: Tail regenerates were found at various levels behind 71/72.

Abnormality: (No. 1) Ectal half of one spermathecal duct in vii where a portion is vesiculate. (No. 2) Male pore of one side midway between eq/x and 10/11, porophore entirely in x. (No. 3) Male organs, including male pore as well as genital markings, of one side lacking.

Remarks: A seeming intrasegmental location of some spermathecal pores may result from a slight and asymmetrical eversion of the cetalmost portion of the spermathecal duct. When such eversion is symmetrical an annular tumescence is recognizable.

The male pores develop at 10/11. Copulation may be facilitated by drawing the prostatic capsule deeper into the parietes so as to raise a rather teatlike external protuberance that then can be inserted into the everted portion of the spermathecal duct. Retraction of the capsule away from the epidermis and nearer to the coelomic cavity then presumably would result in disappearance of any protrusion from the male area.

Relationships, as indicated by the spermathecae, presumably are with the *beddardi* group and therein closest with *D. constricta* and *spissata*.

Drawida sp.

1939. *Drawida* sp., Gates, *Jour. Thailand Res. Soc. Nat. Hist. Nat. Hist. Sup.* 12: p. 75.

(Male pores?) Spermathecal pores, at 7/8 and at C. (Genital markings, none.) Setae, $AA < BC$. Color, white (unpigmented). (Segments?) Size, 33 by 1 mm.

Gizzards, 2, in xiv–xv. Sperm ducts, short, each passing directly into anterior face of a prostate (near parietes?). Prostates, club-shaped, erect, capsule, club-shaped and narrowed ectally. Spermathecae, adiverticulate, duct slightly widened ectally and rather conical, the thickening slightly protuberant into coelom just behind 7/8.

Distribution: Muang Pong (Ban Muang, Thailand).

Remarks: The single specimen was macerated and fragmented during examination. Ovisacs (into xv) and the glandular investment on the prostatic capsule indicated sexual maturity though a clitellum was unrecognizable. However, testis sacs were small, firm, and possibly juvenile.

A supposedly filamentous penis is now believed to have perhaps been only the cuticular lining of the prostatic duct thrown out by strong contraction at preservation. The rather conical male porophore may then represent protrusion of a disc or eversion of a small invagination containing the primary male pore but without a definite penis.

The adiverticulate spermathecae and the absence of genital markings distinguished the worm from *D. vulgaris* the only other drawida at present known from Thailand.

Drawida spissata

1930. *Drawida spissata* Gates, *Rec. Indian Mus.* 32: p. 291. (Type locality, on road over Arakan Yomas through Taungup Pass. Types, none.)
1933. *Drawida spissata*, Gates, *ibid.* 35: p. 471.
1962. *Drawida spissata*, Gates, *Bull. Mus. Comp. Zool. Harvard College* 127: p. 361.

Male pores, in median half of BC and on conical, anteriorly directed protuberances of xi. Spermathecal pores, at 7/8 and at C. Genital markings, median in viii, paired and in median portion of BC posteriorly in x. Setae, $AA = $ or slightly $< BC$. Nephropores, at or close to D (in x?). (Clitellum?) Color, none (unpigmented). Segments, 159. Size, 90–130 by 5–7 mm.

Gizzards, 4–5, in xvii–xxi. (Intestinal origin?) Connectives from extra-esophageals, on anterior face of 8/9. (Nephridia of x?) Sperm ducts, very long, each with a thickened portion in a cluster of long loops that may be three times as large as testis sac, passing into ental end of prostate directly. Prostates, 15–25 mm. long and 1 mm. thick, glandular investment lacking near parietes, capsule digitiform. Spermathecae, adiverticulate, ducts markedly muscular, 45–70 mm. long, an ectal portion 35–50 mm. long most thickened. Ovarian chamber, closed off from parietes. (GM glands, none.)

Distribution: Tanyagyi, Patle, Tsalu, on road over Arakan Yomas through Taungup Pass (Sandoway). At elevations of 1,600–2,500 feet.

Remarks: Known only from the original 12 specimens. All probably were aclitellate. Small size of testis sacs and of spermathecal ampullae, as well as the thinness of the glandular investment on the prostates, warrant a suspicion that the types were not fully mature. Indications of postsexual regression were not recognized.

An ectal portion of the spermathecal duct appears to be eversible. On complete retraction of male protuberances the primary male pores may be invaginate in which case secondary male apertures may be fairly large.

Relationships were thought (Gates, 1962), at least as indicated by the spermathecae, to be with *D.constricta* and *rara*.

Drawida tenellula

1933. *Drawida* sp., Gates, *Rec. Indian Mus.* **35**: p. 476.
1962. *Drawida tenellula* Gates, *Bull. Mus. Comp. Zool. Harvard College* **127**: p. 351. (Type locality, Tharrawaddy. Types, none.)

Male pores, minute, transverse slits at ventral ends of short antero-posteriorly flattened but tubular and penislike porophores at 10/11 in median half of *BC*. Sphermathecal pores, small transverse slits, slightly median to *C*. (Genital markings, lacking but margins of x–xi near penes slightly tumescent.) Setae, $AA < BC$. (Nephropores? Clitellum? Segments?) Color, white (unpigmented). Size, 25–26 by 1–2 mm.

Gizzards, 3–4, in xiii–xvi. (Intestinal origin?) Connectives from extra-esophageals, on anterior face of 8/9. Sperm ducts, long, each thickened in x and there looped in a cluster about as large as testis sac, passing to a prostate near ental end directly. Prostates, erect, club-shaped (capsule digitiform or slenderly club-shaped?). Spermathecae, with fairly thick (muscular?) duct and shortly digitiform, thin-walled diverticulum erect in vii. Ovarian chamber, closed off from parietes.

Distribution: Tharrawaddy (Tharrawaddy), Thayetmyo (Thayetmyo).

Remarks: Known only from the description of two aclitellate specimens.

Relationships presumably are with a *gracilis* group of species, characterized by absence of GM glands and a digitiform shape of the spermathecal diverticulum.

Drawida tumida

1929. *Drawida tumida* Gates, *Proc. U. S. Natl. Mus.* **75**, 10: p. 12. (Type locality, Moulmein. Paratype, in the U. S. Natl. Mus.)
1930. *Drawida tumida*, Gates, *Rec. Indian Mus.* **32**: p. 294.
1931. *Drawida tumida*, Gates, *ibid.* **33**: p. 355.
1933. *Drawida tumida*, Gates, *ibid.* **35**: p. 472.
1962. *Drawida tumida*, Gates, *Bull. Mus. Comp. Zool. Harvard College* **127**: p. 360.

Male pores, very small slits on or in line with 10/11, in median part of *BC*, each at ventral end of a rather conical protuberance bearing a small genital marking and containing a small, flask-shaped gland. Spermathecal pores, at or just median to *C*. Genital markings, with indistinct or unrecognizable rim (as on male porophores) in vicinity of spermathecal pores, elsewhere with wide opaque rim, one large and longitudinally elliptical usually present in front of each male porophore, sometimes also median or paired in *AA* of ix–xiii and there smaller, circular. Setae, $AA < BC$, $DD > \frac{1}{2}C$. Nephropores, at or close to *D*, more dorsal on viii (on x?). Clitellum, red, ix/n–xiv/n. Color, white (unpigmented). Segments, 197. Size, 80–153 by 4–6 mm.

Gizzards, 3–5, in xvi–xxii. (Intestinal origin?) Connectives from extra-esophageals, on anterior face of 8–9. (Nephridia of x?) Sperm ducts, rather long, each with a thickened postseptal portion in a compact cluster of loops that is smaller than the testis sac, passing into prostate below ental end and directly. Prostates, erect or bent over mesially, capsule slightly widened entally. Spermathecal diverticula, digitiform, 2–3 mm. long, thin-walled, erect, bent at right angles, or pressed down against parietes, in vii. Ovarian chamber, closed off from parietes. GM glands, with tough wall, elongately ellipsoidal (with longitudinal markings of x) or spheroidal, often protuberant into coelom through the musculature.

Distribution: Tavoy, Myittha and other localities north to the Ye River (Tavoy), Ye, Moulmein Chaungson, Kyaikmaraw, Kya In (Amherst).

Remarks: The longitudinal genital markings usually are on markedly protuberant ridges that bear the male porophores. The marking was lacking in a few individuals at each locality from which the species was secured.

Relationships clearly are with *D. longatria* of which *D. tumida* might have been considered to be a subspecies were it not for presence in the *tumida* range of two or more *longatria* races. Shortness of the spermathecal atria always distinguished *D. tumida* from those races.

Drawida victoriana

1962. *Drawida victoriana* Gates, *Bull. Mus. Comp. Zool. Harvard College* **127**: p. 352. (Type locality, Mount Victoria, east side near path from Kanpetlet, at 3,000 feet. Types, none.)

Male pores, very small transverse slits, in x and slightly lateral to *B*, at ends of rather conical elevations protuberant anteriorly from posterior margin of x. Spermathecal pores, at or just median to C. Genital markings, small, circular, distinctly demarcated, with a translucent area centrally, unpaired, median, and postsetal on vii–xii, paired, about in line with male porophores, in presetal as well as setal annuli of x and postsetal annulus of xi. Setae, posteriorly $AA < BC$, $(DD = ?C)$. Nephropores, at or close to D (in x of adults?). (Clitellum? Unpigmented? Segments?) Size, 64 by 3 mm.

Gizzards, 4, in xvi–xix. (Intestinal origin?) Connectives from extra-esophageals, on anterior face of 8–9. (Nephridia of x?) Sperm ducts, long, a postseptal portion of each thickened and in a cluster of loops as large as testis sac, passing into prostate just below ental end and directly. Prostates, slenderly club-shaped, $2\frac{1}{2}$ mm. long. Spermathecal diverticula, saccular large, ectal half stalklike, erect in vii. Ovarian chamber, closed off from parietes. (GM glands, if present, underneath longitudinal musculature.)

Distribution: Mount Victoria (Pakokku Chin Hills).

Habitat: Soil, in teak forest and grassland.

Regeneration: The type had a 3 mm. long tail regenerate.

Remarks: To permit entry in the key, presence of solid glands is assumed (cf. notes under the key).

The glandular investment of the prostatic capsule was like that of juveniles in other species but associated with presence of ova in the ovisacs.

Sperm were present in an ectal portion of the diverticular stalk!

The species is known only from the original account of a single specimen secured a year or more before Pearl Harbor.

Drawida vulgaris

1930. *Drawida vulgaris* Gates, *Rec. Indian Mus.* **32**: p. 296. (Type locality, Kalewa. Types, none.)
1931. *Drawida vulgaris*, Stephenson, *ibid.* **33**: p. 179.
1933. *Drawida vulgaris*, Gates, *ibid.* **35**: p. 475.
1939. *Drawida vulgaris*(?), Gates, *Jour. Thailand Res. Soc. Nat. Hist. Sup.* **12**: p. 73.
1962. *Drawida vulgaris*, Gates, *Bull. Mus. Comp. Zool. Harvard College* **127**: pp. 317, 361.

Male pores, small transverse slits, at or slightly median to mBC and apparently in x, slightly in front of 10/11. Spermathecal pores, in CD. Genital markings, small, transverse areas of epidermal thickening, each with a greyish translucent center (sometimes doubled), median or closely paired in AA, paired in BC, usually presetal or in setal annulus, in vii–xiv. Setae, AA usually $< BC$, DD ca. $= \frac{1}{2}C$. Nephropores, at or close to D, more dorsal in viii, lacking in x. Clitellum, red, ix/n–xiv/n. Color, white (unpigmented). Segments, 134. Size, 30–50 by $2\frac{1}{2}$–3 mm.

Gizzards, 2–4, in xii–xvi. Intestinal origin, in xxi ($\pm$ 1?). Connectives from extra-esophageals, on

anterior face of 8/9. Nephridia of x, lacking in adults. Sperm ducts, short, each ca. 8 mm. long, an ectal portion (ca. 5 mm. long) thickened in x and then passing into ental end of prostate directly. Prostates, erect or bent over mesially, capsule club-shaped but only very slightly thickened entally, ca. 1 mm. long. Spermathecal diverticula, erect in vii, ca. 1 mm. long, digitiform or slightly flattened, thick-walled. Ovarian chamber, closed off from parietes. (GM glands, none.)

Distribution: Prome, Paukkaung (Prome), Mount Popa (Myingyan), Masein, Kalewa (Upper Chindwin). Chiengmai, Thailand, at 1,000 feet.

Remarks: GM glands are assumed to be absent although no statement to that effect was published.

Male pores usually were on more or less obvious, posteroventral and flap-like protuberances apparently from x.

Relationships, at least in so far as indicated by the anatomy, are with the *gracilis* group of species.

OCNERODRILIDAE

1891. Ocnerodrilidae Beddard, *Trans. Roy. Soc. Edinburgh* **36**: p. 581.
1892. Ocnerodrilidae, Beddard, *Ann. Mag. Nat. Hist.*, ser. 6, **10**: p. 97.
1893. Ocnerodrilidae + Gordiodrilidae, Eisen, *Proc. California Acad. Sci.*, ser. 2, **3**: p. 279.
1894. Ocnerodrilacea (Microscolecinae, Megascolecidae), Michaelsen, *Mitt. Zool. Mus. Hamburg* **38**: p. 56.
1895. *Gordiodrilus + Nannodrilus + Ocnerodrilus* (Cryptodrilidae) + *Kerria* (Acanthodrilidae) + *Ilyogenia* (Geoscolecidae), Beddard, *A Monogr. of the Order of Oligochaeta* (Oxford), pp. 506, 510, 515, 553, 649.
1896. Ocnerodrilini + *Kerria* (Acanthodrilini), Michaelsen, *Deutsch-Ost-Afrika, Berlin* **4**, 12: p. 45.
1897. Ocnerodrilini (Megascolecidae), Michaelsen, *Verhandl. Naturwiss. Ver. Hamburg*, ser. 3, **4**: p. 25.
1900. Ocnerodrilinae, Eisen, *Proc. California Acad. Sci.*, ser. 3, **2**: p. 108. Ocnerodrilinae (Megascolecidae) + *Maheina* (Acanthodrilinae), Michaelsen, *Das Tierreich*, **10**: pp. 368, 143.
1903. Ocnerodrilinae (Megascolecidae) + *Maheina* (Acanthodrilinae), Michaelsen, *Die geographische Verbreitung der Oligochaeten* (Berlin), pp. 118, 74.
1913. Ocnerodrilinae (Megascolecidae) + *Maheina* (Acanthodrilinae), Michaelsen, in Sarasin, F. & Roux, J., *Nova Caledonia* (Zool.) **1**, 3: p. 188.
1920. Ocnerodrilinae (Acanthodrilidae), Michaelsen, *Mitt. Zool. Mus. Hamburg* **38**: p. 56.
1923. Ocnerodrilinae (Megascolecidae), Stephenson, (*The Fauna of British India*), *Oligochaeta*, p. 479.
1928. Ocnerodrilinae (Acanthodrilidae), Michaelsen, in Kuekenthal & Krumbach, *Handbuch der Zool.* **2**, 28: p. 109.
1930. Ocnerodrilinae (Megascolecidae), Stephenson, *The Oligochaeta* (Oxford), p. 852.
1937. Ocnerodrilinae (Megascolecidae), Pickford, *A Monogr. of the Acanthodriline Earthworms of South Africa* (Cambridge, England), p. 85.
1939. Ocnerodrilidae, Gates, *Jour. Thailand Res. Soc. Nat. Hist. Suppl.* **32**: pp. 73, 109.
1942. Ocnerodrilidae, Gates, *Bull. Mus. Comp. Zool. Harvard College* **89**: p. 65.
1957. Ocnerodrilidae, Gates, *ibid.* **117**: p. 443.

1959. Ocnerodrilinae (Megascolecidae), Avel, *Traité de Zool.* (Paris) **5**: p. 457. Ocnerodrilidae, Gates, *Bull. Mus. Comp. Zool. Harvard College* **121**: p. 254. Acanthodrilinae (part) + *Quechua* (Megascolecinae, Megascolecidae) + *Gordiodrilus*, Lee, *New Zealand Dept. Sci. Indust. Res. Bull.* **103**: p. 33.

1966. Ocnerodrilidae, Gates, *Ann. Mag. Nat. Hist.*, Ser. 13, **9**: p. 51.

Digestive system, with a short esophagus, without intestinal caeca. Vascular system, with complete dorsal (single) and ventral trunks, but no subneural, a supra-esophageal in region of viii–xi, paired extra-esophageals (median to segmental commissures?). Hearts, of ix lateral, of x–xi latero-esophageal, the last pair in xi (or homoeotic equivalent?). Nephridia, holoic, avesiculate, (with cells of peritoneal investment much enlarged in a posterior part of the body?). Setae, eight per segment. (Nephropores, in one longitudinal rank on each side of the body?)

Spermathecal, female (always paired?) and male pores, in that antero-posterior order. Spermathecae, in front of testis segments. Clitellum, multilayered, including female (and male?) pore segments. Ovaries, in xiii, fanshaped and with several egg-strings. (Ovisacs?) Ova, not yolky.[1] Seminal vesicles, trabeculate. Prostates, tubular and of ectodermal origin.

Distribution: Tropical America, tropical Africa, possibly also in adjacent subtropical regions, India, and perhaps Burma.

Remarks: Absence of pigment, dorsal pores, intestinal typhlosoles, and supra-intestinal glands, the small somatic sizes, a pre-intestinal region of only 11 segments, formerly thought (Gates, 1942: p. 65) to characterize the entire family are now known to be liable to sporadic exceptions at species level. *Pygmaeodrilus* remains the only genus with extramural spermathecal diverticula.

African species are at or close to the low limit for megadrile size and so would seem to be much more liable to transportation than the larger American endemics. Michaelsen's criterion for distinguishing endemic from peregrine species also seems to favor an American origin for the family rather than the African one assumed by classical authorities.

KEY TO ORIENTAL GENERA OF OCNERODRILIDAE

1. Exteramural calciferous glands lacking 2
   Extramural calciferous glands present 4
2. One gizzard, in vii 3
   Two gizzards, in vi–vii . . . . *Deccania*
3. Quadrithecal, pores at 7/8–8/9 . . . . *Thatonia*
   Bithecal, pores at 8/9 . . . . *Malabaria*
4. Calciferous glands, unpaired 5
   Calciferous glands, paired 6
5. One gizzard in vii, calciferous glands
   in ix–x . . . . *Curgiona*

---

[1] But unusually large relative to soma size?

No gizzard, calciferous gland in ix . . . . *Gordiodrilus*
6. Calciferous glands with thin wall on
   which are longitudinal ridges that
   do not meet centrally . . . . *Ocnerodrilus*
   Calciferous glands with thick wall,
   not so ridged 7
7. One gizzard, in vii . . . . *Eukerria*
   Two gizzards, in vii–viii . . . . *Nematogenia*

MALABARIINAE

1966. Malabariinae Gates, *Ann. Mag. Nat. Hist.*, Ser. 13, **9**: p. 53. (Type genus, *Malabaria* Stephenson, 1924.)

Digestive system, without extramural calciferous glands and internal calciferous lamellae. Vascular system, with extra-esophageal trunks passing in region of viii–ix to esophagus ventrally on which they are continued into region of xi–xii (where they are joined by posterior lateroparietal trunks?). Spermathecae, adiverticulate. Dorsal pores and pigment, lacking.

Distribution: India and perhaps Burma. One species has been transported, (from Burma?) to Hainan.

All Burmese and Hainan material has been of parthenogenetic morphs not found in India. That statement is, however, of little significance because of our ignorance of so much of the Indian megadrile fauna. Nevertheless, until Burmese morphs can be linked with biparental populations in the Indian peninsula, Burmese forms must be regarded as possibly endemic.

Systematics: The large oceanic gap in the distribution and anatomical differences in a systematically very important part of the esophagus (*cf.* Gates, 1966), possibly associated with significant divergences of the vascular system, were believed to need expression in a classification.

Parthenogenesis now seems to be rampant in the subfamily. Some caution, accordingly, is advisable as to erection of species (*cf.* Gates, 1949, *Proc. Indian Acad. Soc.* **33**: p. 279.)

By a *lapsus calami* or by failure to correct a typographical error, *Malabaria*, then known only from the type species, was said in Stephenson's definition (1930: p. 857) to have paired prostates in xvii and xix. A reader, unfamiliar with the original description of the species but aware of the classical emphasis on the evolution of male terminalia, probably would conclude that the genus was being defined by acanthodrilin male terminalia. Seminal furrows are known, in *Malabaria*, only from *M. sulcata* where they are restricted to male porophores. Moreover, the supposed posterior prostates of *paludicola* are prostatelike GM glands and are in xx. Seminal grooves of the classical acanthodrilin sort have been found, in the Malabariinae, only in *Thatonia* where they are disappearing in various parthenogenetic morphs.

The key below suggests that two genera are distinguished, quite in the manner of the classical system, by characters that often vary intragenerically. This is nearly true. Both genera are retained, however, because there is reason for believing that, in the esophagus, microscopic structure is significantly different if not also in the third genus.

Phylogeny: An esoteric classical phylogeny required derivation of *Malabaria*, then known only from the type species, through ocnerodriles with extramural calciferous glands. Four spaces in the thickened ventral wall of the esophagus in ix and x of *M. paludicola*, accordingly, almost had to be interpreted as vestiges of calciferous caeca that had been withdrawn into the gut wall. Nothing is known about development, either ontogenetically or phylogenetically, of esophageal spaces in ix and x, in any of the three malabariine genera. Nor has any informed suggestion been advanced as to function of the spaces. Secretion of calcium may not be involved.

Modification, for metabolite storage, of peritoneal cells in the nephridial investment, thickening of the esophageal wall in x–xi to contain systems of (glandular?) spaces, are the only macroscopically recognizable, structural specializations in malabariine species. Possibly also secondary is presence of the gizzard in vii. Subtracting those specializations leaves a simple anatomy characterized as follows.

Setae, closely paired, in the ventrum. Dorsal pores and pigment, lacking. Digestive system, with a gizzard in v, an intestinal origin in xii, but without calciferous and supra-intestinal glands, caeca, and typhlosoles. Vascular system, with dorsal (single), ventral, supra-esophageal, and extra-esophageal (median to segmental commissures) trunks, but without a subneural vessel, with lateral and latero-esophageal hearts the last pair of which is in xi. Excretory system, holoic. Reproductive system, with all genital apertures minute, superficial, and in region of *AB*, four pairs of seminal vesicles in ix-xii, paired tubular prostates, two pairs of adiverticulate spermathecae opening at 7/8–8/9.

Worms so characterized, with two additions and one specification, would have been able to give rise to all ocnerodrilids. Additions are, testes in ix, ovaries in xii. The specification is to locate the prostates, in ix–xxii. A hypothetical ancestral form, so characterized, is called a protothatonia.

Evolutionary changes already recorded for the few known species of the malabariine subfamily are as follows. Negative, elimination of prostates except in xvii or xix or xvii and xix, of testes and male funnels in ix, of ovaries and female funnels in xii (rather general), elimination of fewer prostates, of one or the other or both pairs of spermathecae, of testes in x or xi or x-xi (sporadic). Positive, displacement of the single gizzard (first into vi and subsequently?) into vii, and also sporadic, development of a second gizzard (in vi or vii?), of an intestinal

typhlosole, of clear glands in genital markings, appearance of dorsal pores, appearance of muscular prostates (1 species of *Malabaria*), migration of spermathecal pores laterally.

Remarks: Species are most likely to be found in wet soils rich in organic matter. One may be common, in appropriate habitats but the collector who was able to secure a thousand or so specimens of *sulcata* never obtained more than a very few of other malabariine species. The first two species were secured because damage to rice plants was being studied. The family may be ancient and able to persist only in rare and scattered niches.

KEY TO GENERA OF THE MALABARIINAE

1. Esophagus, with one gizzard, in vii............ 2
   Esophagus, with two gizzards, in vi–vii[a]..*Deccania*
2. Quadrithecal and with prostatic pores
   in xvii, xix.........................*Thatonia*
   Bithecal and with prostatic pores in xvii.*Malabaria*

[a] *Deccania* Gates, 1949, known only from the original description of *D. alba* (*Proc. Indian Acad. Sci.* **30**) found near Jubbulpore. The species is of especial interest because male funnels of x are retained in metandric worms and also because of the GM glands, exactly like prostates, located intrasegmentally as well as across intersegmental furrows, in front of, as well as, within and behind the clitellum.

Malabaria

1924. *Malabaria* + *Aphanascus* Stephenson, *Rec. Indian Mus.* **26**: p. 356, 360. (Type species, *M. paludicola* and *A. oryzivorus* Stephenson 1924.)

1930. *Malabaria* + *Aphanascus*, Stephenson, *The Oligochaeta* (Oxford), pp. 857, 858.

1938. *Filodrilus* Chen, *Cont. Biol. Lab. Sci. Soc. China*, (*Zool.*) **12**: p. 422. (Type species, *F. levis* Chen 1938.)

1942. *Malabaria*, Gates, *Bull. Mus. Comp. Zool. Harvard College* **89**: p. 90.

Digestive system, with a gizzard in vii, intestinal origin in xii, but without typhlosoles and supra-intestinal glands. Nephridia, (present from?), large, avesiculate(?), slender ducts passing into parietes in region of *B* (throughout?), posteriorly with most cells of the peritoneal investment enlarged. Nephropores, (at?), inconspicuous. Setae, closely paired. Dorsal pores and pigment, lacking. Septa all present from 5–6.

Distribution: Peninsular India only, from the northernmost edge of the Deccan South almost to Cape Comorin. Burma (?).

The range has been extended recently by transport of a (Burmese?) species to the island of Hainan, China.

As the single Burmese species is known only from parthenogenetic morphs, recent importation from India remains a possibility.

KEY TO SPECIES OF *Malabaria*

1. Holandric................................. 2
   Metandric................................. 5
2. Spermathecal pores, at or close to *B*.......... 3
   Spermathecal pores, at m*BC*[a].............*M.* sp.
3. Prostatelike GM glands present, pores
   in xx[b].............................*paludicola*
   Prostatelike GM glands, lacking.............. 4
4. Spermathecae, long enough to reach into xviii.*levis*
   Spermathecae, shorter[e]................*biprostata*
5. Pyriform glands, open into seminal grooves[d].*sulcata*
   Pyriform glands and seminal grooves,
   lacking[e]...........................*oryzivora*

[a] *M.* sp., Bangalore, South India, distinguished from others of the genus by the more lateral location of the spermathecal pores. Unnamed because adult material has not been available.

[b] *M. paludicola* Stephenson, 1924. South Malabar, southern India, has a rather long clitellum, extending through xi-xxi.

[e] *M. biprostata* Aiyer, 1929. Kumili, Travancore, South India, at 1,500 feet.

[d] *M. sulcata* Gates, 1945. North central portion of the Deccan. Michaelsen presumably would have been overjoyed at sight of this species, because of the classical belief that the Eudrilidae evolved from the Ocnerodrilinae. Elongation internally of a pyriform gland, so that junction with a sperm duct would be further entally, seemingly should result in a condition almost identical with the euprostate that supposedly characterizes the Eudrilidae. Muscular prostates also are present in *Tazelaaria africana* Gates, 1962.

Most ocnerodriles are known from few specimens but individual variation in external characters of 833 and in internal anatomy of 150 of *sulcata* was recorded.

[e] *M. oryzivora* (Stephenson, 1924), south Malabar, has a clear gland in the invagination on each side of xvii that contains discrete male and prostatic pores.

Malabaria levis

1938. *Filodrilus levis* Chen, *Cont. Biol. Lab. Sci. Soc. China (Zool.)* 12: p. 423. (Type locality, Po-peng, Hainan. Types, probably lost.)
1942. *Malabaria levis?* Gates, *Bull. Mus. Comp. Zool. Harvard College* 89: p. 92.
1961. *Malabaria levis*, Gates, *Burma Res. Soc. 50th Anniv. Pub.* No. 1, p. 57.
1966. *Malabaria levis*, Gates, *Ann. Mag. Nat. Hist.*, Ser. 13, 9: p. 45.

Bithecal, pores minute and superficial, at or just lateral to *B*, at 8-9. Male and prostatic pores, in parietal invaginations opening to exterior by transversely slitlike apertures centered at *B*, in xvii. Female pores, at or slightly lateral to *B* and behind 13/14. Genital markings, across 8/9, in viii, sometimes in xviii, each marking associated with a small, solid, clear gland. Setae, *a*,*b*/xvii lacking, $AB = CD$, $AA = BC$, $DD = \frac{1}{2}C$. (Clitellum?) Nephropores, at *B*(?). Prostomium, epilobous. Segments, *ca.* 195. Size, to 87 by 1 mm.

Septa, present from 5/6, 6/7-8/9 thickened, 9/10 slightly thickened. Nephridia, behind the clitellum with cells of peritoneal investment enlarged. Holandric. Seminal vesicles, (small? but soft), in xi, xii.

Prostates, long. Spermathecae, long enough to reach back into xxv, often coiled up in viii or ix.

Reproduction: Probably parthenogenetic in most if not all of the morphs that have been seen.

Distribution: Kungyangon, Twante, Kayan (Hanthawaddy), Pyapon, Kyaiklat (Pyapon), Danubyu (Maubin).

Hainan Island, China.

Biology: All Burmese localities are much nearer sea level than the Kawkareik level achieved by *T. gracilis.*

Polymorphism: *M. levis* is known only from parthenogenetic morphs. Eight or more were represented in Burmese collections comprising only 23 worms. Hainan collections contained at least two morphs unrecorded from Burma. Other morphs certainly can be expected. Among the anatomical degradations allowed by the parthenogenesis are, elimination of one or both spermathecae and/or prostates, of genital markings. Among the reversions to former conditions are, reappearance of ovaries in xii (but without oviducts), reappearance of tests in ix (usually with male funnels). The unusual length of the spermathecae may be a parthenogenetic aberration.

Remarks: If the length of the spermathecae is abnormal, as it may well be, relationships may be closest with the south Indian *M. biprostata* from which *levis* now is distinguished by the presence (usually) of clear glands in viii, across 8/9, or less frequently in xviii. Clear glands have been recorded, in Indian species of *Malabaria*, only from the metandric *oryzivora* where they are unknown outside of the male pore invaginations.

Clitellate specimens of *M. levis* have not been seen.

Thatonia

1942. *Thatonia* Gates, *Bull. Mus. Comp. Zool. Harvard College* 89: p. 101. (Type species, *T. gracilis* Gates, 1942).

Digestive system, with a gizzard in vii, a dendritically branched system of spaces (opening into gut lumen?) in thickened ventral wall of esophagus in ix-x, intestinal origin in xii, but without typhlosoles, caeca, and supra-intestinal glands. Nephridia, (avesiculate?), (present from?), ducts passing into parietes at *B* (throughout?). Nephropores, inconspicuous, at *B* (throughout?). Setae, closely paired. Dorsal pores and pigment, lacking. Septa, all present from 4/5.

Distribution: India, north central part of the Deccan to (and just into?) the Gangetic Valley. (Burma?)

Presence of one species, apparently in very small numbers, at two places in the Gangetic Valley, may have resulted from transportation, if not by man, perhaps by rivers running down from the Deccan.

As the single Burmese species is known only from parthenogenetic morphs, recent importation from India remains a possibility.

Remarks: Parthenogenesis may prove to be as rampant in *Thatonia* as in *Malabaria*. If parthenogenesis, especially when rampant, is an indication of phylogenetic senility, these characteristically Indian genera may be moving on toward extinction.

KEY TO SPECIES OF *Thatonia*

1. Spermathecal pores in or close to *AB* 2
   Spermathecal pores of 8/9 at m*BC*[a] . . . . *exilis*
2. Ventral setae of viii–ix especially enlarged,
   prostate-like GM glands lacking . . . . *gracilis*
   Ventral setae of viii–ix not so enlarged,
   prostate-like, GM glands present in xx[b] . . *parva*

[a] *T. exilis* Gates, 1945, known only from 11 juveniles secured in a north-central portion of the Deccan.

[b] *T. parva* Gates, 1945, known only from 16 specimens, most of which were not mature, found in a north-central portion of the Deccan. Only 11 immature individuals were secured during several years intensive collecting in and around Allahabad.

Thatonia gracilis

1942. *Thatonia gracilis* Gates, *Bull. Mus. Comp. Zool. Harvard College* 89: p. 101. (Type locality, Thongwa. Types, none.)

1961. *Thatonia gracilis*, Gates, *Burma Res. Soc. 50th Anniv. Pub. No. 1*, p. 57.

1966. *Thatonia gracilis*, Gates, *Ann. Mag. Nat. Hist.*, Ser. 13, 9: p. 47.

1970. *Thatonia gracilis*, Soota & Julka, *Proc. Zool. Soc. Calcutta* 23: p. 205.

Quadrithecal, pores at *B* and 7/8–8/9. Quadriprostatic, pores at eq/xvii and eq/xix, at termini of seminal grooves. Male pores, in xviii, anteromedian to *a*. Male field, distinctly delimited and with seminal grooves forming an *H*-shaped figure. Female pores, at *B*, just behind 13/14. Clitellum, saddle-shaped, xiii, xiii/n–xxii, xxiii/n, xxiii. Setae, present from ii, *AA* slightly > *BC*, *DD* ca. = ½*C*, *a,b* in some or all of ii–xii enlarged, especially so in viii–ix, present in xviii, lacking in xvii and xix. Prostomium, epilobous. (Segments?) Size, 63–87 by 1 mm.

Septa, 5/6–8/9 thickened. Nephridia, behind clitellum with cells of nephridial investment enlarged. Holandric. Seminal vesicles, in xi,xii. Sperm ducts, without epididymis and posterior widening. Spermathecae, 3+ mm. long, duct short and with thick wall, ampulla thin-walled and tubular.

Reproduction: Spermatozoal iridescence never was observed on male funnels or in spermathecae. If worms are male sterile, as is indicated by retention of testes in a juvenile state at maturity, reproduction obviously must be parthenogenetic. Variations in the male field are of kinds that could be individual developmental aberrations or parthenogenetic degra-

dations but frequency of divergence from what is believed to be the morphological norm is much more suggestive of parthenogenesis. Nevertheless, seminal vesicles of some individuals were so large as to suggest profuse spermatogenesis.

Distribution: Port Blair (Andaman Islands). Ye, Kawkareik (Amherst), Taungzun, Kyaikto (Thaton), Thongwa, Kayan (Hanthawaddy), Kyaiklat (Pyapon), Maubin, Danubyu (Maubin), Henzada, Ingabu, Zalum (Henzada). Pegu (Pegu).

Biology: At Kawkareik, the species reached its highest known elevation. All other localities are much nearer sea level.

Abnormality and Variation: No. 1. Left side with four spermathecae two of which are in one segment. Right side, the male field about 1 mm. behind that of the left side. Information as to other aberrations and as to the metamerism (probably more or less abnormal?) was not recorded at the time and cannot now be mentioned. Halving of embryonic somites, more anteriorly on the left side, probably was involved.

Other abnormalities (1942 and 1966) now are regarded as parthenogenetic degradations.

Polymorphism: Recorded degradations, of the male terminalia, are; loss of smaller or larger parts of the male field and of one or more prostates. Seemingly only one individual showed the supposedly normal sexprostatic state, but prostates were present on both sides of xvii and xix (4 specimens). A portion of the male field and its prostate was lacking in xvii on the left side (2) on the right side (1). All of the field anterior to 17/18, as well as both prostates of xvii, was lacking in 35 of the few specimens that were identified. The male field, in front of xix on the left side was lacking in one worm with prostates and male field normal on the right side of xvii–xix. Prostates of xix alone were present (2) or were associated with a prostate on the left side of xviii (1), left side of xx (1), right side of xx (1), or with a pair in xx (1).

Location of male pores, unfortunately was not determined in the more interesting of the degraded forms.

Evolution in the Burmese populations seems to be mainly toward the classical balantin (posterobiprostatic) state, a condition rare in any of the three megadrile families where its appearance is theoretically possible. Another line, with prostates only in xix and xx may have provided an approximation, except for segments involved, to the state characterizing *Gordiodrilus tenuis* Beddard, 1892.

Remarks: The anatomical diversity, instanced by the few specimens (less than a hundred) studied, proves polymorphism to be rampant in *T. gracilis*. So great a multiplicity of morphs seems consonant with long establishment of parthenogenesis, unless, of course, the accumulation of mutations that is responsible for such morphs is, in that species, unusually rapid. The subsequent finding (Gates, 1968) of prostatelike GM glands in xx of several Burmese individuals requires consideration of relationships with *T. parva* which also has those glands in the 20th metamere. In that Indian species, known only from 11 juvenile, 3 aclitellate, and 2 clitellate specimens, sperm seemingly are exchanged during copulation. Differences of *T. parva* from *T. gracilis* involve shape of male field and spermathecae, no special enlargement of ventral setae in viii-ix, perhaps also location of male pores. Quite striking divergences in appearance of the male field often are due only to differences in growth stage, physiological condition at time of death, or in method of preservation, and can be accepted as evidence of species distinction only in association with much less variable characters. Hypertrophy of the spermathecal ampulla is one of the structural degradations that has been allowed by parthenogenesis in other families and so the greater length of the spermathecal ampulla in *T. gracilis* may have had a similar origin. However, highly localized enlargement of ventral setae, characterizes only one species of the moniligastrid genus *Drawida* and so may prove to be equally distinctive of *T. gracilis* in its genus. Even if *T. parva* and *T. gracilis* are found to be distinct species, available evidence is suggestive of close relationship. Presumably then, the original home of *T. gracilis* should be sought in a northern portion of the Deccan. If so, presence in Burma may be due to transport and by man.

OCNERODRILINAE

1966. Ocnerodrilinae Gates, *Ann. Mag. Nat. Hist.*, Ser. 13, 9: p. 53. (Type genus, *Ocnerodrilus* Eisen, 1878.)

Digestive system, with extramural calciferous glands in region of ix–x. Vascular system, with extra-extra-esophageal trunks that pass into the calciferous glands, branch therein, and after uniting posteriorly emerge to join the supra-esophageal trunk.

Distribution: Africa, America, in tropical and possibly also in adjacent subtropical regions.

Transportation of various species around the world, followed by successful colonizations, along with inadequate characterizations of many forms, including some type species, has caused so much confusion that original homes of species and genera still are unknown.

Phylogeny: African species, as now known, are all small, close to the low limit for megadrile size. Larger species are known only from South America and Hispaniola. In accordance with Michaelsen's size and other criteria for distinguishing endemics from exotics, the Ocnerodrilinae should have originated in the America hemisphere. Such an origin, of course, rules out an ocnerodrile ancestry of the Eudrilidae.

The Ocnerodrilidae, in the classical system, was the "last" (presumably meaning the most recent) of the stems derived from the supposedly ancestral *Notiodrilus*, in Africa, from which there were radiated off, more than once, branches to the east as well as to the west. The final exposition of those views is in Stephenson's monograph (1930: p. 853).

In evolution of the Ocnerodrilinae from a protothatonia stage, almost the only important change required is development of extramural, esophageal outgrowths that are called calciferous glands. These may be paired or unpaired and ventromedian, possibly with intermediate stages trending from each to the other alternative, in ix or in ix and x. Glands may be thin-walled sacs with low and irregular or higher and lamelliform ridges, thick-walled with a central or subcentral lumen that may be quite small, trabeculate or without connective tissue partitions, with two large and parallel canals, with a system of regularly branching canals, "solid" without any macroscopically recognizable canals. Little is known about the finer microscopic anatomy and nothing about embryonic development. Until considerable data have become available, at least about microscopic structure, if not also about development, of organs known, since 1917, to be of major importance for megadrile systematics, further consideration of phylogeny and even of intergeneric relationships is likely to be futile.

Some of the other evolutionary changes in the ocnerodriline subfamily are as follows. Negative, elimination of all prostates, or of all except 2, 4, or more rarely 6 of those in xvii–xix, much more rarely 2 or 4 of those in xx–xxi, elimination of testes and male funnels in ix (general) in x, xi, or x–xi (sporadic), of gonads and gonoducal funnels in xii (general), of one or the other or both pairs of spermathecae (sporadic), of part or all of the male terminalia (sporadic), elimination of the gizzard (8 genera and some species of another now are agiceriate), abortion of nephridia in some postperistomial segments (no other cephalizations in the excretory system recorded). Positive and sporadic, addition of an extra pair of spermathecae in vii, associated in one species with elimination of the original two pairs, appearance of intramural chambers in spermathecal ducts (some gordiodriles), elongation of the chambers into extramural diverticula (or was the direction of evolution the reverse of that just mentioned? *cf.* Jamieson, 1962, *Proc. Zool. Soc. London*, 139: p. 618, fig. 3), enlargement of spermathecal pores, migration of the pores laterally (some species of several genera), invagination of male pores (first?) into parietal (and then later?) or coelomic chambers possibly eversible to serve as intromittent organs, muscularization of posterior portion of male ducts to serve as ejaculatory bulbs, development of clear glands in association with genital markings or malepore invaginations, development of penial setae (1 species), ornamentation of shaft ectally and/or modification in shape of setal tips, enlargement of setae in certain anterior and/or posterior segments, appearance of lymph glands (1 genus) and of pigment (2 species), transfer of gizzard to vi, vii or viii, doubling of the gizzard (2 in vi–vii, 2 in vii–viii), esophageal elongation so that intestine now begins in xiii (several species), xiv (1 species), appearance of an intestinal typhlosole (2 American and 2 African species), sometimes even associated with short lateral typhlosoles anteriorly, appearance of supra-intestinal glands not associated with a typhlosole, enlargement of cells investing nephridia for storage of metabolites.

Some of the more general negative changes may well have been made earlier than many of the quite sporadic, positive modifications in ancestral anatomy.

Systematics: Calciferous glands, in spite of the importance attributed to them in each section of the present contribution, are unlikely to contribute more to the subfamily definition than the two characters mentioned in the précis, i.e., extramural location, and restriction to ix–x or one of those segments. Divergence from either character, except in abnormal specimens, is unknown and is not now expected. The amount of blood within the glands, according to Eisen's observations on *O. occidentalis*, varies alternately and regularly from little or none to much. If similar alternations characterize other taxa considerable intraspecific difference in appearance of internal organization, hitherto tactily assumed to be invariant, perhaps can be anticipated.

The vascular character of the definition is, because of our ignorance of the system in so many genera, an extrapolation but from many more data than usually were available to classical specialists.

Presence of "Fettkügelchen" in the enlarged peritoneal cells investing the nephridia of *O. calwoodi* Michaelsen, 1899 was mentioned by Michaelsen, but whether that record was based on chemical tests or merely on similarity to conditions found by Timm (1882) in worms of another family is unknown. Enlargement of the peritoneal cells appears to be not uncommon in taxa with holoic nephridia and even sometimes in species with meronephridia and, regardless of the chemical nature of the content, may provide a character of some systematic usefulness. Caution in its use may be required until information is available as to possibility of complete disappearance of the content, either seasonally, during periods of inactivity, or as a result of starvation.

Remarks: Organs herein called calciferous glands have been called; chyle sacs, chylous diverticula, esophageal pockets, sacs, or diverticula, organs "with the structure of calciferous glands." Their function is unknown and in absence of such knowledge is assumed to be the same in all genera of the subfamily. Sacs or pockets seemingly are inapplicable to "solid" structures permeated only by microscopic canaliculi. Although Michaelsen, through his last publication in 1937, continued to call the organs "chylustaschen," food absorption therein is unlikely (*cf.* Stephenson, 1930). No mention of seeing chyle in the glands of *O. occidentalis* was made by Eisen who studied, *in*

vivo, flow of blood through the organs. However, that author also did not mention seeing calcareous granules. The single record of presence of such granules in the glands, rather curiously, was that of Michaelsen for *O. calwoodi*. Esophageal diverticula, as well as intramural glands, of similar structure are known to secrete calcium carbonate in other families. Accordingly, until proof to the contrary is provided, the descriptive characterization of calciferous seems preferable, and especially so to naming the glands after a man who did not discover them and whose contribution to our knowledge of the organs "seems practically valueless."

Some chemical component of the blood is modified in its passage through the calciferous glands as is shown, in properly stained microtome sections by a marked difference in color at distal and proximal ends of the organ.

An American origin of the Ocnerodrilinae (*cf.* p. 263) is of no assistance in explaining relationships between the ocnerodrile subfamilies.

KEY TO ORIENTAL GENERA

1. Gut, agiceriate 2
   Gut, giceriate 3
2. Calciferous glands, paired . . . . *Ocnerodrilus*
   Calciferous gland, unpaired,
      ventromedian . . . . *Gordiodrilus*
3. Gizzards, in vi–vii . . . . *Nematogenia*
   Gizzard, in vii 4
4. Calciferous glands, one pair in ix . . . . *Eukerria*
   Calciferous glands, unpaired, ventral
      in ix and x . . . . *Curgiona*

Curgiona

1921. *Curgia* Michaelsen, *Mitt. Naturhist. Mus. Hamburg* **38**: p. 59 (Non *Curgia* Walker 1860, Trichoptera. Non *Curgia* Walker 1864, Lepidoptera. Type species, *Curgia narayani* n. sp.)

1923. *Curgia*, Stephenson, (*The Fauna of British India*), *Oligochaeta*, p. 481.

1941. *Curgiona* Gates, *Rec. Indian Mus.* **43**: p. 497.

Digestive system, with a gizzard in vii, with a ventromedian calciferous gland in each of ix and x, intestinal origin in xii. Calciferous glands, sessile, with small central lumen. Nephropores, inconspicuous(?), in *AB*. Setae, closely paired. Dorsal pores and pigment, lacking. Spermathecae, adiverticulate.

Distribution: The home range, presumably in Africa, is unknown.

Phylogeny: The eudrilid family, in the classical system, evolved from the Ocnerodrilinae, the "last of the stems" derived from the supposed root genus of a group that Michaelsen eventually termed Series Megascolecina. *Curgia*, with its two unpaired, ventromedian calciferous glands, was believed to diminish the distance between the two families. Until more is known about the anatomy of the type species and some information has been acquired about the original home of the genus, further speculation as to relationships is likely to be futile.

Remarks: The genus is known only from the original description of the type species.

Curgiona narayani

1921. *Curgia narayani* Michaelsen, *Mitt. Naturhist. Mus. Hamburg* **38**: p. 59. (Type locality, River Hatti at Madapur, Coorg, South India. Types, if extant, presumably in the Hamburg Mus.)
1923. *Curgia narayani*, Stephenson, (*The Fauna of British India*), *Oligochaeta*, p. 481.

Bithecal, pores at *B* and 8/9. Male pores, (common openings of male and prostatic ducts?), in papillae just behind eq/xvii. Female pores, presetal and at *B* in xiv. (Clitellum?) Setae, *a,b*/xvii lacking, $AA = BC$, $DD = \frac{1}{2}C$. Prostomium, epilobous, tongue open. Segments, 230. Size, 100 by 0.7–0.9 mm.

Septa, 6/7–7/8 somewhat thickened. Metandric. Seminal vesicles, large, in xii. Copulatory chambers, in xvii into which sperm ducts (and prostatic ducts?) open. Prostates, long.

Distribution: Coorg, South India.

Remarks: Types (3?) probably were relaxed (*ca.* 60 by 1+ mm. if strongly contracted?) and certainly were aclitellate. Spermathecae, with little distinction between duct and ampulla, may not have been fully developed.

Eukerria

1892. *Kerria* Beddard, *Proc. Zool. Soc. London*, **1892**: p. 355. (Type, and only species, *K. halophila* Beddard, 1892.)
1893. *Kerria*, Eisen, *Proc. California Acad. Sci.*, Ser. 2, **3**: p. 293.
1895. *Kerria*, Beddard, *A Monogr. of the order of Oligochaeta* (Oxford), p. 553.
1900. *Kerria*, Michaelsen, *Das Tierreich* **10**: p. 369.
1930. *Kerria*, Stephenson, *The Oligochaeta* (Oxford), p. 859.
1931. *Kerria*, Jackson, *Jour. Roy. Soc. Western Australia* **17**: p. 120.
1935. *Eukerria* Michaelsen, *Ann. Mag. Nat. Hist.*, sec. 10, **15**: p. 102. (Nom. nov. for *Kerria* Beddard, 1892, preoccupied by *Kerria* Targioni-Tozzetti, 1884, Coccidae.)
1970. *Eukerria*, Jamieson, *Bull. British Mus. Nat. Hist. (Zool.)* **20**: p. 133.

Calciferous glands, one pair in ix. Setae, closely paired. Prostatic pores, in xvii and xix. Male pores, in xviii. Proandric. Spermathecae, adiverticulate, with pores at 7/8–8/9.

Distribution: Self-acquired, species ranges seemingly are delimitable in subtropical South America and adjacent tropical regions. Two species, if not more, have been widely transported, presumably by man.

Remarks: *Eukerria* is believed by this author to be a congeries with only the above-mentioned characters in common. Revision of the genus has been handicapped by lack of material of the inadequately characterized type species. (*Cf.* characterizations of calciferous glands in Jamieson's definition, p. 134.)

KEY TO BURMESE SPECIES

Spermathecal pores in *AB* . . . .*kukenthali*
Spermathecal pores slightly median to *C* . . . . .*saltensis*

Eukerria kukenthali

1942. *Eukerria peguana* Gates, *Bull. Mus. Comp. Zool.* **79**: p. 67. (Type locality, Rangoon. Types, none.)
1954. *Eukerria peguana*, Gates, *ibid.* **111**: p. 245.
1961. *Eukerria peguana*, Gates, *Burma Res. Soc. 50th Anniv. Pub.* No. 1: p. 57.
1970. *Eukerria kukenthali*, Jamieson, *Bull. British Mus. Nat. Hist. (Zool.)* **20**: p. 144. *Eukerria peguana*, Soota & Julka, *Proc. Zool. Soc. Calcutta* **23**: p. 204.

Quadrithecal, pores about at m*AB*, at 7/8–8/9. Prostatic pores, minute, on roofs of transversely slit-like parietal invaginations with transverse apertures in *AB* and in protuberant anterior and posterior ends of paired, longitudinal, dumbbell-shaped porophores. A solid "clear gland" with short stalk protrusible from each prostate-pore invagination. Male pores, at eq/xviii and just lateral to *b* which is slightly displaced mesially. Genital marking, a transversely placed area of epidermal thickening in xxi, with a pore on each side just lateral to *b*. Setae, present from ii, $AB = CD$, $AA = BC$, $DD = \frac{1}{2}C$, *a,b*/xvii, xix present(?) but unrecognizable externally in adults. Clitellum, annular but thinner and colorless in *AA*, xiii–xx. Nephropores, (at?), inconspicuous. (Prostomium, prolobous?) Dorsal pores and pigment, lacking. Segments, 105–142. Size, 20–70 by 0.75–1.0 mm.

Septa, 5/6–8/9 muscular. Gizzard, in vii. Calciferous glands, with a small central lumen triangular in section and a thick wall (without canals but presumably with canaliculi?), with (low?) longitudinal ridges almost in contact centrally. Intestinal origin, in xii. Intestinal typhlosoles and supra-intestinal glands, lacking. Nephridia, (lacking in ii–xi? ducts passing into parietes at?, avesiculate?). Seminal vesicles, large, in ix and xi. (Sperm ducts, without epididymis and posterior thickening.) Prostates, long, ducts 1½–1 mm. long. Spermathecal duct, about as long as ampulla, and nearly as wide but with slightly thicker wall and transversely slitlike lumen, moniliform and bound to parietes. GM glands, tubular, stalked, coelomic, in macroscopic appearance like the prostates.

Reproduction: Presumably biparental as sperm are exchanged during copulation.

Distribution: Maya Bundar, Port Blair (Andaman Islands). Car Nicobar. Moulmein (Amherst), Myaungmya (Myaungmya), Rangoon, Kungyangon, Thongwa (Hanthawaddy), Wanetchaung (Insein), Pegu (Pegu), Pyinmana (Yamethin). Selangor (Malay Peninsula). Christmas Island (near Java). Puerto Rico, St. Thomas. Brazil.

The original home of the species may well be in southern South America as long believed.

Variation and Abnormality: Genital marking lacking, one of a number of mature Burmese specimens, one of two Puerto Rican worms. A half of the marking lacking, on right or left side, two Burmese specimens. Genital markings, two, in xx and in xxi, four Burmese worms.

Only one calciferous gland present, on the right side, a Burmese worm. The single gland was, however, almost twice the normal size and, as a result of some developmental accident, perhaps two originally paired rudiments became united into one asymmetrical gland of double size.

Seminal furrows, between equators of xvii and xix, may be present actually or perhaps only potentially. In the latter condition they are represented by ungrooved, longitudinal bands of epidermal translucence. The longitudinal dumbell-shaped areas of the male field seemed quite characteristic of the author's strongly contracted, clitellate material, probably at height of sexual activity. In other stages and with other methods of preservation, appearances well could be so different as to be suggestive of another species.

Remarks: Recognition of genital markings on the types of *E. selangorensis* and of clear glands in prostate-pore invaginations of Michaelsen's Christmas Island specimens by Jamieson, enabled his placing *peguana* as well as two other species names in the synonymy of *E. kukenthali*.

Eukerria saltensis

1895. *Kerria saltensis* Beddard, *Proc. Zool. Soc. London* 1895: p. 225. (Type locality, Salto, Valparaiso, Chile. Types, in the British Mus.)

1942. *Eukerria saltensis*, Gates, *Bull. Mus. Comp. Zool. Harvard College* 89: p. 73.

1961. *Eukerria saltensis*, Gates, *Burma Res. Soc. 50th Anniv. Pub.* No. 1; p. 57.

1970. *Eukerria saltensis*, Jamieson, *Bull. British Mus. Nat. Hist.* (*Zool.*). 20: p. 132.

REPRODUCTION: Gavrilov, 1952, *Acta Zool. Lilloana, Inst. Miguel Lillo, Tucuman, Argentina* 10: p. 673.

Quadrithecal, pores, superficial but not minute and with tumescent margins, at 7/8–8/9, somewhat median to C. Prostatic pores, minute, superficial, at equatorial ends of seminal grooves (slightly lateral to B) containing about at eq/xviii minute and superficial male pores. Female pores, at or near B and anteriorly in xiv. (Genital markings, lacking.) Clitellum, annular but thinner ventrally, intersegmental furrows obliterated, (setae retained or only in part?), xiii, xiii/n, xiv–xix, xx/n, xx. Setae, present from ii, AA ca. $= BC$ (or difference slight), $DD < \frac{1}{2}C$, a,b/xiv–xvi enlarged (?). Nephropores, (at?), inconspicuous. Prostomium, epilobous. (Dorsal pores and pigment, lacking.) Segments, 112–135. Size, 25–100 by 1–2.4 mm.

Septa, present from 4/5, 6/7–8/9 thickened, 5/6 and 9/10–10/11 less so. Gizzard, in vii. Calciferous glands, with a horizontally slitlike, central lumen and a thick wall (without canals but presumably with numerous canaliculi?). Intestinal origin, in xii. Intestinal typhlosoles and supra-intestinal glands, lacking. Nephridia, lacking in ii–iv, v (as well as several additional segments? or rudimentary in several more posterior segments? ducts passing into parietes at?), avesiculate. (Sperm ducts, without epididymis and not enlarged posteriorly.) Seminal vesicles, in ix and xi. Spermathecae, adiverticulate.

Reproduction: Presumably biparental usually (in Burma) as sperm, matured profusely, are exchanged during copulation. However, in Argentina Gavrilov (1952) obtained, by isolation experiments, four generations of uniparental offspring. Whether self-fertilization was involved, and if so whether accomplished before or after laying, was not determined. Some facts were thought to favor internal fertilization but optional parthenogenesis may be implicated.

Distribution: Yamethin (Pyinmana). Mandalay (Mandalay). Monywa (Lower Chindwin). Australia (New South Wales, Western Australia). New Caledonia. Juan Fernandez Island, United States (Oregon, Texas,[2] Georgia,[2] Florida,[2] North Carolina[2]). Easter Island. Chile (Salto, Valparaiso, Coquimbo, Quillota). Argentina (near Bella Vista, Tucuman Province). Brazil (Minas Gerais). South Africa (Cape Province, Natal, Transvaal).

Transport, presumably by man since 1500 A.D., explains most of such a distribution. The original home of the species seemingly should be sought in subtropical South America.

Polymorphism: Male sterile individuals now have been recognized in American collections and for them reproduction must be obligatorily parthenogenetic. That could permit considerable degradation of genital anatomy and also provide an explanation for some of the considerable variation in appearance of the male field, in spermathecal size and shape and, perhaps also, for retention of seminal furrows in a juvenile condition as ungrooved bands of epidermal translucence between prostatic pores of a side.

Systematics: Relationships of this species as well as of *E. kukenthali* with their congeners are not determinable presently. Ten of the species of *Eukerria*, including the type species, are known only from the original material and from descriptions that do not provide necessary data as to structure of calciferous glands now known to be of major systematic importance. Such little information as is now available seems indicative of considerable anatomical heterogeneity—presence or absence of connective tissue partitions, thick (and glandular?) or thin (and non glandular?) walls, presence or absence of large canals, presence or absence of canaliculi, ridged or glabrous lining of the lumen, presence or absence of central lumen, etc.

---

[2] New record.

Remarks: Migration of adult oocytes, previously unknown in megadriles, was recorded by Gavrilov. The gametes were found in each of xii–xix and in several posterior segments including xxxv but more often in xii, xi, less frequently in xvii.

Gordiodrilus

1892. *Gordiodrilus* (part) Beddard, *Ann. Mag. Nat. Hist.*, Ser. 6, 10: p. 93.
1895. *Gordiodrilus* (part), Beddard, *A Monogr. of the Order of Oligochaeta* (Oxford), p. 506.
1900. *Gordiodrilus* (part), Michaelsen, *Das Tierreich* 10: p. 373.
1913. *Gordiodrilus* (part), Michaelsen, *Zoologica* 68: p. 3.
1923. *Gordiodrilus* (part), Stephenson (*The Fauna of British India*), *Oligochaeta*, p. 482.
1930. *Gordiodrilus* (part), Stephenson, *The Oligochaeta* (Oxford), p. 863.
1958. *Gordiodrilus* (part), Omodeo, *Mem. Inst. Français Afrique Noire* 53: p. 21.
1962. *Gordiodrilus*, Gates, *Rev. Zool. Bot. Africaine* 66: p. 351.
1963. *Gordiodrilus* (part), Jamieson, *Bull. British Mus. Nat. Hist.* 9: p. 305. (*Gordiodrilus elegans* Beddard, 1892 designated as type species.)

Digestive system, agiceriate, with a ventro-median calciferous gland having a thick wall without large canals (but with canaliculi?) and a small central lumen opening dorsally without a stalk through floor of gut posteriorly in ix, intestinal origin in xii, without typhlosoles and supra-intestinal glands. Nephridia, (present from?), avesiculate, ducts passing into parietes in or close to *CD* (throughout?). Nephropores, inconspicuous, in or close to *CD* (throughout?). Setae, closely paired. Dorsal pores and pigment, lacking. Septa, all present from 5/6.

Spermathecae, adiverticulate.

Distribution: Tropical Africa. Transportation, presumably by man, is responsible for presence of one (or more?) species in other parts of the world, Burma, India, Dominica, and perhaps non-tropical portions of Africa.

Remarks: The definition above was carefully labeled (1962) tentative primarily because all material available to the author had been of parthenogenetic morphs in which extent of anatomical degradation was uncertain. The species recently designated as the generic type cannot yet be properly characterized as biparental populations have not been studied.

Gordiodrilus elegans (?)

1892. *Gordiodrilus elegans* Beddard, *Ann. Mag. Nat. Hist.*, Ser. 6, 10: p. 84. (Type locality, Kew, soil supposedly received from Lagos, West Africa. Types, in the British Museum, probably of little if any value.)
1910. *Gordiodrilus travancorensis* Michaelsen, *Abhandl. Naturwiss. Ver. Hamburg* 19, 5: p. 98. (Type locality, Nedumangad, Travancore, India. Types, several, in the Hamburg Mus.?)
1931. *Gordiodrilus unicus* Stephenson, *Proc. Zool. Soc. London* 1931: p. 79. (Type locality, Bhamo. Type, in the British Mus.)

1942. *Gordiodrilus peguanus* Gates, *Bull. Mus. Comp. Zool. Harvard College*, 89: p 85. (Type locality, Rangoon, Burma. Types, destroyed during World War II.)
1945. *Gordiodrilus peguanus*, Gates, *Proc. Natl. Acad. Sci. India* 15: p. 47.
1961. *Gordiodrilus peguanus*, Gates, *American Midland Nat.* 65: p. 41. *Gordiodrilus peguanus + G. unicus*, Gates, *Burma Res. Soc. 50th Anniv. Pub.* No. 1: p. 57.
1962. *Gordiodrilus* (*elegans?*), Gates, *Rev. Zool. Bot. Africaines* 66: p. 349.
1970. *Gordiodrilus paski*, Soota & Julka, *Proc. Zool. Soc. Calcutta* 23: p. 205.

Quadrithecal, pores at 7/8–8/9. Seminal grooves, in or close to *AB*. Reproductive apertures, all at or close to *B*. Clitellum, annular, xiii/n–xviii, xix/n, xix, xx/n. Setae, *AB ca. = CD*, *AA* slightly < *BC*, *DD ca.* = $\frac{1}{4}$C. Prostomium, epilobous. Segments, 80–98. Size, 26–47 by 1–1$\frac{1}{2}$ mm.

Holandric. Seminal vesicles, in ix and xii. Spermathecae, with intramural seminal chambers in a middle portion of the duct.

Reproduction: Probably parthenogenetic in anthropochorous morphs.

Distribution: Maya Bundar (Andaman Islands). Ye, Moulmein, Mupun (Amherst), Boyagyi, Taungzun, Kyaikto, Sittang (Thaton), Kyauktan, Syriam, Kungyangon, Rangoon (Hanthawaddy), Bassein, Coomzamu (Bassein), Hmawbi (Insein), Pegu (Pegu), Minbu (Minbu), Indaw Lake (Katha), Bhamo (Bhamo).

India, Africa, Puerto Rico, Bonaco Island, Dominica(?).

Although the home of the species must be somewhere in tropical Africa, the only precise African record is of two specimens from the shore of Lake Tanganyika at Kigoma. Extra-African distribution is attributable to transportation by man and presumably since 1500 A.D.

Polymorphism: The *peguanus* morphs, with prostate pores in xvii–xviii, male pores about at 17/18 in seminal grooves between eq/xvii and eq/xviii, differ from each other as to presence or absence of seminal vesicles in ix and as to degree of development in ix (when present) and in xii. The *travancorensis* morphs have prostate pores in xviii–xix and seminal grooves between eq/xviii and eq/xix. The *unicus* morph, known only from one specimen secured at Bhamo, presumably with a sexprostatic ancestry (pores in xvii–xix), already has lost one of the prostates of xviii and parts of the seminal grooves.

A wider distribution in Burma, than any other ocnerodrile, is indicated by the published records.

Regeneration: Regenerative capacity is expected to be similar to that of *O. occidentalis* but the only published record (Gates, 1961a) is of a heteromorphic tail regenerate at 16/17.

Abnormality: Two specimens (Gates, 1942, 1962) show homoeoses of the kind that result from halving of embryonic somites. Other aberrations, mostly organ losses, are such as appear more often in parthenogenetic morphs than in interbreeding populations.

Remarks: The history of the type specimen of *unicus*, unfortunately, cannot now be learned, but the worm could have been obtained from African plants sent to an orchid dealer in Bhamo as megadriles occasionally are found in shipments of orchids. Subsequent collecting at or near Bhamo, on several occasions, provided no specimens of any gordiodrile morph.

Material of no gordiodrile biparental population has been available to this author. Whether any one has studied individuals of an interbreeding population cannot be determined from the literature. How much anatomical degradation has resulted from the parthenogenesis, in the morphs that were available, cannot now be determined. Yet, merely to have a specific name for the worms now under consideration, a decision had to be made with regard to the extent of degradation.

Fortunately, information as to analogous situations now is available. Seminal grooves, in two families (Ocnerodrilidae, Acanthodrilidae), can be variously shortened or eliminated after reproduction becomes parthenogenetic. One, some or all spermathecae often have been eliminated in parthenogenetic morphs (3 families, at least). Prostates, 1, 2, 3, or 4, likewise have disappeared after reproduction became uniparental (3 families, at least). For existence of a fairly recent, sex-prostatic, ancestral stage in ocnerodrile evolution quite some evidence has been made available. Some data even have been recorded that suggest a somewhat more remote polyprostatic ancestry (reviewed in, Gates, 1966).

Accordingly, it has been assumed for the time being, that the interbreeding biparental population of *G. elegans* is quadrithecal and sexprostatic. From an H morph so characterized, by loss of the posterior pair of prostates, the *zanzibaricus, habessinus, paski, peguanus,* and *bonacanus* morphs could have evolved. By loss of posterior prostates along with one pair of spermathecae, or both pairs, the *ditheca* and *dominicensis* morphs, respectively, could have been derived. By loss of the anterior prostates, the *travancorensis* and *elegans* morphs could have resulted. By loss of one prostate and some shortening of the seminal grooves the *unicus* morph could have arisen. An alternative possibility is that biparental populations (H morphs) are quadriprostatic (1962) in which case two species presumably have been involved in the origin of the anthropochorous and parthenogenetic morphs.

Nematogenia

1900. *Ocnerodrilus (Nematogenia)* Eisen, *Proc. California Acad. Sci.,* Ser. 3, **2**: p. 112. (Type species, *Pygmaeodrilus lacuum* Beddard, 1893.)
1900. *Nematogenia,* Michaelsen, *Das Tierreich* **10**: p. 376.

1923. *Nematogenia,* Stephenson (*The Fauna of British India*), *Oligochaeta,* p. 483.
1930. *Nematogenia,* Stephenson, *The Oligochaeta* (Oxford), p. 862.
1957. *Nematogenia* (part), Gates, *Bull. Mus. Comp. Zool. Harvard College* **117**: p. 428. (Excluding, Hispaniolan autochthones.)

Digestive system, with weak (?) gizzards in vi, vii, paired calciferous glands in ix, intestinal origin in xii, but without typhlosoles and supra-intestinal glands. Calciferous glands, each with fairly thick wall (no large canals) and a central, vertically slitlike lumen, a long and slender stalk bound in a *U*-loop against the esophagus. Nephridia (present from?), avesiculate (?), ducts passing into parietes in or close to *CD* (throughout?). Nephropores, inconspicuous, in or close to *CD* (throughout ?). Setae, closely paired. Pigment, lacking. Septa, all present from 4/5. Spermathecae, adiverticulate.

Distribution: Guinea to Congo, tropical Africa.

The range as thus indicated has been extended around the world, in the tropics, by human transport of one species.

Systematics: The type species unfortunately is known only from the inadequate original description and there are no types. Specimens that Michaelsen first (1922) recorded as *lacuum* were later placed in *panamaensis,* and he still was uncertain (Gates, 1962: p. 264) as to which of the two names should be borne by the last worms he saw. Only the anthropochorous species is adequately characterized for present needs.

Nematogenia panamaensis

1900. *Ocnerodrilus (Nematogenia) lacuum* var. *panamaensis* Eisen, *Proc. California Acad. Sci.,* Ser. 3, **2**: p. 127. (Type locality, Panama. Types, probably none.) *Nematogenia panamaensis,* Michaelsen, *Das Tierreich* **10**: p. 376.
1903. *Nematogenia panamaensis,* Michaelsen, *Ark. Zool. Stockholm* **1**: p. 163. *Die geographische Verbreitung der Oligochaeten* (Berlin), p. 120.
1904. *Nematogenia panamaensis,* Michaelsen, *Sitzber. Böhmischen Ges. Prag 1903,* **40**: p. 16. *N. josephina* Cognetti, *Boll. Mus. Zool. Univ. Torino* 1904, 478: p. 3. (Type locality, San José, Costa Rica. Types, in the Torino Museum?)
1906. *Nematogenia josephina,* Biolley, *Bol. Soc. Nac. Agric. Costa Rica,* 1906: p. 39. Cognetti, *Mem. Acad. Sci. Torino* **56**: p. 55.
1908. "*Nematogonia*" *josephina,* Cognetti, *Ann. Civ. Mus. Genova,* Ser. 3, **4**: p. 33.
1910. *Nematogenia panamaensis,* Michaelsen, *Mitt. Naturhist. Mus. Hamburg* **27**: p. 114. *Abhandl. Naturwiss. Ver. Hamburg* **19,** 5: p. 13.
1915. *Nematogenia panamaensis,* Michaelsen, *Ergeb. 2 Deutschen Zentral-Afrika-Exp. 1910–11,* **1** (Zool. 1): p. 222.
1916. *Nematogenia panamaensis,* Michaelsen, *Ark. Zool. Stockholm* **10,** 9: p. 20.
1921. *Nematogenia panamaensis,* Michaelsen, *Ark. Zool.* **13,** 19: p. 18.
1922. *Nematogenia lacuum,* Michaelsen, *Cap. Zool.* **1,** 3: p. 21.
1923. *Nematogenia panamaensis,* Stephenson (*The Fauna of British India*), *Oligochaeta,* p. 483.
1930. *Nematogenia josephina,* Stephenson, *The Oligochaeta* (Oxford), p. 862.

1935. *Nematogenia panamaensis*, Michaelsen, *Rev. Zool. Bot. Africaine* 12: p. 69.
1942. *Nematogenia panamaensis + N. josephina*, Gates, *American Midland Nat.* 27: p. 98.
1957. *Nematogenia panamaensis*, Gates, *Bull. Mus. Comp. Zool. Harvard College* 117: p. 442.
1958. *Nematogenia panamaensis*, Omodeo, *Mem. Inst. Français Afrique Noire* 53: p. 21.
1962. *Nematogenia panamaensis*, Gates, *Rev. Bot. Zool. Africaines* 65: p. 257.

Bithecal, pores minute and superficial, at or close to *B*, at 8/9. Male porophores, small transversely elliptical discs, at center of each a transversely slitlike prostatic pore. Male pores, just behind prostatic pores. Male field, often slightly raised, occasionally delimited by a circumferential furrow, reaching 16/17, 17/18 and well into *BC*. Female pores, at *B* and slightly nearer eq/xiv. Clitellum, saddle-shaped, reaching down to or nearly to *A*, xiii, xiii/n, xiv–xxi, xxii/n, xxii, xxiii/n, xxiv/n, xxiv. Setae, *a,b*/xvii lacking, *AB ca.* = *CD*, *AA* < *BC*. Dorsal pores, beginning at or behind 8/9. Prostomium, epilobous, tongue small, narrowed posteriorly and often depressed. Segments, 95–124 (usually 103–110?). Size, 40–55 by 1–2 mm.

Septa, present from 4/5, 5/6–9/10 muscular, 6/7–8/9 or 9/10 especially so, 8/9 frequently inserted on parietes in front of intersegmental furrow 8/9. Nematocytes, in coelomic fluid. Metandric. Seminal vesicles, in xii, large. Male gonoducts, without epididymis, slightly thickened in xvii. Prostates, long, ducts slightly narrower, with slight sheen, about as long as two segments. Spermathecae, large enough to reach dorsal parietes, ducts rather slender but with thick wall and narrow central lumen, usually sinuous or in one or two short loops.

Reproduction: Presumably biparental as sperm are exchanged during copulation.

Distribution: Guinea, Liberia, Nigeria, Cameroons, Congo.

Fernando Po, Ceylon, Philippine Islands, Costa Rica, Panama, Venezuela, Bahamas. At sea level, to elevations of 450 m. (Fernando Po), 1,160 m. (Costa Rica).

Remarks: Primarily a lowland, tropical species. Of the supposed difference between *N. panamaensis* and *N. lacuum* as recently stated (Gates, 1962), only shortness of the clitellum and the discrete male pores (male and prostatic ducts not united) then seemed to be of any importance. Individual variation in clitellar extent already recorded for *N. panamaensis* is considerable, xiii, xiii/n, xiv–xxi, xxii/n, xxii, xxiii/n, xxiv/n, xxiv. The clitellum of *N. lacuum* supposedly extended through xiii–xxvi. Further data as to individual variation in the home region of the species are needed.

Ocnerodrilus

1878. *Ocnerodrilus* Eisen, *Nova Acta R. Soc. Sci. Upsaliensis*, Ser. 3, 10, 4: p. 1. (Type species, *O. occidentalis* Eisen 1878.)

1888. *Ocnerodrilus*, Eisen, *Mem. California Acad. Sci.* 2, 1: p. 5.
1889. *Ocnerodrilus*, Vaillant, *Hist. Nat. Anneles, Paris* 3, 1: p. 204.
1891. *Ocnerodrilus*, Beddard, *Trans. Roy. Soc. Edinburgh* 36: p. 581.
1893. *Ocnerodrilus*, Eisen, *Proc. California Acad. Sci.*, Ser. 2, 3: p. 272.
1895. *Ocnerodrilus* (part), Beddard, *A Monogr. of the Order of Oligochaeta* (Oxford), p. 510. (Excluding, *eiseni*, *quilimanensis*, *bukobensis*, *affinis*, and *lacuum*.)
1896. *Ocnerodrilus*, Eisen, *Mem. California Acad. Sci.* 2: p. 172. *Ocnerodrilus* (part), Michaelsen, *Deutsch-Ost-Afrika*, Berlin 4, 12: p. 42. (Excluding, *Pygmaeodrilus* and *Ilyogenia*.)
1900. *Ocnerodrilus* (part) Eisen, *Proc. California Acad. Sci.*, Ser. 3, 2: p. 110. (Excluding, *Leiodrilus*, *Pygmaeodrilus*, *Ilyogenia*, *Nematogenia*, *Haplodrilus*, *Enicmodrilus sanxavieri*.) Michaelsen, *Das Tierreich* 10: p. 377. (Excluding, *Leiodrilus*, *Ilyogenia sanxavieri*, *taste*, *tepicensis*, and *africanus*, *Haplodrilus*.)
1923. *Ocnerodrilus* (part), Stephenson (*Fauna of British India*), *Oligochaeta*, p. 484.
1930. *Ocnerodrilus* (part), Stephenson, *The Oligochaeta* (Oxford), p. 860. (Excluding, *Liodrilus* and part of *Ilyogenia*.) *Ocnerodrilus*, Gates, in MS.

Digestive system, agiceriate, without typhlosoles and supra-intestinal glands. Calciferous glands, one pair, in ix, each with a large lumen into which protrude longitudinal ridges of variable height none of which meet centrally, opening through a short stalk into esophagus laterally just in front of 9/10. Vascular system, with paired extra-esophageals (median to segmental commissures?) that pass into distal ends of calciferous glands where they are divided into several groups of vessels that run longitudinally without anastomoses and after reuniting posteriorly emerge to join the supra-esophageal. Nephridia, ducts passing into parietes at *B*. Nephropores, inconspicuous, at or near *B*. Setae, closely paired. Dorsal pores and pigment, lacking. Septa, all present from 4/5.

Distribution: Tropical America and tropical Africa, possibly also adjacent subtropical portions of the Americas. Presence of species elsewhere is due to accidental introduction presumably by man. Transport is unlikely always to have been away from America or Africa and the extent of intracontinental carriage and subsequent colonization remains to be determined.

Remarks: The type species and others were erected on more or less defective parthenogenetic morphs. Biparental populations have not been studied and until they are adequately characterized, the status of various "species," as also in other ocnerodrilid genera, will remain dubious.

Ocnerodrilus occidentalis

1878. *Ocnerodrilus occidentalis* Eisen, *Nova Acta R. Soc. Sci. Upsaliensis*, Ser. 3, 10, 4: p. 10. (Type locality, Fresno County, California. Specimens collected at Fresno Prairie, April, 1879, and identified by Eisen are in the U. S. Natl. Mus.)
1916. *Ocnerodrilus* (*Ocnerodrilus*) *occidentalis*, Stephenson, *Rec. Indian Mus.* 12: p. 348. (Under flower pots, Ross Island.)

1923. *Ocnerodrilus (Ocnerodrilus) occidentalis*, Stephenson (*Fauna of British India*), *Oligochaeta*, p. 484.
1942. *Ocnerodrilus occidentalis*, Gates, *Bull. Mus. Comp. Zool. Harvard College* **89**: p. 99.
1945. *Ocnerodrilus tenellulus* Gates, *Proc. Indian Acad. Sci.* **21**: p. 223. (Type locality, Allahabad. Types in the Indian Museum.)
1961. *Ocnerodrilus occidentalis*, Gates, *Burma Res. Soc. 50th Anniv. Pub.* No. 1: p. 57.
1970. *Ocnerodrilus occidentalis*, Soota & Julka, *Proc. Zool. Soc. Calcutta* **23**: p. 205.

BIOLOGY: Gates, 1945, *Proc. Natl. Acad. Sci. India* **15**: p. 47. CALCIFEROUS GLANDS: Stephenson & Prashad, 1919, *Trans. Roy. Soc. Edinburgh* **52**: pp. 457, 463. PROSTATES (development of): Stephenson & Ram, 1919, *ibid.*, p. 451. REGENERATION (heteromorphic): Gates, 1961, *American Midland Nat.* **65**: p. 41.

Male porophores, in xvii. (Genital markings, none.) Clitellum, annular, xiii/2, xiii/n, xiv–xix, xx/n, xx. Prostomium, epilobous, tongue usually open but may be closed or even unrecognizable. Setae, *a,b*/xvii lacking, $AA = BC$, $DD = \frac{1}{2}C$. Nephropores, at or near $B(?)$. Segments, 70–84. Size, 12–30 by 1 mm. Septa, 7/8–10/11 somewhat strengthened. Holandric. Sperm ducts, not thickened posteriorly. (Athecal, avesiculate.)

Reproduction: Sperm are matured, in some morphs, probably not in others. Biparental reproduction seems unlikely in absence of spermathecae and of adhesive spermatophores. Parthenogenesis certainly is to be expected in male sterile morphs and is anticipated even in various morphs that still mature sperm. Existence of one or more biparental populations is suspected.

Distribution: Ross Island, etc. (Andaman Islands). Car Nicobar. Rangoon, Kokine, Kayan (Hanthawaddy). Taikkyi, Wanetchaung (Insein). Bassein (Bassein). Pegu (Pegu). Toungoo (Toungoo). Pyinmana (Yamethin). Mount Popa (Myingyan).

India, Pakistan, Ceylon. Singapore. China, Hainan, Japan, Philippines, British Solomon and New Hebrides Islands, California, Arizona, Mexico, St. Thomas. Denmark, Italy, Greece, Cape Verde Islands. Rhodesia, Southwest Africa, Great Comoro. Israel, Lebanon, Central Asia Basin (SSSR). Denmark, Greece, Lebanon, and Japan are included, though the species has not yet been found there, because specimens were intercepted (Gates, 1964) in soil from those countries. One individual was found on potatoes supposedly from Germany in a ship's stores but existence in that country outside greenhouses now seems unlikely.

O. occidentalis was obtained, until after 1893 only at Eisen's vineyard, in a garden plot 100 feet square, 6 miles east of Fresno in the San Joaquin Valley of California. That garden site, five years previous to first finding of the species, was desert—no permanent water within 20 miles. Believed to be native, the worms at first were thought to have been brought down in irrigation water but repeated searches of

canals and mountain springs provided no specimens. Importation seemed unlikely as Eisen knew of no introduced plants in the region. However, by 1900, Eisen already had secured specimens from Mexico, Arizona (near irrigation ditches), and soil with potted plants brought from China. Plant Quarantine interceptions show that the species still is being carried around the world by man. The original home of the species, still unknown, may be somewhere between the United States and South America.

Habitats: Soil. Algal mats at bottom of water tanks. Municipal dump. Banks of ditches (soil pH of 7.5) draining waste effluents of human habitations. Presence in Denmark and Germany outside of greenhouses, is believed to be unlikely.

Biology: Activity is uninterrupted at Burmese and Indian sites where moisture is adequate during the dry season. Elsewhere, quiescence, at unknown depths and in an unknown state, is imposed during the drought. Worms were at the surface, in the type locality, only during irrigation. Saturation may be required for normal activity.

Breeding, from July to May at Allahabad, was indicated by condition of the clitellum.

The distribution hints that climate may be restricting the species to lowlands of tropical and adjacent subtropical regions.

Regeneration: Head regeneration, in an anterior direction, is possible as far back as 12/13 and is probable at more posterior levels. Maximum number of segments in head regenerates at 12/13 was ten. At 19/20, all regenerates on posterior substrates were monstrous though usually in part cephalic but at 21/22 some were monstrous and others were heteromorphic tails. In a posterior direction, regeneration at 20/21–34/35 resulted in normal tails but at 19/20 in normal tails, cephalocaudal monstrosities as well as heteromorphic heads (Gates, MS).

Polymorphism: *O. occidentalis* still is certainly known only from anatomically degraded, parthenogenetic morphs without seminal vesicles and spermathecae. Male terminalia, in many morphs, are disappearing. An AR morph was seen once. An ARZ morph is anticipated as in one worm (Gates, MS) "testes, male funnels and gonoducts, as well as prostates and male pores, could not be found." *O. beddardi, guatemalae, hendrici, sonorae* Eisen, 1893, *O. calwoodi* Michaelsen, 1898, and *O. mexicanus* Eisen, 1900, are believed (*ibid.*) to be morphs of *occidentalis* rather than species.

Systematics: A species is to be defined by characters of an interbreeding population. *O. occidentalis* today is known, with certainty from avesiculate and athecal morphs in which male terminalia already are more or less degraded. The literature does, however, contain disappointingly inadequate descriptions of less degraded or possibly normal morphs that seemingly could have been derived, together with the classical morphs, from a common interbreeding population. Specimens of such morphs have not been available, and most of the types no longer are extant. Accordingly, in the précis above, supposedly specific characters such as number, size, and location of spermathecal pores, situation of seminal vesicles, shape of spermathecae are omitted.

Nevertheless, descriptions of the above-mentioned, supposed species allow an assumption that the amphimictic population from which the classical *O. occidentalis* was segregated by its parthenogenesis may have been characterized as follows: Bithecal, pores large and at 8/9. Male and prostate pores, discrete, in paired parietal invaginations or copulatory chambers eversible to function as penes to transfer sperm directly into spermathecal apertures. Seminal vesicles, in ix and xii. Such a form, perhaps with spermathecal pores centered at *B*, already would have undergone considerable evolutionary specialization in the genitalia while retaining a rather simple organization of the calciferous sacs.

OCTOCHAETIDAE

1900. Octochaetinae + Trigastrinae + *Trinephrus* (part) + *Notoscolex* (part, Megascolecinae, Megascolecidae), Michaelsen, *Das Tierreich* 10: pp. 319, 330, 184, 187. (Exclusions in each instance under this family heading being of species with holoic nephridia or racemose prostates.)

1903. Octochaetinae + Trigastrinae + *Trinephrus* (part) + *Notoscolex* (part), Megascolecidae, Michaelsen, *Die geographische Verbreitung der Oligochaeten* (Berlin), pp. 108, 109, 89, 90.

1909. Octochaetinae + Trigastrinae (Megascolecidae), Michaelsen, *Mem. Indian Mus.* 1: pp. 122, 203.

1910. Octochaetinae + Trigastrinae, Michaelsen, *Abhandl. Naturwiss. Ver. Hamburg* 19: p. 25.

1915. Octochaetinae, Stephenson, *Mem. Indian Mus.* 6: p. 103.

1917. Octochaetinae, Stephenson, *Rec. Indian Mus.* 13: p. 359.

1921. Octochaetinae, Michaelsen, *Mitt. Naturhist. Mus. Hamburg* 38: p. 36. Stephenson, *Proc. Zool. Soc. London* 1921: p. 103.

1923. Octochaetinae + *Spenceriella* + *Megascolides* (Megascolecinae) + Diplocardiinae (part, Megascolecidae), Stephenson, (*Fauna of British India*), *Oligochaeta*, pp. 362, 190, 192, 468.

1930. Octochaetinae + *Megascolides* (part) + *Spenceriella* (part) + Diplocardiinae (meronephric taxa only), Stephenson, *The Oligochaeta* (Oxford), pp. 841, 835, 849.

1937. Octochaetinae, Pickford, *A Monogr. of the Acanthodriline Earthworms of South Africa* (Cambridge, England), pp. 98, 605.

1958. Benhaminae (part, excluding *Neogaster, Wegeneriella, Pickfordia,* Acanthodrilidae), Omodeo, *Mem. Inst. Français Afrique Noire,* No. 53.

1959. Octochaetidae, Gates, *Bull. Mus. Comp. Zoology, Harvard College* 121: p. 254. Acanthodrilinae (part) + Megascolecinae (part, Megascolecidae), Lee, *The Earthworm Fauna of New Zealand* (Wellington), p. 32.

Intestinal origin, behind xiii. Vascular system, with hearts behind xi. Excretory system, meroic. Dorsal pores, present (?).

Prostates, tubular and of ectodermal origin. Seminal vesicles, trabeculate. Spermathecae, pregonadal

and diverticulate. Clitellum, multilayered. Ovaries, in xiii, (fan-shaped and with several egg strings?) Ova, not yolky. (Ovisacs, small and lobed?).

Distribution: Burma, Australia, New Zealand, a northern part of the Pacific coastal region of the United States, Mexico, Central America, West Indies. Central Africa, Madagascar. India.

The vast oceanic discontinuities in the distribution almost guarantee that the family is polyphyletic.

Dichogaster is, of course, represented in the Orient as well as in Europe and the Pacific areas only by anthropochorous species. The congeries, masquerading as a genus under the name *Howascolex,* comprises species supposedly endemic in America, Africa, and India. Otherwise, all oriental genera are confined, in so far as self-acquired ranges are known, to India or India and Burma. Records indicative of colonization by oriental octochaetids, after transoceanic carriage, are very rare.

Burmese octochaetines belong to a group of genera that now appears to have originated and had most of its development in peninsular India—an ancient land mass that never has been submerged. One of the more specialized genera, *Lennogaster,* which appears to have arisen in a northern part of the peninsula, has extended its range, presumably overland, into Chittagong and southern Burma where two species are known. *Eutyphoeus* presumably the last in one of the lines of evolution is believed (Gates, 1958c) to have arisen in the region of the Indo-Burma mountain wall from whence it spread through the Gangetic plain, into the Himalayas and across western and middle portions of Burma to the margin of the Shan Plateau. In so far then as these octochaetines are concerned the Indian contribution to the native Burmese fauna appears to be relatively recent, and aside from *Eutyphoeus,* rather unimportant.

Remarks: The family, no longer defined phylogenetically but rather by morphological characters common to all of its genera, now comprises (Gates, 1959) *Spenceriella* and *Megascolides* of the classical Megascolecinae, the neoclassical *Wegeneriona* and *Neogaster,* in addition to *Barogaster, Celeriella, Lennogaster, Priodochaeta, Priodoscolex, Rillogaster, Scolioscolides, Travoscolides,* oriental genera defined in a more modern manner and so by reference to as much of somatic anatomy as present knowledge allows. Although morphologically defined, the family is not claimed to be monophyletic as of course it was (*cf.* the "phylogenetic tree," in Stephenson, 1930: p. 843) in the classical system. Indeed, there are reasons for suspecting that oriental octochaetids are related neither to those of New Zealand nor to those of Africa-America. However, until very much more is learned about somatic anatomy in all sections of the complex, any formal change in the system can be based on little more than guesses as to results of future research.

Key to Oriental Genera of the Octochaetidae

1. Vascular system, with a subneural
    trunk[a].........................*Priodoscolex*
    Vascular system, without a subneural trunk. . 2
2. Esophageal gizzard, single................. 3
    Esophageal gizzard, doubled............... 14
3. Discrete calciferous glands, none............ 4
    Discrete calciferous glands, present.......... 6
4. Intestinal origin, anterior to xvi........*Ramiella*
    Intestinal origin, behind xv............... 5
5. Calciferous lamellae, in xvi or xvi–xvii[b]
                                            Howascolex
    Calciferous tissues, if any, in xiii–xiv....*Celeriella*
6. Setae, 8 per segment..................... 7
    Setae, more than 8 per segment............ 13
7. Calciferous glands, all behind xiii.......... 8
    Calciferous glands, not or not all behind xiii. . 9
8. Calciferous glands, one pair, usually asym-
    metrical, with short stalk to gut at or close
    to insertion of 15/16............*Octochaetona*
    Calciferous glands, not so characterized[c]
                                            Octochaetoides
9. Calciferous glands, intramural.............. 10
    Calciferous glands, extramural.............. 12
10. Calciferous glands, in xii.................. 11
    Calciferous glands, in xi, xii[d]............*Bahlia*
11. Intestinal typhlosole, not rudimentary
                                            Eutyphoeus
    Intestinal typhlosole, rudimentary..*Scolioscolides*
12. Excretory system, enteronephric[e]...*Travoscolides*
    Excretory system, exonephric[f]........*Calebiella*
13. Calciferous glands, behind xiii[g].....*Priodochaeta*
    Calciferous glands, not behind xiii[h].*Hoplochaetella*
14. Calciferous glands, behind xiii.............. 15
    Calciferous glands or tissues, not behind xiii. . 16
15. Calciferous glands, with one vertical lobe in
    each of xv, xvi, xvii..............*Dichogaster*
    Calciferous glands not so lobed or placed[i]
                                            Octochaetoides
16. In viii, no gizzard......................... 17
    In viii, 1 gizzard[j]................*Octochaetoides*
17. Calciferous glands, 4 pairs, in x–xiii........ 18
    Calciferous glands, less than 4 pairs......... 19
18. Gizzards, in v–vi[j]..................*Pellogaster*
    Gizzards, in vi–vii[k]..................*Rillogaster*
19. Calciferous glands, in x–xii........*Lennogaster*
    Calciferous glands, in xi, xii................ 20
20. Intestinal roof with a gridlike thickening at
    posterior end of typhlosole[l].......*Barogaster*
    Intestinal roof with equatorially separated,
    supra-intestinal glands at end of typhlosole[m]
                                            Eudichogaster

[a] *Priodoscolex* Gates, 1940, known only from the original description of 2 specimens of the type species from South India, has
2 pairs of calciferous glands in each of xv–xvii, an unusually interesting excretory system that needs much further study.

[b] *Howascolex* Michaelsen, 1901, with species supposedly in
India, Africa, and America, is another example of a classical con-

geries. Only characters shared by Indian species were considered
in constructing the key. Some notes were published on excretory
organs of a presumably Indian species called *Howascolex digaster*.
No such species has been erected. *Howascolex* always has been
defined as unigiceriate. The species accordingly does not belong
even in such a waste basket.

[c] *Octochaetoides* Michaelsen, 1922 is another systematic waste
basket, now comprising 5 species, each of which may have to go,
when adequately characterized, into some other genus.

[d] *Bahlia* Gates, 1945, known only from a few specimens of a
species secured in a northern part of the Deccan and an adjacent
portion of the Gangetic plain. Calciferous glands, except for
number, are as in *Eutyphoeus* and *Scolioscolides*.

[e] *Travoscolides* Gates, 1940, with 4 inadequately characterized
species, was found in Travancore and Cochin, South India.

[f] *Calebiella* Gates, 1940, known only from specimens of the type
species obtained in the Gangetic Valley where it may be exotic.

[g] *Priodochaeta* Gates, 1940, known only from the type species,
secured in South India.

[h] *Hoplochaetella* Michaelsen, 1900, with several species in
western peninsular India. Quadriprostatic species have 4 male
pores.

[i] The balance of the genus that did not key out at "8."

[j] *Pellogaster* Gates 1939, found in a northeastern portion of
peninsular India, from Jubbulpore to Orissa and Bengal.

[k] *Rillogaster* Gates 1939, from a part of western India near
Bombay. No supra-esophageal trunk was recognized though
hearts of x–xii bifurcate dorsally.

[l] *Barogaster* Gates 1939, from peninsular India. A gridlike
thickening of intestinal roof at the posterior end of the typhlosole
may have functions like supra-intestinal glands.

[m] *Eudichogaster* Michaelsen 1903, western India, now is fairly
well characterized though much remains to be learned about each
species.

A recent key to the same genera (Gates, 1958: p.
622) used only macroscopically recognizable characters of the calciferous section in the esophagus.
The key was so constructed to show how very useful
a short portion of the gut is to megadrile systematics.
Other important characters will be provided by
studies of microscopic anatomy in the same region.
As a further demonstration that somatic anatomy
should not be derogated as in the past, genital anatomy again has been ignored and characters provided
by other organs are used.

Celeriella

1958. *Celeriella* Gates, Ann. Mag. Nat. Hist., Ser. 13, 1: p. 612.
    (Type species, *Spenceriella duodecimalis* Michaelsen,
    1907.)

Digestive system, without typhlosoles, caeca, supra-
intestinal and calciferous glands, with a gizzard in vi,
esophagus widened in xiii–xiv and there with thickly
lamelliform but rather low longitudinal ridges of
variable height, intestinal origin behind xv. Vascular
system, with unpaired dorsal, ventral and supra-
esophageal trunks but no subneural, paired extra-
esophageals median to the hearts, paired posterior
lateroparietal trunks into xiii. Hearts, of x–xii
(latero?)-esophageal. Excretory system, meroic, all
nephridia small, avesiculate, exoic (? except in the
pharyngeal region and there enteroic?), biramous,
astomate (except the medianmost on each side in

posterior segments?). Septa, all present from 5/6. Setae, more than eight per segment (a,b/xviii lacking?). Dorsal pores, present. (Unpigmented?)

Biprostatic, pores (common apertures of male and prostatic ducts?) in xviii. Clitellum, annular, intersegmental furrows obliterated, dorsal pores occluded, setae retained. Holandric. Seminal vesicles, in xi, xii. Prostates, bound in zigzag loops to parietes through several segments. (Junction of male and prostatic ducts?) Metagynous (ovaries fan-shaped?).

KEY TO SPECIES OF *Celeriella*

1. Setae paired in an anterior part of the body at
   least 2
   Setae not paired, *ca.* 50 per segment
                              kempi (Stephenson, 1924)
2. Quadrithecal . . . .*ditheca* (Stephenson, 1924)
   Bithecal 3
3. Spermathecal pores at or just lateral to B 4
   Spermathecal pores at or close to D.*quadripapillata*
4. Setae in 12 regular ranks throughout the body
                              regularis (Stephenson, 1924)
   Only a and b setae in regular ranks throughout
   the body, behind the middle others irregularly
   placed and number increased to 14–17 per
   segment   *duodecimalis* (Michaelsen, 1907)

Distribution: Indian, Palni Hills, at elevations above 6,000 feet. Domicile in Burma, especially in the plains, seems unlikely.

Remarks: Translocation of one pair of prostates into xviii as assumed in classical phylogeny and in a discussion of derivation of *Celeriella* (Gates, 1958) no longer seems necessary.

Very little is known about this Indian taxon which is of especial interest because of the intrageneric evolution with reference to number and location of the setae.

Celeriella quadripapillata

1924. *Spenceriella duodecimalis* f. *quadripapillata* Stephenson, *Rec. Indian Mus.* 26: p. 331. (Type locality, Kodaikanal, South India, at an elevation of 6,850–7,000 feet. Types, in the Indian Mus.)
1958. *Celeriella quadripapillata*, Gates, *Ann. Mag. Nat. Hist.*, Ser. 13, 1: p. 613.
1961. *Celeriella quadripapillata*, Gates, *Burma Res. Soc. 50th Anniv. Pub.* No. 1: p. 57.

Bithecal, pores minute, superficial, at D and 7/8. Male pores, about at B, between 2, slight, conical protuberances from indistinctly delimited male fields. Female pores, slightly anteromedian to a. Setae, 12 per segment, in regular longitudinal ranks throughout the body, e,f in the dorsum, AB slightly smaller than intervals of other pairs, $FF < AA$. Clitellum, reaching into xiii and xvii (?). First dorsal pore, at 3/4–5/6. Prostomium, epilobous, tongue open. Segments, 81–85 (+?). Size, 27–38 by 1.5–1.75 mm.

Intestinal origin, in xvi.▼ Prostates, in xxiii–xxviii or xxix, ducts thick and straight in xviii–xxi, slenderer and looped in xxii. Spermathecae, large, duct slender and shorter than ampulla. Diverticulum digitiform, longer than main axis, from anterior face of duct at parietes, seminal chamber represented only by slight widening of lumen entally.

Reproduction: Presumably biparental as sperm are matured profusely.

Distribution: India, Kodaikanal, Palni Hills. The Burma record never was confirmed and the single worm obtained (Gates, 1958) "may have been brought from India in soil with plants that were being repotted at the time the collector was in Tharrawaddy."

Dichogaster Beddard, 1888

1958. *Dichogaster*, Gates, *Ann. Mag. Nat. Hist.*, Ser. 13, 1: p. 617.

The 200 species of this classical genus have in common, aside from family characters, only the following anatomy. Digestive system with two gizzards[1] in front of the testis segments and with calciferous glands behind xiii. Setae, lumbricin. The type species is inadequately characterized for present needs and the paucity of information about somatic anatomy obviates any satisfactory revision of the complex until properly preserved material, beginning with *D. damonis* Beddard, 1888, can be studied. As already noted (Gates, 1958) a *bolaui* group, comprising Asiatic species, can be characterized as follows.

Digestive system, with two gizzards in front of 8/9 (segmental allocations?), with a pair of calciferous glands, an intestinal origin in xix, a lamelliform typhlosole beginning in region of xxii–xxiii and lateral typhlosoles through some or all of xxiii–xxix, but without caeca and supra-intestinal glands. Calciferous glands, trilobed, a vertically reniform lobe in each of xv–xvii, the common duct (short and slender) of the three lobes opening into gut dorsolaterally in xvi. Vascular system, with complete dorsal (single) and ventral trunks, a supra-esophageal but no subneural, paired extra-esophageals median to the hearts and with connectives to the supra-esophageal anteriorly in each of x–xii (posterior continuations?), posterior lateroparietal trunks that pass to the supra-esophageal (and/or extra-esophageals?) posteriorly in xiii. Hearts, of ix and anteriorly lateral, of x–xii lateroesophageal. Excretory system, meroic, nephridia exoic (except in ii–iv and there enteroic?), astomate and tubular back to region of xx–xxi, posteriorly discoidal, in several longitudinal rows, astomate except (behind the typhlosole?) where the median nephridium on each side has a preseptal funnel. Discoidal nephridia with cells of peritoneal investment enlarged (and filled with fats or other reserve food substances?). Setae, 8 per segment, closely paired,

---

[1] However, some species have been thought, perhaps mistakenly, to have only one gizzard!

AA $ca.=BC$, $DD > \frac{1}{2}C$, a and b of prostatic segments penial. Dorsal pores, present.

Genital apertures, minute, superficial, male and prostatic pores in seminal grooves, spermathecal pores at or close to 7/8–8/9. Clitellum, intersegmental furrows obliterated, dorsal pores occluded, setae retained. Metagynous, ovaries fan-shaped and with several egg-strings. Ovisacs, paired, in xiv.

Remarks: Discordant determinations of segmental location of organs, in presumably uniform material, have plagued others than Stephenson (1924, *Rec. Indian Mus.* 26: p. 132). He has, however, in the reference just cited, called attention to some of the conditions that cause trouble, even in microtome sections, such as thinness, apposition, incompleteness, and absence of certain septa as well as dislocation of their parietal insertions. Delicate septa may be so completely destroyed as to be unrecognizable merely by pinning out a specimen to permit examination of the internalia.

Individuals of some species have been thought to show stages in uniting the two gizzards to produce a single large organ as in *Monogaster*. However, strong contraction now seems more likely to have so approximated discrete gizzards as to conceal the soft annulus between them.

Calciferous glands often have been said to be three pairs. Presence of only one pair of ducts and of only one pair of apertures into the gut permits an assumption that one pair of glands is evaginated from the gut in that segment in which the ducts open and that during development penetration into the two adjacent segments is associated with demarcation into three more or less reniform lobes. However, correct characterization of the glands may require knowledge of their embryonic development.

The rudimentary seminal vesicles found by Stephenson in segment x suggest that in *Dichogaster*, as in so many other genera, an octovesiculate stage - like that still existing in certain species of the lumbricid genus *Allolobophora*—had been involved in the ancestry.

Species of the *bolaui* group can be derived from the hypothetical protoramiella by appearance of a gizzard in vi, extension of esophagus posteriorly, evagination of extramural calciferous glands behind xiii (in xvi?), development of typhlosoles and acquisition by prostates of ability to induce formation of penial setae in adjacent ventral follicles. As to the cephalization that has been evolving in the group, little is known except that segmental allocation of the gizzards is uncertain because of the thinning and/or abortion of septa along with some dislocation of parietal insertions. Demarcation externally of segments i and ii from each other seems to be disappearing and even in best preservation 1/2 may be unrecognizable or only faintly indicated near mD so that the first setae appear to be in the peristomium. Further continuation of the evolution seemingly under way, by elimination of the setae of ii along with the last traces of 1/2, would result in a regional anterior homoeosis of all organs such as characterizes *Nelloscolex* and *Tonoscolex*.

Each of the Asiatic species undoubtedly is exotic in India and Burma but whether their original home is in the tropics of Africa or America is unknown.

KEY TO BURMESE SPECIES OF *Dichogaster*

1. Quadriprostatic.......................... 2
   Biprostatic.............................*saliens*
2. Female pore (single), at mV.............*bolaui*
   Paired female pores present................ 3
3. Median genital markings usually present....*affinis*
   Genital markings lacking.................... 4
4. Spermathecal duct bulbous, penial setae of one
      kind tapering to a hairlike process...*modiglianii*
   Spermathecal duct not bulbous, penial setae
      without a hairlike ectal termination...*curgensis*

Note: Two species recorded from India are omitted above. *D. parva* (Michaelsen, 1896) is inadequately characterized but appears to be like Stephenson's variety of *curgensis*. *D. travancorensis* (Fedarb, 1898), if allocation of the calciferous glands to xiv–xvi was correct, does not belong in the *bolaui* group even if it belongs in *Dichogaster*.

Dichogaster affinis

1890. *Benhamia affinis* Michaelsen, *Mitt. Naturhist. Mus. Hamburg* 7: p. 9. (Type locality, Quilimane, Zanzibar. Type, in the Hamburg Museum.)
1917. *Dichogaster affinis*, Stephenson, *Rec. Indian Mus.* 13: p. 413. (Tale Sap.)
1931. *Dichogaster sinuosus* Stephenson, *Proc. Zool. Soc. London* 1931: p. 74. *Rec. Indian Mus.* 33: p. 200. (Types, in the British Mus.)
1942. *Dichogaster affinis*, Gates, *Bull. Mus. Comp. Zool. Harvard College* 89: p. 128.
1958. *Dichogaster affinis*, Gates, *Ann. Mag. Nat. Hist.*, Ser. 13, 1: p. 618.
1961. *Dichogaster affinis*, Gates, *Burma Res. Soc. 50th Anniv. Pub. No. 1*: p. 57.

Quadriprostatic, pores at ends of seminal grooves in AB between equators of xvii and xix, male pores at eq/xviii. Spermathecal pores, at or close to A. Female pores, just in front of a. Genital markings (occasionally lacking), median, in AA, across 1–4 of levels 7/8–11/12. Clitellum, annular, though often thinner in AA, xiii, xiv–xxi, xxii. First dorsal pore, at 5/6. Setae, all present in ii, ventral couples usually lacking in xviii. Prostomium, epilobous, tongue narrowing posteriorly (and reaching 1/2?), intersegmental furrow 1/2 often lacking or indistinct. (Unpigmented.) Segments, 105–140. Size, 27–60 by 1–2 mm.

Septa, present from 7/8. Gizzards, in vi, vii (if not v, vi). (Typhlosole, ends in region of lxviii–lxxvi?). Holandric (testis sacs?). Seminal vesicles, vestigial, in xi, xii. Spermathecae, duct longer than ampulla from which it is constricted. Diverticulum, with spheroidal to ovoidal seminal chamber and short stalk passing to ental portion of duct. Penial setae, 0.29–0.43 mm. long, 4–7 μ thick entally, 3–6 μ at midshaft, straight, slightly bowed or slightly sigmoid (two opposite curvatures), sinuous ectally. Tip, bluntly rounded, knobbed or truncate. Ornamentation, of scalelike markings or of teeth in the sinuosities. (GM glands?)

Reproduction: No information now available.

Distribution: Labaw (Mergui). Maungmagaun (Tavoy). Kawkareik (Amherst). Pyapon (Pyapon). Thaton, Bilin, Duyinzeik, Kyaikto (Thaton). Rangoon (Hanthawaddy). Bassein (Bassein). Jungle (Pegu). Toungoo (Toungoo). Thanbula (Thayetmyo). Pyinmana (Yamethin). Magwe, Taungdwingyi (Magwe). Minbu (Minbu). Mount Popa (Myingyan). Dwehla (Kyaukse). Myotha (Sagaing). Taungyi, Maymyo, Kyaukme, Lashio, Namkham (Shan States). Kyaukmyaung (Shwebo). Ingyindaung (Lower Chindwin). Tiangzup (Myitkyina). Thailand. New Caledonia. Mexico. El Salvador. Colombia. French Guiana. Brazil. Haiti. St. Thomas. Cape Verde Islands. Southwest Africa. Madagascar. Zanzibar. Comoro Island. Ceylon. India: Jubbulpore and the Gangetic Plain, Bombay, Baroda, Travancore.

In Salvador at elevations of 450–1,400 m. but in Burma down to sea level.

Habitats: Soil. Rotten wood.

Variation: Genital markings, in Burma, were located as follows; at 7/8 (9 specimens), 8/9 (156), 9/10 (64), 10/11 (6), 11/12 (3).

Regeneration: Head regeneration (homomorphic) apparently is possible at all levels back to 8/9 but number of segments regenerated is unknown. The homoeosis Stephenson found in a Ceylon specimen could have resulted from regeneration of one segment less than had been amputated from the anterior end.

Remarks: Seminal vesicles also have been recorded from x in this species.

The original home of the species is unknown.

Dichogaster bolaui

1891. *Benhamia bolavi* Michaelsen, *Mitt. Naturhist. Mus. Hamburg* 8: p. 307. (Type locality, Bergedorf, Hamburg. Types, in the Hamburg Mus.)

1917. *Dichogaster bolaui*, Stephenson, *Rec. Indian Mus.* 13: p. 413.

1923. *Dichogaster bolaui*, Stephenson (*The Fauna of British India*), *Oligochaeta*, p. 472.

1924. *Dichogaster bolaui*, Stephenson, *Rec. Indian Mus.* 26: p. 132.

1931. *Dichogaster bolaui*, Stephenson, *Proc. Zool. Soc. London* 1931: 1: p. 64. *Rec. Indian Mus.* 33: p. 195.

1936. *Dichogaster bolaui*, Hla Kyaw & Gates, *Jour. Roy. Asiatic Soc. Bengal (Sci.)* 2: p. 166.

1942. *Dichogaster bolaui*, Gates, *Bull. Mus. Comp. Zool. Harvard College* 89: p. 129.

1958. *Dichogaster bolaui*, Gates, *Ann. Mag. Nat. Hist.*, Ser. 13, 1: p. 618.

1961. *Dichogaster bolaui*, Gates, *Burma Res. Soc. 50th Anniv. Pub.* No. 1: p. 57.

1970. *Dichogaster bolaui*, Soota & Julka, *Proc. Zool. Soc. Calcutta* 23: p. 204.

TYPHLOSOLE: Hertling, 1923, *Zeitschr. Wiss. Zool.* 120: p. 190.

Quadriprostatic, pores at ends of seminal grooves at A between equators of xvii and xix, male pores at eq/xviii. Spermathecal pores, at or very close to A. Female pore, equatorial, at mV. (Genital markings, none.) Clitellum, annular though often thinner in AA, xiii, xiv–xviii, xix, xx, xxi/n. First dorsal pore, in region of 5/6–6/7. Prostomium, epilobous, tongue narrowing posteriorly, intersegmental furrow 1/2 often lacking or indistinct. Color, often lacking, red, restricted to dorsum in some or all of i–xii. Segments, 70–98. Size, 20–40 by 1–3 mm.

Septa, present from 7/8. Gizzards, in vi, vii (if not v, vi). Typhlosole, ends in region of lxviii–lxxvi. Holandric, (testis sacs?). Seminal vesicles, (vestigial?), in xi, xii. Spermathecae, duct rather barrel-shaped and of about same size as the ampulla, narrowed in the parietes. Diverticulum, small, digitiform to slightly pyriform from ental portion of duct, directed ventrally. Penial setae, 0.27–0.4 mm. long, 3.5–7.5 μ thick at midshaft. Tip hooked or widened and then scalpel-, spatula-, oar-, or spoon-shaped. Ornamentation of several triangular teeth with point raised away from the shaft.

Reproduction: Presumably biparental as sperm are exchanged during copulation. Cocoon deposition may take place all year round. Worms were sexual at Allahabad from July to May.

Distribution: Car Nicobar. Mergui (Mergui). Maungmagaun (Tavoy). Kyaikto, Sittang (Thaton). Rangoon (Hanthawaddy). Bassein (Bassein). Pegu (Pegu). Tharrawaddy, Myagyaung (Tharrawaddy). Henzada (Henzada). Toungoo (Toungoo). Prome (Prome). Thanbula (Thayetmyo). Taungdwingyi (Magwe). Minbu (Minbu). Pyinmana (Yamethin). Mahlaing (Meiktila). Mount Popa, Taungtha (Myingyan). Dwehla (Kyaukse). Mandalay, Kyaukkyone (Mandalay). Sagaing, Myotha (Sagaing). Kyaukmyaung (Shwebo). Katha (Katha). Wasat Kha, Tingpai (Myitkyina). Taungyi, Maymyo, Lashio, Namkham (Shan States). Ingyindaung (Lower Chindwin). Tiddim (Chin Hills).

Rangamati (Chittagong Hill Tracts, Bangla Desh). Siju Cave (Garo Hills, Assam, India).

Malay Peninsula. Annam. Hainan. Sumatra. Java. Christmas Island. Krakatau. Borneo. Philippine Islands. New Caledonia. New Ireland. New Hebrides. Caroline, Marianna, Palau, Loyalty and Solomon Islands. Australia (Queensland). Hawaiian Islands. Easter Island. California, Florida. Mexico, Panama, Colombia, French Guiana, Venezuela, Bolivia, Brazil. Haiti, Jamaica, St. Thomas, Dominica, St. Vincent, Trinidad, Curacão, Bonaco. Germany.

Cape Verde Islands, Annobon. Togo, Nigeria, Cameroon, Congo, Uganda, Rhodesia, Mozambique, Nyasaland, Natal. Madagascar. Comoro Islands. Ceylon. India: Saharanpur, Dehra Dun, Gangetic Plain, Calcutta, Bombay, Bangalore, Tinnevelly, Cochin, Travancore. Pakistan: Lahore, Kathiawar. **Habitats:** In earth around roots of potted plants (Burma, India). Greenhouses (California, Florida). Fermenting bark of a tannery (Germany). In plumbing of bathtub (Allahabad). Soil, with pH of 7.5 in banks of ditches draining waste effluents of human habitations (Allahabad). In trees (Burma). Found among betel leaves in a bazaar stall (India).

Biology: Activity is uninterrupted at Burmese and Indian sites where moisture is adequate during the dry season. At Allahabad activity was recorded for months of July to May.

Abnormality: Some of the aberrations in three specimens (Gates, 1958) doubtless resulted from suppression of an embryonic somite and from halving of an embryonic somite on one side only. Other abnormalities could not have been produced merely by halving mesoblastic somites.

Absence of the spermathecal diverticulum as well as doubling of it has been recorded from this species.

Regeneration: Head regeneration (homomorphic) is possible back to 8/9 at least. As many as seven excised segments may be replaced. Tail regenerates (homomorphic) were seen at various levels between 33/34 and 62/63. Regeneration of 48 segments at 44/45 suggests a possibility of nearly complete restoration of the lost part in appropriate circumstances (*cf.* segment number in précis).

Remarks: Rudimentary seminal vesicles that "certainly would not have been visible in a dissection" were found by Stephenson (1924) in segment x.

Dichogaster curgensis

1921. *Dichogaster curgensis* Michaelsen, *Mitt. Naturhist. Mus. Hamburg*, **38**: p. 54. (Type locality, Moonad, Coorg, South India. Types, in the Hamburg. Mus.)
1931. *Dichogaster curgensis* var. *unilocularis* Stephenson, *Proc. Zool. Soc. London* 1931: p. 69. *Rec. Indian Mus.* **33**: p. 197. (Types, in the British Mus.)
1961. *Dichogaster curgensis*, Gates, *Burma Res. Soc. 50th Anniv. Pub. No.* 1: p. 57.

Quadriprostatic, pores at ends of seminal grooves between equators of xvii and xix, male pores at eq/xviii(?). Spermathecal pores, at or near A. Female pores, at sites of missing a setae (Michaelsen) or anteromedian to a (Stephenson). (Genital markings, none.) Clitellum, annular, though often thinner in AA, xii, xiii–xx. First dorsal pore, in region of 4/5–5/6. Setae, all present in ii(?), ventral couples of xviii lacking(?). Prostomium, epilobous (tongue narrowing posteriorly?). (Unpigmented.) Segments, 90–110. Size, 21–75 by 1–2 mm.

Septa, 6/7–7/8 present and thickened(?). Gizzards, in vi, vii(?). (Typhlosole?) Holandric (testis

sacs?). Seminal vesicles, lacking or vestigial and then in xi, xii or in xii only. Spermathecae, rather slender, ampulla much shorter than but not especially thicker than nor externally distinguishable from the duct. Diverticulum, pendent from ental end of the duct, with one, three, or four seminal chambers at end of a short stalk or lumen of the stalk may be widened and like a chamber. Penial setae, 0.73–1.0 mm. long, 9 μ thick entally, 3.5–6.5 μ at midshaft, slightly bowed, irregularly sinuous ectally. Tip, tapering to a fine point or rounded (as a result of erosion?). Ornamentation, a toothlike projection at ental side of a sinuosity or a double series of scars the proximal border of which is a single tooth.

Reproduction: Presumably biparental in specimens referred to var. *unilocularis* as sperm were present on the male funnels and in the spermathecae. However, if testis sacs and seminal vesicles are absent, sperm seemingly would have to be matured in the testes a condition that is known to be associated with parthenogenesis but that has not been recorded from forms with obligatory biparental reproduction.

Distribution: Lashio, Kutkai (Northern Shan States).

India (Moonad and Bhagamanola in Coorg.)

The species obviously is exotic in India as well as in Burma.

Remarks: The spermathecal ampulla seems to be poorly developed in this species. Incomplete or aberrant development of the organ is likely if reproduction is parthenogenetic. The diverticulum joins the ampulla (Stephenson) or the duct (Michaelsen). The difference may amount to little more than one of definition but parthenogenetic degradation that is so often illustrated by spermathecae may have been involved.

The Burmese variety was distinguished, as the name suggests by quantitative differences.

Michaelsen thought this species might be the same as Fedarb's inadequately (and perhaps erroneously) characterized *D. travancorensis*. Stephenson, admitting that some of the differences between the two could be attributed to errors in Fedarb's account, was inclined to believe that the spermathecae obviated identity.

Dichogaster modiglianii

1896. *Benhamia modiglianii* Rosa, *Ann. Mus. Sto. Nat. Genova* **36**: p. 510. (Type locality, Padang, Sumatra. Type, in the Genoa Mus.)
1931. *Dichogaster modiglianii*, Stephenson, *Proc. Zool. Soc. London* 1931: p. 65. *Rec. Indian Mus.* **33**: p. 198.
1942. *Dichogaster modiglianii*, Gates, *Bull. Mus. Comp. Zool. Harvard College* **89**: p. 130.
1958. *Dichogaster modiglianii*, Gates, *Ann. Mag. Nat. Hist.*, Ser. 13, **1**: p. 620.
1961. *Dichogaster modiglianii*, Gates, *Burma Res. Soc. 50th Anniv. Pub. No.* 1: p. 57.

Quadriprostatic, pores at ends of seminal grooves at A and between equators of xvii and xix, male pores at eq/xviii. Spermathecal pores, at or very close to *A*. Female pores, just median or posteromedian to *a* setae. (Genital markings, none.) Clitellum, annular though often thinner in *AA*, xiii–xx. First dorsal pore, in region of 4/5–5/6. Prostomium, epilobous, tongue narrowed posteriorly and (as a median groove?) reaching 1/2. (Unpigmented.) Segments, 76–120. Size, 22–60 by 1–2 mm.

Septa, 7/8 as well as 5/6–6/7 lacking (?). Gizzards, in vii, viii(?). Typhlosole, ends in region of lxxviii–lxxxi. Holandric, testis sacs unpaired, that of xi formed by peripheral apposition of 10/11–11/12, that of x ventral. Seminal vesicles, lacking or vestigial and then in xii. Spermathecal duct slightly bulbous and longer than the ampulla from which it is constricted. Diverticulum, with a small, spheroidal to shortly ellipsoidal seminal chamber connected by a very short stalk to middle of the duct. Penial setae, 0.31–0.42 mm. long, 5–9 µ thick entally, straight or slightly bowed. Tip, slightly thickened or truncate or narrowed to a short filament and then straight or recurved (hooked). Ornamentation, of scalelike markings near slight constrictions.

Reproduction: Presumably biparental as sperm are exchanged during copulation.

Distribution: Port Blair, Haddo, Pahargaon (Andaman Islands). Mergui (Mergui). Maungmagaun (Tavoy). Pyapon (Pyapon). Kyaikto, Sittang (Thaton). Twante, Kayan, Rangoon (Hanthawaddy). Bassein (Bassein). Pegu, Nyaunglebin (Pegu). Tharrawaddy (Tharrawaddy). Toungoo, Leiktho (Toungoo). Magwe, Taungdwingyi (Magwe). Pyinmana (Yamethin). Mahlaing (Meiktila). Mount Popa (Myingyan). Sagaing (Sagaing). Taungyi, Lashio (Shan States). Naba, nearby hills (Katha).

Malay Peninsula. Sumatra. Sumba. Philippine Islands. New Caledonia. New Britain. New Hebrides. Banks Island. Colombia, French Guiana, Brazil.

(In Africa, where its original home presumably is to be sought, the species never has been recognized.) India, Calcutta. Pakistan, Lahore.

Habitats: Soil, especially if rich in humus. Trees.

Dichogaster saliens

1893. *Microdrilus saliens* Beddard, *Proc. Zool. Soc. London* 1892: p. 683. (Type locality, undesignated? Types, supposedly from Penang, Singapore, and Java, but obtained from earth in Wardian cases at the Kew Gardens, probably have been lost.
1931. *Dichogaster saliens*, Stephenson, *Proc. Zool. Soc. London* 1931: p. 65. *Rec. Indian Mus.* 33: p. 199.
1942. *Dichogaster saliens*, Gates, *Bull. Mus. Comp. Zool. Harvard College* 89: p. 134.
1958. *Dichogaster saliens*, Gates, *Ann. Mag. Nat. Hist.*, Ser. 13, 1: p. 620.
1961. *Dichogaster saliens*, Gates, *Burma Res. Soc. 50th Anniv. Pub.* No. 1: p. 57.

Biprostatic, pores at anterior ends of seminal grooves on *A* and extending from eq/xvii to 17/18, male pores at posterior ends of grooves which are on posterior faces of protuberances from a transversely, diamond-shaped field. Spermathecal pores, at or close to *A*. Female pores, just median or slightly posteromedian to *a* setae. Genital marking (often lacking), in *AA*, across 15/16. Clitellum, annular though often thinner in *AA*, xiii, xiii/n–xix/n, xix, xx/n. First dorsal pore, in region of 3/4–6/7. Setae, all present in ii, ventral couples of xviii usually lacking. Prostomium, epilobous, tongue narrowed posteriorly and (only as a median groove?) reaching site of 1/2, intersegmental furrow 1/2 indistinct or lacking. (Unpigmented.) Segments, 73–120. Size, 17–70 by 1½–2½ mm.

Septa, 7/8 as well as 5/6–6/7 lacking(?). Gizzards, in vii, viii(?). Typhlosole, ends in region of lxxxi–lxxxviii. Holandric, (testis sacs?). Seminal vesicles, lacking or vestigial and then in xi, xii or xii. Spermathecal duct, slightly bulbous and longer than the ampulla from which it is constricted. Diverticulum, with a small spheroidal to shortly ellipsoidal seminal chamber and a very short stalk to anterior face of duct entally. Penial setae, 0.4–0.71 mm. long, 6–13 µ thick, nearly straight entally, or slightly bowed, more or less obviously sinuous ectally. Tip, knob-shaped, tapering to a point or to a short filament. Ornamentation, of one or more scalelike markings in each sinuosity or of faint ridges. GM gland, within parietes, enclosed in a thin capsule.

Reproduction: Presumably biparental as sperm are exchanged during copulation.

Distribution: Rangoon (Hanthawaddy). Toungoo, Thandaung (Toungoo). Taungyi, Maymyo, Lashio, Kutkai, Namkham (Shan States). Wasat Kha, Tingpai, Sumprabum, Mythonkha (Myitkyina). Penang. Malay Peninsula. Java. Christmas Island. Australia. California. El Salvador. Panama. Congo, Uganda, South Africa. Ceylon, India: Darjiling.

Burmese localities, except Rangoon and Toungoo, are in the hills. Salvador specimens were collected at 1,000 m.

Habitats: Soil. Manure. In earth around roots of potted plants in greenhouses.

Eutyphoeus

1883. *Typhoeus* Beddard, *Ann. Mag. Nat. Hist.*, Ser. 5, 12: p. 219. (Type and only species, *T. orientalis* n. sp.)
1888. *Typhaeus*, Beddard, *Quart. Jour. Micros. Sci.* 28: pp. 403; 29: pp. 111, 117.
1889. *Typhaeus*, Vaillant, *Hist. Nat. Anneles, Paris* 3: 1, p. 182.
1890. *Typhaeus*, Benham, *Quart. Jour. Micros. Sci.* 31: p. 122.
1895. *Typhaeus*, Beddard, *A Monogr. of the Order of Oligochaeta* (Oxford), p. 472.
1900. *Eutyphoeus* Michaelsen, *Das Tierreich* 10: p. 322. (Nom. nov. pro *Typhoeus* Beddard, 1883, preoccupied by *Typhoeus* Leach, 1815, Coleoptera.)
1901. *Typhoeus*, Beddard, *Proc. Zool. Soc. London* 1901: p. 205.
1909. *Eutyphoeus*, Michaelsen, *Mem. Indian Mus.* 1: p. 216.

1922. *Eutyphoeus*, Michaelsen, *Mitt. Naturhist. Mus. Hamburg* **38**: p. 37.
1923. *Eutyphoeus*, Stephenson (*The Fauna of British India*), *Oligochaeta*, p. 420.
1930. *Eutyphoeus*, Gates, *Rec. Indian Mus.* **32**: p. 327. Stephenson, *The Oligochaeta* (Oxford), p. 847.
1933. *Eutyphoeus*, Gates, *Rec. Indian Mus.* **35**: p. 560.
1938. *Eutyphoeus*, Gates, *ibid.*, **40**: p. 60.
1939. *Eutyphoeus*, Gates, *ibid.*, **41**: p. 213.
1958. *Eutyphoeus*, Gates, *ibid.*, **53**: p. 93. Omodeo, *Mem. Inst. Français Afrique Noire*, No., **53**: p. 22.
1959. *Eutyphoeus*, Gates, *Bull. Mus. Comp. Zool. Harvard College* **121**: p. 247.
1961. *Eutyphoeus*, Gates, *Ann. Mag. Nat. Hist.*, Ser. 13, **3**: p. 653.

CASTINGS: Bahl, 1934, *Quart. Jour. Micros. Sci.* **76**. Hla Kyaw & Gates, 1937, *Jour. Asiatic Soc. Bengal (Sci.)* **2**. Nijhawan & Kanwar, 1952, *Indian Jour. Agric. Sci.* **22**. Roy, 1958, *Proc. Zool. Soc. Calcutta* **10**. EXCRETORY SYSTEM: Bahl, 1942, *Quart. Jour. Micros. Sci.* **83**. LUMINESCENCE: Gates, 1925, *Rec. Indian Mus.* **27**. Gates, 1944, *Current Sci.* **13**. SUPRA-INTESTINAL GLANDS: Bahl & Lal, 1933, *Quart. Jour. Micros. Sci.* **76**. Thapar, 1934, *Ann. Mag. Nat. Hist.* **13**. Gates, 1938, *Rec. Indian Mus.* **40**: p. 53. PHYLOGENY: Stephenson, 1923 (*The Fauna of British India*), *Oligochaeta*, p. 363. Gates, 1938, *Rec. Indian Mus.* **40**: pp. 44–47. Gates, 1958, *Idem* **53**: pp. 205–208. WATER RELATIONSHIPS: Bahl, 1934, *Quart. Jour. Micros. Sci.* **76**.

Digestive system, with a gizzard (belonging to vi?) in the space between 5/6 and 8/9, one pair of intramural calciferous glands in xii, intestinal origin in xv, typhlosole terminating posteriorly with a short series of supra-intestinal glands, unpaired, anteriorly directed, small midventral caeca one each in a number of consecutive segments in front of supra-intestinal glands. Calciferous glands, longitudinally hemi-ellipsoidal with flat faces mesially, with numerous transverse vertical partitions and interlamellar spaces communicating dorsally with the esophageal lumen which in xii is T-shaped in cross section. Vascular system, with dorsal (single), ventral (complete) and supraesophageal trunks but no subneural, paired lateroparietal trunks from the anal region passing to hind ends of calciferous glands, paired extra-esophageal trunks median to hearts and passing to front ends of calciferous glands. Hearts, of x–ix and anteriorly lateral, of xi–xiii latero-esophageal. Excretory system, meroic, all nephridia small and avesiculate, numerous astomate biramous and Y-shaped nephridia of iii opening into pharynx, remainder of the system exoic and comprising astomate, biramous and Y-shaped, parietal nephridia which are numerous from v through clitellar segments but posteriorly are in longitudinal ranks, the median nephridium of each side behind the supra-intestinal glands discoidal and with preseptal funnel. Septa, 4/5–5/6 muscular, 6/7–7/8 aborted, 8/9–10/11 thickened and crowded together behind their normal locations, 11/12 approximated to 10/11. Setae, four pairs per segment. Dorsal pores, present. Segments, more than 150.

Biprostatic, pores minute, in region of *AB*, near eq/xvii. Male pores, minute, near to but behind prostatic pores and like them never superficial in

adults. Bithecal, pores superficial, never minute, at 7/8. Female pores, minute, in xiv. Clitellum, annular, extending beyond xiv and xvi, intersegmental furrows obliterated, dorsal pores occluded, setae retained. Subterminal portion of sperm duct thickened as a bulbus ejaculatorius. Ovaries, fan-shaped and with several egg-strings. (Ovisacs, lacking?) Spermathecal diverticula, to ental end of short ducts.

Reproduction: Sperm are exchanged during copulation by nearly all species known from mature material. Genital aberrations of the kinds associated with parthenogenetic polymorphism have been found but rarely. Reproduction probably is biparental, possibly obligatorily so, throughout the genus. Additional mention of reproduction for each species, accordingly seems unnecessary.

Breeding, in plains of Burma, is restricted to a single period centered in September, but may be somewhat later at higher elevations. At Allahabad, breeding extended from a latter part of August into or through October and at a few sites even into November.

Distribution: Burma, from Tenasserim division and western margin of Shan Plateau into the Arakan Yomas. India and Pakistan, from the Burma border into the Gangetic plain and west through the Himalayas beyond Nepal, from sea level to 15,500 feet.

The proper generic range as thus indicated has been extended somewhat, presumably quite recently and as a result of transportation by man of two or three Indian species.

A record for Ceylon almost certainly is erroneous. Colonizing ability may well be rather limited.

Habitats: Soil. A thin layer of mud on lichen-covered rocks at elevations of 13,000–15,500 feet in Nepal was occupied by one species.

Biology: All species are inactive (probably deep down in the soil) during much of the year and in Burma appear to pass into that state at end of a breeding season regardless of continuation or cessation of rainfall. Indian species, according to Bahl (1927), feed and copulate on the surface but such activity never was recorded from Burma. All forms that were studied *in vivo*, those at Rangoon and Allahabad (Gates, 1925, 1944), are luminescent, ejecting through dorsal pores on appropriate stimulation a fluid that soon glows more or less brilliantly.

Castings: All species, so far as could be determined from *in vivo* observations and examination of intestinal contents, are geophagous. Much of the intestinal ejecta is deposited on the surface of the ground in tower-like castings. Attribution (Roy, 1958) of all large castings in Bengal and Bihar to a single species (*E. waltoni*) probably is erroneous.

Abnormality: Abnormalities of the sort that are common in parthenogenetic polymorphism are rare in *Eutyphoeus*. Among many more than 20,000

Burmese and Indian specimens only 2 athecal individuals were found. Four specimens lacked one or the other of the two spermathecae. Male terminalia appeared from external examination to be entirely lacking in one Indian specimen (Gates, 1958) in which, quite unexpectedly, rudiments of prostates and penisetal follicles were found internally. Male terminalia were lacking on one side in one specimen each of 5 species. These statistics together with the data *re* sperm exchange during copulation warrant a belief that parthenogenesis, if at all present in the genus, is rare.

Many of the abnormalities that were found are explainable as results of suppression or halving of mesoblastic somites during embryogenesis.

The above-mentioned individual (Gates, 1958) of *E. nicholsoni* had copulated with a presumably normal partner which would not have received any male gametes during the process. Absence of sperm in spermathecae of a fully mature earthworm cannot always be proof that copulation had not taken place!

Regeneration: Some evidence is available to indicate ability to replace the first two cephalic segments. Although posterior amputees were common, no tail regenerate was ever found during examination of many more than 20,000 specimens. Posterior amputation seemingly is followed by enteroparietal healing and some remodeling of the terminal segment rather than by regeneration of a pygomere. All capacity for tail regeneration may have been lost throughout the genus.

Systematics: Gross anatomy of digestive, circulatory, and excretory systems is fairly uniform throughout the genus in environments ranging from semidesert to rainforest and from tropical sea level to Himalayan heights in the temperate zone. Intrageneric differences in those three systems concern: (1) the median typhlosole, (2) the ventral intestinal caeca, (3) the supra-intestinal glands, (4) the lateral intestinal typhlosoles, (5) the lateral intestinal caeca (when present), (6) the dorsal trunk and associated blood vessels in the first seven segments of the body, (7) nephridia of the postclitellar portion of the body. Knowledge of No. 7 is inadequate for systematic use, though there is some variation as to number of longitudinal ranks. The number of segments with stomate nephridia may be the same as the number of atyphlosolate segments or as the number behind the last supra-intestinal gland. No. 6 provides only three characters; unaborted, lacking in front of hearts of vi, lacking in front of hearts of vii. Exceptional specimens in several species pose difficult questions as to whether the unusual condition is merely an unmodified persistence of an ancestral condition in a small and laggard minority or an individual reversion to that same ancestral condition. Nos. 1–5 provide characters that presently are of little systematic use. The lateral caeca seemingly are just now evolving in some species but even when well established they are in the same segment as the lateral typhlosoles and the beginning of the major typhlosole. Location of supraintestinal glands, in some species, may be useful for characterizing local populations.

Intrageneric differences in the genitalia involve every organ except the ovaries and prostates. Among the more important are the following: Testes present in x–xi (holandry) or only in x (pro-andry) or only in xi (metandry). Presence (with or without part or all of its duct) or absence of male funnels in a segment where testes are aborted. Right oviduct, functional or functionless. Testis sac, lacking, annular, U-shaped and containing hearts of the segment, subesophageal and then with or without secondary extensions dorsally that do not include the hearts. Spermathecae, uni-, bi-, or multi-diverticulate. Spermathecal diverticula, uniloculate to multiloculate, discrete or united in one way or another. Male terminalia, avestibulate, uni- or bi-vestibulate, without or with intromittent organs that may be porophores capable of temporary elevation, eversible invaginations, protrusible annular or tubular penes, without or with penial setae and in the latter case with or without size graded reserves. Genital markings, lacking or present and then paired or unpaired, bi-segmental or segmental and then pre- or post equatorial, without or with definite glands that may or may not be encapsulated. Accordingly the systematics of *Eutyphoeus* must be based mainly on genital characters and especially on those that are macroscopically recognizable, i.e., without greater magnification than is provided by the binocular dissecting microscope, until such future time as histological, cytological, neurological, biochemical, or other substitutes become available.

The male pores, apertures of the sperm ducts, according to Beddard (1888), are one pair, opening in common with or close to the prostatic ducts. The same author, defining the genus in 1895, merely stated that the male pores are in xvii. Whether "each" is to be understood in Michaelsen's Tierreich definition (1900), "Männlichen Poren und Prostataporen 1 Paar am 17 Segm.," is not clear, at least to the writer. Subsequently the character involved was stated (Stephenson, 1923, 1930) in the following way, "Sexual apparatus purely microscolecine (conjoined pores of prostates and male ducts on xvii...)." Although the wording of each of those three authors may be defensible, the condition of the male terminalia certainly was misunderstood as we now know that in the genus *Eutyphoeus* the microscolecin reduction never was completed. Sperm and prostatic ducts, in partial agreement with Beddard, can be said to open in common to the exterior but only through the intermediation of fissures and/or vestibula. Spermiducal pores, when first visible in young juveniles of *Eutyphoeus*, are superficial and behind eq/xvii. Prostatic pores, also originally superficial, at first appearance are in front of eq/xvii. As growth continues apertures of the *a* follicles move close to those of the *b* follicles and both disappear into a slight cleft in an outer portion of the body wall. The prostate pore then moves back to acquire its definitive position on the anterior wall of the slit as the spermiducal pore moves forward to assume its definitive position on the posterior wall. The slit or "fissure" with the permanently discrete follicle apertures on its roof is not, as once was thought, a rudimentary vestibulum.

Around the fissure aperture an annular tumescence sometimes is recognizable in circumstances that did not allow a decision as to whether the annulus was an ephemeral protuberance or a definite organ. Any permanent protrusion from general surface of the body is unlikely in soil-boring megadriles and in taxa with a definitive annulus the latter is within a vestibulum. Elongation of the annulus together with deepening of the vestibulum produces a tubular penis that warrants calling its evolutionary precursor an annular penis. Penisetal follicle apertures remain discrete on the roof of the penial lumen as do the spermiducal and prostatic pores on its posterior and anterior walls. The opening at ventral end of the penis allows protrusion of penial setae at the same time sperm with the prostatic secretion are extruded and so is a male pore though because of past evolution only in a secondary sense. The penes, being more or less deeply withdrawn into vestibula are protruded during copulation through the superficial vestibular apertures which are male pores in a tertiary sense.

Among other taxonomically useful genital characters also ignored or unrecognized by classical specialists are those provided by the testis sac. The coelomic cavity of xi always is small, often because of close approximation and eventually apposition, except of course ventrally, of its bounding septa. By changes that cannot now be precisely characterized but that seem to involve mainly the integrity of septum 11/12 the organs of xi become enclosed in an annular testis sac surrounding the gut. Closing off a dorsomedian portion of the annulus in other species leaves a U-shaped sac with the hearts of xi in the limbs alongside the gut. Those limbs eventually are closed to sperm so as to leave a ventromedian sac. Hearts of xi in species thus characterized are bound down to the gut by delicate tissue that may be remnants of the dorsal and lateral portions of septum 11/12. A ventral sac sometimes has become enlarged by outgrowth from the sac of dorsally directed limbs so that once again the sac approximates more or less closely the previous U-shape. Proof that such shape is secondary and a reacquisition is provided by the hearts of xi still bound down to the gut outside of those dorsally directed limbs.

For critiques of characters formerly believed to have greater systematic importance previous publications (Gates, 1938: pp. 47–60 and 1958: pp. 96–98) may be consulted.

Although somatic and genital characters, previously unused, have enabled some synonymizing, greater precision in generic definition and in some species characterizations, the validity of certain species names remains dubious for reasons such as paucity of material (a grave handicap when individual variation is so "extraordinary" as was noted by Stephenson, 1923: p. 428), poor preservation or even immaturity of all available specimens.

Evolution, seemingly still under way and perhaps rapidly, also has posed systematic problems that remain unsolved. Characters which in many parts of the generic range seem to be stable and species definitive either are now undergoing or recently have undergone modification in a few individuals or in very small, local populations. Some of those changes are of standard sorts that already had been made elsewhere at various times and places in the history of the genus. Erecting species of numerically small size and very restricted distribution seemingly could be avoided while writing the last treatise on this genus (Gates, 1958) only by arbitrary delimitation of several species. As a result each of several small local populations probably were segregated into two species though at certain sites some of the common anatomy was different from that of either species elsewhere.

Information available as to number and location of ventral intestinal caeca and the supra-intestinal glands has been condensed into two brief formulae of the species précis. The caeca always are in consecutive segments and those mentioned in the formulae are the anteriormost and the posteriormost in which the pockets were found. The supra-intestinal glands also are always in consecutive segments. A formula such as "xcv, cxxxiv–ciii, cxl" indicates that the anteriormost gland of the individuals examined was in a region including the 95th to 134th segments and that the posteriormost gland of the same specimens was in a region including the 103rd to 140th segments. Information as to individual and geographical variation will be found in the author's publications on the genus. The supra-intestinal glands have been characterized as two pairs per segment, with slight anteroposterior increase in size, occasionally only one pair present in the first gland segment. Bahl & Lal (1933), however maintained that in spite of the apparent lobing only a single gland is present and that it is continuous not only from one side of a segment to the other but also from one segment to the next. As the organ supposedly arises as paired outgrowths of intestinal epithelium there must then have been median and anteroposterior fusions but so as to leave dorsal lobing as evidence of the process. The gut behind the glands, being without cilia and rodlets present in the epithelium anteriorly, was called a rectum. (Relationships to the proctodaeum?)

The median nephridium on each side of the body in the post-typhlosolar region is, according to Bahl, sac-shaped. That characterization is likely to be misunderstood as there is no saccular or vesicular widening of the slender excretory tubule. Enlarged cells of the investing peritoneum, presumably distended with metabolites, conceal all loops so that the organ appears, at first glance, to be a dorsoventrally flattened thin, white bit of tissue. Accordingly, characterization as discoidal is preferred to saccular.

Two diverticula, one on the median and the other on the lateral side of the spermathecal duct characterize so many species that a bidiverticulate condition was believed to be primitive. Specializations, the nature of some not determinable from external inspection, are various. Elimination of the median or of the lateral diverticulum. Division of a distal portion of a diverticulum into a number of seminal chambers to produce sometimes a berrylike appearance. Bending both diverticula posteriorly, sometimes until in contact with each other, then to be followed by union so as to seem to be a single posteriorly directed diverticulum. Shortening of the diverticular stalk until seminal chambers are sessile, in a median and/or a lateral group, or even in a horizontal band on the posterior face of the duct. In the latter case, two openings into the duct, one on either side of the vertical ridge on the posterior wall of the duct internally, testifies to a double origin.

Phylogeny: Calciferous glands of *Eutyphoeus* are the same as in the Himalayan *Scolioscolides* Gates, 1937, known only from five types of the single species, and as in the supposedly Gangetic *Bahlia* Gates, 1945, also known at present only from a few specimens of the type species. The three genera are so much more closely related than to any other Asiatic octochaetid that a separate derivation from the hypothetical protoramiella seemed (Gates, 1958) to be necessary. The evolution thus envisioned required shifting the gizzard from v into vi, elongating the esophagus so that the intestinal origin is in xv, and, perhaps somewhere in an area comprising northern India and Burma, acquiring calciferous glands by developments suggested by present conditions in certain species of *Ramiella* and *Pellogaster* Gates, 1939. They involved lengthening bifurcations of a ventromedian esophageal typhlosole in xi–xii (if not also in x and xiii) and their union with median margins of transverse calciferous lamellae that reach nearly to the top of the gut. Sooner or later there were acquired in the ancestry of the *Bahlia* group, a pair of hearts in xiii, short lateral and a long median intestinal typhlosoles as well as supra-intestinal glands. From some such a form *Bahlia* diverged, presumably in India, by eliminating the spermathecae of viii as well as all prostates except those in xvii.

Meanwhile, in the *Eutyphoeus* line, calciferous glands of xi and spermathecae of ix were lost, ventral intestinal caeca were acquired and septa 6/7–7/8 got into the habit of aborting ontogenetically. Segregation of *Scolioscolides* involved elimination of lateral typhlosoles and of all prostates except those in xviii, near disappearance of the median typhlosole and passage of spermiducts into ental ends of the prostatic ducts. Meanwhile, all prostates except those of xvii were being eliminated in the immediate ancestry of *Eutyphoeus*. The center of *Eutyphoeus* evolution may have been in the mountain wall between India and Burma and on such an assumption subsequent developments and distribution were explained (Gates, 1958), from data available in 1945, as follows.

From the Indo-Burmese mountain wall an ancestral proto-eutyphoeus passed across Burma into a region comprising the western edge of the Shan plateau as well as the southern hills down through Amherst district to the Ye River where it became segregated into species of the mero-andric *levis* group. In the opposite direction the proto-eutyphoeus penetrated into the Indian lowlands where, without loss of holandry, it became differentiated into *E. incommodus*, *annandalei*, and *quadripapillatus*. To the north there were segregated off species of the Himalayan equivalent of the *levis* group. Subsequently, migration to the east from the mountain wall brought into Burma a race that became *E. hastatus* and which gave rise in the central yomas to species of the *constrictus* group. The oldest of these has hardly escaped from the Pegu Yomas, but another, *E. peguanus*, was able to cross the Sittang into *levis* group territory to get down only to the mouth of the Salween, but elsewhere and perhaps more recently was able to spread through the deltas region. The same ancestor of *E. hastatus*, or a race with a similar evolutionary capacity, passing westward gave rise to the trans-Arakan equivalent of the *constrictus* group (*E. scutarius*, *comillahnus* and others, all of uncertain status). Finally, from that same wall, a proto-foveatus, passing east into the Chindwin-Irrawaddy axis, was segregated into species of the *foveatus* group, due south in the Arakan Yomas into species of the *gigas* group, and to the west into the *gammiei* complex, possibly still further on and in the Indian lowlands, into part or all of the *orientalis* group.

Remarks: Cephalization in *Eutyphoeus* at the generic level involved elimination of the nephridia in ii, considerable specialization of the nephridia in iii, and abortion of septa 6/7–7/8 (instead of 5/6–6/7 as in *Octochaetona*). Further evolution intragenerically resulted in abortion of the dorsal blood vessel and its branches in an increasing number of segments, the climax reached in species such as *E. gigas* and *plenus* in which that portion of the trunk anterior to hearts of vii is lacking. Nothing is known as to other changes in the circulatory system consequent on the abortions.

A vessel from each posterior lateroparietal trunk passes to the supra-intestinal glands. Rhythmic contraction waves, 21–25 per minute, were recorded by Bahl & Lal (1933) as blood flowed into, through and then out of the glands into the dorsal blood vessel. The glands, according to the same authors, are hepatopancreatic, secreting a tryptic enzyme and storing glycogen. Presumably *Eutyphoeus*, as well as certain other octochaetid genera, has two portal systems, one associated with the supra-intestinal glands, the other with the calciferous glands.

The pharyngeal nephridia of contracted specimens cover the parietes in iii except at a midventral region. Occasionally the tubules on each side of the body can be seen to join a single zigzagged cord that might be expected to open into the gut by a single aperture. However, those nephridia according to Bahl (1942: p. 436) open into the pharynx in ii–iii "through several well defined apertures."

Eutyphoeus, according to one of Bahl's extrapolations from observations on a single species (1934), is a surface feeder. Burmese species and those studied in India by the author are geophagous and much if not most of the earth deposited above ground in castings seems to have been swallowed as the worms burrowed their way through the soil. Macroscopically recognizable organic matter, such as bits of leaves, twigs, seeds, etc., were not found in the gut of animals or in the castings examined by the author.

KEY TO SPECIES OF *Eutyphoeus*

34. Testis sac annular. . . . .*bifovis*
    Testis sac ventral. 35
35. Lateral intestinal caeca[d] lacking. 36
    Lateral intestinal caeca present, in xxviii. . . . 41
36. Penes annular. 37
    Penes elongate. 39
37. Genital markings present in xv and usually
    also in xvi. . . . .*orientalis*
    Genital markings not present in xv–xvi. 38
38. Genital markings preclitellar only. . . . . .*aborianus*
    Genital markings in part postclitellar. . . .*callosus*
39. Genital markings across 15/16 only. . . .*nicholsoni*
    Genital markings not so restricted. 40
40. Genital markings pre- intra- and post-clitellar
    waltoni
    Genital markings postclitellar or lacking. . .*kempi*
41. Penial setae present. . . . .*rarus*
    Penial setae lacking. . . . .*quinquepertitus*

ᵃ *E. pharpingianus* is assumed, perhaps without justification, to have an uninterrupted dorsal blood vessel but to lack lateral intestinal caeca, and to be metandric.

ᵇ *E. assamensis* drops out here. Types, possibly of two species, may have been too immature to permit adequate characterization of male terminalia.

ᶜ *E. turaensis* drops out here, if not before. The dorsal blood vessel may be uninterrupted.

ᵈ *E. kempi* is assumed, to permit inclusion in the key, to lack lateral intestinal caeca.

E. annandalei, nainianus, nepalensis, pharpingianus Michaelsen, 1907, *lippus* Gates, 1934, are among the 13 Indian species still known today 30–55 years after erection, only from the original specimens. *E. orientalis* (Beddard, 1883), *incommodus* and *nicholsoni* (Beddard, 1901), *quadripapillatus* and *waltoni* Michaelsen, 1907, also illustrate the paucity of native studies on the systematics of Indian megadriles.

Eutyphoeus aborianus

1914. *Eutyphoeus aborianus* Stephenson, *Rec. Indian Mus.* 8: p. 406. (Type locality, Kobo, Abor Country, in the eastern Himalayas. Type, in the Indian Mus.)
1923. *Eutyphoeus aborianus*, Stephenson (*The Fauna of British India*), *Oligochaeta*, p. 428.
1938. *Eutyphoeus aborianus*, Gates, *Rec. Indian Mus.* 40: p. 62.
1958. *Eutyphoeus aborianus*, Gates, *ibid.* 53: p. 204.

Bivestibulate and penile: vestibula small, vestibular apertures centered about at B, penes annular. Female pore, on left side. Spermathecal pores, just lateral to B. Genital markings, paired, centers slightly median to B, across 9/10. First dorsal pore, at 11/12(?). (Color? Segments?) Size, 210 by 6 mm. (Typhlosole? Lateral typhlosoles? Lateral intestinal caeca, none? Ventral intestinal caeca? Supra-intestinal glands?) Dorsal blood vessel, aborted in front of hearts of vii(?). Metandric, testis sac ventral. Seminal vesicles, between 10/11 and 12/13 which is bulged back. GM glands, none. Spermathecal diverticula, median and lateral (? but posteriorly directed and united externally?).

Distribution: Kobo, the Abor country of the eastern Himalayas in Assam.

Remarks[1]**:** *E. aborianus* is known only from the holotype which, in part because of poor condition, has not provided information indicative of relationships. Available data, at least for the present, contra-indicate synonimization. Thus, the paired vestibula distinguish *E. aborianus* from *E. gammiei*, the genital markings from *E. orientalis* and *callosus*. The species is keyed on assumptions that the dorsal blood vessel is aborted in front of the hearts of vii and that there are no lateral intestinal caeca.

Eutyphoeus annulatus

1931. *Eutyphoeus annulatus* Gates, *Rec. Indian Mus.* 33: p. 418. (Type locality, Sagaing. Types, None.)
1933. *Eutophoeus annulatus* var. *typicus*, Gates, *ibid.*, 35: p. 562.
1958. *Eutyphoeus annulatus*, Gates, *ibid.*, 53: p. 157.

Bivestibulate and penile: (male and prostatic pores? apertures of penisetal follicles?), penes about ½ mm. long, conical, filling well-like vestibula with apertures in AB, vestibular roofs slightly protuberant into coelomic cavity. A U-shaped ridge around each vestibular aperture, with open end mesially. Female pores, both present. Spermathecal pores, centered in BC. Genital markings, paired, in region of AC, postsetal, usually in xv but may be present in any of vii–xxi except viii, xiii and xvii. Clitellum, lacking or very thin at mV. First dorsal pore, at 11/12. Color, brown. Segments, 92(+?). Size, 50–90 by 2–5 mm.

Typhlosole, present from xxiv, xxv. Lateral typhlosoles, in xxiv–xxv or xxv–xxvi. (Lateral intestinal caeca, none.) Ventral caeca, 3–7, in xxviii–xxxv. Supra-intestinal glands, in 4–5 of lxii, lxv–lxvi, lxix. Dorsal blood vessel, complete, with hearts in v–vi. Metandric, testis sac ventral to U-shaped but limbs of the sac do not include hearts of xi. Seminal vesicles, of xii push 12/13 and one or more other septa backwards. Penial setae, 6–8 per battery, 1.95–2.95 mm. long, 35–50 μ thick at base, 27–39 μ at midshaft, 12–25 μ at tip, bowed, tip red and pointed, ornamentation of closely crowded, unbroken circles of spines. GM glands, sessile on parietes. Spermathecal diverticula, median and lateral.

Parasites: Gregarine protozoa were present, in large masses, in coelomic cavities of some or all of preclitellar segments or were distributed throughout the soma. Other protozoa, in one specimen, filled the ovaries. Nematodes also were present in coelomic cavities. The fact that pseudonavicellae cysts and nematode ova can be passed out of the body in coelomic fluid ejected from the dorsal pores needs to be taken into consideration in studying life histories of parasites living in earthworms with such pores.

[1] Penial setae (to 3.3 mm. long by 32 μ thick and with a "gentle S-shaped curve") were softened, sometimes without a spoon-shaped concavity at the tip (Gates, 1938), and the one figured by Stephenson (1923: p. 429) already had undergone some distal fibrillar disintegration.

Remarks: Although the various apertures of the male terminalia were not found, all of them are likely to be in the penial lumen which is believed to be the equivalent of the fissure cavity.

E. annulatus not only is the northernmost of the small-range, *levis* group of species but also the only one of the group west of the Irrawaddy Sittang axis. However, nothing is known of the earthworm fauna along the east bank of the Irrawaddy north of Madaya.

Distribution: Sagaing, Kaungmudaw (Sagaing). Kin-U, Kyaukmyaung (Shwebo).

Eutyphoeus assamensis

1926. *Eutyphoeus assamensis* Stephenson, *Rec. Indian Mus.* **28**: p. 262. (Type locality, Katlicherra, South Cachar, Assam. Types, in the Indian Mus.)
1934. *Eutyphoeus assamensis* Gates, *ibid.*, **36**: p. 271.
1938. *Eutyphoeus assamensis*, Gates, *ibid.*, **40**: p. 65.
1958. *Eutyphoeus assamensis*, Gates, *ibid.*, **53**; p. 204.

(Avestibulate? Apenile? Male and prostatic pores?) Female pores, both present. Spermathecal pores, at or just lateral to *B*. Genital markings, paired, in *AB*, postsetal in xvi. First dorsal pore, at 11/12. (Color?) Segments, 255. Size, 185–245 by 4 mm.

(Typhlosole? Lateral typhlosoles? Lateral intestinal caeca, none?) Ventral caeca, 10–11 in xxxvi–xlvi. Supra-intestinal glands, in 4–6 of ciii, cviii–cvii, cxii. Dorsal blood vessel, complete(?), with hearts in v, vi(?). Metandric, testis sac annular. Seminal vesicles, in xii. Penial setae, to 11 per battery, 2 mm. long, 40 μ thick, 50 μ at thickened tip, rather sickle shaped, ornamentation of irregular circles (often and irregularly interrupted) of spines. GM glands, none (?). Spermathecal diverticula, median and lateral.

Distribution: Kathlicherra (South Cachar, Assam).

Remarks: Known only from aclitellate types in which male terminalia may not have reached definitive development. The dorsal blood vessel suggests that relationships should be sought with species such as *E. scutarius, festivus, lippus, nainianus, nepalensis,* and *pharpingianus.*

Eutyphoeus bifovis

1929. *Eutyphoeus bifovis* Gates, *Proc. U. S. Natl. Mus.* **75**, 10: p. 25. (Type locality, Mandalay. Holotype, lost. Paratypes, in the British Mus., Indian Mus. U. S. Natl. Mus.)
1930. *Eutyphoeus bifovis*, Gates, *Rec. Indian Mus.* **32**: p. 329.
1933. *Eutyphoeus bifovis*, Gates, *ibid.*, **35**: p. 565.
1958. *Eutyphoeus bifovis*, Gates, *ibid.*, **53**: p. 179.

Bivestibulate and penile; penes annular, anterior portion thickened and transversely ellipsoidal; vestibular apertures longitudinal and in *AB*. Female pore, on left side. Spermathecal pores, centered at levels from *C* to *D*. Genital markings, median, in *BB*, across 15/16 and 18/19. First dorsal pore, at (10/11), 11/12. Color, brown (pigmentation dense). Segments, 197. Size, 140–245 by 4–9 mm.

Typhlosole, present from (xxvii) xxviii (xxix). Lateral typhlosoles, represented by longitudinal rows of 1–6 oval red patches in first one or two typhlosolar segments. (Lateral intestinal caeca, none.) Ventral caeca, 4–10, in xxxiii–xli. Supra-intestinal glands, in 5–9, of lxxxvi, xciv–xci, c. Dorsal blood vessel, aborted in front of hearts of vii. Metandric, testis sac annular. Seminal vesicles, in xii. Penial setae, 7–12 per battery, 1.85–3.06 mm. long, 23–51 μ thick at base, 14–22 μ (+?) at midshaft, 18–36 μ at tip. Shaft, straight except for slight curvature of tip region, or sigmoid and then occasionally with a more or less marked spiral twisting. Ornamentation, of small, thorn like to triangular teeth in more or less regular circles. (GM glands, none.) Spermathecal diverticula, median and lateral.

Distribution: Taundwingyi, Magwe, Nyaungbinywa (Magwe). Meiktila, Thazi (Meiktila). Myingyan, Kyaung-gone, Tonbo (Mandalay). Sagaing, Kaungmudaw, Tada-U (Sagaing). Kin-U (Shwebo).

Parasites: Numerous nematodes from the testis sacs and from coelomic cavities were sent to Dr. Cobb but so far as is known they have not yet been studied.

Remarks: A greenish coloration, that gradually disappeared after preservation, could not be traced to any particular tissue or to discrete granules. The color was still recognizable in the parietes internally even after several weeks of formalin preservation.

Relationships presumably are with the *foveatus* group of species. The various characters shared with species of that group require a common ancestry but *E. bifovis* does not appear to be close to any of them. The length (north-south) of the *bifovis* range is much greater than that of any other species of the group. Solving the relationship problem was postponed by placing origin of the species in the north where the fauna is largely unknown and from whence migration to the south may have taken place before junction of the Irrawaddy with the Chindwin.

Eutyphoeus bullatus

1933. *Eutyphoeus bullatus* (part) Gates, *Rec. Indian Mus.* **35**: p. 566. (Type locality, Tiddim. Types, none.)
1958. *Eutyphoeus bullatus*, Gates, *ibid.*, **53**: p. 112.

Bivestibulate and apenile: each male pore on a large tumescence of posterior margin of a fissure on roof of an eversible(?) vestibulum with transversely slit-shaped aperture in *AB*, (prostatic pore anterolateral to fissure?), vestibular roof thickened and protuberant into coelomic cavities as a longitudinally ellipsoidal bulb. Female pores, both present. Spermathecal pores, centered in *AB* or at *B*. Genital markings, small, just in front of and just behind spermathecal pores, in or near *AB* of xvi (postsetal), xviii–xix (presetal), across 17/18, 18/19, unpaired and median in *AA* of viii (presetal) and across 16/17. First dorsal pore, at 10/11. Color, slate(?) in dorsum anteriorly (pigment?). (Segments?) Size, to 210 by 7 mm.

Typhlosole, present from xxvii. (Lateral typhlosoles, in xxvii?) Lateral intestinal caeca, in xxvii. Ventral caeca, 16–24, in xxxii–lv. (Supra-intestinal glands?) Dorsal blood vessel, aborted in front of hearts of vi. Metandric, but with male funnels in x, testis sac annular. Seminal vesicles, of xii bulge 12/13 back against 13/14, in ix rudimentary or lacking. Penial setae, 2–4 per battery, 1.6–2.7 mm. long, 33–50 μ thick at base, 27–36 μ at midshaft, tip flattened and tapering to a point, ornamentation of thornlike spines singly and scattered or in short transverse rows. GM glands, flask-shaped, (with muscular capsule?), protuberant into coelom. Spermathecal diverticula, median and lateral.

Distribution: Tiddim (Chin Hills).

Parasites: Gregarines of fairly large size, perhaps related to *A. singularis*, were present in coelomic cavities anteriorly. Very small nematodes, *Iponema minor* Timm & Maggenti, 1966, were numerous in postclitellar as well as preclitellar segments.

Remarks: Constant presence of male funnels in x, and more especially occasional presence of rudimentary seminal vesicles in ix, are believed to be indicative of recent attainment of metandry. Although metandric, *E. bullatus* seems closer to the holandric *E. marmoreus* than to any of the metandric species now known. The species seemingly can be derived from *E. marmoreus* by the same two changes by which *E. peguanus* can be derived from *E. hastatus*.

Eutyphoeus callosus

1938. *Eutyphoeus callosus* Gates, *Rec. Indian Mus.* **40**: p. 67. (Type locality, Dumpep, Khasi Hills, Assam. Types, in the Indian Mus.)
1958. *Eutyphoeus callosus*, Gates, *ibid.*, **53**: p. 204.

Bivestibulate(?), penes annular (male and prostatic pores?). Female pore, on left side. Spermathecal pores, just lateral to *B*. Genital markings, paired, circular in *A*-m*BC* across 10/11–12/13, rectangular and marginally united(?) at m*V* across 18/19–21/22. First dorsal pore, at 11/12. (Color? Segments?) Size, 247–320 by 7–8 mm.

Typhlosole, present from xxvii. (Lateral typhlosoles? Lateral intestinal caeca, none?) Ventral caeca, 12–30, in xxxvii–lxvi. Supra-intestinal glands, in 5–6 of xciv,xcvii–xcviii,ci. Dorsal blood vessel, aborted in front of hearts of vii. Metandric, testis sac ventral. Seminal vesicles, between 10/11 and 12/13 which with 13/14 is bulged back against 14/15. Penial setae, 6 per battery, 2.–2.5 mm. long, 38–40 μ thick, curved ectally, tip spoon-shaped but not thickened, ornamentation of irregularly interrupted circles of fine spines. GM glands, none (?). Spermathecal diverticula, median and lateral (in contact or united posteriorly but with only two openings into duct?).

Distribution: Dumpep (Khasi Hills, Assam).

Remarks: Known only from the original description of five specimens some of which were not fully mature. Closest relationships may be with *E. aborianus* also of dubious status.

Eutyphoeus cochlearis

1931. *Eutyphoeus cochlearis* Gates, *Rec. Indian Mus.* **33**: p. 422. (Type locality, Pakokku. Types, none.)
1933. *Eutyphoeus cochlearis* Gates, *ibid.*, **35**: p. 570.
1958. *Eutyphoeus cochlearis* Gates, *ibid.*, **53**: p. 178.

Univestibulate and apenile(?): apertures of male deferent and prostatic ducts as well as of penisetal follicles in a fissure on each side of vestibular roof, margin of fissure occasionally slightly tumescent, vestibular aperture transversely slitlike, extending into *BC*. Female pore, on left side. Spermathecal pores, in median portion of *BC*. Genital marking, median, across 15/16, reaching *B* or further laterally and equators of xv–xvi. First dorsal pore, at 11/12. Color, brown (pigmentation dense). Segments, 211. Size, 144–230 by 6–9 mm.

Typhlosole, present from xxvii–xxviii. Lateral typhlosoles, represented by longitudinal rows of 3–6 oval red patches in first typhlosolar segment. (Lateral intestinal caeca, none) Ventral caeca, 4–7, in xxxiv–xlii. Supra-intestinal glands, in 5–6 of xciii, xcv–xcviii,xcix. Dorsal blood vessel, aborted in front of hearts of vii. Metandric, testis sac annular to *U*-shaped. Seminal vesicles, in xii but pushing 12/13 into contact with 13/14. Penial setae, 11–13 per battery, 3–4.4 mm. long, 40–50 μ thick at base, 30–40 μ at mid-shaft, 20–25 μ at neck, 35–39 μ at tip, shaft curved in an *S*-shape, tip more or less spoonshaped, ornamentation of circles of spines. (GM glands, none.) Spermathecal diverticula, median and lateral.

Distribution: Pakokku (Pakokku). Anidaung (Lower Chindwin). Possibly also at Monywa on eastern bank of Chindwin River.

Remarks: Closest relationships of the species are with *E. excavatus* from which it is distinguished by about the same sort of differences as those between *E. foveatus* and *spinulosus*. *E. cochlearis* presently seems to belong primarily to the west of the Chindwin-Irrawaddy axis while the *excavatus* range is to the east of that axis.

Eutyphoeus comillahnus

1907. *Eutyphoeus comillahnus* Michaelsen, *Mitt. Naturhist. Mus. Hamburg* **24**: p. 187. (Type locality, Comillah, Chittagong district., East Bengal. Types, in the Indian Mus., the Hamburg Mus.)
1909. *Eutyphoeus comillahnus*, Michaelsen, *Mem. Indian Mus.* **1**: pp. 111, 219, 242.
1910. *Eutyphoeus comillahnus*, Michaelsen, *Abhandl. Naturwiss. Ver. Hamburg* **19**, **5**: p. 12.
1923. *Eutyphoeus comillahnus*, Stephenson, (*The Fauna of British India*), Oligochaeta, p. 432.

1938. *Eutyphoeus comillahnus*, Gates, *Rec. Indian Mus.* 40: p. 69.
1958. *Eutyphoeus comillahnus*, Gates, *ibid.*, 53: p. 218.

Avestibulate and apenile: male and prostatic pores in a fissure within a small porophore in *A B*. (Female pores?) Spermathecal pores, at or just lateral to *A*. Genital markings, unpaired, in *BB*, postsetal in xii, xiii. First dorsal pore, at 11/12. (Color?) Segments, 240. Size, 65–90 by 3–4 mm. (Typhlosole? Lateral typhlosoles? Lateral intestinal caeca, none? Ventral caeca?) Supra-intestinal glands, in lxxxv–lxxxviii. Dorsal blood vessel, complete (?), with hearts in v,vi. Metandric, testis sac ventral. Seminal vesicles, between 10/11 and 12/13. Penial setae, 2 mm. long, 40 μ thick, tip slightly broadened and hollowed on one side, ornamentation of irregular transverse rows or circles of triangular teeth. (GM glands, none.) Spermathecal diverticula, median and lateral.

Distribution: Comillah (Chittagong district, E. Bengal).

Remarks: The species is known only from the types now softened or macerated. *E. comillahnus*, with *E. scutarius* and other species of uncertain status, were thought (Gates, 1958) to constitute a trans-Arakan equivalent of the *constrictus* group in Burma.

Eutyphoeus compositus

1933. *Eutyphoeus annulatus* var. *compositus* Gates, *Rec. Indian Mus.* 35: p. 563. (Type locality, Tonbo. Types, none.)
1958. *Eutyphoeus compositus*, Gates, *ibid.*, 53: p. 155.

Bivestibulate and penile: (male and prostatic pores? apertures of penisetal follicles?), penes less than ½ mm. long, conical, filling well-like vestibula with apertures at *B*, slightly thickened vestibular roofs scarcely protuberant into coelomic cavity. Female pores, both present. Spermathecal pores, centered at or near *C*. Genital markings, paired, postsetal in *BC* of x,xi, one (rather indistinct but with gland) just in front of and just behind each vestibular aperture. Clitellum, lacking or very thin at mV. First dorsal pore, at 11/12. Color, brown. (Segments?) Size, 45–85 by 2–4 mm.

Typhlosole, present from xxiv or xxv. Lateral typhlosoles, low and simple but folded lamellae, in xxiv–xxv,xxvi. Lateral intestinal caeca, lacking or rudimentary, then variously shaped and directed, in xxiv. Ventral caeca, 3–6, in xxviii–xxxiv. Supra-intestinal glands, in 3–5 of lviii, lx–lx, lxiii. Dorsal blood vessel, complete, with hearts in v,vi. Metandric, testis sac *U*-shaped but including only ventral portions of hearts of xi. Seminal vesicles, in xii. Penial setae, 5–7 per battery, 1.2–1.65 mm. long, 28–43 μ thick at base, 24–33 μ at midshaft, 10–18 μ at tip, bowed, ornamentation of very fine spines in closely crowded circles. GM glands, sessile on parietes. Spermathecal diverticula, median and lateral.

Distribution: Tonbo, Kyaukkyone (Mandalay).

Remarks: Although the apertures of the male terminalia were not found, all of them are likely to be in the penial lumen which is believed to be the equivalent of the fissure cavity.

East of the Irrawaddy-Sittang axis *E. compositus* is the northernmost of the small-range, *levis* group but unlike others of that group except the trans-Irrawaddy *E. annulatus* has the same brown pigment that characterizes certain Burmese and Indian groups.

Eutyphoeus constrictus

1929. *Eutyphoeus constrictus* Gates, *Proc. U. S. Natl. Mus.* 75, 10: p. 28. (Type locality, Meiktila. Holotype, lost. Paratypes, in the U. S. Natl. Mus., Indian Mus., British Mus.)
1933. *Eutyphoeus constrictus*, Gates, *Rec. Indian Mus.* 35: p. 570.
1952. *Eutyphoeus constrictus*, Gates, *American Mus. Nat. Hist. Novitates*, No. 1555: p. 1.
1958. *Eutyphoeus constrictus*, Gates, *Rec. Indian Mus.* 53: p. 113.

Avestibulate and apenile: male pores on posterior wall, prostatic pores on anterior wall, apertures of penisetal follicles on roof of a shallow fissure each in a discrete male porophore. Female pores, both present, at or near *A*. Spermathecal pores, in *BD*. Genital markings, lacking or paired and then usually intrasegmental, in one or more of xiii–xvi, xviii–xx. First dorsal pore, in region of 10/11–11/12. Color, white (unpigmented). Segments, 222–224. Size, 96–230 by 3–7 mm.

Typhlosole, present from (xxv–xxvi)xxvii–xix. Lateral typhlosoles, in first 1 to 3 typhlosolar segments. Lateral intestinal caeca, lacking or in one of xxv–xxix. Ventral caeca, 9–28, beginning in xxxi–xxxvii and ending in xl–lx. Supra-intestinal glands, in 4–7 of lxvii–cix. Dorsal blood vessel, complete, with hearts in v, vi. Metandric, usually with male funnels in x, testis sac annular and including hearts of xi. Seminal vesicles, in xii. Penial setae, 5–14 per battery, 1.40–3.06 mm. long, 36–70 μ thick at base, 35–53 μ at neck, 51–90 μ at blade. Ornamentation, of circles of fine spines. (GM glands, none.) Spermathecal diverticula, median and lateral.

Distribution: Twante (Hanthawaddy). Hlegu (Insein). Padaukchaung (Bassein). Daylo Stream and Pa Taw Lo in Karen Hills, Pegu Yomas (Toungoo). Prome, Labu, Paukkaung (Prome). Thanbula (Thayetmyo). Ywadaw, Pyinmana (Yamethin). Magwe, Natmauk (Magwe), Meiktila (Meiktila). Myingyan, Chappea, Kyaukpadaung (Myingyan). Mount Victoria (Pakokku). Mansum (Myitkyina).

To be expected perhaps in the hills of the entire area between the Chindwin and Irrawaddy rivers as well as hills throughout Myitkyina district if not also further north into India.

Habitats: Soil, of teak and pine forests, of evergreen jungle, of dense bamboo rain forest, grassland, away from towns. Was found on trees, under bark, in beetle holes five feet from the ground. At elevations from sea level to 7,800 feet.

Variation and Abnormality: Absence of genital markings, according to data now available, could characterize a northern race with a range extending from Meiktila and Myingyan districts well into Myitkyina district. More variation was found in the southern portion of the species range, about half of Hanthawaddy and Insein worms ornate. Markings were in xviii of at least 39 specimens mostly from areas in which *E. hastatus* (with similar markings in the same metamere) also was present. Bisegmental markings (across 21/22) were recorded only from one locality where *E. hastatus* also was similarly ornate at the same level.

Presence or absence of lateral intestinal caeca and their location as well as conformation when present were responsible for a suggestion (1958) that those enteric evaginations as yet of little if any apparent function now are being evolved by some populations in one or another of several consecutive metameres.

Male funnels of x seem to have disappeared completely in a Bassein population.

In some 302+ specimens, 603 spermathecae were bidiverticulate. One spermatheca was adiverticulate. Several had an extra small diverticulum on the anterior face of the spermathecal duct. Only two athecal individuals were found.

GM glands, seemingly lacking most everywhere, may have been present in a Daylo population.

Eight small specimens that appeared to be fully mature in so far as clitellar tumescence and ovaries were concerned had obviously juvenile testes and were anandric. Seminal vesicles, prostates and spermathecae were rudimentary. At the time those conditions were noted (some twenty-five or more years ago) development to maturity of the juvenile organs was thought to have been inhibited by substances given off by gregarine parasites that filled coelomic cavities from xix posteriorly. Subsequent discoveries in other genera have made it necessary to ask whether male sterility of a genetic rather than a parasitic nature had not enabled development of such large numbers of protozoa with no apparent effect on function and reproduction of the hosts other than stunting of soma size.

Remarks: In at least three localities, genital markings, male porophores, and/or supra-intestinal glands of *E. constrictus* and *hastatus* are similar and less like those structures elsewhere in their own species ranges. To avoid multiplication of specific names for geographically restricted and quite small populations, the boundary between *E. hastatus* and *constrictus* was drawn arbitrarily some twenty years ago according to the andry. Considerable additional material might have enabled recognition, in a united *E. hastatus-*

montanus, of races without reference to number of testes.

Eutyphoeus excavatus

1929. *Eutyphoeus excavatus* Gates, *Proc. U. S. Natl. Mus.* 75, 10: p. 30. (Type locality, Meiktila. Holotype, lost. Paratypes, in the U. S. Natl. Mus., British Mus., Indian Mus.)
1930. *Eutyphoeus excavatus*, Gates, *Rec. Indian Mus.* 32: p. 329.
1933. *Eutyphoeus excavatus*, Gates, *ibid.*, 35: p. 572.
1958. *Eutyphoeus excavatus*, Gates, *ibid.*, 53: p. 175.

Univestibulate and apenile(?): apertures of male deferent and prostatic ducts as well as of peniseta] follicles in fissures just lateral to *B* on vestibular roof, margin of fissure often more or less markedly protuberant in an annulus, vestibular aperture transversely slitlike, extending laterally to or beyond m*BC*, anterior and posterior lips in contact only in *AB*. Female pore, on left side. Spermathecal pores, centered at or median to *B*. Genital markings, median, in *BB*, across 14/15 and 15/16. First dorsal pore, at 11/12. Color, brown (pigmentation dense). Segments, 254. Size, 110–280 by 4–10 mm.

Typhlosole, present from (xxvii),xxviii. Lateral typhlosoles, represented by longitudinal rows of 2–5 small red patches. (Lateral intestinal caeca, none.) Ventral caeca, 2–11 (but usually 5–7), in xxxiii–xlii. Supra-intestinal glands, in 5–7 of xcii,ci–xcvii,cv. Dorsal blood vessel, aborted in front of hearts of vii. Metandric, testis sac almost annular. Seminal vesicles, push 12/13 back against 13/14 or both against 14/15. Penial setae, 5–17 per battery, 2.95–4.95 mm. long, 40–75 μ thick at base, 20–50 μ at neck, 27–63 μ at tip, straight except for curvature ectal to neck or doubly curved in an *S*-shape, ornamented with spinelike teeth irregularly scattered or more entally in transverse rows and then broken circles. Tip widened, flattened, occasionally concave like bowl of a spoon. (GM glands, none.) Spermathecal diverticulum, lateral.

Distribution: Meiktila, Thazi (Meiktila). Myingyan, Chappea, Taungtha (Myingyan). Kyaukse, Dwehla (Kyaukse). Sagaing, Myotha, Tada-U (Sagaing).

Remarks: Closest relationships of the species are with *E. cochlearis* from which it is distinguished by about the same sort of differences as those between *E. foveatus* and *spinulosus*. *E. excavatus* is the eastern counterpart of *E. cochlearis* and its distribution was thought to warrant a suggestion that *E. excavatus* may have gotten into its present range before the Irrawaddy lost its outlet through the present Sittang River and joined the Chindwin River.

Eutyphoeus ferinus

1958. *Eutyphoeus ferinus* Gates, *Rec. Indian Mus.* 53: p. 146. (Type locality, Myitkyina. Types, none.)

Avestibulate and apenile: male and prostatic pores as well as apertures of penisetal follicles in transversely slitlike fissures, each within a porophore in *BC*. Female pore, on left side. Spermathecal pores, well lateral in *BC* or in *CD*. Genital markings, paired, postsetal in *AA* of xiv–xvi, in *AC* and across 18/19, 19/20, 202/1. First dorsal pore, at 11/12. Color, white (unpigmented). (Segments?) Size, to 270 by 7 mm.

Typhlosole, present from xxvii. Lateral typhlosoles, folded, in xxviii–xxx. (Lateral intestinal caeca, lacking.) Ventral caeca, 2–3, in xxxvii–xxxix. Supraintestinal glands, in four of xciv–ciii. Dorsal blood vessel, aborted in front of hearts of vi. Metandric, with male funnels in x, testis sac ventral. Seminal vesicles, between 10/11 and 12/13. Penial setae, *ca.* 7 per battery, 1.6–2.0 mm. long, shaft curved ectally, ornamentation of transverse rows of fine spines. (GM glands, none?) Spermathecal diverticulum, a horizontal ridge on posterior face of duct with (at least?) two openings into duct lumen.

Distribution: Myitkyina (Myitkyina).

Remarks: Relationships seemingly are closest with *E. constrictus, peguanus,* and *plenus* from each of which *E. ferinus* is distinguished by abortion of the right female pore, abortion of the dorsal blood vessel in front of hearts of vi, reduction of coelomic cavity of xi to a ventral testis sac and possibly also by external union of two diverticula, as in *E. aborianus* and *callosus,* into a single horizontal ridge on posterior face of the spermathecal duct. Present knowledge of distribution requires derivation directly from *E. constrictus* or from a common ancestor of the two species.

Eutyphoeus festivus

1938. *Eutyphoeus festivus* Gates, *Rec. Indian Mus.* **40**: p. 71. (Type locality, Dumpep, Assam. Type, in the Indian Mus.)
1958. *Eutyphoeus festivus,* Gates, *ibid.,* **53**: p. 204.

Avestivulate and apenile: male pores in fissures at centers of small porophores in *AB*. Female pores, both present. Spermathecal pores, in *AB*. Genital markings, in *BB*, across 19/20. First dorsal pore, at 11/12. (Color? Segments?) Size, 146 by 6 mm.

Typhlosole, present from xxv. (Lateral typhlosoles? Lateral intestinal caeca, none.) Ventral caeca, 6 in xxxvii–xliii. Supra-intestinal glands, in xcii–xcv. Dorsal blood vessel, complete (?), with hearts in v,vi. Metandric, testis sac ventral. Seminal vesicles, between 10/11 and 12–13 which with 13/14 is bulged back against 14/15. Penial setae, (several per battery), 1.05–1.15 mm. long, 20–25 μ thick, tip spoon-shaped, ornamentation of irregularly interrupted circles of fine spines. (Genital marking glands?) Spermathecal diverticula median and lateral, but posteriorly directed.

Distribution: Dumpep, (Khasi Hills, Assam).

Remarks: The species is known only from the holotype. *E. festivus,* along with other species of uncertain status, were thought (Gates, 1958) to be a trans-Arakan equivalent of the *constrictus* group in Burma.

Eutyphoeus foveatus

1890. *Typhaeus foveatus* Rosa, *Ann. Civ. Mus. Sto. Nat. Genova* **29**: p. 389. (Type locality, Rangoon. Types, in poor condition, in the Genoa Mus. Specimens from the type locality, collected in the twenties, in the U. S. Natl. Mus.)
1895. *Typhaeus foveatus,* Beddard, *A Monogr. of the Order of Oligochaeta* (Oxford), p. 475.
1900. *Eutyphoeus foveatus,* Michaelsen, *Das Tierreich* **10**: p. 323.
1901. *Typhoeus foveatus,* Beddard. *Proc. Zool. Soc. London* 1901: p. 206.
1903. "*Eutyphaeus*" *foveatus,* Michaelsen, *Die geographische Verbreitung der Oligochaeten* (Berlin), p. 108.
1909. *Eutyphoeus foveatus,* Michaelsen, *Mem. Indian Mus.* **1**: pp. 111, 219.
1923. *Eutyphoeus foveatus,* Stephenson, (*The Fauna of British India*), *Oligochaeta*, pp. 18, 433.
1925. *Eutyphoeus foveatus,* Gates, *Ann. Mag. Nat. Hist.,* Ser. 9, **16**: p. 571. *Rec. Indian Mus.* **27**: p. 472.
1926. *Eutyphoeus foveatus,* Gates, *Jour. Burma Res. Soc.* **15**: p. 215. *Jour. Bombay Nat. Hist. Soc.* **31**: pp. 183, 184, 185. *Biol. Bull.* **51**: p. 403. *Rec. Indian Mus.* **28**: p. 163.
1930. *Eutyphoeus foveatus,* Gates, *ibid.,* **32**: p. 329.
1931. *Eutyphoeus foveatus,* Gates, *ibid.,* **33**: p. 425.
1933. *Eutyphoeus foveatus,* Gates, *ibid.,* **35**: p. 575.
1937. *Eutyphoeus foveatus,* Hla Kyaw & Gates, *Jour. Asiatic Soc. Bengal,* (*Sci.*) **2**: p. 165.
1942. *Eutyphoeus foveatus,* Bahl, *Quart. Jour. Micros. Sci.* **83**: p. 431.
1958. *Eutyphoeus foveatus,* Gates, *Rec. Indian Mus.* **53**: p. 161.
1961. *Eutyphoeus foveatus,* Gates, *American Midland Nat.* **66**: p. 62.

Univestibulate and penile: apertures of male deferent and prostatic ducts as well as of penisetal follicles in tubular penes 1+ mm. long, vestibulum deep with thick wall but thin roof and circular aperture in *BB*, penes protrusible through a median vestibular aperture but vestibulum not (?) eversible. Female pore, on left side. Spermathecal pores, centered at or close to *B*. Genital marking, median, in *BB*, between equators of xiv and xvi margin incised at equator of xv to exclude *b* or also *a*. First dorsal pore, at 11/12. Color, brown (pigmentation dense). Segments, 200–225. Size, 140–380 by 4–10 mm.

Typhlosole, present from xxviii. Lateral typhlosoles, represented by longitudinal rows of one to five small red patches in first one or two typhlosolar segments. Lateral intestinal caeca, lacking, rudimentary or small and then usually in xxviii. Ventral caeca, 17–27, in xxxiii–lx. Supra-intestinal glands, in 5–7 of lxxxv,xcv–xc,ci. Dorsal blood vessel, aborted n front of hearts of vii. Metandric, testis sac ventral, hearts of xi bound down to gut. Seminal vesicles, between 10/11–12/13. Penial setae, 8–16 per battery, 2–5 mm. long, 30–51 μ thick at base, 22–46 μ at midshaft, 11–30 μ at tip, with spiral curvature ectally, tip flattened but not widened, ornamented with spinelike teeth in circles or transverse rows. (GM glands, none.) Spermathecal diverticula, median and lateral.

Distribution: Moulmein, Mupun (Amherst). Kyaikto, Boyagyi (Thaton). Rangoon, Syriam, Kyauktan, Twante (Hanthawaddy). Maubin, Yandoon, Danubyu, Pantanaw (Maubin). Taikkyi, Wanetchaung, Insein, Taukkyan, Hmawbi, Hlegu, Damsite, Hlawga (Insein). Pegu, Thanatpin (Pegu). Thonze, Tharrawaddy, Ngapugale (Tharrawaddy). Henzada, Ingabu, Zalun (Henzada). Toungoo, Pyu (Toungoo). Prome, Paukkaung, Nyaunggyi, Okshitbin, Pegyin, Padaung, Labu (Prome). Thayetmyo, Allanmyo, Thanbula (Thayetmyo). Mandalay (Mandalay). Naba (Katha). From sea level to elevations of at least a thousand feet in the Arakan Yomas west of Prome.

Recent transport by man to Mandalay and from Prome to Naba is believed to be responsible for presence of *E. foveatus* at those two localities. The colony at Mandalay, never represented in subsequent collections, may have become extinct. The Moulmein-Mupun colony probably was established recently and presumably as a result of human transport. Thaton colonies may have been established less recently and as a result of riverine transport. The range includes a southern sector of the Arakan yomas and recent alluvium of the deltas region except in Myaungmya and Pyapon districts which had not been invaded when the Burma survey was ended.

Habitats: Soil. Earth around roots of potted plants (small juveniles only). Very common in open areas under grass or grass and mimosa. Much less common in shade of trees where grass does not grow. Has been found in accumulations of decaying leaves, and commonly in flower beds, gardens, and lawns of Rangoon and various other towns.

Castings: Intestinal ejecta of this species are deposited on the surface of the ground as towers which may be open at the top, capped, or capped and with internal passage more or less completely plugged. Castings of *E. foveatus* having a dry weight of 16.5 and 11.5 pounds, respectively, were secured from 2 plots 10 feet square during 20 days but many more had been washed away by heavy rains before the daily collection was made. Dry weight of castings for the species was estimated to be 13.62 tons per acre for a season of 100 days, with an average of 1.917 pounds per individual.

Abnormality: The abnormalities of specimen No. 1 (1958: p. 165) resulted from halving of a left embryonic somite at 17th level. Abnormalities of No. 2 (*idem*) resulted from halving of right somite at some level in front of the 8th.

Parasites: Monoeystid gregarines of three kinds were seen in seminal vesicles. A large branched monocystid, *Aikinetocystis singularis* Gates, 1926, is usually present in coelomic cavities of a region includ-

ing iii–xii, along with other monocystids, nematodes and occasionally larvae of a small fly. Unbranched gregarines somewhat smaller than *A. singularis* may be present in numbers in the testis sac. Ciliates were seen in the intestinal lumen.

Remarks: The species is most closely related to *E. spinulosus, planatus, cochlearis, excavatus,* and *bifovis.* Characters in common of such a *foveatus* group (Gates, 1958) are; absence of dorsal pores in front of 11/12, presence of a brown pigment, a row of flat, red patches in place of each lateral typhlosole, restriction of supraintestinal glands throughout all six species to a region comprising 21 segments (lxxxv–cv), dorsal blood vessel and its branches aborted in front of the hearts of vii, right female pore aborted, GM pattern stabilized with reference to number (including presence or absence) as well as location (always unpaired, median, in BB), absence of coelomic GM glands, invagination of male fissures into a common vestibule (except in the bivestibulate *E. bifovis*), metandry, association with each functional penial seta of a battery of reserves in various stages of development. Acquisition of metandry was so long ago that the male funnels of x have disappeared.

During evolution of *E. foveatus,* genital markings were restricted to two at 14/15 and 15/16, then subsequently enlarged and finally united though often showing at eq/xv more or less obvious evidence of a double origin. Additionally tissues temporarily protrusible as a sort of annular penis around each male fissure became a permanent tubular penis protrusible from the vestibulum for insertion into the spermathecal duct during copulation. Lateral intestinal caeca, lacking in all other species of the group, now are appearing as rudimentary pockets in some of the *foveatus* populations.

Mutant individuals, with one or another of the *spinulosus* characters (Gates, 1958), were found at five localities within the *foveatus* range, two of the sites deep into that range.

Eutyphoeus gammiei

1888. *Typhaeus gammii* Beddard, *Quart. Jour. Micros. Sci.* **29:** p. 111. (Type locality, Darjiling. Types, probably none but see Gates, 1938, p. 73.)

1895. *Typhaeus gammii,* Beddard, *A Monogr. of the Order of Oligochaeta* (Oxford), p. 473.

1900. *Eutyphoeus gammiei,* Michaelsen, *Das Tierreich* **10:** p. 323.

1901. *Typhoeus gammii,* Beddard, *Proc. Zool. Soc. London* **1901:** p. 205.

1903. "*Eutyphaeus*" *Gammiei,* Michaelsen, *Die geographische Verbreitung der Oligochaeten* (Berlin), p. 108.

1907. *Eutyphoeus chittagongianus* Michaelsen, *Mitt. Naturhist. Mus. Hamburg* **24:** p. 181. (Type locality, Comillah, Chittagong district, East Bengal. Types, in the Indian Mus., the Hamburg Mus.)

1909. *Eutyphoeus chittagongianus* + *E. gammiei,* Michaelsen, *Mem. Indian Mus.* **1:** pp. 111, 218, 219, 231.

1910. *Eutyphoeus chittagongianus* + *E. gammiei,* Michaelsen, *Abhandl. Naturhist. Mus. Hamburg* **19:** 5, p. 12.

1914. *Eutyphoeus koboensis* + *E. magnus* Stephenson, *Rec. Indian Mus.* 8: pp. 404, 408. (Type localities, of *koboensis* Kobo, Abor country, Assam, of *magnus* Upper Rotung, Assam. Types, in the Indian Mus.)
1920. *Eutyphoeus chittagongianus*, Stephenson, *Mem. Indian Mus.* 7: p. 241.
1923. *Eutyphoeus gammiei* (part), Stephenson, (*The Fauna of British India*), *Oligochaeta*, p. 434. (Excluding, *E. kempi.*)
1925. *Eutyphoeus gammiei*, Stephenson, *Rec. Indian Mus.* 27: p. 72.
1926. *Eutyphoeus gammiei*, Stephenson, *ibid.*, 28: p. 264.
1938. *Eutyphoeus gammiei*, Gates, *ibid.*, 40: p. 72.
1958. *Eutyphoeus gammiei*, Gates, *ibid.*, 53: p. 199.

Univestibulate and penile(?); penis (when present) an annular tumescence around a fissure, vestibular aperture with straight posterior and lobed anterior margin. Female pore, on left side. Spermathecal pores, centered at or slightly lateral to B. Genital markings, in BB, unpaired, paired and then in contact or marginally united at mV, across some of the following intersegmental furrows; 9/10–13/14, 19/20–23/24. First dorsal pore, at 11/12. Color, brown in dorsum (?), green elsewhere(?). Segments, 195–263. Size, 182–405 by 7–10 mm. (Typhlosole? Lateral typhlosoles? Lateral intestinal caeca, none?) Ventral caeca, 32–75, in xxv–xcix. Supra intestinal glands, in 4–6 of xcv, cxxxiv–ciii,cxl. Dorsal blood vessel, aborted in front of hearts of vii. Metandric, testis sac ventral. Seminal vesicles, between 10/11 and 12/13, the latter with some following septa bulged backward. Penial setae, several per battery, 2–5 mm. long, 26–40 μ thick, shaft with a slight *S*-curve, tip thickened and spoon-shaped, ornamentation of closely crowded circles of fine spines. (GM glands, none.) Spermathecal diverticula, median and lateral, discrete but posteriorly directed or united and then stalkless.

Distribution: Comilla, (Chittagong district, East Bengal). Katlicherra (South Cachar). Tura (Garo Hills, Assam). Amingaon (Assam Valley). Abor Country, Lokra in the Balipara Frontier Tract, Darjiling district, in the eastern Himalayas.

Remarks: The olive color mentioned by previous authors may well be about the same and similarly produced as the green recorded from the Burmese *E. planatus* and *bifovis*. A characteristic brown pigment may have been leeched from most of the Indian specimens before they were studied.

Some of the difficulties that arose in connection with this species were due to immaturity, maceration as well as variation, and perhaps also, though but rarely, to abnormality. The range is large, extending from the Bay of Bengal into the Himalayas from Darjiling to the easternmost frontier of India. Variation, in a range with such climates, seems, however, to be limited mostly to differences in location of genital markings, intestinal caeca and supra-intestinal glands. Many more than the thirty-odd specimens that have been available will be needed to enable determination

of relationships with Burmese species and any correlation of organ location with geographical distribution. *E. gammiei* clearly is the trans-Arakan equivalent of the *foveatus* group and with relationships, as indicated by the male terminalia, probably closer to *E. cochlearis* and *excavatus*.

Eutyphoeus gigas

1917. *Eutyphoeus gigas* Stephenson, *Rec. Indian Mus.* 13: p. 408. (Type locality, Rangamati, Chittagong Hill Tracts, Bengal. Type, in the Indian Mus.)
1919. *Eutyphoeus gigas*, Stephenson & Prashad, *Trans. Roy. Soc. Edinburgh* 52: p. 465.
1923. *Eutyphoeus gigas*, Stephenson (*The Fauna of British India*), *Oligochaeta*, p. 436.
1930. *Eutyphoeus longiseta* Gates, *Rec. Indian Mus.* 32: p. 332. (Type locality, Sandoway. Types, none.)
1933. *Eutyphoeus longiseta*, Gates, *ibid.*, 35: p. 580.
1934. *Eutyphoeus gigas*, Gates, *ibid.*, 36: p. 272.
1938. *Eutyphoeus gigas*, Gates, *ibid.*, 40: p. 82.
1958. *Eutyphoeus gigas*, Gates, *ibid.*, 53 p. 185.

Avestibulate and penile: apertures of penisetal follicles in transversely elliptical tubercles each surrounded by a large annular penis the ventral portion of which may be variously extended, male pore on posterior portion of annulus, prostatic pore on anterior annulus at or lateral to center of a porophore (disc of parietal thickening) that is variously depressed or folded so as to bring penes below level of body surface. Female pore, on left side. Spermathecal pores, in median half of BC or extending into AB. Genital markings, always present: paired or unpaired in AA, in AB or BC, presetal or postsetal in some of viii–xvi, unpaired in AA and crossing an intersegmental furrow, some of 11/12–15/16, 18/19–21/22. First dorsal pore, at 11/12. Setae, in postclitellar segments ornamented with transverse ridges of fine teeth. Color, brown (pigmentation dense.) Segments, 163–212. Size, 175–290 by 7–11 mm.

Typhlosole, present from xxviii. Lateral typhlosoles, lacking or represented only by a dark red area. Lateral intestinal caeca, usually ventrally directed, in xxviii. Ventral caeca, 24–28, beginning in xxxii–xxxiv and ending in lv–lxi. Supra–intestinal glands, in 4–6 of lxxxii, xc–lxxxv,xcv, aborted in front of hearts of vii. Metandric, testis sac ventral, hearts of xi bound down to gut. Penial setae, 6–12 per battery, 5.0–8.25 mm. long, 50–90 μ thick at base, 52–61 μ at midshift, 24–38 μ at tip, nearly straight, slightly bowed or rather sigmoid and with slight spiral curvature, tip tapering to a spine, ornamentation of very small spine-like teeth in transverse rows. GM glands, sessile on parietes. Spermathecal diverticula, usually with elongate stalk, median and lateral.

Distribution: Sandoway, Taungup, Kyauktaga, Tanywagyi (Sandoway). Myebon (Kyaukpyu). Myohaung, Kyauktaw, Akyab, Buthidaung, Maungdaw (Akyab). Paletwa (Arakan Hills). Rangamati (Chittagong Hill Tracts, Bengal).

At elevations from sea level to 1,600 feet.

Parasites: Elongate gregarines of about the same size as *A. singularis* but unbranched were in anterior coelomic cavities of worms from several localities. Coelomic nematodes were secured at several localities but have not been studied.

Remarks: Closest relationships appear to be with *E. quinquepertitus* and *rarus*. Like those species *E. gigas* has the following characters: Brown pigment present. First dorsal pore at 11/12. Right female pore aborted. Lateral intestinal caeca present in xxviii. Ventral caeca numerous. Dorsal blood vessel and hearts aborted in front of hearts of vii. Metandric. Coelomic cavity of xi reduced to a subesophageal testis sac. Rarity of male funnels in x and absence of their ducts when funnels are present was believed to indicate that the metandry is older than (for instance) in *E. constrictus* and *peguanus*. Six of the listed characters are shared with the *foveatus* group but the two groups were thought to have diverged before the ancestral protofoveatus had lost or reduced the lateral intestinal typhlosoles. Only after the divergence were those typhlosoles reduced in *E. gigas*.

Male terminalia of *E. gigas* seem to have evolved somewhat beyond the stage characterizing *E. pius* of the *levis* group but some of the rather nondescript differences may be apparent only because of the larger size. Male terminalia of *E. rarus* are almost exactly the same as those of *E. macer* (also of the *levis* group) but the terminalia of *E. quinquepertitus* seemingly are more like those of the holandric *E. marmoreus*.

Eutyphoeus hastatus

1929. *Eutyphoeus hastatus* Gates, *Proc. U. S. Natl. Mus.* **75**, 10: p. 32. (Type locality, Thayetmyo. Holotype, lost. Paratypes, in the U. S. Natl. Mus., British Mus., Indian Mus.)

1930. *Eutyphoeus hamatus* + *E. hastatus*, Gates, *Rec. Indian Mus.* **32**: p. 329, 332. (Type locality of *hamatus*, Kalewa. Types, none.)

1933. *Eutyphoeus montanus* + *E. hamatus* + *E. hastatus* + *E. sp.*, Gates, *ibid.*, **35**: pp. 587, 578, 579, 600. (Type locality of *montanus*, Pegu Yomas east of Letpadan. Types, none.)

1958. *Eutyphoeus hastatus*, Gates, *ibid.*, **53**: p. 99.

Avestibulate and apenile(?): apertures of penisetal follicles on roofs of fissures, male pores on posterior walls of fissures or on the anterior face of a posterior lip of an annular tumescence around a fissure (prostatic pores?), each fissure within a more or less definitely delimited porophore, an area of parietal thickening. Female pores, both present. Spermathecal pores, variously located in *AD*. Genital markings, paired but sometimes united at mV, in xvi and/or one or two post-clitellar segments. First dorsal pore, in region of 10/11–11/12. Color, white (unpigmented.) Segments, 188–211. Size, 60–180 by 3–5½ mm.

Typhlosole, present from region of xxvi–xxix. Lateral typhlosoles, in first one or two typhlosolar segments. Lateral intestinal caeca, usually lacking, when present rudimentary or small, in xxviii. Ventral caeca, 6–29, in xxx–lxi. Supra-intestinal glands, in 4–7 of lxix,xcv–lxxvi,ciii. Dorsal blood vessel, complete, with hearts in v,vi. Holandric, testis sac annular. Seminal vesicles, of ix may penetrate into viii, those of xii may penetrate back as far as into xx. Penial setae, 6–12 per battery, 1–4 mm. long, shaft straight except for ectal curve, tip usually thickened, often red, ornamented with thornlike spines in transverse rows or circles. Spermathecal diverticula, median and lateral.

Distribution: Pegu Yomas opposite Letpadan (Tharrawaddy). Prome (Prome). Thayetmyo (Thayetmyo). Pyinmana (Yamethin). Taungdwingyi, Magwe (Magwe). Anidaung, Thindaw, Okma (Lower Chindwin). Laungbyin, Mingin, Kalewa, Masein (Upper Chindwin). Tiddim (Chin Hills).

Parasites: Gregarine trophozoites and cysts were found in spermathecal ampullae of numerous specimens, as well as in coelomic cavities of preclitellar and clitellar segments. Anterior seminal vesicles sometimes were filled.

Remarks: *E. hastatus* shares with *E. annandalei, incommodus, manipurensis, marmoreus,* and *quadripapillatus,* two primitive characters, holandry and presence in v,vi of paired hearts associated with a complete (or at least nearly complete?) dorsal blood vessel. Nevertheless, the 6 species have specialized differently and do not now appear to be closely related to each other. Indeed, in one Burmese area *E. hastatus* appears to be more closely related to *E. constrictus* (q. v.) than to certain conspecific populations of its own range. Because of the great importance attached to andry in the classical system, the boundary between *E. hastatus* and *constrictus* was drawn some twenty years ago according to presence or absence of testes in x. Much evidence has become available since then to show that genital structure is modified phylogenetically much more rapidly than the somatic organization.

An annular penis seemed to be developing around each fissure, in some populations, but whether the annulus was temporary or permanent could not be determined.

Eutyphoeus kempi

1914. *Eutyphoeus kempi* Stephenson, *Rec. Indian Mus.* **8**: p. 401. (Type locality, Kobo, Abor country, Assam. Type, in the Indian Mus.)

1920. *Eutyphoeus kempi*, Stephenson, *Mem. Indian Mus.* **7**: p. 242.

1923. *Eutyphoeus gammiei* (part), Stephenson, (*The Fauna of British India*), *Oligochaeta*, p. 434. (Excluding all except *E. kempi*.)

1938. *Eutyphoeus kempi*, Gates, *Rec. Indian Mus.* **40**: p. 87.

Bivestibulate and penile: penes elongate and within deep, well-like vestibula with apertures in AB. Female pore, on left side. Spermathecal pores, in median part of BC. (Genital markings, unpaired and across 21/22–22/23?) Setae, lacking in ii–iv (and?), $AA = 3–4AB$, $AB < CD < BC < AA < DD < \frac{1}{2}C$. First dorsal pore, at 11/12. Color, light olive green ventrally, slate dorsally (because of presence of pigment?). Segments, 254. Size, 230 by 6 mm. (Typhlosole? Lateral typhlosoles? Lateral intestinal caeca? Ventral caeca?) Supra-intestinal glands, in cxiii–cxviii. Dorsal blood vessel, aborted in front of hearts of vii. Metandric, testis sac ventral. Seminal vesicles, in xii but pushing 12/13–13/14 posteriorly. Penial setae, (per battery?), 2.3–3.9 mm. long, 34 μ thick, shaft curved into an S-shape, tip somewhat spoon-shaped, ornamentation of irregularly and frequently interrupted circles of very fine teeth. (GM glands, none?) Spermathecal diverticula, numerous but joined in a continuous semicircular ridge around duct posteriorly and with two groups of 5–6 apertures into duct lumen.

Distribution: Kobo (Abor country, Assam).

Remarks: Known only from the type. Indistinctness of the supposed genital markings suggests that the worm may not have been fully mature. Relationships presumably are closest with *E. gammiei* of which Stephenson (1920: p. 242) believed it to be a synonym because differences in genital markings and penial setae were insufficient to justify the separation. Even if Stephenson was correct with regard to those characters the suggested synonymy was believed (Gates, 1938) to be contra-indicated by differences in length of penes, in number of vestibula and of diverticular openings into the spermathecal duct. Separation of diverticular apertures, by a vertical ridge on the inner wall posteriorly of the spermathecal duct, into a median and a lateral group of five or six each shows that *E. kempi* had a bidiverticulate ancestry. The grouping of diverticular apertures, also characterizing several other species, may have arisen, after union posteriorly of paired polyloculate diverticula, by abortion of the stalks.

Eutyphoeus levis

1890. *Typhaeus laevis* Rosa, *Ann. Mus. Sto. Nat. Genova* 29: p. 388. (Type locality, Cobapo. Holotype, in the Genoa Mus., even in 1890 was considered to be in too poor condition for study of internal anatomy.)

1895. *Typhaeus laevis* (part), Beddard, *A Monogr. of the Order of Oligochaeta* (Oxford), p. 475. (Excluding *T. laevis* Rosa 1891.)

1900. *Eutyphoeus levis* (part), Michaelsen, *Das Tierreich* 10: p. 323. (Excluding, Ceylon distribution based on *T. laevis* Rosa 1891.)

1901. *Typhoeus laevis*, (part), Beddard, *Proc. Zool. Soc. London* 1901: p. 206. (Excluding, specimens with genital markings.)

1903. "*Eutyphaeus*" *levis* (part), Michaelsen, *Die geographische Verbreitung der Oligochaeten* (Berlin), p. 108. (Excluding, the Ceylon distribution.)

1909. *Eutyphoeus laevis* (part), Michaelsen, *Mem. Indian Mus.* 1: pp. 111, 219. (Excluding, the Ceylon distribution.)

1910. *Eutyphoeus laevis* (part), Michaelsen, *Abhandl. Naturwiss. Ver. Hamburg* 19, 5: p. 12. (Excluding, the Ceylon distribution.)

1923. *Eutyphoeus levis*, Stephenson, (*The Fauna of British India*), *Oligochaeta*, p. 425.

1933. *Eutyphoeus falcifer* Gates, *Rec. Indian Mus.* 35: p. 574. (Type locality, Leiktho Circle. Types, none.)

1958. *Eutyphoeus levis*, Gates, *ibid.*, 53: p. 148.

Avestibulate and apenile: apertures of penisetal follicles on roofs of slight transverse fissures each at B and at center of a porophore reaching to or towards A, male pores on posterior walls of the fissures(?), or on a rudimentary annulus around the fissure, (prostatic pores?). Female pores, both present. Spermathecal pores, centered at or close to B. (Genital markings, none.) Clitellum, lacking or very thin at mV. First dorsal pore, at 11/12. Setae, d chaetae gradually more dorsal posteriorly until DD may be smaller than AA. Color, white (unpigmented). Segments, 152–180. Size, 35–62 by 2–3 mm.

Typhlosole, present from xxi or xxii. Lateral typhlosoles, low and continuous, simple lamellae, in xxi–xxii or xxii–xxiii. (Lateral intestinal caeca, none.) Ventral caeca, 1–2, in xxviii–xxx. Supra-intestinal glands, in 3 of lvii, lviii–lix,lx. Dorsal blood vessel, complete, with hearts in v,vi. Metandric, but male funnels (with ducts) in x, testis sac annular and including hearts of xi. Seminal vesicles, in xii. Penial setae, 2 per battery, 0.82–1.3 mm. long, 10–15 μ thick at tip, 24–30 μ at midshaft, 24–49 μ at base. Shaft, slightly curved or sickle-shaped (under cover glass). Tip, bluntly rounded, narrowed to a point, or occasionally slightly widened and spoonshaped. Ornamentation, of fine spines in closely crowded and almost unbroken circles. Spermathecal diverticulum, digitiform to pyriform, with one seminal chamber or rarely bifid distally, lateral.

Distribution: Thaton, Kyaiktyo, Kyaiktyo Hill (Thaton). Cobapo, Leiktho Circle (Karen Hills of Toungoo district). From sea level, at Thaton, to 1,000–1,400 meters at Cobapo. The southern part of the range may extend from the sea and the Sittang River to the Salween River.

Remarks: Even hearts of iv were still present in one of the few specimens that were studied.

E. levis, along with *E. annulatus*, *compositus*, *macer*, *pius*, *pusillulus*, *sejunctus*, and *strigosus*, provide the eastern, and for Burma, the southern boundary of the *Eutyphoeus* range. Each of that group has retained the dorsal blood vessel in its primitive completeness (along with the hearts of v,vi, sometimes even of iv) and both female pores but has lost all dorsal pores in front of 11/12 and has undergone a meroandric reduction. Each species has a small soma and a small range in which it usually has been hard to find and so may be uncommon if not actually rare. That situation along with invariability of certain characters led to a suggestion (Gates, 1958) that stabilization had been achieved at a cost of adaptability. Phylogenetic filiation of species in the group seemed possible only through a common ancestral form that in passing across Burma from an evolutionary center in the Indo-Burman mountain wall had been segregated at different latitudes into sister species each with its own peculiarities no one of which is unknown elsewhere in the genus.

If, as was suggested, the retained funnels of x indicate a recent acquisition of metandry, the youngest species of group is *E. levis* in which the least progress in reduction of coelomic cavity of xi seems to have been made.

Eutyphoeus macer

1933. *Eutyphoeus macer* Gates, *Rec. Indian Mus.* **35**: p. 582. (Type locality, Thaton, Types, none.)
1958. *Eutyphoeus macer*, Gates, *ibid.*, **53**: p. 152.

Bivestibulate and penile: penes annular and with the posterior half thickened and continued ventrally as two pointed flaps, male and prostatic pores as well as apertures of penisetal follicles in the penes (?). Vestibula, small, well-like, nearly filled by penes, with apertures centered about at *B*, eversible to form with penes columnar intromittent organs. Female pores, both present. Spermathecal pores, centered at or median to *C*. Genital markings, paired, postsetal in *AB* of xvi. Setae, posteriorly *d* chaetae at or slightly dorsal to mL but directed laterally. First dorsal pore, at 11/12. Color, white (unpigmented). Segments, 161. Size, 50–82 by 2–4 mm.

Typhlosole, present from xxv. Lateral typhlosoles, simple low lamellae, in xxv. Lateral intestinal caeca, in xxiv but opening into gut in xxv. Ventral caeca, 4–6, in xxviii–xxxiv. Supra-intestinal glands, in 3 of lxiii, lxiv–lxv, lxvi. Dorsal blood vessel, complete, with hearts in v, vi. Metandric, testis sac *U*-shaped to annular and including hearts of xi. Seminal vesicles, in xii or xii–xiii. Penial setae, 2 per battery, 2–3 mm. long, 19–27 μ thick at base, 13–18 μ at neck, 16–22 μ at blade, 11–16 μ at tip, shaft with slight spiral curvature ectally, tip flattened, occasionally narrowing to a sharp point or a short spine, ornamentation of irregular transverse rows of spines. GM glands, sessile on parietes. Spermathecal diverticula, usually digitiform, posteriorly directed but median and lateral.

Distribution: Thaton, Bilin, Kyaikto (Thaton).

Remarks: Posterior approximation of spermathecal diverticula originally lateral and median seems to be one stage in development of a uniditerticulate condition of the sort that characterizes *E. ferinus*. Relationships of *E. macer* are with the *levis* group.

Eutyphoeus manipurensis

Avestibulate and apenile: apertures of penisetal follicles in a small tubercle on the roof of a fissure in *AB*, male pore in a tumescence from posterior margin

of fissure, prostatic pore anterolaterally (near or in a tumescence of anterior margin of fissure?), each fissure within a distinctly delimited male porophore, an area in *AC* of parietal thickening. Female pores, both present. Spermathecal pores, centered between *A* and m*BC*. Genital markings, small, paired, or unpaired and median, variously located, in vii–ix, xvi–xix. First dorsal pore, in region of 10/11–11/12. Color, white (unpigmented).

Lateral intestinal caeca, and lateral typlosoles, present. Dorsal blood vessel, complete(?), with hearts in v–vi. Holandric, testis sac annular. Seminal vesicles, of ix displace 8/9 anteriorly or penetrate into viii, of xii penetrate into xiii and push 13/14 into contact with 14/15. Penial setae, usually two per battery. GM glands, sessile on parietes. Spermathecal diverticulum, lateral.

Remarks: Samples, totaling only 40 specimens from 3 sites, in a range at least 150 miles long, cannot be expected to provide, in the present circumstances, sufficient data for definitive determination of systematic status. The two subspecies were recognized primarily to call attention to the differences while at the same time emphasizing similarities.

Eutyphoeus manipurensis manipurensis

1921. *Eutyphoeus manipurensis* Stephenson, *Rec. Indian Mus.* **22**: p. 763. (Type locality, Thonga Island, Loktak Lake, Manipur. Types, in the Indian Mus.)
1923. *Eutyphoeus manipurensis*, Stephenson, (*The Fauna of British India*), *Oligochaeta*, p. 441.
1934. *Eutyphoeus manipurensis*, Gates, *Rec. Indian Mus.* **36**: p. 276.
1938. *Eutyphoeus manipurensis*, Gates, *ibid.*, **40**: p. 89.
1958. *Eutyphoeus manipurensis manipurensis*, Gates, *ibid.*, **53**: p. 110.

Spermathecal pores, in *AB*. Genital markings, in viii–ix, xvi–xvii. Segments, 162. Size, 120 by 5 mm.

Typhlosole, present from xxvii. Lateral typhlosoles, in xxvii(?). Lateral intestinal caeca, in xxvii. (Ventral caeca?). Supra-intestinal glands, in 3–5 of lxxiii, lxxiv–lxxvi, lxxviii. GM glands, each with a strong but transparent capsule. Penial setae, 1.5 mm. long, shaft straight, tip slightly curved and tapering but not sharply pointed, ornamentation of fine triangular teeth.

Distribution: Thonga Island, Loktak Lake, Manipur.

Habitat: Swamps.

Remarks: Nothing was recorded as to lateral typhlosoles or as to origin of the major typhlosole but the typhlosoles–caeca correlation of other species warrants a prediction as above.

One lateral intestinal caecum of one of the types is in xxvi (as in *E. chinensis*) instead of xxvii as on the other side of the body. The location, apparently abnormal for a Thonga worm, could have resulted from abortion of one embryonic somite in the region between the 18th and 26th levels.

Eutyphoeus manipurensis chinensis

1933. *Eutyphoeus manipurensis*, Gates, *Rec. Indian Mus.* **35**: p. 383.
1958. *Eutyphoeus manipurensis chinensis* Gates, *ibid.*, **53**: p. 110. (Type locality, Falam, Chin Hills, Types. lost.)

Spermathecal pores, centered at or close to m*BC*. Genital markings, in vii–viii, xvi–xix, often 1–3 in front of and/or behind a spermathecal pore. (Segments?) Size, to 145 by 4 mm. Typhlosole, present from xxvi. Lateral typhlosoles, in xxvi. Lateral intestinal caeca, in xxvi. Ventral caeca, 3–6, in xxxi–xxxiv, xxxvi. Supraintestinal glands, in 4–5 of lxv, lxviii–lxviii, lxxi. GM glands, without a capsule(?) and of different texture within the parietes(?). Penial setae, 1.04–2.0 mm. long, 33–62 μ thick at base, 26–41 μ at midshaft. Ornamentation, variable, of scattered triangular teeth, of smaller spines in transverse rows or irregular circles and then rarely or frequently broken, or almost unrecognizable.

Distribution: Falam to Tiddim, Chin Hills.

Habitat: Cannot now be stated but probably not swamps.

Eutyphoeus marmoreus

1933. *Eutyphoeus marmoreus* + *E. bullatus* (part, 3 holandric specimens with uninterrupted dorsal blood vessel) Gates, *Rec. Indian Mus.* **35**: pp. 585, 569. (Type locality, Tiddim, Chin Hills. Types, none.)
1958. *Eutyphoeus marmoreus*, Gates, *ibid.*, **53**: p. 110.

Bivestibulate and apenile; male pores on small ovoidal tags protuberant from posterior margins of small fissures on roofs of eversible vestibula with apertures in *AB*, the lateral wall markedly thickened and also the roof which protrudes well into coelomic cavity. Female pores, both present. Spermathecal pores, centered at or just lateral to *B*. (Genital markings, none.) First dorsal pore, at 10/11. Color, white (unpigmented). (Segments?) Size, to 230 by 6 mm. Typhlosole, present from xxix. Lateral typhlosoles, in xxix–xxx. Lateral intestinal caeca, in xxix. Ventral caeca, 15–31, in xxxiii–lxix. Supra-intestinal glands, in 3–4 of lxxxv, lxxxvi–lxxxviii. Dorsal blood vessel, complete(?), with hearts in v, vi. Holandric, testis sac of xi annular. Seminal vesicles, of ix may reach into viii, of xii may reach into region of xiv–xxiii. Penial setae, 2–4 per battery, 1.2–1.95 mm. long, 25–45 μ thick at base, 23–40 μ at neck, 29–45 μ at blade, tip flattened and slightly widened, occasionally narrowing to a spine, ornamentation of thorn-like teeth in short transverse rows or in regular circles. Spermathecal diverticula, median and lateral.

Distribution: Tiddim (Chin Hills).

Remarks: Although holandric, relationships seem to be closest to *E. bullatus*, which perhaps could have been derived from *E. marmoreus* by abortion of the dorsal blood vessel in front of hearts of vi and by

abortion of the testes in x. The relationship of *E. marmoreus* and *bullatus* to each other at the only place from which they are now known may be like that of *E. hastatus* and *constrictus* to each other at certain more easterly sites.

Eutyphoeus peguanus

1930. *Eutyphoeus peguanus*, Gates, *Rec. Indian Mus.* **32**: p. 336.
1931. *Eutyphoeus peguanus*, Gates, *ibid.*, **33**: p. 426.
1933. *Eutyphoeus peguanus*, Gates, *ibid.*, **35**: p. 589.
1958. *Eutyphoeus peguanus*, Gates, *ibid.*, **53**: p. 129.
1961. *Eutyphoeus peguanus*, Gates, *Burma Res. Soc. 50th Anniv. Pub. No. 1*: p. 57.

Avestibulate and apenile: apertures of penisetal follicles on roofs of shallow, transverse fissures each in a distinctly delimited male porophore, male pore on a funnel-shaped protuberance from posterior wall of fissure, prostatic pore on anterior wall of fissure sometimes on a smaller protuberance. Female pores, both present. Spermathecal pores, in *BD*. Genital markings, lacking or paired and segmental in *AC*, of xvi and/or one of xix, xx, xxi, xxii. Ventral setae, of some segments ornamented ectally. First dorsal pore, at 11/12. Color, white (unpigmented). Segments, 229–310. Size, 150–240 by 4–10 mm.

Typhlosole, present from xxviii. Lateral typhlosoles, in xxviii–xxix. Lateral intestinal caeca, present in (xxvii) xxviii (xxix) or lacking. Ventral caeca, 10–31, in xxxii–lxx. Supra-intestinal glands, in 4–6 of lxxxiv, cvii–lxxxviii, cxii. Dorsal blood vessel, aborted in front of hearts of vi. Metandric, but usually with male funnels in x, testis sac annular to ventral, hearts of xi entirely or only in part included. Seminal vesicles, in xii. Penial setae, 6–10 per battery, 1.0–2.4 mm. long, 35–75 μ thick at base, 30–60 μ at neck, 32–75 μ at blade, 10–40 at tip, shaft straight except for ectal curve, tip sometimes widened and flattened, ornamentation of transverse rows of fine spine. Spermathecal diverticula, usually with but one seminal chamber, median and lateral.

Distribution: Labaw (Mergui). Tavoy and adjacent areas into the Kamaungthwe River region (Tavoy). Ye, Mupun, Moulmein, Chaungson, Kya-In, Kyundo, Kawkareik (Amherst). Myaungmya, Wakema (Myaungmya). Thaton, Duyinzeik, Kyaik-to, Kyaiktiyo, Kinmunsakhan, Taungzun, Naung-gala (Thaton). Rangoon, Syriam, Kyauktan (Hanthawaddy). Insein, Hmawbi, Taukkyan, Wanetchaung (Insein). Pantanaw (Maubin). Bassein (Bassein). Nyaunglebin, Pazunmyaung (Pegu). Thonze, Pegu Yomas east of Letpadan (Tharrawaddy). Toungoo, Shwegyin, Kyaukkyi, Blachi (Toungoo). Nyaungyo (Prome). Ywadaw, Pyigyaung, Pyinmana (Yamethin). Mandalay (Mandalay). Chaukma (Upper Chindwin). Myitkyina (Myitkyina).

From sea level to elevations of *ca.* 3,500 feet.

Remarks: Annular penial tumescences (around apertures of male pore fissures) such as seem to be appearing in some populations of *E. hastatus* and *constrictus* have not been seen in *E. peguanus*. Instead, the male pore often is located on a special protuberance, funnel-shaped or otherwise, from the posterior wall of the fissure. A protuberance from the anterior wall bearing the prostate pore, when recognizable, is of less definite shape. Those porophores are small and being rather delicate are easily damaged or rendered unrecognizable.

Genital markings of *E. peguanus* apparently are always intra-segmental, not crossing intersegmental furrows as in various other species of the genus. Although there is some variation as to segmental location a marking was found only once in xviii (but see *E. plenus* and *constrictus*) and then on the left side of an abnormal specimen.

Lateral intestinal caeca usually are present but form and location have not yet become stabilized in *E. peguanus*. Demarcation from the intestine often is partial or indistinct. A mere bulge that in other circumstances would attract no attention sometimes seemingly is to be regarded as an early rudiment. A distinctly demarcated caecum is apt to be overlooked when completely retracted into the intestinal lumen.

Male funnels usually are present in x. They may be plicate and about as large as the funnels of xi, mere rounded knobs or of various intermediate sizes and degrees of vestigialization. The sperm duct from a funnel of x may be complete, lacking behind the funnel septum, or in various intermediate conditions.

The coelomic cavity of xi in *E. peguanus* has been reduced to an annular testis sac and then sometimes to a *U*-shaped sac or finally to a subesophageal chamber. The sac more often is annular in the delta of the Irrawaddy but east of the Sittang is more apt to subesophageal.

The previous discussion of *E. peguanus* (1958) was based on that part of data secured during the period between 1932 and 1941 not destroyed in the World War II sack of Rangoon. Recognition of two subspecies did seem allowable as the article was typed. Each contained externally identifiable, geographically isolated or seemingly true-breeding races that, in accordance with long classical practice, had been provided with Latin varietal names. Infra-subspecific taxa have been disallowed by the International Rules. Distributions of such local races are separately indicated below to provide a basis for a future solution of the problem. When new material is available from Lower Burma, especially from the Pegu Yomas, a western species, *E. peguanus*, and an *eastern*, *E. similis*, each with subspecies, perhaps can be recognized.

Eutyphoeus peguanus peguanus

1925. *Eutyphoeus peguanus* Gates, *Ann. Mag. Nat. Hist.*, Ser. 9, 15: p. 323. (Type locality, Rangoon. Holotype, lost.

Paratypes, in U. S. Natl. Mus., Indian Mus.) *Rec. Indian Mus.* 27: p. 472.

1926. *Eutyphoeus peguanus*, Gates, *Rec. Indian Mus.* 28: p. 163. *Jour. Bombay Nat. Hist. Soc.* 31: p. 183. *Jour. Burma Res. Soc.* 15: p. 216. *Biol. Bull.* 51: p. 403.

1929. *Eutyphoeus peguanus*, Gates, *Proc. U. S. Natl. Mus.* 75, 10: p. 39.

1930. *Eutyphoeus peguanus* vars. *typicus, promotus, simplex*, Gates, 1930. *Rec. Indian Mus.* 32: p. 336.

1931. *Eutyphoeus peguanus* vars. *typicus, promotus, simplex, postremus*, Gates, *idem* 33: p. 426.

1936. *Eutyphoeus peguanus*, Hla Kyaw & Gates, *Jour. Roy. Asiatic Soc. Bengal*, (Sci.) 2: p. 170.

1958. *Eutyphoeus peguanus peguanus*, Gates, *Rec. Indian Mus.* 53: p. 142.

1961. *Eutyphoeus peguanus*, Gates, *American Midland Nat.* 66: p. 62.

Male porophores and genital markings, all of about the same large size. Fissures, eccentrically located on male porophores. Genital markings restricted to one of segments xix–xxii.

Distribution: The Irrawaddy delta region, including the following districts: Pyapon, Myaungmya, Bassein, Hanthawaddy, Maubin, Insein, Tharrawaddy (lowlands only), Henzada. As indicated below, the subspecies has been found outside the delta but because of transportation presumably by man.

Habitats: Soil, of various kinds including that of bare ground where dense shade seemingly prevented growth of grass and most other vegetation. Gardens, flower beds around and near houses. Accumulations of decaying leaves.

Castings: Intestinal ejecta, deposited on the surface of the ground, have rather tower-like shapes. A laboratory steward, K. John, who had much experience in collecting earthworms, recognized the places where *E. peguanus* could be secured by the castings. Those checked by the author usually were thicker (and perhaps shorter?) than castings of *E. foveatus* but other seemingly distinguishing characteristics cannot now be stated.

Variation and Abnormality: In one sampling (250 specimens), 498 spermathecae had each a median and a lateral diverticulum. One spermatheca of one worm lacked the median diverticulum. One spermatheca (of another worm), certainly adiverticulate, was quite abnormal being represented only by 2 vesicles sessile on the parietes. Athecal individuals were not found.

The extra testis and male funnel in Abnormal No. 1 (1958) may have resulted from halving of the left mesoblastic somite at the eleventh level.

Parasites: Large branched gregarines, presumably *Aikinetocystis singularis*, occasionally were found in coelomic cavities of preclitellar segments. Nothing has been published as yet about nematodes from coelomic cavities of *E. peguanus*.

Remarks: Luminescence of *E. peguanus* is more brilliant (1925b) than that of any other Rangoon species.

Male porophores appear to be structurally identical, except for inclusion of male pore fissures, with the genital markings. Porophores as well as genital markings of this subspecies may function as suckers during copulation.

1

1930. *Eutyphoeus peguanus* var. *simplex*, Gates, 1930, *Rec. Indian Mus.* 32: p. 340.

1958. *Eutyphoeus peguanus peguanus* f. *simplex*, *ibid.* 53: p. 141.

Genital markings, none. Penial setae, tips often scarcely thickened.

Distribution: Mergui and Tavoy districts, Mandalay, Chaukma (Upper Chindwin) and Nyaungyo (Prome) where the species is represented only by this form. Presumably transported there from the delta region of the Irrawaddy. Also present at Myitkyina to which it presumably was transported along with two other forms from Rangoon.

In the delta region was recorded from the vicinity of Syriam, from Kyauktan, Taukkyan, and Wanetchaung where it is the commonest form.

2

1930. *Eutyphoeus peguanus* var. *promotus* Gates, *Rec. Indian Mus.* 32: p. 339.

1958. *Eutyphoeus peguanus peguanus* f. *promotus*, Gates, *ibid.*, 53: p. 141.

Genital markings, in xix.

Distribution: Rangoon, Kyauktan, Hmawbi, Wanetchaung, Myitkyina.

Remarks: Presumably transported to Myitkyina from the Irrawaddy delta region.

3

1958. *Eutyphoeus peguanus peguanus* f. *intermedius* Gates, *Rec. Indian Mus.* 53: p. 141.

Genital markings, in xx.

Distribution: Hlegu (the only form), Kyauktan (the commonest form), Hmawbi (common), Hlegu (the commonest form but associated with 1), Wanetchaung (except for 1, the commonest form).

Remarks: Never found in Rangoon during the period of nearly twenty years in which earthworms were collected there.

4

1930. *Eutyphoeus peguanus* var. *typicus*, Gates, *Rec. Indian Mus.* 32: p. 337.

1958. *Eutyphoeus peguanus peguanus* f. *typicus*, Gates, *ibid.*, 53: p. 141.

Genital markings, in xxi.

Distribution: Rangoon and vicinity, Myaungmya and Wakema (Myaungmya). Myitkyina.

Remarks: Presumably transported to Myitkyina, perhaps from Rangoon and vicinity.

5

1931. *Eutyphoeus peguanus* var. *postremus* Gates, *Rec. Indian Mus.* 33: p. 427.

1958. *Eutyphoeus peguanus peguanus* f. *postremus*, Gates, *ibid.*, 53: p. 141.

Genital markings, in xxii.

Distribution: Syriam and vicinity where this form was repeatedly secured. Several specimens each were recorded from Kyauktan and Wanetchaung.

Eutyphoeus peguanus similis

1929. *Eutyphoeus similis* Gates, *Proc. U. S. Natl. Mus.* 75, 10: p. 37. (Type locality, Kawkareik. Type, lost.)

1930. *Eutyphoeus peguanus* var. *tumidus* + var. *similis*, Gates, *Rec. Indian Mus.* 32: pp. 341, 342.

1931. *Eutyphoeus peguanus* var. *tumidus* + var. *similis* + var. *praecox*, Gates, *idem.* 33: pp. 426, 428.

1958. *Eutyphoeus peguanus similis*, Gates, *ibid.*, 53: p. 142.

Male porophores small, fissures usually central. Genital markings, larger than the porophores.

Variation: Each of 627 spermathecae (of 315 specimens) had a lateral and a median diverticulum. Three spermathecae lacked one or the other of the two diverticula as follows; median diverticulum of left spermatheca, median diverticulum of right spermatheca, lateral diverticulum of left spermatheca.

Distribution: Amherst district from Ye north through Thaton district and hills at western margin of the Shan Plateau to Toungoo.

Remarks: The distributions were thought (1958) to provide some justification for retention of the three varieties until further studies can solve some of the problems that are involved.

6

1958. *Eutyphoeus peguanus similis* f. *similis*, Gates, *Rec. Indian Mus.* 53: p. 142.

Genital markings, in xvi and usually also in xxi.

Distribution: Ye, Mupun, Moulmein, Chaungson, Kya In, Kyundo, Kawkareik (Amherst district).

Remarks: Towards the Siamese border in the vicinity of Kya In, Kyundo, and Kawkareik genital markings of xxi often are lacking.

7

1931. *Eutyphoeus peguanus* var. *praecox* Gates, *Rec. Indian Mus.* 33: pp. 426, 428.

1958. *Eutyphoeus peguanus similis* f. *praecox*, Gates, *ibid.*, 53: p. 142.

Genital markings, in xvi and xx.

Distribution: Thaton district and north to Shwegyin, Pazunmyaung, and Nyaunglebin in Toungoo district.

Remarks: At four of the Thaton district localities markings of xvi are often or perhaps always lacking.

8

1930. *Eutyphoeus peguanus* var. *tumidus* Gates, *Rec. Indian Mus.* 32: p. 341.
1958. *Eutyphoeus peguanus similis* f. *tumidus*, Gates, *ibid.*, 53: p. 142.

Genital markings, in xix.
Distribution: Toungoo and vicinity.
Remarks: North of Toungoo into Yamethin district genital markings are much like or the same as those of the *peguanus* forms.

Eutyphoeus pius

1958. *Eutyphoeus pius* Gates, *Rec. Indian Mus.* 53: p. 150. (Type locality, Kyaiktiyo. Types, none.)

Avestibulate and penile: penes annular just lateral to *B* and in a regularly concave depression within an indistinctly delimited porophore in *A*-m*BC*, male and prostatic pores as well as apertures of penisetal follicles in the penes(?). Female pores, both present. Spermathecal pores, centered at or near *C*. Genital markings, paired, postsetal in *AB* of xiii–xv. Clitellum, lacking or very thin at m*V*. Setae, posteriorly *d* chaetae at or slightly dorsal to m*L*. First dorsal pore, at 11/12. Color, white (unpigmented). (Segments?) Size, 42 by 2½–3 mm.
Typhlosole, present from xxvi. Lateral typhlosoles, low, simple lamellae, in xxvi. Lateral intestinal caeca, in xxv but opening into gut in xxvi. Ventral caeca, 3, in xxxi–xxxiv. Supra-intestinal glands, in 2–3 of lxv–lxvi, lxvii. Dorsal blood vessel, complete(?), with hearts in v–vi. Metandric, testis sac *U*-shaped with limbs reaching shortly along gut and including ventral portions of hearts of xi. Seminal vesicles, in xii–xiii. Penial setae, 2 per battery, 1.45–1.5 mm. long, 51–70 μ thick at base, 40 μ at mid-shaft, 29–30 μ at neck, curved in a spiral fashion, tip thickened and with or without a spoon-shape, ornamentation of slightly irregular and closely crowded circles of very fine spines. GM glands, sessile on parietes. Spermathecal diverticula, digitiform and looped in a shortly zigzag manner, median and lateral.
Distribution: Kyaiktiyo (Thaton).
Remarks: The species is known only from the original description of the two clitellate types both of which may be amputees. Apertures of the male terminalia are likely to be in the penes which seemed to be permanent structures, each in the concavity of a watch-crystal-shaped porophore probably of some parietal thickening. Whether the entire porophore with the penis at its tip or only the penis is protruded during copulation was not learned.
Relationships of *E. pius* are with species of the *levis* group.

Eutyphoeus planatus

1929. *Eutyphoeus planatus* Gates, *Proc. U. S. Natl. Mus.* **75**: 10, p. 35. (Type locality, Thayetmyo. Holotype, lost. Paratypes, in the U. S. Natl. Mus. and the Indian Mus.)

1933. *Eutyphoeus planatus*, Gates, *Rec. Indian Mus.* 35: p. 592.
1958. *Eutyphoeus planatus*, Gates, *ibid.*, 58: p. 169.

Univestibulate and apenile (?): apertures of male deferent and prostatic ducts as well as of penisetal follicles in small transverse fissures (margins of which may be temporarily tumescent as small annuli?) on roof of vestibulum, vestibular aperture in *BB*, transversely slitlike to *Y*-shaped. Female pore, on left side. Spermathecal pores, centered at, close to, or lateral to *B*. Genital marking, median, in *BB*, postsetal in xiii (or across 13/14?). First dorsal pore, at 11/12. Color, brown (pigmentation dense). Segments, 149. Size, 100–240 by 5–9 mm.
Typhlosole, present from xxviii–xxix. Lateral typhlosoles, represented by longitudinal rows of one to five red patches. (Lateral intestinal caeca, none.) Ventral caeca, 4–7, in xxxiii–xli. Supra-intestinal glands, in 5–7 (rarely 8–9) of lxxxv, xcix–xc, civ. Dorsal blood vessel, aborted in front of hearts of vii. Metandric, testis sac ventral or annular, hearts of xi bound down to gut or included. Seminal vesicles, between 10/11 and 12/13 but pushing 12/13–13/14 back against 14/15. Penial setae, 4–11 per battery, 1–3 mm. long, 28–50 μ thick at base, 12–30 μ at neck, 28–60 μ at blade, shaft curved only ectally or slightly bowed, tip widened and flattened, scoop to spade-shaped, ornamented with transverse rows or circles of fine spines. (GM glands, none?) Spermathecal diverticula, median and lateral.
Distribution: Prome (Prome). Thayetmyo, Allanmyo (Thayetmyo). Magwe (Magwe), Minbu (Minbu).
Parasites: Large monocystid gregarines, with terminal suckerlike or digitiform appendages, were present in region of iii, singly or in pairs, trios or quartettes. Other gregarines were numerous in postprostatic coelomic cavities of juveniles. Nematodes, from testis sacs and coelomic cavities, *Iponema major* Timm & Maggenti, 1966, Prome.
Remarks: A greenish coloration, that gradually disappeared after preservation, could not be traced to any particular tissue or to discrete granules.
Some six characters, in a small range with a north-south length of about a hundred miles, apparently could distinguish a northern from a southern subspecies.
Relatively recent transport across the Irrawaddy is indicated by the restricted distribution on the east bank and by the similarity of worms from opposite sides of the river at the same latitude.
Relationships of *E. planatus* are with species of the *foveatus* group. An origin from the common ancestor of the group in the eastern hills of the Arakan yomas north of the *foveatus* range is anticipated.

Eutyphoeus plenus

1958. *Eutyphoeus plenus* Gates, *Rec. Indian Mus.* **53**: p. 142. (Type locality, Pyinmana. Types, none.)

Avestibulate and apenile: male and prostatic pores as well as apertures of penisetal follicles in transverse fissures each centrally located about at B in a porophore that is smaller than the genital markings. Female pores, both present. Spermathecal pores, in BD. Genital markings, paired, in AC, in two or more of xiv–xvi, xix–xxi. First dorsal pore, at 11/12. Color, white (unpigmented). (Segments?) Size, 210–335 by 7–10 mm.

Typhlosole, present from xxviii(?). Lateral typhlosoles, present but small (in xxviii?). Lateral intestinal caeca, usually in xxviii but occasionally in xxvii or xxix. Ventral caeca, 40–44, in xxxiv–lxxix. Supra-intestinal glands, in 5–7 of civ, cx–cx, cxv. Dorsal blood vessel, aborted in front of hearts of vii. Metandric, but with male funnels in x, testis sac annular (hearts of xi included?). Seminal vesicles, in xii. Penial setae, 8–15 per battery, 2–3 mm. long, 36–75 μ thick at base, 40–70 μ at neck, 60–115 μ at blade, 25–45 μ at tip, straight except for an ectal curve, tip widened and flattened, ornamented with fine, thornlike spines closely crowded or in irregularly interrupted circles. (GM glands, none?) Nerve cord sheath, very thickly muscular anteriorly. Spermathecal diverticula, median and lateral.

Distribution: Pyinmana and vicinity (Yamethin).

Variation: Genital markings were present in xv, xvi and xx of 219 (out of 265 adult) specimens. A single, asymmetrical marking in xviii was present in two specimens. Usual absence of markings in xviii is shared with *E. peguanus*. Frequently distribution of segmental location of supra-intestinal glands in 69 Pyinmana specimens was as follows; cvi–12, cvii–33, cviii–51, cix–62, cx–69, cxi–65, cxii–52, cxiii–35, cxiv–15, cxv–5. Lateral intestinal caeca were lacking only in one of 290 specimens.

Remarks: The male pore, when first recognizable in juveniles, is superficial and about halfway between eq/xvii and 17/18. Migration of the male pore towards the prostatic pore during growth was also noted in other species but in none of them do the male pores unite with the prostatic pores.

Relationships appear to be with *E. hastatus*, *constrictus* and *peguanus*, the metandry suggestive of closer relationships with *E. constrictus* and *peguanus*. Derivation directly from *E. constrictus* would involve abortion of all of the dorsal blood vessel in front of the hearts of vii at one step as one specimen of *E. hastatus* (one of the types of *E. montanus*) seems to indicate is possible. Derivation indirectly from *E. constrictus* through a *peguanus* stage would involve, in a first step, abortion of the trunk and associated vessels in front of the hearts of vi as well as the portion of the trunk still remaining in front of the hearts of vii. Derivation directly from *E. hastatus* would have required simultaneously with abortion of the dorsal trunk an abortion of testes in x.

Eutyphoeus pusillulus

1931. *Eutyphoeus pusillulus* Gates, *Rec. Indian Mus.* **33**: p. 248. (Type locality, Ye. Types, none.)
1933. *Eutyphoeus pusillulus*, Gates, *ibid.*, **35**: p. 593.
1958. *Eutyphoeus pusillulus*, Gates, *ibid.*, **53**: p. 150.

Avestibulate and apenile: apertures of penisetal follicles on roofs of very small fissures each within an oval and slightly diagonal porophore in AB. (Male and prostatic pores, in usual positions?) Female pores, both present. Spermathecal pores, at B. (Genital markings, none?) Clitellum, thinner at mV. First dorsal pore, at 10/11. Setae, posteriorly the d setae in the dorsum(?). Color, white (unpigmented). Segments, 80 (+?). Size, 32(+?) by 3 mm.

Typhlosole, present from xxii. Lateral typhlosoles, low, simple lamellae in xxii–xxiii. (Lateral intestinal caeca, lacking.) Ventral caecum, in xxix. Supraintestinal glands, in lvi–lviii. Dorsal blood vessel, complete, with hearts in (v?)vi. Proandric, but with coelomic cavity of xi reduced to an annular testis sac containing only the hearts of xi. Seminal vesicles, in ix. Penial setae, 2 per battery(?), *ca.* 0.65 mm. long and 20 μ thick, curved ectally, ornamented with broken circles of very fine spines. Spermathecal diverticulum, lateral.

Distribution: Ye, at southern boundary of Amherst district.

Remarks: Known only from the holotype which may have been a posterior amputee. Subsequent collecting in the immediate neighborhood of the type locality provided no additional specimens but nothing is known of the earthworm fauna from the vicinity of Ye all the way to the Siamese border. The type of this enigmatical form is the only specimen of a *levis* group species to be found below the Thaton district border and then some hundred miles away at the southern border of Amherst district.

Although absence of male funnels in xi indicates early ontogenetic abortion of gonads of that segment and a proandry that is older (for example) than the metandry of *E. constrictus* the coelomic cavity still is reduced (ontogenetically) to a functionless testis sac.

E. pusillulus appears to be closest to *E. levis* from which it is distinguished by the proandry that is unknown elsewhere in the genus.

Eutyphoeus quinquepertitus

1930. *Eutyphoeus quinquepertitus* Gates, *Rec. Indian Mus.* **32**: p. 343. (Type locality Nyaungyo. Types, none.)
1933. *Eutyphoeus quinquepertitus*, Gates, *ibid.*, **35**: p. 594.
1958. *Eutyphoeus quinquepertitus*, Gates, *ibid.*, **53**: p. 189.

Bivestibulate and apenile(?): male and prostatic pores discrete, on roofs of eversible(?) vestibula with transversely slit-like apertures reaching beyond A and B, thick roofs of vestibula protuberant into coelomic cavity. Female pore, on left side. Spermathecal pores, centered at or just lateral to B. Genital markings: transversely elliptical, one just anterior and one just posterior to each vestibular aperture, a larger one in BB between equators of xviii–xix. Setae, ventral follicles of xvii lacking in adults. First dorsal pore, at 11/12. Color, brown but pigment sparse. Segments, 120 (+?). Size, 156–190(+?) by 9 mm.

Typhlosole, present from xxviii. Lateral typhlosoles, in xxviii. Lateral intestinal caeca, in xxviii. Ventral caeca, 20–24, in xxxii–lv. Supra-intestinal glands, in 5 of lxxix, lxxx–lxxxiii, lxxxiv. Dorsal blood vessel, aborted in front of hearts of vii. Metandric, testis sac ventral. Seminal vesicles, between 10/11 and 12/13 which with 13/14 is pushed back into contact with 14/15. Glandular (?) tissue in the parietes associated with the large genital marking. (Penial setae, none.) Spermathecal diverticula, median and lateral, each with short stalk and an acinous mass of seminal chambers.

Distribution: Nyaungyo, Nyaungyi, Sanbot (Prome), at elevations of 2,500–3,500 feet in the Arakan Yomas.

Remarks: Only eight specimens were secured and each may have been a posterior amputee. Supposed penes were later thought to be early stages of vestibular eversion involving, at moment of preservation, only the region immediately around male and prostatic pores. Fissures were not found and presumably are not to be expected as they develop primarily in connection with apertures of penisetal follicles. Loss of penial setae, in this species., does not involve reversion to an ancestral condition but rather abortion of a and b follicles of xvii, presumably well before maturity, as in the Himalayan E. nainianus. An earlier stage in such evolution was thought to be shown by E. quadripapillatus in which penisetal follicles are retained into maturity though the secretory cells now deposit chaetal substance in discontinuous bits of various sizes and shapes. The apenisetal state like so many other characters has appeared independently in India and in Burma.

Relationships of E. quinquepertitus were believed to be with E. gigas and rarus.

Eutyphoeus rarus

1925. Eutyphoeus rarus Gates, Ann. Mag. Nat. Hist., Ser. 9, 16: p. 59. (Type locality, Rangoon. Type, lost. Paratypes, in the U. S. Natl. Mus., the Indian Mus.) Rec. Indian Mus. 27: p. 473.

1926. Eutyphoeus rarus, Gates, Jour. Bombay Nat. Hist. Soc. 31: p. 182; Jour. Burma Res. Soc. 15: p. 216; Biol. Bull. 51: p. 403; Rec. Indian Mus. 28: p. 164.

1928. Eutyphoeus rarus, Cobb, Jour. Washington Acad. Sci. 18: p. 200.

1930. Eutyphoeus rarus, Gates, Rec. Indian Mus. 32: p. 345.

1933. Eutyphoeus rarus, Gates, ibid., 35: p. 595.

1958. Eutyphoeus rarus, Gates, ibid., 53: p. 190.

1961. Eutyphoeus rarus, Gates, American Midland Nat. 66: p. 62.

Bivestibulate and penile: apertures of penisetal follicles in transversely elliptical tubercles each surrounded by an annular penis with posterior margin continued into two pointed flaps, male pore on posterior portion of annulus(?) (prostatic pore on anterior portion?), vestibula small, well-like, nearly filled by penes. Female pore, on left side. Spermathecal pores, with centers at or median to mBC. Genital markings, usually present, most frequently in xv–xvi, in region of AB, BC or AC, pre-setal and postsetal in vii–viii and xiv–xviii or across 18/19–22/23; in AA and unpaired or if paired in contact or marginally united at mV, presetal and postsetal in vii–viii, xiv–xvii (or across intersegmental furrows?), across 18/19. First dorsal pore at 11/12. Color, brown (pigment sparse). Segments, 145–187. Size, 90–205 × 3–7 mm.

Typhlosole, present from xxviii. Lateral typhlosoles, in xxviii. Lateral intestinal caeca, in xxviii but opening into gut at 28/29. Ventral caeca, 12–24, in xxxii–lvi. Supra-intestinal glands, in 4–6 of lxxiii, lxxxi–lxxvi, lxxxvi. Dorsal blood vessel, aborted in front of hearts of vii. Metandric, testis sac subesophageal and including only ventral portions of hearts of xi. Penial setae, 5–11 per battery, 3–6 mm. long, 30–80 μ thick at base, 22–47 μ at midshaft, 38–51 μ near tip, straight except for an ectal curve, occasionally much thickened ectally, tip tapering to a spine, ornamentation of very small thornlike teeth in irregularly interrupted circles. GM glands, sessile on parietes. Spermathecal diverticula, median and lateral.

Distribution: Rangoon, Twante, Kungyangon (Hanthawaddy). Insein, Taukkyan, Wanetchaung (Insein). Bassein, Kochi (Bassein). Thonze, Myagyaung (Tharrawaddy). Kyangin (Henzada). Prome, Laboo, along road over Arakan Yomas to Taungup (Prome). Sandoway, Taungup, (Sandoway).

Habitat: Soil.

Regeneration: The first segment and prostomium, had been replaced by the only individual of the genus Eutyphoeus to show evidence of cephalic regeneration.

Parasites: SPOROZOA: Large branched monocystids, presumably Aikinetocystis singularis occasionally were found in coelomic cavities of preclitellar segments. NEMATODA: Ungella secta Cobb, 1928, Adieronema eutyphoei Timm, 1967.

Remarks: Penes are protrusible from the vestibula but whether a vestibulum is eversible was not learned. Penes and vestibula are like those of the not closely related E. macer. Like E. quinquepertitus, E. rarus has retained the lateral typhlosoles that E gigas seems to have lost. However, E. rarus and gigas (q. v.) may be more closely related to each other than either is to E. quinquepertitus.

Eutyphoeus scutarius

1907. Eutyphoeus scutarius Michaelsen, Mitt. Naturhist. Mus. Hamburg 24: p. 186. (Type locality, Comillah, Chittagong district, East Bengal. Types, in the Indian Mus., the Hamburg Mus.)

1909. *Eutyphoeus scutarius*, Michaelsen, *Mem. Indian Mus.* 1: p. 240.
1910. *Eutyphoeus scutarius*, Michaelsen, *Abhandl. Naturwiss. Ver. Hamburg* 19, 5: p. 13.
1923. *Eutyphoeus scutarius*, Stephenson (*The Fauna of British India*), *Oligochaeta*, p. 452.
1938. *Eutyphoeus scutarius*, Gates, *Rec. Indian Mus.* 40: p. 108.
1958. *Eutyphoeus scutarius*, Gates, *ibid.*, 53: p. 218.

Avestibulate and apenile: each male pore in a fissure within a small porophore centered at *B*. Female pores, both present. Spermathecal pores, at or near m*BC*. Genital marking, in *BB* between equators of xv and xvi. First dorsal pore, at 11/12. (Color?) Segments, 290. Size, 140–180 by 5 mm. (Typhlosole? Lateral typhlosoles? Lateral intestinal caeca, none? Ventral caeca?) Supra-intestinal glands, in 5 of cxxvi, cxxxiii–cxxx, cxxxvii. Dorsal blood vessel, complete(?), with hearts in v, vi. Metandric, testis sac annular(?). Seminal vesicles, in xii but bulging several septa backwards. Penial setae, 6 per battery, 2 mm. long, 80–95 μ thick, slightly curved, ornamented with irregular and interrupted circles of fine spines. (GM glands, none.) Spermathecal diverticula, median and lateral.

Distribution: Comillah (Chittagong district, E. Bengal).

Remarks: Seminal vesicles, in specimens examined by the author, did not extend behind 12/13.

The species is known only from the three types. *E. scutarius*, with *E. comillahnus* and other species of uncertain status were thought (Gates, 1958) to be a trans-Arakan equivalent of the *constrictus* group in Burma.

Eutyphoeus sejunctus

1930. *Eutyphoeus sejunctus* Gates, *Rec. Indian Mus.* 32: p. 349. (Type locality, Thandaung. Types, none.)
1931. *Eutyphoeus sejunctus*, Gates, *ibid.*, 33: p. 430.
1933. *Eutyphoeus sejunctus*, Gates, *ibid.*, 35: p. 596.
1958. *Eutyphoeus sejunctus*, Gates, *ibid.*, 53: p. 153.

Bivestibulate and penile: penes about ½ mm. long, conical, filling well-like vestibula with apertures in *AB* and with thickened roofs markedly protuberant into coelomic cavity, male and prostatic pores as well as apertures of penisetal follicles in the penes(?). Females pores, both present. Spermathecal pores, centered at or close to *B*. Genital markings, paired, postsetal in *AB* of xi. Clitellum, lacking or very thin at m*V*. Setae, posteriorly ornamented with short transverse rows of fine spines, *d* setae (below m*L* anteriorly) gradually become more dorsal posteriorly until *DD* < *CD* as well as *AA*. First dorsal pore, at 11/12. Color, white (unpigmented). Segments, 178. Size, 64–146 by 3–4 mm.

Typhlosole, present from xxvii. Lateral typhlosoles, low simple lamellae, in xxvii. Lateral intestinal caeca, in xxvi but opening into gut at 26/27. Ventral caeca, 7–13, in xxxii–xliv. Supra–intestinal glands,

in 2–3 of lxiii, lxvi–lxv, lxviii. Dorsal blood vessel, complete, with hearts in v, vi. Metandric, testis sac ventral and dumb-bell-shaped to *U*-shaped, occasionally *U*-shaped but dorsal limbs not including hearts of xi. Seminal vesicles, push 12/13 and one or more other septa back. Penial setae, to 9 per battery, 2.46–2.48 mm. long, 48–50 μ thick at base, 45 μ at midshaft, 38–40 μ at tip, spirally curved, ornamentation of fine spines in unbroken circles ectally and scattered rows more entally. GM glands, sessile on the parietes. Spermathecal diverticula, median and lateral.

Distribution: Thandaung, Leiktho, Pelachi, Shwenyaungbin, Dawpakho (Karen Hills of Toungoo district).

Remarks: To have obtained its present range, *E. sejunctus* or its ancestors must have crossed lowlands at the site of the present Chindwin-Irrawaddy axis, climbed the Pegu Yomas, crossed the Sittang Valley in the Toungoo region (where it never was found) and then up into the Karen Hills east of what is now the edge of the Shan Plateau.

Relationships of *E. sejunctus* are with species of the *levis* (*q.v.*) group.

Eutyphoeus spinulosus

1926. *Eutyphoeus spinulosus* Gates, *Rec. Indian Mus.* 28: p. 164. (Type locality, Bassein. Types, none.) *Biol. Bull.* 51: p. 403.
1933. *Eutyphoeus spinulosus*, Gates, *Rec. Indian Mus.* 35: p. 597.
1958. *Eutyphoeus spinulosus*, Gates, *ibid.*, 53: p. 168.

Univestibulate and penile: apertures of male deferent and prostatic ducts as well as of penisetal follicles in tubular penes 1 + mm. long, vestibulum deep with thick wall but thin roof and circular aperture in *BB*, (penes protrusible through vestibular aperture but vestibulum not eversible?). Female pore, on left side. Spermathecal pores, centered at or median to *B*. Genital marking, median, in *BB*, between equators of xiv and xvi but portion in front of eq/xv usually much smaller, or smaller and separated, or lacking. First dorsal pore, at 11/12. Color, brown (pigmentation dense). (Segments?) Size, 170–230 by 4–9 mm.

Typhlosole, present from xxviii. Lateral typhlosoles, represented by longitudinal rows of one to five small red patches in first one or two typhlosolar segments. (Lateral intestinal caeca, none.) Ventral caeca in xxxiv, xxxvi–lx, lxvi. Supra-intestinal glands, in 4–6 of lxxxvii, xci–xci, xcvi. Dorsal blood vessel, aborted in front of hearts of vii. Metandric, testis sac ventral, hearts of xi bound down to gut. Seminal vesicles, between 10/11–12/13. Penial setae, 6–8 per battery, 3.0–4.37 mm. long, 32–55 μ thick at base, 20–40 μ at midshaft, 31–40 μ at neck, 45–59 μ at blade, straight except for slight curve ectally, tip widened and flattened or with membranous expansions, ornamented with thornlike teeth in more or less regular circles. (GM glands, none.) Spermathecal diverticulum, lateral.

Distribution: Bassein, Coomzamu, Kokya, Padaukchaung (Bassein).

Parasites: Large, branched monocystids, possibly *Aikinetocystis singularis*, occasionally were found in coelomic cavities of preclitellar segments.

Remarks: Relationships are closest with *E. foveatus*. Similarities are such that direct derivation from *E. foveatus* once seemed possible, by modifications such as eliminating one of the spermathecal diverticula. The species is more variable than others of the *foveatus* group but is known only from Bassein and vicinity which may be at the eastern boundary of the range.

Eutyphoeus strigosus

1933. *Eutyphoeus strigosus* Gates, *Rec. Indian Mus.* 35: p. 598. (Type locality, Shoko. Types, none.)
1958. *Eutyphoeus strigosus*, Gates, *ibid.*, 53: p. 154.

Bivestibulate and penile: penes small, shortly conical and filling small, well-like, eversible vestibula with apertures about at *B*. Male and prostatic pores as well as apertures of penisetal follicles in the penes (?). Female pores, both present. Spermathecal pores, centered at or near *C*. Genital marking, postsetal in *AA* of xii. (Clitellum, thin at mV(?). Setae, posteriorly *d* chaetae at or slightly dorsal to mL but laterally directed. First dorsal pore, at 11/12. Color, white (unpigmented). Segments, 189. Size, 139 by 5 mm.

Typhlosole, present from xxvii. Lateral typhlosoles, low, simple lamellae, in xxvii. Lateral intestinal caeca, in xxvii but opening into gut at 27/28. Ventral caeca, 11–14, in xxxiii–xlvii. Supra-intestinal glands, in 3 of lxix, lxx–lxxi, lxxii. Dorsal blood vessel, complete, with hearts in v, vi. Metandric, testis sac *U*-shaped but entered only below gut by hearts of xi. Seminal vesicles, push 12/13 and one or more other septa back. Penial setae, bowed but less strongly than in *sejunctus*. GM gland, sessile on the parietes. Spermathecal diverticula, median and lateral.

Distribution: Shoko, Blachi, Ko Haw Der (Karen Hills of Toungoo district).

Remarks: A deformation of the gizzard shown by so many of the types is not now believed to be characteristic of the species. *E. strigosus* is related to members of the *levis* (*q.v.*) group.

Eutyphoeus turaensis

1920. *Eutyphoeus turaensis* Stephenson, *Mem. Indian Mus.* 7: p. 244. (Type locality, Tura, Garo Hills, Assam. Types, in the Indian Mus.)
1923. *Eutyphoeus turaensis*, Stephenson (*The Fauna of British India*), *Oligochaeta*, p. 453.
1938. *Eutyphoeus turaensis*, Gates, *Rec. Indian Mus.* 40: p. 110.
1958. *Eutyphoeus turaensis*, Gates, *ibid.*, 53: p. 204.

Avestibulate(?) and apenile (?); male and prostatic pores as well as apertures of penisetal follicles in

fissures (or in penes?). (Female pores?) Spermathecal pores, just lateral to *B*. Genital markings, paired, centered at *A* or in *AB*, postsetal in xiv–xvi, presetal in xviii. First dorsal pore, at 11/12. (Color?) Segments, 171. Size, 100 by 3.5 mm.

(Typhlosole? Lateral typhlosoles? Lateral intestinal caeca, none? Ventral caeca?) Supra-intestinal glands, in 3–7 of lxv, lxvi–lxvii, lxxi. Dorsal blood vessel (aborted in front of 5/6?), hearts of vi present. Metandric (testis sac?). Seminal vesicles, between 10/11 and 12/13 which with 13/14 is bulged back against 14/15. Penial setae (several per battery), to 1.5 mm. long, 35 μ thick, shaftly slightly curved, tip thickened, pointed and claw-shaped, ornamentation of frequently interrupted irregular circles of fine spines. (GM glands, none?) Spermathecal diverticula, median and lateral.

Distribution: Tura (Garo Hills, Assam).

Parasites: Monocystids like *Aikinetocystis* but unbranched were present in coelomic cavities. This is the northernmost record for these large gregarines.

Remarks: Known only from the types which are juvenile.

E. turaensis, if the dorsal trunk ends with the hearts of vi, will be the only trans-Arakan representative of a *peguanus* stage in cephalization of the circulatory system.

Lennogaster

1939. *Lennogaster* Gates, *Rec. Indian Mus.* 41: p. 183. (Type species, *Eudichogaster yeicus* Stephenson, 1931.)
1958. *Lennogaster*, Gates, *Ann. Mag. Nat. Hist.*, Ser. 13, 1: p. 616. *Rec. Indian Mus.* 53: p. 210.
1959. *Lennogaster*, Gates, *Bull. Mus. Comp. Zool. Harvard College* 121: p. 254. Lee, *Bull. New Zealand Dept. Sci. Indus. Res.* 130: p. 33.
1961. *Lennogaster*, Gates, *Ann. Mag. Nat. Hist.*, Ser. 13, 3: p. 653. EXCRETORY SYSTEM: Bahl, 1942, *Quart. Jour. Micros. Sci.* 83: p. 446. PHYLOGENY: Gates, 1958, *Rec. Indian Mus.* 53: p. 208.

Digestive system, without intestinal caeca, with gizzards in v–vi, paired calciferous glands in x–xii, intestinal origin in xv (lateral typhlosoles?), intestinal typhlosole enlarged at its posterior end but without definite supra-intestinal glands. Calciferous glands, vertically ovoidal, opening through dorsal poles without distinct stalks into longitudinal grooves in lateral walls of esophagus, each groove bounded mesially by a thick ridge and in communication dorsally with esophageal lumen there *T*-shaped in cross section. Vascular system, with unpaired dorsal, ventral, and supra-esophageal trunks but no subneural, paired extra-esophageal trunks (median to hearts) connected by transverse vessels near 5/6–6/7 and passing posteriorly in esophagus within the ridges covering apertures to calciferous glands, paired lateroparietal trunks from posterior end of body passing to gut in xiii. Hearts, of ix lateral, of x–xii lateroesophageal. Excretory system, meroic, all nephridia small, avesiculate, biramous(?), astomate numerous nephridia of iii (opening into pharynx?) in vertical bands, remainder of the system exoic and comprising astomate nephridia in transverse parietal rows of three to six on each side in iv–xvii but with additional tubules mesially in xiv–xvii, from xviii posteriorly in three to five longitudinal parietal ranks on each side, the median nephridium of each side posteriorly (behind the typhlosole?) larger and with preseptal funnel. Septa, present from 4/5. Setae, four pairs per segment, ventral couples of xvii penial. Dorsal pores, present. (Unpigmented.)

Reproductive apertures, all minute and superficial, prostatic and male pores in seminal grooves. Clitellum, annular, in xiii–xvii, intersegmental furrows obliterated, dorsal pores occluded, setae retained. (Genital markings, lacking.) Metagynous (ovaries fanshaped?). Spermathecae, diverticulate.

Distribution: Burma. Bangla Desh and India from the Burmese border through the Gangetic valley and into the northern part of the peninsula.

The natural range as thus indicated may have been extended, presumably by man, to include the island of Sumba (south of Flores) from which an Indian species, *L. barkudensis*, once was recorded.

Remarks: Characters by which classical genera often were defined, such as hol-, pro- and met-andry, bi- and quadri-thecal, bi- and quadri-prostatic, are unavailable in the present instance because of intrageneric variation in number of testes, spermathecae and prostates.

Evagination of those portions of the esophageal wall that bear calciferous lamellae in *Eutyphoeus* and *Bahlia* conceivably could give rise to the lennogaster glands. However, further speculation on phylogenetic relationships (*cf.* Gates, 1958: p. 616) should await acquisition of information as to internal anatomy of the evaginated sacs.

The excretory system, according to Bahl (1942) closely resembles that of *Eutyphoeus*. Unfortunately, the extent of the similarities was not indicated nor any of the differences which probably do exist.

L. elongatus Gates, 1945, *falcifer* and *trichochaetus* (Stephenson, 1920), with seminal vesicles in ix and xii, four each of spermathecae, testes, and prostates, clearly are more primitive than the other species. In a northwestern portion of the Indian peninsula inhabited by those primitive species is to be sought the original home of *Lennogaster*. Four species, including *pusillus* and *barkudensis* (Stephenson, 1920), have lost the seminal vesicles and in addition one pair of testes, spermathecae, or prostates. Loss of gonads, whether in x or xi, was so recent that associated male funnels and ducts still are present. Absence of seminal vesicles warrants a suspicion of parthenogenesis in the four eastern species.

Two Burmese species are tentatively regarded as endemic in the area from which they are now known.

1. Holandric[a] 2
   Meroandric[a] 4
2. Male field confined to xvii–xix 3
   Male field extends posteriorly at least to 20/21[b] .... *elongatus*
3. Penial setae 2 + mm. long, *a* and *b* of viii–ix copulatory[c] .... *trichochaetus*
   Penial setae less than 2 mm. long, *a* of vii copulatory .... *falcifer*
4. Quadriprostatic .... *yeicus*
   Biprostatic 5
5. Metandric[a] .... *barkudensis*
   Proandric 6
6. The *a* setae of viii copulatory ...... *chittagongensis*
   No copulatory setae in viii[d] .... *pusillus*

[a] Andry of *L. barkudensis* is unknown. Testes apparently were found by Stephenson in xi but not in x. If testes are present in x the species is distinguished from other holandric forms by the single pair of spermathecae.

[b] Extension of seminal grooves behind eq/xix in *L. elongatus*, is now known not to be an individual abnormality of the type as the condition has since been found in some fourteen further specimens.

[c] Copulatory setae of *L. trichochaetus* and *falcifer* are like those of *Bahlia, Calebiella, Pellogaster* and *Octochaetoides*. Similarity in shape and ornamentation is not believed to indicate close relationships.

[d] *L. pusillus*, by a *lapsus calami*, was credited (*Proc. Indian Acad. Sci.* 21: (1945) p. 252) with an extra pair of spermathecae. Accordingly, "and ix," sixth line from the bottom of the page should be deleted.

Dichogaster parvus Fedarb, 1898, so inadequately (if not also erroneously) characterized as to be generically unidentifiable, may have been a lennogaster.

Lennogaster chittagongensis

1917. *Eudichogaster chittagongensis* Stephenson, *Rec. Indian Mus.* 13: p. 411. (Type locality, Rangamati, Chittagong Hill Tracts, now Bangla Desh. Types, in the Indian Mus.)
1923. *Eudichogaster chittagongensis*, Stephenson (*The Fauna of British India*), Oligochaeta, p. 411.
1931. *Eudichogaster chittagongensis*, Stephenson, *Proc. Zool. Soc. London*, 1931: p. 78; *Rec. Indian Mus.* 33: p. 192.
1939. *Lennogaster chittagongensis*, Gates, *ibid.*, 41: p. 192.
1955. *Lennogaster chittagongensis*, Gates, *ibid.*, 52: p. 75.
1958. *Lennogaster chittagongensis*, Gates, *Ann. Mag. Nat. Hist.*, Ser. 13, 1: p. 615.
1961. *Lennogaster chittagongensis*, Gates, *Burma Res. Soc. 50th Anniv. Publ.* No. 1: p. 57.

Biprostatic, pores (common apertures of prostatic duct and penisetal follicles?) at anterior ends of seminal grooves, male pores at posterior ends. Grooves in *AB*, with anterior ends at *A* and eq/xvii, posterior ends at *B* and 17/18, both grooves in a transverse field reaching laterally in xvii to or well towards *C*. Bithecal, pores in *AB*, at or just in front of eq/viii. First dorsal pore, at 11/12. Setae, *a*, *b* of xviii–xix sigmoid and ornamented ectally by several transverse rows of fine teeth, *a* of viii copulatory. Prostomium, epilobous, tongue narrowing posteriorly and nearly reaching 1/2. Segments, 121. Size, 24–78 by 1–2 mm.

Proandric, but male funnels with ducts present in xi, testes and anterior funnels in paired testis sacs that may extend well dorsally but without inclusion of the hearts. (Seminal vesicles, lacking.) Penial setae, 0.5–0.64 mm. long, 3–6 μ thick at base and midshaft, 2–3 μ at tip, variously curved or bent, margins of ectal third to half serrate, tip truncate or narrowed to a short spine, ornamented with 12–17 circles(?) of fine spines. Copulatory setae, 0.22–0.3 mm. long, 11–14 μ thick, tip flattened, ornamentation of scattered triangular teeth or longitudinal rows of gouges with jagged ental margins. Spermathecal duct, much longer than ampulla, a short ectal portion with thick, smooth wall and narrow lumen, remainder irregularly constricted, bulged or narrowed and with larger lumen. Diverticulum, pendent anteriorly from base of ental portion of duct to or nearly to parietes, shortly digitiform.

Reproduction: Sperm are matured but none of the records still available mentions presence of sperm in the spermathecae. Parthenogenesis is suspected, primarily because of absence of seminal vesicles, the organs in which sperm usually are matured.

Distribution: Ye, Moulmein, Kya-In (Amherst). Kyaikto (Thaton). Rangoon (Hanthawaddy). Wanetchaung, Taikkyi (Insein). Bassein, Coomzamu (Bassein). Pegu (Pegu). Henzada (Henzada). Sandoway (Sandoway). Toungoo, Shwegyin (Toungoo), Paukkaung (Prome). Thanbula (Thayetmyo). Ramree (Kyaukpyu). Mount Popa (Myingyan). Kalewa, Paungbyin, Kindat, Masein, Pantha, Mawlaik (Upper Chindwin). Kengtung, Lashio (Shan States). Katha, Indaw Lake, Naba (Katha). Rangamati (Chittagong Hill Tracts, Bangla Desh). Seemingly endemic in middle and western Burma as well as Chittagong. Possibly introduced recently to the two localities on the Shan plateau.

Remarks: Although definite genital markings were not found, the region around apertures of *a* and *b* follicles of viii may be tumescent.

Lennogaster yeicus

1931. *Eudichogaster yeicus* Stephenson, *Rec. Indian Mus.* **33**: p. 193. (Type locality, Chaungson. Types, in the British Mus.)
1939. *Lennogaster yeicus*, Gates, *ibid.*, **41**; p. 186.
1955. *Lennogaster yeicus*, Gates, *ibid.*, **52**: p. 75.
1958. *Lennogaster yeicus*, Gates, *Ann. Mag. Nat. Hist.*, Ser. 13, **1**: p. 615.
1961. *Lennogaster yeicus*, Gates, *Burma Res. Soc. 50th Anniv. Publ.* No. 1: p. 57.

Quadriprostatic, pores in *AB*, in ventrally concave, bracket-shaped grooves at or just median to *B* and between eq/xvii and eq/xix, male pores at eq/xviii, margins of grooves especially swollen along the mesially directed end portions. Quadrithecal, pores in *AB*, at or in front of eq/viii, at or just behind 8/9.

First dorsal pore, at 11/12. Setae, ventral couples of xviii lacking, of xvii and xix penial, none copulatory. Prostomium, epilobous. Segments, 150. Size, 24–52 by 1–2 mm.

Lateral typhlosoles, in xvii–xx, interrupted. Proandric, but male funnels with ducts present in xi, testes and anterior funnels in paired testis sacs that may extend well dorsally but without inclusion of hearts. (Seminal vesicles, lacking.) Penial setae, 0.3–0.46 mm. long, 4–6 μ thick at base, 3–5 μ at midshaft, 2 μ at neck, 4 μ at tip, nearly straight, tip flattened and narrowing to a sharp point or widened, bifid and webbed or truncate, ornamentation of slight marginal serrations in ectal half. Spermathecal duct, much longer than ampulla, a short ectal portion with thick, smooth wall and narrow lumen, remainder irregularly constricted. Diverticulum, pendent anteriorly from base of ental portion of duct to or nearly to parietes, shortly digitiform.

Reproduction: Presumably biparental as sperm are exchanged in copulation.

Distribution: Ye, Chaungson, Kya In (Amherst). Thaton, Kinmunsakhan (Thaton). Rangoon, Kokine (Hanthawaddy). Doedaung Hill (Sandoway). Presumably endemic in Burma.

Remarks: Seminal vesicles apparently were found in ix, by Stephenson, perhaps only in one specimen. Conditions in the two Burmese species of *Lennogaster* would seem to indicate that testis sacs make seminal vesicles unnecessary if reproduction is not parthenogenetic.

Penial setae usually are truncate, the tip presumably having been worn or broken off at the neck.

Recorded differences as to relative sizes of major intersetal intervals require confirmation or correction. For that reason the characters are omitted from the précis of *L. chittagongensis* as well as of *L. yeicus*.

Octochaetona

1962. *Ochochaetona* Gates, *Ann. Mag. Nat. Hist.*, Ser. 13, **5**: p. 211.
(Type species, *Octochaetus surensis* Michaelsen, 1910.)

CALCIFEROUS GLANDS: Stephenson & Prashad, 1919, *Trans. Roy. Soc. Edinburgh* **52**. EXCRETORY SYSTEM: Gates, 1939, *Rec. Indian Mus.* **41**: p. 213. Bahl, 1942, *Quart. Jour. Micros. Sci.* **83**: p. 447. Matthew, 1950, *Jour. Zool. Soc. India* 2: p. 143.

Digestive system, with a small esophageal gizzard (belonging to vi?) in a coelomic space where septa are abortive, one pair of extramural, shortly and slenderly stalked, calciferous glands opening into esophagus in region of insertion of septum 15/16, an intestinal origin behind xvi, a ventrally bifid intestinal typhlosole, but without intestinal caeca and supra-intestinal glands. Vascular system, with unpaired dorsal, ventral and supra-esophageal trunks but no subneural, paired subesophageal trunks in region of vi–xiv, paired extra-esophageals median to the hearts and opening into the supra-esophageal in viii, paired lateroparietal trunks from the posterior end that open into the supra-esophageal in xiii. Hearts, of x–xii latero-esophageal. Excretory system, meroic, all nephridia small, avesiculate, astomate, and biramous, a pair of large clusters of nephridia anterior to 4/5 (with ducts opening into pharynx?), posteriorly only exoic parietal nephridia. Septa, 5/6–6/7 (at least) abortive. Setae, 4 pairs per segment, ventral setae of xvii and xix penial. Dorsal pores, present.

Quadriprostatic, pores at termini of seminal grooves extending from eq/xvii to eq/xix. Male pores, like the prostatic minute, in the seminal grooves, at eq/xviii. Quadrithecal, pores at or behind 7/8 and 8/9. Female pores, in xiv. Clitellum, annular, female pore segment included, intersegmental furrows obliterated, dorsal pores occluded, setae retained, always extending beyond xiv and xvi. Metagynous, ovaries fan-shaped and with several egg-strings. Ovisacs, present in xiv.

Distribution: India, peninsular portion from the northern margin down at least to South Arcot but not including the western coast.

The proper generic range as thus indicated has been extended over land and sea, presumably in large part through transport by man of two species, to include the western coast of India, Indo-Gangetic Valleys, Nepal, Assam, Burma, the Malay Peninsula, and the Philippines.

Remarks: Segments, according to published records, range from 100 to 246 but some of the counts now seem likely to have been from amputees. Numbers, in most if not all species, may normally be above 150. Pigment, when present, appears to be of the same sort as in *Eutyphoeus* though usually not deposited as densely. Dorsal pores never have been found in front of 11/12.

Structure of the calciferous glands has been studied in one species. The gland is said (Stephenson & Prashad, 1919: p. 464) to have a great number of thin lamellae arising from the wall and projecting into the interior, ending in a free edge near the place where the lumen of the gland opens by a narrow duct into the esophagus. Calcareous matter of the glands is in fine granules. Concretions, two mm. thick, seemingly too large to pass through the slender gland stalks, were found in the esophageal lumen of xvi.

Octochaetona has been evolving one kind of cephalization that may be characteristic but which can be described as yet almost only by reference to the septa. Stages shown by species already studied are as follows. I. Septum 4/5 thickly muscular, 5/6–6/7 membranous, 7/8–12/13 muscular. II. Septa 5/6–6/7 aborted. III. Septa 5/6–6/7 aborted, 7/8 membranous. IV. Septa 5/6–7/8 aborted. V. Septa 5/6–8/9 aborted, 9/10 membranous. VI. Septa 5/6–9/10 aborted, 10/11–11/12 still muscular but displaced posteriorly and united peripherally. Associated with that cephalization is some elongation

of the esophagus to bring the intestinal origin, perhaps at first in xv or xvi, back into xvii, then xviii. In contrast with those intrageneric modifications in somatic anatomy is the conservatism of the genitalia, the ancestral 4 prostates and 4 spermathecae having been retained throughout and the 4 testes in all but one of species.

Too little is known about the endemic forms to warrant discussion of intrageneric and intergeneric relationships. *Octochaetona* does now seem likely to be terminal in one line of nephridial evolution.

KEY TO BURMESE SPECIES OF *Octochaetona*

Spermathecal pores in *AA*, unpigmented,
  metandric............................*beatrix*
Spermathecal pores in *AB*, pigmented,
  holandric............................*surensis*

Note: For a key to known species of the genus *cf.* Gates, 1962.

Octochaetona beatrix

1902. *Octochaetus Beatrix* Beddard, *Ann. Mag. Nat. Hist.*, Ser. 7, 9: p. 456. (Type locality, Calcutta, India. Types, none.)
1929. *Octochaetus lunatus* Gates, *Proc. U. S. Natl. Mus.* 75, 10: p. 24. (Type locality, Mandalay, Burma. Types, in the Indian, British and U. S. Natl. Mus.)
1930. *Octochaetus lunatus*, Gates, *Rec. Indian Mus.* 32: p. 325.
1931. *Octochaetus lunatus*, Gates, *ibid.*, 33: p. 417. *Octochaetus (Octochaetoides) fermori*, Stephenson, *ibid.*, 33: p. 192.
1933. *Octochaetoides fermori*, Gates, *ibid.*, 35: p. 559.
1955. *Octochaetoides beatrix*, Gates, *ibid.*, 52: p. 76.
1961. *Octochaetoides beatrix*, Gates, *Burma Res. Soc. 50th Anniv. Pub. No. 1*: p. 57.
1962. *Octochaetona beatrix*, Gates, *Ann. Mag. Nat. Hist.*, Ser. 13, 5: p. 213.

ECOLOGY: Gates, 1945, *Proc. Natl. Acad. Sci. India* 15: p. 48. NEPHRIDIA: Bahl, 1947, *Biol. Rev.* 22: p. 132. Matthew, 1950, *Jour. Zool. Soc. India* 2: p. 143. TYPHLOSOLE: Hertling, 1923, *Zeitschr. Wiss. Zool.* 120: p. 185.

Spermathecal pores, minute, at or slightly in front of eq/viii and eq/ix, in *AA*. Seminal grooves, in *AA*. (Genital markings, lacking. No differentiation of a special male field.) Setae, $AB < CD < BC < AA$, $DD > \frac{1}{2}C$, $a,b/$viii–ix modified slightly, apertures of *a* penisetal follicles discrete (*b* follicles and prostatic ducts opening to exterior through common apertures?). Female pores, in *AA*. Clitellum, xiii–xvii, xviii. First dorsal pore, at (11/12), 12/13. Prostomium, with posteriorly narrowed tongue. (Unpigmented.) Segments, to 196. Size, 50–100 by 2–5 mm.

Septa, 4/5 thickly muscular, 5/6–7/8 lacking, 8/9–11/12 muscular, 8/9–10/11 displaced posteriorly and crowded together, 10/11–11/12 adherent to each other peripherally. Intestinal origin, in xvii. Typhlosole, ending abruptly in region of civ–cxii. Last hearts, in xiii, latero-esophageal. Metandric, but male funnels of x retained, testis sac *U*-shaped, sub-esophageal. Seminal vesicles, small, in xii. Penial setae, 0.50–

0.56 mm. long and 0.015–0.020 mm. thick, with sparse ornamentation of triangular teeth. Spermathecae, small, beneath the gut, duct shorter than ampulla. Diverticulum, from anterior face of duct near parietes, spheroidal, shortly pyriform, ellipsoidal or flattened and shelflike, often with a very short and slender stalk. Ovisacs, in xiv, finely acinous. Setae of ventral couples in viii–ix, "slightly modified."

Reproduction: Presumably biparental as sperm are matured and exchanged in copulation. Breeding may be possible throughout periods of activity. Several eggs may be present in an ovisac.

Distribution: Moulmein, Ye, Chaungson, Kyaikmaraw (Amherst). Thaton, Duyinzeik, Bilin, Kyaikto, Taungzun, (Thaton). Bassein, Coomzamu, Thinbawgyin (Bassein). Shwegyin (Pegu). Thonze (Tharrawaddy). Henzada (Henzada). Sandoway, Ngapoli, Patle (Sandoway). Toungoo, Shwegyin (Toungoo). Pyinmana (Yamethin). Magwe (Magwe). Minbu (Minbu). Meiktila, Mondine (Meiktila). Mount Popa, Kyappea (Myingyan). Pakokku (Pakokku). Akyab, Padali, Buthidaung, Maungdaw (Akyab). Namkham, Lashio, Maymyo (Northern Shan States). Kyaukse (Kyaukse). Mandalay, Tonbo, Kyaukkyone (Mandalay). Sagaing, Kaungmudaw, Tada-U, Myotha (Sagaing). Monywa, Ingyindaung, Powindaung, Thindaw, Anidaung (Lower Chindwin). Ye-U, Kin-U, Kyaukmyaung (Shwebo). Indaw Lake, Wuntho, Naba (Katha). Mingin, Chaukma, Kalewa, Kindat, Masein, Paungbyin, Mawleik (Upper Chindwin). Bhamo (Bhamo). Myitkyina, Mohnyin, Weshi (Myitkyina).

India, from Trivandrum in Travancore north to Kassauli, Hoshiarpur, Dehra Dun, and Raniganj. Nepal, Katmandu, and Gowchar.

Malay Peninsula (Kuala Lumpur). Philippine Islands (three localities).

Paucity of records for southern India suggests that this uniquely metandric species of the genus arose in some northern portion of the peninsula.

Presence in the Philippines clearly is a result of transportation. Burma, Kuala Lumpur, and Himalayan Nepal doubtless were reached passively and even much of the Indian distribution may be due to man.

Habitats: Soil. Earth around roots of potted plants (Burma, India). Saturation with kitchen drainage apparently resulted in disappearance of the species at one Allahabad site.

Biology: The species is geophagous and is believed to copulate as well as feed underground. Activity probably is uninterrupted at favorable sites where moisture is adequate during the dry season. Elsewhere, quiescence is imposed during drier portions of the year.

The species has been able to maintain itself, after introduction, in three widely separated localities on the Shan Plateau as well as in Himalayan Nepal, at

various places in the dry zone of central Burma, in the coastal strips of Arakan and Tenasserim where annual rainfall may be as much as 200 inches. Although introduction to Rangoon would seem more likely than to most any other place in Lower Burma, the species never was found in Hanthawaddy and Insein districts.

Abnormality: No. 1. Posterior prostates, lacking. No. 2. Right anterior spermatheca, lacking. No. 3. Three spermathecae in viii on left side, each with own pore. Ducts of the posterior spermathecae unite to open by a common median pore.

Regeneration: Tail regenerates were found at various levels from 85/86 to 173/174.

Parasites: Insect larvae, probably of *Sarcophaga* sp., emerged from anterior ends of several individuals. Protozoa occasionally were found in the coelom of postprostatic segments but further information no longer is available.

Remarks: The prostomium has been said to be pro-, proepi- or epilobous but none of the standard classical terms adequately characterize the organ as seen in most well-preserved individuals. A tongue obviously is present but it is narrowed posteriorly. The laterally demarcating grooves often seem to unite posteriorly but close convergence associated with a median depression sometimes is responsible for an appearance as of a single median groove. The latter usually does not reach 1/2. At level of the anterior margin of i a transverse groove sometimes demarcates the tongue from the real prostomium.

As the young mature, apertures of the ventral follicles of xvii and xix are moved mesially. Setae in the ventral follicles of viii–ix retain some of their sigmoid shape but usually acquire ectally a sparse ornamentation of rather triangular teeth.

Although the paired subesophageal trunks, in well-preserved specimens, are easily traced almost to the gizzard, connections with other trunks were not seen.

A shelflike spermathecal diverticulum flattened against the parietes often is unrecognizable until after removal of concealing tissues. Failure to dissect off such tissues was responsible for records of adiverticulate spermathecae.

Octochaetona surensis

1910. *Octochaetus surensis* Michaelsen, *Abhandl. Naturwiss. Ver. Hamburg* **19**, 5: p. 88. (Type locality, Sur Lake, Puri district, Orissa, India. Types, none.)
1925. *Octochaetus (Octochaetoides) birmanicus* Gates, *Ann. Mag. Nat. Hist.*, Ser. 9, **16**: p. 55. (Type locality, Rangoon. Types, in the Indian, British and U. S. Natl. Mus.)
1926. *Octochaetus birmanicus*, Gates, *Rec. Indian Mus.* **28**: p. 162; *Jour. Bombay Nat. Hist. Soc.* **31**: p. 183; *Jour. Burma Res. Soc.* **15**: p. 214.
1930. *Octochaetus birmanicus*, Gates, *Rec. Indian Mus.* **32**: p. 325.
1931. *Octochaetus birmanicus*, Gates, *ibid.* **33**: p. 417.
1933. *Octochaetoides birmanicus*, Gates, *ibid.*, **35**: p. 557.
1934. *Octochaetoides birmanicus*, Gates, *ibid.*, **36**: p. 269.
1937. *Octochaetoides birmanicus*, Hla-Kyaw & Gates, *Jour. Roy. Asiatic Soc. Bengal*, (Sci.) **2**: p. 166.

1955. *Octochaetoides surensis*, Gates, *Rec. Indian Mus.* 52: p. 78.
1961. *Octochaetoides surensis*, Gates, *American Midland Nat.* 66: p. 62. *Burma Res. Soc. 50th Anniv. Pub.* No. 1: p. 57.
1962. *Octochaetona surensis*, Gates, *Ann. Mag. Nat. Hist.*, Ser. 13, 5: p. 213.

Spermathecal pores, minute, at or near equators of viii–ix, in *AB*. Seminal furrows, about at *B*, within a distinct male field extending into xvi and xx and laterally well towards *C*, usually with two deep transverse depressions involving a postsetal portion of xvii and a presetal portion of xix. Genital markings, areas of slight epidermal modifications, in *AA*, median or paired, postsetal in some of xviii–xxii (or at intersegmental levels?). Setae, $AB < CD < BC < AA$, $DD > \frac{1}{2}C$ anteriorly to $= \frac{1}{2}C$ posteriorly, *BC* becoming smaller posteriorly, *a,b*/viii–ix copulatory and follicle apertures (associated with indistinctly delimited, irregular tumescences) often displaced from *A* and *B* meridians, *a,b*/xvii and xix penial (prostatic duct and two associated penisetal follicles with a common external aperture?). Clitellum, xiii–xvi, xvii. First dorsal pore, at 12/13. Prostomium, with posteriorly narrowed tongue. Color, brown. Segments, to 180. Size, 60–140 by 2.5–6 mm.

Septa, 4/5 thickly muscular, 5/6–7/8 lacking, 8/9–10/11 displaced posteriorly and crowded together. Pigment, in or immediately under the epidermis. Intestinal origin, in xvii. Typhlosole, beginning in xxii–xxiii, deeply bifid ventrally to region of xxx–xl, much less so posteriorly, ending abruptly in region of ci–cxv (atyphlosolate segments, to 67). Last hearts, in xiii, latero-esophageal. Holandric, testis sacs cylindrical. Seminal vesicles, in ix and xii. Sperm ducts, loosely looped on the parietes, several times as long as distance between funnel septa and xviii, uniting only within parietes of xviii. Penial setae, 1.2–1.8 mm. long, 0.040–0.045 mm. thick at base, 0.025–0.030 mm. at mid-shaft, tip claw-shaped (?), ornamentation of longitudinal rows of triangular teeth. Spermathecae, rather small, beneath the gut, duct barrel-shaped ectally. Diverticulum, multilocular, with very short stalk from narrow ental part of duct. Copulatory setae, 0.85–1.2 mm. long, 0.026–0.030 mm. thick at base, 0.020–0.025 mm. at mid-shaft, tip slightly claw-shaped, ornamentation of longitudinal rows of spike or thorn-like protuberances.

Reproduction: Presumably biparental as sperm are matured and exchanged during copulation. Breeding may be possible throughout periods of activity. An ovisac may contain as many as 40–50 ova.

Distribution: Labaw (Mergui). Tavoy (Tavoy). Moulmein, Martaban, Kyaikmaraw, Mupun (Amherst). Pyapon, Thanchitaw, Thameintaw, Kyaiklat, Dedaye, Bogale (Pyapon). Wakema (Myaungmya). Kyaikto, Sittang, Mokpalin (Thaton). Rangoon, Thongwa, Syriam, Twante, Kungyangon, Kyauktan (Hanthawaddy). Insein, Hmawbi, Taikkyi, Wanet-chaung, Hlawga (Insein). Yandoon, Danubyu, Pantanaw (Maubin). Bassein, Coomzamu, Kochi, Thinbawgyin, Padaukchaung (Bassein). Pegu, Shwegyin, Nyaunglebin, Pazunmyaung, Kyauktan, Thanatpin (Pegu). Letpadan, Thonze (Tharrawaddy). Henzada (Henzada). Toungoo, Tantabin (Toungoo). Pyinmana (Yamethin). Kyaukse (Kyaukse). Naba (Katha). Bhamo (Bhamo).

India, Sur Lake, Sambalpur(?), Cuttack (Orissa). Robertsganj (United Provinces). Safraha, Baraila, Gaurighat, Jubulpore (Central Provinces). Amingaon, (Assam).

The original home of *surensis* now appears to have been in a northwestern part of peninsular India, between the Orissa coast and the Jubbulpore section of the Central Provinces. Absence in the Arakan yomas and coastal areas (also no records for Chittagong) requires over sea transport to Burma. Over land transport to the Assam locality seems probable for similar reasons. The time since introduction to Burma has been long enough to allow attainment of a more or less continuous distribution through the south central part of the country. Isolated colonies at a single locality in each of Mergui, Tavoy, Katha, and Bhamo districts represent much more recent introductions presumably from the area of earlier colonization.

Habitats: Soil, in and around villages, towns and cities (Burma). Earth around roots of potted plants (Burma).

Biology: The species is geophagous and is believed to copulate as well as feed underground. Activity probably is uninterrupted at favorable sites where moisture is adequate during the dry season. Elsewhere, quiescence is imposed during drier parts of the year. Breeding may be possible throughout a considerable portion of periods of activity but data as to laying at Rangoon no longer are available.

Although *surensis* has maintained itself after introduction into two dry zone localities, it does seem to do better in alluvial soils deposited by the Irrawaddy and Sittang rivers. Although obtained from so-called jungles at Labaw (Mergui) and Pegu, the species never was found in rain forests east of Toungoo, on the Shan Plateau or in the Pegu and Arakan yomas. Prevalence in Rangoon at the gardens of the Agri-Horticultural Society should have provided many opportunities for transport to areas where the species has not colonized. Restriction to the lowlands seemingly is indicated.

O. surensis was found along with *O. beatrix* in twelve localities of nine districts. Annual rainfall in four of those districts may be as much as 100 inches.

Castings: Intestinal ejecta of *O. surensis* are believed to be deposited, at least occasionally, on the surface of the ground and then in small piles of more or less irregularly constricted cords. However, deposition on the surface was not obtained in laboratory experiments.

Variation: An *a* seta of vii in one specimen also had become copulatory.

Oviducts had united within the parietes to open by a single pore at mV in 16 of 141 specimens in an Assam series.

Abnormality: No. 1. An extra prostate on one side of xx was associated with penial setae and a posterior continuation of the seminal groove. (Location of male pore on the abnormal side is unknown.) Halving of a mesoblastic somite at the 19th level presumably could have been responsible for the abnormality.

No. 2. Female pores, paired, in xiv–xv. Ovaries, paired, in xiii–xiv. Posterior half of male field serially repeated in each of xxi–xxviii. (Location of male pores, unknown.) Both mesoblastic somites at the 13th level presumably had been halved. Some halving of somites may have been involved posteriorly but cannot be invoked to explain all of the serial repetitions.

Regeneration: Head regeneration (homomorphic) is possible back to 7/8 at least. As many as 6 excised segments may be replaced. Tail regenerates were seen at various levels from 66/67 to 141/142.

Parasites: Cysts, presumably of protozoan parasites, have been found, in large numbers, in coelomic cavities, but only rarely.

Remarks: Color of the worms resembles that of the pigmented species of *Eutyphoeus*.

Epidermal thickenings behind spermathecal pores, though obvious, are indistinctly delimited like genital tumescences in the Lumbricidae but the shape usually is more irregular.

As the young mature, usually the *b* setae of xviii are lost and later the *a* setae though one or more of the four may be retained for some time. Meanwhile apertures of the *a* follicles in xvii and xix are moved nearer to openings of the *b* follicles.

Ramiella

1921. *Ramiella* Stephenson, *Proc. Zool. Soc. London* 1921: p. 109.
1922. "*Ramella*," Michaelsen, *Mitt. Naturhist. Mus. Hamburg* **38**: p. 37. (*Octochaetus bishambari* Stephenson, 1914, designated as type species.)
1923. *Ramiella*, Stephenson (*The Fauna of British India*), Oligochaeta, p. 397.
1930. *Ramiella*, Stephenson, *The Oligochaeta* (Oxford), p. 845.
1939. *Ramiella*, Gates, *Rec. Indian Mus.* **41**: p. 212.
1945. *Ramiella*, Gates, *Proc. Indian Acad. Sci.* **21**: p. 229.
1957. *Ramiella* (part), Gates, *Breviora*, No. 75: p. 6. (Excluding, American species.)
1958. *Ramiella*, Gates, *Ann. Mag. Nat. Hist.*, Ser. 13, 1: p. 611; *Rec. Indian Mus.* **53**: p. 207. Omodeo, *Mem. Inst. Français Afrique Noire* **58**: p. 22.

Digestive system, without caeca, supra-intestinal and calciferous glands, with one esophageal gizzard and an intestinal origin behind xiii. Vascular system, with complete dorsal (single) and ventral trunks, a supra-esophageal trunk but no subneural, paired extra-esophageals median to the hearts, paired posterior lateroparietal trunks that join extra-esophageals in region of xii–xiii. Hearts, of ix–x lateral (?), of xi (latero?)-esophageal, of xii latero-esophageal. Excretory system, meroic, all nephridia small, avesiculate and exoic (? none opening into pharynx?), compact, discoidal and in longitudinal ranks behind the clitellum, astomate except in posterior segments where the medianmost on each side has a preseptal funnel. Septa, all present from 4/5. Setae, four pairs per segment. Dorsal pores, present. (Unpigmented.)

Quadriprostatic, pores equatorial in xvii and xix. Male pores, in xviii. Seminal grooves, between equators of xvii and xix. Quadrithecal, pores at or behind 7/8–8/9. Clitellum, extending into xiii and xvii, intersegmental furrows obliterated, dorsal pores occluded, setae retained. Holandric. Spermathecae, diverticulum opening into duct entally. Metagynous, ovaries fan-shaped and with several egg strings. Ovisacs, in xiv.

Distribution: India, a western portion of the Gangetic plain south through western part of the peninsula to Coorg.

The proper generic range, which presently can be characterized only as above, has been extended, presumably by man, as a result of transportation of the type species.

Remarks: Two American species belong (Gates, 1957) in *Ramiella* as defined in the classical system. Subsequent study of additional American material[1] warrants a belief that somatic organization, as in other instances of classical polyphyly, will enable recognition of monophyletic taxa.

A hypothetical protoramiella (Gates, 1958) lacking those specializations that have appeared in known Indian species, is primitive enough to have given rise to Indian octochaetids. Derivation of the protoramiella from a primitive ocnerodrilid (without calciferous glands) requires only transverse fragmentation of nephridial anlage such as does take place during embryogenesis in species of *Eutyphoeus*, *Octochaetus*, and *Pheretima*.

KEY TO SPECIES OF *Ramiella*

1. Ventral setae of xvii–xix present, unmodified. Spermathecae adiverticulate . . . . *parva*
   Ventral setae of xviii lacking, of xvii and xix penial. Spermathecae diverticulate 2
2. Spermathecal pores at intersegmental furrows, in lateral part of *BC* . . . . *heterochaeta*ᵃ
   Spermathecal pores intrasegmental and at or median to *B* 3
3. Intestinal origin in xiv, typhlosole lacking or rudimentary . . . . *bishambari*
   Intestinal origin behind xv, typhlosole lamelliform 4

---

[1] Gates, 1961, *Proc. California Acad. Sci.*, Ser. 4, 31.

4. Spermathecal pores at sites of missing *a* setae, oviducts discrete . . . . *pallida*[b]

Spermathecal pores presetal, *a* and *b* of viii–ix present, oviducts unite to open by a single pore at mV . . . . *nainiana*[c]

[a] *R. heterochaeta* Michaelsen, 1921, from Somavarpatana in Coorg, still has the gizzard in v but an intestinal typhlosole has appeared.

[b] *R. pallida* (Stephenson, 1920), from Panchgani and Mahableshwar, in Bombay Presidency, also has a typhlosole but the gizzard now is in vi.

[c] *R. nainiana* Gates, 1945, in a western sector of the Gangetic Valley and in adjacent portions of the Indian Peninsula, also has a typhlosole and the gizzard in vi.

Ramiella bishambari

1914. *Octochaetus bishambari* Stephenson· *Rec. Indian Mus.* 10: p. 347. (Type locality, Saharanpur, India. Types, none.)

1931. *Ramiella cultrifera* Stephenson, *ibid.*, 33: p. 187. (Type locality, Rangoon. Types may have been lost as they are not mentioned in lists supplied by the British and Indian Museums.)

1942. *Ramiella cultrifera*, Gates, *Bull. Mus. Comp. Zool. Harvard College* 89: p. 122.

1955. *Ramiella bishambari*, Gates, *Rec. Indian Mus.* 52: p. 75.

1958. *Ramiella bishambari*, Gates, *Ann. Mag. Nat. Hist.*, Ser. 18, 1: p. 609.

1961. *Ramiella bishambari*, Gates, *Burma Res. Soc. 50th Anniv. Pub. No. 1*: p. 57.

ECOLOGY and DISTRIBUTION: Gates, 1945, *Proc. Indian Acad. Sci.* 21: p. 230 and *Proc. Natl. Acad. Sci. India* 15: p. 48; 1956, *Proc. Natl. Acad. Sci. India* 26: pp. 146, 152. EXCRETORY ORGANS: Bahl, 1942, *Quart. Jour. Micros. Sci.* 83: p. 441.

Spermathecal pores, just behind intersegmental furrows, at *B*. Prostatic pores (conjoined openings of prostatic ducts and penisetal follicles?), in *AB*. Male pores, in seminal grooves midway between prostatic pores. Female pores, anteromedian to *a*. Genital markings, small, circular to shortly and transversely elliptical, variable in number and location; in or near *AB*, presetal in vii, viii, ix, xvii, xx, postsetal in vii, viii, x, xi; unpaired and median, postsetal in xix or at 19/20. Setae, present from ii, *AB ca.* = *CD*, *AA ca.* = *BC*, ventral couples of xvii and xix penial. Clitellum, annular (always?), xiii, xiii/n–xvii/n, xvii. First dorsal pore, in region of 6/7–10/11. Prostomium, epilobous, tongue open. Segments, 82–87. Size, 20–35 by 1–1.2 mm.

Gizzard, in vi. Intestinal origin, in xiv. Typhlosole, represented only by a low, rounded ridge. Nephridia, of postclitellar segments in 2 or 3 longitudinal ranks on each side. Seminal vesicles, in (xi) xii. Penial setae, two per follicle, ribbonlike, rolled so as to appear solid, 0.50–0.95 mm. long, 0.020–0.036 mm. wide, tip narrowed (solid?) and hooked, ornamentation of 7–15 transverse rows of triangular teeth. Spermathecal duct, slender, longer than ampulla. Diverticulum, spheroidal to ellipsoidal, sessile at ental end of duct. (GM glands?)

Reproduction: Presumably biparental as sperm are exchanged during copulation. However, no sperm were recognizable in two clitellate specimens from the Philippines. Functional metandry was suspected in bivesiculate Burmese specimens because of absence of vesicles in xi and of iridescence on male funnels of x. Breeding is possible, in favorable conditions, throughout the year.

Distribution: Aberdeen, Port Blair (South Andaman Island). Rangoon (Hanthawaddy), Hmawbi (Insein), Toungoo (Toungoo), Mount Popa (Myingyan), Kalewa (Lower Chindwin).

India, Gangetic plain in United Provinces from Allahabad to Saharanpur, south into Central India and Central Provinces. Christmas Island (Java). Philippine Islands. (China, Amoy in Fukien Province?)[2]

The species presumably originated somewhere in sub-Himalayan India. Presence in Burma, Andaman, Christmas and Philippine Islands is attributable to transportation, presumably by man.

Habitats: Soil, especially when saturated and with plenty of organic matter. Soil, with pH of 7.5, banks of drainage ditches saturated with waste effluents from human habitations. Municipal dump.

Biology: Activity is uninterrupted at Burmese and Indian sites where moisture is adequate during the dry season. Elsewhere, quiescence is imposed during drier parts of the year.

Regeneration: Head regeneration (homomorphic) is possible at all levels back to 12/13 at least and as many as eleven excised segments may be replaced. Tail regeneration (homomorphic) is possible at least back from 34/35.

Remarks: The clitellum often appears to be saddle-shaped and sometimes may actually be so but further information is needed.

Occasional presence of seminal vesicles in xi suggests that *R. bishambari* is in a different line of descent from other Indian species of *Ramiella* all of which have vesicles in ix and xii. A record of vesicles in x has not been confirmed but if correct would hint at an octovesiculate ancestral state for *Ramiella*.

[2] The queried Chinese record is of *Howascolex sinicus* Chen, 1936 (*Cont. Biol. Lab. Sci. Soc. China, Zool.* 11: p. 113). This species is known only from the holotype which was sectioned and is distinguishable from *R. bishambari* only as follows: (1) By presence of gizzard in vii, the segment where Stephenson thought it was located in *R. cultrifera*. Determination of the gizzard location often is not easy because of the delicacy of the septa and their close apposition to each other. (2) By the supposed intestinal origin in xx. Chen's fig. 28 (p. 115) however is suggestive of an intestinal origin in xiv unrecognized because of relaxation of the esophageal valve. (3) By presence of "hearts" in xiii. Supposed hearts however are connectives to dorsal trunk from the extraesophageal vessels. Similar connectives, in xii of various lumbricids, also have been mistaken for hearts. (4) By the penial setae. The rolled structure could have been unrecognized or overlooked as in case of all specimens examined by Stephenson as well as Michaelsen. Recognition of ornamentation in ordinary microtome sections seems unlikely.

Ramiella parva

1924. *Ramiella parva* Stephenson, *Rec. Indian Mus.* 26: p. 344. (Type locality, White Crow Stream, Yaungwhe, Shan States. Types, in the Indian Mus.)
1942. *Ramiella parva*, Bahl, *Quart. Jour. Micros. Sci.* 83: p. 441.
1945. *Ramiella parva*, Gates, *Proc. Indian Acad. Sci.* 21: pp. 229.
1958. *Ramiella parva*, Gates, *Ann. Mag. Nat. Hist.*, Ser. 13, 1: p. 610.

Spermathecal pores, behind the intersegmental furrows, (in or near *AB*?). Prostate pores, near *A*. Males pores, near *A*. (Female pores? Genital markings? Dorsal pores? Clitellum? Segments?) Setae, *AB = CD*, *AA > BC*, (*DD* = ?), *a* and *b* of xvii and xix more closely paired than in xviii but unmodified, anteriorly from ix *a* and *b* enlarged. Prostomium, proepilobous. Size, 30–35 by 1 mm.

(Gizzard? Intestinal origin? Typhlosole?) Seminal vesicles, in xii. Spermathecae, adiverticulate, duct slender, sinuous, longer than ampulla.

Distribution: Yaungwhe, Southern Shan States, Burma. The species was "presumably introduced accidentally into Burma by man" (Gates, 1958) and if *parva* really belongs as a valid species in *Ramiella* the unknown original home is to be sought in peninsular India.

Remarks: The spermathecae, as figured by Stephenson, seemingly are like those of certain ocnerodrilids. The only contra-indications to such an identification, in the original description, are presence of hearts in xii and absence of nephrostomes. Nephridia are, however, stomate posteriorly according to Bahl (1942) but, so far as can be guessed from his meager characterization, the excretory system is meroic. Although ovaries were said to be large the worms were aclitellate and may have been immature in which case typical genital characteristics (including spermathecal diverticula?) may not have been recognizable as might also be the case if postsexual regression had been nearly completed. Another possibility now has to be taken into consideration, that types of *R. parva* are of a uniparental clone of *R. bishambari* in which parthenogenetic degradation has resulted in loss of spermathecal diverticula and of the ability to develop penial setae.

If the last explanation is found to be correct, parthenogenesis has completely reversed evolution with respect to two different organs.

Scolioscolides

1937. *Scolioscolides* Gates, *Rec. Indian Mus.* 39: p. 305. (Type species, *Megascolides bergtheili* Michaelsen, 1907).
1958. *Scolioscolides*, Gates, *ibid.*, 53: p. 205.

Digestive system, with a gizzard (belonging to vi?) in the space between 5/6–8/9, one pair of intramural calciferous glands in xii, intestinal origin in xv, paired lateral intestinal caeca(?), unpaired anteriorly directed, small midventral caeca one each in a number of consecutive segments in front of a short series of supra-intestinal glands, a rudimentary typhlosole terminating posteriorly with the supra-intestinal glands. Calciferous glands, longitudinally hemi-ellipsoidal with flat faces mesially, with numerous transverse vertical partitions and interlamellar spaces communicating dorsally with the esophageal lumen which in xii is *T*-shaped in cross section. Vascular system, with dorsal (single), ventral (complete) and supra-esophageal trunks but no subneural, paired latero-parietal trunks from anal region passing to hind ends of calciferous glands, paired extra-esophageal trunks median to hearts and passing to front ends of calciferous glands. Hearts, of x and anteriorly lateral, of xi–xiii latero-esophageal. Excretory system meroic, all nephridia small and avesiculate, numerous astomate biramous and *Y*-shaped nephridia of iii opening into pharynx, remainder of the system exoic and comprising astomate biramous and *Y*-shaped parietal nephridia which are numerous from v through clitellar segments but posteriorly are in longitudinal ranks, the median nephridium of each side behind the supra-intestinal glands discoidal and with preseptal funnel. Septa, 4/5–5/6 muscular, 6/7–7/8 aborted, 8/9–10/11 thickened and crowded together behind their normal locations, 11/12 approximated to 10/11. Setae, four pairs per segment. Dorsal pores, present. (Segments, more than 150?)

Biprostatic, male pores, in xviii. Bithecal, pores superficial, not minute. Female pores, minute, in xiv. Clitellum, annular, intersegmental furrows obliterated, dorsal pores occluded, setae retained. Sperm ducts, join prostatic ducts entally. Ovaries, fan-shaped and with several egg-strings(?). (Ovisacs, lacking?). Spermathecal diverticula, to ental end of a short duct.

Distribution: Possibly in the Himalayas east of Darjiling.

Remarks: The prostatic anlage, from their more posterior location in this genus, exert their attracting influence on the sperm ducts before the latter have grown back through xviii and so effectively that the sperm ducts pass straight into the developing glands, presumably about at parietal level. Subsequent growth of the prostates seemingly carries level of union away from the parietes so that in adults the sperm duct disappears into the ental end of the prostatic duct. That coelomic union of sperm and prostatic ducts to open to the exterior by a single pair of apertures in segment xviii constituted, in the classical system, the only character-in-common of the group ranked as a subfamily (Megascolecinae) by Stephenson or as a family (Megascolecidae) by Michaelsen. Megascolecin terminalia are now known to be not uncommon in the Octochaetidae—already recorded from *Priodochaeta pellucida* (Bourne, 1894), *Priodoscolex montana* Gates, 1940, probably characterizing all species of *Barogaster* Gates, 1939, and *Travoscolides* Gates, 1940.

Scolioscolides is closely related to *Eutyphoeus* (*q.v.*) and *Bahlia* Gates, 1945. Indeed, anatomy of *Eutyphoeus* and *Scolioscolides* is so much alike as to warrant some question at least as to generic, if not also as to subgeneric, distinctness.

Scolioscolides bergtheili

1907. *Megascolides bergtheili* Michaelsen, *Mitt. Naturhist. Mus. Hamburg* 24: p. 150. (Type locality, Sandakphu, British Sikkim. Types, in the Indian and the Hamburg Museums.)

1909. *Megascolides bergtheili*, Michaelsen, *Mem. Indian Mus.* 1: p. 159.

1910. *Megascolides bergtheili*, Michaelsen, *Abhandl. Naturhist. Ver. Hamburg* 19, 5: p. 9.

1916. *Megascolides bergtheili*, Michaelsen, *Mjöbergs Australian Exped.*, p. 48.

1923. *Megascolides bergtheili*, Stephenson (*Fauna of British India*), *Oligochaeta*, p. 196.

1937. *Scolioscolides bergtheili*, Gates, *Rec. Indian Mus.* 39: p. 307.

1942. *Scolioscolides berghtheili*, Bahl, *Quart. Jour. Micros. Sci.* 83: p. 446.

1958. *Scolioscolides bergtheili*, Gates, *Rec. Indian Mus.* 53: p. 206.

Male pores (apertures of united sperm and prostatic ducts), small, transversely crescentic, each within a distinctly demarcated and rather penis-like but small tubercle in a porophore that reaches into *BC* and *AA*. Female pores, both present. Spermathecal pores, in *AB*. Genital markings, unpaired and median, primarily presetal, reaching into *BC*, in some of xii, xiii, xx, xxi. Clitellum, annular, xiii–xvii. Setae, rather small, the ventral closely paired, the lateral almost separated, *a,b*/xviii lacking. First dorsal pore, at 11/12. Prostomium, tanylobous. Color, white (unpigmented?). Segments, 146–175. Size, 100–120 by 4.5–5 mm.

(Lateral typhlosoles, lacking?) Lateral caeca, in xxi. Ventral caeca, in xxiv–xxviii. Supra-intestinal glands, in lix, lx–lx, lxi. Dorsal blood vessel, complete, with hearts in iv–v. Holandric. Testis sacs, (of x U-shaped? of xi annular?). Seminal vesicles, in ix and xii. Prostates, in xviii–xix, ducts 1¾ mm. long, looped entally. (Penial setae, none. GM glands, none.) Spermathecal diverticula, median and lateral.

Reproduction: Presumably biparental as sperm are exchanged during copulation.

Distribution: Sandakphu, British Sikkim, Darjiling District, Bengal.

Remarks: Known only from five specimens secured at the type locality.

SPARGANOPHILIDAE

1918. Sparganophilinae (Lumbricidae), Michaelsen, *Zool. Jahrb. Syst.* 41: p. 301.

1921. Sparganophilidae (Lumbricina), Michaelsen, *Arch. Naturgesch.* 86, A: p. 141.

1928. Sparganophilidae (Lumbricina), Michaelsen, *Handbuch der Zoologie, Berlin* 2, 2–8: p. 107.

1930. Sparganophilinae (Glossoscolecidae), Stephenson, *The Oligochaeta* (Oxford), p. 899.

1959. Sparganophilidae, Gates, *Bull. Mus. Comp. Zool. Harvard College* 121: p. 255.

Digestive system, without gizzard, calciferous glands or lamellae, caeca, typhlosoles and supra-intestinal glands, with an intestinal origin in front of the testis segments. Vascular system, with complete dorsal and ventral trunks, two pairs of anterior lateroparietal trunks one of which passes to the dorsal vessel and the other to the ventral vessel in xiv, but without subneural and supra-esophageal trunks. Hearts, lateral, moniliform, in vii–xi. Blood glands, protuberances from capillaries in septal glands. Nephridia, holoic, aborted at maturity in first 12 or more segments, avesiculate, peritoneal cells investing postseptal portions enlarged, ducts without muscular thickening passing into parietes in *AB*. Nephropores, inconspicuous, in *AB*. Setae, eight per segment. Dorsal pores and pigment, lacking. Prostomium, zygolobous. Anus, dorsal.

Reproductive apertures, all minute and superficial, female pores in xiv, spermathecal pores in front of testis segments. Clitellum, multilayered, including male pore segment. Seminal vesicles, trabeculate. Ovaries, in xiii, each terminating in a single eggstring. Ova, not yolky. Ovisacs, in xiv, small and lobed. Spermathecae, adiverticulate.

Distribution: Central and North America. (England? France?)

S. eiseni Smith, 1895, has been recorded from trans-Mississippi states of Iowa, Missouri, Louisiana, and several eastern states, including Wisconsin, Florida, and Massachusetts. The species just gets into Canada at one site each on the northern shores of Lakes Erie and Ontario. One subspecies of *S. benhami* Eisen, 1896, for each of Guatemala, Mexico, and Iowa, was erected by Eisen but *S. benhami* usually has been regarded as a synonym of *S. eiseni*. If a dubious and unconfirmed record (based on a juvenile) for Oregon is found to have been correct, *S. eiseni* may be present throughout most if not all of the family range. The species is confined to the muddy banks of streams, rivers, ponds, and lakes. Distinguishing a self-acquired portion of the species range seems likely to become increasingly difficult as time goes by, but discontinuities in the glaciated portion of United States indicate transportation of some sort may have been involved.

The type species certainly has been present in England but was reported only from the Thames and Lake Windermere. Introduction from America usually has been assumed, and the two records for that country do not contra-indicate a recent arrival there. Absence of any American distribution has little meaning at present because of the paucity of our knowledge of the earthworm faunas of Mexico, Central America and even much of the United States.

France was included in the *Sparganophilus* range by Cernosvitov when he placed *Pelodrilus cuenoti*

Tetry, 1934 in the synonymy of *S. tamesis* Benham, 1892. Tetry did not say where genital apertures were located, but in a figure she showed the sperm ducts passing straight through the parietes to male pores which seemingly are shown to be about at eq/xi, xii. If those locations are correct, *cuenoti* belongs to the genus in which Tetry placed it, as in *Sparganophilus* the ducts turn posteriorly underneath the epidermis (a character supposedly unique in megadriles) and there pass back to region of 18/19 before opening to the exterior.

Present knowledge of the family accordingly provides support neither for a North Atlantic bridge nor for a Wegenerian approximation of continental land masses.

Systematics: The family is monogeneric which enables some simplification of the following discussion. Two species only have been well characterized.

Benham defined his new genus by genitalia and in addition by seven somatic characters of which four were negative; absence of gizzard, intestinal typhlosole, intestinal caeca, subneural blood vessel. Beddard (1895) omitted from his definition of the genus somatic characters except that of the calciferous gland and the zygolobous shape of the prostomium. Later definitions of the genus and subfamily included some of the negative characters, along at times with one or more of three other somatic characters; setae closely paired, nephropores in region of *AB*, last hearts in xi.

Benham's definition contained extrapolations from data concerning a few specimens of a single species, but two pairs of anterior lateroparietal trunks (instead of one pair of extra-esophageals) later were found in *S. eiseni*. The character is unknown elsewhere and now, with somewhat more justification, is regarded as definitive.

The simplicity of the digestive tract parallels that of the Haplotaxidae but of course can be secondary rather than primitive. Little evidence is available for either alternative. Retention of ten or eleven pairs of direct connectives, in i–xi, between dorsal and ventral trunks, does seem indicative of less cephalization in that part of the vascular system than in most megadriles.

Now regarded as one of the most important indicators of affinity, especially at higher levels in the hierarchy, is shape of the ovaries and that character suggests a common ancestry of the Sparganophilidae with the Lumbricidae. Differences in so much somatic anatomy are such as to suggest only a distant relationship.

Remarks: *S. eiseni* has paired glands, improperly called prostates and spermiducal glands, that are invaginated from the epidermis near the ventral setae. Finding those organs in so many of the first thirty metameres led Eisen (*Mem. California Acad. Sci.* 2, 5: p. 152) as long ago as 1896, to suggest it is "not unlikely that originally this genus possessed many more pairs of spermiducal glands, perhaps one in every somite." Ocnerodrilid glands, similarly invaginated but in a shorter portion of the main axis, clearly are the same as tubular prostates of the classical system and likewise seemed to require a polyprostatic or, more accurately, a polyglandular ancestry.

The longitudinal musculature of *S. eiseni*, according to Harman (*in lit.*), is fasciculate.

Eisen thought he had evidence showing that "a large number of species of this genus will soon be found on the American continent." The "soon" of the prediction was incorrect as Eisen did not anticipate a subsequent derogation of systematics and in particular of the Oligochaeta. Existence, in North America, of two, three or even more species additional to *S. eiseni*, is now anticipated.

APPENDIX

MICRODRILI

The author's survey of Burmese oligochaetes was limited to terrestrial forms, including glyphidriles that live in mud under water. Several species in each of three microdrile families have been reported from Burma and adjacent areas as listed below. Some of records resulted from collecting in certain Asiatic lakes by officers of the Zoological Survey of India. Other records are casual. Although never identified, aeolosomatids are known to be common, as are also various naids.

Enchytraeidae

Hemienchytraeus rangoonensis (Stephenson, 1931), originally recorded as *Enchytraeus rangoonensis*, Rangoon. Possibly the same as the Assam species according to Dr. C. Overgaard Nielsen (*in lit.* Nov. 4, 1963).

Hemienchytraeus stephensoni (Cognetti, 1927), originally recorded as *Enchytraeus cavicola* Stephenson, 1924 (preoccupied by *Enchytraeus cavicola* Joseph, 1840), Siju Cave, Assam.

Naididae

Chaetogaster diastrophus (Gruithuisen, 1828), originally recorded as *C. annandalei* Stephenson, 1917, Inle Lake, Shan Plateau.

Chaetogaster limnaei Baer, 1827, originally recorded as *C. bengalensis* Annandale, 1905, Inle Lake.

Perhaps *C. bengalensis* should be recognized as a subspecies.

Slavina appendiculata (Udekem, 1855), Andaman Islands.

Aulophorus furcatus (Oken, 1815), Rangoon.

Nais communis Piguet, 1906, originally recorded as var. *caeca* Stephenson, 1910, Loktak Lake, Manipur, Assam.

Tubificidae

Bothrioneuron iris Beddard, 1901, Peninsular Thailand, Kurseong (Eastern Himalayas, India). Also recorded from Palni Hills and Trivandrum in South India.

Branchiura sowerbyi Beddard, 1892, Inle Lake (Shan Plateau), Indawgyi Lake (Myitkyina), Mongyai (Shan Plateau), Loktak Lake, Manipur (Assam).

Limnodrilus hoffmeisteri Claparede, 1862, originally recorded as *L. socialis* Stephenson, 1912, Rangoon.

Limnodrilus grandisetosus Nomura, 1932, Indawgyi Lake (Myitkyina). This record may have been the basis for Chen's (1940, Cont. Biol. Lab. Sci. Soc. China, Zool. 14, p. 9) presumably erroneous listing of *L. chacoensis* from "Upper Burma."

Note: Inclusion of Burma in the distribution of *Tubifex blanchardi* as in certain recent publications is erroneous according to K. V. Naidu (*in lit.*, July 20, 1966).

GLOSSARY

ABBREVIATIONS

a, the medianmost seta on each side of a segment. In a phrase like "anteromedian to *a*," reference is to the seta if protuberant to the exterior or visible within the parietes or, otherwise, to the aperture of its follicles.

A, in parthenogenetic polymorphism, an abbreviation for any "athecal morph."

A, a meridian of longitude passing anterioposteriorly along apertures of the *a* setal follicles.

AA, median space ventrally between the two *A* meridians.

AB, space between *A* and *B* meridians.

aq, anterior to a segmental equator.

AR, abbreviation for athecal, anarsenosomphic morphs.

ARZ, abbreviation for any athecal, anarsenosomphic parthenogenetic morph without testes.

b, the seta next lateral to *a*.

B, a meridian of longitude passing along apertures of *b* follicles.

BC, space between meridians of longitude *B* and *C*.

c, the seta next lateral to *b*.

C, a meridian of longitude passing along apertures of the *c* follicles.

C, abbreviation of circumference, in German publications = U, abbreviation for Umfang.

CD, the space between the *C* and the *D* meridians of longitude.

d, the seta next lateral to *c*, and when there eight per segment, the lateral-most on each side.

D, a meridian of longitude passing along apertures of the *d* follicles. D, dorsal.

DD, the space dorsally between the two *D* meridians.

e, the fifth setae from mV or the one lateral to *d* when setal arrangement is not lumbricin.

E, the meridian of longitude passing across apertures of the *e* follicles.

EF, the space between the *E* and *F* meridians of longitude.

eq., abbreviation for equator or equatorial.

GM, genital marking. Usually referring to more distinctly delimited epidermal areas than the lumbricid genital tumescences.

GS, genital setae, in the lumbricidae modified setae, not associated primarily with the spermathecae or the male terminalia, but in follicles opening through genital tumescences.

H, a morph with biparental reproduction of a species with parthenogenetic morphs.

H*p*, a hermaphroditic parthenogenetic morph in which organs such as spermathecae, seminal vesicles, testes, and prostates remain juvenile.

i, ii, iii, iv, v. These Roman numerals indicate the segments of the body in anteroposterior direction beginning with the peristomium. Some authors, a hundred years ago, did not count the peristomium as a segment and, occasionally, instances of counting the prostomium as a segment also may cause confusion. Some editors insist on capitalizing the numerals thus, I, II, III, IV, V, etc. Thus I is understood to mean segment one, II, segment two.

xii/2, xiii/2, xviii/2, etc., refers to the equator of the segment indicated by the roman numeral.

I, parthenogenetic morphs that show intermediate stages of anatomical degradation or modification. With a subscript 1 understood, intermediate between H and A morphs.

I₂, intermediate stages between H and R morphs.

I₃, intermediate stages between R and AR.

I₄, intermediate stages between H and AR.

I₅, intermediate stages between A and AR.

ibid, *ibidem*, in the same place.

i.e., *id est*, that is.

m, mid, as in, at m*BC*, at the middle of *BC*, and as in, mD = middorsal.

mD, mid-dorsal.

mL, mid-lateral.

mm, millimeters.

mV, mid-ventral.

n, a fraction or portion of, usually with reference to a segment, as xiii/n–xviii/n, which indicates that a clitellum reaches slightly into, or to varying extents into, segments xiii and xviii.

pq, posterior to the segmental equator, postsetal.

R, a parthenogenetic morph without male terminalia.

TP, of or pertaining to the tuberculum pubertatis (plural, tubercula pubertatis).

U, abbreviation of Umfang meaning circumference, in German contributions.

V, ventral, as in mV, meaning midventral.

x, the seta next lateral in the dorsum to *y* when the arrangement is perichaetin.

X, the meridian of longitude along apertures of *x* follicles when setal arrangement is perichaetin.

y, the dorsal seta next lateral to *z*, when setae are perichaetin.

YZ, the space between the longitudinal meridians Y and Z when setae are perichaetin.

z, the medianmost setae in the dorsum when setae are perichaetin, regardless of the number involved.

Z, the meridian of longitude passing across apertures of the *z* follicles.

Z, the designation of a parthenogenetic morph without testes.

CONVENTIONS

¼, ½, ¾, when referring to a prostomium indicates the extent to which the tongue of the prostomium penetrates through the first segment. If furrows demarcating the tongue laterally reach 1/2 the prostomium is said to be tanylobous.

1/2, 2/3, 3/4, 4/5, 5/6, numbers so printed, never as ½, ¾, etc., designate the intersegmental furrows (*q.v.*), not grooves, bounding the segments. Thus, 1/2 marks the boundary of segments i and ii.

1 – 1 – 1, or any similar set of three figures indicates from left to right the number of juvenile, aclitellate and clitellate specimens.

1 – 1 – 1 – 1, indicates in addition the number of postsexual, aclitellate individuals in which the clitellum had regressed, here = 1.

– 1 – 1, or – 1 – 1 – 1, indicates separating immature individuals into age classes was impractical from external examination alone.

The above conventions *re* developmental stages precisely indicate that which is quickly determinable from external inspection and involve no possibly false assumptions as to semi, near-, or other degree of maturity. Other stages at present can only be determined by dissection and/or microtome sections. Such determinations are unavailable to experimenters who must keep their animals alive and to those concerned with rapid counting during population studies.

Juveniles have no readily recognizable rudiments of genital markings such as tubercula pubertatis, papillae, tumescences, clitellum, etc. Further subdivision of the category sometimes is feasible as size differences allow recognition at least of early and late juveniles. Aclitellate refers to all remaining stages without a clitellum on which other genital markings obviously are developing. Individuals hav-

317

ing no trace of clitellar differentiation frequently have been found with mature sperm aggregated on the male funnels and presumably are able to copulate. Such worms then are fully mature in so far as maleness is concerned. Female maturity is not yet indicated by the ovaries or ovisacs when the latter are present. Hermaphrodites in which sperm are matured before the ova are protandric (protandrous), which is not to be confused with pro-andric. Aclitellate individuals with sperm in their spermathecae demonstrate that copulation sometimes does take place before female maturity even though special studies of the subject report copulation only between clitellate individuals. Clitellate stages need no explanation except to note that included are those with the slightest definitely recognizable epidermal tumescence on the appropriate segments. A well-developed clitellum need not always indicate full maturity of the ovaries.

The fourth stage usually is distinguished from the pre-sexual aclitellate condition by more or less obvious discoloration at sites of clitellum and of other genital markings. If those indications do disappear, distinction of postsexual from presexual aclitellates may be impossible even after examination of internal genitalia.

DEFINITIONS

Aborted, vestigial and functionless, or lacking and having become so during individual development.

Acanthodrilin, in the classical system, acanthodriline, with reference to the male terminalia, having prostatic pores in xvii and xix, male pores in xviii, all pores often in seminal furrows = grooves between eq/xvii and eq/xix. Sometimes applied to homoeotic equivalents of segments xvii–xix, as in various species of the American genus *Diplocardia*.

Aclitellate, without a clitellum, adult or nearly so but still without clitellar tumescence of the epidermis. The second number in the set of three or four of the usually recognized growth stages.

Adiverticulate, without diverticula, of spermathecae.

Agiceriate, gizzardless, without a gizzard.

Amphigonic, biparental reproduction (*cf.* amphimictic).

Amphimictic, amphimixis, reproduction involving fertilization of an ovum by a sperm. In megadriles, with the connotation of biparental.

Ampulla, ental portion of an adiverticulate spermatheca in which spermatozoa are stored temporarily or the widened ental portion of the main axis of diverticulate spermatheca and then without such a storage function. Also, when qualified by dorsal or primary, referring to a distal constricted-off portion of a seminal vesicle.

Amphodynamous, having a diapause or not, according to circumstances.

Anal, having to do with the anus, the posterior opening of the digestive tract, or referring to the region in which that aperture is located.

Anandric, anandry, without testes.

Andry, referring to testis containing segments.

Anarsenosomphic, without male terminalia as in parthenogenetic morphs, some cephalic regenerates, abnormal individuals.

Anterobiprostatic, referring to the male terminalia of parthenogenetic morphs in which the posterior prostates of an acanthodrilin set are lacking.

Anthropochorous, transported by man, usually unintentionally.

Aprostatic, without prostates.

Arsenosomphic, with male terminalia.

Asetal, without setae, as in the peristomium and pygomere.

Astomate, with reference to nephridia, "closed," without a nephrostome.

Athecal, without spermathecae.

Atrial, glandular tissue associated with a cleft or coelomic invagination containing the male pore in the Lumbricidae. Atrial sac, referring in species of the moniligastrid *Drawida*, to the spermathecal diverticulum.

Atrium (plural atria), a diverticulum of a spermatheca in two moniligastrid genera, *Drawida* and *Moniligaster*, in older contributions of Beddard, Benham, etc., may mean a tubular prostate as in the Acanthodrilidae, Ocnerodrilidae, and Octochaetidae or a capsular prostate as in moniligastrids, also certain genital organs in microdriles.

Atyphlosolate, without a typhlosole.

Autochthonous, native, endemic.

Autogamy, reproduction involving fertilization of an animal's ova by its own sperm.

Autotomy, autotomize, self-amputation, in megadriles the process of breaking off a portion of the tail, often induced by picking up a worm near its hind end.

Avesiculate, without seminal vesicles when referring to genital system, without a bladder when referring to a nephridium.

Balantin, in classical terminology balantine, with male and prostatic pores in xix. In Avel, 1959: p. 326, fig. 261, mislabeled megascolecine, and showing a union of prostatic and male ducts that may be lacking.

Bidiverticulate, with two diverticula (of spermathecae).

Bigiceriate, with two gizzards.

Biprostatic, with two prostates.

Bithecal, with two spermathecae.

Blood glands, follicles clustered in region of the pharynx, some species of *Pheretima*, or in a collar on esophagus just behind the gizzard, some species of *Pheretima*. Function, supposedly production of haemoglobin and of blood corpuscles.

Brown bodies, spheroidal, ellipsoidal or discoidal bodies, free in coelomic cavities, filled with corpuscles or brown debris along with setae and foreign bodies of various sorts such as cysts of parasites, nematodes and their ova.

Cephalization, loss of metameric uniformity at anterior end of body involving some or all of the following; abortion of septa beginning with 1/2, of nephridia and setal follicles beginning with ii, of segmental hearts or connectives between dorsal and ventral blood vessels. Sometimes involving total or almost complete abortion of one or more segments. (See Glossoscolecidae and *Tonoscolex-Nelloscolex*.)

Chaeta, chaetae, see seta.

Classical, of or pertaining to the "Classical System" of the Oligochaeta (see next definition).

Classical System, was defined (Gates, 1959, *Bull. Mus. Comp. Zool. Harvard* 121: p. 235) as the classification of the Oligochaeta, initially presented in vol. 10 of *Das Tierreich* by Michaelsen in 1900 and expressed in its final form by Stephenson (1930) in "The Oligochaeta." That system crystallized (Gates, 1959, *idem*) around two suppositions. (1) Genera, even subfamilies and families, can be defined and arranged in straight-line phylogenetic sequences of a mother-daughter-granddaughter sort by a very few and simple or generalized characters such as lumbricin and perichaetin, micronephric and meganephric, presence or absence of gizzards and calciferous glands. (2) Other structure, including much of the digestive and excretory as well as all of the vascular and nervous systems, is phylogenetically meaningless and hence of little or no systematic importance. The present author, on the contrary, has attempted to consider each group *de novo*, without preconceptions other than certain general assumptions that were recorded, in 1960 (*Bull. Mus. Comp. Zool.* 123: p. 281). The modern system that is emerging slowly from studies of individual, anomalous, and regenerative variation, as well as of genital and geographic polymorphism, involves much greater emphasis than in the past on somatic characters as well as on various, previously neglected genital characters.

Clitellate, having a clitellum, the age or stage during which the worm has a clitellum.

Clitellum, a regional tumescence of the epidermis, the gland cells of which secrete material to form a cocoon.

Composite, with reference to certain stalked glands, each of which comprises several similar units, in some pheretimas.

Copulatory chamber, an invagination, containing the male pore, that reaches through the body wall into the coelom. Bourses copulatrices, copulatory chambers or pouches of some publications prior to 1900. Previous authors usually did not distinguish between invaginations confined to the parietes and those that reached more or less deeply into the coelomic cavity.

Copulatory pouches, spermathecae (some older publications).

Copulatory setae or chaetae, those in the same segment as, also near, spermathecae. Occasionally refers to similar setae in an adjacent but athecal segment.

Cosmopolitan, a word that should have little use in scientific discussions because it may mean, found in every country, or only, present in more than one country.

Crop, a widened portion of the digestive system that lacks the muscularity of a gizzard. In the lumbricidae, at beginning of the intestine and in front of the gizzard.

Cysticercoid, larval stage of a cestode (tapeworm).

Decathecal, with ten spermathecae, usually in five pairs.

Definitive, capable of being used in defining a taxon.

Diagnostic, uniquely characterizing a taxon.

Diapause, originally characterizing a stage in insect development and perhaps with certain connotations inapplicable to megadriles but now usually meaning a state in which gut is empty and the worm is tightly coiled in a closed off cell. The state has been characterized as obligatory, internally controlled, and also as optional.

Distal, away from place of attachment, as in a regenerate, an organ on a septum, the gut, or body wall. A quite different meaning, in some European publications, with respect to the body as a whole, is peripheral.

Diverticulate, having a diverticulum, an outgrowth of some sort from the main axis of an organ.

Duodecathecal, having 12 spermathecae, usually in 6 pairs.

Ectal, ectally, near to or towards the body wall.

Endemic, indigenous, native.

Ental, entally, away from the body wall, toward the center of the body.

Enteroic, when referring to the excretory system, opening into gut lumen.

Enterosegmental organs, paired, segmentally repeated structures of unknown function close to mD on dorsal face of the postgizzard gut in moniligastrids. Each is a bundle of glandular tubes bound together by a delicate connective tissue investment. A tube may be erect, with one end free, or curved into a horseshoe shape. Some tubes are continued ventrally within the lateral wall of the gut nearly to mV but others open directly into gut lumen. Size of the cluster and length of the units decreases posteriorly. The tubes, in preserved material, often are translucent and red, looking like thin-walled blood vessels.

Epilobous, epilobic, referring to a prostomium that is continued by a tongue into the peristomium but without reaching 1/2.

Epimorphic, epimorphosis. Regeneration that results in addition of new tissue at the level of amputation.

Equator, a central meridian of latitude of a segment often or usually equivalent to a circumferential line passing across apertures of setal follicles.

Equatorial, at, of or pertaining to a central meridian of latitude in a segment.

Equimeric, equimery, with reference to regenerates, having the same number of segments as had been amputated, the state of being such.

Esophageal, when referring to the digestive system, that portion of the gut between the pharynx and the intestine, ending posteriorly in an esophageal valve. When referring to the circulatory system, a heart that opens dorsally into the supra-esophageal trunk and beneath the gut into the ventral trunk.

Estivation, a state of rest or inactivity during unfavorable summer conditions.

Evaginate, evagination, to grow out from, an outgrowth of, as calciferous sacs of the lumbricid esophagus.

Exoic, when referring to the excretory system, opening to the exterior through the epidermis, in contrast to enteroic (q.v.).

Exotic, imported, foreign, alien, in contrast to native, endemic, and autochthonous.

Family, a taxon usually comprising a number of genera, with a name ending in ae, as Lumbricidae.

Female ducts, female gonoducts. Oviducts (q.v.).

Female funnel, enlargement of ental end of an oviduct to facilitate entry of ova on their way to the exterior.

Female pores, external apertures of the female ducts.

Fissure, an epidermal crevice containing, in species of Eutyphoeus, discrete male and prostatic pores, as well as apertures of penisetal follicles.

Genital setae, vide seta.

Genital tumescences, in the Lumbricidae, areas of modified epidermis without distinct boundaries and through which follicles of genital setae open.

Giceriate, having one or more gizzards in the digestive system.

Gonad, gonadal, a testis, ovary, or an organ simultaneously or consecutively producing sperm and ova. Of or pertaining to a gonad.

Gonoducts, male, female, ducts or passages that carry gametes from coelomic funnels to or towards the exterior. (cf. Sperm ducts, oviducts.)

-gyny, the characterization of ovarian location along the main axis, cf. holo-, meta- and pro-gyny.

Haemerophilic, not averse to culture or some human interference with the environment.

Haemerophobic, averse to culture and human interference with the environment.

Hearts, enlarged, segmental pulsating connectives in an anterior region of the body between the ventral and one or two other longitudinal trunks, the dorsal and supra-esophageal. According to the Perrier-Bourne-Gates terminology those opening into the dorsal trunk are lateral, into the supra-esophageal are esophageal, into both are latero-esophageal.

Heterodynamous, development indirect because interrupted by a period of rest called diapause.

Heteromorphic, with reference to regenerates, a head regenerated instead of a tail or a tail instead of an amputated head.

Hibernation, the state of rest or inactivity during unfavorable winter conditions.

Hibernestivation, the state of rest or inactivity during unfavorable conditions, in the monsoon tropics, that extends through both the cool and the hot seasons.

Holandric, a classical term that now means no more than, testes restricted to x–xi, or a homoeotic equivalent. Andric and gynous characterizations of the past have been applicable only to conditions of, or those derived by reduction from, a supposedly octogonadal megadrile ancestor with testes in x–xi and ovaries in xii–xiii. The terminology was retained, whether unwittingly or not by Stephenson (1930) though his ancestral oligochaete with testes in x–xii and ovaries in xiii–xiv nullified the meanings. As there is still need for such terms, they are retained with the meaning implicit in Stephenson's later usage which is now precisely stated. Illustrations of homoeotic equivalents, resulting from suppression of a segment anteriorly, are provided by *Nelloscolex* and *Tonoscolex*. These are *holandric* and *metagynous* though testes are in ix–x and ovaries are in xii. Holandry, the state or condition of being holandric.

Hologynous, a classical term that now means only, ovaries restricted to xii and xiii or a homoeotic equivalent. Hologyny, the state or condition of being hologynous.

Holoic, referring to an excretory system, having a pair of stomate, exoic nephridia in each segment of the body except the first and last, referring to a nephridium, having a preseptal nephrostome or funnel opening into the coelomic cavity, a post-septal looped tubule opening to the exterior through a single epidermal nephropore and derived without fragmentation from a single embryonic rudiment.

Holonephridial, an unnecessarily long term with same meaning as holoic which avoids tautologies such as "holonephridial nephridia."

Homodynamous, development direct, i.e., not interrupted by a diapause.

Homoeosis, was defined (Gates, 1949: p. 134) as ordinarily used with reference to megadriles as follows; Presence of an organ, or pairs of organs, or a series of organs, in a segment or series of segments, other than that or those, in which usually or normally found. Reference primarily is to intraspecific variation, secondarily to phylogenetic variation, for a species or genus may be homoeotic with reference to related species or genera. In case of individual homoeosis, the dislocation may involve one or both organs of a segmental pair. The former is asymmetrical, the latter symmetrical homoeosis. Homoeotic, the state of being such.

Homomorphic, of regenerates, of the same cephalic or caudal nature as the part that was amputated.

Hoplochaetellin, of male terminalia in which one pair of sperm ducts open together with the prostatic ducts of xvii or close to the prostatic pores, the other pair of sperm ducts similarly associated with the prostates of xix.

Hyperandric, having testes additional to those of x–xi. Hyperandry, the state of being such.

Hypergynous, having ovaries additional to those of xii–xiii. Hypergyny, the state of being such.

Hypermeric, of regenerates, having more segments than had been removed prior to the regeneration. Hypermery, the state of being hypermeric.

Hyperplasia, in oligochaetes, having more than the usual number of organs in a set or battery. Hyperplasic, the state of being such.

Hypomeric, of regenerates, with fewer segments than had been removed.

Hypomery, the state of being such.

Intersegmental furrow, the boundary between two consecutive segments, almost in a geometrical sense, but actually the level at which the epidermis is thinnest and where color is lacking in pigmented species.

Intersegmental groove, a circumferential depression of strongly contracted specimens that contains the intersegmental furrow. A common failure, in classical writings, to distingtish between groove and furrow resulted in lack of precision with reference to systematically useful characterizations.

Invagination, an ingrowth, as of the epidermis into the parietes, or of the whole body wall into the coelom.

Juvenile, referring to young from time of hatching till appearance of seminal furrows or grooves, genital tumescences, markings and/or pores. The first of the 3 or 4 stages usually recorded when indicating number of specimens from a site or locality.

Lateral, on, of, or pertaining to the sides of a body or of an organ but in connection with the vascular system, a heart joining the ventral trunk below the gut and the dorsal blood vessel above the gut. Also any segmental commissure with the same relationship to the 2 major trunks.

Latero-esophageal, with reference to the vascular system, a heart or other vessel joining the ventral trunk below the gut but bifurcating above the gut, with one branch to the supra-esophageal trunk, the other to the dorsal trunk.

Lumbricin, having 4 setae per segment.

Lymph glands, organs on the anterior faces of septa and associated with the dorsal blood vessel, in the intestinal regions of pheretimas (perhaps of some other genera?), supposedly functioning in production of phagocytes.

Macroic, large, with reference to excretory organs, a substitute for the classical meganephridium, which resulted in tautological characteriziations such as meganephridial excretory or nephridial systems.

Macroscopic, larger than microscopic and capable of being studied with unaided eye, a magnifying glass or a dissecting binocular.

Male ducts, male gonoducts. Sperm ducts (*q.v.*).

Male funnel, a funnel or rosette-shaped enlargement of ental end of a sperm duct, with central aperture through which sperm pass into lumen of the duct on their way to the exterior. Sperm, prior to entering the ducts in many species, temporarily aggregate on the funnels in such a way as to reveal their presence by iridescence.

Male pores, primarily openings to the exterior of the male ducts. The pores may be superficial, be invaginated into chambers confined to the parietes or reaching more or less extensively into the coelomic cavities, or the apertures of such chambers may be withdrawn into a depression that can be closed off by apposition of its margins. Union of male and prostatic ducts also introduces other complications, i.e., the ducts may unite just beneath the epidermis, deeper within the body wall, or within a chamber invaginated into the coelom which may or may not have a penis or some sort of a porophore in which case male and prostatic pores may still be discrete.

Manicate, glove-shaped, usually referring to an intestinal caecum of certain pheretimas in which the organ comprises several anteriorly directed secondary caeca.

Meganephridia, meganephridial, meganephridium, large excretory organs, in the classical system often meaning organs now called holoic (*q.v.*).

Megascolecin, in the classical system megascolecine, indicating that the single pair of prostates, tubular or racemose, opened to the exterior in xviii, along side of or together with the sperm ducts. In Avel 1959: p. 326, fig. 261, mislabeled microscolecine and showing sperm ducts passing into prostate glands far entally which if at all known must be very rare. Megascolecin terminalia supposedly arose by deletion of prostates of xvii and dislocation into xviii of prostates from xix.

Megascolecoid, of or pertaining to worms or taxa of the classical family Megascolecidae which really was at least equivalent of a superfamily.

Meroic, divided, with reference to the excretory system, nephridial tubules formed by longitudinal or transverse fragmentation of the original single pair of embryonic rudiments of each segment. When parietal, often numerous and then said to be "in forests" almost covering the body wall. Used in place of the classical meronephridial which involved tautologies such as meronephridial nephridia.

Meronephric, meronephridial, see meroic.

Metagynous, metagyny, classical terms now meaning only, ovaries restricted to xiii or a homoeotic equivalent, the state of being such.

Metamere, a segment. Metameric, pertaining to segmentation.

Metandric, metandry, classical terms now meaning only, testes restricted to xi, the state of being such.

Metagynous, having ovaries only in xiii or a homoeotic equivalent.

Metagyny, the state of being such.

Microic, smaller than macroic, substituting for the classical micronephridial, a term often applied to nephridia as large as, or even larger than meganephridia.

Micronephridia, in classical systematics often with meaning of meroic (*q.v.*).

Microscopic, smaller than macroscopic, usually requiring an inverting microscope for elucidation of structure or recognition of whatever characters are involved.

Microscolecin, in classical terminology—microscolecine, provided with a pair of tubular prostates opening to exterior in xvii alongside or together with the sperm ducts. In Avel, 1959: p. 326, fig. 261, mislabeled balantine and showing union of male and prostatic ducts which often is lacking.

Monophyletic, with a single common ancestry.

Monothecal, having only one spermatheca.

Morph, with reference to parthenogenesis, a group of individuals that reproduce parthenogenetically or that have the option to do so, and which share a common anatomy as a result of the degradations, deletions, or other changes from structure of the ancestral amphimictic population as a result of isolation because of the uniparental reproduction.

Morphallatic, morphallaxis, a regenerative process in which the new parts are reorganized from the old in situ instead of being formed anterior or posterior to the level of amputation.

Octoprostatic, having 8 prostates.

Octothecal, having 8 spermathecae.

Ontogenetic, ontogeny, having to do with individual development.

Oviducts, ducts or passages carrying female gametes, usually from a coelomic funnel to or toward the exterior.

Parthenogenesis, uniparental reproduction in which the ova develop without fertilization by spermatozoa. Parthenogenetic of or pertaining to that manner of reproduction.

Penial setae, refer to setae.

Peptonephridia, classical term for organs, supposedly modified nephridia, opening into the buccal cavity or pharynx.

Peregrine, exotic, foreign. In the past often with the implication of widely wandering but now usually more accurately characterized as anthropochorous.

Perichaetin, location of the setae, when more than eight per segment, in a more or less complete circle around the equator of a segment.

Periproct, preferred to pygomere (*q.v.*) by some because of similarity to peristomium.

Peristomium, anteriormost portion of the body, around the mouth and like the anus, lacking major characteristics of a segment though counted as one.

Phylogenetic, phylogeny, having to do with past evolutionary development, as distinct from ontogenetic.

Poly-andric, having testes in more segments than x–xi.

Polydiverticulate, with reference to spermathecae, having more than two diverticula.

Polygiceriate, with several gizzards.

Polygonadal, having more than four gonads.

Polyloculate, referring to a spermathecal diverticulum, having several seminal chambers.

Polymorphic, of or pertaining to polymorphism. The latter, with reference to megadriles, is of several kinds, of which the most important for systematics are geographical and parthenogenetic.

Polyphyletic, of mixed evolutionary origin, not derived from a common ancestor.

Polyprostatic, having more than six prostates in three segments or more than eight in two segments.

Polystomate, having many mouths, referring to nephridia, with several nephrostomes.

Polytesticulate, having more than two pairs of testes.

Polythecal, having more than one or two pairs of spermathecae per segment.

Porophore, any area, protuberance or special structure bearing a pore, usually that of a spermatheca, ovi- or sperm-duct.

Posterobiprostatic, with prostates in xix after loss of a pair in xvii of an acanthodrilin set.

Precis, a brief summary of sytematically important characters as presently known, sometimes with mention of characters about which needed information is not available. Used rather than diagnosis (a term that seems inapplicable to such summaries) and definition inasmuch as available data are insufficient to allow characterization to be finally "definitive."

Proandric, proandry, classical terms that now mean only, testes restricted to x or a homoeotic equivalent, the state of being such.

Progynous, a classical term now meaning only, ovaries restricted to xii or a homoeotic equivalent. Progyny, the state of being such.

Prolobic, prolobous, characterizing a prostomium demarcated from and without a tongue in i.

Prostate, originally with a meaning in human anatomy that is inapplicable to oligochaete organs of a different nature. Moreover, the word has been used in classical texts, for many different kinds of glands, some of which are not even remotely associated with the male ducts. Now commonly used for tubular glands that open beside the male pores, in acanthodrilids, ocnerodrilids, and octochaetids. Gland ducts and sperm ducts may unite in the coelom or within the parieties. The very same glands may, however, open to the exterior in the segment next in front, or next behind, that of the male pores or even at sites several segments away from the male pores. Exact equivalents of the moniligastrid capsular prostates, and of the eudrilid "euprostates," but not associated with sperm ducts, also have been found. The function of each of the various kinds of "prostates" is unknown. Usage, in the present text is restricted to stalked glands opening to the exterior in a segment containing the male pores, and, as a relict of classical usage, to the same sort of gland opening to the exterior in a segment next in front of or next behind that of the male pore. Lumbricid "prostates," without stalks are characterized as "atrial glands."

Prostomium, a protuberance anteriorly and above the mouth from the first segment. The megadrile equivalent of an elephant's trunk.

Protandric, protandrous, proterandrous, maturing sperm before the ova. Protandry, the state of being such.

Proximal, near to, towards, place of attachment, as in a regenerate, an organ on a septum, the gut, or body wall.

Pseudogamic, of or pertaining to pseudogamy, a method of reproduction in which entrance of the sperm stimulates development of the egg but without involving biparental heredity.

Pseudovesicles, structures on posterior faces of 12/13 or 13/14 that are serially homologous with seminal vesicles. Usually retained from an embryonic or early juvenile state. Function unknown.

Pygomere, the terminal portion of the body, sometimes called the anal segment but lacking some of the characters of a metamere.

Quadriprostatic, with four prostates.

Quadrithecal, with four spermathecae.

Quincunx, a pattern involving location of setae of three consecutive segments in a group of five, with one centrally as here ⦂∴

Racemose, from the latin racemus, meaning bunch, as perhaps of grapes, long used to characterize the lobular kind of prostate present in *Pheretima* and related genera. Lobulation may or may not be obvious superficially but within those glands a prostatic duct branches repeatedly, the subdivisions usually unrecognizable macroscopically. The finest ductules end in microlobules. Prostatic glands with a central lumen, from which short branches pass out, were believed by classical authorities (*cf.* Stephenson, 1930: p. 369), to be intermediate between the simpler tubular kind (without such branches) and the polylobular glands (without a central or macroscopically recognizable lumen). Intermediate stages between a purely mesodermal organ (Stephenson, 1930: p. 368) and one arising as an ectodermal invagination seem improbable if not impossible.

re, (Latin, ablative of *res*), in the matter of, regarding, with reference to.

Regularization, anatomical adjustments involved in reducing the asymmetry resulting from unilateral splitting of mesoblastic somites.

Segment, a portion of the body, along the anteroposterior axis, between 2 consecutive intersegmental furrows and the associated septa.

Seminal, literally of or pertaining to seeds, but characterizing structures in which sperm are involved, as follows.

Seminal furrows or grooves, except in the Lumbricidae, referring to distinct and permanent markings in the epidermis that are associated with male, and sometimes also prostatic, pores, and through which sperm and/or prostatic secretions move at time of copulation.

Seminal receptacles, formerly used for spermathecae or occasionally even for seminal vesicles.

Seminal reservoirs, in some older publications-seminal vesicles.

Seminal vesicles, pockets from a septum in which sperm are matured. A German phrase for testis sacs was consistently mistranslated (Michaelsen, 1909) as seminal vesicles. The German for seminal vesicles was translated as sperm sacs. Supposed seminal vesicles in x, as well sometimes as in xi, species of *Pheretima*, have been testis sacs, of ocnerodrilid genera have been hypertrophied testes. The septal outgrowth may be simple, tubular and then is posteriorly directed as in Group I megadriles. Vesicles in Group II megadriles are lobed, with connective tissue partitions internally, not posteriorly directed primarily though secondary growth may result in posterior elongation, in ix and x on anterior faces of the septa, in xi and xii on posterior faces of the septa.

Seta, setae, from Latin meaning bristle, hence more appropriate than chaeta from Greek meaning hair or mane. Also note, setole (Italian), soie (French), cedas (Spanish), cerda (Portuguese), but Borste, Borsten, meaning bristle (German). Solid rods, secreted by cells at ental end of a tubular epidermal ingrowth, the setal follicle. Follicles are provided with protractor and retractor muscles so that the seta can be partially protruded beyond the epidermis or retracted. A normal, unspecialized seta has a slight double curvature providing a shape called sigmoid, a pointed outer end called the tip, a thickening somewhere near the middle of the shaft called the nodulus, and a blunt inner end called the base. Specialized setae, usually no longer sigmoid, often ornamented in one or more of several ways, if associated more or less closely with male or prostatic pores are called penial, but copulatory if associated with spermathecae. Modified setae associated with genital tumescences and/or with special glands but not especially with the male, female, prostatic or spermathecal pores are designated only as genital. Setae often are variously modified in shape and may be sculptured (ornamented) ectally in numerous ways. Ornamentation by circles of fine spines or teeth thus may characterize enlarged setae at either end of the body.

Sexprostatic, with 6 prostates in 3 consecutive segments.

Sexthecal, with 3 pairs of spermathecae.

Salivary gland, in classical writings a term for organs, opening into buccal cavity or the pharynx, that sometimes were thought to be modified excretory organs = peptonephridia.

Somatic, of or pertaining to any portion of the anatomy except the reproductive organs.

Spermatheca, spermathecal, an organ in which sperm received from a copulatory partner are stored until extrusion during laying.

Sperm ducts, ducts or tubes that carry sperm from the male funnels to or towards the exterior.

Spermiducal glands, spermiducal pores, of older texts now are called prostates, prostatic or male pores.

Sperm sacs, in writing of some classical authorities, seminal vesicles, or sometimes referring to testis sacs or spermathecae.

Stomate, having a mouth, referring to a nephridium, with a funnel. A nephridium with a funnel sometimes is said to be "open."

Sulcate, having seminal furrows or grooves.

Tanylobous, characterizing a prostomium with a tongue that reaches all the way through segment i to 1/2.

Taxon, any unit in a system of classifying plants or animals.

Testicular chamber, testis sac.

Testicular vesicle, testis sac or some part of it.

Testicular sac, sometimes mistaken, when enlarged in parthenogenetic morphs for seminal vesicles.

Testis sac, usually a closed off coelomic space containing one or both testis and male funnels of a segment.

Thecal, having spermatheca.

Trabeculate, characterizing megadrile seminal vesicles that develop as connective tissue proliferations from a septum so as to have numerous irregular spaces that remain inconsiderable until spermatogonia begin to enter.

Troglophile, cave-loving, but sometimes used to characterize animals living permanently and reproducing in caves. Various species of earthworms can be called troglophile but none of them are confined to caves, i.e., they are not obligatory troglophiles.

Trogloxene, cave guest, sometimes used to characterize animals that do not complete all of the life cycle in caves.

Typhlosolate, having a typhlosole, usually with reference to a segment or segments of the intestine or the intestinal region of the body.

Typhlosole, any longitudinal fold of gut wall, especially if projecting into gut lumen from the roof at mD or the floor at mV. Lateral typhlosoles (in the intestine) usually are rudimentary.

Unidiverticulate, having one diverticulum, as of spermathecae.

Uniloculate, having only one seminal chamber, as of spermathecal diverticulum.

Vas deferens, plural, vasa deferentia. Sperm duct(s).

Vesicle, referring to the excretory system, the bladder, referring to the reproductive system, anteriorly or posteriorly directed pockets of a septum in which male germ cells mature.

Vesiculate, with reference to a nephridium—provided with a bladder, with reference to a reproductive system—having seminal vesicles, with reference to tissue or organ structure—having small spaces.

Vestibulate, having one or two vestibula.

Vestibulum, plural vestibula, an invagination in species of *Eutyphoeus* that contains a penis or male porophore, regardless of whether it is confined to the body wall or reaches more or less conspicuously into the coelom. Not homologous with but containing the male pore fissure.

Yoma, local Burmese name for long, backbonelike mountain ridges.

Zygolobous, a prostromium that is not in any way demarcated from the first segment.

BIBLIOGRAPHY OF BURMESE OLIGOCHAETA

BAHL, K. N. 1941. "The enteronephric Type of Nephridial System in the Genus *Tonoscolex* (Gates)." *Quart. Jour. Micros. Sci.* **82**: pp. 443–466.

—— 1942. "Studies on the Structure, Development and Physiology of the Nephridia of Oligochaeta. Part I. General Introduction, and the Nephridia of the Subfamily Octochaetinae. Part II. Multiple Funnels of the Nephridia." *Idem* **83**: pp. 423–449.

BEDDARD, F. E. 1883. "Note on Some Earthworms from India." *Ann. Mag. Nat. Hist.*, Ser. 5, **12**: pp. 213–224.

—— 1895. *A Monograph of the Order of Oligochaeta* (Oxford).

CHAPMAN, M. 1939. "Notes on the Circulatory System in Burmese Species of *Drawida* (Oligochaeta)." *Rec. Indian Mus.* **41**: pp. 117–120.

GATES, G. E. 1925a. "Some New Earthworms from Burma. *Ann. Mag. Nat. Hist.*, Ser. 9, **15**: pp. 316–328.

—— 1925b. "Some Notes on *Pheretima anomala* Mich., and a Related Species New to India and Burma." *Ibid.*, pp. 538–550.

—— 1925c. "On Some New Earthworms from Burma. II." *Ibid.*, Ser. 9, **16**: pp. 49–64.

—— 1925d. "Note on an Abnormal Specimen of *Pheretima posthuma.*" *Rec. Indian Mus.* **27**: pp. 237–240.

—— 1925e. "Notes on Rosa's Rangoon Earthworms, *Pheretima peguana* and *Eutyphoeus foveatus.*" *Ann. Mag. Nat. Hist.*, Ser. 9, **16**: pp. 566–577.

—— 1925f. "Note on a New Species of *Drawida* from Rangoon." *Ibid.*, pp. 660–664.

—— 1925g. "Note on Luminescence in the Earthworms of Rangoon." *Rec. Indian Mus.* **27**: pp. 471–473.

—— 1926a. "Note on a New Species of *Pheretima* from Rangoon." *Ann. Mag. Nat. Hist.*, Ser. 9, **17**: pp. 411–415.

—— 1926b. "Notes on the Rangoon Earthworms. The Peregrine Species." *Ibid.*, pp. 439–473.

—— 1926c. "Notes on the Seasonal Occurrence of Rangoon Earthworms." *Jour. Bombay Nat. Hist. Soc.* **31**: pp. 180–185.

—— 1926d. "The Earthworms of Rangoon." *Jour. Burma Res. Soc.* **15**: pp. 196–221.

—— 1926e. "Notes on Earthworms from Various Places in the Province of Burma, with Descriptions of Two New Species." *Rec. Indian Mus.* **28**: pp. 141–170.

—— 1927a. Note on *Perichaeta campanulata* Rosa and *Pheretima houlleti* (E. Perrier). *Ann. Mus. Civ. Sto. Nat. Genova* **52**, pp. 227–231.

—— 1927b. "Note on a New Species of *Notoscolex* with a List of the Earthworms of Burma." *Ann. Mag. Nat. Hist.*, Ser. 9, **19**: pp. 609–615.

—— 1927c. "Regeneration in a Tropical Earthworm, *Perionyx excavatus* E. Perr." *Biol. Bull.* **53**: pp. 351–364.

—— 1929. "A Summary of the Earthworm Fauna of Burma with Descriptions of Fourteen New Species." *Proc. U. S. Natl. Mus.* **75**: 10, pp. 1–41.

—— 1930. "The Earthworms of Burma. I." *Rec. Indian Mus.* **32**: pp. 257–356.

—— 1931. "The Earthworms of Burma. II." *Ibid.* **33**: pp. 327–442.

—— 1932. "The Earthworms of Burma. III." *Ibid.* **34**: pp. 357–549.

—— 1933. "The Earthworms of Burma. IV." *Ibid.* **35**: pp. 413–606.

—— 1934. "Notes on Some Earthworms from the Indian Museum." *Ibid.*, **36**: pp. 233–277.

—— 1936. "The Earthworms of Burma. V." *Ibid.* **38**: pp. 377–468.

—— 1937. "Indian Earthworms. II." *Scolioscolides. Ibid.* **39**: pp. 305–310.

—— 1938a. "Indian Earthworms. III. The genus *Eutyphoeus.*" *Ibid.* **40**: pp. 39–119.

—— 1938b. "Indian Earthworms. IV. The Genus *Lampito* Kinberg. V. *Nellogaster*, gen. nov., with a Note on Indian Species of *Woodwardiella.*" *Ibid.*, pp. 423–429.

—— 1939a. "Indian Earthworms. VI. *Nelloscolex*, gen. nov." *Ibid.*, **41**: pp. 37–44.

—— 1939b. "Indian Earthworms. VII. Contribution to a Revision of the Genus *Eudichogaster.*" *Ibid.*, pp. 151–218.

—— 1939c. "Thai Earthworms." *Jour. Thailand Res. Soc. Nat. Hist. Suppl.* **12**: pp. 65–114.

—— 1941. "Further Notes on Regeneration in a Tropical Earthworm, *Perionyx excavatus* (E. Perrier, 1872)." *Jour. Exp. Zool.* **88**: pp. 161–185.

—— 1942. "Notes on Various Peregrine Earthworms." *Bull. Mus. Comp. Zool. Harvard College* **89**: pp. 63–144.

—— 1943a. "On Some American and Oriental Earthworms." *Ohio Jour. Sci.* **43**: pp. 87–116.

—— 1943b. "Some Further Notes on Regeneration in *Perionyx excavatus* (E. Perrier)." *Proc. Natl. Acad. Sci. India, Allahabad* **13**: pp. 168–179.

—— 1945. "On Some Indian Earthworms. II." *Jour. Roy. Asiatic Soc. Bengal, (Sci.)* **11**: pp. 54–91.

—— 1948. "On Segment Formation in Normal and Regenerative Growth of Earthworms." *Growth* **12**: pp. 165–180.

—— 1952. "On Some Earthworms from Burma." *American Mus. Nat. Hist. Novitates*, No. 1555; pp. 1–13.

—— 1954a. "On Some Earthworms from Northeast Burma." *Ark. Zool. Stockholm* 6, **20**: pp. 433–439.

—— 1954b. "On the Evolution of an Oriental Earthworm Species, *Pheretima anomala.* (Michaelsen, 1907)." *Breviora, Mus. Comp. Zool. Harvard College*, No. 37: pp. 1–8.

—— 1955. "The Earthworms of Burma. VI." *Rec. Indian Mus.* **52**: pp. 55–93.

—— 1956. "Reproductive Organ Polymorphism in Earthworms of the Oriental Megascolecine Genus *Pheretima* Kinberg, 1867." *Evolution* **10**: pp. 213–227.

—— 1958a. "On Burmese Earthworms of the Megascolecid Subfamily Octochaetinae." *Ann. Mag. Nat. Hist.*, Ser. 13, **1**: pp. 609–624.

—— 1958b. "On Indian and Burmese Earthworms of the Genus *Glyphidrilus.*" *Rec. Indian Mus.* **53**: pp. 53–66.

—— 1958c. "The Earthworms of Burma. VII. The Genus *Eutyphoeus*, with Notes on Several Indian Species." *Ibid.*, pp. 93–222. (Notification of acceptance of the manuscript for publication was dated December 15, 1954. Page proof was returned to the editor July 16, 1956. When reprints were received July 18, 1959, the text was found to have many of the original typographical errors. Furthermore new typographical errors had been introduced. Presumably some, if not all, of the pages had been reset. Although publication date supposedly was February 20, 1959, the journal number was not available to permanent subscribers in this country until much later. A few reprints with many of the errors corrected by hand are available.)

—— 1960. "On Burmese Earthworms of the Family Megascolecidae." *Bull. Mus. Comp. Zool. Harvard College* **123**: pp. 203–282.

—— 1961a. "Normal Heteromorphism in Earthworms." *American Midland Nat.* **65**: pp. 40–43.

—— 1961b. "Ecology of Some Earthworms with Special Reference to Seasonal Activity." *Ibid.* **66**: pp. 61–86.

—— 1961c. "Earthworms of Burma." *Burma Res. Soc. 50th Anniv. Pub.* No. 1: pp. 51–58.

—— 1961d. "On Some Burmese and Indian Earthworms of the Family Acanthodrilidae." *Ann. Mag. Nat. Hist.*, Ser. 13, **4**: pp. 417–429.

—— 1962a. "Miscellanea Megadrilogica. VI. Reproduction in a Tropical Earthworm." *Proc. Biol. Soc. Washington* **75**: pp. 140–143.

—— 1962b. "On Some Burmese Earthworms of the Moniligastrid Genus *Drawida.*" *Bull. Mus. Comp. Zool. Harvard College* **127**: pp. 297–373.

—— 1962c. "Contributions to a Revision of the Earthworm Family Ocnerodrilidae. IV. On a Species of the African Genus *Gordiodrilus.*" *Rev. Zool. Bot. Africaine* **66**: pp. 344–350.

—— 1966. "Contributions to a Revision of the Earthworm Family Ocnerodrilidae. VII–VIII. Part VII. Final Notes on Some Indian and Burmese Species. Part VIII. On the Family Ocnerodrilidae." *Ann. Mag. Nat. Hist.*, Ser. 13, **9**: pp. 45–53.

GATES, G. E., & MAUNG HLA KYAW. 1937. "The Clitellum and Sexual Maturity in the Megascolecinae." *Jour. Roy Asiatic Soc. Bengal (Sci.)* **2**: pp. 123–125.

HLA KYAW, MAUNG, and G. E. GATES. 1937. "On Earthworm Populations and the Formation of Castings in Rangoon, Burma." *Ibid.* **2**: pp. 165–170.

MICHAELSEN, W. 1900. "Oligochaeta." *Das Tierreich* **10**: p. 575.

—— 1907. "Neue Oligochäten von Vorderindien, Ceylon, Birma und den Andaman-Inseln." *Mitt. Naturhist. Mus. Hamburg* **24**: pp. 143–188.

—— 1909. "The Oligochaeta of India, Nepal, Ceylon, Burma and the Andaman Islands." *Mem. Indian Mus.* **1**: pp i–ii, 103–253. (In this English translation, testicular vesicles (Moniligastridae) refer to testis sacs. Sperm sacs refer to seminal vesicles. Seminal vesicles (*Pheretima* and *Eutyphoeus* sp.) refer to testis sacs.)

—— 1910. "Die Oligochätenfauna der vorderindisch-ceylonischen Region." *Abhandl. Naturwiss. Ver. Hamburg* **19**, 5: pp. 1–108.

ROSA, D. 1888a. "Lombrichi della Birmania, del Tenasserim e dello Scioa." *Boll. Mus. Zool. Univ. Torino* **3**, 50: pp. 1–2.

—— 1888b. "Viaggio di Leonardo Fea in Birmania e Regioni vicini. V. Perichetidi." *Ann. Mus. Civ. Sto. Nat. Genova* 26: pp. 155–167.

—— 1890a. "Viaggio di Leonardo Fea in Birmania e Regioni vicini. XXV. Moniligastridi, Geoscolecidi ed Eudrilidi." *Ibid.* 29: pp. 368–400.

—— 1890b. "Viaggio di Leonardo Fea in Birmania e Regioni vicini. Perichetidi." *Ibid.* 30: pp. 107–122.

—— 1890c. "Terricolas ex Birmania et ex austral America." *Boll. Mus. Zool. Univ. Torino* **5**, 93: pp. 1–3.

SOOTA, T. D., & J. M. JULKA. 1970. "Notes on Earthworms of the Andaman and Nicobar Islands, India." *Proc. Zool. Soc. Calcutta* **23**: pp. 201–206.

STEPHENSON, J. 1912. "Contributions to the fauna of Yunnan. VIII. Earthworms." *Rec. Indian Mus.* **7**: pp. 273–278.

—— 1916. "On a Collection of Oligochaeta Belonging to the Indian Museum." *Ibid.* **12**: pp. 294–354.

—— 1917. "On a Collection of Oligochaeta from Various Parts of India and Further India." *Ibid.* **13**: pp. 353–416.

—— 1918. "Aquatic Oligochaeta of the Inle Lake." *Ibid.* **14**: pp. 9–18.

—— 1923. *Oligochaeta. The Fauna of British India including Ceylon and Burma* (London), pp. xxiv + 518.

—— 1924a. "Oligochaeta of the Siju Cave, Garo Hills, Assam." *Rec. Indian Mus.* 26: pp. 127–135.

—— 1924b. "On some Indian Oligochaeta with a description of two new genera of Ocnerodrilinae." *Ibid.* pp. 317–365.

—— 1925. "On Some Oligochaeta Mainly from Assam, South India and the Andaman Islands." *Ibid.* **27**: pp. 43–73.

—— 1929. "The Oligochaeta of the Indawgyi Lake (Upper Burma)." *Ibid.* **31**: pp. 225–229.

—— 1930. *The Oligochaeta* (Oxford).

—— 1931a. "Oligochaeta from Burma, Kenya and other parts of the world." *Proc. Zool. Soc. London* 1931: pp. 33–92.

—— 1931b. "Descriptions of India Oligochaeta. II." *Rec. Indian Mus.* **33**: pp. 173–202.

—— 1933. "Oligochaeta from Australia, North Carolina and Other Parts of the World." *Proc. Zool. Soc. London* 1932: pp. 899–941.

INDEX OF SPECIES NAMES

This index does not include names of species mentioned in keys but not otherwise in the text.

www.ingramcontent.com/pod-product-compliance
Lightning Source LLC
Chambersburg PA
CBHW081338190326
41458CB00018B/6042